Transition Metals for Organic Synthesis
Volume 2

Edited by
M. Beller and C. Bolm

Further Reading from Wiley-VCH:

R. Mahrwald (Ed.)

Modern Aldol Reactions
2 Vols.

2004, ISBN 3-527-30714-1

de Meijere, A., Diederich, F. (Eds.)

Metal-Catalyzed Cross-Coupling Reactions
2nd Ed., 2 Vols.

2004, ISBN 3-527-30518-1

Krause, N., Hashmi, A. S. K. (Eds.)

Modern Allene Chemistry
2 Vols.

2004, ISBN 3-527-30671-4

Cornils, B., Herrmann, W. A. (Eds.)

Aqueous-Phase Organometallic Catalysis
2nd Ed.

2004, ISBN 3-527-30712-5

Transition Metals for Organic Synthesis

Building Blocks and Fine Chemicals

Second Revised and Enlarged Edition

Volume 2

Edited by
M. Beller and C. Bolm

WILEY-VCH Verlag GmbH & Co. KGaA

Edited by

Professor Dr. Matthias Beller
Leibniz-Institute for Organic Catalysis
University of Rostock
Buchbinderstraße 5–6
18055 Rostock
Germany

Professor Dr. Carsten Bolm
Department of Chemistry
RWTH Aachen
Professor-Pirlet-Straße 1
52056 Aachen
Germany

Library of Congress Card No.: applied for

British Library Cataloguing-in-Publication Data
A catalogue record for this book is available from the British Library.

Bibliographic information published by Die Deutsche Bibliothek
Die Deutsche Bibliothek lists this publication in the Deutsche Nationalbibliografie; detailed bibliographic data is available in the Internet at <http://dnb.ddb.de>

Printed in the Federal Republic of Germany
Printed on acid-free paper

Composition K+V Fotosatz GmbH, Beerfelden
Printing Strauss GmbH, Mörlenbach
Bookbinding Litges & Dopf Buchbinderei GmbH, Heppenheim

ISBN 3-527-30613-7

Preface to the Second Edition

Is there really a need for a second edition of a two-volume book on the use of Transition Metals in Organic Synthesis after only 6 years? How will the community react? Are there going to be enough interested colleagues, who will appreciate the effort (and spend their valuable money in times of shortened budgets)? Do we, the editors, really want to invest into a project, which, for sure, will be most time-consuming? All of these questions were asked about three years ago, and together with Wiley/VCH we finally answered them positively. *Yes*, there has been enough progress in the field. *Yes*, the community will react positively, and *yes*, it is worth spending time and effort in this project, which once more will show und underline the strength of modern transition metal chemistry in organic synthesis.

The Nobel Prize in Chemistry 2001, which was awarded to K. Barry Sharpless, Ryoji Noyori (who both are authors in this book), and William S. Knowles for their contributions in asymmetric catalysis, nicely highlighted the area and demonstrated once more the high synthetic value of the use of transition metals for both small-scale laboratory experiments and large-scale industrial production.

During the past six years the field has matured and at the same time expanded into areas, which were rather unexplored before. Taking this development into account we decided to pursue the following concept: On the one hand the authors of the first edition were asked to up-date their original chapters, and most of them kindly responded positively. In a few cases the contributions of the first edition were reused and most often up-dated by an additional chapter written by another author. Some fields are now covered by other authors, which proved most interesting, since the same topic is now presented from a different perspective. New research areas have been summarized by younger active colleagues and leading experts.

It should be clearly stated that the use of transition metals in organic synthesis can not be fully covered even in a two-volume set. Instead, the present book presents a personal selection of the topics which we believe are the most interesting and actual ones. In general, the focus of the different contributions is on recent research developments since 1998. Literature up to mid – sometimes end – of 2003 has been taken into account. Hence, we believe the new book complements nicely the first more general edition of this book.

Transition Metals for Organic Synthesis, Vol. 2, 2nd Edition.
Edited by M. Beller and C. Bolm

ISBN: 3-527-30613-7

Most importantly, as editors we thank all contributors for their participation in this project and, in some case, for their patience, when it took longer than expected. We also acknowledge the continuous stimulus by Elke Maase from Wiley/VCH, who did not push but challenged. It remains our hope that the readers will enjoy reading the new edition and discover aspects, which will stimulate their own chemistry and create ideas for further discoveries in this most timely and exciting area of research and science.

Aachen, June 2004 *Carsten Bolm*
Rostock, June 2004 *Matthias Beller*

Contents

Transition Metals for Organic Synthesis, Vol. 2, 2nd Edition.
Edited by M. Beller and C. Bolm

ISBN: 3-527-30613-7

List of Contributors

ELISABETTA ALBERICO
CNR –
Instituto di Chimica Biomolecolare
sezinone di Sassari, Regione Baldinca
07040 Li Punti (Sassari)
Italy

ALEXANDRE ALEXAKIS
Dépt. de Chimie Organique
Univ. de Genève
30, quai Ernest Ansermet
1211 Genève 4
Switzerland

HOWARD ALPER
Department of Chemistry
University of Ottawa
10 Marie Curie
Ottawa, Ontario
K1N 6N5
Canada

ISABEL W. C. E. ARENDS
Laboratory of Organic Chemistry
and Catalysis
Delft University of Technology
Julianalaan 136
2628 BL Delft
The Netherlands

TERRY T.-L. AU-YEUNG
Department of Applied Biology
and Chemical Technology
The Hong Kong Polytechnic University
Hong Kong
China

JAN-E. BÄCKVALL
Department of Organic Chemistry
Arrhenius Laboratory
Stockholm University
106 91 Stockholm
Sweden

J.-M. BASSET
Laboratoire COMS CPE Lyon –
UMR CNRS 9986
43 bd du 11 novembre 1918
69616 Villeurbanne Cedex
France

OLIVER BECKMANN
Institut für Organische Chemie
der RWTH Aachen
Professor-Pirlet-Str. 1
52056 Aachen
Germany

MATTHIAS BELLER
Institut für Organische
Katalyseforschung
an der Universität Rostock e.V.
Buchbinderstraße 5–6
18055 Rostock
Germany

FRANK BIENEWALD
Ciba Spezialitätenchemie AG
Postfach
Klybeckstrasse 141
4002 Basel
Switzerland

Transition Metals for Organic Synthesis, Vol. 2, 2nd Edition.
Edited by M. Beller and C. Bolm

ISBN: 3-527-30613-7

Hans-Ulrich Blaser
Solvias AG
Catalysis Research, WRO-1055.628
P.O. Box
4002 Basel
Switzerland

Carsten Bolm
Institut für Organische Chemie
der RWTH Aachen
Professor-Pirlet-Str. 1
52056 Aachen
Germany

Helmut Bönnemann
MPI für Kohlenforschung
Kaiser-Wilhelm-Platz 1
45470 Mülheim an der Ruhr
Germany

Armin Börner
Institut für Organische
Katalyseforschung
an der Universität Rostock e.V.
Buchbinderstr. 5–6
18055 Rostock
Germany

Stephen M. Brown
Zeneca Process Technology
Development
Huddersfield Works
P.O. Box A38, Leeds Road
Huddersfield HD2 1FF
United Kingdom

Werner Brijoux
MPI für Kohlenforschung
Kaiser-Wilhelm-Platz 1
45470 Mülheim an der Ruhr
Germany

Albert L. Casalnuovo
The Dupont Company
Central Research and Development
Department
P.O. Box 80328
Experimental Station
Wilmington, Delaware 19880-0328
USA

Albert S.C. Chan
Department of Applied Biology and
Chemical Technology
The Hong Kong Polytechnic University
Hong Kong
China

Shu-Sun Chan
Department of Applied Biology
and Chemical Technology
The Hong Kong Polytechnic University
Hong Kong
China

Isabelle Chellé-Regnault
Laboratoire de Chimie Organique
Place Louis Pasteur, 1
1348 Louvain-la-Neuve
Belgium

Pedro Cintas
Departamento de Quimica Organica
Facultad de Ciencias-UEX
06071 Badajoz
Spain

Paul J. Comina
School of Chemistry
University of Reading
PO BOX 224, Whiteknights,
Reading RG6 6AD
UK

Rosenildo Correa da Costa
Institut für Organische Chemie
Universität Erlangen-Nürnberg
Henkestraße 42
91054 Erlangen
Germany

Florian Dehmel
Institut für Organische Chemie
Universität Köln
Greinstraße 4
50939 Köln
Germany

Kanae Doda
Laboratoire de Chimie Organique
Place Louis Pasteur, 1
1348 Louvain-la-Neuve
Belgium

KARL HEINZ DÖTZ
Institut für Organische Chemie
und Biochemie
Universität Bonn
Gerhard-Domagk-Str. 1
53121 Bonn
Germany

RAPHAL DUMEUNIER
Laboratoire de Chimie Organique
Place Louis Pasteur, 1
1348 Louvain-la-Neuve
Belgium

RUDOLF O. DUTHALER
Novartis Pharma AG
WSJ-507.109
Postfach
4002 Basel
Switzerland

PETER EILBRACHT
FB Chemie, Org. Chemie I
der Universität Dortmund
Otto-Hahn-Straße 6
44227 Dortmund
Germany

BASSAM EL ALI
Chemistry Department
King Fahd University of
Petroleum & Minerals
QF 31261 Dhahran
Saudi-Arabia

B. L. FERINGA
University of Groningen
Department of Chemistry
Nijenborgh 4
9747 AG Groningen
The Netherlands

RICHARD W. FISCHER
Anorganisch-Chemisches Institut
TU München
Lichtenbergstr. 4
85747 München
Germany

GIANCARLO FRANCIÒ
Institut für Technische Chemie und
Makromolekulare Chemie
Bereich Technische Chemie
Worringerweg 1
52074 Aachen
Germany

MATHIAS U. FREDERIKSEN
Chemistry Department
Stanford University
Stanford, CA 94305-5080
USA

GREGORY C. FU
MIT Department of Chemistry
77 Massachusetts Avenue
Room 18-411
Cambridge, MA 02139
USA

ALOIS FÜRSTNER
MPI für Kohlenforschung
Kaiser-Wilhelm-Platz 1
45470 Mülheim an der Ruhr
Germany

ARNAUD GAUTIER
Laboratoire de Chimie Organique
Place Louis Pasteur, 1
1348 Louvain-la-Neuve
Belgium

PAUL R. GILES
Laboratoire de Chimie Organique
Place Louis Pasteur, 1
1348 Louvain-la-Neuve
Belgium

SERAFINO GLADIALI
Dipartimento di Chimica
Università di Sassari
Via Vienna 2
07100 SASSARI
Italy

JOHN A. GLADYSZ
Institut für Organische Chemie
Universität Erlangen-Nürnberg
Henkestraße 42
91054 Erlangen
Germany

Dirk Gördes
Institut für Organische Katalyseforschung an der Universität Rostock e.V.
Buchbinderstraße 5–6
18055 Rostock
Germany

Helena Grennberg
Department of Organic Chemistry
Arrhenius Laboratory
Stockholm University
106 91 Stockholm
Sweden

Andreas Hafner
Ciba Specialties Chemical Ltd.
Additive Research
P.O. Box 64
1723 Marly 1
Switzerland

Tamio Hayashi
Kyoto University
Department of Chemistry
Faculty of Science
Sakyo, Kyoto 606-01
Japan

Horst Hennig
Universität Leipzig
Fakultät für Chemie und Mineralogie
Institut für Anorganische Chemie
Johannisallee 29
04103 Leipzig
Germany

Wolfgang A. Herrmann
Anorganisch-Chemisches Institut
TU München
Lichtenbergstr. 4
85747 München
Germany

Andreas Heumann
Faculté de St.-Jérome
ENSSPICAM, UMR-CNRS 6516
Université d'Aix-Marseille
13397 Marseille Cedex 20
France

Dennis J. Hlasta
Johnson+Johnson
Pharmaceutical Research & Development
Welsh and McKean Roads
PO Box 776
Spring House, PA 19477
USA

Lukas Hintermann
Institut für Organische Chemie der RWTH Aachen
Professor-Pirlet-Str. 1
52056 Aachen
Germany

David M. Hodgson
The Dyson Perrins Laboratory
University of Oxford
South Parks Road
Oxford OX1 3QY
UK

Jens Holz
Institut für Organische Katalyseforschung an der Universität Rostock e.V.
Buchbinderstr. 5–6
18055 Rostock
Germany

Amir Hoveyda
Department of Chemistry
Merkert Chemistry Center
Boston College
Chestnut Hill, Ma 02067
USA

Yasushi Imada
Department of Chemistry
Graduate School of Engineering Science
Osaka University
1–3, Machikaneyama, Toyonaka
Osaka 560-8531
Japan

Henri B. Kagan
Universite Paris-Sud
Institut de Chimie Moleculaire d'Orsay
Laboratoire des Reactions Organiques
Selectives
91405 Orsay Cedex
France

Tsutomu Katsuki
Department of Molecular Chemistry
Graduate School of Science
Kyushu University
Hakozaki, Higashi-ku
Fukuoka 812-8581
Japan

Wilhelm Keim
Institut für Technische Chemie und
Petrolchemie der Technischen
Hochschule Aachen
Templergraben 55
52074 Aachen
Germany

Hans-Joachim Knölker
Institut für Organische Chemie
Technische Universität Dresden
Bergstraße 66
01069 Dresden
Germany

Shū Kobayashi
Grad. School of Pharmaceutical
Sciences, University of Tokyo
Hongo, Bunkyo-ku
Tokyo 113-0033
Japan

Hartmuth Kolb
Department of Chemistry, SP-227
The Scripps Research Institute
10550 N. Torrey Pines Rd
La Jolla, CA 92037
USA

Fritz E. Kühn
Anorganisch-Chemisches Institut
TU München
Lichtenbergstr. 4
85747 München
Germany

Kamal Kumar
Institut für Organische
Katalyseforschung
an der Universität Rostock e.V.
Buchbinderstraße 5–6
18055 Rostock
Germany

Jung Lee
Johnson+Johnson
Pharmaceutical Research
& Development
Welsh and McKean Roads
PO Box 776
Spring House, PA 19477
USA

Walter Leitner
Institut für Technische Chemie
und Makromolekulare Chemie
Bereich Technische Chemie
Worringerweg 1
52074 Aachen
Germany

Jacques Le Paih
Institut für Organische Chemie
der RWTH Aachen
Professor-Pirlet-Str. 1
52056 Aachen
Germany

T. O. Luukas
L'Oréal
Advanced Research
Aulnay sous Bois
France

Ilan Marek
Department. of Chemistry
Technion – Israel Institute of
Technology, Technion City
Haifa 32000
Israel

István E. Markó
Laboratoire de Chimie Organique
Place Louis Pasteur, 1
1348 Louvain-la-Neuve
Belgium

Ana Minatti
Institut für Organische Chemie
und Biochemie
Universität Bonn
Gerhard-Domagk-Str. 1
53121 Bonn
Germany

Christian Mössner
Institut für Organische Chemie
der RWTH Aachen
Professor-Pirlet-Str. 1
52056 Aachen
Germany

Kilian Muñiz-Fernandez
c/o K. H. Dötz
Kekule-Institut für Organische Chemie
und Biochemie
Gerhard-Domagk-Str. 1
53121 Bonn
Germany

Shun-Ichi Murahashi
Department of Applied Chemistry
Okayama University of Science
Ridai-cho 1-1, Okayama 700-0005
Japan

Jean-Luc Mutonkole
Laboratoire de Chimie Organique
Place Louis Pasteur, 1
1348 Louvain-la-Neuve
Belgium

Helfried Neumann
Institut für Organische
Katalyseforschung
an der Universität Rostock e.V.
Buchbinderstraße 5–6
18055 Rostock
Germany

Ronny Neumann
Department of Organic Chemistry
Weizmann Institute of Science
Rehovot
Israel 76100

Hisao Nishiyama
School of Material Science
Toyohashi University of Technology
Tempaku-cho
Toyohashi 441
Japan

Ryoji Noyori
Research Center for Materials Science
Nagoya University
Chikusa, Nagoya 464-8602
Japan

Takeshi Ohkuma
Research Center for Materials Science
Nagoya University
Chikusa, Nagoya 464-8602
Japan

Sei Otsuka
Tokushima Research Institute
Otsuka Pharmaceutical Co. Ltd.
Kawauchi-cho
Tokushima 771-01
Japan

Chiara Palazzi
Institut für Organische Chemie
der RWTH Aachen
Professor-Pirlet-Str. 1
52056 Aachen
Germany

Nicos A. Petasis
Department of Chemistry
University of Southern California
Los Angeles, CA 90089-1661
USA

Andreas Pfaltz
Institut für Organische Chemie
der Universität Basel
St.-Johanns-Ring 19
4056 Basel
Switzerland

Freddi Philippart
Laboratoire de Chimie Organique
Place Louis Pasteur, 1
1348 Louvain-la-Neuve
Belgium

T.V. RajanBabu
Department of Chemistry
The Ohio State University
100 W. 18th Avenue
Columbus, OH 43210
USA

Oliver Reiser
Institut für Organische Chemie
Universität Regensburg
Universitätsstr. 31
93053 Regensburg
Germany

Thomas H. Riermeier
Degussa AG
Projekthaus Katalyse
Industriepark Höchst, G 830
65926 Frankfurt am Main
Germany

Karola Rück-Braun
Institut für Organische Chemie
TU Berlin
Straße des 17. Juni 135
10623 Berlin
Germany

Hiroaki Sasai
ISIR
Osaka University
8-1 Mihogaoka, Ibaraki-Shi
Osaka 567
Japan

Gunther Schlingloff
Ciba Specialities
Köchlinstr. 1
79630 Grenzach-Wyhlen
Germany

Hans-Günther Schmalz
Institut für Organische Chemie
Universität Köln
Greinstraße 4
50939 Köln
Germany

Axel M. Schmidt
FB Chemie, Org. Chemie I
der Universität Dortmund
Otto-Hahn-Straße 6
44227 Dortmund
Germany

Jaysree Seayad
Institut für Organische
Katalyseforschung
an der Universität Rostock e.V.
Buchbinderstraße 5–6
18055 Rostock
Germany

K. Barry Sharpless
Department of Chemistry
The Scripps Research Insitute
10666 North Torrey Pines Road
La Jolla, CA 92037
USA

Roger A. Sheldon
Laboratory of Organic Chemistry
and Catalysis
Delft University of Technology
Julianalaan 136
2628 BL Delft
The Netherlands

M. Shibasaki
Graduate School
of Pharmaceutical Sciences
University of Tokyo
7-3-1 Hongo, Bunkyo-ku
Tokyo 113-0033
Japan

Georgiy B. Shul'pin
Semenov Institute
of Chemical Physics
Russian Academy of Sciences
ul. Kosygina, dom 4,
Moscow 119991
Russia

Denis Sinou
Universite Claude-Bernard Lyon 1
Lab. de Synthese Asymmetrique
U.M.R.
U.C.B.L. CNRS 5622
C.P.E. Lyon – Batiment 1918
69622 Villeurbanne Cedex
France

Barry B. Snider
Department of Chemistry
Brandeis University
415 South Street, MS 015
Waltham, MA 02454-9110
USA

Felix Spindler
Solvias AG
Catalysis Research, WRO-1055.6
P.O. Box
4002 Basel
Switzerland

Heinz Steiner
Solvias AG
Catalysis Research, WRO-1055.628
P.O. Box
4002 Basel
Switzerland

Dirk Strübing
Institut für Organische
Katalyseforschung
an der Universität Rostock e.V.
Buchbinderstraße 5–6
18055 Rostock
Germany

Martin Studer
Solvias AG
Catalysis Research, WRO-1055.6.17
P.O. Box
4002 Basel
Switzerland

Kenneth S. Suslick
School of Chemical Sciences
University of Illinois
at Urbana-Champaign
Chemical & Life Sciences Laboratory
600 S. Mathews Av.
Urbana, Illinois 61801
USA

Kazuhide Tani
Department of Chemistry
Osaka University
Toyonaka
Osaka 560-8531
Japan

Daniella Banon-Tenne
Department of Chemistry Technion
Israel Institute of
Technology, Technion City
Haifa 32000
Israel

Oliver R. Thiel
Amgen Inc.
Small Molecule Process Research
One Amgen Center Drive, Thousand
Oaks California 91320
USA

Werner R. Thiel
Institut für Anorganische Chemie
Technische Universität Chemnitz
Straße der Nationen 62
09107 Chemnitz
Germany

Annegret Tillack
Institut für Organische
Katalyseforschung
an der Universität Rostock e.V.
Buchbinderstraße 5–6
18055 Rostock
Germany

Masao Tsukazaki
Laboratoire de Chimie Organique
Place Louis Pasteur, 1
1348 Louvain-la-Neuve
Belgium

Barry M. Trost
Chemistry Department
Stanford University
Stanford, CA 94305-5080
USA

Christopher J. Urch
Zeneca Agrochemicals
Jealott's Hill Research Station
Bracknell
Berkshire RG42 6ET
United Kingdom

Axel Jacobi von Wangelin
Institut für Organische
Katalyseforschung
an der Universität Rostock e.V.
Buchbinderstraße 5–6
18055 Rostock
Germany

Peter Wasserscheid
Institut für Technische Chemie und
Makromolekulare Chemie
Bereich Technische Chemie
Worringerweg 1
52074 Aachen
Germany

Thomas Weskamp
Anorganisch-Chemisches Institut
TU München
Lichtenbergstr. 4
85747 München
Germany

Keiji Yamamoto
Kyoto University
Department of Chemistry
Faculty of Science
Sakyo, Kyoto 606-01
Japan

Naoki Yoshikawa
Graduate School
of Pharmaceutical Sciences
University of Tokyo
7-3-1 Hongo, Bunkyo-ku
Tokyo 113-0033
Japan

Alexander Zapf
Institut für Organische
Katalyseforschung
an der Universität Rostock e.V.
Buchbinderstraße 5–6
18055 Rostock
Germany

1

Reductions

Transition Metals for Organic Synthesis, Vol. 2, 2nd Edition.
Edited by M. Beller and C. Bolm

ISBN: 3-527-30613-7

1.1 Homogeneous Hydrogenations

1.1.1 Olefin Hydrogenations

Armin Börner and Jens Holz

1.1.1.1 Various Applications

The addition of hydrogen to olefins occupies an important position in transition metal-mediated transformations [1]. Historically, the field has been dominated by heterogeneous catalysts for a considerable time [2]. However, for the past few decades, soluble metal complexes have also emerged as indispensable tools in laboratory-scale synthesis as well as in the manufacturing of fine chemicals. Homogeneous hydrogenation catalysts offer distinct advantages, such as superior chemo-, regio- and stereoselectivity compared with their heterogeneous counterparts.

A multitude of transition metal complexes (including organolanthanides and organoactinides) are known to reduce olefins, which are generally more readily hydrogenated than any other functional groups with the exception of triple bonds [3]. In particular, metals of subgroup VIII of the periodical table have seen broad application. Unfortunately, scaled-up application is often hampered by considerable costs and fluctuation of the metal prices on the world market. Appropriate ligands associated with the metal, which are capable of retrodative π-bonding, such as phosphines, cyanide, carbonyl, or cyclopentadienyl (Cp), facilitate the activation of molecular hydrogen and stabilize catalytically active metal hydrides. Prominent and widely applied soluble catalysts are the Ziegler-type systems, which are carbonyl complexes, e.g., $(Cp)_2Ti(CO)_2$, $(arene)Cr(CO)_3$, $Co_2(CO)_8$, $Fe(CO)_5$, and the water-soluble $[Co(CN)_5]^{2-}$.

One of the most versatile metal catalysts for double-bond saturation in the homogeneous phase is $RhCl(PPh_3)_3$ [4] (commonly referred to as Wilkinson complex) and its ruthenium(II) analog $RuCl_2(PPh_3)_3$. The cationic iridium(I) complex $[Ir(COD)(PCy_3)(py)]PF_6$ (COD = *cis,cis*-cycloocta-1,5-diene) discovered by Crabtree is similarly useful, but is less selective in the hydrogenation of polyolefins [5]. Related

Transition Metals for Organic Synthesis, Vol. 2, 2nd Edition.
Edited by M. Beller and C. Bolm

ISBN: 3-527-30613-7

ruthenium or rhodium catalysts, based on chelating chiral diphosphines, can reliably discriminate between diastereo- or enantiotopic faces of functionalized olefins [6]. Recently, even rhodium(I) catalysts based on monodentate *P*-ligands have been shown to be highly enantioselective [7]. Rhodium(I) phosphine complexes are commonly applied as cationic complexes or generated *in situ* by mixing ligand and transition metal complex in the hydrogenation solvent [8]. For the stabilization of these complexes, diolefins (COD, NBD = norborna-2,5-diene) are coordinated (precatalyst). To obtain the benefit of the whole amount of the catalyst, sufficient time has to be allowed to generate the catalytically active species from the precatalyst [9]. This holds true particularly when fast reacting substrates are subjected to hydrogenation.

In general, olefin hydrogenation can easily be carried out with Wilkinson-type catalysts. The reaction proceeds with good rates under mild conditions. In many cases, atmospheric H_2 pressure is sufficient. Since aromatic nuclei are inert, the reaction can be performed in aromatic solvents. However, it should be noted that an aromatic solvent may coordinate to the metal, inhibiting the catalytic reaction [10]. Highly useful for hydrogenations are alcohols, THF, and acetone. The catalysts are, with the exception of carbonyl groups (decarbonylation of aldehydes!), compatible with a variety of functional groups (hydroxy, ester, carboxy, azo, ether, nitro, chloro). Even olefins bearing sulfur groups (e.g., thiophene) which generally poison heterogeneous catalysts can be reduced cleanly at a pressure of 3–4 atm [11].

1.1.1.1.1 **Hydrogenation of Mono- and Polyolefins**

Monoolefins are preferably hydrogenated with heterogeneous catalysts if there are no special requirements respecting regio- and stereoselectivity. However, soluble catalysts, e.g., the Vaska complex *trans*-$[Ir(PPh_3)_2(CO)Cl]$ [12] or chloroplatinic acid in the presence of stannous chloride [13], are also capable of transforming simple olefins (e.g., ethylene, propylene, but-1-ene, hex-1-ene, fumaric acid) into alkanes.

Aromatic compounds (benzene, naphthalene, phenol, xylenes) can be efficiently converted into fully saturated cycloalkanes with the remarkably active Ziegler catalyst $Et_3Al/Ni(2\text{-ethylhexanoate})_2$ at elevated temperatures (150–210 °C) and moderate pressures (about 75 atm) [14]. Particularly attractive is the synthesis of cyclohexane by reduction of benzene. This method is used for the large-scale production of adipic acid, a major intermediate in the production of nylon. In the so-called IFP process (Institut Français du Pétrole), a Ziegler system based on triethylaluminum, nickel and cobalt salts is employed for the hydrogenation [15]. The volatility of cyclohexane facilitates the separation of the product from the homogeneous catalyst. Because of its efficiency, this approach is likely to replace the conventional process with Raney nickel.

The hydrogenation of a conjugated double bond can occur either by 1,2- or 1,4-addition of the first molecule of hydrogen. The 1,4-product may also be formed by 1,2-addition followed by isomerization. Regioselective saturation of conjugated polyolefins is a domain of arene chromium tricarbonyl complexes (arene, e.g., benzene, methylbenzoate, toluene). In general, these complexes are air-stable. Monoolefins are not reduced. By this method, a variety of important acyclic and cyclic monoolefins such as hex-2-ene, cyclohexene, and cyclooctene are available starting from the appropriate polyolefins [16]. Cyclooctene provides an essential feedstock for the production of 1,9-decadiene in the Shell FEAST (Further Exploitation of Advanced Shell Technology) metathesis process with ethene [17]. Self-metathesis of cyclooctene in a ring-opening polymerization gives an elastomer known as *trans*-polyoctenamer produced on a multi-ton scale by Degussa (Vestenamer®) for application in blends with other rubbers [18].

The ability of Cr catalysts to reduce 1,3-dienes via 1,4-addition to *cis*-monoolefins is interesting (e.g., synthesis of *cis*-pent-2-ene from penta-1,3-diene) [16]. In the presence of $CpCrH(CO)_3$, isoprene can be converted into 2-methylbut-2-ene in excellent selectivity by reaction in benzene [19]. In general, for Cr catalysts, elevated temperatures (40–200 °C) and H_2-pressure (30–100 atm) are required to reach completion.

The rate of addition of hydrogen to olefins, which occurs in a strictly *cis* manner [20], depends upon the steric bulk of the groups surrounding the double bond. In polyenes the less-hindered double bond is always reduced best. Conjugated double bonds react more slowly than terminal olefins. The *cis* configuration facilitates the hydrogenation in comparison to the *trans* arrangement. Generally, tri- and tetrasubstituted olefins react only under more severe conditions.

A critical step in Merck's semi-synthetic approach to the broad-spectrum antiparasitic agent ivermectin is the selective hydrogenation of the bacterial metabolite avermectin B_{1a} (Fig. 1, compound **1**) [21]. By means of the Wilkinson com-

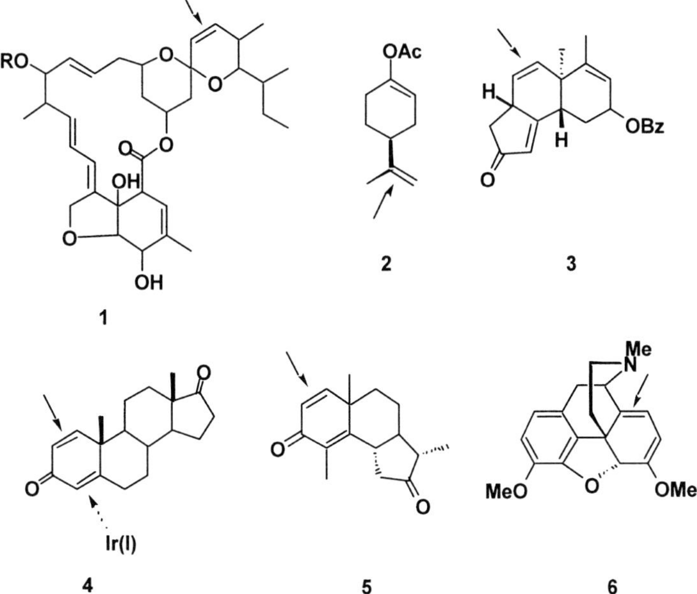

Fig. 1 Regioselective hydrogenation of polyolefins with Wilkinson-type complexes (arrows indicate where reaction takes place).

plex, the C-22/C-23 double bond of the precursor is reduced regioselectively in toluene at 25 °C and 1 atm H_2, producing a yield of 85%. The macrolide antibiotic, originally developed for application in veterinary medicine, is now successfully administered in the treatment of onchocerciasis ("African river blindness"), a disease afflicting several million people every year in Africa and Central America.

On the way to the natural product noroxopenlanfuran, isolated from the marine sponge *Dysidea fragilis* (which is native to the North Brittany Sea), the aim is the regioselective reduction of an exocyclic double bond [22]. By treatment of diolefin **2** with hydrogen in the presence of the Wilkinson complex the isopropylidene group is selectively hydrogenated. Another useful application of the same catalyst concerns the regioselective reduction of the triolefin **3** [23]. The desired diolefin can be obtained in a yield of 92%. After various subsequent steps, a diterpenoid of the cyathin family results, in nature produced by bird's nest fungi of the genus *Cyathus*.

Selectivity in enone hydrogenation has been copiously exemplified in steroidal chemistry, and only a glimpse of this large area can be given here. Thus, with $[Ir(COD)(PCy_3)(py)]PF_6$, both double bonds of androsta-1,4-diene-3,17-dione (**4**) are affected, affording androstane-1,4-dione, whereas with the less reactive Wilkinson complex the sterically more crowded double bond resists hydrogenation [24]. Regioselective hydrogenation of a fused cyclohexa-1,4-diene-3-one ring is also the aim in the hydrogenation of α-santonin (**5**) to 1,2-dihydro-α-santonin [25]. Applying the Wilkinson complex, this crucial intermediate in the synthesis of (+)-arbusculin B, representing a sesquiterpene of potential biological activity, can be obtained fairly quantitatively.

Heteroatom substituents may advantageously impede attack of hydrogen on an adjacent olefin. As an example, reference is made to the hydrogenation of the opium alkaloid thebaine (**6**) to 8,14-dihydrothebaine with the Wilkinson complex in benzene [26]. Although both olefinic bonds are trisubstituted, the methoxy substituent is more inhibitory than the alkylene group.

1.1.1.1.2 **Diastereoselective Hydrogenation**

In cyclic and acyclic systems, several functionalities, e.g., alcoholate, hydroxy, ether, carboxy, or amide groups, if properly situated, chelate onto the catalytically active metal and can thus direct hydrogenations, providing a high degree of selectivity (*anchor effect*) [27]. Results obtained with soluble transition metals often contrast advantageously with those of heterogeneous catalysts, which invariably lead to mixtures containing appreciable quantities of undesired isomers [28].

Diastereoselective hydrogenation of trisubstituted homoallyl alcohols is of considerable importance in the synthesis of structural features of polyether and macrolide antibiotics. The preparation of the C10-C19 fragment in the Merck total synthesis of FK-506, an immunosuppressant isolated from *Streptomyces tsukubaensis*, involves two consecutive hydrogenations of a galactopyranoside-derived precursor catalyzed with $[Rh(NBD)(1,4\text{-bis(diphenylphosphino)butane})]BF_4$, affording the saturated polyether in a yield of 90% (Tab. 1, Entry 1) [29]. Hydroxy group-direc-

Tab. 1 Functional group-directed hydrogenation

Entry	*Substrate*	*Product*
1	MeOOC, OH, OBz, MeO MeO	MeOOC, OH, OBz, MeO MeO
2	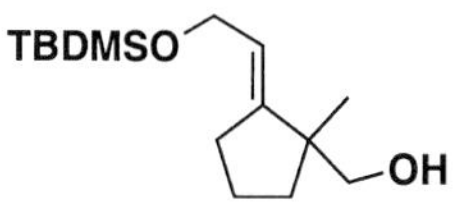 TBDMSO, OH	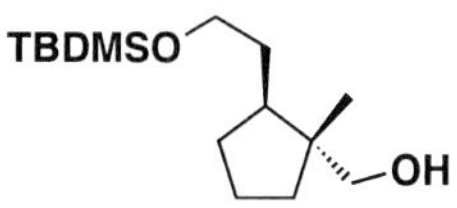 TBDMSO, OH

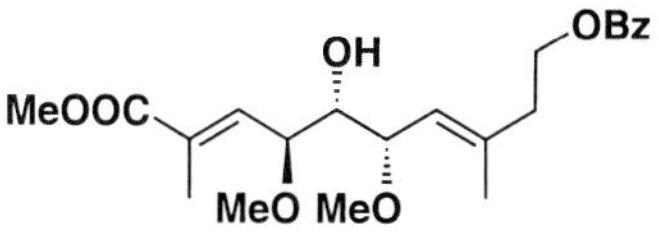

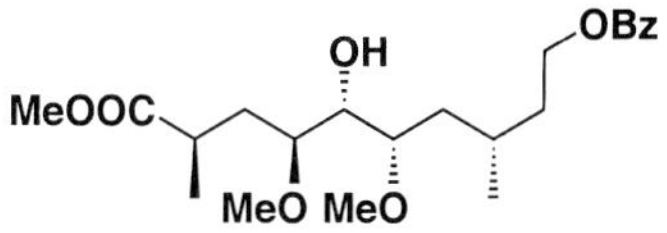

ted hydrogenation of a functionalized alkylidene cyclopentane (Entry 2) with Crabtree's complex $[Ir(COD)(PCy_3)(py)]PF_6$ furnished the desired epimer with excellent yield and selectivity (93%, diastereomeric ratio >99:1) [30]. The trisubstituted cyclopentane, which can be prepared in a medium-scale approach, is an important building block (C-ring) in the total synthesis of ophiobolane sesquiterpenes, which have been isolated from phytopathogenic fungi and protective wax secreted by scale insects, respectively.

1.1.1.1.3 **Asymmetric Hydrogenation**

Enantiomerically pure compounds show a rapidly growing potential in the pharmaceutical, agrochemical, and cosmetics industries because in several applications only one of the enantiomers exhibits the desired biological activity (*eutomer* [31]), while the optical antipode is inactive or may even cause the reverse effect (*diastomer, isomeric ballast* [31]) [32]. In general, for commercial use, catalysts inducing a selectivity exceeding 90% *ee* are desired. Sometimes, the optical purity of enantiomerically enriched hydrogenation products may be enhanced by consecutive crystallizations.

Among the vast number of chiral catalysts, Rh(I) and Ru(II) diphosphine complexes have been revealed to be the most efficient for asymmetric reduction of functionalized olefins. In particular, ruthenium(II) catalysts based on the atropisomeric ligands (*R*)- and (*S*)-BINAP (Fig. 2, ligand **7**) discovered by Noyori and Takaya play a pivotal role in asymmetric scale-up hydrogenations [33]. Enzyme-like enantioselectivities matching the requirements of natural product synthesis were also reported with Rh(I) complexes based on (*R*)-BICHEP (**8**) [34], (*S*,*S*)-BDPP (**9**) [35], and (*S*,*S*)-DuPHOS (**10**) [36]. By electronic and steric "tuning" of chiral parent diphosphines such as Kagan's DIOP or Achiwa's BPPM, the ligands (*R*,*R*)-MOD-DIOP (**11**) [37] and (–)-phenyl-CAPP (**12**) [38] resulted, which show similarly superior enantioface discriminating abilities in the hydrogenation of the olefins considered here. With the aminoalkyl-substituted ferrocenyldiphosphine **13**

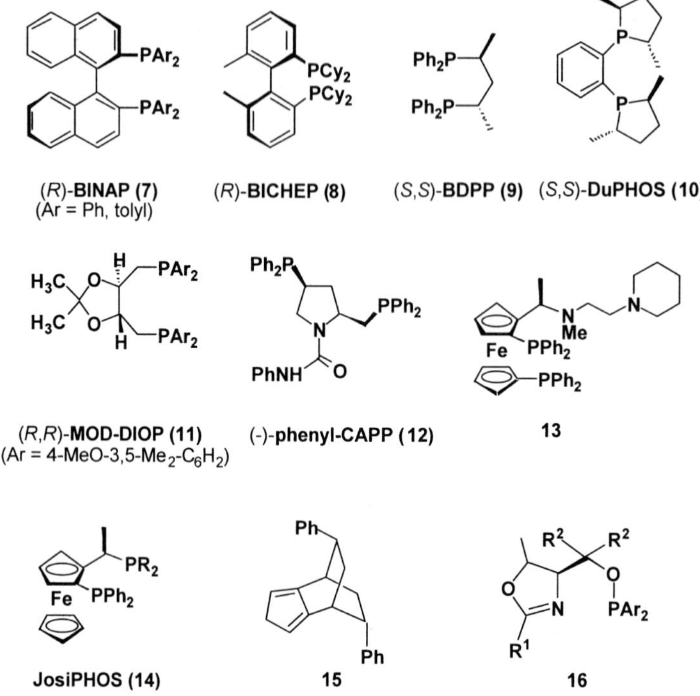

Fig. 2 Chiral ligands utilized as ligands for efficient asymmetric olefin hydrogenation.

associated with Rh(I), even tri- and tetra-substituted acrylic acids are stereoselectively hydrogenated at 50 atm, benefiting from an attractive interaction build-up between ligand and substrate [39].

A high level of selectivity (> 95% *ee*) in the reaction with unfunctionalized di- and tri-substituted prochiral olefins can be achieved with chiral titanocene complexes based on cyclopentadienyl ligands such as **15** [40]. Related catalysts such as chiral homogeneous Ziegler-Natta systems [41] and organolanthanide complexes [42] also effect the asymmetric reduction of unfunctionalized olefins with good stereoselectivity. Ir catalysts bearing a PHOX-ligand of type **16** can induce excellent enantioselectivities in the hydrogenation of nonfunctionalized olefins [43]. Full conversions were obtained with catalyst loadings as low as 0.02 mol%.

With these catalysts available, there is a great potential to displace cumbersome optical resolutions still being operated on a large scale in the production of fine chemicals. Unfortunately, only a few of the ligands, generally synthesized via multi-step sequences, have been commercialized. Among the so-called "privileged ligands", which means commercially available ligands with high stereodiscriminating abilities for a range of different metal catalyzed reactions, BINAP (**7**), DuPHOS (**10**) and JosiPHOS (**14**) [44] attract the most attention.

Ru(II)-BINAP-catalyzed hydrogenation of a wide range of α/β-unsaturated carboxylic acids (esters are poor substrates) proceeds with excellent selectivity [45].

Depending on the substitution pattern of the substrate, an appropriate H_2 pressure has to be applied in order to achieve high enantioface selection [46]. Thus, tiglic acid, geranic acid, and atropic acid can be converted to chiral saturated carbonic acids at 4, 101 and 112 atm, respectively. The optical purity of the products ranging from 87 to 92% *ee* can be increased by recrystallization of the corresponding salts. The chiral acids find widespread application as building blocks in organic synthesis.

R³ COOH / R² R¹ — 4-112 atm H_2, Ru(II)-BINAP, MeOH, 15-30 °C, 12-24 h → R³ COOH / R² R¹ (*) (1)

85 - 99 % ee

R^1 = alkyl, aryl, hydroxycarbonylalkyl
R^2 = H, hydroxyalkyl
R^3 = H, alkyl

There is considerable latitude in the choice of substituents. For example ω-hydroxyalkyl-2-en carboxylic acids are reduced in 93–95% *ee* to methyl-substituted γ- and δ-lactones, which are important intermediates in the synthesis of natural products.

The nonsteroidal anti-inflammatory compound (*S*)-2-(6′-methoxynaphth-2′-yl)-propanoic acid commercialized as Naproxen® is one of the largest-selling prescription drugs. There is a strict need for the selective production of the enantiopure (*S*)-isomer, because the (*R*)-enantiomer is a liver toxin. In general, the agent is produced by kinetic resolution. Alternatively, the Ru[(*S*)-BINAP]Cl_2-catalyzed hydrogenation process may become another practical route, since the patent concerned expired in 1993. After reduction of α-naphthylacrylic acid, available by a two-step synthesis including an electrochemical reduction of acetylnaphthalene with CO_2, Naproxen® is obtained in more than 92% yield and in 97% *ee* [45]. The same hydrogenation protocol has also been used as a key step in the preparation of a core unit of HIV protease.

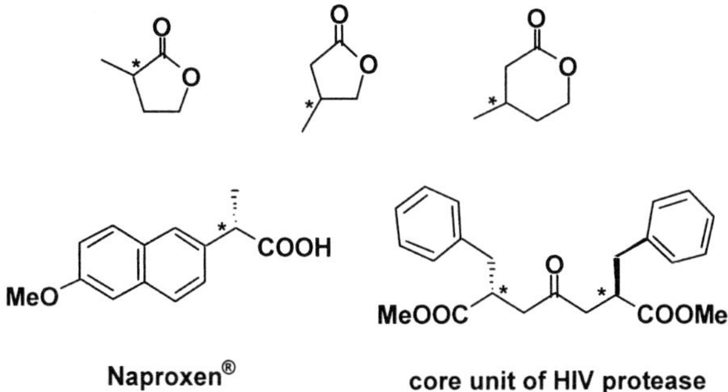

Fig. 3 Useful products by hydrogenation with Ru(II)-BINAP complexes (the asterisk indicates the newly created asymmetric carbon atom).

The rhodium(I) complex of MOD-DIOP is a competent catalyst for the reduction of alkylidene succinic acids [37]. The products are applied for the assemblage of naturally occurring or modified cytotoxic lignans (podophyllotoxin) and serve as precursors to clinical antitumor reagents (etoposide, teniposide). For example, treatment of *α*-piperonylidene succinic acid half-ester with hydrogen in the presence of a precatalyst prepared *in situ* by mixing (*S*,*S*)-MOD-DIOP with $[Rh(COD)Cl]_2$ furnished the product in 93% *ee*. A single crystallization step afforded the enantiopure (*R*)-configurated piperonyl-succinic acid half-ester.

COOMe COOH —H_2, $[Rh(COD)Cl]_2$, (*S*,*S*)-MOD-DIOP→ COOMe COOH (2)

A technical process for the large-scale manufacture of the fragrance (+)-*cis*-methyl dihydrojasmonate is based on the reduction of the relevant tetrasubstituted cyclopentene substrate [47]. JosiPHOS and DuPHOS coordinated to a newly developed Ru precursor gives the best performance in this hydrogenation, which proceeds with a substrate/catalyst ratio of 2000.

COOMe O —H_2, [HRu(P-P*)]→ COOMe O (3)

P-P*: DuPHOS or JosiPHOS

Similarly, the regio- and enantioselective hydrogenation of substituted allylic alcohols with Ru(II)-BINAP at 30 atm initial hydrogen pressure proceeds effectively, giving rise to chiral terpene alcohols [48]. The products are widely used as fragrances in perfume design and production. Using Ru(II)-(*S*)-BINAP, geraniol is reduced to (natural) (*R*)-citronellol, which is a rose scent component, with up to 99% *ee*. The C6-C7 double bond is not attacked under these conditions. Noteworthy is the extremely high substrate/catalyst mole ratio of 50000 applied. In addition, the catalyst is easily recovered by distillation of the product. It can be used for further runs without loss of efficiency.

OH —H_2, $Ru[(S)\text{-BINAP}](OCOCF_3)_2$→ OH (4)

geraniol 99 % ee **(*R*)-citronellol**

The enantioselectivity of the hydrogenation is dependent upon the reaction pressure. Under reduced pressure (low hydrogenation rate), the isomerization of geraniol to *γ*-geraniol comes into play as a serious competing reaction [49]. Unfortunately, *γ*-geraniol is hydrogenated with Ru(II)-(*S*)-BINAP to (*S*)-citronellol. Therefore, depending upon the degree of isomerization, a loss of enantioselectivity is observed. The pressure effect may be masked by insufficient mixing of the reaction solution [50]. As a result, the diffusion of hydrogen becomes the rate-limiting step, and preequilibria responsible for high enantioselection are disturbed.

Unnatural citronellol can be produced by reduction of geraniol with the (*R*)-BINAP complex [48]. The isomeric allylic alcohol (nerol) can be equally utilized as substrate. A similar hydrogenation protocol was followed in the synthesis of (3*R*,7*R*)-3,7,11-trimethyldodecanol, representing a key intermediate in the production of vitamin E (α-tocopherol) and vitamin K_1.

The diastereoselective hydrogenation of an allylic alcohol linked to a chiral azetidinone with Ru[(*S*)-(tolBINAP)](OAc)$_2$ under atmospheric pressure has been suggested for the creation of a new class of carbapenem antibiotics which exhibit enhanced metabolic and chemical stability in comparison to related antibiotics such as thienamycin [51].

tert-BuMe$_2$SiO ... H ... OH ... N H ... O —— H_2, Ru[(*R*)-tolBINAP)](OAc)$_2$ ——→ *tert*-BuMe$_2$SiO ... H ... OH ... N H ... O (5)

99.9 %

Asymmetric hydrogenation of racemic allylic alcohols with Ru(II)-BINAP complexes affords a high level of kinetic enantiomer selection [52]. Using this method, (*R*)-4-hydroxy-2-cyclopentenone can be produced by treatment of the racemic mixture. The reaction can be carried out in a multi-kilogram scale and is used in the industrial three-component prostaglandin synthesis.

O ... HO (±) —— H_2, Ru(II)-(*S*)-BINAP / 68 % conversion ——→ O ... HO + O ... HO (6)

98 % ee

The efficiency of Ru(II) complexes based on BINAP and related atropisomeric ligands [53] was also shown in the synthesis of a variety of naturally ubiquitous isoquinoline alkaloids by reduction of (*Z*)-2-acyl-1-benzylidene-1,2,3,4-tetrahydroisoquinolines [54].

(RO)$_m$ ... NCOR ... (OR)$_n$ —— H_2, Ru(II)-BINAP ——→ (RO)$_m$ ... NCOR * ... (OR)$_n$ (7)

95 - 100 % ee

This procedure can be applied for the stereoselective synthesis of naturally occurring and artificial alkaloids based on the morphinane skeleton [55]. Several compounds of this class exhibit important analgesic effects (morphine) or bronchodilating activity (dextromethorphan).

References

1 P.A. Chaloner, M.A. Esteruelas, F. Joó, L.A. Oro, *Homogeneous Hydrogenation*, Kluwer Academic Publishers, Dordrecht, **1994**, p. 119.

2 P.N. Rylander, *Catalytic Hydrogenation in Organic Syntheses*, Academic Press, New York, **1979**, p. 31. S. Nishimura, *Handbook of Heterogeneous Catalytic Hydrogenation for Organic Synthesis*, Wiley, New York, **2001**, p. 64.

3 B.R. James, *Homogeneous Hydrogenation*, John Wiley & Sons, New York, **1973**. H. Pracejus, *Koordinationschemische Katalyse organischer Reaktionen*, Theodor Steinkopff, Dresden, **1977**, Chapter 2, A.F. Noels, A. J. Hubert in *Industrial Applications of Homogeneous Catalysis* (Eds.: A. Mortreux, F. Petit), D. Reidel Publishing Company, Dordrecht, **1988**, p. 65.

4 J.A. Osborn, F.H. Jardine, J.F. Young, G. Wilkinson, *J. Chem. Soc.* (A) **1966**, 1711.

5 R.H. Crabtree, H. Felkin, T. Fillebeen-Khan, G.E. Morris, *J. Organomet. Chem.* **1979**, *168*, 183.

6 H. Takaya, T. Ohta, R. Noyori in *Catalytic Asymmetric Synthesis* (Ed.: I. Ojima), VCH, Weinheim, **1993**, p. 1.

7 I. Komarov, A. Börner, *Angew. Chem.* **2001**, *113*, 1237; *Angew. Chem. Int. Ed.* **2001**, *40*, 1197.

8 H. Brunner in *Applied Homogeneous Catalysis with Organometallic Compounds* (Eds.: B. Cornils, W.A. Herrmann), VCH, Weinheim, **1996**, Vol. 1, p. 209.

9 D. Heller, J. Holz, S. Borns, A. Spannenberg, R. Kempe, U. Schmidt, A. Börner, *Tetrahedron: Asymmetry* **1997**, *8*, 213. A. Börner, D. Heller, *Tetrahedron Lett.* **2001**, *42*, 233.

10 D. Heller, H.-J. Drexler, A. Spannenberg, B. Heller, J. You, W. Baumann, *Angew. Chem.* **2002**, *114*, 814; *Angew. Chem. Int. Ed.* **2002**, *41*, 777.

11 P.D. Clark, N.M. Irvine, P. Sarkar, *Can. J. Chem.* **1991**, *69*, 1011.

12 L. Vaska, J.W. DiLuzio, *J. Am. Chem. Soc.* **1961**, *83*, 2784.

13 R.D. Cramer, E.L. Jenner, R.V. Lindsey Jr., U.G. Stolberg, *J. Am. Chem. Soc.* **1963**, *85*, 1691.

14 S.J. Lapporte, W.R. Schuett, *J. Org. Chem.* **1963**, *28*, 1947.

15 G.W. Parshall, S.D. Ittel, *Homogeneous Catalysis*, 2nd edn., John Wiley, New York, **1992**, p. 180.

16 E.N. Frankel, *J. Org. Chem.* **1972**, *37*, 1549 and references therein.

17 P. Chaumont, C.S. John, *J. Mol. Catal.* **1988**, *46*, 317.

18 G.W. Parshall, W.A. Nugent, *Chem-Tech* **1988**, 314.

19 A. Miyake, H. Kondo, *Angew. Chem.* **1968**, *80*, 663; *Angew. Chem., Int. Ed. Engl.* **1968**, *7*, 631.

20 F.J. McQuillin, *Homogeneous Hydrogenation in Organic Chemistry*, D. Reidel Publishing Company, Dordrecht, **1976**, p. 22.

21 G.W. Parshall, W.A. Nugent, *Chem-Tech* **1988**, 184.

22 M. Kato, M. Watanabe, B. Vogler, Y. Tooyama, A. Yoshikoshi, *J. Chem. Soc., Chem. Commun.* **1990**, 1706.

23 D.E. Ward, *Can. J. Chem.* **1987**, *65*, 2380.

24 J.W. Suggs, S.D. Cox, R.H. Crabtree, J. M. Quirk, *Tetrahedron Lett.* **1981**, *22*, 303.

25 A.E. Greene, J.-C. Muller, G. Ourisson, *J. Org. Chem.* **1974**, *39*, 186.

26 A.J. Birch, K.A.M. Walker, *J. Chem. Soc.* [C] **1966**, 1894.

27 J.M. Brown, *Angew. Chem.* **1987**, *99*, 169; *Angew. Chem., Int. Ed. Engl.* **1987**, *26*, 190. A.H. Hoveyda, D.A. Evans, G.C. Fu, *Chem. Rev.* **1993**, *93*, 1307.

28 P. N. Rylander, *Catalytic Hydrogenation over Platinum Metals*, Academic Press, New York, **1967**.

29 A. Villalobos, S. J. Danishefsky, *J. Org. Chem.* **1990**, *55*, 2776.

30 W. G. Dauben, A. M. Warshawsky, *J. Org. Chem.* **1990**, *55*, 3075.

31 E. J. Ariens, *Med. Res. Rev.* **1986**, *6*, 451. E. J. Ariens in *Metabolism of Xenobiotics*, (Eds.: J. W. Gorrod, H. Oelschläger, J. Caldwell), Tayler & Francis, London, **1988**, p. 325.

32 I. W. Wainer, D. E. Drayer, *Drug Stereochemistry*, Marcel Dekker Inc., New York, **1988**. A. N. Collins, G. N. Sheldrake, J. Crosby, *Chirality in Industry*, John Wiley & Sons, Chichester, **1992**. J. S. Millership, A. Fitzpatrick, *Chirality* **1993**, *5*, 573.

33 R. Noyori, *Acta Chem. Scand.* **1996**, *50*, 380.

34 T. Chiba, A. Miyashita, H. Nohira, H. Takaya, *Tetrahedron Lett.* **1991**, *32*, 4745.

35 P. Bissel, R. Sablong, J.-P. Lepoittevin, *Tetrahedron: Asymmetry* **1995**, *6*, 835.

36 M. J. Burk, *J. Am. Chem. Soc.* **1991**, *113*, 8518.

37 T. Morimoto, M. Chiba, K. Achiwa, *Tetrahedron Lett.* **1990**, *31*, 261.

38 H. Jendralla, *Tetrahedron Lett.* **1991**, *32*, 3671.

39 T. Hayashi, N. Kawamura, Y. Ito, *J. Am. Chem. Soc.* **1987**, *109*, 7876.

40 R. L. Halterman, K. P. C. Vollhardt, M. E. Welker, D. Bläser, R. Boese, *J. Am. Chem. Soc.* **1987**, *109*, 8105. R. L. Halterman, K. P. C. Vollhardt, *Organometallics* **1988**, *7*, 883. R. D. Broene, S. L. Buchwald, *J. Am. Chem. Soc.* **1993**, *115*, 12569. See also: L. A. Paquette, J. A. McKinney, M. L. McLaughlin, A. L. Rheingold, *Tetrahedron Lett.* **1986**, *27*, 5599.

41 R. Waymouth, P. Pino, *J. Am. Chem. Soc.* **1990**, *112*, 4911 M. V. Troutman, D. H. Appella, S. L. Buchwald, *J. Am. Chem. Soc.* **1999**, *121*, 4916.

42 V. P. Conticello, L. Brard, M. A. Giardello, Y. Tsuji, M. Sabat, C. L. Stern, T. J. Marks, *J. Am. Chem. Soc.* **1992**, *114*, 2761.

43 J. Blankenstein, A. Pfaltz, *Angew. Chem.* **2001**, *113*, 4577; *Angew. Chem. Int. Ed.* **2001**, *40*, 4445. F. Menges, A. Pfaltz, *Adv. Synth. Catal.* **2002**, *344*, 40.

44 H.-U. Blaser, W. Brieden, B. Pugin, F. Spindler, M. Studer, A. Togni, *Top. Catal.* **2002**, *19*, 3.

45 T. Ohta, H. Takaya, M. Kitamura, K. Nagai, R. Noyori, *J. Org. Chem.* **1987**, *52*, 3174.

46 R. Noyori, *Asymmetric Catalysis in Organic Synthesis*, John Wiley & Sons, New York, **1994**, p. 32.

47 D. A. Dobbs, K. P. M. Vanhessche, E. Brazi, V. Rautenstrauch, J.-Y. Lenoir, J.-P. Genêt, J. Wiles, S. H. Bergens, *Angew. Chem.* **2000**, *112*, 2080; *Angew. Chem. Int. Ed.* **2000**, *39*, 1992.

48 H. Takaya, T. Ohta, N. Sayo, H. Kumobayashi, S. Akutagawa, S.-I. Inoue, I. Kasahara, R. Noyori, *J. Am. Chem. Soc.* **1987**, *109*, 1596.

49 Y. Sun, J. Wang, C. LeBlond, R. N. Landau, J. Laquidara, J. R. Sowa Jr., D. G. Blackmond, *J. Mol. Catal. A: Chemical* **1997**, *115*, 495.

50 Y. Sun, R. N. Landau, J. Wang, C. LeBlond, D. G. Blackmond, *J. Am. Chem. Soc.* **1996**, *118*, 1348.

51 M. Kitamura, K. Nagai, Y. Hsiao, R. Noyori, *Tetrahedron Lett.* **1990**, *31*, 549.

52 M. Kitamura, I. Kasahara, K. Manabe, R. Noyori, H. Takaya, *J. Org. Chem.* **1988**, *53*, 708.

53 B. Heiser, E. A. Broger, Y. Crameri, *Tetrahedron: Asymmetry* **1991**, *2*, 51.

54 R. Noyori, M. Ohta, Y. Hsiao, M. Kitamura, T. Ohta, H. Takaya, *J. Am. Chem. Soc.* **1986**, *108*, 7117.

55 M. Kitamura, Y. Hsiao, R. Noyori, H. Takaya, *Tetrahedron Lett.* **1987**, *28*, 4829.

1.1.2 Unnatural α-Amino Acids via Asymmetric Hydrogenation of Enamides

Terry T.-L. Au-Yeung, Shu-Sun Chan, and Albert S. C. Chan

1.1.2.1 Introduction

By definition, the term "unnatural amino acids" embraces all amino acid derivatives but excludes the 22 genetically encoded α-amino acids commonly found in all living organisms [1]. Much of the use of unnatural amino acids is linked to drug discovery and synthesis in macromolecular systems such as protein engineering [2], peptidomimetics [3], or glycopeptides synthesis [4]. Sometimes, even structurally simple α-amino acids may exhibit interesting biological properties [5]. Given that the natural abundance of free unnatural α-amino acids is limited, chemical synthesis may provide a viable solution to increase their availability.

Hailed as one of the most efficient, cleanest and economical technologies, transition metal-catalyzed stereoselective hydrogenation is the ideal methodology for the synthesis of an enormous number of chiral compounds. Unnatural amino acids, too, can be obtained via hydrogenation of the respective enamides (Scheme 1). In fact, the hydrogenation of acetamidocinnamic acid has been serving as a testing platform for the evaluation of the performance of many newly designed ligands in asymmetric catalysis. Nevertheless, industrial use of the latter is still in general overshadowed by the more conventional biocatalysis and classical resolution, primarily because of the need for using relatively high catalyst loading. This section highlights a continual worldwide effort, mainly through the design and synthesis of new ligands and the study of mechanistic details, that has contributed to our understanding and the practical applications of asymmetric hydrogenation of a broad spectrum of enamides.

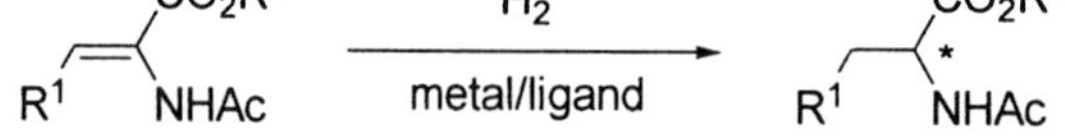

CO_2R^2 H_2 CO_2R^2
R^1 NHAc — metal/ligand → R^1 * NHAc

Scheme 1

1.1.2.2 Metals

The most popular and efficacious metal catalyst precursors used in the asymmetric hydrogenation of enamides are still rhodium(I)-based compounds in conjunction with a chiral ligand (see below). Most often used are $[Rh(diene)_2]^+X^-$, where diene = cyclooctadiene (COD) or norbornadiene (NBD) and X = non-coordinating or weakly coordinating anion such as ClO_4^-, BF_4^-, PF_6^-, OTf^-, etc. Other transition metal systems are occasionally employed, but will not be discussed here.

1.1.2.3 Ligands

Since the catalyst system is relatively invariant, an intensive search for chiral ligands has become a prevailing model for identifying efficient rhodium catalysts. Consequently, a plethora of bidentate phosphorus ligands have been synthesized in the past few decades, and a comprehensive review of these has become almost impossible within the limit of a book chapter. Although many of these ligands have been established to be excellent chiral inducers in the asymmetric hydrogenation of acetamidocinnamic acid (ACA) and acetamidoacrylic acid (AAA) or their methyl esters (MAC, MAA), information on their performance on a wide range of other substrates is lacking [6]. Thus, this chapter is by no means exhaustive. Rather, we only intend to include representatives of each class of ligands which have been demonstrated to show a relatively broad substrate scope with a respectable turnover number (TON) and turnover frequency (TOF), novel chirality features, or unusual donor properties. Apart from these criteria, some of the more important trends that have emerged in recent years warrant special attention.

1.1.2.4 Other Reaction Parameters

Solvent: The choice of solvent can sometimes have a dramatic effect on selectivity. Solvents ranging from protic or non-protic organic solvents to environmentally benign solvents such as water [7], ionic liquids [8], or even supercritical fluid [9] can be used. Their effect, however, is unpredictable, and one usually has to discover the best solvent for a particular ligand by trial and error.

Temperature: The temperature at which hydrogenation is carried out is often ambient. Usually, lower reaction temperature does not give better *ee* according to mechanistic considerations, although sometimes a reduction of temperature may be conducive to *ee* enrichment but at the cost of activity.

Pressure of H_2: For simple substrates, a low hydrogen gas pressure normally suffices. With more difficult substrates, high pressures are sometimes required.

1.1.2.5 Asymmetric Hydrogenation of Enamides

1.1.2.5.1 Diphospholane Derivatives

Burk and co-workers introduced the excellent modular ligands DuPHOS and BPE, and the corresponding rhodium complex worked highly efficiently for the stereoselective, regioselective, and chemoselective hydrogenation of functionalized C=C bonds with 95–99% *ee* at a very low catalyst loading (max. TON = 50000, TOF = 5000 h^{-1}) [10]. The ingenious design of DuPHOS has been proved to with-

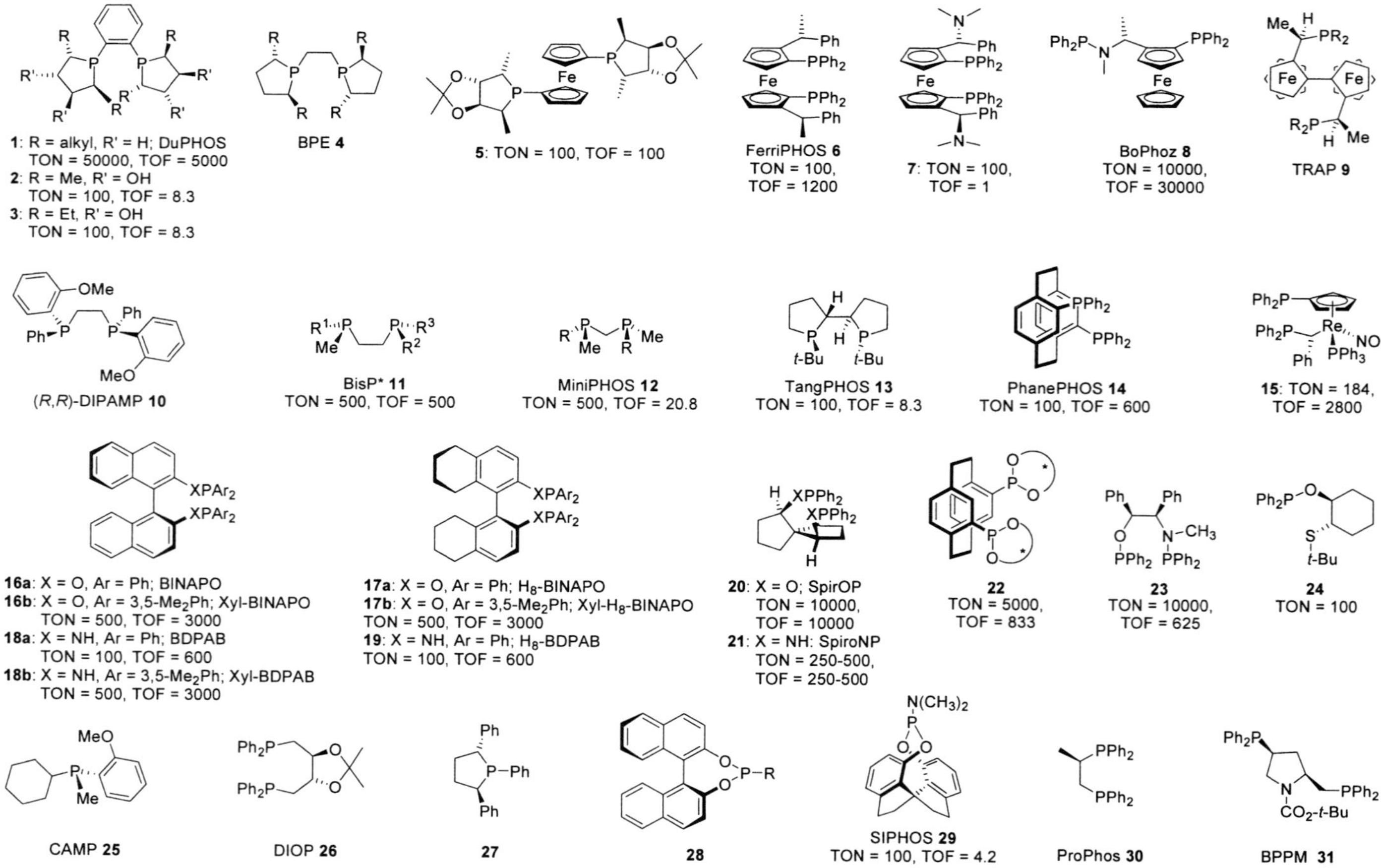

Fig. 1 Chiral phosphorus ligands for asymmetric hydrogenation.

Tab. 1 Enantioselective hydrogenation of (*Z*)-aryl or (*Z*)-alkyl-amidoacrylic acids and their methyl esters with various ligands

$$R^1\text{-CH=C(NHAc)}CO_2R^2 \xrightarrow[H_2]{L^*/Rh(I)} R^1\text{-CH}_2\text{-C}^*H(NHAc)CO_2R^2$$

Substrate		*L**																							
R^1	***R^2***	**1**	**2**	**3**	**5**	**6**	**7**	**8**	**11**	**12**	**13**	**14**	**16a**	**16b**	**17a**	**17b**	**18a**	**18b**	**19**	**20**	**21**	**22**	**23**	**24**	**28**	**29**
H	Me	>99	98	>99	99.9	98	97	99	>99.9	>99.9		99.6		94	81[e]	97	93	97	97	99	95	99			>99	97
	H		>99	>99	99.4	98		96	42	>99.9				86		90	94	98	99	>99.9	78[a]	97	95		>99	
Me	Me	>99										94												98		99
Ph	Me	>99	>99	>99	99.5	99	98	99.1	99.9	>99	99	98	64[b]	92	84	94	90	98	96	96	99	99	98	97	97	98
	H		>99	>99		97		99.4	99	98	>99		18[c]		74[f]		90	98	94	98	>98	99	97		97	
4-MePh	Me	>99												92	84	94	90	98	94	96	98		97			98
4-MeOPh	Me	>99	98	>99	98						>99			91		94		98	93	96	>99		97	98	94	96
	H		98	99							>99															
4-BrPh	Me	>99													96			98	96	96	>99		98			
3-BrPh	Me	>99		>99	99.3						>99													95		
	H			99	99						>99															
4-ClPh	Me						>99							91	81	93	88	98	94	94	>99		98		94	99
3-ClPh	Me												55[d]		78		90	97	94		>99		95			
2-ClPh	Me			98	98						>99			90	85	93	90	97	97		96		98			97
	H			98	99.3						>99						90		94	97	96					
4-FPh	Me	>99	98	>99	98		99				>99							98	93	96	>99		97	94	96	
	H		99	>99	99						>99														93	
4-NO_2Ph	Me													89		91		96	91		94		98		95	99
4-AcO-3-MeOPh	Me								99.8	95													98		96	

Tab. 1 (cont.)

Substrate		*L**																								
R^1	R^2	1	*2*	*3*	*5*	*6*	*7*	*8*	*11*	*12*	*13*	*14*	*16a*	*16b*	*17a*	*17b*	*18a*	*18b*	*19*	*20*	*21*	*22*	*23*	*24*	*28*	*29*
	H								97	95											94					
2-Naph-thyl	Me	>99		>99	99	98	96				>99															
Furyl	Me													89	64	92		98	91	97	>99		91			
2-Thienyl	Me	>99	>99	>99							>99															
	H		>99	>99							>99															

a) 15% conv.;
b) 86% conv.;
c) 71% conv.;
d) 70% conv.;
e) 49% conv.;
f) 82% conv..

stand the challenge of a variety of structurally diverse substrates (see below). In the light of this significant success, it is therefore not surprising to see the appearance of other phospholane ligands bearing a structural resemblance to DuPHOS. Several functionalized DuPHOS-type ligands have been synthesized from low-cost, commercially available D-mannitol, and many of them have shown enantioselectivity toward the standard substrates similar to that of the parent DuPHOS. Of particular interest is the presence of two free hydroxyl groups at the 3,4-positions of the phospholane units (**2**,**3** R=alkyl, R′=OH), which allows for secondary interactions between ligand and substrates [11]. These ligands hydrogenate both acid and methyl ester substrates, giving up to >99% *ee* regardless of change in electronic and steric properties in the substrate. However, the *ee* decreased gradually with the size of R beyond the size of Et. When the hydroxyl groups are masked by a ketal and the two phospholanes are scaffolded by ferrocene (**5**), the same high enantioselectivities can be obtained [12]. Unfortunately, only relatively low levels of TON and TOF have been recorded so far with these new ligands.

1.1.2.5.2 Ferrocene-based Diphosphines

As a result of its chemical robustness, modifiability, crystallizability, and highly electron-donating nature, ferrocene-bridged diphosphines have become popular targets. These compounds often possess an assortment of center and planar chirality. FERRIPHOS, a C_2-symmetrical ferrocenyl diphosphine, reduces dehydroamino acids derivatives with a remarkable activity even at low temperatures [13]. The corresponding diamino analog **7**, with the replacement of the two methyl groups by dimethylamino, afforded comparable *ee*, albeit with lower reaction activity [14]. Being readily prepared from cheap reagents involving non-pyrophoric and non-air-sensitive intermediates, BoPhoz has shown tremendous potential, as an extraordinary TOF of 30000 h^{-1} has been observed with a TON as high as 10000 under a low pressure of H_2 [15].

1.1.2.5.3 *P*-Chiral Diphosphines

Despite the commercial success of DIPAMP, leading to the first industrial production of chiral fine chemicals almost three decades ago [16], this type of diphosphines was less pursued in the ensuing twenty-year development, probably because of a shortage of sophisticated methods for preparing these compounds. However, they have made a recent comeback, thanks to the advent of much ameliorated synthetic methodologies [17]. Unaffected by the possible δ-λ conformational equilibrium in the ethylene bridge, C_2-symmetric and electron-rich *t*-BuBisP* (**11**, $R^1=R^2=$*t*-Bu, $R^3=$Me) performed admirably in the enantioselective Rh(I)-catalyzed hydrogenation of α-dehydroamino acids, with completion within 1 h [18]. Unsymmetrical BisP* also gave results comparable to those with BisP* in the hydrogenation of MAC. *t*-Bu-MiniPHOS (**12**, R=*t*-Bu), in which two stereogenic phosphorus atoms are connected by a methylene group only, forms a four-membered ring with Rh(I). This unusually highly strained metallacycle gave similar results to those with BisP*, yet with lower activity (ca. 24 h). However, it gave better enantioselectivity in the case of AAA than BisP* (99.9% *ee* vs 42% *ee*). Another conformationally rigid ligand, TangPhos, also provided almost immaculate enantioselectivity for a wide array of substrates [19]. Two common and distinctive features of these ligands are that (i) the presence of two stereochemically disparate substituents on the phosphorus, and (ii) their electron-rich character appear to be critical for attaining good results. Nonetheless, in the syntheses of BisP*, MiniPHOS, and TangPHOS, the precursors containing two identical enantiotopic or diastereotopic groups are desymmetrized by a *sec*-BuLi-(–)-sparteine complex, and the apparent shortcoming is therefore that only one enantiomeric form of the ligand is accessible.

1.1.2.5.4 Miscellaneous Diphosphines

Unlike the ferrocene-type ligands, PhanePHOS is the first effective planar chiral diphosphine ligand devoid of any other form of chirality. Its extraordinary activity permits reactions to be carried out at very low temperatures without sacrificing yield and selectivity [20]. A unique cyclopentadienyl-rhenium-based diphosphine (S_{Re},R_C)-**15** having a metal chiral center was shown to be effective, with a turnover frequency reaching 2800 h^{-1} [21].

1.1.2.5.5 Bidentate Phosphorus Ligands Containing One or More P-O or P-N Bonds

We have seen above that a handful of diphosphines are highly effective for the enantioselective hydrogenation of enamides; however, many of their syntheses are either not trivial (e.g., DuPHOS, PhanePHOS) or restrictive to the access of their antipodes (e.g., BisP*, TangPhos). Although the applications of diphosphinites, diphosphinamidites, and related ligands in asymmetric hydrogenation have been known for a long time, their full potential has not been realized until recently. The attractive attributes of these types of ligands are the ubiquity of chiral diols, diamines, and amino alcohols and the ease of operation associated with the ligand synthesis. We found that by partially hydrogenating BINOL or BINAM to H_8-BINOL or H_8-BINAM, respectively, and subsequently preparing the corresponding BINAPOs and BDPABs [22], the enantioselectivities can be much boosted in the case of **16a** vs **17a** or in the case of **18a** vs **19**. The boost in *ee* can also be induced by replacing Ph with 3,5-Me_2Ph (**16a** vs **16b**, **17a** vs **17b**, **18a** vs **18b**). In our other findings, the rigidity of SpirOP and SpiroNP also led to desirable, highly stereoselective and complementary outcomes [23]. It should be noted that a TON of 10000 and a TOF of 10000 h^{-1} with the use of SpirOP have been observed. Diphosphonate **22** (diol=(*R*)-BINOL) [24] and phosphinite-phosphinamidite **23** [25] gave consistently high levels of enantioselectivity with reasonably good turnover numbers. Finally, a sulfur-containing chelating phosphinite (**24**) was also found to effect Rh-catalyzed enantioselective hydrogenation of α-dehydroamino acids [26].

1.1.2.5.6 Chiral Monodentate Phosphorus Ligands

One of the first examples of chiral monodentate ligands, employed by Knowles et al. in the asymmetric hydrogenation of dehydroamino acids and dating back to 1968, utilized a rhodium catalyst containing monophosphane CAMP and yielded *N*-acetylphenylalanine in up to 88% *ee* [27]. However, since the inception of DIOP in 1972 [28], the development of chiral ligands has abandoned monodentate *P*-ligands in favor of bidentate phosphorus compounds. Sharing a similar fate with *P*-chirogenic ligands, a resurgence of interest in monodentate ligands has taken place recently. Fiaud (27), Pringle (**28a**, R=alkyl=phosphonites), Feringa (**28b**, R=dimethylamino=phosphoramidite) and Reetz (**28c**, R=alkoxy=phosphites) independently rediscovered the latent effectiveness of monodentate phosphorus ligands by showing their rhodium complexes to be capable of hydrogenating *α*-dehydroamino acids and their derivatives with *ee* >90% [29]. The better-performing structures **28** are all composed of the common 2,2′-dihydroxy-bi-1-naphthyl backbone. That the latter is cheap but efficient, and that variation of the R group is convenient, render these compounds attractive targets for low-cost ligand optimization. Further, the faster hydrogenation rate exhibited by the relatively less basic **28** challenges the notion that electron-rich phosphines are the *sine qua non* for achieving enhanced rate [30]. Another monodentate phosphoramidite, SIPHOS, also demonstrated similar competence [31]. Reetz et al. have taken the use of the monodentate ligands a step further by introducing a rather special and intriguing concept. They first synthesized a library of monodentate phosphonites (**28**, R=alkyl or aryl) and phosphites (**28**, R=alkoxy or aryloxy). Subsequently, they combined different pairs of monodentates with a rhodium complex in a 1:1:1 ratio to generate a high-throughput screening system. This idea of heterocombination based on a molecular self-assembly motif to produce the most efficient transition metal catalyst(s) was proved to be more effective than the analogous homocombination [32].

1.1.2.6 Cyclic Substrates

Hydrogenation of enamide substrates containing an endocyclic C=C bond furnishes heterocyclic amino acids. This reaction type was rarely investigated in the past, but sporadic reports have appeared in recent years. The successful development of this reaction is apparently fruitful, as a lot of chiral alkaloid structures with an *α*-carboxylic acid functionality become accessible. Although this process is still under development, a few examples are shown in Tab. 2 to illustrate the current status of the art. The enantioselective hydrogenation of tetrahydropyrazine **32** (entry 1), whose product is an important intermediate of the HIV protease inhibitor Indinavir, was promoted by Et-DuPHOS under forcing conditions with mediocre *ee* [33a]. In contrast, whilst under much milder conditions, PhanePHOS delivered good *ee* in a much shorter time [20]. Respectable *ee* was obtained with *i*-Bu-TRAP (**9** R=*i*-Bu) [33b] but at the expense of yield. Utilizing the DuPHOS [34] or the TRAP ligands [35], 1-aza-2-cycloalkene-2-carboxylates **33** (entry 2) could be hy-

Tab. 2 Asymmetric hydrogenation of cyclic substrates

Substrate	*Ligand* [a]	*TON*	*Time (h)*	*Temp. (°C)*	*pH_2 (atm)*	*Solvent*	*Yield (%)*	*ee (%)*	*Ref.*
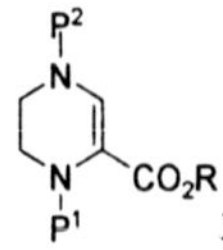 32	Phanephos	33	6	–40	1.5	MeOH	100 [d]	86	20
	Et-DuPHOS	33	18	40	70	TFE	97	50	33a
	i-BuTRAP	50	24	50	1.0	$(ClCH_2)_2$	52 [d]	92	33b
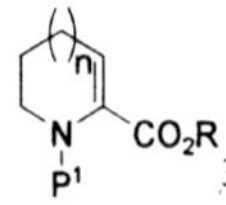 33 [b), c)]	Et-DuPHOS	17	24	rt	6.1	MeOH	84–97	0–97	34
	Ph-TRAP	100	24	50	1.0	$(ClCH_2)_2$	100 [d]	73–93	35
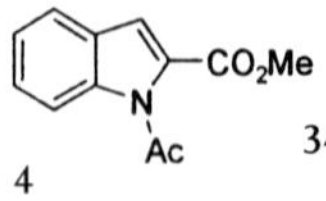 34	Ph-TRAP	100	0.5	60	1	*i*-PrOH	95	95	36

a) Rhodium catalyst precursor was used unless otherwise stated.
b) n=0–4, 8, 11 for Et-DuPHOS.
c) n=1 for *i*-PrTRAP.
d) Conversion.

drogenated with good to excellent *ees*, except for substrates where n=0 or 1 with DuPHOS. Partial dearomatization of the fused five-membered indole ring **34** (entry 3) has been accomplished with high stereoinduction via hydrogenation with Ph-TRAP in only 30 min [36].

1.1.2.7
β,*β*-Disubstituted Enamides

Whilst Me-BPE has previously been the privileged ligand for the asymmetric hydrogenation of this substrate class [37], BisP* [38] and phosphinite-thioether **24** [26] have lately made successful entries into this category. This reaction is typically slower than the hydrogenation of *β*-monosubstituted amidoacrylic esters because of the presumably poorer coordination as a consequence of an increase in steric bulk. When the two *β*-substituents are the same, only one stereocenter is obtained upon hydrogenation. Yet, when the two *β*-substituents are non-equivalent, two asymmetric centers can be created. Moreover, multifunctional *α*-amino acids are attainable when the *β*- or farther positions are substituted by heteroatoms. In this regard, PrTRAP was found to be particularly effective [39]. It is worthy of note that the hydrogen atoms are added in the *cis*-fashion, as is borne out by the stereochemistry of the products.

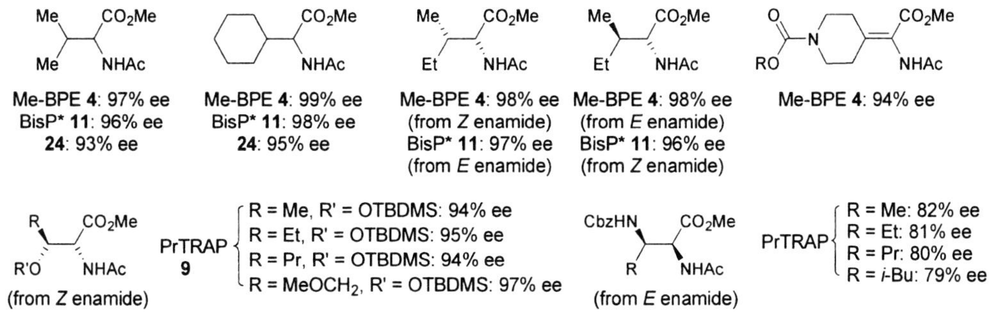

Me-BPE **4**: 97% ee
BisP* **11**: 96% ee
24: 93% ee

Me-BPE **4**: 99% ee
BisP* **11**: 98% ee
24: 95% ee

Me-BPE **4**: 98% ee (from *Z* enamide)
BisP* **11**: 97% ee (from *E* enamide)

Me-BPE **4**: 98% ee (from *E* enamide)
BisP* **11**: 96% ee (from *Z* enamide)

Me-BPE **4**: 94% ee

(from *Z* enamide)
PrTRAP **9**
R = Me, R' = OTBDMS: 94% ee
R = Et, R' = OTBDMS: 95% ee
R = Pr, R' = OTBDMS: 94% ee
R = $MeOCH_2$, R' = OTBDMS: 97% ee

(from *E* enamide)
PrTRAP
R = Me: 82% ee
R = Et: 81% ee
R = Pr: 80% ee
R = *i*-Bu: 79% ee

Fig. 2 *β,β*-Disubstituted *α*-amino acid derivatives via asymmetric hydrogenation.

1.1.2.8 Selected Applications

Aside from the synthesis of simple fine chemicals, the scope and importance of asymmetric hydrogenation is further underscored by its applications in the synthesis of valuable building blocks for the construction of complex molecules of

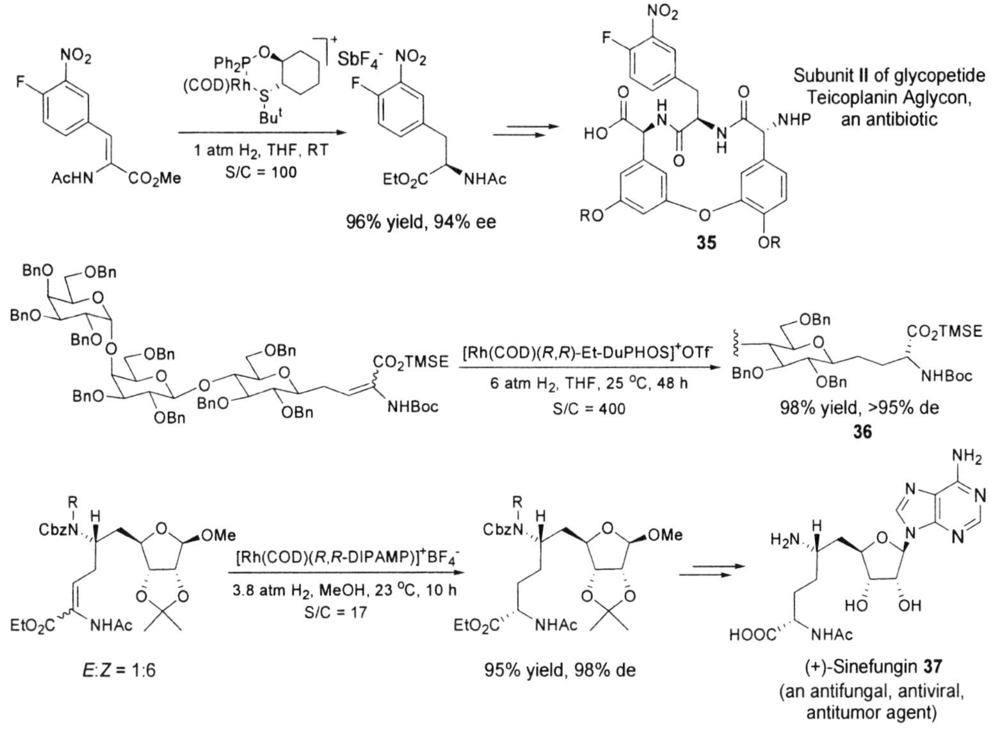

1 atm H_2, THF, RT
S/C = 100
96% yield, 94% ee
Subunit **II** of glycopetide Teicoplanin Aglycon, an antibiotic
35

[Rh(COD)(*R,R*)-Et-DuPHOS]$^+$OTf$^-$
6 atm H_2, THF, 25 °C, 48 h
S/C = 400
98% yield, >95% de
36

E:*Z* = 1:6
[Rh(COD)(*R,R*-DIPAMP)]$^+$$BF_4^-$
3.8 atm H_2, MeOH, 23 °C, 10 h
S/C = 17
95% yield, 98% de
(+)-Sinefungin **37** (an antifungal, antiviral, antitumor agent)

Scheme 2 Stereoselective hydrogenation of selected biologically active compounds or their fragments.

medical or biological significance through judicious design of mimetics of biological molecules or variations of natural product structures. A few selected examples serve to illustrate the versatile uses of unnatural amino acids via enantioselective or diastereoselective hydrogenation (Scheme 2) [4b, 40, 41].

1.1.2.9
Mechanistic Studies – New Developments

It is indisputable that a thorough understanding of a mechanism can lead to a better design of ligand or catalytic system. Previously, Halpern et al. [42] and Brown et al. [43] elucidated the mechanism of asymmetric hydrogenation with *cis*-chelating bis(alkyldiarylphosphine)-rhodium complex (the so-called unsaturated mechanism). The essential steps of this mechanism (Fig. 3, left-hand cycle) involve the pre-coordination of the enamide, the minor isomer **39**, prior to the rate-determining oxidative addition of dihydrogen, although the putative dihydride species **40** has never been observed. With the advent of PhanePHOS, Bargon and Brown managed to detect an agostic hydride (**41**) prior to the formation of the alkyl hydride species **42**, thus allowing a peek at the events during which the dihydride is transformed to the alkyl hydride for the first time [43].

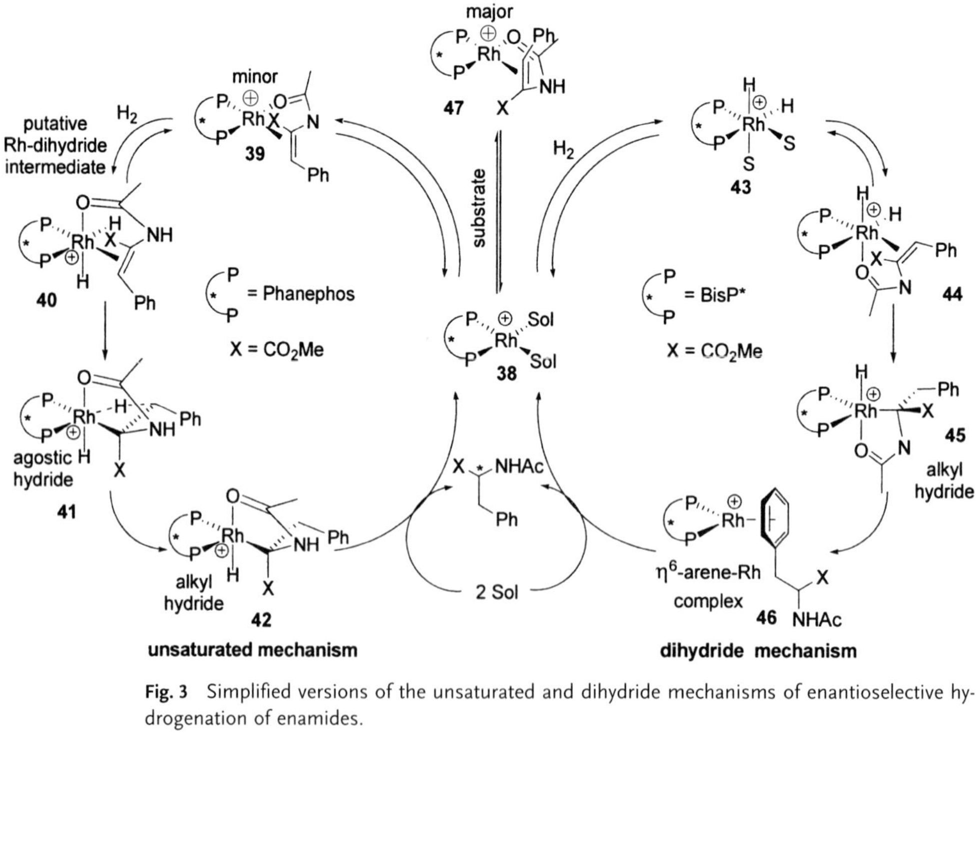

Fig. 3 Simplified versions of the unsaturated and dihydride mechanisms of enantioselective hydrogenation of enamides.

Recently, Imamoto and Crépy published results on the study of the asymmetric hydrogenation mechanism with their electron-rich BisP* (Fig. 3, right-hand cycle) [18]. Detailed (PHIP) NMR and kinetic studies led Imamoto and co-workers to conclude that dihydrogen is initially oxidatively added to the solvated Rh-diphosphine complex to form the stable dihydride **43**, a rudimentary step that constitutes the dihydride mechanism. Upon coordination of the substrate, the unstable dihydride **44** thus formed undergoes migratory insertion to give the alkyl hydride **45**. After reductive elimination of the latter, an η^6-arene-Rh species (**46**) was observed before extrusion of the product to regenerate the catalytically active **38**.

It is appropriate at this point to mention that the minor catalyst-substrate complex predicts the correct stereochemistry of the product in the unsaturated mechanism. This is in part general for a C_2-symmetric disphosphine ligand. However, when it comes to ligands with donors having distinctly differentiated *trans*-influence, such as **24**, the product configuration appears to be originated through the major interaction between the catalyst and the enamide **47** [26].

With regard to the monodentate phosphorus ligands, kinetic and mechanistic studies of the MonoPHOS series have been initiated [44, 45]. A predominant tetra-coordinated Rh-complex with four MonoPHOS molecules (cf. Imamoto's $[bis(MiniPHOS)Rh]^+$ [18]) has been confirmed by X-ray crystallography [44, 45]. Whilst it might be premature to suggest that the homoleptic cationic rhodium complex cannot be a catalyst itself as asserted by Feringa et al., more data are undoubtedly required to draw such a conclusion.

1.1.2.10
Catalyst Recycle [46]

From an industrial perspective, when the TON and TOF of enamide hydrogenations are low, there is a need to recycle the expensive catalysts. Several strategies have evolved over the years, and they are briefly described below.

Homogeneous catalysts can be anchored to a number of supported materials such as aluminum, carbon, lanthana, or montmorillonite K by using heteropoly acid. For instance, anchored rhodium catalysts containing DIPAMP, ProPhos **30**, Me-DuPHOS, BPPM **31** have been examined in the asymmetric hydrogenation of MAA. The reaction rate and the product *ee* were found to be comparable to the corresponding homogeneous catalyst. In some cases even better results were obtained after recycling the catalyst. In the case of Rh(DIPAMP) supported on phosphotungstic acid-treated montmorillonite K, the catalyst could be reused fifteen times without loss of activity and enantioselectivity [47]. Alternatively, [(*R*,*R*)-Me-DuPHOS-Rh(COD)]OTf, non-covalently immobilized on mesoporous MCM-41 by the interaction of the triflate counter ion and the surface silanols of the silica support, showed better enantioselectivity than the homogeneous catalyst in the asymmetric hydrogenation of the three dehydroamino esters tested. This catalyst could be reused without loss of activity or enantioselectivity [48].

Fan et al. showed that ACA could be hydrogenated with a polymeric rhodium catalyst associated with a MeO-PEG-supported (3*R*,4*R*)-Pyrphos **48** (Fig. 4), and that *ees* in

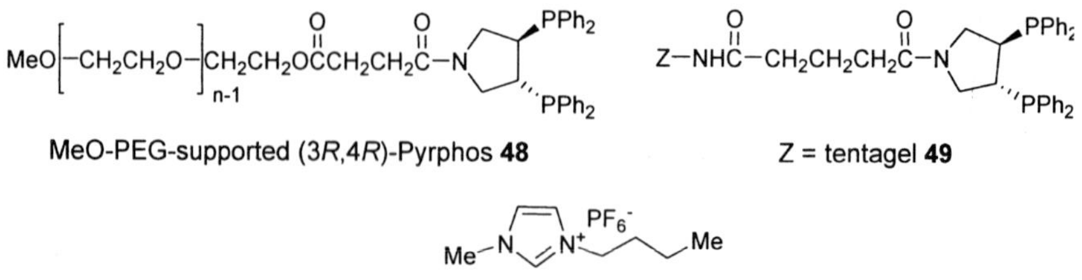

MeO-PEG-supported (3*R*,4*R*)-Pyrphos **48**

Z = tentagel **49**

$[BMIM][PF_6]$ **50**

Fig. 4

the range 87–96% could be obtained [49]. The catalyst was reused at least three times without loss of enantioselectivity. In contrast, the insoluble polyethylene oxide-grafted polystyrene matrix (TentaGel)-supported analog **49** could be reused only once [50].

Geresh and co-workers reported the asymmetric hydrogenation of MAA in water catalyzed by Rh-Me-DuPHOS occluded in polydimethylsiloxane [51]. Up to 97% *ee* was achieved by increasing the silica content to 20 wt%. A slight diminution in *ee* was observed after reuse of the occluded catalyst. Ionic liquid, $[BMIM][PF_6]$ **50**, another environmentally friendly solvent, provided extra stability to air-sensitive Rh-Me-DuPHOS in the asymmetric hydrogenation of MAA and MAC. Similar enantioselectivities were obtained for both substrates, comparing well with the homogeneous catalyst, but gradually decreasing catalytic activities were found for MAC after successive reuse of the catalyst [8].

Rh-Et-DuPHOS may be recovered using nanofiltration techniques. Thus, asymmetric hydrogenation of MAA has been performed continuously with the reaction mixture filtered through a nano-membrane, which permeates the product while retaining the catalyst for recycling. However, the activity and the enantioselectivity of the catalyst decline over time [52].

1.1.2.11
Conclusion

Notwithstanding the enduring success of DuPHOS and BPE, recent results have indicated that virtually all neutral phosphorus(III) compounds, whether bidentate or monodentate, combined with various chirality features, stand a chance of inducing high enantioselectivities. Only the curiosity of scientists, coupled with the ever-expanding applications of α-amino acids, will reveal what is still in store for us to discover in this ostensibly mature field.

Acknowledgement

We thank The Hong Kong Research Grants Council Central Allocation Fund (Project ERB003), The University Grants Committee Area of Excellence Scheme in Hong Kong (AoE P/10-01), and The Hong Kong Polytechnic University Area of Strategic Development Fund for financial support of this study.

References

1 (a) J.F. Atkins, R. Gesteland, *Science* **2002**, *296*, 1409. (b) G. Srinivasan, C. M. James, J.A. Krzycki, *Science* **2002**, *296*, 1459. (c) B. Hao, W. Gong, T.K. Ferguson, C.M. James, J.A. Krzycki, M.K. Chan, *Science* **2002**, *296*, 1462.

2 J.L. Cleland, C.S. Craik, *Protein Engineering: Principles and Practice*, Wiley-Liss, New York, **1996**.

3 W. M. Kazmierski, *Peptidomimetics Protocols*, Humana Press, New Jersey, **1999**.

4 For leading examples, see (a) J.R. Allen, C.R. Harris, S.J. Danishefsky, *J. Am. Chem. Soc.* **2001**, *123*, 1890, (b) D.A. Evans, J.L. Katz, G.S. Peterson, T. Hintermann, *J. Am. Chem. Soc.* **2001**, *123*, 12411. (c) Y. Zou, N.E. Fahmi, C. Vialas, G.M. Miller, S.M. Hecht, *J. Am. Chem. Soc.* **2002**, *124*, 9476.

5 (a) M. Adamczyk, S.R. Akireddry, R.E. Reddy, *Org. Lett.* **2001**, *3*, 3157. (b) M. Adamczyk, S.R. Akireddry, R.E. Reddy, *Tetrahedron* **2002**, *58*, 6951.

6 The hydrogenation of these substrates with new ligands has been amply covered in a recent review: H.-U. Blaser, C. Malan, B. Pugin, F. Spindler, H. Steiner, M. Studer, *Adv. Synth. Catal.* **2003**, *345*, 103.

7 D. Sinou, *Adv. Synth. Catal.* **2002**, *344*, 219.

8 S. Guernik, A. Wolfson, M. Herskowitz, N. Greenspoon, S. Geresh, *Chem. Commun.* **2001**, 2314.

9 M.J. Burk, S. Feng, M.F. Gross, W. Tumas, *J. Am. Chem. Soc.* **1995**, *117*, 8277.

10 M.J. Burk, *Acc. Chem. Res.* **2000**, *33*, 363 and references therein.

11 (a) W. Li, Z. Zhang, D. Xiao, X. Zhang, *Tetrahedron Lett.* **1999**, *40*, 6701. (b) W. Li, Z. Zhang, D. Xiao, X. Zhang, *J. Org. Chem.* **2000**, *65*, 3489.

12 D. Liu, W. Li, X. Zhang, *Org. Lett.* **2002**, *4*, 4471.

13 J.J. Almena Perea, A. Börner, P. Knochel, *Tetrahedron Lett.* **1998**, *39*, 8073.

14 J.J. Almena Perea, M. Lotz, P. Knochel, *Tetrahedron: Asymmetry* **1999**, *10*, 375.

15 N.W. Boaz, S.D. Debenham, E.B. Mackenzie, S.E. Large, *Org. Lett.* **2002**, *4*, 2421.

16 W.S. Knowles, M.J. Sabacky, B.D. Vineyard, D.J. Weinkauff, *J. Am. Chem. Soc.* **1975**, *97*, 2567; B.D. Vineyard, W.S. Knowles, G.L. Bachman, D.J. Weinkauff, *J. Am. Chem. Soc.* **1977**, *99*, 5946.

17 (a) K.M. Pietrusiewicz, M. Zablocka, *Chem. Rev.* **1994**, *94*, 1375. (b) J.M. Brunel, B. Faure, M. Maffei, *Coord. Chem. Rev.* **1998**, *178-180*, 665. (c) O.I. Kolodiazhnyi, *Tetrahedron: Asymmetry* **1998**, *9*, 1279. (d) M. Ohiff, J. Holz, M. Quirmbach, A. Börner, *Synthesis* **1998**, 1391. (e) B. Carboni, L. Monnier, *Tetrahedron* **1999**, *55*, 1197.

18 K.V. L. Crépy, T. Imamoto, *Adv. Synth. Catal.* **2003**, *345*, 79 and references therein.

19 W. Tang, X. Zhang, *Angew. Chem. Int. Ed.* **2002**, *41*, 1612.

20 P.J. Pye, K. Rossen, R.A. Reamer, N.N. Tsou, R.P. Volante, P.J. Reider, *J. Am. Chem. Soc.* **1997**, *119*, 6207.

21 K. Kromm, P.L. Osburn, J.A. Gladysz, *Organometallics* **2002**, *21*, 4275.

22 T.T.-L. Au-Yeung, S.S. Chan, A.S.C. Chan, *Adv. Synth. Catal.* **2003**, *345*, 537.

23 (a) A.S.C. Chan, W. Hu, C.-C. Pai, C.-P. Lau, Y. Jiang, A. Mi, M. Yan, J. Sun, R. Lou, J. Dang, *J. Am. Chem. Soc.* **1997**, *119*, 9570. (b) C.W. Lin, Ph. D. Thesis, The Hong Kong Polytechnic University, **1999**.

24 A. Zanotti-Gerosa, C. Malan, D. Herzberg, *Org. Lett.* **2001**, *3*, 3687.

25 R. Lou, A. Mi, Y. Jiang, Y. Qin, Z. Li, F. Fu, A.S.C. Chan, *Tetrahedron* **2000**, *56*, 5857.

26 D.A. Evans, F.E. Michael, J.S. Tedrow, R. Campos, *J. Am. Chem. Soc.* **2003**, *125*, 3534.

27 (a) W.S. Knowles, M.J. Sabacky, *J. Chem. Soc. Chem. Commun.* **1968**, 1445. (b) L. Horner, H. Siegel, H. Bthe, *Angew. Chem. Int. Ed. Engl.* **1968**, *7*, 942.

28 H.B. Kagan, T.P. Dang, *Chem. Commun.* **1971**, 481; H.B. Kagan, T.P. Dang, *J. Am. Chem. Soc.* **1972**, *94*, 6429.

29 F. Lagasse, H.B. Kagan, *Chem. Pharm. Bull.* **2000**, *48*, 315; I.V. Komarov, A. Brner, *Angew. Chem. Int. Ed.* **2001**, *40*, 1197 and references therein.

30 (a) K. Inoguchi, S. Sakuraba, K. Achiwa, *Synlett* **1992**, 169. (b) D. Peña, A. J. Minnaard, A. H. M. de Vries, J. G. de Vries, B. L. Feringa, *Org. Lett.* **2003**, *5*, 475.

31 Y. Fu, J.-H. Xie, A.-G. Hu, H. Zhou, L.-X. Wang, Q.-L. Zhou, *Chem. Commun.* **2002**, 480.

32 M.T. Reetz, T. Sell, A. Meiswinkel, G. Mehler, *Angew. Chem. Int. Ed.*. **2003**, *42*, 790.

33 (a) K. Rossen, S.A. Weissman, J. Sager, R.A. Reamer, D. Askin, R.P. Volante, P.J. Reider, *Tetrahedron Lett.* **1995**, *36*, 6419. (b) R. Kuwano, Y. Ito, *J. Org. Chem.* **1999**, *64*, 1232.

34 K.C. Nicolaou, G.-Q. Shi, K. Namoto, F. Bernal, *Chem. Commun.* **1998**, 1757.

35 R. Kuwano, D. Karube, Y. Ito, *Tetrahedron Lett.* **1999**, *40*, 9045.

36 R. Kuwano, K. Sato, T. Kurokawa, D. Karube, Y. Ito, *J. Am. Chem. Soc.* **2000**, *122*, 7614.

37 (a) M.J. Burk, M.F. Gross, J.P. Martinez, *J. Am. Chem. Soc.* **1995**, *117*, 9375. (b) M.J. Burk, M.F. Gross, T. Gregory, P. Harper, C.S. Kalberg, J.R. Lee, J.P. Martinez, *Pure Appl. Chem.* **1996**, *68*, 37.

38 A. Ohashi, S.-I. Kikuchi, M. Yasutake, T. Imamoto, *Eur. J. Org. Chem.* **2002**, 2535.

39 (a) R. Kuwano, S. Okuda, Y. Ito, *J. Org. Chem.* **1998**, *63*, 3499. (b) R. Kuwano, S. Okuda, Y. Ito, *Tetrahedron: Asymmetry* **1998**, *9*, 2773.

40 S.D. Debenham, J. Cossrow, E.J. Toone, *J. Org. Chem.* **1999**, *64*, 9153.

41 A.K. Ghosh, W. Liu, *J. Org. Chem.* **1996**, *61*, 6175.

42 (a) J. Halpern, *Acc. Chem. Res.* **1982**, *15*, 332. (b) J. Halpern, *Pure Appl. Chem.* **1983**, *55*, 99. (c) C.R. Landis, J. Halpern, *J. Am. Chem. Soc.* **1987**, *109*, 1746.

43 (a) J.M. Brown in Comprehensive Asymmetric Catalysis (Ed.: E.N. Jacobsen, A. Pfaltz, H. Yamamoto), Springer, Berlin, **1999**, *1*, 124–137. (b) R. Giernoth, H. Heinrich, N.J. Adams, R.J. Deeth, J. Bargon, J. M. Brown, *J. Am. Chem. Soc.* **2000**, *122*, 12381.

44 X. Li, A.S.C. Chan, unpublished results.

45 M. van den Berg, A.J. Minnaard, R.M. Haak, M. Leeman, E.P. Schudde, A. Meetsma, B.L. Feringa, A.H.M. de Vries, C.E.P. Maljaars, C.E. Willans, D. Hyett, J.A.F. Boogers, H.J.W. Henderickx, J.G. de Vries, *Adv. Synth. Catal.* **2003**, *345*, 308.

46 Q.-H. Fan, Y.-M. Li, A.S.C. Chan, *Chem. Rev.* **2002**, *102*, 3385.

47 R. Augustine, S. Tanielyan, S. Anderson, H. Yang, *Chem. Commun.* **1999**, 1257.

48 F.M. de Rege, D.K. Morita, K.C. Ott, W. Tumas, R.D. Broene, *Chem. Commun.* **2000**, 1797.

49 Q.-H. Fan, G.-J. Deng, C.-C. Lin, A.S.C. Chan, *Tetrahedron: Asymmetry* **2001**, *12*, 1241.

50 U. Nagel, J. Keipoid, *Chem. Ber.* **1996**, *129*, 815.

51 A. Wolfson, S. Janssens, I. Vankelecom, S. Geresh, M. Gottlieb, M. Herskowitz, *Chem. Commun.* **2002**, 388.

52 K.D. Smet, S. Aerts, E. Ceulemans, I.F.J. Vankelecom, P.A. Jacobs, *Chem. Commun.* **2001**, 597.

1.1.3
Carbonyl Hydrogenations

Takeshi Ohkuma and Ryoji Noyori

1.1.3.1
Introduction

The hydrogenation of carbonyl compounds is one of the most important synthetic reactions. Molecular hydrogen is catalytically activated by appropriate metals or metal complexes and delivered to the C=O functionality to give the corresponding reduction products. This transformation has been not only of academic interest but also of industrial significance because of its simplicity, environmental friendliness, and economics. Practical hydrogenation catalysts are required to have high activity, selectivity, and stability. The ideal system hydrogenates the organic substrates quantitatively with a small amount of the catalyst under mild conditions within a short period. In organic synthesis of fine chemicals, hydrogenation should be accomplished with high chemo-, enantio- and diastereoselectivity. The applicability to a wide variety of substrates is obviously desirable, particularly in research into biologically active substances and advanced functional materials. Historically, hydrogenation of carbonyl compounds was accomplished mainly by heterogeneous catalysts such as Ni, Pd, and PtO_2 [1]. Until recently, highly active homogeneous catalysts for carbonyl hydrogenation remained undeveloped. This chapter presents recent topics in this important field.

1.1.3.2
Ketones and Aldehydes

1.1.3.2.1 Simple Ketones and Aldehydes

Reactivity

The discovery of the late transition metal complexes with phosphine ligands, such as $RhCl[P(C_6H_5)_3]_3$ and $RuCl_2[P(C_6H_5)_3]_3$, by Wilkinson [2] has led to a great advancement in homogeneous hydrogenation of olefinic and acetylenic compounds [2, 3]. Reaction of these complexes with H_2 efficiently produces the active metal-hydride species under mild conditions. The remarkable advantage of the homogeneous catalysts is their ability to be designed rationally by considering their activity and stereoselectivity.

Homogeneous hydrogenation of simple ketones has remained difficult to achieve even under a high hydrogen pressure and at high temperature. Up to now, only a limited number of Rh and Ru complexes have shown good catalytic activity in the hydrogenation of ketonic substrates possessing no functionality adjacent to the carbonyl group. Some Rh-catalyzed reactions effected under atmospheric pressure of H_2 and at ambient temperature are represented in Fig. 1. The bipyridine-based complexes such as $[RhCl_2(bipy)_2]Cl$ and $Rh_2Cl_2(OCOCH_3)_2(bipy)_2$ show high activity un-

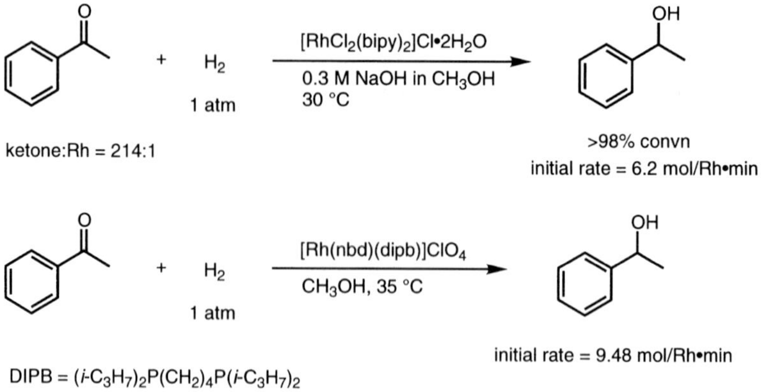

+ H_2 1 atm; $[RhCl_2(bipy)_2]Cl \cdot 2H_2O$; 0.3 M NaOH in CH_3OH, 30 °C

ketone:Rh = 214:1

>98% convn
initial rate = 6.2 mol/Rh•min

+ H_2 1 atm; $[Rh(nbd)(dipb)]ClO_4$; CH_3OH, 35 °C

initial rate = 9.48 mol/Rh•min

DIPB = $(i\text{-}C_3H_7)_2P(CH_2)_4P(i\text{-}C_3H_7)_2$

Fig. 1 Hydrogenation of acetophenone catalyzed by Rh complexes.

der basic conditions [4]. A phosphine complex system, $RhCl(cod)[P(C_6H_5)_3]_3$-$NaBH_4$, requires an addition of 45 equivalents of KOH [5]. Cationic complexes with basic alkylphosphine ligands such as $[RhH_2\{P(C_6H_5)(CH_3)_2\}_2L_2]X$ (L=solvent, X=PF_6 or ClO_4), $[Rh(nbd)(dipb)]ClO_4$ (NBD=norbornadiene, DIPB=1,4-bis(diisopropylphosphino)butane), and $[Rh(cod)(dipfc)]OSO_2CF_3$ (DiPFc=1,1′-bis(diisopropylphosphino)ferrocene) [6] also effect catalytic hydrogenation of ketones. The basic ligands increase electron density on the central metal so that the oxidative addition of H_2 can be accelerated [7].

Ru catalysts [8, 9] are less active than Rh complexes. Notably, an anionic complex, $K_2[Ru_2H_4P(C_6H_5)_2\{P(C_6H_5)_3\}_3] \cdot 2O(CH_2CH_2OCH_3)_2$, shows a much higher reactivity than other Ru complexes so far reported (Fig. 2) [10]. Although the high reactivity was ascribed to the anionic property of the complex the real active species was recently proposed to be a neutral hydride complex, $RuH_4[P(C_6H_5)_3]_3$ [11]. The trinuclear Ru complex, $[RuHCl(dppb)]_3$ (DPPB=1,4-bis(diphenylphosphino)butane), catalyzes the hydrogenation of acetophenone at atmospheric pressure [12].

Although $RuCl_2[P(C_6H_5)_3]_3$ is not very active for the hydrogenation of ketones, the catalytic activity is remarkably enhanced when small amounts of $NH_2(CH_2)_2NH_2$ and KOH are added to this complex (Fig. 3a) [13]. Acetophenone can be hydrogenated quantitatively at 1 atm of H_2 and at room temperature in 2-propanol with a high initial rate. At 50 atm of H_2, the turnover frequency (TOF), defined as moles of product per mole of catalyst per h or s, reaches up to 23000 h^{-1}. The presence of both diamine and inorganic base as well as the use of 2-propanol as solvent are crucial to achieving the high catalytic activity. The activity of the in situ prepared catalyst is increased by more than 20 times when a preformed complex *trans*-$RuCl_2[P(C_6H_4\text{-}4\text{-}CH_3)_3]_2[NH_2(CH_2)_2NH_2]$ and $(CH_3)_3COK$ is used as a catalyst (Fig. 3b) [14, 15]. Cyclohexanone is quantitatively reduced in the presence of the catalyst with a substrate/catalyst mole ratio (S/C) of 100000 at 60 °C under 10 atm of H_2 to give cyclohexanol. The initial TOF is reached at 563000 h^{-1} or 156 s^{-1}. A combination of RuHCl(diphosphine)(1,2-diamine) and a strong base also shows high cat-

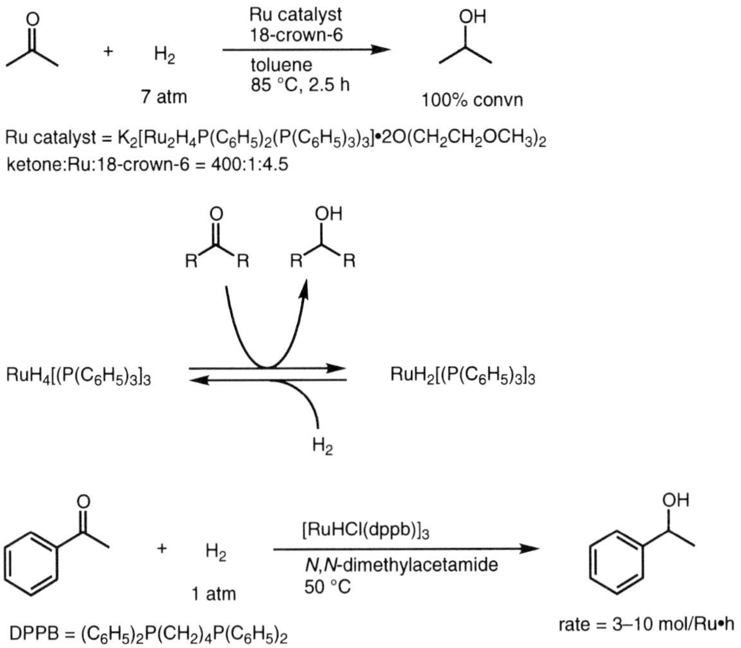

Fig. 2 Hydrogenation of ketones catalyzed by Ru complexes.

alytic activity [16]. $RuH(\eta^1\text{-}BH_4)$(diphosphine)(1,2-diamine) [17] (see below) as well as the RuH_2 complexes [18] do not require an additional base to catalyze this transformation. A *trans*-$RuCl_2$(diphosphine)(pyridine)$_2$ promotes the hydrogenation of acetophenone in the presence of $(CH_3)_3COK$ [19].

Historically, hydrogenation of ketones has been recognized to proceed through a $[\sigma 2 + \pi 2]$ transition state consisting of a carbonyl group and a metal hydride [6, 10, 20]. However, the phosphine/1,2-diamine–Ru catalyst is supposed to promote the transformation by an entirely different mechanism [14]. Fig. 4 illustrates a summary of the proposed mechanism. The preformed complex $RuCl_2(PR_3)_2[NH_2(CH_2)_2NH_2]$ (**A**) is not the real catalytic species. It is converted to $RuHX(PR_3)_2[NH_2(CH_2)_2NH_2]$ (**B**; X = H, OR, etc.) in the presence of two equiv. of an alkaline base and a hydride source, H_2, and a trace of 2-propanol. The added base primarily operates to neutralize HCl liberated from **A**. The 18-electron species **B**, which has no vacant site to interact directly with substrates, immediately hydrogenates a ketone through a pericyclic six-membered transition state $\mathbf{TS_1}$ to afford the 16-electron complex **C** and a product alcohol. Collaboration of the charge-alternating $H^{\delta-}$–$Ru^{\delta+}$–$N^{\delta-}$–$H^{\delta+}$ arrangement with the $C^{\delta+}$=$O^{\delta-}$polarization notably stabilizes $\mathbf{TS_1}$. The 16-electron species **C** heterolytically cleaves the H_2 molecule to restore **B** through the four-membered $\mathbf{TS_2}$ or the six-membered $\mathbf{TS_3}$ promoted by a hydrogen-bonded alcohol molecule. An alternative pathway to regenerate **B** via species **D** and **E** is possible. Protonation of **C** in 2-propanol media

a

$RuCl_2[P(C_6H_5)_3]_3$
$NH_2(CH_2)_2NH_2$, KOH
$(CH_3)_2CHOH$
28 °C

acetophenone + H_2 → 1-phenylethanol (OH)

>99% yield

H_2, atm	ketone:Ru:diamine:KOH	TOF, mol/Ru•h
1	500:1:1:2	880
3	5000:1:1:20	6700
50	10,000:1:1:40	23,000

b

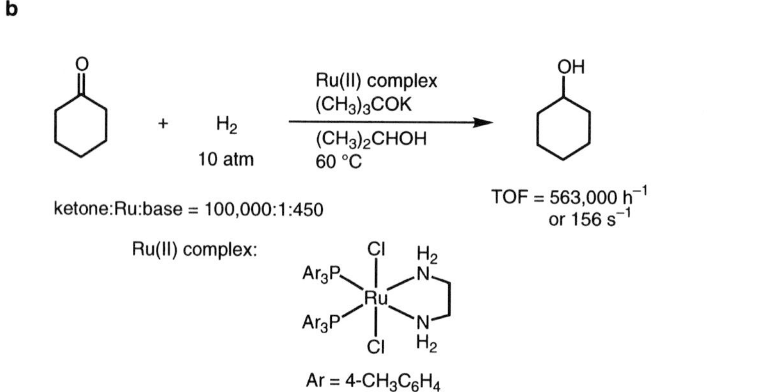

Fig. 3 High-speed hydrogenation of simple ketones.

gives the cationic species **D** followed by H_2-molecule binding on the Ru center, resulting in **E**. A base assisting cleavage of H_2 on **E** completes the catalytic cycle. The non-classical metal–ligand difunctional mechanism has been supported by both experimentally (structures and kinetics [21]) and theoretically (*ab initio* MO and DFT [22, 23]) in the closely related transfer hydrogenation of ketones catalyzed by Ru complexes in 2-propanol [24]. Other transition state models have been also proposed [25, 26].

A copper(I) complex prepared from $[CuH\{P(C_6H_5)_3\}]_6$ and an excess amount of $P(CH_3)_2C_6H_5$ is also active for the hydrogenation of 4-*tert*-butylcyclohexanone [27], in which a highly hydridic complex, $[CuH\{P(CH_3)_2C_6H_5\}]_n$, may be the active species. A Mo(II) complex, $[MoCp(CO)_2\{P(cyclo\text{-}C_6H_{11})_3\}(\eta^1\text{-3-pentanone})]B[C_6H_3\text{-3,5-}(CF_3)_2]_4$ (Cp = cyclopentadienyl) catalyzes the hydrogenation of 3-pentane with an S/C of 10–12 at 23 °C under 4 atm of H_2 [28]. The TOF was determined as 2 h^{-1}.

An alkaline base-catalyzed hydrogenation of aromatic ketones without any transition metals was reported in 1961 [29]. Recently, this reduction has been reinvestigated. Under vigorous conditions (135 atm H_2, 210 °C), benzophenone is reduced in the presence of 0.2 equiv. of $(CH_3)_3COCs$ to give benzhydrol contami-

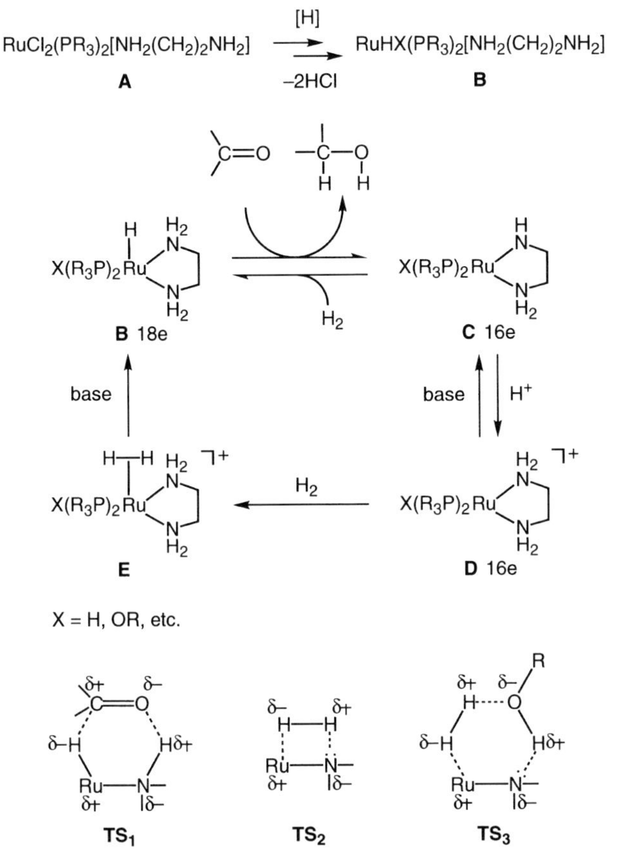

$RuCl_2(PR_3)_2[NH_2(CH_2)_2NH_2]$ (A) —[H], −2HCl→ $RuHX(PR_3)_2[NH_2(CH_2)_2NH_2]$ (B)

X = H, OR, etc.

TS$_1$ TS$_2$ TS$_3$

Fig. 4 Proposed mechanism for hydrogenation of ketones with diphosphine/diamine–Ru complexes.

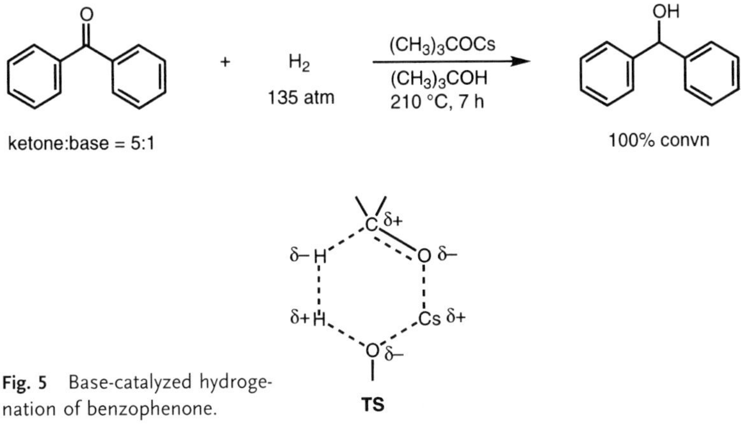

ketone:base = 5:1 + H_2 135 atm —$(CH_3)_3COCs$, $(CH_3)_3COH$, 210 °C, 7 h→ 100% convn

TS

Fig. 5 Base-catalyzed hydrogenation of benzophenone.

nated with about 1% of diphenylmethane (Fig. 5) [30]. Other alkaline alcoholates are also usable: The efficiency of metals decreases in the order Cs > Rb ~ K ≫ Na ≫ Li. The hydrogenation is supposed to proceed via the six-membered transition state described in Fig. 5 that closely resembles that of a phosphine/diamine–Ru-catalyzed hydrogenation (see $\mathbf{TS_1}$ of Fig. 4).

Carbonyl selectivity

Unsaturated carbonyl compounds are classified into two types, nonconjugated and conjugated. Isolated C=O and C=C linkages are distinctly different. The carbonyl carbon of a simple ketone normally reacts with nucleophiles, whereas an isolated olefinic bond reacts with electrophiles. With α,β-unsaturated carbonyl compounds, a nucleophile can react with both the carbonyl carbon and β-carbon, while an electrophile reacts only at the C=C bond. Chemical differentiation between such C=O and C=C (or C≡C) moieties is important. Most existing heterogeneous and homogeneous catalysts using molecular hydrogen preferentially saturate carbon–carbon multiple bonds over carbonyl groups [1a, 31]. This selectivity is conceived to arise from the easier interaction of the metal center with an olefinic or acetylenic π bond than with a carbonyl linkage.

Certain catalyst systems, however, exhibit notable carbonyl selectivities. Figs. 6 and 7 illustrate competitive hydrogenation between isolated carbonyl and olefinic bonds in favor of carbonyl saturation. Under basic conditions, $[RhCl_2(bipy)_2]Cl$ selectively hydrogenates ketones in the presence of olefins [4a]. The $[CuH\{P(C_6H_5)_3\}]_6$–$P(CH_3)_2C_6H_5$ system converts 4-cycloocten-1-one to the unsaturated alcohol [27]. The copper catalyst system hydrogenates carbonyl groups preferentially over acetylenic bonds. A catalyst system consisting of a Ru(II) phosphine complex, diamine, and inorganic base shows an excellent carbonyl selectivity over an olefinic or acetylenic function [32]. The combined effects of $NH_2(CH_2)_2NH_2$ and KOH decelerate olefin hydrogenation catalyzed by $RuCl_2[P(C_6H_5)_3]_3$ and accelerate carbonyl hydrogenation. Thus, a competition experiment using a mixture of heptanal and 1-octene with $RuCl_2[P(C_6H_5)_3]_3$ reveals that the terminal olefin is hydrogenated 250 times faster than the aldehyde. However, when $NH_2(CH_2)_2NH_2$ and KOH are present, heptanal is hydrogenated 1500 times faster than 1-octene. 1,2,3,6-Tetrahydrobenzaldehyde is hydrogenated by $Ir(ClO_4)(CO)[P(C_6H_5)_3]_2$ to give an enol without olefin hydrogenation [33]. Chromium-modified Raney nickel, Raney cobalt, or cobalt black shows carbonyl selectivity limited over tri- or tetrasubstituted olefins [34].

The Ir complexes depicted in Fig. 8 serve as catalysts for selective hydrogenation of α,β-unsaturated ketones to the allylic alcohols. Hydrogenation of benzalacetone with a catalyst system consisting of $[Ir(OCH_3)(cod)]_2$ and 10 equivalents of $P(C_6H_5)_2C_2H_5$ gives 4-phenyl-3-penten-2-ol with a 97% selectivity [35]. The use of a large phosphine ligand with a cone angle ranging from 135 to 150° is required to obtain >90% selectivity. $IrH_3[P(C_6H_5)_2C_2H_5]_3$ was proposed as a real active species. Hydrogenation of 2-cyclohexen-1-one with an $[Ir(OCH_3)(cod)]_2$–(*S,S*)-DIOP (see Fig. 13) system gives (*R*)-2-cyclohexen-1-ol (25% *ee*) with 95% selectivity at 65% conversion [36]. Benzala-

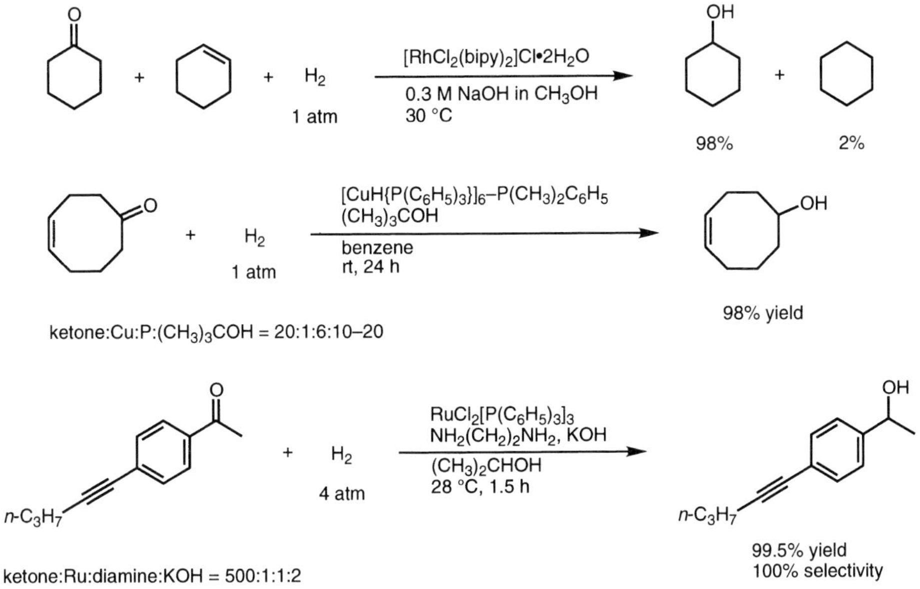

+ + H_2 (1 atm) → $[RhCl_2(bipy)_2]Cl{\bullet}2H_2O$, 0.3 M NaOH in CH_3OH, 30 °C → OH (98%) + (2%)

+ H_2 (1 atm) → $[CuH\{P(C_6H_5)_3\}]_6$–$P(CH_3)_2C_6H_5$, $(CH_3)_3COH$, benzene, rt, 24 h → OH (98% yield)

ketone:Cu:P:$(CH_3)_3COH$ = 20:1:6:10–20

O + H_2 (4 atm) → $RuCl_2[P(C_6H_5)_3]_3$, $NH_2(CH_2)_2NH_2$, KOH, $(CH_3)_2CHOH$, 28 °C, 1.5 h → OH, n-C_3H_7 (99.5% yield, 100% selectivity)

n-C_3H_7

ketone:Ru:diamine:KOH = 500:1:1:2

Fig. 6 Carbonyl-selective hydrogenation of nonconjugated unsaturated ketones.

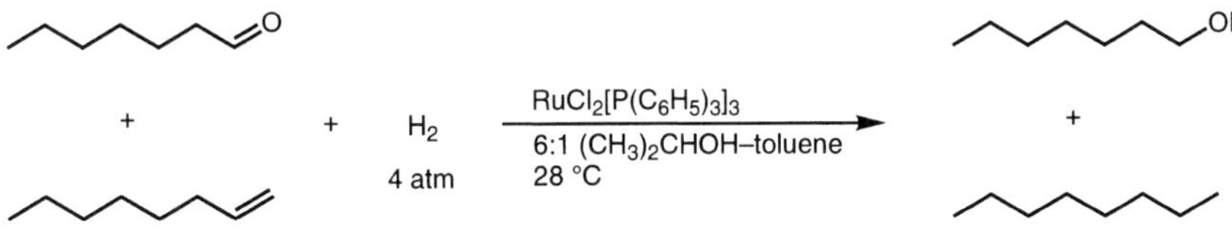

O + + H_2 (4 atm) → $RuCl_2[P(C_6H_5)_3]_3$, 6:1 $(CH_3)_2CHOH$–toluene, 28 °C → OH +

conditions	time, min	1-heptanol:octane
no base (aldehyde:Ru = 500:1)	150	1:250
$NH_2(CH_2)_2NH_2$ and KOH (aldehyde:Ru:diamine:KOH = 500:1:1:2)	10	1500:1

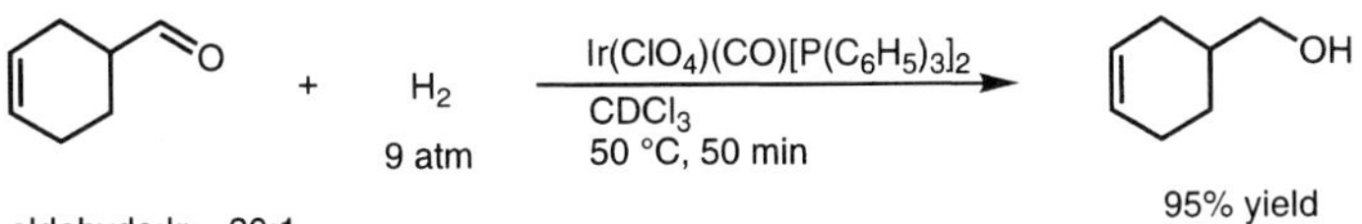

O + H_2 (9 atm) → $Ir(ClO_4)(CO)[P(C_6H_5)_3]_2$, $CDCl_3$, 50 °C, 50 min → OH (95% yield)

aldehyde:Ir = 30:1

Fig. 7 Carbonyl-selective hydrogenation of nonconjugated unsaturated aldehydes.

cetone is hydrogenated in the presence of $[Ir(cod)\{(R)\text{-binap}\}]BF_4$ (BINAP, see Fig. 13) and 1.5 equivalents of *o*-dimethylaminophenyldiphenylphosphine to afford the *R* allylic alcohol (65% *ee*) with a 97% carbonyl selectivity at 72% conversion [37]. High electron density on Ir caused by the aminophosphine ligand is supposed to promote preferential reduction of the carbonyl group. The $[CuH\{P(C_6H_5)_3\}]_6$–

+ H_2 30 atm

$[Ir(OCH_3)(cod)]_2$–$P(C_6H_5)_2C_2H_5$

toluene, 100 °C, 10 h

OH

ketone:Ir:P = 500:1:10

97% selectivity at 99% conv

O

+ H_2 48 atm

$[Ir(cod)\{(R)\text{-binap}\}]BF_4$–P,N ligand

THF, 60 °C, 47 h

OH

ketone:Ir:P,N ligand = 120:1:1.5

97% selectivity at 72% conv *R*, 65% ee

P,N ligand = $P(C_6H_5)_2$ $N(CH_3)_2$

O

+ H_2 34 atm

$[CuH\{P(C_6H_5)_3\}]_6$–$C_6H_5P(CH_3)C_2H_5$ $(CH_3)_3COH$

benzene rt, 21 h

OH

ketone:Cu:P:$(CH_3)_3COH$ = 120:1:6:40

>98% selectivity 95% yield

Fig. 8 Carbonyl-selective hydrogenation of α,β-unsaturated ketones.

O

+ H_2 4 atm

$RuCl_2[P(C_6H_5)_3]_3$ $NH_2(CH_2)_2NH_2$, KOH

$(CH_3)_2CHOH$ 28 °C, 18 h

OH

ketone:Ru:diamine:KOH = 10,000:1:1:2

>99.9% selectivity at 100% conv

OH

99.6% selectivity 98.2% conv

OH

100% selectivity >99% conv

OH

>99.9% selectivity 99.8% conv

Fig. 9 Carbonyl-selective hydrogenation of α,β-unsaturated ketones.

$C_6H_5P(CH_3)C_2H_5$ catalyst system converts β-ionone to β-ionol with >98% selectivity in 95% isolated yield [27 c]. The selectivity is slightly better than that obtained using $P(CH_3)_2C_6H_5$ instead of $C_6H_5P(CH_3)C_2H_5$. As illustrated in Fig. 9, a range of α,β-unsaturated ketones except 2-cyclohexen-1-one are hydrogenated in the presence of the $RuCl_2[P(C_6H_5)_3]_3$–$NH_2(CH_2)_2NH_2$–KOH combined catalyst system to give the allylic alcohols [32, 38, 39]. The carbonyl selectivity is almost perfect. An Ru/C catalyst can be used for the hydrogenation of ketones conjugated with trisubstituted olefinic bonds [40].

An $[Ir(OCH_3)(cod)]_2$–$P(C_6H_5)_2C_2H_5$ catalyst system shows perfect carbonyl selectivity in the hydrogenation of cinnamaldehyde at 96% conversion (Fig. 10) [41]. The bulkiness of phosphine ligands largely affects the selectivity. An $[Ir(cod)\{P(CH_2OH)_3\}_3]Cl$ with five equivalents of the phosphine ligand also shows high selectivity in a biphasic medium [42]. A water-soluble Ru catalyst, prepared *in situ* from $RuCl_3$ and tris(*m*-sulfonyl)phosphine trisodium salt (TPPTS), effects efficient two-phase hydrogenation of α,β-unsaturated aldehydes, leading to allylic alcohols with excellent selectivity [43]. This procedure has been extended to an industrial use [44]. The selectivity highly depends on the pH of the reaction media, which controls the equilibrium distribution of hydride complexes: a RuHCl complex selectively reducing the C–C double bond of enals dominantly exists at pH≤3.3, while the RuH_2 species preferably hydrogenating carbonyl group exclusively exists at pH≥7 [45]. Hydrogen pressure also affects the equilibrium [46]. Hydrogenation of citral in the presence of $RuHCl[P(C_6H_5)_3]_3$ and five equivalents of HCl gives the corresponding alcohol with 99% selectivity, where the addition of

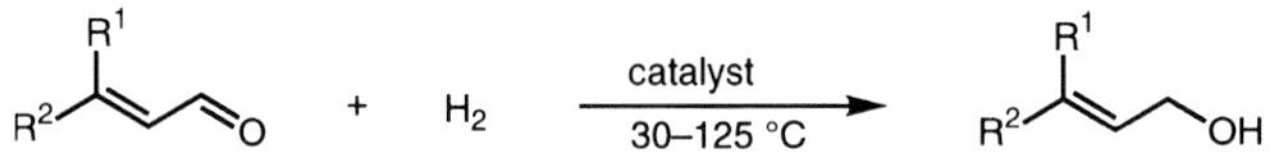

R^1, R^2 (enal) + H_2 —catalyst, 30–125 °C→ R^1, R^2 (allylic alcohol, OH)

R^1	R^2	Catalyst	Solvent	H_2 (atm)	% conv.	% selec. [a]
H	C_6H_5	$[Ir(OCH_3)(cod)]_2$–$P(C_6H_5)_2C_2H_5$	$(CH_3)_2CHOH$	30	96	100
H	C_6H_5	$[Ir(cod)\{P(CH_2OH)_3\}_3]Cl$	C_6H_6–H_2O	90	97	97
H	C_6H_5	$RuCl_2[P(C_6H_5)_3]_3$–EN [b]+ KOH	$(CH_3)_2CHOH$–toluene	4	99.7	99.8
H	C_6H_5	$[CuH\{P(C_6H_5)_3\}]_6$–$(C_6H_5)P(CH_2)_4$ [c]	C_6H_6–$(CH_3)_3COH$	5	89	98.8
H	CH_3	$RuCl_3\cdot 3H_2O$–TPPTS [d]	Toluene–H_2O	20	95	99
CH_3	$(CH_3)_2C{=}CH(CH_2)_2$	$RuCl_3\cdot 3H_2O$–TPPTS [d]	Toluene–H_2O	49	96	98
CH_3	$(CH_3)_2C{=}CH(CH_2)_2$	$RuHCl[P(C_6H_5)_3]_3$+HCl	Toluene–C_2H_5OH	6	99	99
CH_3	$(CH_3)_2C{=}CH(CH_2)_2$	$RuCl_2[P(C_6H_5)_3]_3$–EN [b]+ KOH	$(CH_3)_2CHOH$–toluene	4	92	100
H	C_6H_5	Pt–Ge/Nylon 66	C_2H_5OH	1	>99	90–95
H	C_6H_5	Pt/graphite	$(CH_3)_2CHOH$–H_2O	39	50	98
H	C_6H_5	Co/SiO_2	C_2H_5OH	10	90	96
CH_3	$(CH_3)_2C{=}CH(CH_2)_2$	Rh–Sn/SiO_2	Heptane	80	100	98

a) Carbonyl selectivity.
b) EN = $NH_2(CH_2)_2NH_2$.
c) Phenylphospholane.
d) TPPTS = $P(m\text{-}C_6H_4SO_3Na)_3$.

Fig. 10 Carbonyl-selective hydrogenation of α,β-unsaturated aldehydes.

HCl effectively increases both activity and selectivity [47]. The ternary catalyst system consisting of $RuCl_2I P(C_6H_5)_3]_3$, $NH_2(CH_2)_2NH_2$, and KOH shows excellent activity and carbonyl selectivity with α,β-unsaturated aldehydes as well [32]. The $[CuH\{P(C_6H_5)_3\}]_6$–phenylphospholane combined catalyst reduces cinnamaldehyde with 98.8% selectivity [27c].

Heterogeneous carbonyl-selective hydrogenation of α,β-unsaturated aldehydes has been studied mainly by using group VIII metal catalysts [1a]. The first selective hydrogenation of cinnamaldehyde to cinnamyl alcohol was achieved by the use of an unsupported Pt–Zn–Fe catalyst [48]. The activity and selectivity of the catalyst are highly affected by the metal, support, and additive as well as preparation conditions. Pt is the most frequently used metal. A variety of catalyst properties are exhibited by metal supports such as Al_2O_3, graphite, Nylon, SiO_2, and zeolite. Addition of 1 to 7 atomic% of Ge to a Nylon-supported Pt catalyst leads to an improved selectivity of up to 95% at >90% conversion [49]. A Pt/graphite catalyst which has large, faceted metal particles exhibits a higher selectivity than a catalyst which has small particles [50]. Both activity and selectivity of a Co/SiO_2 catalyst prepared from $CoCl_2$ are dependent on the amount of remaining chlorine [51]. Cinnamyl alcohol is obtained from cinnamaldehyde with 96% selectivity at around Cl/Co = 0.2. A Ru catalyst supported on nanometer-scale carbon tubules was reported to afford cinnamyl alcohol with up to 92% selectivity at 80% conversion [52]. This selectivity is much better than that using an Al_2O_3- or carbon-supported catalyst. A polymer-bound Rh catalyst prepared from aminated polystyrene and $Rh_6(CO)_{16}$ shows up to 96% carbonyl selectivity in the hydrogenation of phenyl-substituted α,β-unsaturated aldehydes [53]. Hydrogenation of citral with a bimetallic $Rh–Sn/SiO_2$ catalyst gives the allylic alcohol with 98% selectivity at complete conversion [54]. The Sn/Rh ratio of 0.95 is crucial to achieve high carbonyl selectivity.

Diastereoselectivity

Homogeneous hydrogenation of 4-*tert*-butylcyclohexanone catalyzed by $[Rh(nbd)(dppb)]ClO_4$ (DPPB = 1,2-bis(diphenylphosphino)butane) [6b], $[RhH_2\{P(CH_3)_2C_6H_5\}_2L_2]X$ (L = solvent, X = PF_6 or ClO_4) [6a], $[Rh(cod)(dipfc)]OSO_2CF_3$ [6d], and $[CuHP(C_6H_5)_3]_6$–$P(CH_3)_2C_6H_5$ [27] gives *trans*-4-*tert*-butylcyclohexanol and the *cis* isomer in a 99:1, 86:14, 86:14, and 74:26 ratio, respectively. The hydrogenation to this conformationally anchored ketone occurs preferentially from the axial direction to form the *trans* alcohol. In contrast, a Ru(II) catalyst *in situ* formed from $RuCl_2[P(C_6H_5)_3]_3$, $NH_2(CH_2)_2NH_2$, and KOH in 2-propanol [13, 32] tends to hydrogenate the same substrate from the less crowded direction to give a 98:2 mixture of the *cis* and *trans* alcohols (Fig. 11) [55]. The stereoselectivity of the reaction of other 4-substituted cyclohexanones is controlled basically by the population of the equatorial and axial conformers, leading to a predominance of the *cis* alcohols. Hydrogenation of 3-methylcyclohexanone affords quantitatively a 96:4 mixture of the *trans* and *cis* alcohols. In a similar manner, 2-methyl- and 2-isopropylcyclohexanone are hydrogenated to afford the corresponding *cis* alcohols with

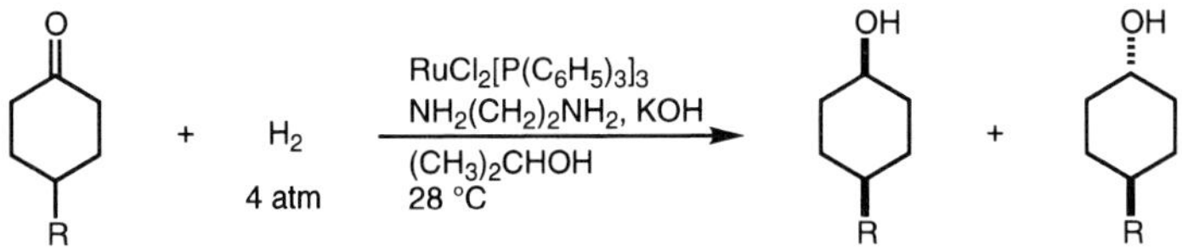

RuCl$_2$[P(C$_6$H$_5$)$_3$]$_3$
NH$_2$(CH$_2$)$_2$NH$_2$, KOH
(CH$_3$)$_2$CHOH
28 °C

+ H$_2$ 4 atm

ketone:Ru:diamine:KOH = 500:1:1:2

R	% yield	cis:trans
t-C$_4$H$_9$	>99	98:2
C$_6$H$_5$	>99	96:4
CH$_3$	97	92:8
OH	>99	83:17

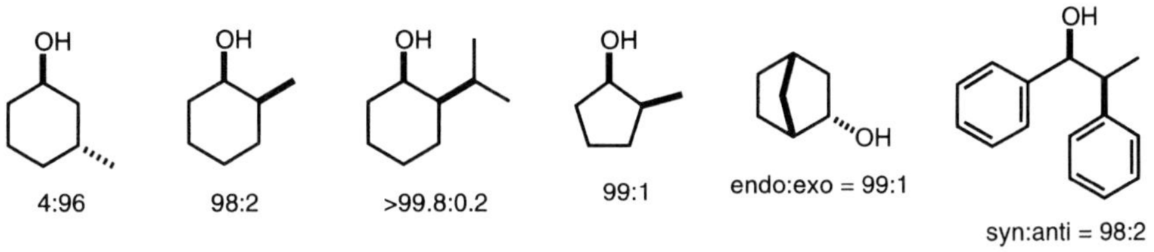

4:96 · 98:2 · >99.8:0.2 · 99:1 · endo:exo = 99:1 · syn:anti = 98:2

Fig. 11 Diastereoselective hydrogenation of ketones.

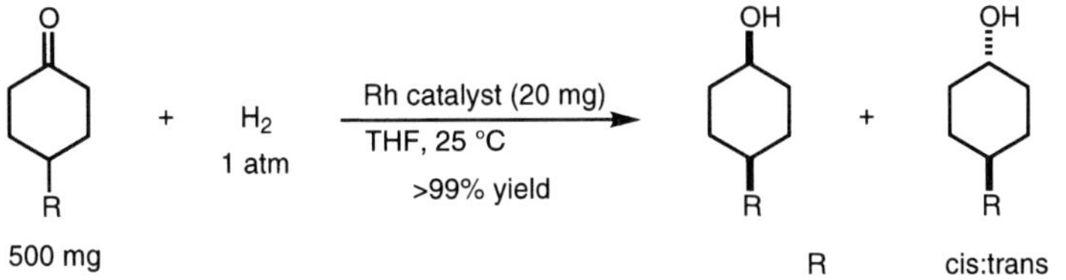

500 mg + H$_2$ 1 atm

Rh catalyst (20 mg)
THF, 25 °C
>99% yield

R	cis:trans
t-C$_4$H$_9$	99:1
CH$_3$	90:10

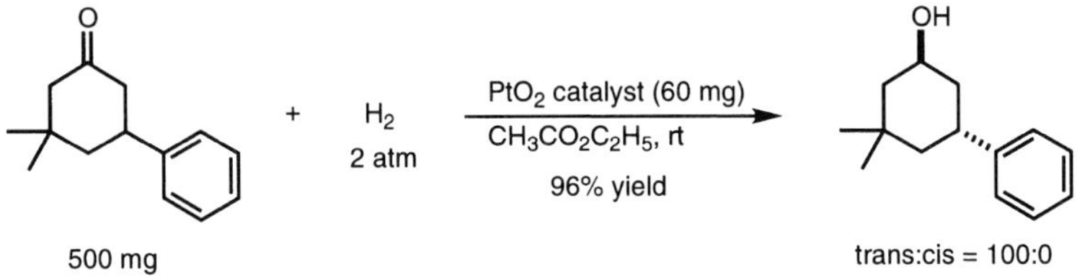

500 mg + H$_2$ 2 atm

PtO$_2$ catalyst (60 mg)
CH$_3$CO$_2$C$_2$H$_5$, rt
96% yield

trans:cis = 100:0

Fig. 12 Diastereoselective hydrogenation of substituted cyclohexanones.

98:2 and >99.8:0.2 selectivity. 2-Methylcyclopentanone is converted to the *cis* alcohol with 99:1 selectivity, whereas bicyclo[2.2.1]heptan-2-one gives a 99:1 mixture of the *endo* and *exo* alcohols. Reaction of conformationally flexible 1-phenylethyl ketones displays a high Cram selectivity. The degrees of the kinetic diastereoface

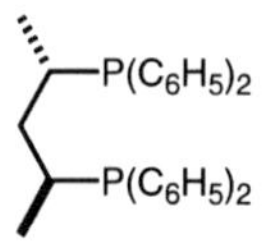

$P(C_6H_5)_2$
$P(C_6H_5)_2$
(*S,S*)-BDPP
(SKEWPHOS)

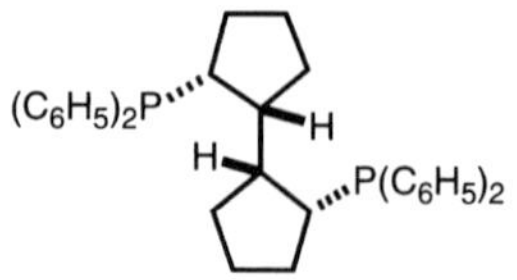

$(C_6H_5)_2P$
H
H
$P(C_6H_5)_2$
(*R,R*)-BICP

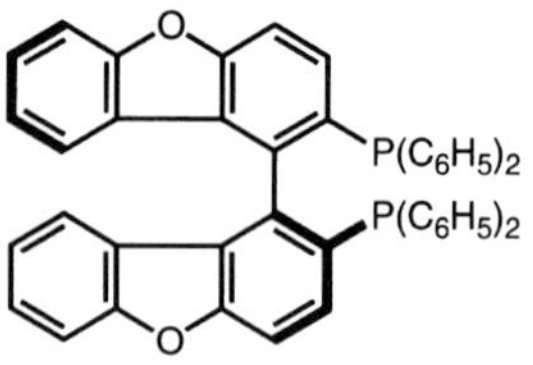

O
$P(C_6H_5)_2$
$P(C_6H_5)_2$
O
(*S*)-BIFAP

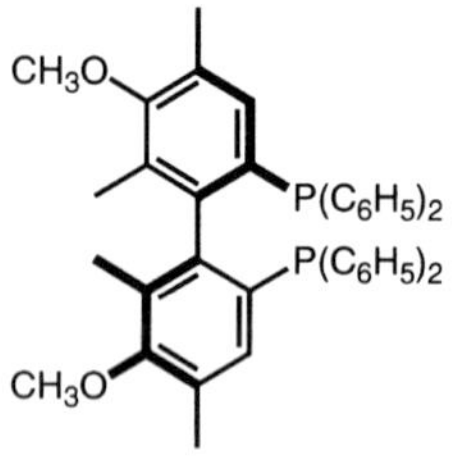

CH_3O
$P(C_6H_5)_2$
$P(C_6H_5)_2$
CH_3O
(*R*)-BIMOP

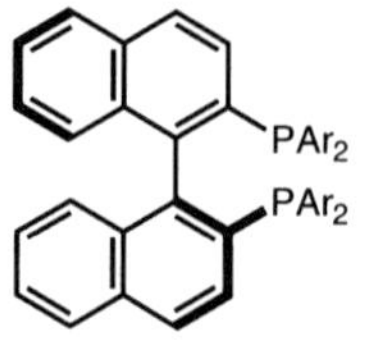

PAr_2
PAr_2
(*S*)-BINAP
BINAP: Ar = C_6H_5
TolBINAP: Ar = 4-$CH_3C_6H_4$
XylBINAP: Ar = 3,5-$(CH_3)_2C_6H_3$
DTBBINAP: Ar = 3,5-$(t\text{-}C_4H_9)_2C_6H_3$
(absolute configuration unknown)

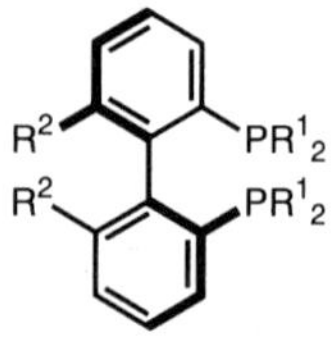

R^2
R^2
PR^1_2
PR^1_2
(*S*)-BIPHEMP
BIPHEMP: $R^1 = C_6H_5$; $R^2 = CH_3$
MeO-BIPHEP: $R^1 = C_6H_5$; $R^2 = CH_3O$
BICHEP: R^1 = *cyclo*-C_6H_{11}; $R^2 = CH_3$

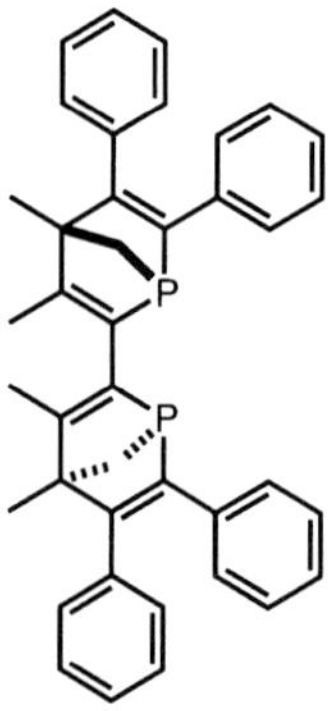

P
P
(*R,R*)-BIPNOR

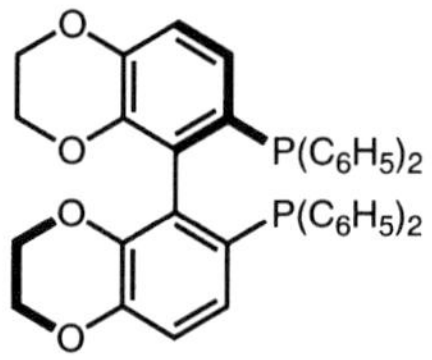

O
O
$P(C_6H_5)_2$
$P(C_6H_5)_2$
O
O
(*R*)-BisbenzodioxanPhos

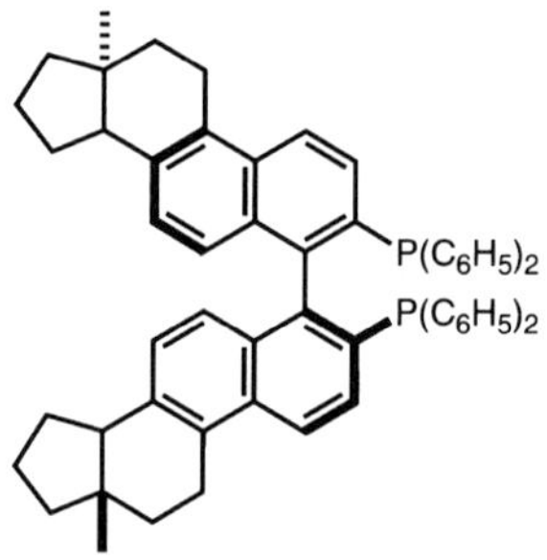

$P(C_6H_5)_2$
$P(C_6H_5)_2$
(*S*)-bis-steroidal phosphine

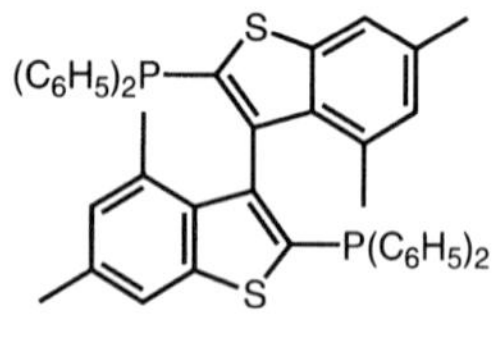

$(C_6H_5)_2P$
S
S
$P(C_6H_5)_2$
tetraMe-BITIANP
(absolute configuration unknown)

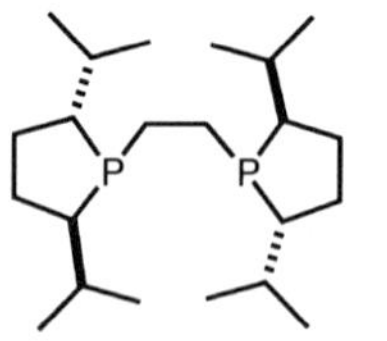

P
P
(*R,R*)-*i*-Pr-BPE

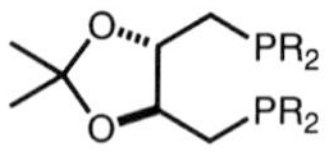

O
O
PR_2
PR_2
(*S,S*)-DIOP
DIOP: R = C_6H_5
CyDIOP: R = *cyclo*-C_6H_{11}

Fig. 13 C_2-chiral diphosphine ligands (in alphabetical order).

PAr_2
PAr_2

(*R*)-Xyl-HexaPHEMP
Ar = 3,5-$(CH_3)_2C_6H_3$

$P(C_6H_5)_2$
$P(C_6H_5)_2$

(*S*,*S*)-NORPHOS

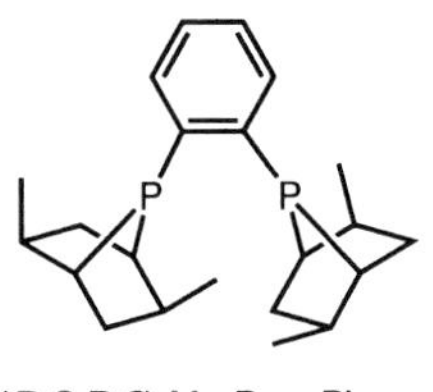

(*R*,*S*,*R*,*S*)-Me-PennPhos

PAr_2
PAr_2

(*S*)-[2.2]PHANEPHOS

PHANEPHOS: Ar = C_6H_5
xylyl-PHANEPHOS: Ar = 3,5-$(CH_3)_2C_6H_3$

OCH_3
N
CH_3O PAr_2
CH_3O PAr_2
N
OCH_3

(*R*)-P-Phos

P-Phos: Ar = C_6H_5
Tol-P-Phos: Ar = 4-$CH_3C_6H_4$
Xyl-P-Phos: Ar = 3,5-$(CH_3)_2C_6H_3$

O
O PAr_2
O PAr_2
O

(*R*)-SEGPHOS

SEGPHOS: Ar = C_6H_5
DTBM-SEGPHOS:
Ar = 4-CH_3O-3,5-(t-$C_4H_9)_2C_6H_2$
(absolute configuration unknown)

O $P(C_6H_5)_2$
O $P(C_6H_5)_2$

(*R*)-C4TunaPhos

Fig. 13 (cont.)

discrimination compare well with those accomplished by stoichiometric reduction using Selectride reagents [56].

The study on diastereoselective hydrogenation of simple ketones using heterogeneous catalysts was almost complete by the middle of the 1980s, as detailed in the reviews [1a, 1b, 57]. Excellent selectivity was reported for hydrogenation of substituted cyclohexanones having an anchored conformation (Fig. 12). For example, hydrogenation of 4-*tert*-butylcyclohexanone in the presence of a Rh catalyst gives the *cis* and *trans* alcohols in a 99:1 ratio [58]. The 4-methyl analog also shows a good *cis* selectivity. This procedure is applicable to diastereoselective reduction of steroids [58b]. 3,3-Dimethyl-5-phenylcyclohexanone is converted by PtO_2 catalyst to the *trans* alcohol in a pure form [59].

Enantioselectivity

Chiral ligands

Nowadays, a wide variety of optically active organic ligands are available [60, 61]. A suitable combination of a metal species and chiral ligand is the key to preparing high-performance catalysts for asymmetric hydrogenation [62]. Commonly used C_2-chiral diphosphine ligands are listed in Fig. 13. Figs. 14 and 15 show diphosphines without C_2 symmetry and amido- or aminophosphines. Chiral amines and amino alcohols are indicated in Figs. 16 and 17. Recently reported immobilized chiral diphosphine ligands are shown in Fig. 18.

Alkyl aryl ketones

Asymmetric hydrogenation of simple ketones has remained difficult to realize. Only a few catalysts enable unfunctionalized chiral alcohols with high optical purity to be produced [62g–j]. Some cationic or neutral transition metal catalysts with monodentate or bidentate chiral phosphine ligands were developed [60], but the

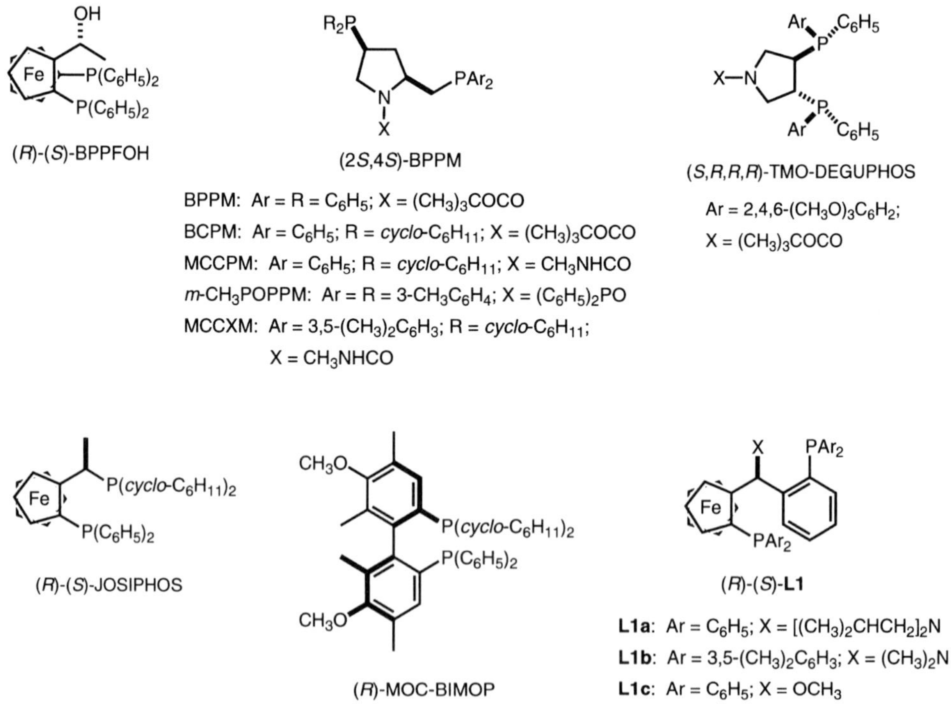

(*R*)-(*S*)-BPPFOH

(2*S*,4*S*)-BPPM

BPPM: Ar = R = C_6H_5; X = $(CH_3)_3COCO$
BCPM: Ar = C_6H_5; R = *cyclo*-C_6H_{11}; X = $(CH_3)_3COCO$
MCCPM: Ar = C_6H_5; R = *cyclo*-C_6H_{11}; X = CH_3NHCO
m-CH_3POPPM: Ar = R = 3-$CH_3C_6H_4$; X = $(C_6H_5)_2PO$
MCCXM: Ar = 3,5-$(CH_3)_2C_6H_3$; R = *cyclo*-C_6H_{11}; X = CH_3NHCO

(*S*,*R*,*R*,*R*)-TMO-DEGUPHOS

Ar = 2,4,6-$(CH_3O)_3C_6H_2$; X = $(CH_3)_3COCO$

(*R*)-(*S*)-JOSIPHOS

(*R*)-MOC-BIMOP

(*R*)-(*S*)-**L1**

L1a: Ar = C_6H_5; X = $[(CH_3)_2CHCH_2]_2N$
L1b: Ar = 3,5-$(CH_3)_2C_6H_3$; X = $(CH_3)_2N$
L1c: Ar = C_6H_5; X = OCH_3

Fig. 14 Diphosphine ligands without C_2 chirality.

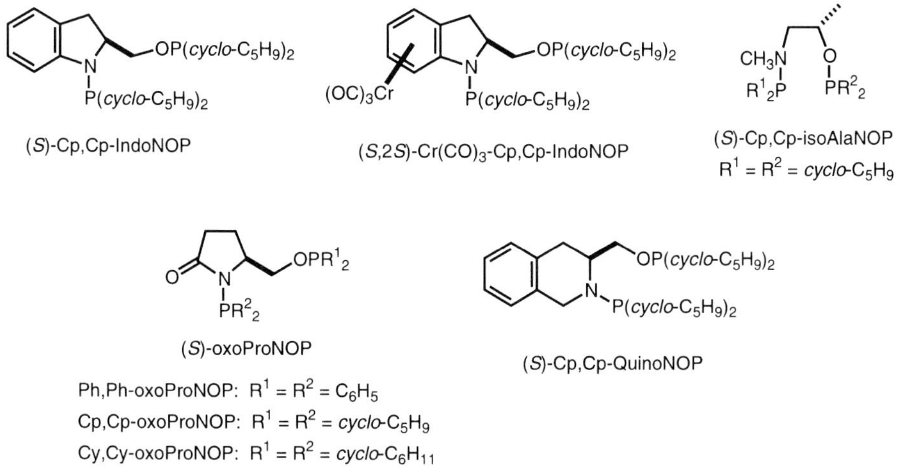

(*S*)-Cp,Cp-IndoNOP

(*S*,2*S*)-Cr(CO)$_3$-Cp,Cp-IndoNOP

(*S*)-Cp,Cp-isoAlaNOP
$R^1 = R^2 =$ *cyclo*-C_5H_9

(*S*)-oxoProNOP

Ph,Ph-oxoProNOP: $R^1 = R^2 = C_6H_5$
Cp,Cp-oxoProNOP: $R^1 = R^2 =$ *cyclo*-C_5H_9
Cy,Cy-oxoProNOP: $R^1 = R^2 =$ *cyclo*-C_6H_{11}

(*S*)-Cp,Cp-QuinoNOP

Fig. 15 Amido- or aminophosphine ligands (in alphabetical order).

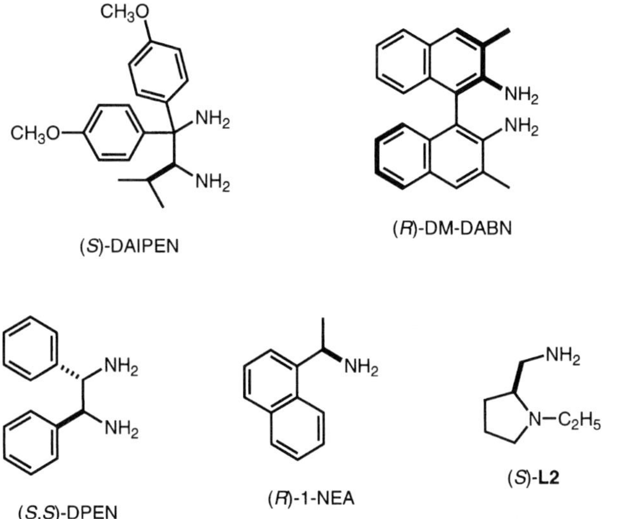

Fig. 16 Amine ligands.

enantioselectivity remained unsatisfactory. However, a breakthrough has been provided by the development of Ru catalysts, which have BINAP as a chiral diphosphine and a chiral 1,2-diamine [14].

The discovery of Ru catalysts consisting of $RuCl_2$(diphosphine)(1,2-diamine) and alkaline base has achieved high-speed and practical asymmetric hydrogenation of simple ketones [14]. For example, acetophenone (601 g) is completely converted to (*R*)-1-phenylethanol in 80% *ee* with only 2.2 mg of *trans*-$RuCl_2$[(*S*)-tolbi-

cinchona alkaloids

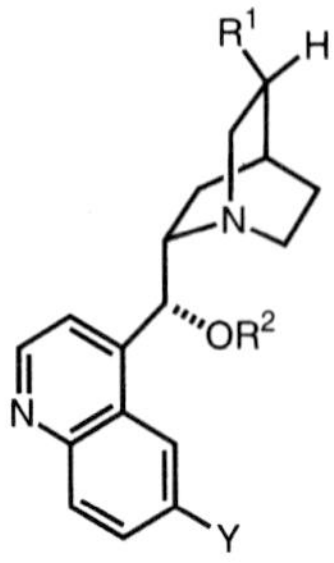

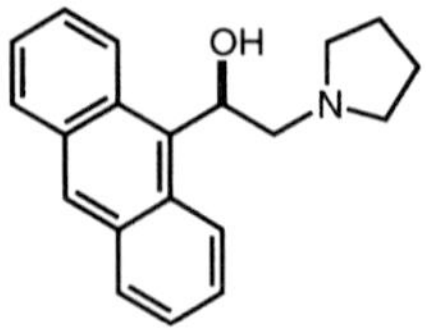

(*R*)-**L3**

cinchonidine: $R^1 = CH_2{=}CH$; $R^2 = H$; Y = H

HCd: $R^1 = C_2H_5$; $R^2 = H$; Y = H

quinine: $R^1 = CH_2{=}CH$; $R^2 = H$; $Y = CH_3O$

MeOHCd: $R^1 = C_2H_5$; $R^2 = CH_3$; Y = H

Fig. 17 Amino alcohols and esters.

nap][(*S,S*)-dpen] at 30 °C and under 45 atm of H_2 (Fig. 19) [15]. The turnover number (TON), defined as moles of product per mole of catalyst, reaches 2 400 000, while the TOF at 30% conversion was 228 000 h^{-1} or 63 s^{-1}. 2′-Methylacetophenone and 1′-acetonaphthone are hydrogenated in the presence of TolBINAP/1,2-diamine–Ru catalysts, **C1** and **C2**, with an S/C ratio of 100 000 to give the corresponding chiral alcohols in 99% and 98% *ee*, respectively (Fig. 20) [15]. However, no single chiral catalyst can be universal because of the structural diversity of ketonic substrates. Only limited kinds of ketones are reduced in sufficiently high enantioselectivity with catalysts **C1** and **C2**.

The use of more sterically hindered XylBINAP as a chiral diphosphine ligand has greatly expanded the scope of this reaction [39]. A wide variety of alkyl aryl ketones are hydrogenated with *trans*-$RuCl_2$[(*S*)-xylbinap][(*S*)-daipen] (or the *R*/*R* combination) and $(CH_3)_3COK$, resulting in chiral alcohols with a consistently high optical purity, while the reactivity slightly decreases. For example, acetophenone is reduced in the presence of the (*S*)-XylBINAP/(*S*)-DAIPEN–Ru catalyst (*S,S*)-**C3** with an S/C of 100 000 under 8 atm of H_2 to give (*R*)-1-phenylethanol in 99% *ee* quantitatively. 3′-Methyl- and 4′-methoxyacetophenone are reduced with 100% optical yield. DPEN is also usable as a diamine ligand. The hydrogenation tolerates many functional groups on the aromatic ring, including F, Cl, Br, I, CF_3, OCH_3, $CO_2CH(CH_3)_2$, NO_2, and NH_2. The influence of electric and steric character of substituents on enantioselectivity is rather small. Propiophenone, isobutyrophenone, cyclopropyl phenyl ketone, and 1′- and 2′-acetonaphthone are also reduced in excellent optical yield. This reaction is applied to the asymmetric synthesis of a potent therapeutic agent for prostatomegary, TF-505 [63]. *trans*-RuHCl[(*S*)-binap][(*S,S*)-cydn] with $(CH_3)_3COK$ also showed high catalytic activity [16]. Ru catalysts with biaryl diphosphine ligands, Xyl-HexaPHEMP [64] and Xyl-P-Phos [65], instead of XylBINAP, give similar results. Combination of (*R*)-Xylyl-Phanephos and (*S,S*)-DPEN also provided a high

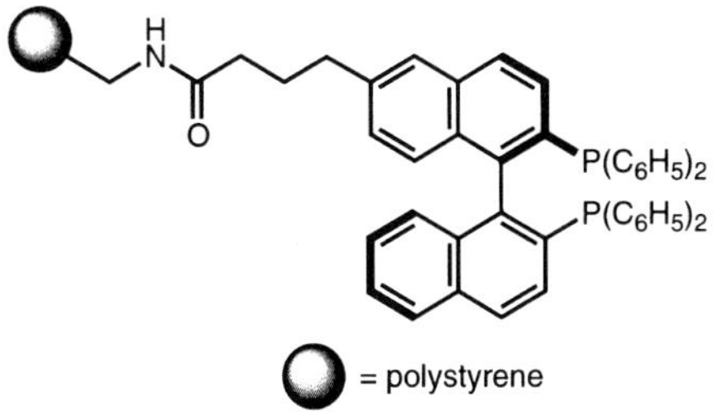

(*R*)-APBBINAP

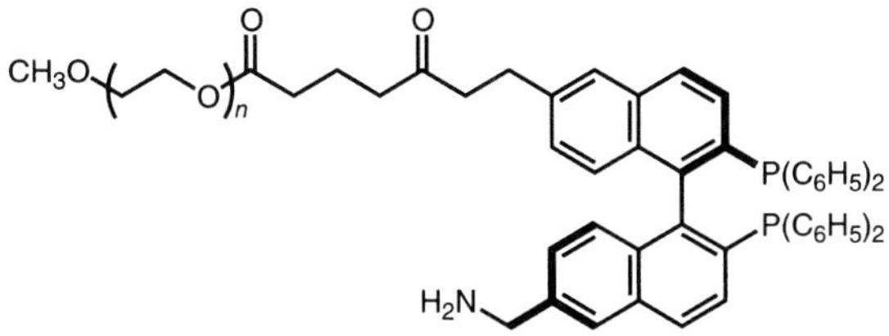

PEG-(*R*)-Am-BINAP

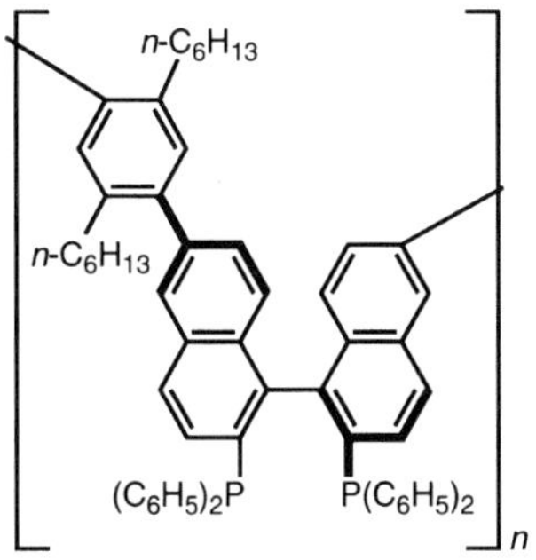

(*R*)-poly(BINAP)

Fig. 18 Immobilized BINAP ligands (in alphabetical order).

level of enantioselection [66]. Several aromatic ketones are converted to the chiral alcohols in 99% *ee*. Hydrogenation of acetophenone promoted by (*S,S*)-BDPP/(*S,S*)-DPEN–Ru catalyst gives (*R*)-1-phenylethanol in 84% *ee* [19]. *In situ*-generated catalyst from $RuBr_2$[(*R,R*)-bipnor], (*S,S*)-DPEN, and KOH mediates reduction of 2′-acetonaphthone with an optical yield of 81% [67]. Pivalophenone, a sterically hindered aromatic ketone, is hydrogenated with a ternary catalyst system consisting of RuClCp*(cod) (Cp* = pentamethylcyclopentadienyl), chiral diamine (*S*)-**L2**, and

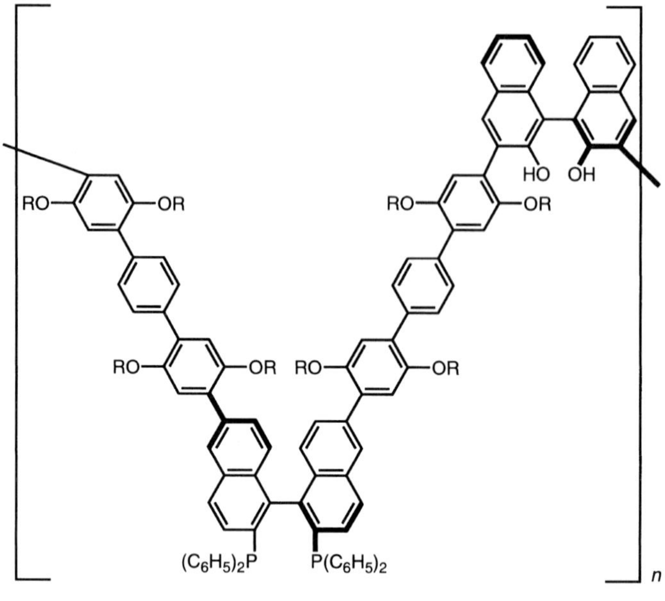

$R = n\text{-}C_6H_{13}$

(*R*,*R*)-poly(BINOL-BINAP)

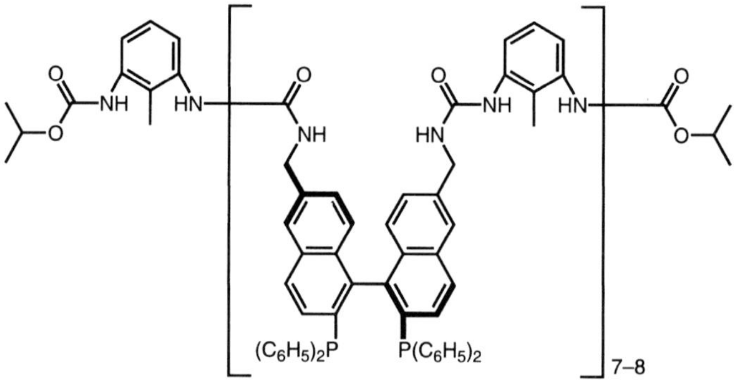

(*R*)-poly-NAP

Fig. 18 (cont.)

KOH to afford the *R* alcohol in 81% *ee* [68]. Hydrogenation of 2′-halo-substituted acetophenones with $[NH_2(C_2H_5)_2][\{RuCl[(S)\text{-tolbinap}]\}_2(\mu\text{-}Cl)_3]$ under 85 atm of H_2 resulted in the halogenated alcohols in up to >99% *ee* [69]. The reaction was supposed to proceed via a stable six-membered intermediate constructed by the chelation of carbonyl oxygen and halogen at the 2′ position to the Ru metal [62c].

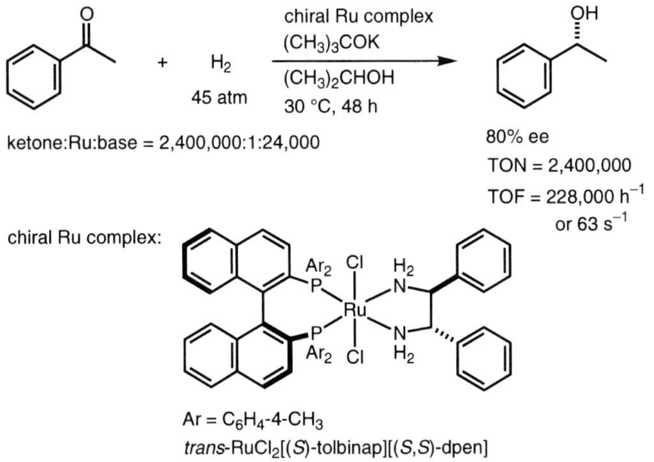

chiral Ru complex
$(CH_3)_3COK$
$(CH_3)_2CHOH$
30 °C, 48 h

+ H_2 45 atm

ketone:Ru:base = 2,400,000:1:24,000

80% ee
TON = 2,400,000
TOF = 228,000 h^{-1}
or 63 s^{-1}

chiral Ru complex:

Ar = C_6H_4-4-CH_3
trans-$RuCl_2$[(*S*)-tolbinap][(*S*,*S*)-dpen]

Fig. 19 Practical asymmetric hydrogenation of simple ketones.

$RuCl_2$(diphosphine)(1,2-diamine) type complexes require an addition of strong base to generate catalytic species for hydrogenation of ketones mainly for neutralization of releasing HCl, as shown in Fig. 4. Therefore, highly base-sensitive ketonic substrates cannot be reduced with the catalyst systems. A newly devised *trans*-RuH(η^1-BH_4)[(*S*)-xylbinap][(*S*,*S*)-dpen], which is prepared from the corresponding $RuCl_2$ complex and excess amount of $NaBH_4$, generates active species in the absence of an additional base [17]. Acetophenone is completely hydrogenated using the *S,SS* complex with an S/C of 100 000 under 8 atm of H_2 within 7 h to give the *R* alcohol in 99% *ee* (Fig. 20). The reaction is even accelerated by an addition of base, so that the substrate is completely converted in 45 min with the same optical yield under otherwise identical conditions. The base-free procedure is successfully applied to hydrogenation of several ketonic substrates containing a base-sensitive substituent [17]. For example, hydrogenation of (*R*)-glycidyl 3-acetylphenyl ether in the presence of the *S,SS* catalyst results in the *R,R* product in 99% *de* quantitatively, leaving the base-labile epoxy ring intact (Fig. 20). Ethyl 4-acetylbanzoate is reduced with the same catalyst to afford ethyl (*R*)-4-(1-hydroxyethyl)benzoate in 99% *ee* as a sole product without any detectable transesterification.

Early attempts at asymmetric hydrogenation of simple ketones were done with chiral Rh catalysts. Acetophenone and 1′-acetonaphthone are hydrogenated with a catalyst system prepared from $[RhCl(nbd)]_2$, (*S*,*S*)-DIOP, and $(C_2H_5)_3N$ at an S/C of 200 under 69 atm of H_2 to give the corresponding alcohols in 80% and 84% *ee*, respectively (Fig. 20) [70, 71]. $RhH[P(C_6H_5)_3]_3$ is supposed to be a reactive species. When (*S*,*S*)-BDPP is used as a chiral ligand, acetophenone is hydrogenated with an 82% optical yield [72]. Several aromatic ketones are hydrogenated with an (*R*,*S*,*R*,*S*)-Me-PennPhos–Rh complex (S/C=100) in the presence of 2,6-lutidine and KBr under 30 atm of H_2 to give the *S* alcohols in high yield and with up to 95% *ee* [73]. The additives play key roles in increasing both reactivity and enantio-

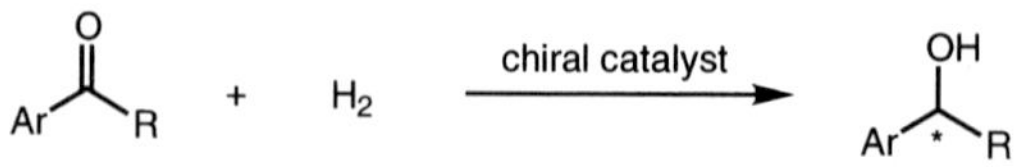

$$\text{ArC(O)R} + H_2 \xrightarrow{\text{chiral catalyst}} \text{ArC}^*\text{H(OH)R}$$

chiral catalyst:

trans-$RuCl_2$[(*S*)-tolbinap][(*S*)-daipen] + $(CH_3)_3COK$; (*S*,*S*)-**C1**
trans-$RuCl_2$[(*S*)-tolbinap][(*S*,*S*)-dpen] + $(CH_3)_3COK$; (*S*,*SS*)-**C2**
trans-$RuCl_2$[(*S*)-xylbinap][(*S*)-daipen] + $(CH_3)_3COK$; (*S*,*S*)-**C3**
trans-$RuCl_2$[(*S*)-xylbinap][(*S*,*S*)-dpen] + $(CH_3)_3COK$; (*S*,*SS*)-**C4**
trans-RuHCl[(*S*)-binap][(*S*,*S*)-cydn] + $(CH_3)_2CHOK$; (*S*,*SS*)-**C5**
trans-RuH(η^1-BH_4)[(*S*)-xylbinap][(*S*,*S*)-dpen]; (*S*,*SS*)-**C6**
trans-RuH(η^1-BH_4)[(*S*)-xylbinap][(*S*,*S*)-dpen] + $(CH_3)_3COK$; (*S*,*SS*)-**C7**
trans-$RuCl_2$[(*S*)-xyl-hexaphemp][(*S*)-daipen] + $(CH_3)_3COK$; (*S*,*S*)-**C8**
trans-$RuCl_2$[(*R*)-xyl-p-phos][(*R*,*R*)-dpen] + $(CH_3)_3COK$; (*R*,*RR*)-**C9**
trans-$RuCl_2$[(*R*)-xylyl-phanephos][(*S*,*S*)-dpen] + $(CH_3)_3COK$; (*R*,*SS*)-**C10**
$RuCl_2$[(*S*,*S*)-bdpp][(*S*,*S*)-dpen] + $(CH_3)_3COK$; (*SS*,*SS*)-**C11**
$RuBr_2$[(*R*,*R*)-bipnor]–(*S*,*S*)-DPEN + KOH; (*RR*,*SS*)-**C12**
[$NH_2(C_2H_5)_2$][{RuCl[(*S*)-tolbinap]}$_2$(μ-Cl)$_3$]; (*S*)-**C13**
RuClCp*(cod)–(*S*)-**L2** + KOH; (*S*)-**C14**
[RhCl(nbd)]$_2$–(*S*,*S*)-DIOP + $(C_2H_5)_3N$; (*S*,*S*)-**C15**
[RhCl(nbd)]$_2$–(*S*,*S*)-BDPP + $(C_2H_5)_3N$; (*S*,*S*)-**C16**
[RhCl(cod)]$_2$–(*R*,*S*,*R*,*S*)-Me-PennPhos + 2,6-lutidine + KBr; (*R*,*S*,*R*,*S*)-**C17**
[Ir{(*S*)-binap}(cod)]BF_4–P[2-N$(CH_3)_2C_6H_4]_2C_6H_5$; (*S*)-**C18**

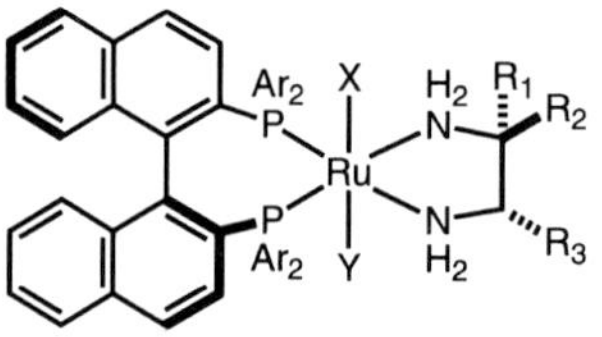

trans-$RuCl_2$[(*S*)-tolbinap][(*S*)-daipen]: X = Y = Cl,
Ar = 4-$CH_3C_6H_4$, R^1 = R^2 = 4-$CH_3OC_6H_4$, R^3 = $(CH_3)_2CH$

trans-$RuCl_2$[(*S*)-tolbinap][(*S*,*S*)-dpen]: X = Y = Cl,
Ar = 4-$CH_3C_6H_4$, R^1 = H, R^2 = R^3 = C_6H_5

trans-$RuCl_2$[(*S*)-xylbinap][(*S*)-daipen]: X = Y = Cl,
Ar = 3,5-$(CH_3)_2C_6H_3$, R^1 = R^2 = 4-$CH_3OC_6H_4$, R^3 = $(CH_3)_2CH$

trans-$RuCl_2$[(*S*)-xylbinap][(*S*,*S*)-dpen]: X = Y = Cl,
Ar = 3,5-$(CH_3)_2C_6H_3$, R^1 = H, R^2 = R^3 = C_6H_5

trans-RuH(η^1-BH_4)[(*S*)-xylbinap][(*S*,*S*)-dpen]: X = H, Y = η^1-BH_4,
Ar = 3,5-$(CH_3)_2C_6H_3$, R^1 = H, R^2 = R^3 = C_6H_5

Fig. 20 Asymmetric hydrogenation of aromatic ketones.

R	Ar	Catalyst	S/C[a)]	H_2 (atm)	Temp (°C)	% yield	% ee	Confign
CH_3	C_6H_5	(*S,S*)-**C3**	100 000	8	45	100	99	*R*
CH_3	C_6H_5	(*S,SS*)-**C5**	5000	3	20	100	88	*R*
CH_3	C_6H_5	(*S,SS*)-**C6**	100 000	8	45	100	99	*R*
CH_3	C_6H_5	(*S,SS*)-**C7**	100 000	8	45	100	99	*R*
CH_3	C_6H_5	(*S,S*)-**C8**	3000	8	rt	>99	99	*R*
CH_3	C_6H_5	(*R,RR*)-**C9**	100 000	34	25–28	99.7	99	*S*
CH_3	C_6H_5	(*R,SS*)-**C10**	20 000	8	18–20	>99	99	*R*
CH_3	C_6H_5	(*SS,SS*)-**C11**	500	2	rt	100	84	*R*
CH_3	C_6H_5	(*S,S*)-**C15**	200	69	50	64	80	–
CH_3	C_6H_5	(*S,S*)-**C16**	100	69	50	72	82	*S*
CH_3	C_6H_5	(*R,S,R,S*)-**C17**	100[b)]	30	rt	97	95	*S*
CH_3	C_6H_5	(*S*)-**C18**	100	54–61	60	63	54	*S*
CH_3	2-$CH_3C_6H_4$	(*S,S*)-**C1**	100 000	10	28	94	99	*R*
CH_3	3-$CH_3C_6H_4$	(*S,S*)-**C3**	10 000	10	28	98	100	*R*
CH_3	4-$CH_3C_6H_4$	(*R,RR*)-**C4**	2000	4	28	100	98	*S*
CH_3	4-*n*-$C_4H_9C_6H_4$	(*R,RR*)-**C4**	2000	4	28	100	98	*S*
CH_3	2,4-$(CH_3)_2C_6H_3$	(*R,R*)-**C3**	2000	4	28	99	99	*S*
CH_3	2-FC_6H_4	(*S,S*)-**C3**	2000	8	28	100	97	*R*
CH_3	2-FC_6H_4	(*S*)-**C13**	1300	85	35	21	>99	–
CH_3	3-FC_6H_4	(*R,RR*)-**C4**	2000	4	28	99	98	*S*
CH_3	4-FC_6H_4	(*R,R*)-**C3**	2000	4	28	100	97	*S*
CH_3	2-ClC_6H_4	(*R,RR*)-**C4**	2000	4	28	99.5	98	*S*
CH_3	2-BrC_6H_4	(*R,R*)-**C1**	10 000	10	28	100	98	*S*
CH_3	2-BrC_6H_4	(*R,R*)-**C3**	2000	4	28	99	96	*S*
CH_3	2-BrC_6H_4	(*S*)-**C13**	950	85	35	95	97	*S*
CH_3	3-BrC_6H_4	(*R,R*)-**C3**	2000	4	28	100	99.5	*S*
CH_3	4-BrC_6H_4	(*S,S*)-**C3**	20 000	8	28	99.9	99.6	*R*
CH_3	4-BrC_6H_4	(*S,S*)-**C3**	500	1	28	99.7	99.6	*R*
CH_3	4-BrC_6H_4	(*R,SS*)-**C10**	3000	8	18–20	>99	99	*R*
CH_3	4-IC_6H_4	(*S,S*)-**C3**	2000	8	28	99.7	99	*R*
CH_3	2-$CF_3C_6H_4$	(*R,R*)-**C3**	2000	4	28	99	99	*S*
CH_3	3-$CF_3C_6H_4$	(*R,R*)-**C3**	2000	4	28	100	99	*S*
CH_3	3-$CF_3C_6H_4$	(*R,SS*)-**C10**	3000	8	18–20	>99	99	*R*
CH_3	4-$CF_3C_6H_4$	(*S,S*)-**C3**	10 000	10	28	100	99.6	*R*
CH_3	2-$CH_3OC_6H_4$	(*R,R*)-**C3**	2000	4	28	100	92	*S*
CH_3	3-$CH_3OC_6H_4$	(*R,R*)-**C3**	2000	4	28	99	99	*S*
CH_3	4-$CH_3OC_6H_4$	(*S,S*)-**C3**	2000	10	28	100	100	*R*
CH_3	4-$CH_3OC_6H_4$	(*R,S,R,S*)-**C17**	100	30	rt	83	94	*S*
CH_3	3-(*R*)-glycidyl-oxyphenyl	(*S,SS*)-**C6**	2000	8	25	99	99	*R,R*
CH_3	4-(C_2H_5OCO)-C_6H_4	(*S,SS*)-**C6**	4000	8	25	100	99	*R*
CH_3	4-$[(CH_3)_2CH$-$OCO]C_6H_4$	(*S,S*)-**C3**	2000	8	28	100	99	*R*
CH_3	4-$NO_2C_6H_4$	(*S,S*)-**C3**	2000	8	28	100	99.8	*R*
CH_3	4-$NH_2C_6H_4$	(*S,S*)-**C3**	2000	8	28	100	99	*R*
CH_3	1-naphthyl	(*S,SS*)-**C2**	100 000	10	28	99.5	98	*R*

Fig. 20 (cont.)

R	*Ar*	*Catalyst*	*S/C*[a)]	*H₂, atm*	*Temp (°C)*	*% yield*	*% ee*	*Confign*
CH_3	1-naphthyl	(*R,RR*)-**C4**	2000	4	28	99	99	*S*
CH_3	1-naphthyl	(*S,S*)-**C15**	200	69	50	100	84	–
CH_3	2-naphthyl	(*R,RR*)-**C4**	2000	4	28	99	98	*S*
CH_3	2-naphthyl	(*RR,SS*)-**C12**	500	5	28	65	81	*R*
C_2H_5	C_6H_5	(*R,RR*)-**C4**	2000	4	28	100	99	*S*
C_2H_5	C_6H_5	(*S,RR*)-**C10**	3000	5.5	18–20	>99	98	*S*
C_2H_5	C_6H_5	(*R,S,R,S*)-**C17**	100	30	rt	95	93	*S*
C_2H_5	4-FC_6H_4	(*R,RR*)-**C4**	2000	4	28	99	99	*S*
C_2H_5	4-ClC_6H_4	(*S,S*)-**C3**	20 000	8	28	99.9	99	*R*
$(CH_3)_2CH$	C_6H_5	(*R,R*)-**C3**	10 000	8	28	99.7	99	*S*
$(CH_3)_2CH$	C_6H_5	(*S*)-**C18**	200	54–61	90	78	84	*R*
cyclo-C_3H_5	C_6H_5	(*S,S*)-**C3**	2000	8	28	99.7	99	*R*
$(CH_3)_2$-$CHCH_2$	C_6H_5	(*S*)-**C14**	100	10	30	98	95	*R*
$(CH_3)_3C$	C_6H_5	(*S*)-**C14**	100	10	30	99	81	*R*

a) Substrate/catalyst mole ratio.
b) Without addition of KBr.

Fig. 20 (cont.)

selectivity. A cationic BINAP–Ir(I) complex combined with an aminophosphine is successfully used for the hydrogenation of cyclic aromatic ketones (Fig. 21) [74]. Although the reaction requires H_2 pressures up to 57 atm and temperatures as high as 90 °C, this was the first example in which >90% optical yield is achieved in the hydrogenation of simple ketones. Enantioselectivity in the hydrogenation of alkyl phenyl ketones with the Ir catalyst is highly dependent on the bulkiness of the alkyl groups (Fig. 20) [75].

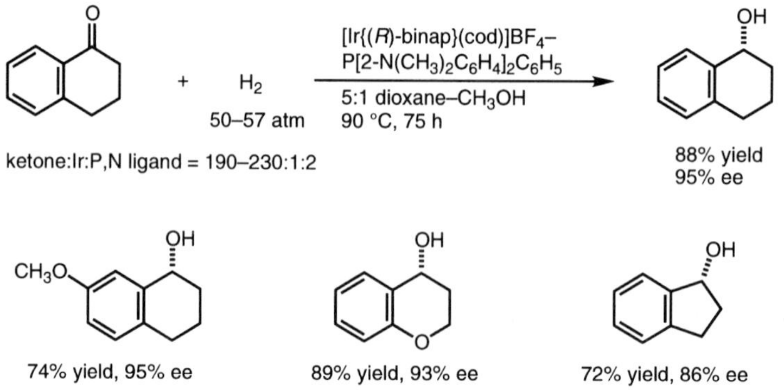

+ H_2 50–57 atm

[Ir{(*R*)-binap}(cod)]BF_4–P[2-N$(CH_3)_2C_6H_4]_2C_6H_5$
5:1 dioxane–CH_3OH
90 °C, 75 h

ketone:Ir:P,N ligand = 190–230:1:2

88% yield
95% ee

CH_3O

74% yield, 95% ee

89% yield, 93% ee

72% yield, 86% ee

Fig. 21 Asymmetric hydrogenation of aromatic ketones with a BINAP–Ir complex.

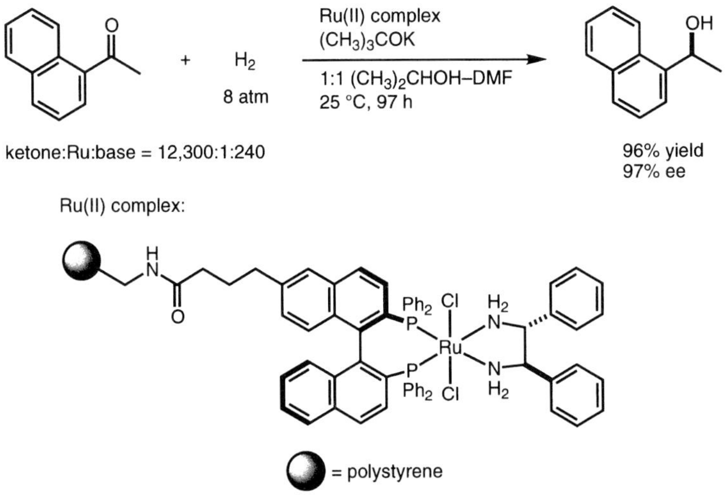

Ru(II) complex
$(CH_3)_3COK$
+ H_2 8 atm
1:1 $(CH_3)_2CHOH$–DMF
25 °C, 97 h
ketone:Ru:base = 12,300:1:240
96% yield
97% ee
Ru(II) complex:
= polystyrene

Fig. 22 Asymmetric hydrogenation of ketones with a polymer-bound BINAP/diamine–Ru catalyst.

Immobilized catalysts on solid supports have inherent benefits, such as easy separation from products and facility for recycling use. These catalysts are expected to be useful for combinatorial synthesis. The use of a polystyrene-bound BINAP as the chiral diphosphine ligand enabled immobilization of the BINAP/DPEN–Ru catalyst [76]. Hydrogenation of 1′-acetonaphthone in the presence of the immobilized complex (beads) and $(CH_3)_3COK$ with an S/C of 12 300 in a 2-propanol–DMF mixture (1:1 v/v) under 8 atm of H_2 affords (*S*)-1-(1-naphthyl)ethanol in 98% *ee* and 96% yield (Fig. 22). The polymer-bound catalyst is separated simply by a filtration. When the reaction is conducted with an S/C of 2470 under otherwise identical conditions, the catalyst can be used 14 times without loss of enantioselectivity, achieving a total TON of 33 000. Several BINAP-incorporated polymers have been used for the same purpose. Hydrogenation of 1′-acetonaphthone promoted by the (*S*)-Poly-Nap/(*S,S*)-DPEN–Ru catalyst with an S/C of 1000 under 40 atm of H_2 gave the *R* alcohol in 96% *ee* quantitatively [77]. The reaction could be repeated four times without loss of optical yield. Acetophenone is completely converted to (*S*)-1-phenylethanol with an optical yield of 84% in the presence of (*R,R*)-poly(BINOL–BINAP)/(*R,R*)-DPEN–Ru catalyst with an S/C of 4900 under 12 atm of H_2 [78]. The same catalyst efficiency was achieved by the use of poly(BINAP) as a diphosphine ligand [79].

Diaryl ketones

Generally, asymmetric hydrogenation of pro-chiral diarylketones is difficult because it requires differentiation of two electrically and sterically similar aryl groups. Furthermore, the produced diaryl methanols are easily converted to the corresponding diaryl methanes under regular hydrogenation conditions. However,

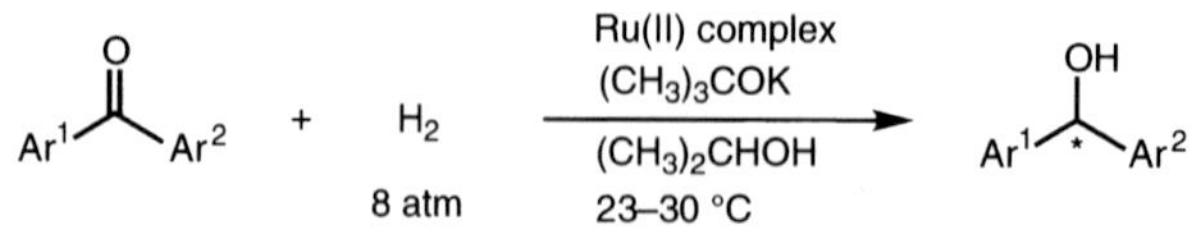

$$Ar^1C(=O)Ar^2 + H_2 \text{ (8 atm)} \xrightarrow[\text{(CH}_3)_2\text{CHOH, 23–30 °C}]{\text{Ru(II) complex, (CH}_3)_3\text{COK}} Ar^1\overset{*}{C}H(OH)Ar^2$$

Ru(II) complex = *trans*-RuCl$_2$[(*S*)-xylbinap][(*S*)-daipen]

Ar[1]	*Ar*[2]	*S/C*[a)]	*% yield*	*% ee*	*Confign*
2-$CH_3C_6H_4$	C_6H_5	2000	99	93	*S*
2-$CH_3OC_6H_4$	C_6H_5	2000	100	99	*S*
2-FC_6H_4	C_6H_5	2000	99	97	*S*
2-ClC_6H_4	C_6H_5	20000	99	97	*S*
2-BrC_6H_4	C_6H_5	2000	99	96	*S*
2-BrC_6H_4	4-$CH_3C_6H_4$	2000	99	98	*S*
4-$CH_3OC_6H_4$	C_6H_5	2000	95	35	*R*
4-$CF_3C_6H_4$	C_6H_5	2000	99	47	*S*
4-$CH_3OC_6H_4$	4-$CF_3C_6H_4$	2000	97	61	–
Ferrocenyl[b)]	C_6H_5	2000	100	95	*S*

a) Substrate/catalyst mole ratio.
b) *trans*-RuCl$_2$[(*S*)-tolbinap][(*S*)-daipen] + $(CH_3)_3COK$ is used as a catalyst.

Fig. 23 Asymmetric hydrogenation of diaryl ketones.

2-substituted benzophenones are hydrogenated with *trans*-RuCl$_2$[(*S*)-xylbinap][(*S*)-daipen] and $(CH_3)_3COK$ at an S/C as high as 20000 under 8 atm of H_2 to give quantitatively the corresponding diaryl methanol in up to 99% *ee* (Fig. 23) [80]. No diaryl methane derivative is detected. Substrates having an electron-donating and an electron-attracting group such as CH_3, CH_3O, F, Cl, or Br are reduced with consistently high optical yield. Chiral alcohols derived from the reduction of 2-methyl- and 2-bromo-4′-methylbenzophenones are key intermediates for the synthesis of antihistaminic (*S*)-orphenadrine and (*R*)-neobenodine, respectively [80]. Benzophenones substituted at 3 or 4 position are hydrogenated with an only moderate optical yield. Hydrogenation of benzoylferrocene with *trans*-RuCl$_2$[(*S*)-tolbinap][(*S*)-daipen] and a base resulted in the *S* alcohol in 95% *ee*.

Hetero-aromatic ketones

General asymmetric hydrogenation of hetero-aromatic ketones has been realized by the use of a XylBINAP/DAIPEN–Ru(II) catalyst. A variety of chiral alcohols connecting an electron-rich and an electron-deficient hetero-aromatic group at the chiral center are prepared with consistently high optical purity (Fig. 24) [81]. 2-Acetylfurane is hydrogenated in the presence of *trans*-RuCl$_2$[(*R*)-xylbinap][(*R*)-daipen] and $(CH_3)_3COK$ with an S/C of 40000 under 50 atm of H_2 to afford (*S*)-1-(2-furyl)ethanol in 99% *ee* leaving the furan ring intact. Hydrogenation of 2- and 3-acetylthiophene with the same catalyst under 1–8 atm of H_2 results in the chiral alcohols in >99% *ee* quantitatively. The sulfur-containing group does not affect the

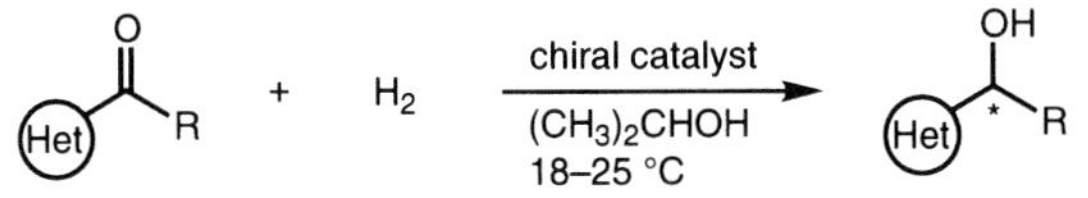

$$\text{(Het)}-C(=O)-R + H_2 \xrightarrow[\text{18–25 °C}]{\text{chiral catalyst, }(CH_3)_2CHOH} \text{(Het)}-C^{*}H(OH)-R$$

Het	*R*	*Catalyst*[a] *(S/C*[b]*)*	*H_2 (atm)*	*% yield*	*% ee*	*Confign*
2-furyl	CH_3	(*R,R*)-**C3** (40 000)	50	96	99	*S*
2-furyl	CH_3	(*R,SS*)-**C10** (3000)	5.5	>99	96	*R*
2-furyl	CH_3	(*R,S,R,S*)-**C17** (100)	30[c]	83	96	*S*
2-furyl	*n*-C_5H_{11}	(*R,R*)-**C3** (2000)	8	100	98	*S*
2-thienyl	CH_3	(*R,R*)-**C3** (5000)	8	100	99	*S*
2-thienyl	CH_3	(*S,S*)-**C3** (1000)	1	100	99	*R*
2-thienyl	CH_3	$RuCl_2$[(*R,R*)-bicp]-(tmeda)–(*R,R*)-DPEN + KOH (500)	4[d]	100	93	*S*
3-thienyl	CH_3	(*R,R*)-**C3** (5000)	8	100	99.7	*S*
3-thienyl	CH_3	(*S,RR*)-**C10** (3000)	5.5	>99	98	*S*
2-(1-methyl)pyrrolyl	CH_3	(*S,S*)-**C3** (1000)	8	61	97	–
2-[1-(4-toluene-sulfonyl)]pyrrolyl	CH_3	(*R,R*)-**C3** (1000)	8[e]	93	98	*S*
2-thiazolyl	CH_3	(*R,R*)-**C3** (2000)[f]	8	100	96	*S*
2-pyridyl	CH_3	(*R,R*)-**C3** (2000[f])	8	99.7	96	*S*
2-pyridyl	$(CH_3)_2CH$	(*R,R*)-**C3** (2000)	8	100	94	*S*
3-pyridyl	CH_3	(*R,R*)-**C3** (5000)	8	100	99.6	*S*
3-pyridyl	CH_3	(*R,SS*)-**C10** (1500)	8	>99	99	*R*
4-pyridyl	CH_3	(*R,R*)-**C3** (5000)	8	100	99.8	*S*
2,6-diacetylpyridine		(*R,R*)-**C3** (10 000)	8	99.9	100	*S,S*

a) See Fig. 20.
b) Substrate/catalyst mole ratio.
c) Reaction in methanol.
d) At –30 °C.
e) Reaction in 1:10 DMF–2-propanol.
f) $B[OCH(CH_3)_2]_3$ is added. Ketone/B = 100.

Fig. 24 Asymmetric hydrogenation of hetero-aromatic ketones.

rate. Hydrogenation of the 2-(1-methyl)pyrrolyl ketone does not complete, although the optical yield is high. The 1-(4-toluenesulfonyl)pyrrolyl analog is completely converted to the alcohol in 98% *ee* (93% isolated yield). Hydrogenation of 2-acetylthiazol and 2-acetylpyridine does not complete under regular conditions. This may be because of the high binding ability of the alcoholic products to the metal. This problem is resolved by an addition of $B[OCH(CH_3)_2]_3$ (ketone:Ru:borate = 2000:1:20), which is known as an efficient agent for trapping amino alcohols [81]. Hydrogenation of the isopropyl 2-pyridyl ketone proceeds smoothly without addition of borate. Reduction of 3- and 4-acetylpyridine under the standard conditions gives the corresponding alcohols with an excellent optical yield. Double hydrogenation of 2,6-diacetylpyridine with the (*R*)-XylBINAP/(*R*)-DAIPEN–Ru(II) catalyst results in the optically pure *S,S* diol as a sole product.

The (*R*)-Xylyl-Phanephos/(*S*,*S*)-DPEN–Ru(II) catalyst is also an excellent catalyst for this purpose [66]. Hydrogenation of 3-acetylpyridine with an S/C of 1500 under 8 atm of H_2 gives the *R* alcohol in 99% *ee* quantitatively. An *in situ*-prepared catalyst from $RuCl_2$[(*R*,*R*)-bicp](tmeda), (*R*,*R*)-DPEN, and KOH catalyzes reduction of 2-acetylthiophene to afford the *S* alcohol in 93% *ee* [82]. Hydrogenation of 2-acetylfuran with a Rh catalyst consisting of $[RhCl(cod)]_2$, (*R*,*S*,*R*,*S*)-Me-PennPhos, 2,6-lutidine, and KBr gives the *S* alcohol in 96% *ee* [73].

Fluoro ketones

In recent years, much attention has been directed toward the synthesis of a variety of chiral fluorinated compounds. Asymmetric hydrogenation of fluorinated ketones provides a reliable method to give the fluorinated chiral alcohols. Hydrogenation of 2,2,2-trifluoroacetophenones in the presence of *trans*-$RuCl_2$[(*S*)-xylbinap][(*S*)-daipen] and $(CH_3)_3COK$ gives the corresponding *S* alcohols with an optical yield of 94–96% (Fig. 25) [39]. Substitution of an electron-attracting and an electron-donating group at the 4′ position of the ketonic substrates has little effect on the optical yield. The sense of the enantioface selection is the same as that observed in the hydrogenation of acetophenones (see Fig. 20).

A chiral Rh catalyst $[Rh(OCOCF_3)\{(S)\text{-cy,cy-oxopronop}\}]_2$ effectively differentiates trifluoromethyl and pentafluoroethyl groups from alkyl groups (Fig. 26) [83]. For example, *n*-octyl trifluoromethyl ketone is hydrogenated using the *S* catalyst with an S/C of 200 under 20 atm of H_2 to afford the *R* alcohol in 97% *ee* quantitatively. Benzyloxymethyl trifluoromethyl ketone is also converted to the chiral alcohol in 86% *ee*.

Dialkyl ketones

To achieve high enantioselectivity in the hydrogenation of pro-chiral dialkyl ketones is still a challenging scientific subject. Currently reported Rh catalyst consisting of $[RhCl(cod)]_2$, (*R*,*S*,*R*,*S*)-Me-PennPhos, 2,6-lutidine, and KBr shows good enantioselectivity in the reaction of *n*-alkyl methyl ketones [73]. Hydrogenation of 2-hexanone with this catalyst at an S/C of 100 under 30 atm of H_2 results in (*S*)-2-hexanol in 75% *ee* (Fig. 27). 4-Methyl-2-pentanone was converted to the *S* alcohol

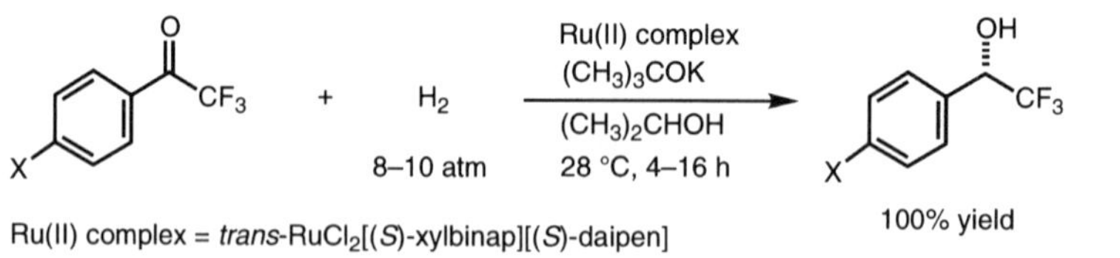

CF3 + H2 (8–10 atm) → Ru(II) complex, (CH3)3COK / (CH3)2CHOH, 28 °C, 4–16 h

OH, CF3, X

100% yield

Ru(II) complex = *trans*-$RuCl_2$[(*S*)-xylbinap][(*S*)-daipen]
S/C = 2000–11000

X	% ee
H	96
Cl	94
Br	94
CH_3O	96

Fig. 25 Asymmetric hydrogenation of 2,2,2-trifluoroacetophenones.

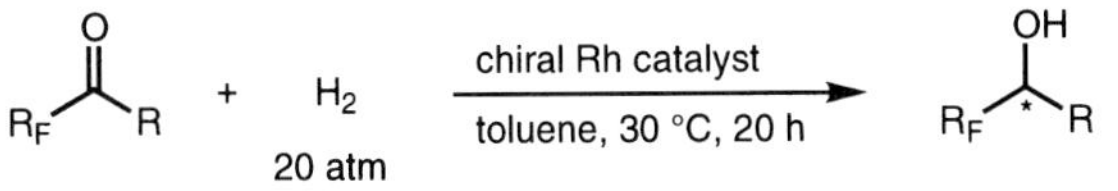

$$R_F\text{C(O)R} + H_2 \text{ (20 atm)} \xrightarrow[\text{toluene, 30 °C, 20 h}]{\text{chiral Rh catalyst}} R_F\overset{*}{\text{C}}\text{H(OH)R}$$

Rh catalyst = $[Rh(OCOCF_3)\{(S)\text{-cy,cy-oxopronop}\}]_2$
S/C = 200

R_F	*R*	*% yield*	*% ee*	*Confign*
CF_3	C_6H_5	93	73	*R*
CF_3	*cyclo*-C_6H_{11}	90	97	*R*
CF_3	*n*-C_8H_{17}	99	97	*R*
CF_3	$C_6H_5CH_2OCH_2$	100	86	–
C_2F_5	*n*-C_9H_{19}	100	97	*R*

Fig. 26 Asymmetric hydrogenation of fluoroketones.

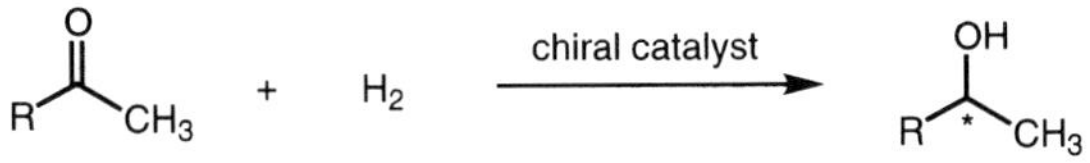

$$\text{RC(O)CH}_3 + H_2 \xrightarrow{\text{chiral catalyst}} \text{R}\overset{*}{\text{C}}\text{H(OH)CH}_3$$

R	*Catalyst* [a]	*S/C* [b]	*Solvent*	*H_2 (atm)*	*% yield*	*% ee*	*Confign*
n-C_4H_9	(*R,S,R,S*)-**C17**	100	CH_3OH	30	96	75	*S*
$(CH_3)_2$-$CHCH_2$	(*R,S,R,S*)-**C17**	100	CH_3OH	30	66	85	*S*
$(CH_3)_2CH$	(*R,S,R,S*)-**C17**	100	CH_3OH	30	99	84	*S*
cyclo-C_3H_5	(*S,S*)-**C3**	11 000	$(CH_3)_2CHOH$	10	96	95	*R*
cyclo-C_6H_{11}	(*R,S,R,S*)-**C17**	100	CH_3OH	30	90	92	*S*
cyclo-C_6H_{11}	(*S,S*)-**C3**	10 000	$(CH_3)_2CHOH$	8	99	85	*R*
$(CH_3)_3C$	(*R,S,R,S*)-**C17**	100	CH_3OH	30	51	94	*S*
$(CH_3)_3C$	[Rh{(*S,R,R,R*)-tmo-deguphos}-(cod)]BF_4	1000	$(CH_3)_2CHOH$	73	30	84	*S*
1-methyl-cyclopropyl	(*S,S*)-**C3**	500	$(CH_3)_2CHOH$	4	96	98	–

a) See Fig. 20.
b) Substrate/catalyst mole ratio.

Fig. 27 Asymmetric hydrogenation of aliphatic ketones with homogeneous catalysts.

in 85% *ee*. Reaction of 3-methyl-2-butanone and cyclohexyl methyl ketone results in optical yields of 84% and 92%, respectively. Pinacolone, a sterically hindered ketone, was hydrogenated to give the *S* alcohol in 94% *ee* and 51% yield. Hydrogenation of cyclopropyl methyl ketone and cyclohexyl methyl ketone in the presence of *trans*-$RuCl_2$[(*S*)-xylbinap][(*S*)-daipen] and $(CH_3)_3COK$ with an S/C of >10 000 under 8–10 atm of H_2 affords the corresponding *R* alcohols in 95% and 85% *ee*, re-

+ H_2 90 atm

Raney Ni–(*R,R*)-tartaric acid
NaBr
10:9 THF–pivalic acid
50–60 °C

>98% yield

80% ee

85% ee

72% ee

+ H_2 90 atm

fine Ni powder–(*R,R*)-tartaric acid
NaBr
10:9 THF–pivalic acid
120 °C

>98% yield
31% ee

Fig. 28 Asymmetric hydrogenation of 2- or 3-alkanones with chirally modified Ni catalysts.

spectively (Fig. 27) [39]. Methyl 1-methylcyclopropyl ketone was hydrogenated using the same catalyst with an optical yield of 98% [14a]. Hydrogenation of pinacolone catalyzed by $[Rh\{(S,R,R,R)\text{-tmo-deguphos}\}(cod)]BF_4$ yielded the *S* alcohol in 84% *ee* [84].

Heterogeneous Ni catalysts modified with tartaric acid and NaBr are effective for the asymmetric hydrogenation of alkanones [85]. Hydrogenation of 2-alkanones in the presence of the modified Raney Ni and an excess amount of pivalic acid gave 2-alkanols quantitatively in up to 85% *ee* (Fig. 28). When the NaBr/tartaric acid ratio is 22, the optical yield of hydrogenation of 2-butanone reaches 72% [85d]. This asymmetric environment enables a distinction of methyl even from ethyl. When Raney Ni is replaced by fine nickel powder, 3-octanone is hydrogenated to give 3-octanol in 31% *ee* [85c].

Amino, hydroxy, methoxy, and phenylthio ketones

Asymmetric hydrogenation of amino ketones is one of the direct and reliable procedures to obtain the corresponding amino alcohols. 2-Aminoacetophenone hydrochloride is hydrogenated by MOC-BIMOP–$[Rh(nbd)_2]ClO_4$ [86] or $[RhX(cy,cy\text{-oxo-pronop})]_2$ (X=Cl, $OCOCF_3$) [87] to give the corresponding amino alcohol in 93% *ee* (Fig. 29). An MCCPM/Rh catalyst has achieved highly reactive and enantioselective hydrogenation of *α*-amino ketone hydrochlorides [88, 89]. 2-(Dimethylamino)acetophenone hydrochloride is hydrogenated with the catalyst at an S/C of 100000 under 20 atm of H_2 to afford the chiral amino alcohol in 96% *ee* [89]. The mono-*N*-benzyl analog is enantioselectively hydrogenated with this complex [88]. Epinephrine hydrochloride with 95% optical purity is synthesized via hydrogenation catalyzed by $[Rh(nbd)(bppfoh)]ClO_4$ with $(C_2H_5)_3N$ [90]. *α*-Dialkylamino ketones are effectively converted to the chiral alcohols with up to 99% *ee* by the reaction catalyzed by BINAP–Ru [91, 92] and DIOP–Rh complexes [93]. Hydrogenation of *β*- and *γ*-amino ketone hydrochlorides with MCCPM–Rh complex gives the corresponding chiral amino alcohols in up to 91% *ee* [94]. Recently, quite high op-

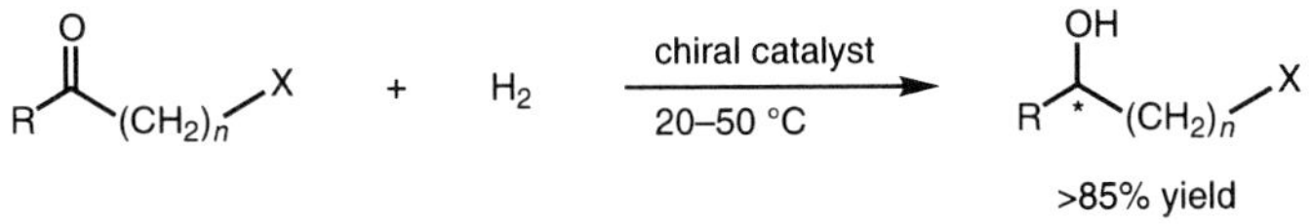

$$RCO(CH_2)_nX + H_2 \xrightarrow[20\text{–}50\ ^\circ C]{\text{chiral catalyst}} RC^{*}H(OH)(CH_2)_nX$$

>85% yield

R	*n*	*X*	*Catalyst* [a] *(S/C* [b]*)*	*H_2, (atm)*	*% ee*	*Config*
C_6H_5	1	$ClNH_3$	$[Rh(nbd)_2]ClO_4$–(*R*)-MOC-BIMOP + $(C_2H_5)_3N$ (1000)	90	93	*R*
C_6H_5	1	$ClNH_3$	[Rh{(*S*)-cy,cy-oxopronop}-(cod)]BF_4 (200)	50	93	*S*
C_6H_5	1	$ClC_6H_5CH_2NH_2$	$[RhCl(cod)]_2$–(2*S*,4*S*)-MC-CPM + $(C_2H_5)_3N$ (1000)	20	93	*S*
3,4-$(OH)_2$-C_6H_3	1	$ClCH_3NH_2$	[Rh{(*R*)-(*S*)-bppfoh}(nbd)]-ClO_4 + $(C_2H_5)_3N$ (100)	50	95	*R*
CH_3	1	$(CH_3)_2N$	[RuI{(*S*)-binap}-(*p*-cymene)]I (1100)	102	99	*S*
CH_3	1	$(CH_3)_2N$	(*R*,*R*)-**C3** (2000)	8	92	*S*
C_6H_5	1	$(CH_3)_2N$	$RuBr_2$[(*S*)-binap] (500)	100	95	*S*
C_6H_5	1	$(CH_3)_2N$	(*R*,*R*)-**C3** (2000)	8	93	*R*
2-naphthyl	1	$(C_2H_5)_2N$	$[RhCl(nbd)]_2$–(*S*,*S*)-DIOP (200)	70	95	+
CH_3	1	$Cl(CH_3)_2NH$	[Rh{(*S*)-cy,cy-oxopronop}-(cod)]BF_4 (200)	50	97	*S*
C_6H_5	1	$Cl(CH_3)_2NH$	[Rh($OCOCF_3$){(*S*)-cp,cp-indonop}]$_2$ (200)	50	>99	*S*
3-ClC_6H_4	1	$Cl(CH_3)_2NH$	[Rh{(*R*)-cy,cy-oxopronop}-(cod)]BF_4 (200)	1	96	*R*
C_6H_5	1	$Cl(C_2H_5)_2NH$	$[RhCl(cod)]_2$–(2*S*,4*S*)-MCCPM + $(C_2H_5)_3N$ (100 000)	20	97	*S*
C_6H_5	1	$C_6H_5CO(CH_3)N$	(*R*,*R*)-**C3** (2000)	8	99.8	*R*
4-C_6H_5-$CH_2OC_6H_4$	1	$C_6H_5CO[3,4\text{-}(CH_3O)_2\text{-}C_6H_3(CH_2)_2]N$	(*R*,*R*)-**C3** (2000)	8	97	*R*
C_6H_5	2	$ClCH_3NH_2$	$[RhCl(cod)]_2$–(2*S*,4*S*)-MCCPM + $(C_2H_5)_3N$ (1000)	30	79.8	*R*
C_6H_5	2	$(CH_3)_2N$	(*S*,*S*)-**C3** (10 000) [c]	8	97.5	*R*
C_6H_5	2	$(CH_3)_2N$	(*S*,*SS*)-**C6** (4000)	8	97	*R*
2-thienyl	2	$(CH_3)_2N$	(*R*,*R*)-**C3** (2000) [c]	8	92	*S*
C_6H_5	2	$Cl(CH_3)_2NH$	[Rh{(*S*)-cy,cy-oxo-pronop}(cod)]BF_4 (200)	50	93 [d]	*R*
C_6H_5	2	$ClC_6H_5CH_2(CH_3)NH_2$	$[RhCl(cod)]_2$–(2*S*,4*S*)-MCCPM + $(C_2H_5)_3N$ (1000)	30	90.8	*R*

Fig. 29 Asymmetric hydrogenation of amino ketones.

R	n	X	Catalyst [a)] (S/C [b)])	H_2, (atm)	% ee	Confign
4-FC_6H_4	3	R′ [e)]	(*S*,*S*)-**C3** (10000)	8	99	*R*
C_6H_5	3	$Cl(CH_3)_2NH$	[Rh{(*S*)-cy,cy-oxopronop}-(cod)]BF_4 (200)	50 [f)]	92	*R*
C_6H_5	3	$ClC_6H_5CH_2(CH_3)NH_2$	$[RhCl(cod)]_2$–(2*S*,4*S*)-MCCPM + $(C_2H_5)_3N$ (250)	50	88.4	*R*
CH_3	1	HO	$RuCl_2$[(*S*)-binap] (2570)	100	92	*S*
CH_3	1	HO	$[NH_2(C_2H_5)_2][\{RuCl[(R)\text{-segphos}]\}_2(\mu\text{-}Cl)_3]$ (3000)	30 [g)]	99.5	*R*
n-C_3H_7	1	HO	$RuCl_2$[(*R*)-binap] (–)	–	95	*R*
C_6H_5	1	CH_3O	(*R*,*R*)-**C3** (2000)	8	95	*R*
CH_3	2	HO	$RuCl_2$[(*R*)-binap] (900)	70	98	*R*
CH_3	2	C_6H_5S	$Ru[\eta^3\text{-}CH_2C(CH_3)CH_2]_2$-(cod)–(*S*)-MeO-BIPHEP + HBr (50)	30	98	*S*
C_2H_5	2	C_6H_5S	$RuBr_2$[(*S*,*S*)-bdpp] (50–100)	30	95	*S*
CH_3	3	C_6H_5S	$Ru[\eta^3\text{-}CH_2C(CH_3)CH_2]_2$-(cod)–(*S*)-BINAP + HBr (50)	115 [f)]	70 [h)]	*S*

a) See Fig. 20.
b) Substrate/catalyst mole ratio.
c) *trans*-$RuCl_2$(xylbinap)(daipen) was treated with $(CH_3)_3COK$ in $(CH_3)_2CHOH$ prior to hydrogenation.
d) Contaminated with 5% of propiophenone.
e) R′ = F–(pyrimidinyl)–N(piperazine)N

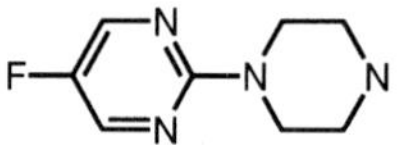

f) At 80 °C.
g) At 60 °C.
h) 70% yield.

Fig. 29 (cont.)

tical yield has been obtained by means of (*S*)-Cy,Cy-oxoProNOP–Rh [95] and (*S*)-Cp,Cp-IndoNOP–Rh [96] catalysts for this purpose. Dimethylaminoacetone and 2-(dimethylamino)acetophenone are converted to the corresponding *S* amino alcohols in 97% and >99% *ee*, respectively (Fig. 29). This method is applied to the synthesis of an atypical *β*-adrenergic phenylethanolaminotetraline agonist SR58611A [97]. The activity and enantioselectivity of these catalysts are lower in the reaction of *β*- and *γ*-amino ketone hydrochlorides [95]. A chiral catalyst consisting of *trans*-$RuCl_2$[(*R*)-xylbinap][(*R*)-daipen] and $(CH_3)_3COK$ efficiently mediates hydrogenation of *α*-, *β*-, and *γ*-amino ketones [98]. Dimethylaminoacetone is hydrogenated with the *R*,*R* catalyst at an S/C of 2000 under 8 atm of H_2 to give the *S* amino alcohol in 92% *ee* (Fig. 29). Interestingly, the sense of enantioselection with the same catalyst reverses in the range of 185% in the reaction of 2-(dimethylamino)acetophenone. These observations indicate that the enantio-directing ability of groups in this hydrogenation decreases in the order phenyl > (dimethylamino)methyl > methyl. Excellent optical yield of as high as 99.8% is achieved in the hydrogenation of acetophenone derivatives, which have an amido group at the α position with the (*R*)-XylBINAP/(*R*)-DAIPEN–Ru catalyst [98]. The procedure can be applied to the synthesis of (*R*)-denopamine, a β_1-receptor agonist used for treating congestive heart failure. β-Amino ketones are difficult substrates to be hydrogenated under basic conditions because of the inherent instability of substrates. 3-(Dimethylamino)propiophenone is hydrogenated with the (*S*)-XylBINAP/(*S*)-DAIPEN–Ru catalyst prepared from the corresponding $RuCl_2$ complex and a minimum amount of $(CH_3)_3COK$ to afford the *R* amino alcohol in 97.5% *ee* and 96% yield accompanied by 2% of 1-phenyl-1-propanol [98]. When hydrogenation of this β-amino ketone is performed in the presence of *trans*-$RuH(\eta^1\text{-}BH_4)$[(*S*)-xylbinap][(*S,S*)-dpen] under base-free conditions, the desired β-amino alcohol in 97% *ee* is obtained quantitatively without any special care [17]. In a similar manner, a 2-thienyl derivative was also hydrogenated selectively [81]. The obtained chiral β-amino alcohols are key building blocks in the synthesis of antidepressants (*R*)-fluoxetine and (*S*)-duloxetine. Hydrogenation of a γ-amino ketone indicated in Fig. 29 in the presence of the (*S*)-XylBINAP/(*S*)-DAIPEN–Ru catalyst with an S/C of 10000 under 8 atm of H_2 resulted in the *R* alcohol in 99% *ee*, which is known as a potent antipsychotic, BMS 181100 [98]. Chiral 1,2-diols with >92% *ee* are obtained by the BINAP–Ru catalyzed hydrogenation of α-hydroxy ketones (Fig. 29) [91, 99]. Hydrogenation of hydroxyacetone with a SEGPHOS–Ru complex gives the diol in 99.5% *ee* [100]. 4-Hydroxy-2-butanone, a β-hydroxy ketone, is converted to the 1,3-diols in the presence of the BINAP–Ru catalyst with an optical yield of 98% [91]. Hydrogenation of β-phenylthio ketones are catalyzed by an Ru complex of BINAP, MeO-BIPHEP, or BDPP to give the chiral thio alcohols in up to 98% *ee* (Fig. 29) [101]. The reaction of a γ-phenylthio ketone, which requires somewhat drastic conditions, gives a moderate optical yield. Hydrogenation of 2-methoxyacetophenone with *trans*-$RuCl_2$[(*R*)-xylbinap][(*R*)-daipen] and $(CH_3)_3COK$ results in the *R* α-methoxy alcohol in 95% *ee* (Fig. 29) [14a]. The sense of enantioface selection is the same as that observed in the reaction of simple acetophenone (see Fig. 20).

Hydrogenation of pyruvic aldehyde dimethylacetal with the *R,R* catalyst gives the *S* alcohol in 98% *ee* quantitatively (Fig. 30) [14 a]. The high level of enantio-directing ability of the dimethoxymethyl group leads to such an excellent optical yield. Heterogeneous asymmetric hydrogenation of α-keto acetals provides the

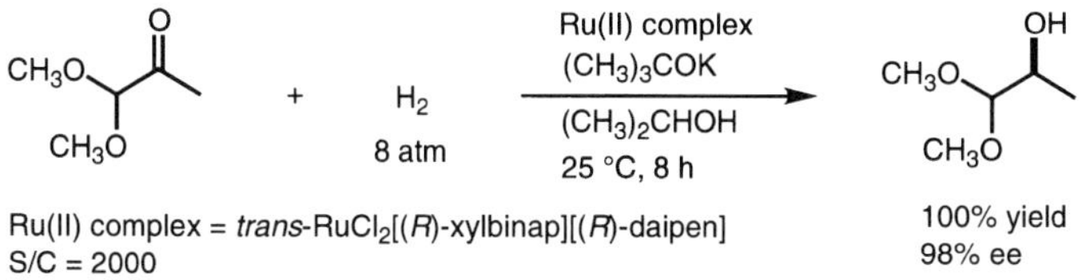

CH_3O, CH_3O (ketone) + H_2 (8 atm) → [Ru(II) complex, $(CH_3)_3COK$, $(CH_3)_2CHOH$, 25 °C, 8 h] → alcohol (OH), 100% yield, 98% ee

Ru(II) complex = *trans*-$RuCl_2$[(*R*)-xylbinap][(*R*)-daipen]
S/C = 2000

Fig. 30 Asymmetric hydrogenation of pyruvic aldehyde dimethylacetal.

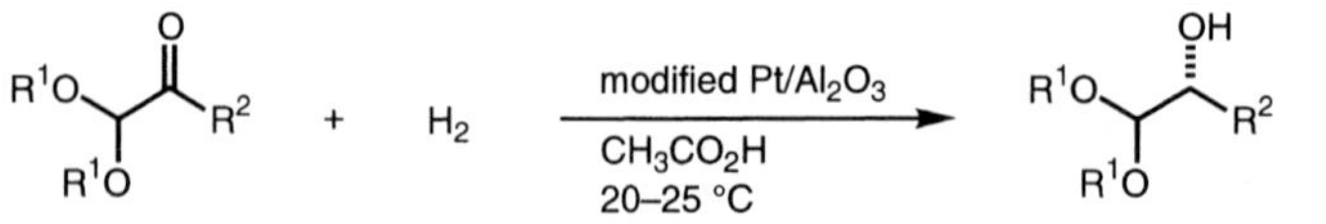

R^1	R^2	Modifier	H_2 (atm)	% ee
CH_3	CH_3	Cinchonidine	1	96.5
CH_3	CH_3	MeOHCd	60	96.5
$-(CH_2)_3-$	CH_3	MeOHCd	60	97
CH_3	C_6H_5	HCd	60	89

Fig. 31 Asymmetric hydrogenation of pyruvic aldehyde acetals with modified Pt/Al_2O_3.

chiral alcohols with excellent enantioselectivity (Fig. 31). Pyruvic aldehyde dimethylacetal is hydrogenated with an optical yield of 96.5% in the presence of Pt/Al_2O_3 catalysts modified by cinchonidine and 10,11-dihydro-*O*-methylcinchonidine (MeOHCd) [102, 103]. The cyclic acetal derivative is converted to the chiral alcohol in 97% *ee*. Hydrogenation of an aromatic α-keto acetal catalyzed by a 10,11-dihydrocinchonide (HCd)-modified Pt/Al_2O_3 gives 89% optical yield.

Fig. 32 illustrates highly enantioselective hydrogenation of bifunctionalized ketones. Hydrogenation of 1-aryloxy-2-oxo-3-propylamine derivatives in the presence of a (2S,4*S*)-MCCPM–Rh complex gives the (*S*)-amino alcohols in up to 97% *ee* [104]. The BINAP–Ru catalyst efficiently distinguishes a hydroxy group from an alkoxy or an aryloxy group, and even a *n*-octadecyl from a triphenylmethoxy group [105].

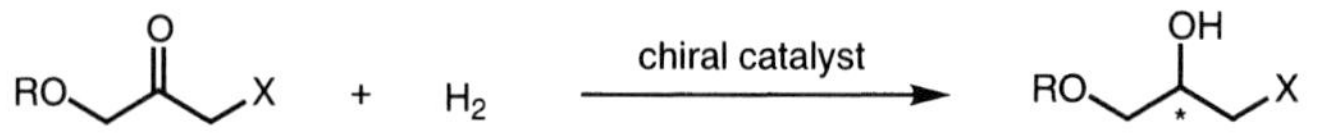

R	X	chiral catalyst	H_2, atm	temp, °C	% yield	% ee	confign
C_6H_5	$ClNH_2CH_2C_6H_5$	(2*S*,4*S*)-MCCPM–Rh	20	50	100	97	*S*
Ar[a]	OH	(*S*)-BINAP–Ru	94	25	86	>95	*R*
$CH_2C_6H_5$	OH	(*S*)-BINAP–Ru	97	20	>98	93	*R*
n-$C_{18}H_{37}$	$OC(C_6H_5)_3$	(*S*)-BINAP–Ru	97	20	>70	>96	*R*

[a] OAr =

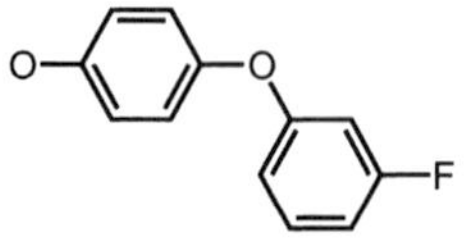

Fig. 32 Asymmetric hydrogenation of bifunctionalized ketones.

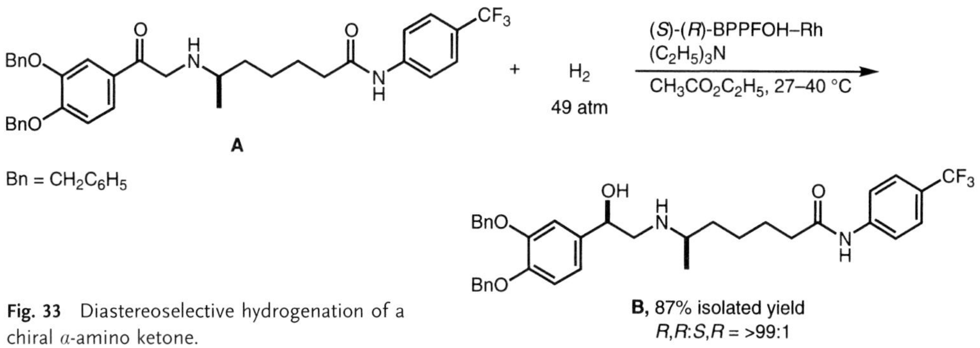

BnO, BnO, A, + H_2 49 atm, (S)-(R)-BPPFOH–Rh, $(C_2H_5)_3N$, $CH_3CO_2C_2H_5$, 27–40 °C

Bn = $CH_2C_6H_5$

B, 87% isolated yield
R,R:S,R = >99:1

Fig. 33 Diastereoselective hydrogenation of a chiral α-amino ketone.

Examples of diastereoselective hydrogenation of chiral amino or hydroxy ketones using a homogeneous optically active catalyst are shown in Fig. 33. Hydrogenation of the (*R*)-amino ketone **A** with a neutral (*S*)-(*R*)-BPPFOH–Rh complex in ethyl acetate gives the (*R,R*)-amino alcohol **B** in >99% purity, whereas reduction in the presence of a cationic Rh complex in methanol gives the *S,R* isomer predominantly [106].

Kinetic resolution of racemic 1-hydroxy-1-phenyl-2-propanone is achieved by means of hydrogenation with $Ru(OCOCH_3)_2$[(*R*)-binap] in the presence of HCl to give the unreacted *R* hydroxy ketone in 92% *ee* (49.5%) and the corresponding 1*S*,2*R* diol in 92% *ee* (50.5%, *syn*:*anti*=98:2) (Fig. 34) [62c]. The extent of enantiomer differentiation, k_{fast}/k_{slow}, is calculated to be 64. Racemic 2-methoxycyclohexanone can be resolved through hydrogenation with *trans*-RuH(η^1-BH_4)[(*S*)-xylbinap][(*S,S*)-dpen] to afford the unreacted *R* ketone in 94% *ee* at 53% conversion accompanied by the 1*S*,2*R* alcohol in 91% *ee* (*cis*:*trans*=100:0) [17]. The k_{fast}/k_{slow} is

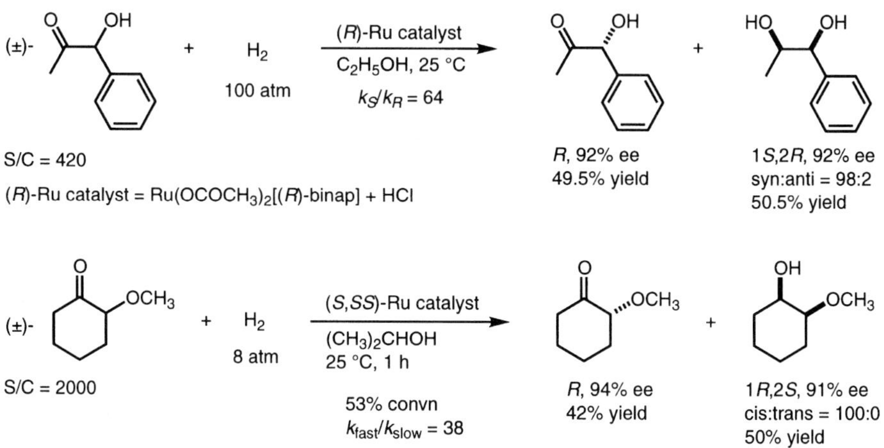

Fig. 34 Kinetic resolution of racemic α-substituted ketones through asymmetric hydrogenation.

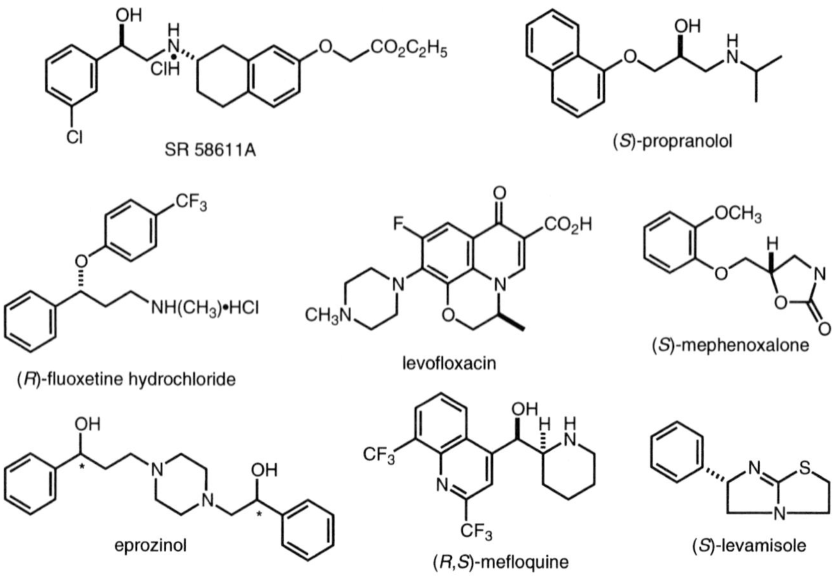

Fig. 35 Examples of biologically active compounds obtainable by homogeneous asymmetric hydrogenation of amino or hydroxy ketones.

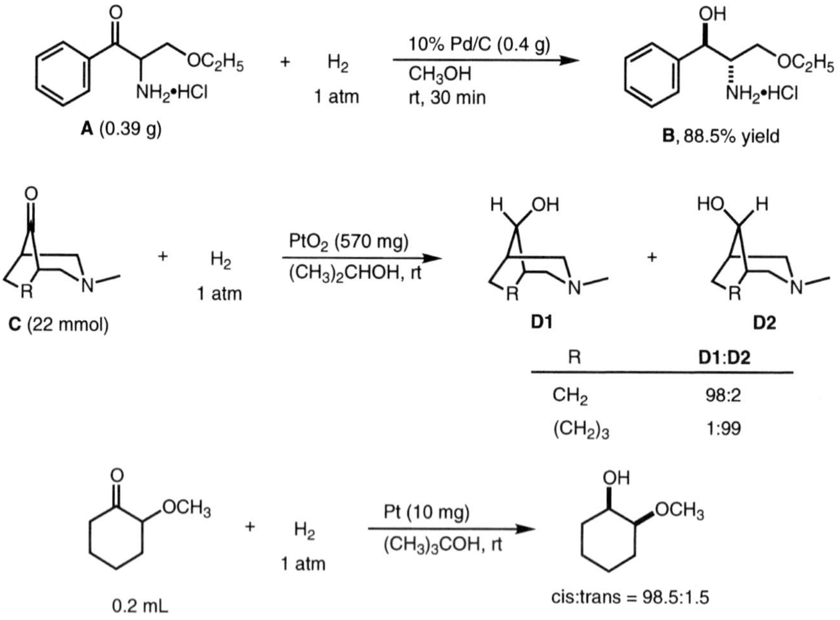

R	**D1:D2**
CH_2	98:2
$(CH_2)_3$	1:99

Fig. 36 Diastereoselective hydrogenation of amino or alkoxy ketones.

determined to be 38. The obtained chiral ketone is a key intermediate for the synthesis of a potent antibacterial sanfetrinem [107].

Chiral amino or hydroxy alcohols obtained via homogeneous asymmetric hydrogenation are used for synthesis of some biologically active compounds. Examples are shown in Fig. 35 [88, 94c, 97, 104, 105c, 108]. (*R*)-1,2-Propanediol obtained by BINAP–Ru-catalyzed hydrogenation of 1-hydroxy-2-propanone (50 tons/year at Takasago International Co.) is now used for commercial synthesis of levofloxacin, an antibacterial agent (Dai-ichi Pharmaceutical Co.).

Diastereoselective hydrogenation of amino and alkoxy ketones in heterogeneous phase has been reviewed [1]. Some examples are depicted in Fig. 36. Hydrogenation of the α-amino-β-ethoxy ketone hydrochloride **A** with a Pd/C catalyst gives the *anti* alcohol **B** as an only detectable product [109]. The bicyclic amino ketones **C** are hydrogenated with high stereoselectivity [110]. When R is CH_2, **D1** is obtained exclusively, whereas in case of $R=(CH_2)_3$, **D2** is a predominant product. The ring size strongly influences the conformation of substrates, reversing the diastereoselectivity. Hydrogenation of 2-methoxycyclohexanone catalyzed by Pt in *tert*-butyl alcohol gives the *cis* alcohol predominantly [111]. The *cis* stereoselectivity is about 5 times higher than that with 2-methylcyclohexanone.

Unsaturated ketones

Asymmetric hydrogenation of unsaturated ketones resulting in chiral unsaturated alcohols is difficult to achieve because most existing hydrogenation catalysts preferentially reduce C=C bonds rather than C=O linkages (vide supra). The long-sought solution for this problem has been achieved by applying BINAP/chiral 1,2-diamine–Ru(II) catalysts [15, 17, 32, 38, 39, 112]. For example, 1-(2-furyl)-4-penten-1-one, an unconjugated enone, is hydrogenated with *trans*-$RuCl_2$[(*S*)-xylbinap]-[(*S*)-daipen] and $(CH_3)_3COK$ in 2-propanol to give quantitatively the *R* unsaturated alcohol in 97% *ee* (Fig. 37) [81]. No saturation of olefinic bond is detected.

Asymmetric hydrogenation of α,β-unsaturated ketones to chiral allylic alcohols with different structural and electronic characteristic is achieved with the BIANAP/1,2-diamine–Ru catalyst system. A variety of the conjugated enones can be converted to the corresponding allylic alcohols with high optical yields in the presence of *trans*-$RuCl_2$[(*S*)-xylbinap][(*S*)-daipen] (or the *R*/*R* combination) and K_2CO_3 [39]. Use of the relatively weak base efficiently prevents the formation of undesired polymeric compounds. For example, hydrogenation of benzalacetone in the presence of the *S*,*S* catalyst with an S/C of 100000 under 80 atm of H_2 results

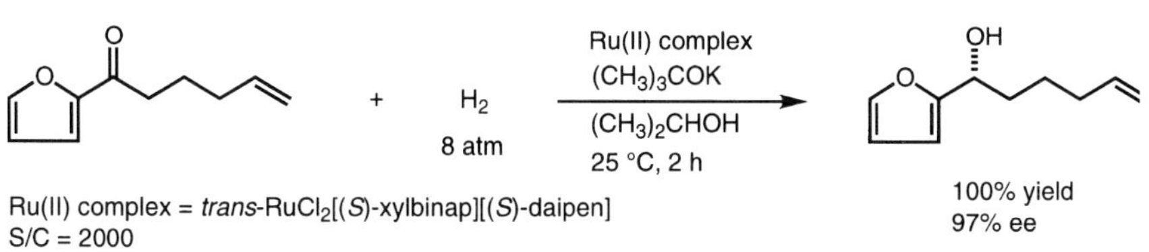

+ H_2 8 atm; Ru(II) complex, $(CH_3)_3COK$, $(CH_3)_2CHOH$, 25 °C, 2 h

100% yield
97% ee

Ru(II) complex = *trans*-$RuCl_2$[(*S*)-xylbinap][(*S*)-daipen]
S/C = 2000

Fig. 37 Asymmetric hydrogenation of an unconjugated enone.

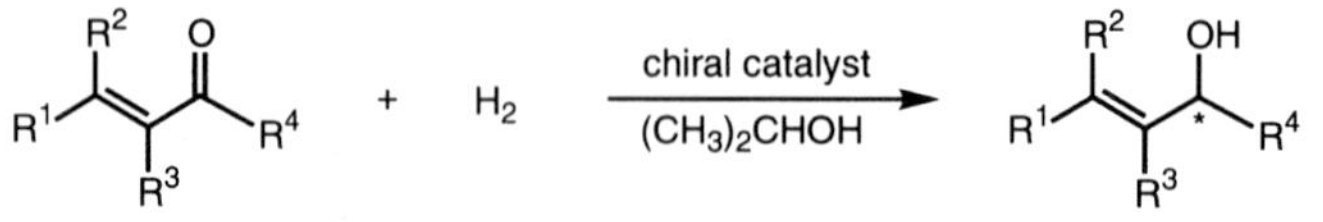

a: $R^1 = C_6H_5$; $R^2 = R^3 = H$; $R^4 = CH_3$

b: $R^1 = C_6H_5$; $R^2 = R^3 = H$; $R^4 = (CH_3)_2CH$

c: R^1 = 2-thienyl; $R^2 = R^3 = H$; $R^4 = CH_3$

d: $R^1 = n\text{-}C_5H_{11}$; $R^2 = R^3 = H$; $R^4 = CH_3$

e: $R^1 = CH_3$; $R^2 = R^3 = H$; $R^4 = (CH_3)_2CHCH_2$

f: $R^1 = R^2 = R^4 = CH_3$; $R^3 = H$

g: $R^1\text{–}R^3 = (CH_2)_4$; $R^2 = H$; $R^4 = CH_3$

h: $R^1\text{–}R^3 = (CH_2)_5$; $R^2 = H$; $R^4 = CH_3$

i: $R^1\text{–}R^3 = (CH_2)_3$; $R^2 = R^4 = CH_3$

j: R^1 = 2,6,6-trimethylcyclohexenyl; $R^2 = R^3 = H$; $R^4 = CH_3$ (β-ionone)

catalyst:

trans-$RuCl_2$[(*S*)-xylbinap][(*S*)-daipen] + K_2CO_3; (*S*,*S*)-**C19**

trans-$RuCl_2$[(*S*)-xylbinap][(*S*,*S*)-dpen] + K_2CO_3; (*S*,*SS*)-**C20**

Substrate	*Catalyst* [a)]	*S/C* [b)]	*H_2 (atm)*	*% yield*	*% ee*	*Confign*
a	(*S*,*S*)-**C19**	100 000	80	100	97	*R*
a	(*S*,*S*)-**C19**	10 000	10	100	96	*R*
a	(*R*,*SS*)-**C10**	3000	5.5	>99	97	*R*
b	(*S*,*S*)-**C19**	2000	8	100	86	*R*
c	(*R*,*R*)-**C19**	5000	8	100	91	*S*
d	(*S*,*S*)-**C3**	2000	8	98	97	*R*
d	(*S*,*SS*)-**C6**	4000	8	95	99	*R*
e	(*R*,*R*)-**C19**	2000	10	100	90	*S*
f	(*S*,*SS*)-**C20**	10 000	8	100	93	*R*
g	(*S*,*S*)-**C3**	10 000	10	99	100	*R*
h	(*S*,*S*)-**C3**	2000	8	99.9	99	*R*
i	(*S*,*S*)-**C3**	13 000	10	100	99	*R*
j	(*S*,*S*)-**C19**	10 000	8	99	94	*R*

a) See also Fig. 20.
b) Substrate/catalyst mole ratio.

Fig. 38 Asymmetric hydrogenation of α,β-unsaturated ketones.

in (*R*)-(*E*)-4-phenyl-3-buten-2-ol in 97% *ee* quantitatively (Fig. 38). Thienyl-substituted ketone is also selectively reduced [39]. Reaction of (*E*)-6-methyl-2-hepten-4-one with the *R,R* catalyst afforded the *S* allylic alcohol in 90% *ee* [39], which is known to be a key intermediate for the synthesis of α-tocopherol (vitamin E) side chain. Hydrogenation of more substituted, less base-sensitive substrates is performed more rapidly and conveniently by the use of a stronger alkaline base $(CH_3)_3COK$ instead of K_2CO_3. 1-Acetylcycloalkenes are hydrogenated with an optical yield as high as 100%. Hydrogenation of highly base-sensitive 3-nonene-2-one with *trans*-$RuCl_2$[(*S*)-xylbinap][(*S*)-daipen] and K_2CO_3 under conditions of high dilution of the substrate (0.1 M) gives the *R* allylic alcohol in 97% *ee* and in high yield [39]. The use of *trans*-RuH(η^1-BH_4)[(*S*)-xylbinap][(*S,S*)-dpen] without addition of base caused the conversion of 3-nonene-2-one under 2.0 M substrate concentration to the *R* alcohol in 99% *ee* and in 95% yield [17]. The (*R*)-Xylyl-PhanePhos/(*S,S*)-DPEN–Ru catalyst also promoted hydrogenation of benzalacetone to give the *R* allylic alcohol in 97% *ee* [66]. A BINAP–Ir catalyst shown in Fig. 8 reduces benzalacetone with a moderate enantioselectivity [35].

2,4,4-Trimethyl-2-cyclohexenone, a cyclic enone, is hydrogenated with *trans*-$RuCl_2$[(*R*)-tolbinap][(*S,S*)-dpen] (not *R*/*R,R*) and $(CH_3)_3COK$ to give quantitatively the *S* allylic alcohol in 96% *ee* (Fig. 39) [38, 112]. In contrast to the hydrogenation of alkyl aryl ketones, the *R*/*R,R* or *S*/*S,S* combination of catalysts gives lower reactivity and enantioselectivity. The obtained cyclic allylic alcohol in both enantiomers is a versatile building block for the synthesis of carotenoid-derived odorants and other bioactive terpenes. Simple 2-cyclohexenone is reduced with an (*S,S*)-DIOP–Ir catalyst to give selectively (*R*)-2-cyclohexenol in 25% *ee* (Fig. 39) [113].

Hydrogenation of (*R*)-carvone, a chiral dienone, requires many selectivity problems to be overcome, that is 1,2- versus 1,4-reduction at the conjugated enone part, chemoselective reduction of conjugated versus unconjugated olefinic bond, and diastereoselective formation of 1,5-*cis* versus 1,5-*trans* alcohol when the carbonyl group is hydrogenated. A (*S*)-BINAP/(*R,R*)-DPEN–Ru catalyst prepared *in situ*

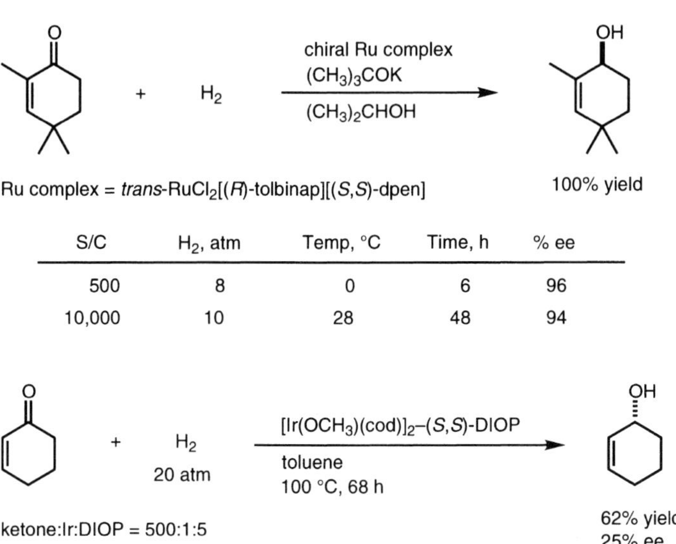

S/C	H_2, atm	Temp, °C	Time, h	% ee
500	8	0	6	96
10,000	10	28	48	94

Fig. 39 Asymmetric hydrogenation of cyclohexenones.

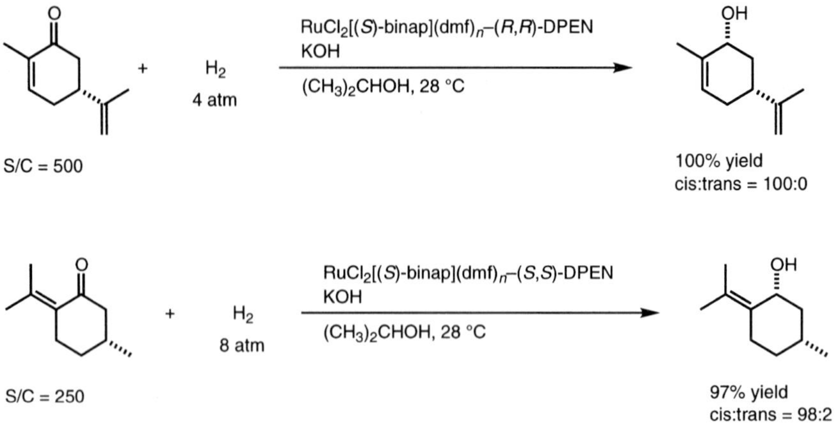

$RuCl_2[(S)\text{-binap}](dmf)_n$–(R,R)-DPEN
KOH
+ H_2 4 atm
$(CH_3)_2CHOH$, 28 °C
S/C = 500
100% yield
cis:trans = 100:0

$RuCl_2[(S)\text{-binap}](dmf)_n$–(S,S)-DPEN
KOH
+ H_2 8 atm
$(CH_3)_2CHOH$, 28 °C
S/C = 250
97% yield
cis:trans = 98:2

Fig. 40 Hydrogenation of (*R*)-carvone and (*R*)-pulegone with chiral Ru(II) catalysts.

as described in Fig. 40 chemo- and diastereoselectively reduces the C=O function of the dienone to give quantitatively (1*R*,5*R*)-carveol (*cis*:*trans*=100:0) [38]. When the reaction is conducted with the (*R*)-BINAP/(*S*,*S*)-DPEN–Ru catalyst, a 34:66 mixture of the *cis* and *trans* products is produced. On the other hand, (*R*)-pulegone, an *s-cis* enone, is most selectively reduced with the (*S*)-BINAP/(*S*,*S*)-DPEN (not *S*/*R*,*R*) combined catalyst to give the 1*R*,5*R* alcohol (*cis*:*trans*=98:2) in 97% yield (Fig. 40) [38]. Racemic carvone can be kinetically resolved through asymmetric hydrogenation catalyzed by the (*S*)-BINAP/(*R*,*R*)-DPEN–Ru catalyst to afford the unreacted *S* substrate in 94% *ee* and the 1*R*,5*R* alcohol in 93% *ee* at 54% conversion (Fig. 41) [38]. The k_{fast}/k_{slow} is calculated to be 33.

Asymmetric activation and deactivation

Hydrogenation of prochiral ketones promoted by racemic catalysts normally provides racemic alcohols. However, a surrounding non-racemic environment sometimes differently affects the catalyst efficiency of two enantiomeric molecules. A racemic metal complex can be activated as a chiral catalyst by an addition of chiral ligand. A racemic $RuCl_2(tolbinap)(dmf)_n$ feebly catalyzes hydrogenation of 2,4,4-trimethyl-2-cyclohexenone. The reaction proceeds smoothly with this complex in the presence of an equimolar amount of (*S*,*S*)-DPEN in a 7:1 2-propanol–toluene mixture containing KOH to afford (*S*)-2,4,4-trimethyl-2-cyclohexenol in 95% *ee* quantitatively (Fig. 42) [112]. The optical yield approaches 96% available in the hydrogenation mediated by an optically pure (*R*)-TolBINAP/(*S*,*S*)-DPEN–Ru catalyst under otherwise identical conditions [38, 112]. The highly enantioselective catalyst cycle generated by the (*R*)-TolBINAP/(*S*,*S*)-DPEN–Ru complex occurs 121 times faster than the diastereomeric catalyst cycle involving the *S*,*SS* species that gives the *R* allylic alcohol in only 26% *ee*. The structures of diphosphine, diamine, and ketonic substrate affect the degree and sense of the resulting enantioselectivity. Hydrogenation of 2′-methylacetophenone, an acyclic aromatic ketone, with the

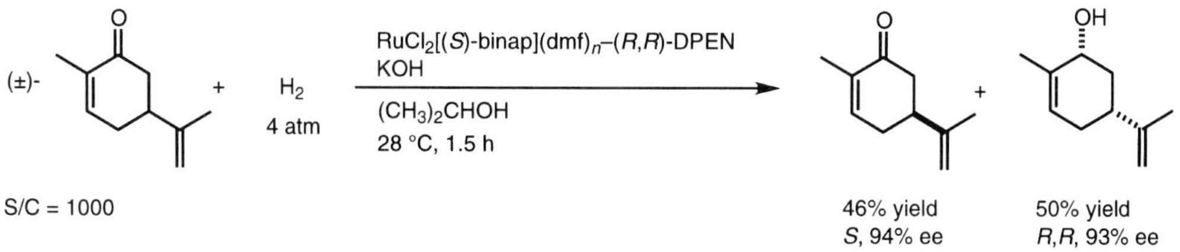

(±)- [carvone] + H_2 4 atm → $RuCl_2[(S)\text{-binap}](dmf)_n$–(R,R)-DPEN, KOH, $(CH_3)_2CHOH$, 28 °C, 1.5 h

S/C = 1000

46% yield, S, 94% ee + 50% yield, R,R, 93% ee

Fig. 41 Kinetic resolution of racemic carvone through asymmetric hydrogenation.

(±)-TolBINAP/(*S,S*)-DPEN–Ru catalyst affords the *R* alcohol in 90% *ee* (Fig. 42) [112]. In this case, the major catalyst cycle involving the *S,SS* species resulting in the *R* alcohol with 97.5% *ee* proceeds 13 times faster than the minor *R,SS* catalyst cycle, giving the *S* alcohol in only 8% *ee*.

DM-BIPHEP is a conformationally flexible diphosphine existing as an *R* and *S* configurated mixture in equilibrium (Fig. 43) [114]. When $RuCl_2$(dm-biphep)-$(dmf)_n$ is mixed with (*S,S*)-DPEN, a 3:1 diastereo-mixture of (*S*)-DM-BIPHEP/(*S,S*)-DPEN–$RuCl_2$ complex and the *R,SS* isomer is obtained. The major *S,SS* species is more reactive and enantioselective in the hydrogenation of acyclic aromatic ketones. Hydrogenation of 1′-acetonaphthone with the mixed Ru complex in 2-propanol containing KOH at −35 °C gives the *R* alcohol in 92% *ee* and in >99% yield.

(*R*)-DM-DABN, a chiral aromatic diamine, preferably interacts with $RuCl_2[(R)$-xylbinap]$(dmf)_n$ rather than with the *S* complex, producing catalytically inactive

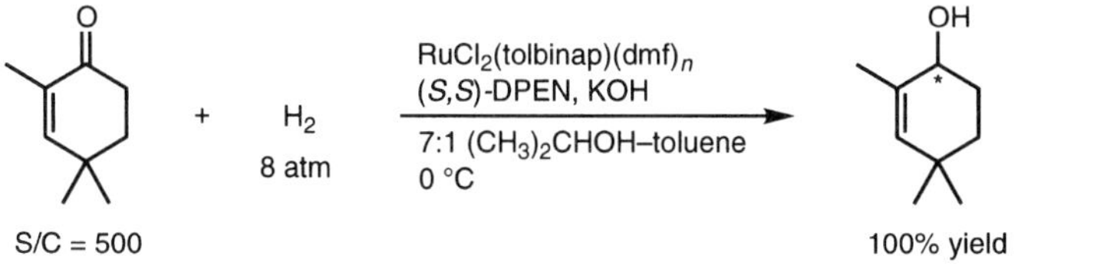

[ketone] + H_2 8 atm → $RuCl_2$(tolbinap)$(dmf)_n$, (S,S)-DPEN, KOH, 7:1 $(CH_3)_2CHOH$–toluene, 0 °C

S/C = 500

100% yield

TolBINAP	% ee	confign
±	95	*S*
R	96	*S*
S	26	*R*

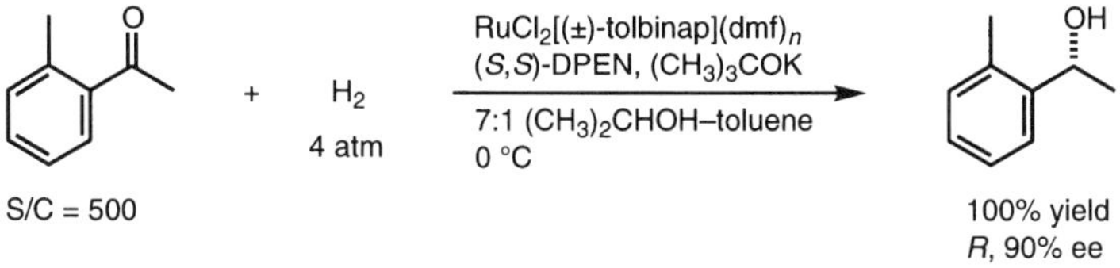

[ketone] + H_2 4 atm → $RuCl_2[(\pm)\text{-tolbinap}](dmf)_n$, (S,S)-DPEN, $(CH_3)_3COK$, 7:1 $(CH_3)_2CHOH$–toluene, 0 °C

S/C = 500

100% yield, R, 90% ee

Fig. 42 Asymmetric hydrogenation of ketones with a racemic TolBINAP–Ru complex and an optically pure DPEN.

$RuCl_2$[(*R*)-xylbinap][(*R*)-dm-dabn] for hydrogenation of aromatic ketones (Fig. 44) [115]. The enantiomer-selective deactivation cooperates well with the asymmetric activation described above (see Fig. 42), giving a highly enantioselective catalyst system using a racemic XylBINAP–$RuCl_2$ complex. Hydrogenation of 1′-acetonaphthone conducted with a catalyst system consisting of $RuCl_2$[(±)-xylbinap]-$(dmf)_n$, (*R*)-DM-DABN, (*S,S*)-DPEN, and KOH in a 1:0.55:0.5:2 ratio affords quantitatively the *R* alcohol in 96% *ee*.

1-Deuteriobenzaldehydes

Asymmetric hydrogenation of 1-deuteriobenzaldehyde and its derivatives in the presence of $Ru(OCOCH_3)_2$[(*R*)-binap] and 5 equivalents of HCl gives 1-deuterio benzyl alcohols in up to 89% *ee* (Fig. 45) [116]. Substrates with a heteroatom at the 1′-position show good enantioselectivity caused by a directing effect of the heteroatom interacting with the Ru catalyst. Hydrogenation of 1-deuterio-1′-methybenzaldehyde with *trans*-$RuCl_2$[(*S*)-tolbinal][(*S*)-daipen] and $(CH_3)_3COK$ in 2-propanol gives the *S* alcohol in 89% *ee* [14a]. Use of XylBINAP instead of TolBINAP results in a lower selectivity. Reaction of 1-deuteriobenzaldehyde gives poor enantioselectivity.

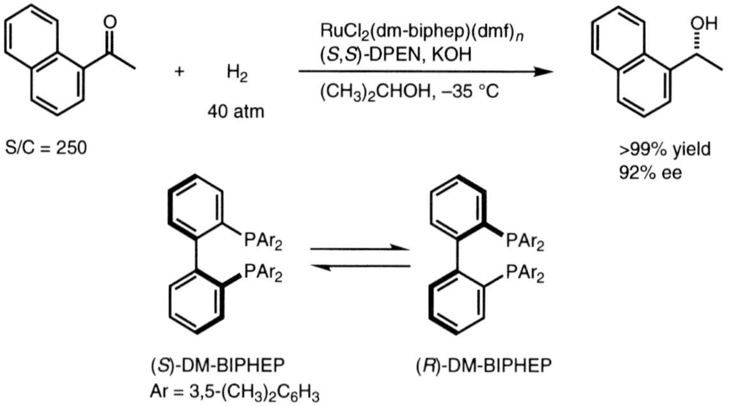

O
+ H_2
40 atm
$RuCl_2(\text{dm-biphep})(\text{dmf})_n$
(*S,S*)-DPEN, KOH
$(CH_3)_2CHOH$, –35 °C
OH
S/C = 250
>99% yield
92% ee
PAr_2
PAr_2
PAr_2
PAr_2
(*S*)-DM-BIPHEP
Ar = 3,5-$(CH_3)_2C_6H_3$
(*R*)-DM-BIPHEP

Fig. 43 Asymmetric hydrogenation of 1′-acetonaphthone with a DM-BIPHEP–Ru complex and (*S,S*)-DPEN.

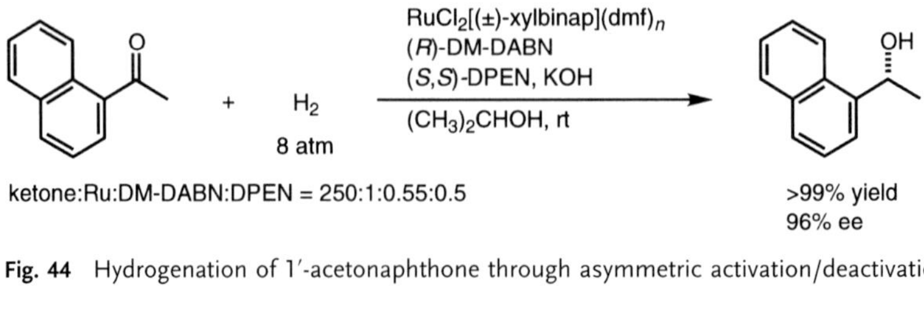

O
+ H_2
8 atm
$RuCl_2[(\pm)\text{-xylbinap}](\text{dmf})_n$
(*R*)-DM-DABN
(*S,S*)-DPEN, KOH
$(CH_3)_2CHOH$, rt
OH
ketone:Ru:DM-DABN:DPEN = 250:1:0.55:0.5
>99% yield
96% ee

Fig. 44 Hydrogenation of 1′-acetonaphthone through asymmetric activation/deactivation.

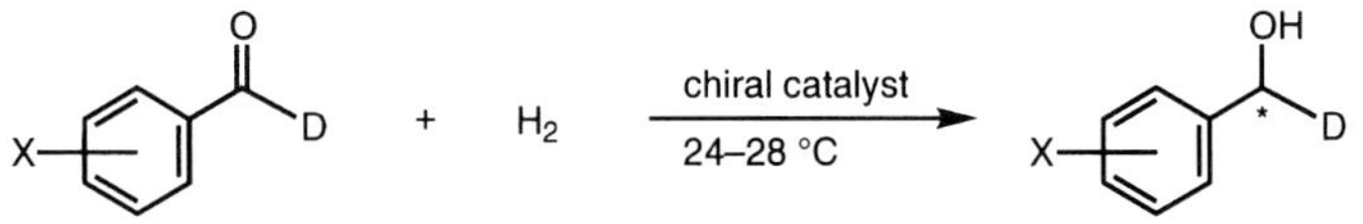

X	Catalyst [a]	S/C [b]	H_2 (atm)	% yield	% ee	Confign
H	$Ru(OCOCH_3)_2$-[(*R*)-binap] + HCl	85–100	11	100	65	*S*
H	(*S*,*S*)-**C1**	250	8	99.8	46	*S*
2-CH_3	(*S*,*S*)-**C1**	250	8	99	89	*S*
2-Br	$Ru(OCOCH_3)_2$-[(*R*)-binap] + HCl	85–100	11	100	89	*S*
3-Cl	$Ru(OCOCH_3)_2$-[(*R*)-binap] + HCl	85–100	11	67	73	–
4-Cl	$Ru(OCOCH_3)_2$-[(*R*)-binap] + HCl	85–100	11	100	70	–

a) See Fig. 20.
b) Substrate/catalyst mole ratio.

Fig. 45 Asymmetric hydrogenation of 1-deuteriobenzaldehydes.

As described above, the BINAP/1,2-diamine–Ru(II) complexes catalyze hydrogenation of a wide variety of simple ketones including aromatic, hetero-aromatic, amino, and unsaturated ketones with excellent chemo-, diastereo- and enantioselectivities. Some aliphatic ketones are also hydrogenated with high stereoselectivity. Kinetic resolution of racemic ketones gives the chiral ketones with high optical yield. The hydrogenation has been used as a key reaction in the synthesis of medicines, perfumes, etc. Fig. 46 lists the examples [14, 17, 32, 38, 39, 63, 80, 81, 107, 117].

1.1.3.2.2 Functionalized Ketones

Keto esters and their derivatives

Asymmetric hydrogenation of ketones which have a heteroatom adjacent to the carbonyl group has recently been a major subject in organic synthesis [62]. The functionality which is capable of interacting with Lewis acidic metals effectively accelerates hydrogenation of the carbonyl moiety and also directs the enantioface differentiation.

Homogeneous asymmetric hydrogenation of α-keto acid derivatives catalyzed by chiral phosphine–Rh complexes exhibits an excellent enantioface selection [118]. As illustrated in Fig. 47, hydrogenation of methyl pyruvate using a complex prepared *in situ* from $[RhCl(cod)]_2$ and MCCPM gives methyl lactate in 87% *ee* [119]. The chirally arranged diphenylphosphino group on the methylene at C2 is proposed to control the enantioselection, and the electron-donating dicyclohexylphosphino function at C4 is proposed to enhance the activity of the catalyst. A Cy,Cy-

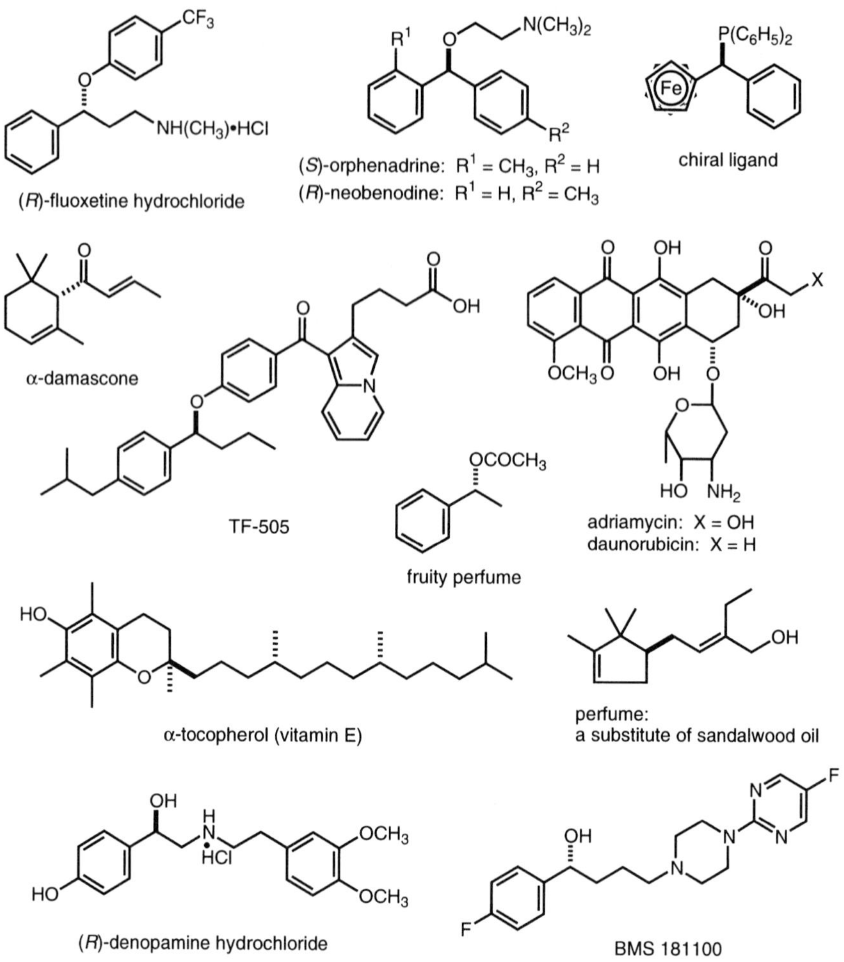

(*R*)-fluoxetine hydrochloride

(*S*)-orphenadrine: $R^1 = CH_3$, $R^2 = H$
(*R*)-neobenodine: $R^1 = H$, $R^2 = CH_3$

chiral ligand

α-damascone

TF-505

fruity perfume

adriamycin: X = OH
daunorubicin: X = H

α-tocopherol (vitamin E)

perfume:
a substitute of sandalwood oil

(*R*)-denopamine hydrochloride

BMS 181100

Fig. 46 Products obtained by routes involving hydrogenation of ketones catalyzed by diphosphine/diamine–Ru(II) complexes.

oxoProNOP–Rh complex can reduce ethyl pyruvate and benzoylformamide derivatives with 95% optical yield [120]. The use of Cp,Cp-QuinoNOP as a ligand achieves >99% optical yield [121]. An Rh catalyst with the $Cr(CO)_3$-complexed Cp,Cp-IndoNOP shows higher enantioselectivity than that with the original ligand (97% *ee* versus 91% *ee*) [96]. A neutral NORPHOS–Rh complex is effective for the hydrogenation of ethyl 2-oxo-4-phenylbutanoate [122]. Phosphine–Ru complexes sometimes work better than Rh catalysts. Although hydrogenation of methyl pyruvate using a neutral (*R*)-BINAP–Ru complex gives the (*R*)-hydroxy ketone in 83% *ee* [91], the cationic complex with aqueous HBF_4 which catalyzes the hydrogenation of methyl 4′-methylbenzoylformate gives 93% optical yield [92]. An Ru complex of BICHEP, an electron-rich biaryl ligand, effects hydrogenation of

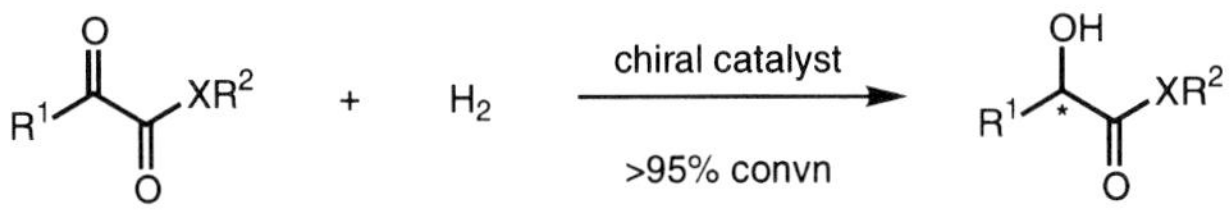

$R^1COCOXR^2 + H_2 \xrightarrow[>95\%\ convn]{chiral\ catalyst} R^1C^*H(OH)COXR^2$

R^1	XR^2	Chiral catalyst (S/C[a])	Solvent	H_2 (atm)	% ee	Confign
CH_3	OCH_3	$[RhCl(cod)]_2$–(2*S*,4*S*)-MCCPM (1000)	THF	20	87	*R*
CH_3	OCH_3	$RuCl_2$[(−)-tetrame-bitianp] (590)	CH_3OH	100	88	*S*
CH_3	OC_2H_5	$[RhOCOCF_3${(*S*)-cy,cy-oxo-pronop}$]_2$ (350)	toluene	50	95	*R*
t-C_4H_9	OC_2H_5	$[NH_2(C_2H_5)_2]$[{RuCl-[(*R*)-segphos]}$_2(\mu\text{-}Cl)_3$] (1000)	C_2H_5OH	50	98.6	*R*
$C_6H_5(CH_2)_2$	OC_2H_5	$[RhCl(nbd)]_2$–(*S*,*S*)-NOR-PHOS (50)	CH_3OH	100	96	*S*
$C_6H_5(CH_2)_2$	OC_2H_5	$[NH_2(C_2H_5)_2]$[{RuCl-[(*R*)-segphos]}$_2(\mu\text{-}Cl)_3$] (1500)	C_2H_5OH	50	95.7	*R*
C_6H_5	OCH_3	Ru[η^3-$CH_2C(CH_3)CH_2]_2$-[(*S*)-meo-biphep] + HBr (100)	CH_3OH	20	86	*S*
C_6H_5	OCH_3	[RuI(*p*-cymene){(*R*)-bichep}]I (100)	C_2H_5OH	5	>99	*S*
4-$CH_3C_6H_4$	OCH_3	[RuCl(*p*-cymene){(*S*)-binap}]Cl + HBF_4 (100)	CH_3OH	100	93	*S*
C_6H_5	$NHCH_2C_6H_5$	[RhCl{(*S*)-cy,cy-oxopronop}$]_2$ (50)	toluene	50	95	*S*
C_6H_5	$NHCH_2C_6H_5$	[RhCl{(*S*)-cp,cp-indonop}$]_2$ (200)	toluene	1	91	*S*
C_6H_5	$NHCH_2C_6H_5$	[RhCl{(*S*,2*S*)-$Cr(CO)_3$-cp,cp-indonop}$]_2$ (200)	toluene	1	97	*S*
C_6H_5	$NHCH_2C_6H_5$	[RhCl{(*S*)-cp,cp-quinonop}$]_2$ (200)	toluene	50	>99	*S*
C_6H_5	$NHCH_2C_6H_5$	[RuCl(*p*-cymene){(*S*)-bichep}]Cl (100)	CH_3OH	40	96	*R*

a) Substrate/catalyst mole ratio.

Fig. 47 Asymmetric hydrogenation of α-keto acid derivatives.

methyl benzoylformate and the amide derivative, giving the corresponding alcohols in up to >99% *ee* [123]. A MeO-BIPHEP–Ru complex is also usable [124]. A Ru complex with (*R*)-SEGPHOS as a ligand effects asymmetric hydrogenation of aliphatic α-keto esters ($R^1 = t\text{-}C_4H_9$, $C_6H_5(CH_2)_2$) with an S/C of >1000, resulting in the *R* alcohols in >95% *ee* [100]. An Ru complex of tetraMe-BITIANP possessing five-membered heteroaromatic rings also shows high selectivity for the hydrogenation of methyl pyruvate [125].

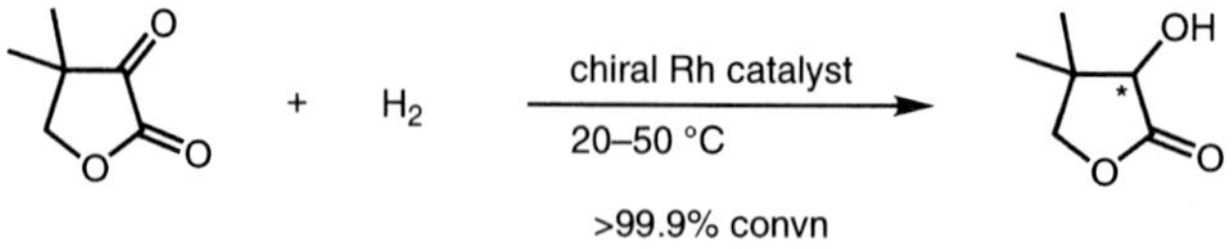

+ H_2 → chiral Rh catalyst, 20–50 °C, >99.9% convn

Chiral Rh catalyst (S/C[a])	*Solvent*	*H_2 (atm)*	*% ee*	*Confign*
$[RhCl(cod)]_2$–(2*S*,4*S*)-BPPM (95–101)	C_6H_6	50	86.7	*R*
$[RhCl(cod)]_2$–(2*S*,4*S*)-BCPM (100)	THF	50	92.0	*R*
$[RhCl(cod)]_2$–(2*S*,4*S*)-*m*-CH_3POPPM (770)	Toluene	12	94.8	*R*
$[RhOCOCF_3\{(S)$-cp,cp-indonop$\}]_2$ (200)	Toluene	1	>99	*R*
$[RhOCOCF_3\{(S)$-cp,cp-oxopronop$\}]_2$ (200)	Toluene	1	98.7	*R*
$[RhOCOCF_3\{(S)$-cp,cp-isoalanop$\}]_2$ (200)	Toluene	1	97.0	*S*

a) Substrate/catalyst mole ratio.

Fig. 48 Asymmetric hydrogenation of ketopantolactone.

As illustrated in Fig. 48, asymmetric hydrogenation of ketopantolactone catalyzed by a Rh complex with (2*S*,4*S*)-BPPM, a pyrrolidine-based diphosphine ligand, gives (*R*)-pantoyl lactone with 86.7% optical purity [126]. The BCPM–Rh complex shows better enantioselection [127]. The reaction with a *m*-CH_3POPPM–Rh catalyst affords the hydroxy lactone in 95% *ee* [128, 129]. When the hydrogenation is conducted with an S/C of 150000, the TOF of 50000 h^{-1} is achieved. A 200 kg batch reaction has been performed (Hoffmann-La Roche, Ltd). The use of $[RhOCOCF_3(cp,cp\text{-}oxopronop)]_2$ gives 98.7% optical yield and a TOF as high as 3300 h^{-1} [130]. The high rate is due to the electron-rich property of the phosphine ligand. The Cp,Cp-IndoNOP–Rh catalyst affords the hydroxy lactone in >99% *ee* [96]. Similarly, Cp,Cp-isoAlaNOP is effective for this purpose [131].

Highly enantioselective hydrogenation of *β*-keto esters is achieved by the use of BINAP–Ru(II). Hydrogenation of methyl 3-oxobutanoate, a representative substrate, catalyzed by (*R*)-BINAP–Ru(II) halide complex gave (*R*)-methyl 3-hydroxybutanoate quantitatively in up to >99% *ee* (Fig. 49) [62c, 132, 133]. Halogen-containing complexes with a formula of RuX_2(binap) (X=Cl, Br, or I; empirical formula with a polymeric form) or $RuCl_2(binap)(dmf)_n$ (oligomeric form) [134] display excellent catalytic performance in the hydrogenation of a wide variety of *β*-keto esters. The reaction can be conducted with an S/C as high as 10000. *β*-Keto amides and thioesters are also hydrogenated with high enantioselectivity [91, 135]. Its remarkable efficiency urged the chemists to develop convenient procedures to prepare active BINAP–Ru species [19, 92, 124, 136]. The reaction is remarkably accelerated under strongly acidic conditions [136b, d]. Other biaryl diphosphines such as BIPHEMP [124], BIMOP [137], MeO-BIPHEP [138], C4TunaPhos [139], BIFAP [140], BisbenzodioxanPhos [141], P-phos [142], tetraMe-BITIANP [125], and bis-steroidal phosphine [143] are also excellent chiral ligands for the hydrogenation of *β*-keto esters. An Ru complex possessing *i*-Pr-BPE, a fully alkylated diphosphine, effectively promotes the reaction under a low pressure [144]. The electron-

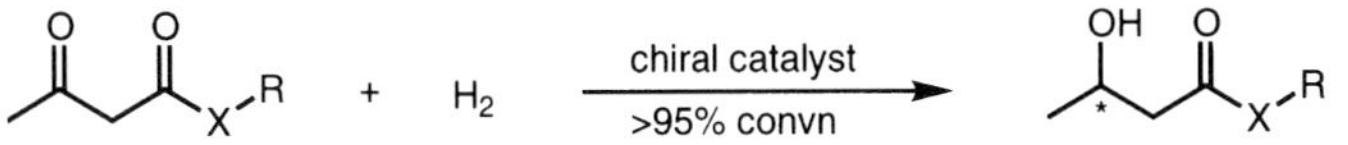

XR	Chiral catalyst (S/C[a])	Solvent	H_2 (atm)	Temp (°C)	% ee	Confign
OCH_3	$RuCl_2$[(*R*)-binap] (2000)	CH_3OH	100	23	>99	*R*
OCH_3	$RuCl_2$[(*R*)-binap](dmf)$_n$ (1960)	CH_3OH	100	25	99	*R*
OCH_3	$RuCl_2$[(*R*)-binap](dmf)$_n$ (2330)	CH_3OH	4	100	98	*R*
OCH_3	[$NH_2(C_2H_5)_2$][{RuCl[(*R*)-binap]}$_2$(μ-Cl)$_3$] (1410)	CH_3OH	100	25	>99	*R*
OCH_3	[RuI{(*S*)-binap}C_6H_6]I (2380)	CH_3OH	100	20	99	*R*
OCH_3	Ru[η^3-$CH_2C(CH_3)CH_2$]$_2$-[(*S*)-binap] + HBr (50)	CH_3OH	1	rt	97[b)]	*S*
OCH_3	*trans*-$RuCl_2$[(*R*)-binap]py$_2$ + HCl (1000)	CH_3OH	3.7	60	99.9	*R*
OCH_3	[$RuCl_2$(cod)]$_n$–(*R*)-BINAP (100)	CH_3OH	4	rt	99	*R*
OCH_3	$RuCl_2$[(*S*)-bis-steroidal phosphine](dmf)$_n$ (1270)	CH_3OH	100	100	99	*S*
OCH_3	$RuBr_2$[(*S*)-biphemp] (200)	CH_3OH	5	50	>99	*S*
OCH_3	[RuI_2(*p*-cymene)]$_2$–(*R*)-BIMOP (2000)	1:1 CH_3OH–CH_2Cl_2	10	30–40	100	*R*
OCH_3	$RuCl_3$–(*S*)-MeO-BIPHEP (100)	CH_3OH	4	50	99	*S*
OCH_3	$RuCl_2$[(*R*)-c4tunaphos]-(dmf)$_n$ (100)	CH_3OH	51	60	99.1	*R*
OCH_3	$RuCl_2$[(*S*)-bifap](dmf)$_n$ (1000)	CH_3OH	100	70	100	*S*
OCH_3	Ru[η^3-$CH_2C(CH_3)CH_2$]$_2$-[(*R,R*)-*i*-pr-bpe] + HBr (500)	9:1 CH_3OH–H_2O	4	35	99.3	*S*
OCH_3	Ru($OCOCF_3$)$_2$[(*S*)-[2.2]-phanephos] + (n-C_4H_9)$_4$NI (125–250)	10:1 CH_3OH–H_2O	3	–5	96	*R*
OCH_3	$RuCl_2$[(*R*)-poly-nap](dmf)$_n$ (1000)	CH_3OH	40	50	99	*R*
OCH_3	Ru[η^3-$CH_2C(CH_3)CH_2$]$_2$-[peg-(*R*)-am-binap] + HBr (10000)	CH_3OH	100	50	99	*R*
OC_2H_5	$RuCl_2$[(*R*)-bisbenzo-dioxanephos](dmf)$_n$ (1000)	C_2H_5OH	3.4	80–90	99.5	*R*
OC_2H_5	$RuCl_2$[(*S*)-p-phos](dmf)$_n$ (400)	10:1 C_2H_5OH–CH_2Cl_2	3.4	70	98.6	–
OC_2H_5	$RuCl_2$[(–)-tetrame-bitianp] (1000)	CH_3OH	100	70	99	*R*

Fig. 49 Asymmetric hydrogenation of β-keto acid derivatives.

XR	Chiral catalyst (S/C[a])	Solvent	H_2 (atm)	Temp (°C)	% ee	Confign
OC_2H_5	Ru[η^3-$CH_2C(CH_3)CH_2$]$_2$-(cod)–(*R*)-(*S*)-**L1a**+HBr (200)	C_2H_5OH	50	50	98.6	*S*
OC_2H_5	[Rh(nbd)$_2$]BF_4–(*R*)-(*S*)-JOSIPHOS (100)	CH_3OH	20	rt	97	*S*
$OC(CH_3)_3$	[$NH_2(C_2H_5)_2$][{RuCl[(*R*)-binap]}$_2$(μ-Cl)$_3$]+HCl (2170)	CH_3OH	3	40	>97	*R*
NHC_6H_5	[$NH_2(C_2H_5)$][{RuCl[(*R*)-binap]}$_2$(μ-Cl)$_3$] (500)	CH_3OH	30	60	>95	*R*
$N(CH_3)_2$	$RuBr_2$[(*S*)-binap] (670)	C_2H_5OH	63	27	96	*S*
SC_2H_5	$RuCl_2$[(R)-binap] (530)	C_2H_5OH	95	27	93[c]	*R*

a) Substrate/catalyst mole ratio.
b) 80% yield.
c) 42% yield.

Fig. 49 (cont.)

donating property is considered to be the origin of the high reactivity. Ru($OCOCF_3$)$_2$([2.2]-phanephos) with (n-C_4H_9)$_4$NI exhibits high catalytic activity under a low temperature and a low H_2 pressure conditions in the absence of strong acids [145]. A Ru complex with chiral 1,5-diphosphinylferrocene **L1a** [146] as well as a JOSIPHOS–Rh complex [147] are also excellent for asymmetric hydrogenation of β-keto esters. Some recyclable catalysts effectively promote the hydrogenation of β-keto esters. An oligomeric (*R*)-Poly-NAP–Ru-catalyzed hydrogenation of methyl 3-oxobutanoate with an S/C of 1000 can be repeated 5 times to give the *R* alcohol in >98% *ee* (Fig. 49) [148]. A PEG-Am-BINAP–Ru complex effects the reaction with an S/C of 10 000 under 100 atm of H_2 [149]. Hydrogenation in water is promoted by a Ru catalyst with a water-soluble diam-BINAP [150]. Immobilized catalysts in a polydimethylsiloxane membrane [151] or on a polystyrene are also usable [152].

Hydrogenation of benzoylacetic acid derivatives with high enantioselectivity has been difficult to achieve. Recently, an (*R*)-SEGPHOS–Ru complex catalyzed the hydrogenation of the ethyl ester with an S/C of 10 000 under 30 atm of H_2, resulting in the *S* alcohol in 97.6% *ee* (Fig. 50) [100]. MeO-BIPHEP [138], Tol-P-Phos [153], and a chiral ferrocenyl diphosphine **L1c** [154] are also excellent ligands for this purpose. Hydrogenation of *N*-methylbenzoylacetamide in the presence of an (*R*)-BINAP–Ru catalyst affords the *S* alcohol in >99.9% *ee*, while the yield is 50% [155].

α,α-Difluoro-β-keto esters are hydrogenated with (*R*)-BINAP–Ru [156] and (*S*)-Cy,Cy-OxoProNOP–Rh [157] complexes under 20 atm of H_2 at an S/C as high as 1000 to give the corresponding *R* alcohols in >95% *ee* (Fig. 51). The sense of enantioselection is the same as that in the reaction of simple β-keto esters (see Fig. 49). Hydrogenation of ethyl 4,4,4-trifluoro-3-oxobutanoate is catalyzed by Cy,Cy-OxoProNOP–Rh complex to give the *R* alcohol in 91% *ee* [157]. A MeOHCd-

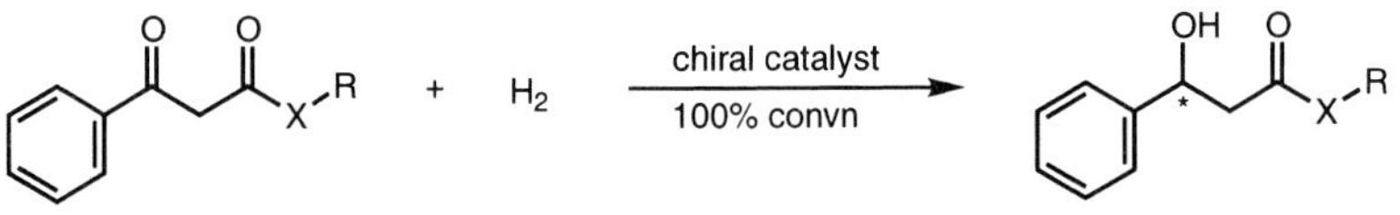

XR	Chiral catalyst (S/C[a])	Solvent	H_2 (atm)	Temp (°C)	% ee	Confign
OC_2H_5	$RuCl_3$–(*S*)-Meo-BIPHEP (100)	CH_3OH	4	80	95	*R*
OC_2H_5	$[NH_2(C_2H_5)_2][\{RuCl[(R)\text{-segphos}]\}_2(\mu\text{-Cl})_3]$ (10 000)	C_2H_5OH	30	80	97.6	*S*
OC_2H_5	$RuCl_2[(S)\text{-tol-p-phos}](dmf)_n$ (800)	1:1 C_2H_5OH–CH_2Cl_2	20	90	96.4	*S*
OC_2H_5	$Ru[\eta^3\text{-}CH_2C(CH_3)CH_2]_2$-(cod)–(*R*)-(*S*)-**L1c** + HBr (200)	C_2H_5OH	50	50	98	*R*
$NHCH_3$	$RuCl_2[(R)\text{-binap}](dmf)_n$ (1800)	CH_3OH	14	100	>99.9[b]	*S*

a) Substrate/catalyst mole ratio.
b) 50% yield.

Fig. 50 Asymmetric hydrogenation of benzoylacetic acid derivatives.

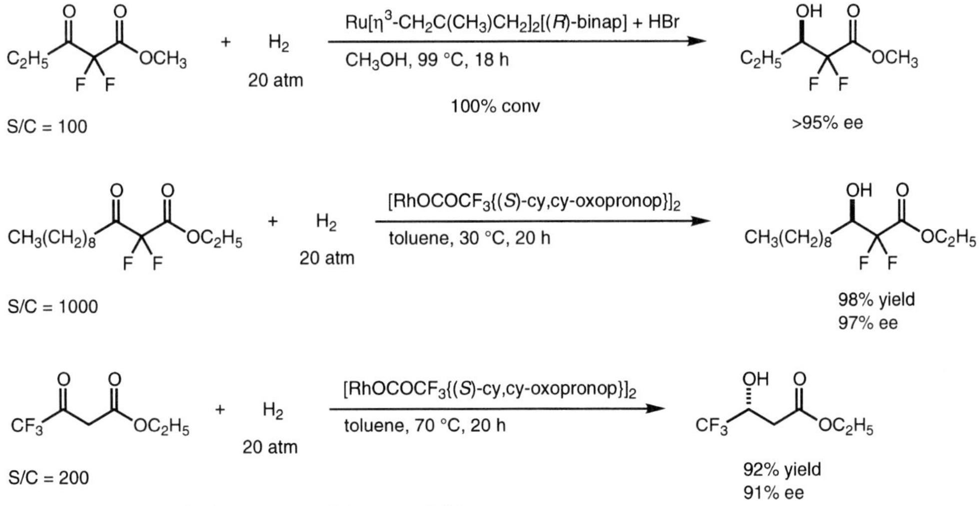

Fig. 51 Asymmetric hydrogenation of fluorinated β-keto esters.

modified Pt/Al_2O_3 catalyst hydrogenates the trifluoroketo ester under 10 atm of H_2 to give the *S* alcohol in 93% *ee* [148 d, 158].

Hydrogenation of γ-keto esters or *o*-acylbenzoic esters catalyzed by a BINAP–Ru complex gives the corresponding γ-lactones or *o*-phthalides with an excellent enantioselectivity (Fig. 52) [159, 160].

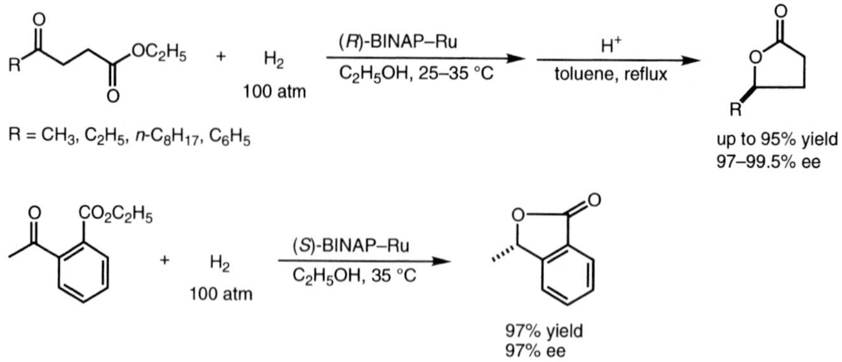

Fig. 52 Asymmetric hydrogenation of γ-keto esters.

Homogeneous asymmetric hydrogenation of α-, β- or γ-keto esters catalyzed by BINAP–Ru(II) complexes is now conveniently used for the synthesis of a wide range of natural and unnatural compounds [133, 161]. Fig. 53 illustrates some examples. Chiral centers induced by the asymmetric reduction are labeled by *R* or *S*.

In asymmetric hydrogenation of bifunctionalized ketones, competitive interaction of the functionalities to the center metal of the catalyst tends to decrease the enantioselectivity, depending on the steric and electronic nature of the coordinative groups. Hydrogenation of methyl 5-benzyloxy-3-oxopentanoate with the BINAP–Ru complex gives the corresponding alcohol with the same degree and sense of enantioface selection as the reaction of methyl 3-oxobutanoate (Fig. 54) [91]. On the other hand, the reaction of 4-benzyloxy- or 4-chloro-3-oxobutanoate is only moderately enantioselective at 100 atm of H_2 and room temperature. Their enantioselectivity, however, is increased to 98 and 97%, respectively, by raising the temperature to 100 °C [162]. The analog possessing a bulky triisopropylsilyloxy group at the C4 position shows high selectivity at room temperature. 4-Trimethylamino chloride derivatives are also reduced with a high enantioselectivity [124]. Similarly, Ru complexes modified by C_2-symmetric chiral diphosphines also exhibit high enantioselectivity in the hydrogenation of ethyl 4-chloro-3-oxobutanoate at higher temperatures [100, 138, 141, 142]. Hydrogenation of methyl 4-methoxy-3-oxobutanoate in the presence of *i*-Pr-BPE–Ru complex gives the corresponding hydroxy ester in 95.5% *ee* at 35 °C, whereas the enantioselectivity is moderate in the reaction of the 4-chloro analog [144]. A Ru complex with Ph,Ph-oxoProNOP catalyzed the hydrogenation of ethyl 4-chloro-3-oxobutanoate with an optical yield of 75% at 20 °C [163]. The 4-dimethylamino hydrochloride derivative is hydrogenated by the MCCXM–Rh complex with good enantioselectivity [164].

Asymmetric hydrogenation of bifunctionalized ketones catalyzed by BINAP–Ru complexes is applicable to enantioselective synthesis of several bioactive compounds

→

Fig. 53 Examples of biologically active compounds obtainable through BINAP–Ru-catalyzed hydrogenation of α-, β-, or γ-keto esters.

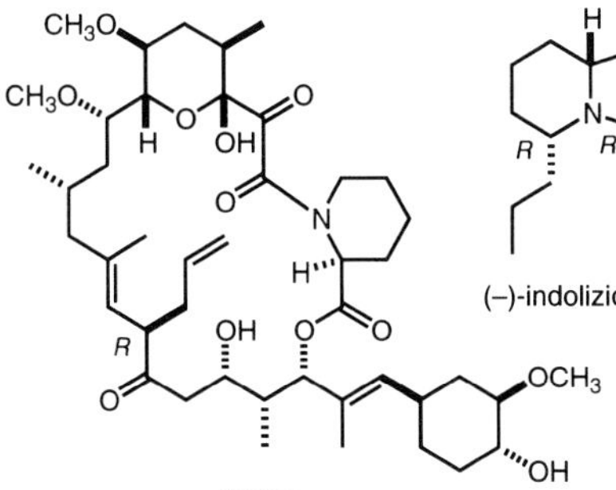

FK506
(–)-indolizidine 223AB 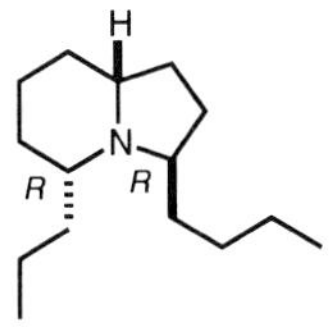
(–)-pyrenophorin 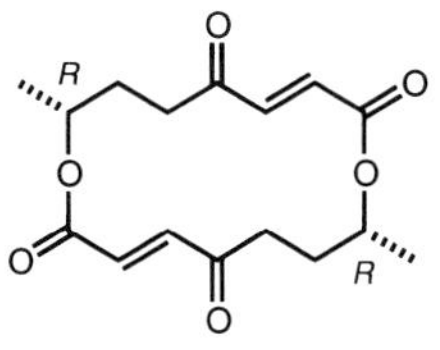
(+)-araguspongine B 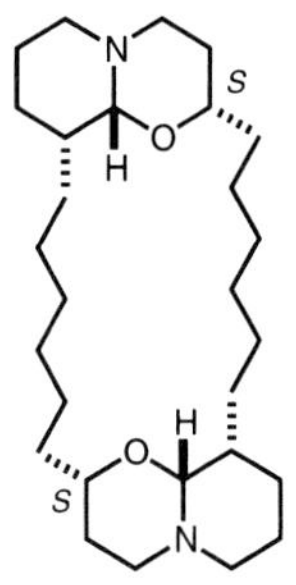
(–)-tetrahydrolipstatin 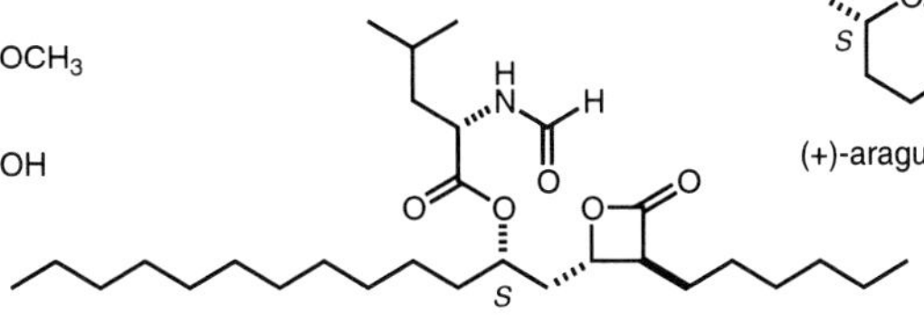
(–)-gloeosporone 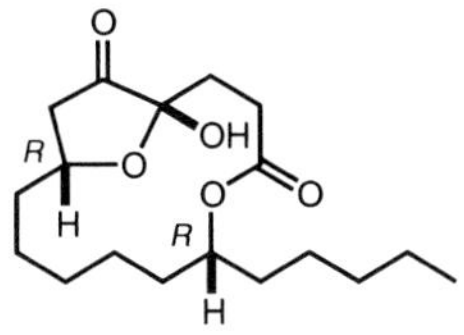
(+)-α-lipoic acid 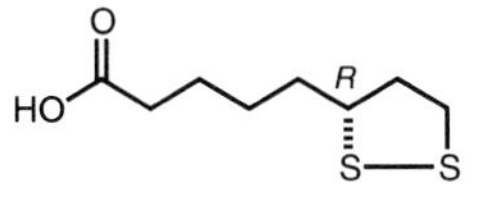
lipid A 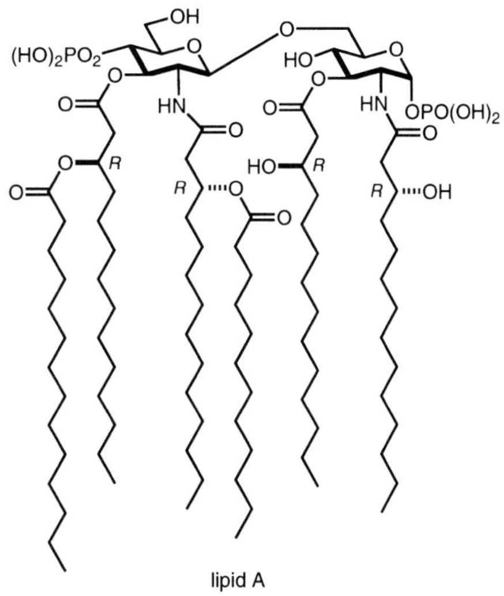
epothilone B 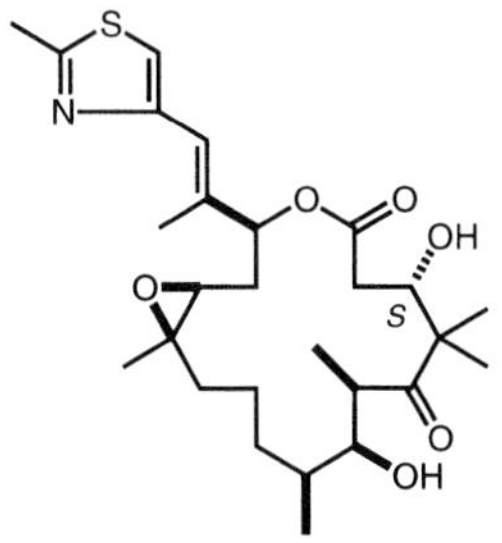
(+)-brefeldin A 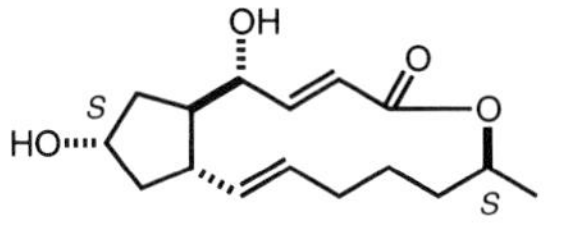
alkaloid 251F 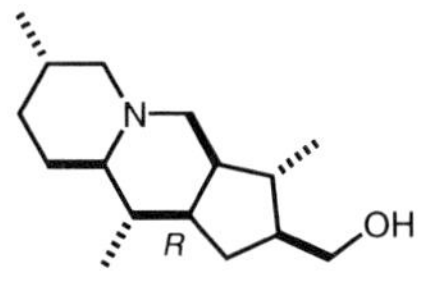
(–)-pateamine A 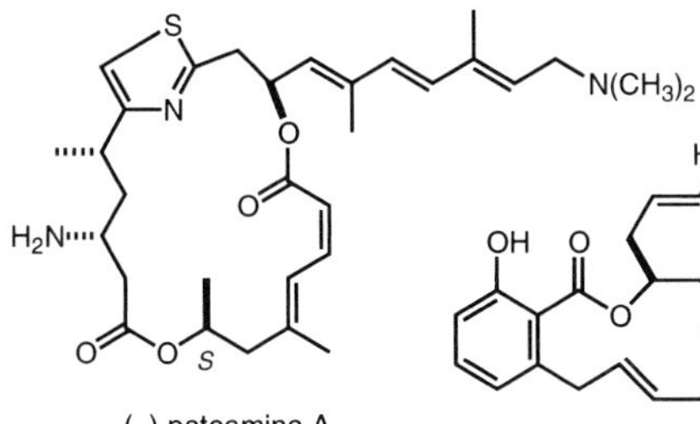
salicylihalamide A 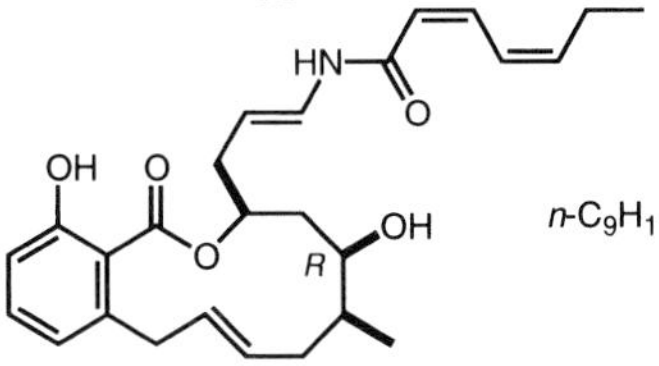
arthrobacilin A 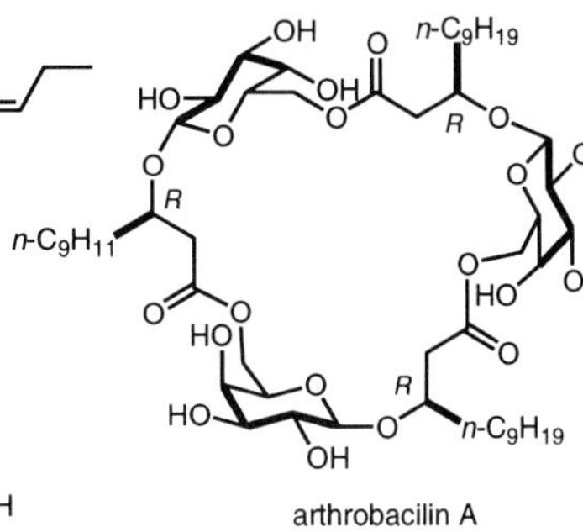
sulfobacin A

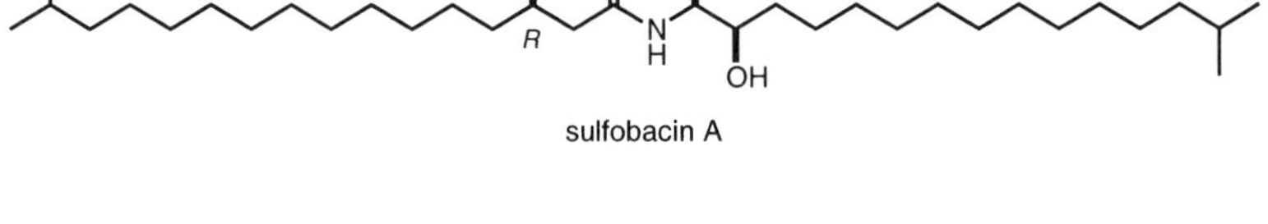

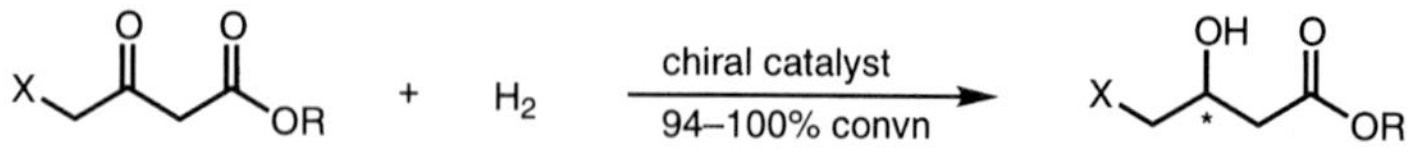

$$\text{X}\!-\!\text{CH}_2\text{C(O)CH}_2\text{C(O)OR} + \text{H}_2 \xrightarrow[\text{94–100\% convn}]{\text{chiral catalyst}} \text{X}\!-\!\text{CH}_2\text{C}^{*}\text{H(OH)CH}_2\text{C(O)OR}$$

X	*R*	*Chiral catalyst (S/C[a])*	*H_2 (atm)*	*Temp (°C)*	*% ee*	*Confign*
$C_6H_5CH_2OCH_2$	CH_3	$RuBr_2$[(*S*)-binap] (370)	50	26	99	*S*
CH_3O	CH_3	Ru[η^3-$CH_2C(CH_3)CH_2$]$_2$-[(*R,R*)-*i*-pr-bpe]+HBr (500)	4	35	95.5	*R*
$C_6H_5CH_2O$	C_2H_5	$RuBr_2$[(*S*)-binap] (560)	100	28	78	*R*
$C_6H_5CH_2O$	C_2H_5	$RuBr_2$[(*S*)-binap] (560)	100	100	98	*R*
$[(CH_3)_2CH]_3SiO$	C_2H_5	$RuBr_2$[(*S*)-binap] (290)	100	27	95	*R*
Cl	C_2H_5	Ru[η^3-$CH_2C(CH_3)CH_2$]$_2$[(*R,R*)-*i*-pr-bpe]+HBr (500)	4	35	76	*R*
Cl	C_2H_5	$RuBr_2$[(*S*)-binap] (1080)	77	24	56	*R*
Cl	C_2H_5	$RuBr_2$[(*S*)-binap] (1300)	100	100	97	*R*
Cl	C_2H_5	$Ru(OCOCH_3)_2$[(*S*)-ph,ph-oxo-pronop] (150)	140	20	75	*S*
Cl	C_2H_5	$RuCl_3$–(*S*)-MeO-BIPHEP (100)	4	120	92	*R*
Cl	C_2H_5	$[NH_2(C_2H_5)_2]$[{RuCl[(*R*)-seg-phos]}$_2$(μ-Cl)$_3$] (2500)	30	90	98.5	*S*
Cl	C_2H_5	$RuCl_2$[(*R*)-bisbenzodioxane-phos](dmf)$_n$ (1000)	3.4	80–90	97	*S*
Cl	C_2H_5	$RuCl_2$[(*S*)-p-phos](dmf)$_n$ (2780)	3.4	80	98	–
$Cl(CH_3)_2NH$	C_2H_5	$[RhCl(cod)]_2$–(2*S*,4*S*)-MCCXM (100)	20	50	85	*S*
$Cl(CH_3)_3N$	C_2H_5	$[NH_2(C_2H_5)_2]$[{RuCl[(*R*)-binap]}$_2$(μ-Cl)$_3$] (–)	100	25	96[b]	*S*

a) Substrate/catalyst mole ratio.
b) 75% yield.

Fig. 54 Asymmetric hydrogenation of bifunctionalized ketones.

as shown in Fig. 55 [62c, 162, 164, 165]. Stereocenters determined by BINAP–Ru-catalyzed reaction are labeled by *R* or *S*.

As illustrated in Fig. 56, diastereoselective hydrogenation of the α-keto amide **A** derived from an (*S*)-amino ester, catalyzed by an (*R,R*)-CyDIOP–Rh complex, preferentially gives the (*S,S*)-hydroxy amide **B** [166]. On the other hand, when the (*S,S*)-catalyst is used, the *R,S* product is obtained selectively. The *N*-Boc-protected (*S*)-γ-amino β-keto esters **C** are converted predominantly to the *syn* alcohols **D** with the (*R*)-BINAP–Ru complex [167]. The use of the *S* catalyst preferentially gives the *anti* isomer. *N*-Acetyl- or *N*-boc-protected γ-amino γ,δ-unsaturated β-keto esters **E** are tandem hydrogenated in the presence of (*S*)-BINAP–Rh and –Ru catalysts to give predominantly (3*R*,4*R*)-**F** in one pot [168]. The BINAP–Rh catalyst preferentially reduces the olefinic function of **E** at a low H_2 pressure, and the BINAP–Ru catalyst then hydrogenates the carbonyl group under high-pressure conditions. Hydrogenation of the *N*-Boc-protected (*S*)-δ-amino β-keto ester **G** followed by cyclization gives the *trans* lactone **H** stereoselectively [169]. The chiral products

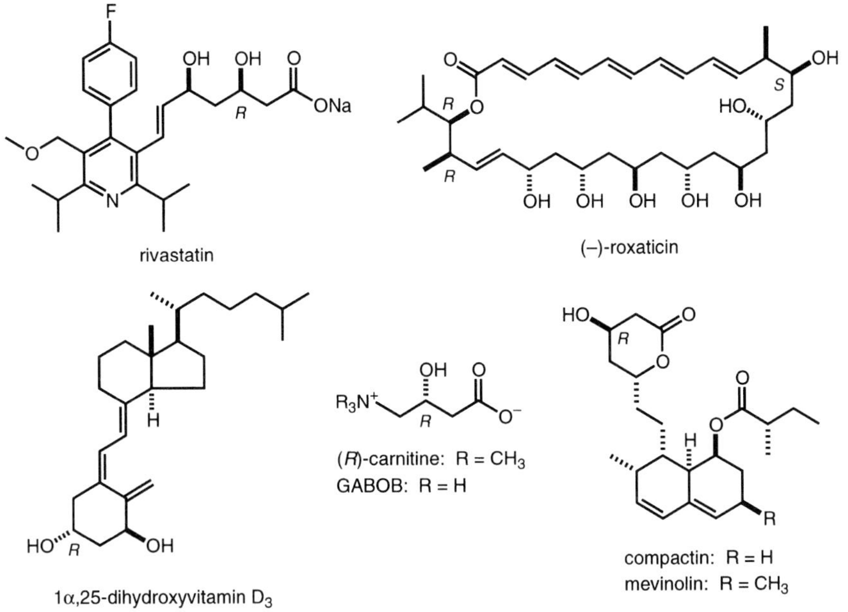

Fig. 55 Examples of biologically active compounds obtainable through BINAP–Ru-catalyzed hydrogenation of difunctionalized ketones.

D and **F** are useful intermediates for a statin series – essential components of aspartic proteinase inhibitors [167, 168]. The product **H** is also a useful building block for the synthesis of theonellamide F, an antifungal agent [169].

Heterogeneous asymmetric hydrogenation of α-keto esters was first achieved with an alkaloid-modified Pt/Al_2O_3 catalyst [62j, 102, 170, 171]. Hydrogenation of methyl pyruvate catalyzed by Pt/Al_2O_3 in the presence of quinine proceeded in benzene to give (*R*)-methyl lactate in 87% *ee* (Fig. 57) [170b]. Ethyl benzoylformate is hydrogenated to give the corresponding hydroxy ester in up to 90% *ee* [170c]. When hydrogenation of ethyl pyruvate is conducted in the presence of a cinchonidine-modified catalyst with ultrasonic pretreatment, the *ee* of the product increases to as high as 97% [172]. The smaller metal particle size (3.9 nm) of the catalyst may cause the high enantioselectivity. The reduction promoted by a cinchonidine-modified catalyst in toluene with the addition of an achiral tertiary amine, quinuclidine, afforded the *R* alcohol in 95% *ee* [173]. The *ee* value is much higher than that in the absence of the achiral amine (78%). An HCd-modified catalyst with appropriate surface conditions ($Pt_{surface}$/modifier = 5–12) reduces the α-keto ester with an optical yield of 94% [174]. High TON values of >28 000 relative to the modifier and a TOF of 4 s^{-1} are achieved. Ethyl 2-oxo-4-phenylbutanoate is also hydrogenated with a high enantioselectivity using this catalyst [174]. Hydrogenation of ethyl benzoylformate with the HCd–Pt/Al_2O_3 catalyst in a 1:1 acetic acid–toluene mixture affords (*R*)-ethyl mandelate in 98% *ee* [175]. The use of MeOHCd-modifier in the hydrogenation of ethyl pyruvate gives an optical yield of

A + H_2 (1 atm) —Rh catalyst→ B

A

Rh catalyst = $[RhCl(C_2H_4)_2]_2$–(*R*,*R*)-CyDIOP

B

S,*S*:*R*,*S* = 86:14

C + H_2 (100 atm) —Ru catalyst→ **D** → statine series

C: R = $CH_2C_6H_5$, $CH_2CH(CH_3)_2$, CH_2-*cyclo*-C_6H_{11}

Boc = OCO-*t*-C_4H_9

Ru catalyst = $RuBr_2$[(*R*)-binap]

D

92–97% yield,
syn:anti = >99:1

statine series

E + H_2 —Rh catalyst, Ru catalyst, $(C_2H_5)_3N$; 10 atm for 24 h then 90 atm for 24 h→ (3*R*,4*R*)-**F**

E: R^1 = $COCH_3$, OCO-*t*-C_4H_9; R^2 = CH_3, C_2H_5;
X = H, Cl

Rh catalyst = [Rh{(*S*)-binap}(cod)]ClO_4; Ru catalyalyst = $RuBr_2$[(*S*)-binap]

(3*R*,4*R*)-**F**

90–99% yield, >95% ee
syn:anti = 100:0

G + H_2 (100 atm) —Ru catalyst→ —H^+→ **H**

G

Ru catalyst = $RuBr_2$[(*R*)-binap](dmf)$_n$

H

85% yield
trans:cis = 96:4

theonellamide F

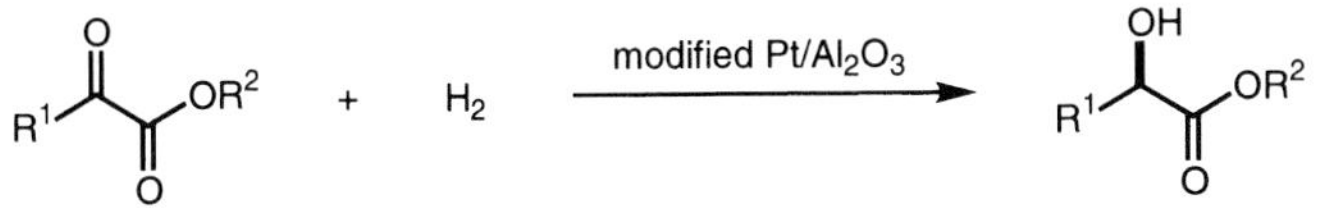

$$R^1C(=O)C(=O)OR^2 + H_2 \xrightarrow{\text{modified } Pt/Al_2O_3} R^1CH(OH)C(=O)OR^2$$

R^1	R^2	Modifier	Solvent	H_2 (atm)	Temp (°C)	% ee
CH_3	CH_3	Quinine	Benzene with quinine	70	rt	86.8
CH_3	C_2H_5	Cinchonidine[a]	CH_3CO_2H	10	25	97.1
CH_3	C_2H_5	Cinchonidine/ Quinuclidine	Toluene	50	10	94.6
CH_3	C_2H_5	HCd	CH_3CO_2H	6	17	94
CH_3	C_2H_5	MeOHCd	CH_3CO_2H	100	20–25	95
CH_3	C_2H_5	(*R*)-1-NEA	CH_3CO_2H	8	9	82
CH_3	C_2H_5	(*R*)-**L3**	CH_3CO_2H	70	10	87
$C_6H_5(CH_2)_2$	C_2H_5	HCd	CH_3CO_2H	6	17	91
$C_6H_5(CH_2)_2$	C_2H_5	MeOHCd	CH_3CO_2H	70	rt	92
C_6H_5	C_2H_5	HCd	1:1 CH_3CO_2H–toluene	25	0	98
$C_2H_5OCO(CH_2)_2$	C_2H_5	MeOHCd	CH_3CO_2H	20	20	96[b]
CH_3	H	MeOHCd	9:1 C_2H_5OH–H_2O	100	20–30	79
C_6H_5	H	MeOHCd	9:1 C_2H_5OH–H_2O	100	20–30	85

a) Ultrasonicated Pt/Al_2O_3 is used.
b) (*R*)-Ethyl 5-oxotetrahydrofuran-2-carboxylate is observed.

Fig. 57 Asymmetric hydrogenation of *α*-keto esters catalyzed by modified Pt/Al_2O_3.

95% [176]. Hydrogenation of ethyl 2-oxo-4-phenylbutanoate affords the chiral alcohol in 92% *ee* [102h, 122]. Ethyl 2-oxoglutarate is converted in this reaction to (*R*)-ethyl 5-oxotetrahydrofuran-2-carboxylate in 96% *ee* [177]. This catalyst is also effective for the asymmetric hydrogenation of *α*-keto acids [178]. High enantioselectivity is also achievable with catalysts modified with simple chiral amines, (*R*)-1-(1-naphthyl)ethylamine [(*R*)-1-NEA] and (*R*)-1-(9-anthracenyl)-2-(1-pyrrolidinyl)ethanol [(*R*)-**L3**] [179, 180]. An extended aromatic *π*-system for binding on the metal surface is crucial to achieve high enantioselectivity. The real parameter affecting the enantioselectivity was proposed to be the concentration of H_2 in the liquid phase [181]. The crucial structural elements are: (1) the tertiary quinuclidine nitrogen, (2) the flat quinoline ring, and (3) the stereogenic center(s) close to the nitrogen [102, 180a, 182, 183]. A colloidal Pt catalyst stabilized by HCd promotes hydrogenation of ethyl pyruvate with 91% optical yield [184].

Ketopantolactone and 1-ethyl-4,4-dimethylpyrrolidine-2,3,5-trione are hydrogenated with a Pt catalyst modified by cinchonidine to give the corresponding alco-

Fig. 56 Diastereoselective hydrogenation of chiral ketones.

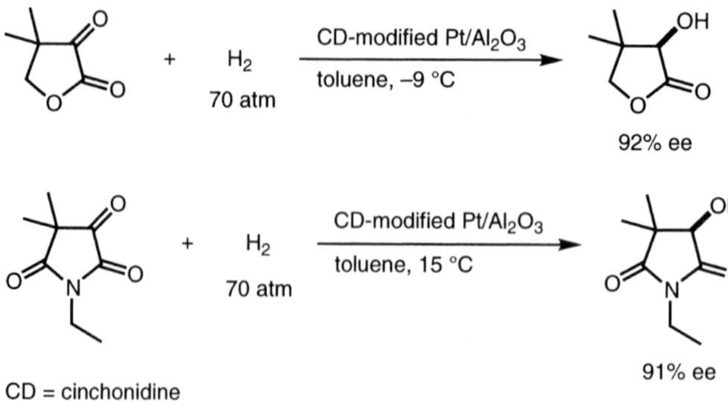

Fig. 58 Asymmetric hydrogenation of α-keto lactone and lactam.

hols in 92% and 91% *ee*, respectively (Fig. 58) [185]. These reactions can be conducted with an S/C as high as 237 000 [185 a].

The Raney Ni catalyst modified by tartaric acid and NaBr is an excellent heterogeneous catalyst for the asymmetric hydrogenation of β-keto esters (Fig. 59) [171 b, 186, 187]. The enantiodiscrimination ability of the catalyst is highly dependent on the preparation conditions. Appropriate pH (3–4), temperature (100 °C), and concentration of the modifier (1%) should be carefully chosen. Addition of NaBr as a second modifier is also crucial. Ultrasonic irradiation of the catalyst leads to even better activity and enantioselectivity up to 98.6% [186 f, g]. The Ni catalyst is considered to consist of a stable, selective and weak, nonselective surface area, while the latter is selectively removed by ultrasonication.

Heterogeneous hydrogenation of α- and β-keto esters is also used for the synthesis of various biologically active compounds [171, 188–190]. Some examples are depicted in Fig. 60. The syntheses of benazepril, an angiotensin-converting en-

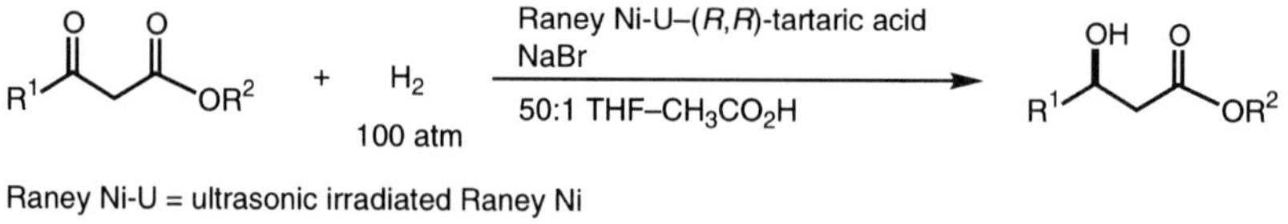

R¹	*R²*	*Temp (°C)*	*Time, h*	*% ee*
CH_3	CH_3	100	4	86
C_2H_5	CH_3	60	34	94
n-C_6H_{13}	CH_3	60	52	90
$(CH_3)_2CH$	CH_3	60	71	96
cyclo-C_3H_5	CH_3	60	48	98.6
CH_3	$(CH_3)_2CH$	60	45	87
CH_3	$(CH_3)_3C$	60	40	88

Fig. 59 Asymmetric hydrogenation of β-keto esters catalyzed by modified Raney Ni.

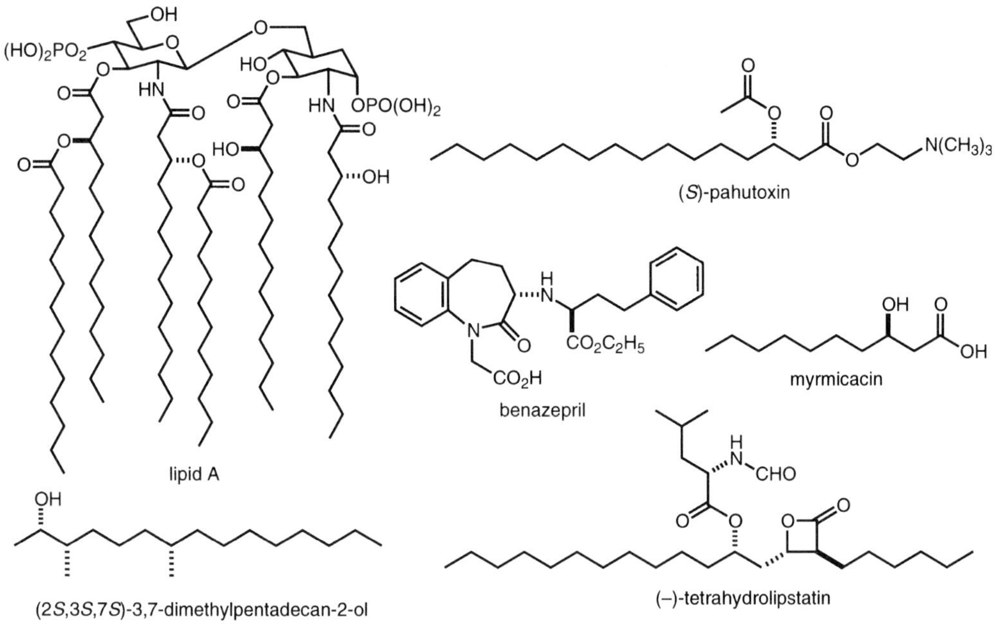

Fig. 60 Examples of biologically active compounds obtainable by asymmetric hydrogenation of α- or β-keto esters catalyzed by modified Raney Ni or Pt/Al_2O_3.

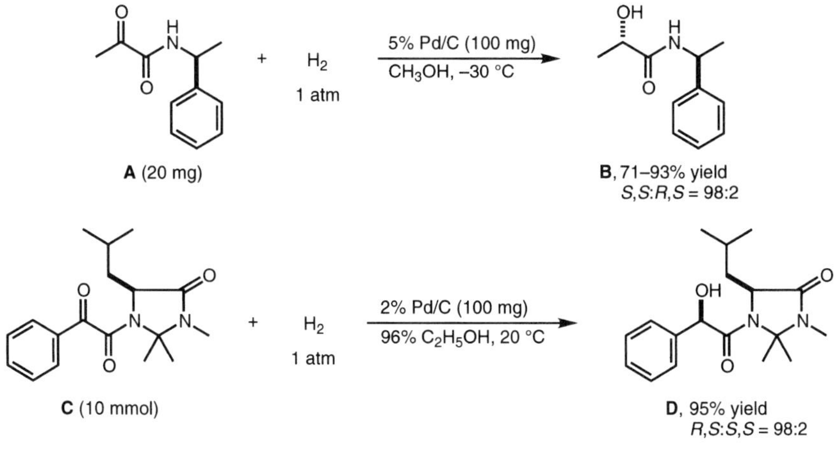

Fig. 61 Diastereoselective hydrogenation of chiral α-keto amides.

zyme inhibitor (Novartis Service AG and Solvias AG) [189], and (–)-tetrahydrolipstatin (orlistat), a pancreatic lipase inhibitor (F. Hoffmann-La Roche AG) [190], are performed on an industrial scale.

Heterogeneous catalysts effect diastereoselective hydrogenation of α-keto acid derivatives without chiral auxiliaries [57]. Thus chiral amides, **A** and **C**, are hydrogenated by Pd/C catalyst in an alcoholic medium to give the (*S,S*)-hydroxy amide **B** and *R,S* product **D**, respectively, in high yield (Fig. 61) [191].

Diketones

Enantioselective hydrogenation of *α*-diketones is rare. However, hydrogenation of benzil in the presence of a quinine–$NH_2CH_2C_6H_5$–Co(dmg)$_2$ catalyst system gives (*S*)-benzoin in 78% *ee* (Fig. 62) [192]. A catalyst with the BDM 1,3-pn ligand also shows a similar selectivity [193]. Double hydrogenation of 2,3-butanedione catalyzed by an (*R*)-BINAP–Ru complex gives a 26:74 mixture of enantiomerically pure (*R,R*)-2,3-butanediol and the *meso* diol (Fig. 63) [91].

Hydrogenation of *β*-diketones gives the corresponding chiral diols with excellent diastereo- and enantioselectivity (Fig. 64). (*R*)-BINAP–Ru-catalyzed hydrogenation of 2,4-pentanedione gives enantiomerically pure (*R,R*)-2,4-pentanediol in 99% yield [91, 194]. Hydrogenation of 5-methyl-2,4-hexanedione and 1-phenyl-1,3-butanedione gives the chiral *anti* diols stereoselectively. Ru complexes containing BIPHEMP [195] and BDPP [196] also show high selectivity. Methyl 3,5-dioxohexanoate is hydrogenated with the BINAP–Ru catalyst to give an 81:19 mixture of the *anti* (78% *ee*) and *syn* dihydroxy esters [197]. The absolute cnofiguration of the product shows that

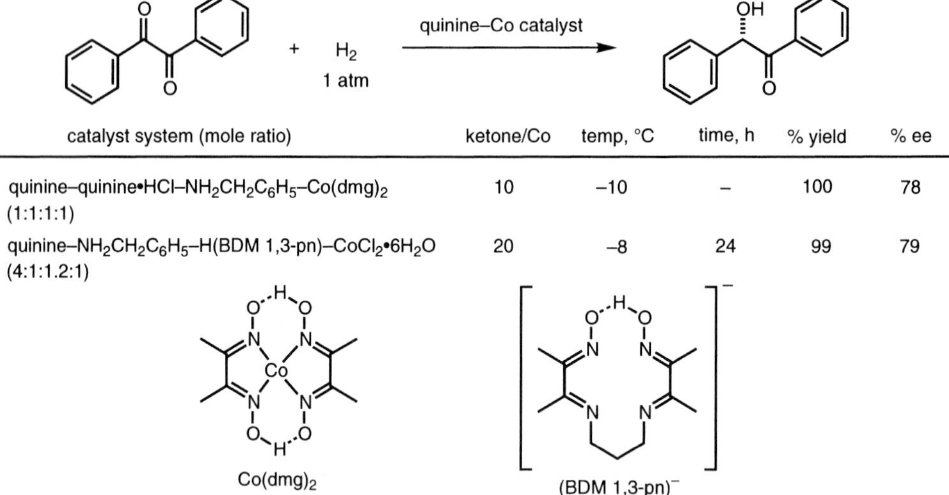

catalyst system (mole ratio)	ketone/Co	temp, °C	time, h	% yield	% ee
quinine–quinine•HCl–$NH_2CH_2C_6H_5$–Co(dmg)$_2$ (1:1:1:1)	10	−10	–	100	78
quinine–$NH_2CH_2C_6H_5$–H(BDM 1,3-pn)–$CoCl_2$•$6H_2O$ (4:1:1.2:1)	20	−8	24	99	79

Fig. 62 Asymmetric hydrogenation of benzyl.

+ H_2 80 atm; $RuCl_2[(R)\text{-binap}]$, C_2H_5OH, 26 °C, anti:syn = 26:74; S/C = 2000; anti 100% ee + syn

Fig. 63 Asymmetric hydrogenation of 2,3-butanedione.

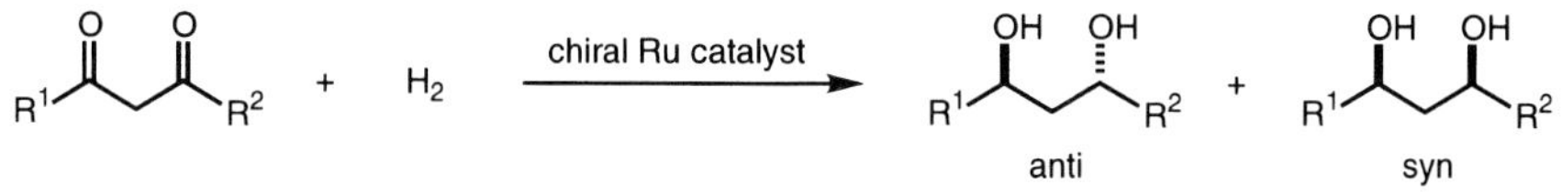

$R^1COCH_2COR^2 + H_2$ —chiral Ru catalyst→ anti + syn

anti

syn

R^1	R^2	Catalyst (S/C[a)])	H_2 (atm)	Temp (°C)	% yield	dr[b)]	% ee[c)]
CH_3	CH_3	$RuCl_2$[(*R*)-binap] (2000)	72	30	100	99:1	100
CH_3	CH_3	RuHCl[(*R*)-biphemp]-[$P(C_6H_5)_3$] + HCl (2000)	100	50	100	99:1	>99.9
CH_3	CH_3	[$RuCl_2(C_6H_6)$]$_2$–(*R*,*R*)-BDPP (1695)	80	80	100	75:25	97
CH_3	$(CH_3)_2CH$	[$NH_2(C_2H_5)_2$][{RuCl[(*R*)-binap]}$_2$(μ-Cl)$_3$] (500)	50	50	92	97:3	98
CH_3	C_6H_5	$RuBr_2$[(*R*)-binap] (360)	83	26	98	94:6	94
CH_3	CH_3OCOCH_2	[$NH_2(C_2H_5)_2$][{RuCl[(*R*)-binap]}$_2$(μ-Cl)$_3$] (–)	100	50	100[d)]	81:19	78
CH_3	C_2H_5OCO	$RuBr_2$[(*S*)-meo-biphep] (200)	100	80	>99[e)]	84:16	98
C_6H_5	C_6H_5	$RuCl_2$[(*R*)-biphemp] (170)	100	40	70	94:6	87
C_6H_5	C_6H_5	Ru[η^3-$CH_2C(CH_3)CH_2$]$_2$-(cod)–(*S*)-(*R*)-**L1c** + HBr (200)	50	50	100	>99.5:0.5	>99
$ClCH_2$	$ClCH_2$	[$NH_2(C_2H_5)_2$][{RuCl[(*R*)-binap]}$_2$(μ-Cl)$_3$] (–)	85	102	–	–	92–94

a) Substrate/catalyst mole ratio.
b) *Anti*:*syn* diastereomer ratio.
c) %*ee* of the *anti* diol.
d) A mixture of diol and δ-lactone.
e) (3*R*,5*S*)-3-Hydroxy-5-methyltetrahydrofuran-2-one.

Fig. 64 Asymmetric hydrogenation of β-diketones.

the C3 carbonyl group is preferably hydrogenated over the C5 carbonyl function. Ethyl 2,4-dioxopentanoate is hydrogenated with an (*S*)-MeO-BIPHEP–Ru catalyst to give (3*R*,5*S*)-3-hydroxy-5-methyltetrahydrofuran-2-one in 98% *ee* and the 3*R*,5*R* isomer in 87% *ee* with an 84:16 ratio after *in situ* cyclization [198]. A Ru complex with a chiral ferrocenyl diphosphine (*S*)-(*R*)-**L1c** exhibits almost perfect diastereo- and enantioselectivity in the hydrogenation of 1,3-diphenyl-1,3-propanedione [154]. A BIPHEMP–Ru catalyst also shows high stereoselectivity [199]. Optically active 1,5-dichloro-2,4-pentanediol, a useful chiral synthon, has been synthesized via BINAP–Ru-catalyzed hydrogenation of the corresponding dione [200]. Hydrogenation of 1-phenyl-1,3-butanedione with [$NH_2(C_2H_5)_2$][{RuCl[(*R*)-binap]}$_2$(μ-Cl)$_3$] under appropriate conditions affords selectively (*R*)-1-phenyl-3-hydroxybutan-1-one (Fig. 65) [194]. The BINAP–Ru-catalyzed hydrogenation of β-diketones is useful for the synthesis of organic compounds with contiguous polyhydroxy groups, as exemplified in Fig. 66 [201]. Hydrogenation of 2,5-hexadione, a γ-diketone, with a BINAP–Ru catalyst under acidic conditions gives optically pure *syn*-2,5-hexanediol in

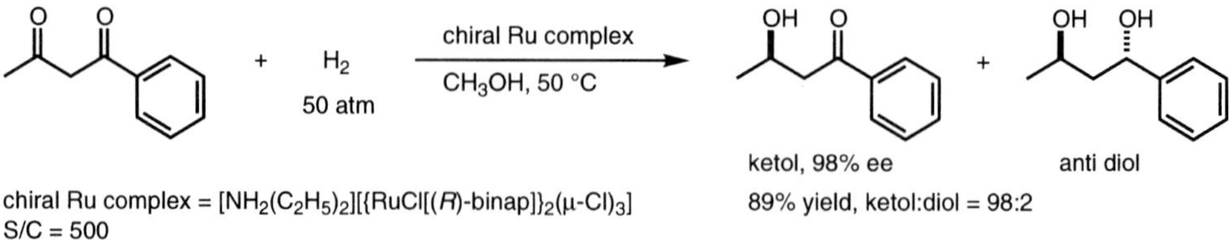

Fig. 65 Asymmetric hydrogenation of 1-phenyl-1,3-butanedione.

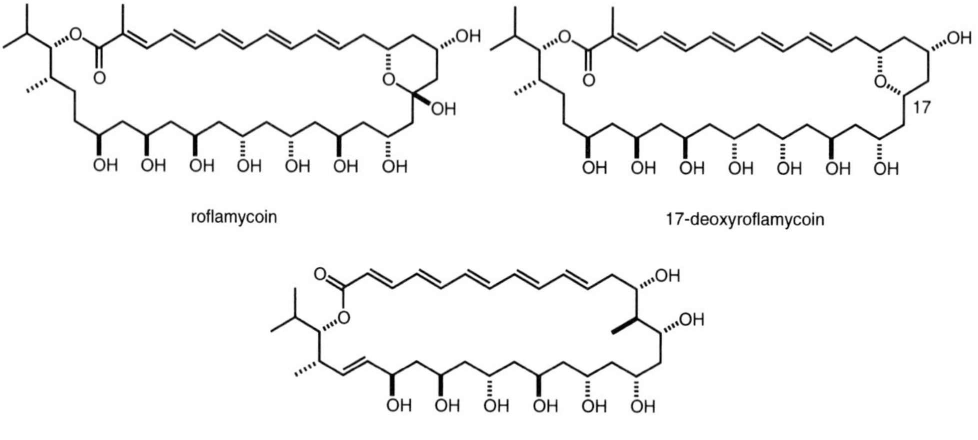

Fig. 66 Examples of bioactive compounds obtainable through BINAP–Ru-catalyzed hydrogenation of β-diketones.

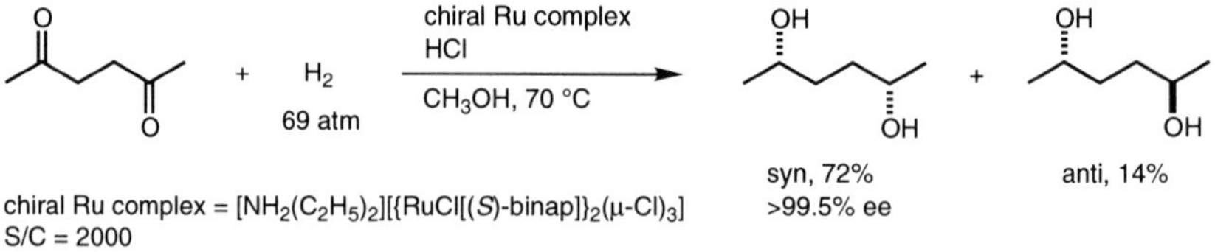

Fig. 67 Asymmetric hydrogenation of γ-diketones.

72% yield (Fig. 67) [202]. Remarkable rate enhancement is observed with the addition of HCl (Ru:HCl=1:4).

Heterogeneous asymmetric hydrogenation of 1,3-diketones is achieved by using a chirally modified Raney Ni catalyst (Fig. 68) [203]. Desired chiral diols are obtained with about 90% *ee*. This procedure is applicable to the synthesis of some natural compounds such as africanol and ngaione [204].

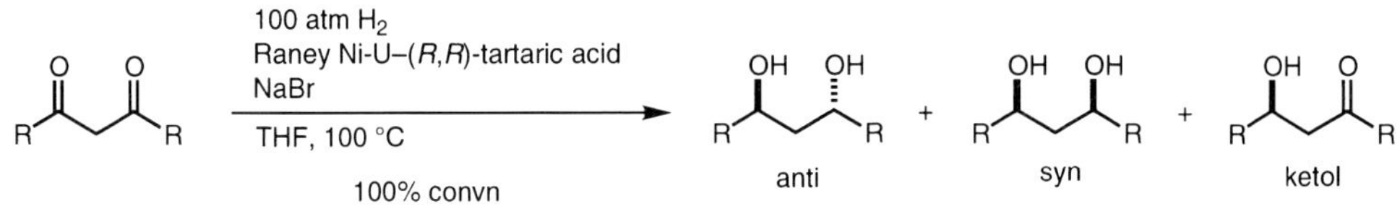

R	anti:syn:ketol	% ee of anti diol
CH_3	86:7:7	91
$CH(CH_3)_2$	72:22:6	90

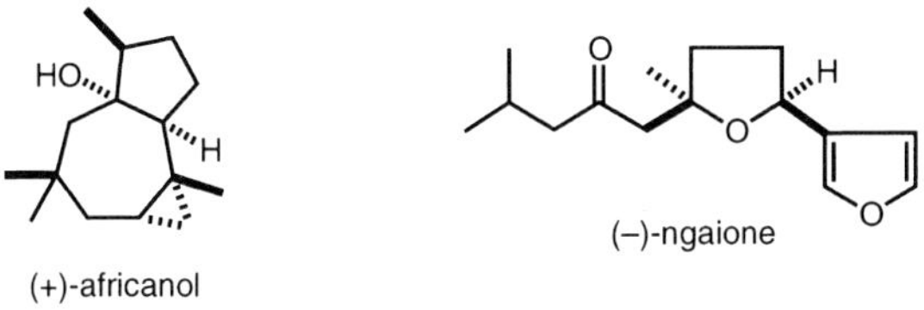

Fig. 68 Asymmetric hydrogenation of β-diketones catalyzed by chirally modified Raney Ni.

Keto phosphonates, sulfonates, sulfones, and sulfoxides

BINAP–Ru complexes effect asymmetric hydrogenation of β-keto phosphonates under mild conditions (1–4 atm of H_2, room temperature) to give the corresponding β-hydroxy phosphonates in up to 99% *ee* (Fig. 69) [205]. The sense of enantioface discrimination is the same as that of hydrogenation of β-keto carboxylic esters (see Fig. 49). A BDPP–Ru complex also shows high enantioselectivity [101 b].

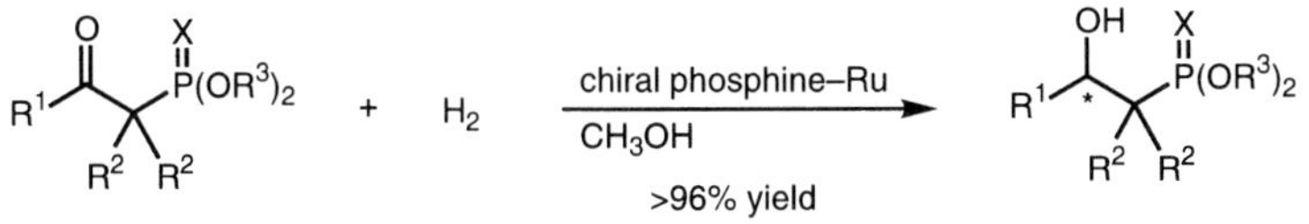

R^1	R^2	R^3	X	Chiral phosphine (S/C[a])	H_2 (atm)	Temp (°C)	% ee	Confign
CH_3	H	CH_3	O	(*R*)-BINAP (1220)	4	25	98	*R*
CH_3	H	C_2H_5	O	(*S*)-BINAP (50)	1	50	99	*S*
CH_3	H	C_2H_5	O	(*R,R*)-BDPP (50–100)	30	rt	95	*R*
CH_3	CH_3	CH_3	O	(*R*)-BINAP (370–530)	4	50	98	*R*
n-C_5H_{11}	H	CH_3	O	(*S*)-BINAP (100)	100	rt	98	*S*
$(CH_3)_2CH$	H	CH_3	O	(*S*)-BINAP (370–530)	4	80	96	*S*
C_6H_5	H	CH_3	O	(*R*)-BINAP (370–530)	4	60	95	*R*
n-C_5H_{11}	H	CH_3	S	(*S*)-MeO-BIPHEP (100)	100	rt	94	*S*
$(CH_3)_2CH$	H	CH_3	S	(*S*)-MeO-BIPHEP (100)	10	rt	93	*S*

a) Substrate/catalyst mole ratio.

Fig. 69 Asymmetric hydrogenation of β-keto phosphonates and thiophosphonates.

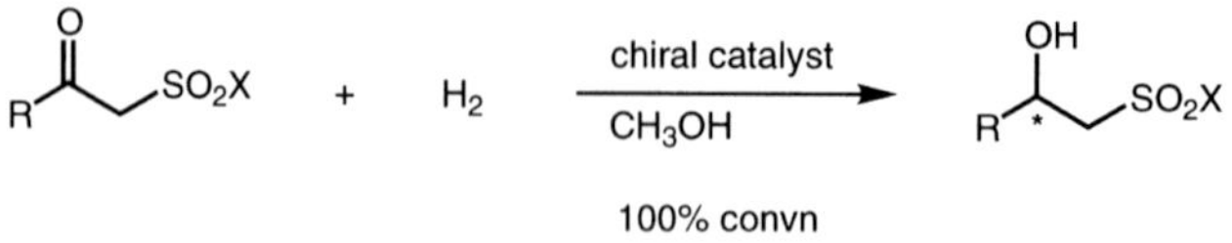

R	X	Chiral catalyst (S/C[a])	H_2 (atm)	Temp (°C)	% ee	Confign
CH_3	ONa	$RuCl_2[(R)\text{-binap}](dmf)_n$ + HCl (200)	1	50	97	*R*
n-$C_{15}H_{31}$	ONa	$RuCl_2[(R)\text{-binap}](dmf)_n$ + HCl (200)	1	50	96	*R*
$(CH_3)_2CH$	ONa	$RuCl_2[(R)\text{-binap}](dmf)_n$ + HCl (200)	1	50	97	*R*
C_6H_5	ONa	$RuCl_2[(R)\text{-binap}](dmf)_n$ + HCl (200)	1	50	96	*R*
CH_3	C_6H_5	$RuBr_2[(R)$-meo-biphep] (100)	1	65	>95	*R*
n-C_5H_{11}	C_6H_5	$RuBr_2[(R)$-meo-biphep] (100)	1	65	>95	*R*
cyclo-C_6H_{11}	C_6H_5	$RuBr_2[(R)$-meo-biphep] (100)	1	65	>95	*R*
C_6H_5	C_6H_5	$RuBr_2[(S)$-meo-biphep] (100)	75	40	>95	*S*

a) Substrate/catalyst mole ratio.

Fig. 70 Asymmetric hydrogenation of *β*-keto sulfonates and sulfones.

In a similar manner, asymmetric hydrogenation of *β*-keto thiophosphonates is achieved by using a MeO-BIPHEP–Ru catalyst [205 b].

BINAP–Ru catalysts are effective for the asymmetric hydrogenation of *β*-keto sulfonates. Sodium *β*-keto sulfonates are hydrogenated with the *R* catalyst to give quantitatively the corresponding *R β*-hydroxy sulfonates in up to 97% *ee* (Fig. 70) [206]. Similarly, several *β*-keto sulfones are hydrogenated with the (*R*)-MeO-BIPHEP–Ru complex to afford the *R* hydroxy sulfones in consistently >95% *ee* [207]. Diastereoselective hydrogenation of *R β*-keto sulfoxides is achievable by the use of

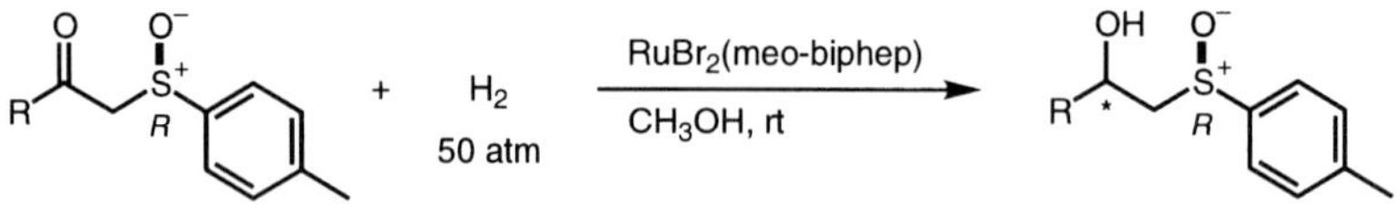

R	MeO-BIPHEP	Time (h)	% yield	S,R:R,R
n-C_6H_{13}	*S*	25	82	>99:1
n-C_6H_{13}	*R*	25	74	6:94
C_6H_5	*S*	63	70	>99:1
C_6H_5	*R*	63	95	10:90

Fig. 71 Asymmetric hydrogenation of chiral *β*-keto sulfoxides.

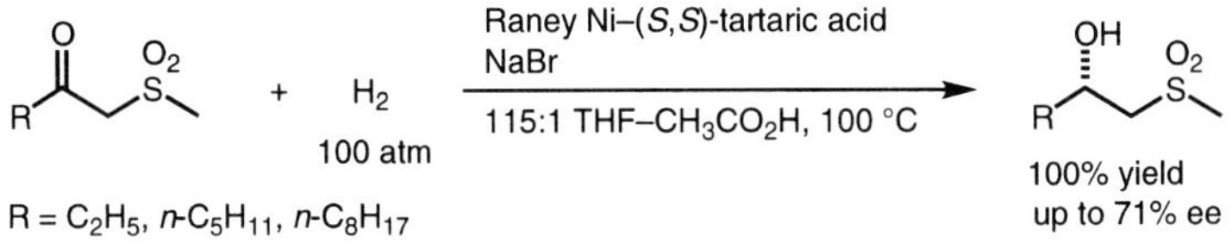

Fig. 72 Asymmetric hydrogenation of β-keto sulfones with a modified Raney Ni catalyst.

Meo-BIPHEP–Ru catalysts [208]. The *R* chiral center of the substrate collaborates well with the *S* configuration of catalyst, resulting in the corresponding *S,R* alcohols predominantly (Fig. 71), whereas use of the *R* catalyst for this reaction gives

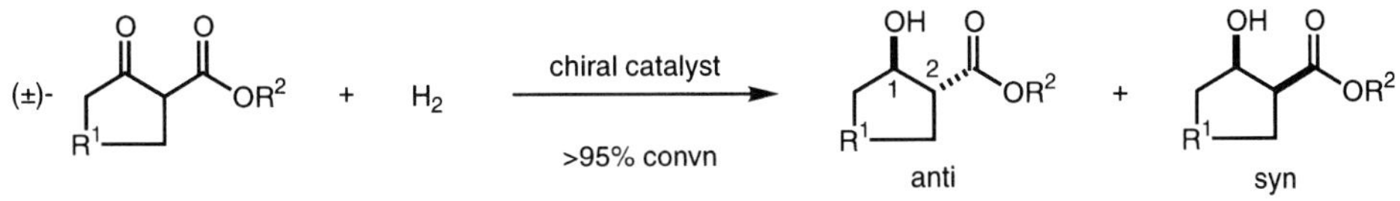

R^1	R^2	Catalyst (S/C[a])	Solvent	H_2 (atm)	dr[b]	Anti alcohol % ee	Anti alcohol Confign
CH_2	CH_3	[RuCl{(*R*)-binap}C_6H_6]Cl (1820)	CH_2Cl_2	100	99:1	92	1*R*,2*R*
CH_2	CH_3	[RuI{(*S*)-binap}-(*p*-cymene)]I (1370)	CH_2Cl[c]	100	99:1	95	1*S*,2*S*
CH_2	CH_3	Ru[η^3-$CH_2C(CH_3)CH_2$]$_2$-[(*R,R*)-*i*-pr-bpe]+HBr (500)	9:1 CH_3OH–H_2O	4	96:4	98.3	1*S*,2*S*
CH_2	CH_3	$RuCl_2$[(+)-tetrame-bitianp] (1000)	CH_3OH	100	93:7	99	1*R*,2*R*
CH_2	C_2H_5	Ru[η^3-$CH_2C(CH_3)CH_2$]$_2$-(cod)–(*R*)-BINAP+HBr (100)	CH_3OH	20	97:3[d]	94	1*R*,2*R*
$(CH_2)_2$	C_2H_5	[RuCl{(*R*)-binap}C_6H_6]Cl (530)	CH_2Cl_2	100	95:5	90	1*R*,2*R*
$(CH_2)_2$	C_2H_5	Ru[η^3-$CH_2C(CH_3)CH_2$]$_2$-(cod)–(*S*)-BINAP+HBr (100)	CH_3OH	20	74:26	91	1*S*,2*S*
$(CH_2)_2$	C_2H_5	Ru[η^3-$CH_2C(CH_3)CH_2$]$_2$-(cod)–(*R*)-(*S*)-**L1b**+HBr (200)	C_2H_5OH	50	92:8	>99	1*R*,2*R*
$(CH_2)_3$	CH_3	[RuCl{(*R*)-binap}C_6H_6]Cl (910)	CH_2Cl_2	100	93:7	93	1*R*,2*R*

a) Substrate/catalyst mole ratio.
b) *Anti*:*syn* diastereomer ratio.
c) Contaminating <1% of water.
d) 50% convn.

Fig. 73 Stereoselective hydrogenation of racemic β-keto esters.

a 6:94–10:90 diastereo mixture of *S,R* and *R,R* alcohols. The stereochemistry of the products is mostly regulated by the configuration of the catalyst. Asymmetric hydrogenation of *β*-keto sulfones catalyzed by an (*S,S*)-tartaric acid-modified Raney Ni gives the corresponding (*S*)-alcohols in up to 71% *ee* (Fig. 72) [209].

Dynamic kinetic resolution

Hydrogenation of *β*-keto esters having an *α*-substituent gives four possible stereoisomeric hydroxy esters. However, since the *α* position is configurationally labile, asymmetric hydrogenation of the racemic substrate can give a single stereoisomer selectively and in high yield by utilizing its *in situ* racemization. In fact, as shown in Fig. 73, hydrogenation of racemic 2-methoxycarbonylcyclopentanone with [RuCl{(*R*)-binap}C_6H_6]Cl gives the corresponding 1*R*,2*R* product with a 99:1 *anti* selection in 92% *ee* [210, 211]. When the ring size of the substrate is increased, the diastereoselectivity is decreased to some extent, while the *ee* of the product is not affected. Ru complexes with *i*-Pr-BPE [144], tetraMe-BITIANP [125], and a chiral ferrocenyl diphosphine **L1b** [146] show excellent stereoselectivity. The *anti* alcohols are obtained in up to >99% *ee*. The success in the asymmetric hydrogenation via dynamic kinetic resolution is based on both catalyst-based intermolecular asymmetric induction and substrate-based intramolecular asymmetric induction as well as suitable kinetic parameters [212]. Computer-aided analysis of hydrogenation of racemic 2-ethoxycarbonylcycloheptanone catalyzed by an (*R*)-BINAP–Ru complex in dichloromethane revealed that the *R* substrate is hydrogenated 9.8 times faster than the *S* isomer and that equilibration between the enantiomeric substrates occurs 4.4 times faster than hydrogenation of the slow-reacting *S* keto ester. Racemic 3-acetyltetrahydrofuran-2-one is hydrogenated with an (*S*)-BINAP–Ru catalyst to give the 3*R*,6*S* isomer exclusively in up to 97% *ee* (Fig. 74) [92, 210b]. A tetraMe-BITIANP–Ru catalyst also shows high stereoselectivity [125].

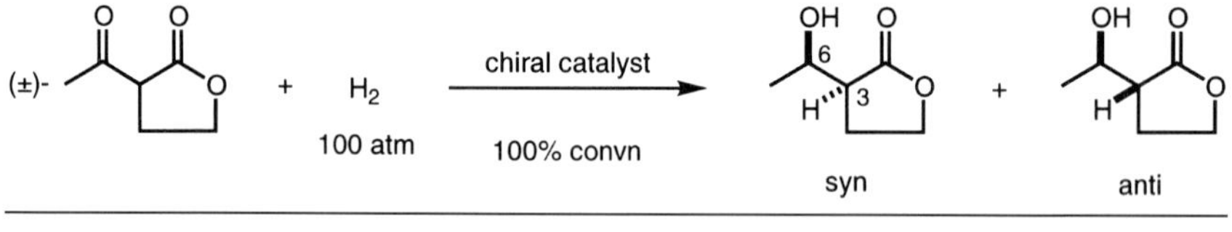

(±)- + H_2 (100 atm) —chiral catalyst, 100% convn→ syn + anti

Catalyst (S/C[a])	Solvent	dr[b]	Syn alcohol % ee	Syn alcohol Confign
[RuCl{(*R*)-binap}C_6H_6]Cl (1350)	C_2H_5OH	98:2	94	3*S*,6*R*
[RuI_2(*p*-cymene)]$_2$–(*S*)-BINAP (770)	3:1 CH_3OH–CH_2Cl_2	99:1	97	3*R*,6*S*
$RuCl_2$[(+)-tetrame-bitianp] (1000)	CH_3OH	96:4	91	3*S*,6*R*

a) Substrate/catalyst mole ratio.
b) *Syn*:*anti* diastereomer ratio.

Fig. 74 Stereoselective hydrogenation of racemic substrates.

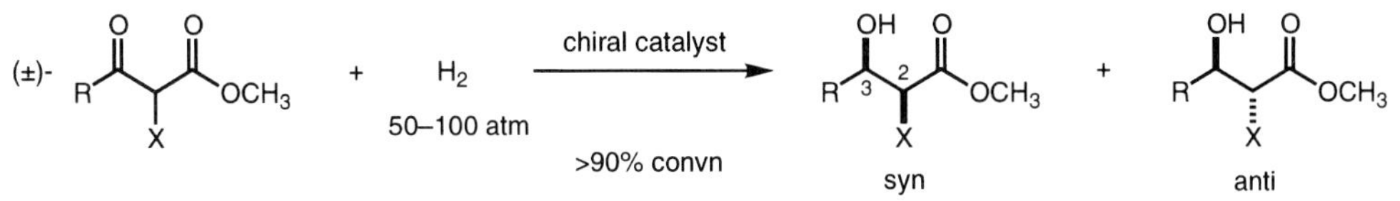

$$(\pm)\text{-}RCOCH(X)COOCH_3 + H_2 \xrightarrow[>90\%\ convn]{chiral\ catalyst} \text{syn} + \text{anti}$$

50–100 atm

R	X	Catalyst (S/C[a])	Solvent	dr[b]	Syn alcohol % ee	Syn alcohol Confign
CH_3	CH_3CONH	$RuBr_2$[(*R*)-binap] (270)	CH_2Cl_2	99:1	98	2*S*,3*R*
CH_3	CH_3CONH	Ru[η^3-$CH_2C(CH_3)CH_2$]$_2$[(*R*)-binap]+HCl (100)	CH_3OH	76:24	95	2*S*,3*R*
CH_3	$(CH_3)_2CHCONH$	Ru[η^3-$CH_2C(CH_3)CH_2$]$_2$[(*R*)-binap]+HBr (100)	CH_3OH	77:23	92	2*S*,3*R*
Ar[c]	CH_3CONH	$RuBr_2$[(*R*)-binap] (265)	CH_2Cl_2	99:1	94	2*S*,3*R*
CH_3	$C_6H_5CONHCH_2$	[$NH_2(C_2H_5)_2$][{RuCl[(*R*)-binap]}$_2$-(μ-Cl)$_3$] (100)	CH_2Cl_2	94:6	98	2*S*,3*R*
CH_3	$C_6H_5CONHCH_2$	[RuI{(*S*)-binap}(*p*-cymene)]I (100)	99.5:0.5 CH_2Cl_2–H_2O	94:6	97	2*R*,3*R*
CH_3	$C_6H_5CONHCH_2$	[RuI_2(*p*-cymene)]$_2$–(+)-DTBBINAP (1000)	1:7 CH_3OH–CH_2Cl_2	99:1[d]	99	2*S*,3*R*
CH_3	$C_6H_5CONHCH_2$	[$NH_2(C_2H_5)_2$][{RuCl[(–)-dtbm-segphos]}$_2$(μ-Cl)$_3$] (–[e])	–[e]	99.3:0.7	99.4	2*S*,3*R*
CH_3	Cl[f]	Ru[η^3-$CH_2C(CH_3)CH_2$]$_2$(cod)–(*R*)-BINAP (200)	CH_2Cl_2	1:99	99[g]	2*R*,3*R*[g]
CH_3	CH_3	Ru[η^3-$CH_2C(CH_3)CH_2$]$_2$[(*R*,*R*)-*i*-pr-bpe]+HBr (500)	9:1 CH_3OH–H_2O[h]	58:42	96	2*R*,3*R*
CH_3	CH_3	[RuCl{(*R*)-binap}C_6H_6]Cl (625)	CH_2Cl_2	32:68	94	2*R*,3*R*

a) Substrate/catalyst mole ratio.
b) *Syn*:*anti* diastereomer ratio.
c) 3,4-Methylenedioxyphenyl.
d) 55% convn.
e) Not reported.
f) Ethyl ester.
g) Value of the *anti* alcohol.
h) 4 atm of H_2.

Fig. 75 Stereoselective hydrogenation of racemic substrates.

This methodology is applicable to the hydrogenation of α-acylamino-, α-amidomethyl-, or α-chloro-substituted β-keto esters (Fig. 75) [92, 210a, 213]. Hydrogenation of the α-acylamino and α-amidomethyl substrates with an (*R*)-BINAP–Ru catalyst gives the corresponding 2*S*,3*R* (*syn*) alcohols in up to 98% *ee* [92, 210a]. Ru complexes with sterically hindered ligands, DTBBINAP and DTBM-SEGPHOS, provide the almost pure *syn* α-amidomethyl β-hydroxy ester [92, 100]. Hydrogenation of the α-chloro analog in the presence of the BINAP–Ru dimethallyl complex

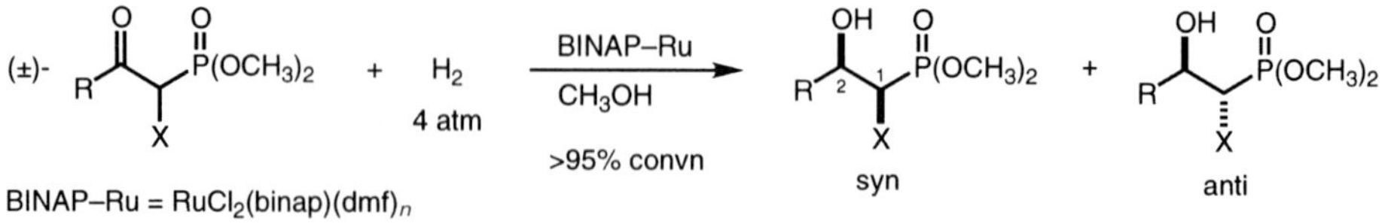

(±)- R–CO–CH(X)–P(O)(OCH₃)₂ + H₂ (4 atm) → [BINAP–Ru, CH₃OH, >95% convn] syn + anti

BINAP–Ru = $RuCl_2(binap)(dmf)_n$

R	X	BINAP (S/C[a])	Temp (°C)	dr[b]	Syn alcohol % ee	Syn alcohol Confign
CH_3	CH_3CONH	*R* (590)	25	97:3	>98	1*R*,2*R*
C_6H_5	CH_3CONH	*R* (100)	45	98:2	95	1*R*,2*R*
CH_3	Br	*S* (590)	25	90:10[c]	98	1*R*,2*S*

a) Substrate/catalyst mole ratio.
b) *Syn*:*anti* diastereomer ratio.
c) Contaminated with 15% of a debrominated compound.

Fig. 76 Stereoselective hydrogenation of racemic substrates.

predominantly gives the *anti* chloro alcohol in 99% *ee* [213 b]. The simple α-methyl analogs are difficult substrates to be hydrogenated with high diastereoselectivity, while the products are obtained with high optical purity [144, 210]. In the same manner, α-acylamino- or α-bromo-substituted β-keto phosphonates are hydrogenated with the BINAP–Ru catalyst, giving the corresponding *syn* alcohols preferentially in up to >98% *ee* (Fig. 76) [205 a, 214]. The sense of enantio- and diastereoselection is the same as that of the reaction of α-substituted β-keto carboxylic esters.

The stereoselective hydrogenation of configurationally unstable α-substituted β-keto carboxylates and phosphonates via dynamic kinetic resolution is widely applicable to the synthesis of useful biologically active compounds as well as some chiral diphosphines [62 c, g–i, 133, 205 a, 210, 215]. Selected examples are given in Fig. 77. The stereogenic center derived from the BINAP–Ru-catalyzed hydrogenation is labeled by *R* or *S*. The asymmetric synthesis of the 2-acetoxyazetidinone, a key intermediate for the synthesis of carbapenems, is now performed on an industrial scale at Takasago International Corporation via stereoselective hydrogenation of methyl 2-benzamidomethyl-3-oxobutanoate (Fig. 78) [62 c, 215 b, 216].

Asymmetric hydrogenation via dynamic kinetic resolution is applicable to simple α-substituted ketones. For example, hydrogenation of racemic 2-isopropylcyclohexanone, a configurationally labile α-substituted ketone, in the presence of a $RuCl_2$[(*S*)-binap]$(dmf)_n$–(*R,R*)-DPEN combined catalyst in 2-propanol containing an excess amount of KOH affords quantitatively the 1*R*,2*R* alcohol in 93% *ee* (*cis*:*trans*=99.8:0.2) (Fig. 79) [55]. Computer-aided analysis shows that the *R* ke-

Fig. 77 Examples of bioactive compounds and chiral diphosphines obtainable through BINAP–Ru-catalyzed hydrogenation via dynamic kinetic resolution.

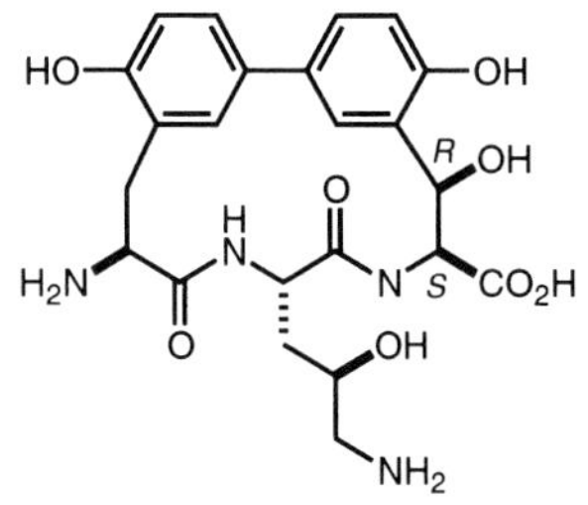

biphenomycin A

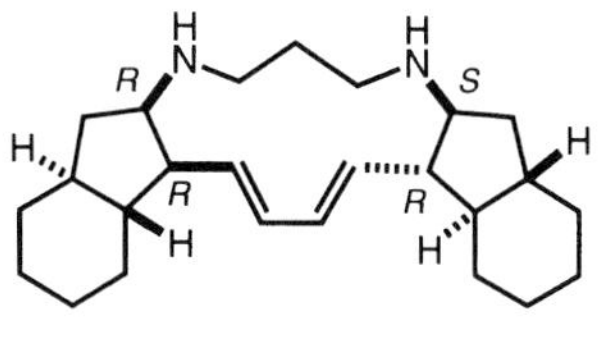

(–)-haliclonadiamine

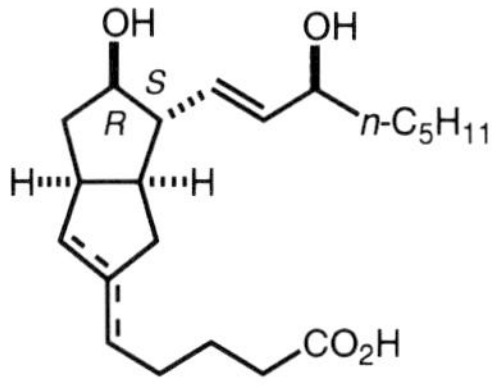

carbacyclins

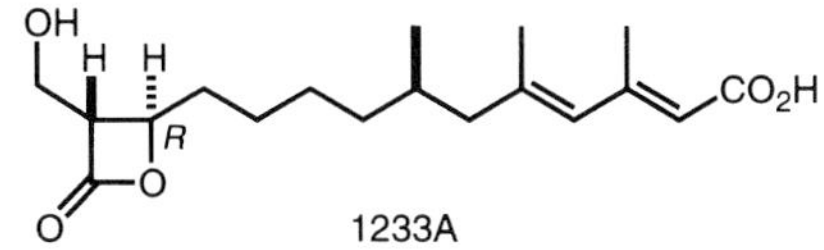

1233A

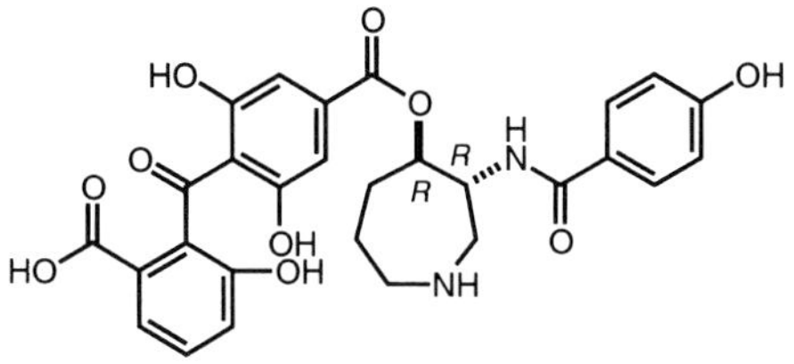

(–)-balanol

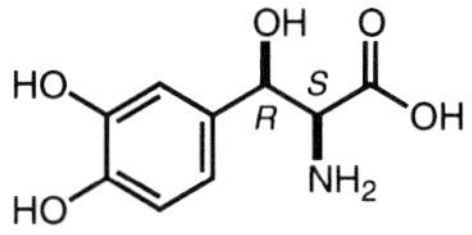

L-DOPS

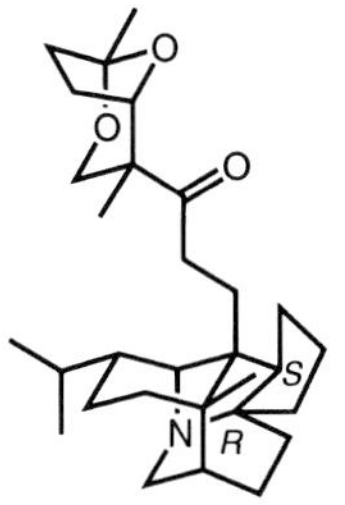

(+)-codaphniphylline

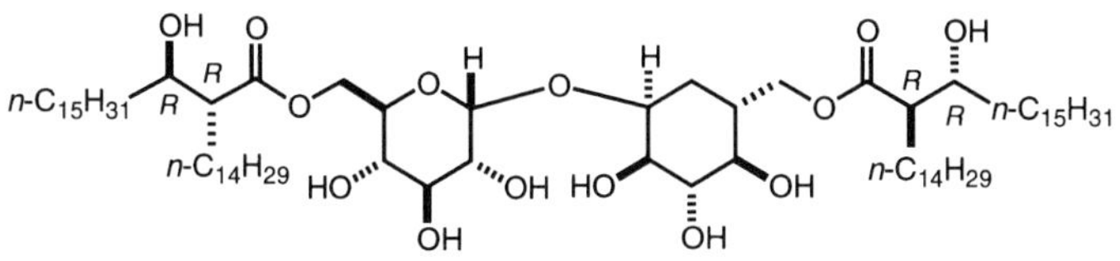

trehalose dicorynomycolate

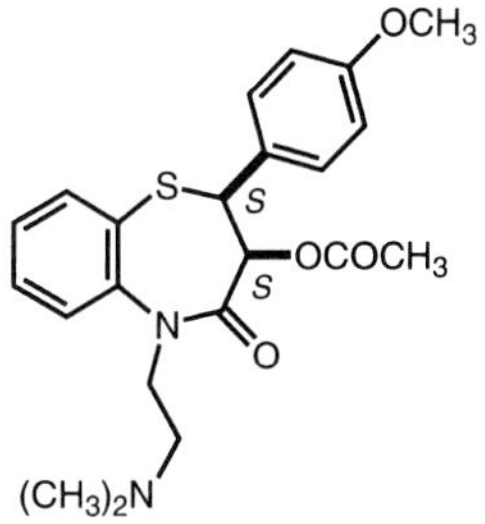

dilthiazem

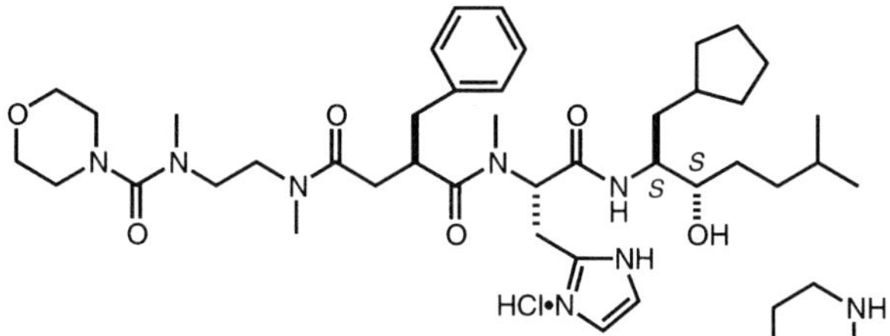

FK906

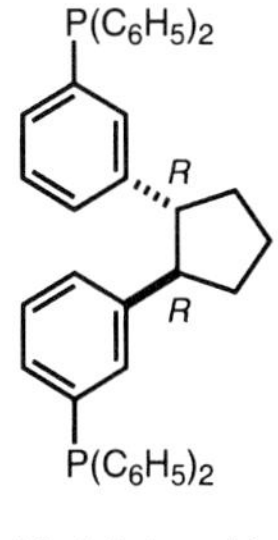

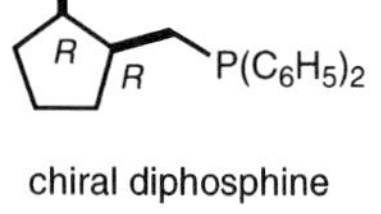

chiral diphosphine

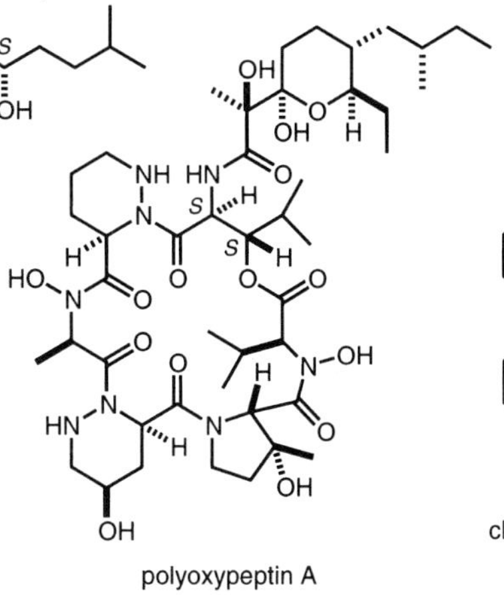

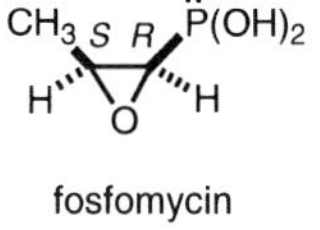

fosfomycin

polyoxypeptin A

chiral diphosphine

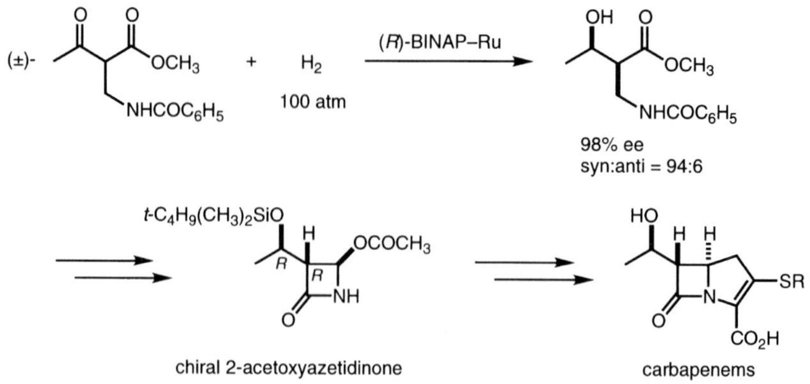

Fig. 78 Industrial synthesis of a carbapenem intermediate by BINAP–Ru-catalyzed hydrogenation.

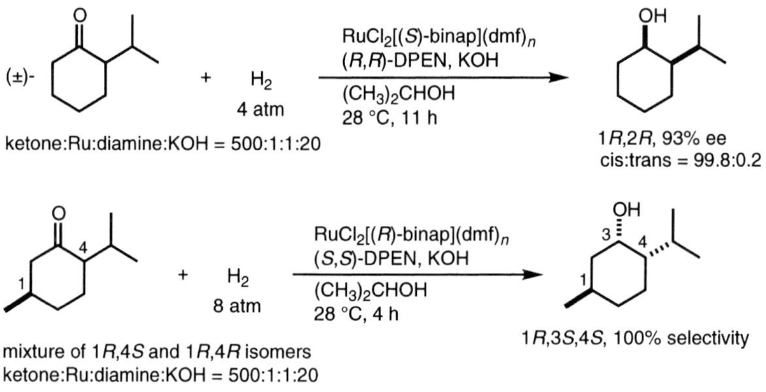

Fig. 79 Asymmetric hydrogenation via dynamic kinetic resolution.

tone substrate is hydrogenated 36 times faster than the *S* ketone. The slow-reacting *S* substrate undergoes *in situ* stereochemical inversion 47 times faster than its hydrogenation, leading to the efficient dynamic kinetic resolution. (–)-Menthone possesses a configurationally stable C1 and an unstable C4 stereogenic center. When a mixture of menthone and its 4*R* epimer is subjected to hydrogenation with an (*R*)-BINAP–(*S*,*S*)-DPEN combined system under the protic, basic conditions, (+)-neomenthol is formed exclusively [55]. On the other hand, reaction of racemic 2-isopropylcyclohexanone with *trans*-RuH(η^1-BH_4)[(*R*)-xylbinap][(*S*,*S*)-dpen] in the absence of an additional base gives the unreacted *S* ketone in 91% *ee* at 53% conversion because of very slow stereo-mutation at the α position [17].

Hydrogenation of racemic 2-methoxycyclohexanone catalyzed by an (*S*)-XylBINAP/(*S*,*S*)-DPEN–Ru complex in the presence of base at 5 °C and under 50 atm of H_2 gives (1*R*,2*S*)-2-methoxycyclohexanol in 99% *ee* (*cis*:*trans*=99.5:0.5) (Fig. 80) [107]. The chiral product is applicable to the synthesis of the potent antibacterial san-

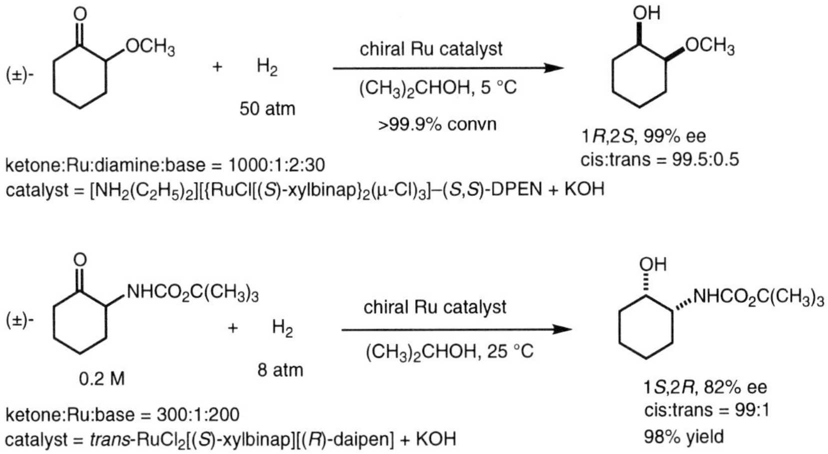

(±)- [2-methoxycyclohexanone] + H_2 (50 atm) → chiral Ru catalyst, $(CH_3)_2CHOH$, 5 °C, >99.9% convn

1*R*,2*S*, 99% ee
cis:trans = 99.5:0.5

ketone:Ru:diamine:base = 1000:1:2:30
catalyst = $[NH_2(C_2H_5)_2][\{RuCl[(S)\text{-xylbinap}]\}_2(\mu\text{-Cl})_3]$–(*S*,*S*)-DPEN + KOH

(±)- [2-($NHCO_2C(CH_3)_3$)cyclohexanone] (0.2 M) + H_2 (8 atm) → chiral Ru catalyst, $(CH_3)_2CHOH$, 25 °C

1*S*,2*R*, 82% ee
cis:trans = 99:1
98% yield

ketone:Ru:base = 300:1:200
catalyst = *trans*-$RuCl_2$[(*S*)-xylbinap][(*R*)-daipen] + KOH

Fig. 80 Asymmetric hydrogenation via dynamic kinetic resolution.

(±)- [2-phenylpropiophenone] (0.2 M) + H_2 (4 atm) → chiral Ru catalyst, $(CH_3)_3COK$, $(CH_3)_2CHOH$, 28 °C

1*R*,2*R*, 96% ee
syn:anti = 99:1
96% yield

ketone:Ru:base = 1000:1:10
catalyst = *trans*-$RuCl_2$[(*S*)-xylbinap][(*S*)-daipen]

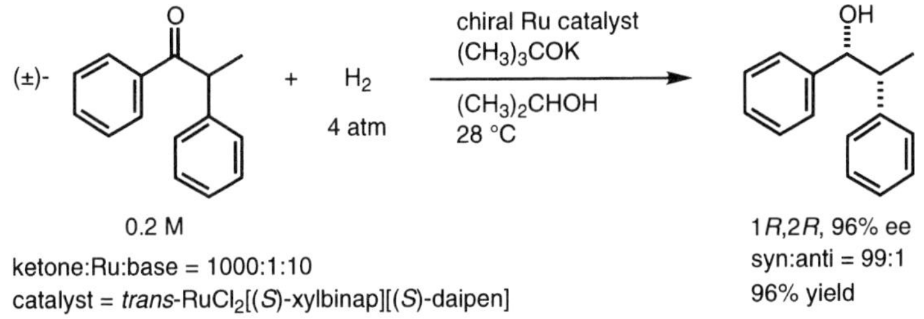

Fig. 81 Asymmetric hydrogenation via dynamic kinetic resolution.

fetrinem after it is oxidized to the chiral ketone. Similarly, hydrogenation of racemic 2-(*tert*-butoxycarbonylamino)cyclohexanone with an (*S*)-XylBINAP/(*R*)-DAIPEN–Ru catalyst under basic conditions affords the 1*S*,2*R* amino alcohol in 82% *ee* (*cis*:*trans*=99:1) [39].

Racemic 2-phenylpropiophenone, an acyclic α-substituted ketone, is hydrogenated with $RuCl_2$[(*S*)-xylbinap][(*S*)-daipen] and $(CH_3)_3COK$ to afford the 1*R*,2*R* alcohol in 96% *ee* (*syn*:*anti*=99:1) (Fig. 81) [14a].

1.1.3.3
Carboxylic Acids and their Derivatives

Hydrogenation of carboxylic acids and their derivatives is an important process. These compounds are less reactive to nucleophiles than aldehydes and ketones, so that drastic reaction conditions are generally required [1a]. For example, hydrogenation catalyzed by Cu chromite, a representative catalyst developed by Adkins, requires 300 atm of H_2 and heating to 250 °C [217]. Recent studies are mainly focused on developing more active hydrogenation catalysts.

1.1.3.3.1 Carboxylic Acids

Hydrogenation of decanoic acid with a Re–Os bimetallic catalyst in the presence of thiophene at 100 atm of H_2 and 100 °C gives 1-decanol with 90% selectivity at 94% conversion contaminated with small amounts of a hydrocarbon and ester [218]. The reaction proceeds at 25 atm of H_2 and 120 °C at a reasonable rate. A bimetallic catalyst system consisting of a Group VIII transition metal and a Group VIB or VIIB metal carbonyl shows high activity for the hydrogenation of carboxylic acids [219]. For example, pentadecanoic acid is hydrogenated effectively in the presence of $Rh(acac)_3$–$Re_2(CO)_{10}$ catalyst in DME under 100 atm of H_2 at 160 °C to afford 1-pentadecanol in 97% yield contaminated with 3% of pentadecane (Fig. 82). No ester formation has been observed. Carboxylic acid is reduced in preference to esters. Arabinoic acid in equilibrium with arabinonolactones is

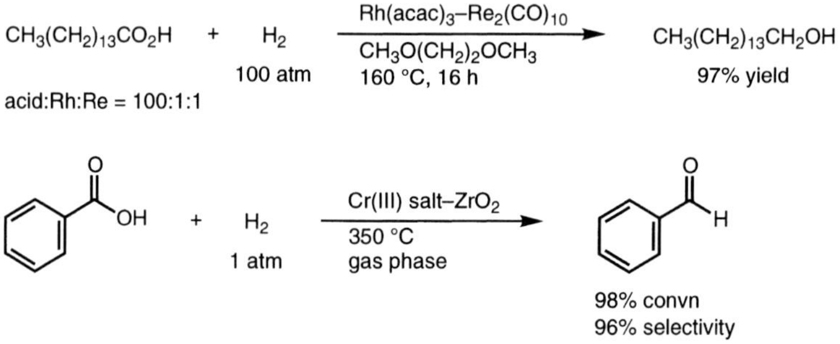

$CH_3(CH_2)_{13}CO_2H$ + H_2 (100 atm) → [$Rh(acac)_3$–$Re_2(CO)_{10}$; $CH_3O(CH_2)_2OCH_3$; 160 °C, 16 h] $CH_3(CH_2)_{13}CH_2OH$, 97% yield

acid:Rh:Re = 100:1:1

Benzoic acid + H_2 (1 atm) → [Cr(III) salt–ZrO_2; 350 °C; gas phase] benzaldehyde, 98% convn, 96% selectivity

Fig. 82 Hydrogenation of carboxylic acids.

RCO_2H + H_2 (30 atm) → [$Pd[P(C_6H_5)_3]_4$; $[(CH_3)_3CCO]_2O$; THF, 80 °C, 24 h] RCHO, up to 99% yield

acid:Pd:anhydride = 100:1:300

n-$C_7H_{15}CHO$ 98% yield; NC–C_6H_4–CHO 99%; levulinaldehyde (CHO) 97%; OHC–$(CH_2)_{10}$–CHO 99%

cinnamaldehyde (CHO) 84% with $Pd(OCOCH_3)_2$–$P(C_6H_5)_3$ catalyst; CH_2=CH$(CH_2)_7$CH$_2$CHO 80%; n-C_7H_{15}CH=CH$(CH_2)_5$CHO 99%

60 atm, 5 h in DMF; Pd:anhydride = 1:150

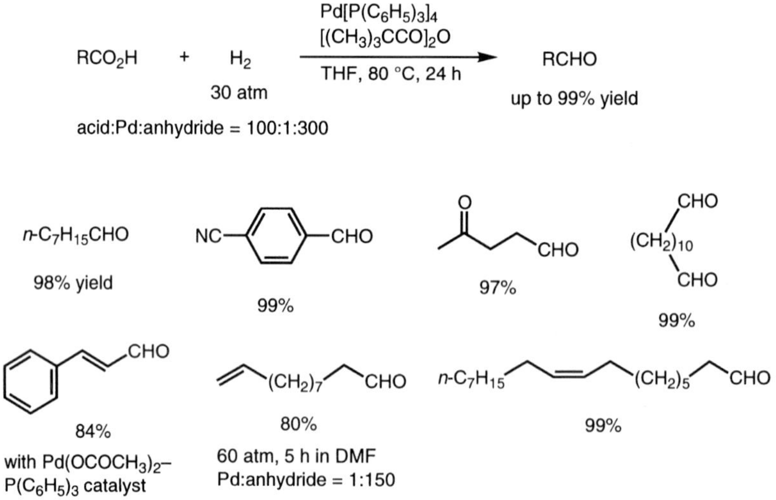

Fig. 83 Hydrogenation of carboxylic acids catalyzed by a Pd complex.

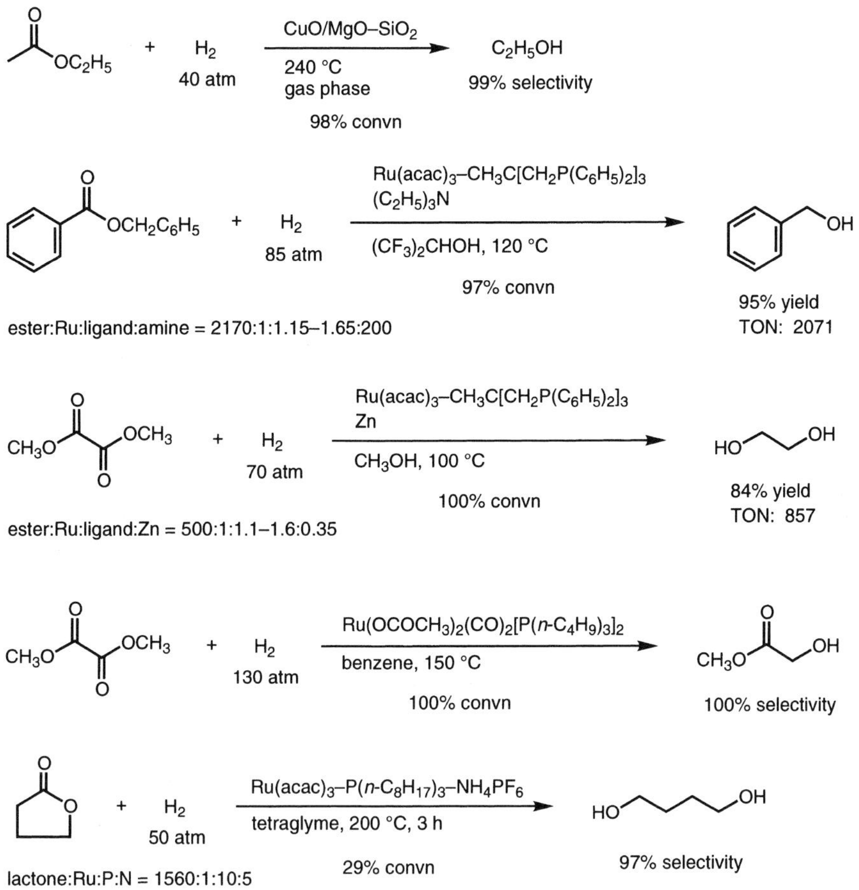

OC_2H_5 + H_2 40 atm; $CuO/MgO–SiO_2$, 240 °C, gas phase; 98% convn → C_2H_5OH 99% selectivity

$OCH_2C_6H_5$ + H_2 85 atm; $Ru(acac)_3$–$CH_3C[CH_2P(C_6H_5)_2]_3$, $(C_2H_5)_3N$, $(CF_3)_2CHOH$, 120 °C; 97% convn → OH 95% yield TON: 2071

ester:Ru:ligand:amine = 2170:1:1.15–1.65:200

CH_3O OCH_3 + H_2 70 atm; $Ru(acac)_3$–$CH_3C[CH_2P(C_6H_5)_2]_3$, Zn, CH_3OH, 100 °C; 100% convn → HO OH 84% yield TON: 857

ester:Ru:ligand:Zn = 500:1:1.1–1.6:0.35

CH_3O OCH_3 + H_2 130 atm; $Ru(OCOCH_3)_2(CO)_2[P(n\text{-}C_4H_9)_3]_2$, benzene, 150 °C; 100% convn → CH_3O OH 100% selectivity

O + H_2 50 atm; $Ru(acac)_3$–$P(n\text{-}C_8H_{17})_3$–NH_4PF_6, tetraglyme, 200 °C, 3 h; 29% convn → HO OH 97% selectivity

lactone:Ru:P:N = 1560:1:10:5

Fig. 84 Hydrogenation of carboxylic acid esters.

hydrogenated with a Ru/C catalyst under 100 atm of H_2 and at 80 °C in aqueous solution to give arabitol with 98.9% selectivity at 98% conversion [220].

Aromatic aldehydes are produced from the corresponding carboxylic acids by gas phase hydrogenation using a Cr salt-doped ZrO_2 catalyst at 1 atm of H_2 and at 350 °C (Fig. 82) [221]. This catalyst is applicable to a variety of aldehydes except for normal alkanals. This process is performed on an industrial scale at Mitsubishi Chemical Corporation. CeO_2 catalyst also gives benzaldehyde with high selectivity in the hydrogenation [222]. A variety of aliphatic, aromatic, and heteroaromatic carboxylic acids are hydrogenated with Pd complexes in the presence of pyvalic anhydride to give the corresponding aldehydes with excellent selectivity. For example, hydrogenation of octanoic acid catalyzed by $Pd[P(C_6H_5)_3]_4$ with the anhydride (acid : Pd : anhydride = 100 : 1 : 300) in THF under 30 atm of H_2 and at 80 °C gives octanal in 98% yield (Fig. 83) [223]. Sterically hindered substrates show lower reactivity. Diacids are also converted to the diformyl compounds. Carbonyl functions of ketones and esters as well as internal olefinic groups are left intact. Chemoselective hydrogenations of α,β-unsaturated acids are achieved by use of a $Pd(OCOCH_3)_2$–$P(C_6H_5)_3$ catalyst system. Substrates with a terminal olefin are hydrogenated selectively to give the unsaturated aldehyde under appropriate conditions. Preferential hydrogenation of a carboxylic acid functionality over a carbon–carbon double bond is achieved by a Ru–Sn/Al_2O_3 catalyst prepared by a sol-gel method [224]. Oleic acid is converted under 55 atm of H_2 at 250 °C to (*E*)- and (*Z*)-9-octadecen-1-ol with 81% selectivity at 81% conversion. Hydrogenation of succinic acid with $Ru_4H_4(CO)_8[P(n\text{-}C_4H_9)_3]_4$ in dioxane under 130 atm of H_2 at 180 °C gives γ-butyrolactone in 100% yield [225].

1.1.3.3.2 **Esters and Lactones**

Hydrogenation of ethyl acetate in the gas phase catalyzed by a CuO/MgO–SiO_2 catalyst at 40 atm of H_2 and 240 °C gives ethanol with 99% selectivity at 98% conversion (Fig. 84) [226]. Benzyl benzoate is hydrogenated in the presence of a $Ru(acac)_3$–$CH_3C[CH_2P(C_6H_5)_2]_3$ catalyst with $(C_2H_5)_3N$ in $(CF_3)_2CHOH$ under 85 atm of H_2 and at 120 °C to afford benzyl alcohol in 95% yield [227]. The TON reaches a value as high as 2071. Use of the fluorinated alcohol as a solvent significantly accelerates the reaction. A bimetallic Ru–Sn/Al_2O_3 catalyst which is effective for hydrogenation of carboxylic acid also promotes hydrogenation of methyl laurate in DME under 97 atm of H_2 at 280 °C to give lauryl alcohol with 96% selectivity at 99% conversion [228]. Contamination of chloride should be avoided to gain high reactivity. This bimetallic catalyst hydrogenates olefinic groups as well. On the other hand, a Ru–Sn–B/γ-Al_2O_3 terdentate catalyst preferentially promotes hydrogenation of ester groups [229]. Methyl 9-octadecenoate is hydrogenated at 43 atm of H_2 and at 270 °C to produce 9-octadecen-1-ol with 77% selectivity at 80% conversion. A potassium hydrido(phosphine)ruthenate complex [10], Rh–Sn/SiO_2 [230], and Cu–Zn/SiO_2 [231] are also known as effective catalysts. A Rh–PtO_2-catalyzed hydrogenation of chiral α-amino acid esters gives the corresponding α-amino alcohols without loss of optical purity [232]. Hydrogenation of dimethyl oxalate using a $Ru(acac)_3$–$CH_3C[CH_2P(C_6H_5)_2]_3$ catalyst with an addition of Zn in methanol under 70 atm of H_2 and at 100 °C gives selectively ethylene glycol in 84% yield (Fig. 84) [233]. The TON reaches 857. When $Ru(OCOCH_3)_2(CO)_2[P(n\text{-}C_4H_9)_3]_2$ is used as a catalyst, dimethyl oxalate is converted predominantly to methyl glycolate [234]. When Raney Cu is employed as catalyst, 1,4-butanediol is obtained selectively from dimethyl succinate [235]. A $Ru(acac)_3$–$P(n\text{-}C_8H_{17})_3$–acidic promoter system effectively converts γ-butyrolactone, δ-valerolactone, and ε-caprolactone to the corresponding diols [236]. NH_4PF_6, H_3PO_4, or its derivative is usable as an acid promoter.

1.1.3.3.3 **Anhydrides**

Benzoic anhydride is hydrogenated with $Pd[P(C_6H_5)_3]_4$ under 30 atm of H_2 and at 80 °C to give benzaldehyde in 99% yield accompanied by benzoic acid in 97% yield (Fig. 85) [237]. Octanoic anhydride, an aliphatic anhydride, is also converted to octanal in 97% yield. Reaction of pivalic anhydride is sluggish. Hydrogenation of succinic anhydride with $RuCl_2[P(C_6H_5)_3]_3$ in toluene under 10 atm of H_2 at 100 °C gives a mixture of γ-butyrolactone and succinic acid [238]. When $Ru_4H_4(CO)_8[P(n\text{-}C_4H_9)_3]_4$ is used, γ-butyrolactone is obtained in 100% yield [225]. A $Ru(acac)_3$–$P(n\text{-}C_8H_{17})_3$-p-TsOH system gives γ-butyrolactone with 98% selectivity at 97% conversion [239]. Under the identical conditions ethyl acetate is obtained with 99% selectivity from acetic anhydride. Maleic anhydride is hydrogenated with a Cu–Cr [240] or Cu–Zn–Al [241] catalyst to give γ-butyrolactone selectively. A Pd/Al_2O_3 catalyst is also effective for the conversion of maleic anhydride to γ-butyrolactone in supercritical CO_2 media [242, 243]. The regioselective hydrogenation of 2,2-dimethylglutaric anhydride using a $RuCl_2$(ttp) catalyst gives 2,2-dimethyl-δ-valerolactone [244]. A similar result is obtainable by using $RuCl_2[P(C_6H_5)_3]_3$ as a catalyst [245]. Asymmetric hydrogenation of 3-substituted glutaric anhydrides with BINAP–Ru(II) or DIOP–Ru(II) gives 3-substituted δ-valerolactone in about 60 and 20% *ee*, respectively [246].

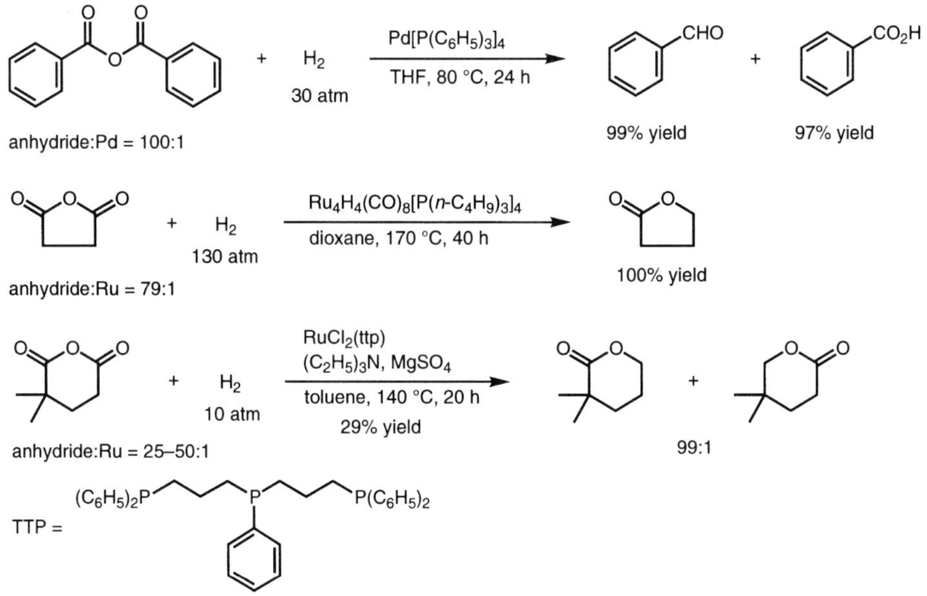

+ H_2 30 atm; $Pd[P(C_6H_5)_3]_4$; THF, 80 °C, 24 h; CHO; 99% yield; + CO_2H; 97% yield; anhydride:Pd = 100:1

+ H_2 130 atm; $Ru_4H_4(CO)_8[P(n\text{-}C_4H_9)_3]_4$; dioxane, 170 °C, 40 h; 100% yield; anhydride:Ru = 79:1

+ H_2 10 atm; $RuCl_2$(ttp); $(C_2H_5)_3N$, $MgSO_4$; toluene, 140 °C, 20 h; 29% yield; 99:1; anhydride:Ru = 25–50:1

TTP = $(C_6H_5)_2P$ ... P ... $P(C_6H_5)_2$

Fig. 85 Hydrogenation of carboxylic anhydrides.

1.1.3.4
Carbon Dioxide

Carbon dioxide (CO_2) fixation is of great interest as a future technology. Well-designed conditions including reaction media and catalysts are crucial for achieving this purpose because of the high thermodynamic stability of CO_2. Hydrogenation of CO_2 to give formic acid/formate anion was first achieved by the use of Raney Ni as a catalyst [247]. Since the first report of homogeneous hydrogenation of CO_2 catalyzed by a Ru complex [248], many effective hydrogenation systems in homogeneous media have been explored [249]. Selected examples are depicted in Fig. 86. Pd [250], Rh [251–253], and Ru [248, 254–256] complexes are successfully used as catalysts. Addition of a base, normally $N(C_2H_5)_3$, is crucial to achieve a high turnover number (TON). It improves the reaction enthalpy, while dissolution of gases improves the entropy [249]. An accelerating effect of a small amount of water has also been observed [248, 254, 257], probably due to a donating interaction between H_2O and the carbon atom of CO_2 [249]. An extremely high catalytic activity is obtained with $RuX_2[P(CH_3)_3]_4$ (X=H or Cl) in the presence of $N(C_2H_5)_3$ and H_2O in supercritical CO_2 (sc-CO_2) [254, 258]. A TON of 7200 and a TOF of 1400 h^{-1} have been

$$CO_2 + H_2 \xrightarrow{\text{catalyst}} HCO_2H$$

Catalyst	*Solvent*	*Additives*	*H_2/CO_2 (atm)*	*Temp (°C)*	*TON*	*TOF*
$PdCl_2$	H_2O	KOH	110/n [a]	160	1580	530
$[RhH(cod)]_4$	DMSO	$N(C_2H_5)_3$+DPPB	total 40	rt	312 [b]	390
$[RhH(cod)]_4$	DMSO	$N(C_2H_5)_3$+DPPB	total 40	rt	2198 [c]	122
$[RhCl(cod)]_4$	DMSO	$N(C_2H_5)_3$+DPPB	total 40	rt	1150	52
$RhCl[P(C_6H_4\text{-}m\text{-}SO_3Na)_3]_3$	H_2O	$NH(C_2H_5)_2$	20/20	rt	3439	287
$RuH_2[P(C_6H_5)_3]_4$	C_6H_6	$N(C_2H_5)_3+H_2O$	25/25	rt	87	4
$RuH_2[P(CH_3)_3]_4$	sc-CO_2 [d]	$N(C_2H_5)_3+H_2O$	85/120	50	1400	1400
$RuCl_2[P(CH_3)_3]_4$	sc-CO_2 [d]	$N(C_2H_5)_3+H_2O$	85/120	50	7200	150
$RuCl(OCOCH_3)\text{-}[P(CH_3)_3]_4$	sc-CO_2 [d]	$N(C_2H_5)_3+C_6F_5OH$	70/120	50	– [a]	95 000
$[RuCl_2(CO)_2]_n$	H_2O–$(CH_3)_2CHOH$	$N(C_2H_5)_3$	81/27	80	396	1300
$TpRuH[P(C_6H_5)_3]\text{-}(CH_3CN)$ [e]	H_2O–THF	$N(C_2H_5)_3$	25/25	100	760	– [a]

a) Not mentioned.
b) 0.8-hour reaction.
c) 18-hour reaction.
d) Supercritical CO_2.
e) Tp=hydrotris(pyrazolyl)borate.

Fig. 86 Hydrogenation of CO_2 to formic acid.

$$CO_2 + H_2 + NHR_2 \xrightarrow{\text{catalyst}} HCONR_2 + H_2O$$

Catalyst	*R*	*Solvent*	*H_2/CO_2 (atm)*	*Temp (°C)*	*TON*	*TOF*
$IrCl(CO)[P(C_6H_5)_3]_2$	H	CH_3OH	50–68/13–17	125	1145	5
$IrCl(CO)[P(C_6H_5)_3]_2$	CH_3	C_6H_6	27/27	125	1200	71
$Pt_2(dppm)_3$ [a)]	CH_3	Toluene	67–94/10–12	75	1460	61
$RuCl_3/dppe/Al(C_2H_5)_3$ [b)]	CH_3	Hexane	29/29	130	3400	567
$RuCl_2[P(CH_3)_3]_4$	CH_3	sc-CO_2	80/130	100	370000	10000
$RuCl_2(dppe)_2$ [b)]	CH_3	sc-CO_2	84/128	100	740000	360000
$RuCl_2L_3$ [c)] cocondensed with $Si(OC_2H_5)_4$	CH_3	sc-CO_2	84/128	133	110800	1860

a) dppm = $(C_6H_5)_3PCH_2P(C_6H_5)$.
b) dppe = $(C_6H_5)_3P(CH_2)_2P(C_6H_5)$.
c) L = $P(CH_3)_2(CH_2)_2Si(OC_2H_5)_3$.

Fig. 87 Hydrogenation of CO_2 with amines to formamides.

achieved. Because sc-CO_2 has densities intermediate between those of liquid and gaseous CO_2, it dissolves a huge amount of H_2 and acts as a good medium for its own hydrogenation [259]. The use of $RuCl(OCOCH_3)[P(CH_3)_3]_4$ as a catalyst with $N(C_2H_5)_3$ and C_6F_5OH, a highly acidic alcohol, in sc-CO_2 achieves even higher reactivity (TOF = 95 000 h^{-1}) [260]. Methyl formate is available by hydrogenation of CO_2 in the presence of CH_3OH. This reaction with a homogeneous catalyst was first achieved by the use of $IrH_3[P(C_6H_5)_3]_3$ as a catalyst [261]. Phosphine complexes with basic cocatalysts such as $RhCl[P(C_6H_5)_3]_3$–1,4-diazabicyclo[2.2.2]octane [262] and $RuCl_2[P(C_6H_5)_3]_3$–basic Al_2O_3 [263] are effective in CH_3OH to achieve a TON of 200 and 470, respectively. $RuCl_2[P(CH_3)_3]_4$ or $RuCl_2(dppe)_2$ (DPPE = 1,2-bis(diphenylphosphino)ethane) with CH_3OH and $N(C_2H_5)_3$ in sc-CO_2 also shows a remarkable activity with TON values of 3500 or 12 900, respectively [264, 265]. A fine $Cu/Zn/Al_2O_3$ catalyst promotes hydrogenation of CO_2 to give methanol [266].

Hydrogenation of CO_2 in the presence of NHR_2 (R = H or CH_3) under appropriate conditions produces a formamide, $HCONR_2$ (Fig. 87). This type of reaction was first achieved by the use of Raney Ni as catalyst [247]. *N,N*-dimethylformamide (DMF) is produced with a high TON of up to 1200 in homogeneous hydrogenation catalyzed by $IrCl(CO)[P(C_6H_5)_3]_2$, $CoH(dppe)_2$, or $CuCl[P(C_6H_5)_3]_3$ [267]. $IrCl(CO)[P(C_6H_5)_3]_2$ is also effective for the formation of formamide [268]. $Pt_2(dppm)$ (DPPM = bis(diphenylphosphino)methane) is an efficient catalyst for the formation of DMF [269]. A Ru complex with DPPE, a bidentate phosphine ligand, is found to be an even more effective catalyst (TON = 3400) in hexane than $RuCl_2[P(C_6H_5)_3]_3$ (TON = 2650) [270]. $RuCl_2[P(CH_3)_3]_4$ shows remarkable efficiency in sc-CO_2 [271]. The TON reaches 370 000. The use of $RuCl_2(dppe)_2$ gives an even higher value, 740 000 [265]. A hybrid material derived from $RuCl_2[P(CH_3)_2(CH_2)_2$-$Si(OC_2H_5)_3]_3$ by cocondensation with $Si(OC_2H_5)_4$ exhibits enough activity for the formation of DMF in sc-CO_2 (TON = 110 800) [272]. The immobilized complex is easily separated from products. $RuCl_2$ and RuH_2 complexes with resin-supported diphosphine ligands are also effective for the hydrogenation in sc-CO_2 [273].

References

1 Reviews: (a) P. Rylander, *Catalytic Hydrogenation in Organic Syntheses,* Academic Press, New York, **1979**. (b) M. Bartók, *Stereochemistry of Heterogeneous Metal Catalysis,* Wiley, Chichester, **1985**, Chapter 7.

2 (a) J.F. Young, J.A. Osborn, F.H. Jardine, G. Wilkinson, *Chem. Commun.* **1965**, 131–132. (b) F.H. Jardine, J.A. Osborn, G. Wilkinson, J.F. Young, *Chem. Ind.* **1965**, 560. (c) D. Evans, J.A. Osborn, F.H. Jardine, G. Wilkinson, *Nature* **1965**, *208*, 1203–1204.

3 (a) B.R. James, *Homogeneous Hydrogenation*, Wiley, New York, **1973**. (b) A.J. Birch, D.H. Williamson, *Organic Reactions* **1976**, *24*, 1–186. (c) B.R. James, *Adv. Organomet. Chem.* **1979**, *17*, 319–405.

4 (a) G. Mestroni, R. Spogliarich, A. Camus, F. Martinelli, G. Zassinovich, *J. Organomet. Chem.* **1978**, *157*, 345–352. (b) H. Pasternak, E. Lancman, F. Pruchnik, *J. Mol. Catal.* **1985**, *29*, 13–18. (c) V. Pénicaud, C. Maillet, P. Janvier, M. Pipelier, B. Bujoli, *Eur. J. Org. Chem.* **1999**, 1745–1748.

5 M. Gargano, P. Giannoccaro, M. Rossi, *J. Organomet. Chem.* **1977**, *129*, 239–242.

6 (a) R.R. Schrock, J.A. Osborn, *Chem. Commun.* **1970**, 567–568. (b) K. Tani, K. Suwa, E. Tanigawa, T. Yoshida, T. Okano, S. Otsuka, *Chem. Lett.* **1982**, 261–264. (c) K. Tani, E. Tanigawa, Y. Tatsuno, S. Otsuka, *J. Organomet. Chem.* **1985**, *279*, 87–101. (d) M.J. Burk, T.G. P. Harper, J.R. Lee, C. Kalberg, *Tetrahedron Lett.* **1994**, *35*, 4963–4966.

7 I.M. Lorkovic, R.R. Duff, Jr., M.S. Wrighton, *J. Am. Chem. Soc.* **1995**, *117*, 3617–3618.

8 M.A. Bennett, T.W. Matheson in *Comprehensive Organometallic Chemistry* (Eds: G. Wilkinson, F.G.A. Stone, E.W. Abel), Pergamon Press, Oxford, **1982**, Vol. 4, Chapter 32.9.

9 T. Naota, H. Takaya, S. Murahashi, *Chem. Rev.* **1998**, *98*, 2599–2660.

10 R.A. Grey, G.P. Pez, A. Wallo, *J. Am. Chem. Soc.* **1981**, *103*, 7536–7542.

11 (a) D.E. Linn, Jr., J. Halpern, *J. Am. Chem. Soc.* **1987**, *109*, 2969–2974. (b) J. Halpern, *Pure Appl. Chem.* **1987**, *59*, 173–180.

12 B.R. James, A. Pacheco, S.J. Rettig, I.S. Thorburn, R.G. Ball, J.A. Ibers, *J. Mol. Catal.* **1987**, *41*, 147–161.

13 T. Ohkuma, H. Ooka, S. Hashiguchi, T. Ikariya, R. Noyori, *J. Am. Chem. Soc.* **1995**, *117*, 2675–2676.

14 (a) R. Noyori, T. Ohkuma, *Angew. Chem. Int. Ed.* **2001**, *40*, 40–74. (b) R. Noyori, M. Koizumi, D. Ishii, T. Ohkuma, *Pure Appl. Chem.* **2001**, *73*, 227–232. (c) R. Noyori, *Angew. Chem. Int. Ed.* **2002**, *41*, 2008–2022. (d) R. Noyori, *Adv. Synth. Catal.* **2003**, *345*, 15–32. (e) R. Noyori, T. Ohkuma, *Pure Appl. Chem.* **1999**, *71*, 1493–1501.

15 H. Doucet, T. Ohkuma, K. Murata, T. Yokozawa, M. Kozawa, E. Katayama, A. F. England, T. Ikariya, R. Noyori, *Angew. Chem. Int. Ed.* **1998**, *37*, 1703–1707.

16 K. Abdur-Rashid, A.J. Lough, R.H. Morris, *Organometallics* **2001**, *20*, 1047–1049.

17 T. Ohkuma, M. Koizumi, K. Muñiz, G. Hilt, C. Kabuto, R. Noyori, *J. Am. Chem. Soc.* **2002**, *124*, 6508–6509.

18 K. Abdur-Rashid, A.J. Lough, R.H. Morris, *Organometallics* **2000**, *19*, 2655–2657.

19 O.M. Akotsi, K. Metera, R.D. Reid, R. Mcdonald, S.H. Bergens, *Chirality* **2000**, *12*, 514–522.

20 (a) R.A. Sanchez-Delgado, J.S. Bradley, G. Wilkinson, *J. Chem. Soc. Dalton Trans.* **1976**, 399–404. (b) C.W. Jung,

P. E. Garrou, *Organometallics* **1982**, 1, 658–666.

21 K.-J. Haack, S. Hashiguchi, A. Fujii, T. Ikariya, R. Noyori, *Angew. Chem. Int. Ed. Engl.* **1997**, *36*, 288–290.

22 (a) M. Yamakawa, H. Ito, R. Noyori, *J. Am. Chem. Soc.* **2000**, *122*, 1466–1478. (b) M. Yamakawa, I. Yamada, R. Noyori, *Angew. Chem. Int. Ed.* **2001**, *40*, 2818–2821. (c) R. Noyori, M. Yamakawa, S. Hashiguchi, *J. Org. Chem.* **2001**, *66*, 7931–7944.

23 D. A. Alonso, P. Brandt, S. J. M. Nordin, P. G. Andersson, *J. Am. Chem. Soc.* **1999**, *121*, 9580–9588.

24 R. Noyori, S. Hashiguchi, *Acc. Chem. Res.* **1997**, *30*, 97–102.

25 (a) K. Abdur-Rashid, M. Faatz, A. J. Lough, R. H. Morris, *J. Am. Chem. Soc.* **2001**, *123*, 7473–7474. (b) K. Abdur-Rashid, S. E. Clapham, A. Hadzovic, J. N. Harvey, A. J. Lough, R. H. Morris, *J. Am. Chem. Soc.* **2002**, *124*, 15104–15118.

26 R. Hartmann, P. Chen, *Angew. Chem. Int. Ed.* **2001**, *40*, 3581–3585.

27 (a) J. F. Daeuble, J. M. Stryker in *Catalysis of Organic Reactions* (Eds: M. G. Scaros, M. L. Prunier), Dekker, New York, **1995**, pp 235–247. (b) J.-X. Chen, J. F. Daeuble, D. M. Brestensky, J. M. Stryker, *Tetrahedron* **2000**, *56*, 2153–2166. (c) J.-X. Chen, J. F. Daeuble, J. M. Stryker, *Tetrahedron* **2000**, *56*, 2789–2798.

28 (a) R. M. Bullock, M. H. Voges, *J. Am. Chem. Soc.* **2000**, *122*, 12594–12595. (b) M. H. Voges, M. Bullock, *J. Chem. Soc. Dalton Trans.* **2002**, 759–770.

29 (a) C. Walling, L. Bollyky, *J. Am. Chem. Soc.* **1961**, *83*, 2968–2969. (b) C. Walling, L. Bollyky, *J. Am. Chem. Soc.* **1964**, *86*, 3750–3752.

30 (a) A. Berkessel, T. J. S. Schubert, T. N. Müller, *J. Am. Chem. Soc.* **2002**, *124*, 8693–8698. (b) A. Berkessel, *Curr. Opin. Chem. Biol.* **2001**, *5*, 486–490.

31 Reviews: (a) R. L. Augustine, *Adv. Catal.* **1976**, *25*, 56–80. (b) M. Hudlicky, *Reductions in Organic Chemistry*, Wiley, New York, **1984**. (c) S. Siegel in *Comprehensive Organic Synthesis* (Eds: B. M. Trost, I. Fleming), Pergamon Press, Oxford, **1991**, Vol. 8, Chapter 3.1. (d) H. Takaya, R. Noyori in *Comprehensive Organic Synthesis* (Eds: B. M. Trost, I. Fleming), Pergamon Press, Oxford, **1991**, Vol. 8, Chapter 3.2. (e) E. Keinan, N. Greenspoon in *Comprehensive Organic Synthesis* (Eds: B. M. Trost, I. Fleming), Pergamon Press, Oxford, **1991**, Vol. 8, Chapter 3.5.

32 T. Ohkuma, H. Ooka, T. Ikariya, R. Noyori, *J. Am. Chem. Soc.* **1995**, *117*, 10417–10418.

33 C. S. Chin, B. Lee, S. C. Park, *J. Organomet. Chem.* **1990**, *393*, 131–135.

34 (a) P. S. Gradeff, G. Formica, *Tetrahedron Lett.* **1976**, 4681–4684. (b) J. Ishiyama, S. Maeda, K. Takahashi, Y. Senda, S. Imaizumi, *Bull. Chem. Soc. Jpn.* **1987**, *60*, 1721–1726.

35 E. Farnetti, J. Kaspar, R. Spogliarich, M. Graziani, *J. Chem. Soc., Dalton Trans.* **1988**, 947–952.

36 R. Spogliarich, S. Vidotto, E. Farnetti, M. Graziani, N. V. Gulati, *Tetrahedron: Asymmetry* **1992**, *3, 1001–1002.*

37 K. Mashima, T. Akutagawa, X. Zhang, H. Takaya, T. Taketomi, H. Kumobayashi, S. Akutagawa, *J. Organomet. Chem.* **1992**, *428*, 213–222.

38 T. Ohkuma, H. Ikehira, T. Ikariya, R. Noyori, *Synlett* **1997**, 467–468.

39 T. Ohkuma, M. Koizumi, H. Doucet, T. Pham, M. Kozawa, K. Murata, E. Katayama, T. Yokozawa, T. Ikariya, R. Noyori, *J. Am. Chem. Soc.* **1998**, *120*, 13529–13530.

40 G. Gilman, G. Cohn, *Adv. Catal.* **1957**, *9*, 733–742.

41 E. Farnetti, M. Pesce, J. Kaspar, R. Spogliarich, M. Graziani, *J. Mol. Catal.* **1987**, *43*, 35–40.

42 A. Fukuoka, W. Kosugi, F. Morishita, M. Hirano, L. McCaffrey, W. Henderson, S. Komiya, *Chem. Commun.* **1999**, 489–490.

43 J. M. Grosselin, C. Mercier, G. Allmang, F. Grass, *Organometallics* **1991**, *10*, 2126–2133.

44 B. Cornils, W. A. Herrmann, R. W. Eckl, *J. Mol. Catal. A: Chemical* **1997**, *116*, 27–33.

45 (a) F. Joó, J. Kovács, A. C. Bényei, Á. Kathó, *Angew. Chem. Int. Ed.* **1998**, *37*, 969–970. (b) F. Joó, J. Kovács, A. C. Bén-

YEI, Á. KATHÓ, *Catal. Today* **1998**, *42*, 441–448.

46 G. PAPP, J. ELEK, L. NÁDASDI, G. LAURENCZY, F. JOÓ, *Adv. Synth. Catal.* **2003**, *345*, 172–174.

47 (a) K. HOTTA, *J. Mol. Catal.* **1985**, *29*, 105–107. (b) K. HOTTA, *Kagaku to Kogyo* **1986**, *60*, 196–205.

48 W.F. TULEY, R. ADAMS, *J. Am. Chem. Soc.* **1925**, *47*, 3061–3068.

49 S. GALVAGNO, Z. POLTARZEWSKI, A. DONATO, G. NERI, R. PIETROPAOLO, *J. Chem. Soc. Chem. Commun.* **1986**, 1729–1731.

50 A. GIROIR-FENDLER, D. RICHARD, P. GALLEZOT, *Catal. Lett.* **1990**, *5*, 175–181.

51 (a) Y. NITTA, Y. HIRAMATSU, T. IMANAKA, *Chem. Express* **1989**, *4*, 281–284. (b) Y. NITTA, Y. HIRAMATSU, T. IMANAKA, *J. Catal.* **1990**, *126*, 235–245. (c) C. ANDO, H. KUROKAWA, H. MIURA, *Appl. Catal. A: General* **1999**, *185*, L181–L183.

52 J.M. PLANEIX, N. COUSTEL, B. COQ, V. BROTONS, P.S. KUMBHAR, R. DUTARTRE, P. GENESTE, P. BERNIER, P.M. AJAYAN, *J. Am. Chem. Soc.* **1994**, *116*, 7935–7936.

53 K. KANEDA, T. MIZUGAKI, *Organometallics* **1996**, *15*, 3247–3249.

54 F. LEFEBVRE, J.-P. CANDY, C.C. SANTINI, J.-M. BASSET, *Top. Catal.* **1997**, *4*, 211–216.

55 T. OHKUMA, H. OOKA, M. YAMAKAWA, T. IKARIYA, R. NOYORI, *J. Org. Chem.* **1996**, *61*, 4872–4873.

56 (a) H.C. BROWN, S. KRISHNAMURTHY, *J. Am. Chem. Soc.* **1972**, *94*, 7159–7161. (b) S. KRISHNAMURTHY, H.C. BROWN, *J. Am. Chem. Soc.* **1976**, *98*, 3383–3384.

57 Reviews: (a) K. HARADA in *Asymmetric Synthesis* (Ed.: J.D. MORRISON), Academic Press, Orlando, **1985**, Vol. 5, Chapter 10. (b) K. HARADA, T. MUNEGUMI in *Comprehensive Organic Synthesis* (Eds: B.M. TROST, I. FLEMING), Pergamon, Oxford, **1991**, Vol. 8, Chapter 1.6.

58 (a) S. MITSUI, H. SAITO, Y. YAMASHITA, M. KAMINAGA, Y. SENDA, *Tetrahedron* **1973**, *29*, 1531–1539. (b) S. NISHIMURA, M. ISHIGE, M. SHIOTA, *Chem. Lett.* **1977**, 963–966.

59 M. BALASUBRAMANIAN, A. D'SOUZA, *Tetrahedron* **1968**, *24*, 5399–5408.

60 (a) H.B. KAGAN in *Asymmetric Synthesis* (Ed: J.D. MORRISON), Academic Press, Orlando, **1985**, Vol. 5, Chapter 1. (b) H. BRUNNER, *Topics in Stereochemistry* **1988**, *18*, 129–247. (c) H.-U. BLASER, *Chem. Rev.* **1992**, *92*, 935–952. (d) H. BRUNNER, W. ZETTLMEIER, Handbook of Enantioselective Catalysis, VCH, Weinheim, **1993**. (e) J. SEYDEN-PENNE, Chiral Auxiliaries and Ligands in Asymmetric Synthesis, Wiley, New York, **1995**. (f) L. SCHWINK, P. KNOCHEL, *Chem. Eur. J.* **1998**, *4*, 950–968. (g) C.J. RICHARDS, A. J. LOCKE, *Tetrahedron: Asymmetry* **1998**, *9*, 2377–2407. (h) K.V.L. CRÉPY, T. IMAMOTO, *Adv. Synth. Catal.* **2003**, *345*, 79–101.

61 D. LUCET, T. LE GALL, C. MIOSKOWSKI, *Angew. Chem. Int. Ed.* **1998**, *37*, 2580–2627.

62 Reviews: (a) R. NOYORI, M. KITAMURA in *Modern Synthetic Methods* (Ed: R. SCHEFFOLD), Springer, Berlin, **1989**, *5*, 115–198. (b) H. TAKAYA, T. OHTA, R. NOYORI in *Catalytic Asymmetric Synthesis* (Ed: I. OJIMA), VCH, New York, **1993**, Chapter 1. (c) R. NOYORI, *Asymmetric Catalysis in Organic Synthesis,* Wiley, New York, **1994**, Chapter 2. (d) I. OJIMA, M. EGUCHI, M. TZAMARIOUDAKI in *Comprehensive Organometallic Chemistry II* (Eds: E.W. ABEL, F.G.A. STONE, G. WILKINSON), Pergamon, Oxford, **1995**, Vol. 12, Chapter 2. (e) H. BRUNNER, *Methods of Organic Chemistry (Houben-Weyl) 4th edn.* **1995**, Vol. E21d, Chapter 2.3.1. (f) J.P. GENET in *Reductions in Organic Synthesis* (Ed: A.F. ABDEL-MAGID), American Chemical Society, Washington, DC, **1996**, Chapter 2. (g) T. OHKUMA, R. NOYORI in *Transition Metals for Organic Synthesis* (Eds: M. BELLER, C. BOLM), Wiley-VCH, Weinheim, **1998**, *2*, 25–69. (h) T. OHKUMA, R. NOYORI in *Comprehensive Asymmetric Catalysis* (Eds: E.N. JACOBSEN, A. PFALTZ, H. YAMAMOTO), Springer, Berlin, **1999**, Vol. 1, Chapter 6.1. (i) T. OHKUMA, M. KITAMURA, R. NOYORI in *Catalytic Asymmetric Synthesis 2nd edn.* (Ed: I. OJIMA), Wiley-VCH, New York, **2000**, Chapter 1. (j) H.-U. BLASER, C. MALAN, B. PUGIN, F. SPINDLER, H. STEINER, M. STUDER, *Avd. Synth. Catal.* **2003**, *345*, 103–151.

63 K. TERASHIMA, T. OHKUMA, R. NOYORI, *Japan Kokai Tokkyo Koho* 2000-26344, **2000**.

64 J.P. HENSCHKE, M.J. BURK, C.G. MALAN, D. HERZBERG, J.A. PETERSON, A.J. WILD-

SMITH, C. J. COBLEY, G. CASY, *Adv. Synth. Catal.* **2003**, *345*, 300–307.

65 J. WU, H. CHEN, W. KWOK, R. GUO, Z. ZHOU, C. YEUNG, A. S. C. CHAN, *J. Org. Chem.* **2002**, *67*, 7908–7910.

66 M. J. BURK, W. HEMS, D. HERZBERG, C. MALAN, A. ZANOTTI-GEROSA, *Org. Lett.* **2000**, *2*, 4173–4176.

67 F. ROBIN, F. MERCIER, L. RICHARD, F. MATHEY, M. SPAGNOL, *Chem. Eur. J.* **1997**, *3*, 1365-1369.

68 M. ITO, M. HIRAKAWA, K. MURATA, T. IKARIYA, *Organometallics* **2001**, *20*, 379–381.

69 R.-X. LI, P.-M. CHENG, D.-W. LI, H. CHEN, X.-J. LI, C. WESSMAN, N.-B. WONG, K.-C. TIN, *J. Mol. Catal. A: Chemical* **2000**, *159*, 179–184.

70 B. HEIL, S. TÖRÖS, J. BAKOS, L. MARKÓ, *J. Organomet. Chem.* **1979**, *175*, 229–232.

71 S. TÖRÖS, B. HEIL, L. KOLLÁR, L. MARKÓ, *J. Organomet. Chem.* **1980**, *197*, 85–86.

72 J. BAKOS, I. TÓTH, B. HEIL, L. MARKÓ, *J. Organomet. Chem.* **1985**, *279*, 23–29.

73 Q. JIANG, Y. JIANG, D. XIAO, P. CAO, X. ZHANG, *Angew. Chem. Int. Ed.* **1998**, *37*, 1100–1103.

74 X. ZHANG, T. TAKETOMI, T. YOSHIZUMI, H. KUMOBAYASHI, S. AKUTAGAWA, K. MASHIMA, H. TAKAYA, *J. Am. Chem. Soc.* **1993**, *115*, 3318–3319.

75 X. ZHANG, H. KUMOBAYASHI, H. TAKAYA, *Tetrahedron: Asymmetry* **1994**, *5*, 1179–1182.

76 T. OHKUMA, H. TAKENO, R. NOYORI, *Adv. Synth. Catal.* **2001**, *343*, 369–375.

77 R. TER HALLE, E. SCHULZ, M. SPAGNOL, M. LEMAIRE, *Synlett* **2000**, 680–682.

78 H.-B. YU, Q.-S. HU, L. PU, *J. Am. Chem. Soc.* **2000**, *122*, 6500–6501.

79 H.-B. YU, Q.-S. HU, L. PU, *Tetrahedron Lett.* **2000**, *41*, 1681–1685.

80 T. OHKUMA, M. KOIZUMI, H. IKEHIRA, T. YOKOZAWA, R. NOYORI, *Org. Lett.* **2000**, *2*, 659–662.

81 T. OHKUMA, M. KOIZUMI, M. YOSHIDA, R. NOYORI, *Org. Lett.* **2000**, *2*, 1749–1751.

82 P. CAO, X. ZHANG, *J. Org. Chem.* **1999**, *64*, 2127–2129.

83 Y. KUROKI, Y. SAKAMAKI, K. ISEKI, *Org. Lett.* **2001**, *3*, 457–459.

84 U. NAGEL, C. ROLLER, *Z. Naturforsch. Ser. B* **1998**, *53*, 267–270.

85 (a) T. OSAWA, *Chem. Lett.* **1985**, 1609–1612. (b) T. OSAWA, T. HARADA, A. TAI, *J. Mol. Catal.* **1994**, *87*, 333–342. (c) T. OSAWA, A. TAI, Y. IMACHI, S. TAKASAKI in *Chiral Reactions in Heterogeneous Catalysis* (Eds: G. JANNES, V. DUBOIS), Plenum, New York, **1995**, pp 75–81. (d) T. HARADA, T. OSAWA in *Chiral Reactions in Heterogeneous Catalysis* (Eds: G. JANNES, V. DUBOIS), Plenum, New York, **1995**, pp 83–88.

86 K. YOSHIKAWA, N. YAMAMOTO, M. MURATA, K. AWANO, T. MORIMOTO, K. ACHIWA, *Tetrahedron: Asymmetry* **1992**, *3*, 13–16.

87 A. ROUCOUX, M. DEVOCELLE, J.-F. CARPENTIER, F. AGBOSSOU, A. MORTREUX, *Synlett* **1995**, 358–360.

88 S. SAKURABA, H. TAKAHASHI, H. TAKEDA, K. ACHIWA, *Chem. Pharm. Bull.* **1995**, *43*, 738–747.

89 H. TAKEDA, T. TACHINAMI, M. ABURATANI, H. TAKAHASHI, T. MORIMOTO, K. ACHIWA, *Tetrahedron Lett.* **1989**, *30*, 363–366.

90 T. HAYASHI, A. KATSUMURA, M. KONISHI, M. KUMADA, *Tetrahedron Lett.* **1979**, 425–428.

91 M. KITAMURA, T. OHKUMA, S. INOUE, N. SAYO, H. KUMOBAYASHI, S. AKUTAGAWA, T. OHTA, H. TAKAYA, R. NOYORI, *J. Am. Chem. Soc.* **1988**, *110*, 629–631.

92 K. MASHIMA, K. KUSANO, N. SATO, Y. MATSUMURA, K. NOZAKI, H. KUMOBAYASHI, N. SAYO, Y. HORI, T. ISHIZAKI, S. AKUTAGAWA, H. TAKAYA, *J. Org. Chem.* **1994**, *59*, 3064–3076.

93 S. TÖRÖS, L. KOLLÁR, B. HEIL, L. MARKÓ, *J. Organomet. Chem.* **1982**, *232*, C17–C18.

94 (a) S. SAKURABA, K. ACHIWA, *Synlett* **1991**, 689–690. (b) S. SAKURABA, N. NAKAJIMA, K. ACHIWA, *Synlett* **1992**, 829–830. (c) S. SAKURABA, K. ACHIWA, *Chem. Pharm. Bull.* **1995**, *43*, 748–753.

95 M. DEVOCELLE, F. AGBOSSOU, A. MORTREUX, *Synlett* **1997**, 1306–1308.

96 C. PASQUIER, S. NAILI, L. PELINSKI, J. BROCARD, A. MORTREUX, F. AGBOSSOU, *Tetrahedron: Asymmetry* **1998**, *9*, 193–196.

97 M. DEVOCELLE, A. MORTREUX, F. AGBOSSOU, J.-R. DORMOY, *Tetrahedron Lett.* **1999**, *40*, 4551–4554.

98 T. Ohkuma, D. Ishii, H. Takeno, R. Noyori, *J. Am. Chem. Soc.* **2000**, *122*, 6510–6511.

99 G. M. R. Tombo, D. Belluš, *Angew. Chem. Int. Ed. Engl.* **1991**, *30*, 1193–1215.

100 T. Saito, T. Yokozawa, T. Ishizaki, T. Moroi, N. Sayo, T. Miura, H. Kumobayashi, *Adv. Synth. Catal.* **2001**, *343*, 264–267.

101 (a) J.-P. Tranchier, V. Ratovelomanana-Vidal, J.-P. Genêt, S. Tong, T. Cohen, *Tetrahedron Lett.* **1997**, *38*, 2951–2954. (b) D. Blanc, J.-C. Henry, V. Ratovelomanana-Vidal, J.-P. Genêt, *Tetrahedron Lett.* **1997**, *38*, 6603–6606.

102 Reviews: (a) A. Baiker, H. U. Blaser in *Handbook of Heterogeneous Catalysis* (Eds: G. Ertl, H. Knözinger, J. Weitkamp), VCH, Weinheim, **1997**, Vol. 5, Chapter 4.14. (b) H.-U. Blaser, H.-P. Jalett, M. Müller, M. Studer, *Catal. Today* **1997**, *37*, 441–463. (c) A. Baiker, *J. Mol. Catal. A: Chemical* **1997**, *115*, 473–493. (d) P. B. Wells, A. G. Wilkinson, *Top. Catal.* **1998**, *5*, 39–50. (e) P. B. Wells, R. P. K. Wells in *Chiral Catalyst Immobilization and Recycling* (Eds: D. E. De Vos, I. F. J. Vankelecom, P. A. Jacobs), Wiley-VCH, Weinheim, **2000**, Chapter 6. (f) A. Baiker in *Chiral Catalyst Immobilization and Recycling* (Eds: D. E. De Vos, I. F. J. Vankelecom, P. A. Jacobs), Wiley-VCH, Weinheim, **2000**, Chapter 7. (g) M. von Arx, T. Mallat, A. Baiker, *Top. Catal.* **2002**, *19*, 75–87. (h) M. Studer, H.-U. Blaser, C. Exner, *Adv. Synth. Catal.* **2003**, *345*, 45–65.

103 (a) B. Török, K. Felöldi, K. Balázsik, M. Bartók, *Chem. Commun.* **1999**, 1725–1726. (b) M. Studer, S. Burkhardt, H.-U. Blaser, *Chem. Commun.* **1999**, 1727–1728.

104 H. Takahashi, S. Sakuraba, H. Takeda, K. Achiwa, *J. Am. Chem. Soc.* **1990**, *112*, 5876–5878.

105 (a) E. Cesarotti, A. Mauri, M. Pallavicini, L. Villa, *Tetrahedron Lett.* **1991**, *32*, 4381–4384. (b) E. Cesarotti, P. Antognazza, M. Pallavicini, L. Villa, *Helv. Chim. Acta* **1993**, *76*, 2344–2349. (c) H.-P. Buser, F. Spindler, *Tetrahedron: Asymmetry* **1993**, *4*, 2451–2460.

106 H. P. Märki, Y. Crameri, R. Eigenmann, A. Krasso, H. Ramuz, K. Bernauer, M. Goodman, K. L. Melmon, *Helv. Chim. Acta* **1988**, *71*, 320–336.

107 T. Matsumoto, T. Murayama, S. Mitsuhashi, T. Miura, *Tetrahedron Lett.* **1999**, *40*, 5043–5046.

108 R. Schmid, E. A. Broger, M. Cereghetti, Y. Crameri, J. Foricher, M. Lalonde, R. K. Müller, M. Scalone, G. Schoettel, U. Zutter, *Pure Appl. Chem.* **1996**, *68*, 131–138.

109 T. Matsumoto, T. Nishida, H. Shirahama, *J. Org. Chem.* **1962**, *27*, 79–84.

110 H. O. House, H. C. Müller, C. G. Pitt, P. P. Wickham, *J. Org. Chem.* **1963**, *28*, 2407–2416.

111 S. Nishimura, M. Katagiri, Y. Kunikata, *Chem. Lett.* **1975**, 1235–1240.

112 T. Ohkuma, H. Doucet, T. Pham, K. Mikami, T. Korenaga, M. Terada, R. Noyori, *J. Am. Chem. Soc.* **1998**, *120*, *1086–1087*.

113 R. Spogliarich, S. Vidotto, E. Farnetti, M. Graziani, N. V. Gulati, *Tetrahedron: Asymmetry* **1992**, *3*, 1001–1002.

114 K. Mikami, T. Korenaga, M. Terada, T. Ohkuma, T. Pham, R. Noyori, *Angew. Chem. Int. Ed.* **1999**, *38*, 495-497.

115 K. Mikami, T. Korenaga, T. Ohkuma, R. Noyori, *Angew. Chem. Int. Ed.* **2000**, *39*, 3707–3710.

116 T. Ohta, T. Tsutsumi, H. Takaya, *J. Organomet. Chem.* **1994**, *484*, 191–193.

117 (a) T. Aida, M. Harada, T. Yamamoto, H. Iwai, A. Amano, T. Yamasaki, *Japan Kokai Tokkyo Koho* 10-147547, **1998**. (b) T. Aida, M. Harada, T. Yamamoto, H. Iwai, A. Amano, T. Yamasaki, US Patent 5994291, **1999**.

118 (a) T. Hayashi, T. Mise, M. Kumada, *Tetrahedron Lett.* **1976**, 4351–4354. (b) I. Ojima, T. Kogure, K. Achiwa, *J. Chem. Soc. Chem. Commun.* **1977**, 428–430. (c) T. Hayashi, M. Kumada, *Acc. Chem. Res.* **1982**, *15*, 395–401. (d) I. Ojima, *Pure Appl. Chem.* **1984**, *56*, 99–110.

119 (a) H. Takahashi, T. Morimoto, K. Achiwa, *Chem. Lett.* **1987**, 855–858. (b) K. Inoguchi, S. Sakuraba, K. Achiwa, *Synlett* **1992**, 169–178.

120 J.-F. Carpentier, A. Mortreux, *Tetrahedron: Asymmetry* **1997**, *8*, 1083–1099.

121 C. Pasquier, S. Naili, A. Mortreux, F. Agbossou, L. Pélinski, J. Brocard, J. Eilers, I. Reiners, V. Peper, J. Martens, *Organometallics* **2000**, *19*, 5723–5732.

122 H.-U. Blaser, H.-P. Jalett, F. Spindler, *J. Mol. Catal. A: Chemical* **1996**, *107*, 85–94.

123 T. Chiba, A. Miyashita, H. Nohira, H. Takaya, *Tetrahedron Lett.* **1993**, *34*, 2351–2354.

124 J.P. Genet, C. Pinel, V. Ratovelomanana-Vidal, S. Mallart, X. Pfister, L. Bischoff, M.C. Cano de Andrade, S. Darses, C. Galopin, J.A. Laffitte, *Tetrahedron: Asymmetry* **1994**, *5*, 675–690.

125 T. Benincori, E. Brenna, F. Sannicolo, L. Trimarco, P. Antognazza, E. Cesarotti, F. Demartin, T. Pilati, *J. Org. Chem.* **1996**, *61*, 6244–6251.

126 (a) I. Ojima, T. Kogure, *J. Organomet. Chem.* **1980**, *195*, 239–248. (b) I. Ojima, T. Kogure, Y. Yoda, *Org. Synth.* **1985**, *63*, 18–25.

127 H. Takahashi, M. Hattori, M. Chiba, T. Morimoto, K. Achiwa, *Tetrahedron Lett.* **1986**, *27*, 4477–4480.

128 (a) E.A. Broger, Y. Crameri, Eur. Patent Appl. 0218970, **1987**. (b) E.A. Broger, Y. Crameri, US Patent 5142063, **1992**. (c) R. Schmid, *Chimia* **1996**, *50*, 110–113.

129 H.-U. Blaser, B. Pugin, F. Spindler in *Applied Homogeneous Catalysis with Organometallic Compounds* (Eds: B. Cornils, W.A. Herrmann), VCH, Weinheim, **1996**, Vol. 2, Chapter 3.3.

130 A. Poucoux, L. Thieffry, J.-F. Carpentier, M. Devocelle, C. Méliet, F. Agbossou, A. Mortreux, A.J. Welch, *Organometallics* **1996**, *15*, 2440–2449.

131 A. Poucoux, M. Devocelle, J.-F. Carpentier, F. Agbossou, A. Mortreux, *Synlett* **1995**, 358–360.

132 R. Noyori, T. Ohkuma, M. Kitamura, H. Takaya, N. Sayo, H. Kumobayashi, S. Akutagawa, *J. Am. Chem. Soc.* **1987**, *109*, 5856–5858.

133 R. Noyori, *Acta Chem. Scand.* **1996**, *50*, 380–390.

134 (a) M. Kitamura, M. Tokunaga, T. Ohkuma, R. Noyori, *Tetrahedron Lett.* **1991**, *32*, 4163–4166. (b) M. Kitamura, M. Tokunaga, T. Ohkuma, R. Noyori, *Org. Synth.* **1993**, *71*, 1–13.

135 P.L. Gendre, M. Offenbecher, C. Bruneau, P.H. Dixneuf, *Tetrahedron: Asymmetry* **1998**, *9*, 2279–2284.

136 (a) T. Ikariya, Y. Ishii, H. Kawano, T. Arai, M. Saburi, S. Yoshikawa, S. Akutagawa, *J. Chem. Soc. Chem. Commun.* **1985**, 922–924. (b) D.F. Taber, L.J. Silverberg, *Tetrahedron Lett.* **1991**, *32*, 4227–4230. (c) B. Heiser, E.A. Broger, Y. Crameri, *Tetrahedron: Asymmetry* **1991**, *2*, 51–62. (d) S.A. King, A.S. Thompson, A.O. King, T.R. Verhoeven, *J. Org. Chem.* **1992**, *57*, 6689–6691. (e) J.B. Hoke, L.S. Hollis, E.W. Stern, *J. Organomet. Chem.* **1993**, *455*, 193–196. (f) J.P. Genet, V. Ratovelomanana-Vidal, M.C. Cano de Andrade, X. Pfister, P. Guerreiro, J.Y. Lenoir, *Tetrahedron Lett.* **1995**, *36*, 4801–4804. (g) H. Doucet, P.L. Gendre, C. Bruneau, P.H. Dixneuf, J.-C. Souvie, *Tetrahedron: Asymmetry* **1996**, *7*, 525–528. (h) S.A. King, L. DiMichele in *Catalysis of Organic Reactions* (Eds: M.G. Scaros, M.L. Prunier), Dekker, New York, **1995**, pp 157–166. (i) T. Ohta, Y. Tonomura, K. Nozaki, H. Takaya, K. Mashima, *Organometallics* **1996**, *15*, 1521–1523. (j) L. Shao, K. Takeuchi, M. Ikemoto, T. Kawai, M. Ogasawara, H. Takeuchi, H. Kawano, M. Saburi, *J. Organomet. Chem.* **1992**, *435*, 133–147. (k) K. Mashima, T. Hino, H. Takaya, *J. Chem. Soc. Dalton Trans.* **1992**, 2099–2107. (l) D.D. Pathak, H. Adams, N.A. Bailey, P.J. King, C. White, *J. Organomet. Chem.* **1994**, *479*, 237–245. (m) P. Guerreiro, M.-C. Cano de Andrade, J,-C. Henry, J.-P. Tranchier, P. Phansavath, V. Ratovelamanana-Vidal, J.-P. Genêt, T. Homri, A. R. Touati, B. B. Hassine, *C.R. Acad. Paris* **1999**, 175–179.

137 M. Murata, T. Morimoto, K. Achiwa, *Synlett* **1991**, 827–829.

138 J. Madec, X. Pfister, P. Phansavath, V. Ratovelomanana-Vidal, J.-P. Genêt, *Tetrahedron* **2001**, *57*, 2563–2568.

139 Z. Zhang, H. Qian, J. Longmire, X. Zhang, *J. Org. Chem.* **2000**, *65*, 6223–6226.

140 A.E.S. Gelpke, H. Kooijman, A.L. Spek, H. Hiemstra, *Chem. Eur. J.* **1999**, *5*, 2472–2482.

141 C.-C. Pai, Y.-M. Li, Z.-Y. Zhou, A. S. C. Chan, *Tetrahedron Lett.* **2002**, *43*, 2789–2792.

142 C.-C. Pai, C.-W. Lin, C.-C. Lin, C.-C. Chen, A. S. C. Chan, W. T. Wong, *J. Am. Chem. Soc.* **2000**, *122*, 11513–11514.

143 V. Enev, C. L. J. Ewers, M. Harre, K. Nickisch, J. T. Mohr, *J. Org. Chem.* **1997**, *62*, 7092–7093.

144 M. J. Burk, T. G. P. Harper, C. S. Kalberg, *J. Am. Chem. Soc.* **1995**, *117*, 4423–4424.

145 P. J. Pye, K. Rossen, R. A. Reamer, R. P. Volante, P. J. Reider, *Tetrahedron Lett.* **1998**, *39*, 4441–4444.

146 T. Ireland, K. Tappe, G. Grossheimann, P. Knochel, *Chem. Eur. J.* **2002**, *8*, 843–852.

147 A. Togni, C. Breutel, A. Schnyder, F. Spindler, H. Landert, A. Tijani, *J. Am. Chem. Soc.* **1994**, *116*, 4062–4066.

148 R. ter Halle, B. Colasson, E. Schulz, M. Spagnol, M. Lemaire, *Tetrahedron Lett.* **2000**, *41*, 643–646.

149 P. Guerreiro, V. Ratovelomanana-Vidal, J.-P. Genêt, P. Dellis, *Tetrahedron Lett.* **2001**, *42*, 3423–3426.

150 T. Lamouille, C. Saluzzo, R. ter Halle, F. Le Guyader, M. Lemaire, *Tetrahedron Lett.* **2001**, *42*, 663–664.

151 (a) D. Tas, C. Thoelen, I. F. J. Vankelecom, P. A. Jacobs, *Chem. Commun.* **1997**, 2323–2324. (b) I. Vankelecom, A. Wolfson, S. Geresh, M. Landau, M. Gottlieb, M. Hershkovitz, *Chem. Commun.* **1999**, 2407–2408.

152 D. J. Bayston, J. L. Fraser, M. R. Ashton, A. D. Baxter, M. E. C. Polywka, E. Moses, *J. Org. Chem.* **1998**, *63*, 3137–3140.

153 J. Wu, H. Chen, Z.-Y. Zhou, C. H. Yeung, A. S. C. Chan, *Synlett* **2001**, 1050–1054.

154 M. Lotz, K. Polborn, P. Knochel, *Angew. Chem. Int. Ed.* **2002**, *41*, 4708–4711.

155 H.-L. Huang, L. T. Liu, S.-F. Chen, H. Ku, *Tetrahedron: Asymmetry* **1998**, *9*, 1637–1640.

156 D. Blanc, V. Ratovelomanana-Vidal, J.-P. Gillet, J.-P. Genêt, *J. Organomet. Chem.* **2000**, *603*, 128–130.

157 Y. Kuroi, D. Asada, K. Iseki, *Tetrahedron Lett.* **2000**, *41*, 9853–9858.

158 M. von Arx, T. Bürgi, T. Mallat, A. Baiker, *Chem. Eur. J.* **2002**, *8*. 1430–1437.

159 T. Ohkuma, M. Kitamura, R. Noyori, *Tetrahedron Lett.* **1990**, *31*, 5509–5512.

160 T. Nishi, M. Kataoka, Y. Morisawa, *Chem. Lett.* **1989**, 1993–1996.

161 (a) S. L. Schreiber, S. E. Kelly, J. A. Porco, Jr., T. Sammakia, E. M. Suh, *J. Am. Chem. Soc.* **1988**, *110*, 6210–6218. (b) M. D. Nakatsuka, J. A. Ragan, T. Sammakia, D. B. Smith, D. E. Uehling, S. L. Schreiber, *J. Am. Chem. Soc.* **1990**, *112*, 5583–5601. (c) S. C. Case-Green, S. G. Davies, C. J. R. Hedgecock, *Synlett* **1991**, 781–782. (d) D. F. Taber, L. J. Silverberg, E. D. Robinson, *J. Am. Chem. Soc.* **1991**, *113*, 6639–6645. (e) J. E. Boldwin, R. M. Adlington, S. H. Ramcharitar, *Synlett* **1992**, 875–877. (f) D. F. Taber, P. B. Deker, L. J. Silverberg, *J. Org. Chem.* **1992**, *57*, 5990–5994. (g) K. Nozaki, N. Sato, H. Takaya, *Tetrahedron: Asymmetry* **1993**, *4*, 2179–2182. (h) S. D. Rychnovsky, R. C. Hoye, *J. Am. Chem. Soc.* **1994**, *116*, 1753–1765. (i) D. M. Garcia, H. Yamada, S. Hatakeyama, M. Nishizawa, *Tetrahedron Lett.* **1994**, *35*, 3325–3328. (j) D. F. Taber, K. K. You, *J. Am. Chem. Soc.* **1995**, *117*, 5757–5762. (k) D. S. Keegan, S. R. Hagen, D. A. Johnson, *Tetrahedron: Asymmetry* **1996**, *7*, 3559–3564. (l) C. Spino, N. Mayes, H. Desfossés, *Tetrahedron Lett.* **1996**, *37*, 6503–6506. (m) A. Balog, C. Harris, K. Savin, X.-G. Zhang, T. C. Chou, S. J. Danishefsky, *Angew. Chem. Int. Ed.* **1998**, *37*, 2675–2678. (n) N. Irako, T. Shioiri, *Tetrahedron Lett.* **1998**, *39*, 5793–5796. (o) J. E. Boldwin, A. Melman, V. Lee, C. R. Firkin, R. C. Whitehead, *J. Am. Chem. Soc.* **1998**, *120*, 8559–8560. (p) D. Romo, R. M. Rzasa, H. A. Shea, K. Park, J. M. Langenhan, L. Sun, A. Akhiezer, J. O. Liu, *J. Am. Chem. Soc.* **1998**, *120*, 12237–12254. (q) T. T. Upadhya, M. D. Nikalje, A. Sudalai, *Tetrahedron Lett.* **2001**, *42*, 4891–4893. (r) A. Fürstner, T. Dierkes, O. R. Thiel, G. Blanda, *Chem. Eur. J.* **2001**, *7*, 5286–5298.

162 M. Kitamura, T. Ohkuma, H. Takaya, R. Noyori, *Tetrahedron Lett.* **1988**, *29*, 1555–1556.

163 F. Hapiot, F. Agbossou, A. Mortreux, *Tetrahedron: Asymmetry* **1997**, *8*, 2881–2884.

164 H. Takeda, S. Hosokawa, M. Aburatani, K. Achiwa, *Synlett* **1991**, 193–194.

165 (a) S. D. Rychnovsky, R. C. Hoye, *J. Am. Chem. Soc.* **1994**, *116*, 1753–1765. (b) B. M. Trast, P. R. Hanson, *Tetrahedron Lett.* **1994**, *35*, 8119–8122. (c) G. Beck, H. Jendralla, K. Kesseler, *Synthesis* **1995**, 1014–1018.

166 (a) K. Tani, E. Tanigawa, Y. Tatsuno, S. Otsuka, *Chem. Lett.* **1986**, 737–738. (b) K. Tani, K. Suwa, E. Tanigawa, T. Ise, T. Yamagata, Y. Tatsuno, S. Otsuka, *J. Organomet. Chem.* **1989**, *370*, 203–221.

167 T. Nishi, M. Kitamura, T. Ohkuma, R. Noyori, *Tetrahedron Lett.* **1988**, *29*, 6327–6330.

168 T. Doi, M. Kokubo, K. Yamamoto, T. Takahashi, *J. Org. Chem.* **1998**, *63*, 428–429.

169 K. Tohdo, Y. Hamada, T. Shioiri, *Synlett* **1994**, 105–106.

170 (a) Y. Orito, S. Imai, S. Niwa, *Nippon Kagaku Kaishi* **1979**, 1118–1120. (b) Y. Orito, S. Imai, S. Niwa, *Nippon Kagaku Kaishi* **1980**, 670–672. (c) S. Niwa, S. Imai, Y. Orito, *Nippon Kagaku Kaishi* **1982**, 137–138.

171 Reviews: (a) H.-U. Blaser, M. Müller in *Heterogeneous Catalysis and Fine Chemicals II* (Eds: M. Guisnet et al.), Elsevier, Amsterdam, **1991**, pp 73–92. (b) G. Webb, P. B. Wells, *Catal. Today* **1992**, *12*, 319–337. (c) H.-U. Blaser, B. Pugin in *Chiral Reactions in Heterogeneous Catalysis* (Eds: G. Jannes, V. Dubois), Plenum, New York, **1995**, pp 33–57.

172 B. Török, K. Felföldi, G. Szakonyi, K. Balázsik, M. Bartók, *Catal. Lett.* **1998**, *52*, 81–84.

173 J. L. Margitfalvi, E. Tálas, M. Hegedûs, *Chem. Commun.* **1999**, 645–646.

174 C. LeBlond, J. Wang, A. T. Andrews, Y.-K. Sun, *Top. Catal.* **2000**, *13*, 169–174.

175 M. Sutyinszki, K. Szöri, K. Felföldi, M. Bartók, *Catal. Commun.* **2002**, *3*, 125–127.

176 H.-U. Blaser, H. P. Jalett, J. Wiehl, *J. Mol. Catal.* **1991**, *68*, 215–222.

177 K. Balázsik, K. Szöri, K. Felföldi, B. Török, M. Bartók, *Chem. Commun.* **2000**, 555–556.

178 H.-U. Blaser, H. P. Jalett in *Heterogeneous Catalysis and Fine Chemicals III* (Eds: M. Guisnet et al.), Elsevier, Amsterdam, **1993**, pp 139–146.

179 B. Minder, M. Schürch, T. Mallat, A. Baiker, T. Heinz, A. Pfaltz, *J. Catal.* **1996**, *160*, 261–268.

180 (a) A. Pfaltz, T. Heinz, *Top. Catal.* **1997**, *4*, 229–239. (b) M. Schürch, T. Heinz, R. Aeschimann, T. Mallat, A. Pfaltz, A. Baiker, *J. Catal.* **1998**, *173*, 187–195.

181 Y. Sun, R. N. Landau, J. Wang, C. LeBlond, D. G. Blackmond, *J. Am. Chem. Soc.* **1996**, *118*, 1348–1353.

182 H.-U. Blaser, H. P. Jalett, D. M. Monti, A. Baiker, J. T. Wehrli in *Structure-Activity and Selectivity Relationships in Heterogeneous Catalysis* (Eds: R. K. Grasselli, A. W. Sleight), Elsevier, Amsterdam, **1991**, pp 147–155.

183 (a) M. Garland, H.-U. Blaser, *J. Am. Chem. Soc.* **1990**, *112*, 7048–7050. (b) O. Schwalm, B. Minder, J. Weber, A. Baiker, *Catal. Lett.* **1994**, *23*, 271–279. (c) K. E. Simons, P. A. Meheux, S. P. Griffiths, I. M. Sutherland, P. Johnston, P. B. Wells, A. F. Carley, M. K. Rajumon, M. W. Roberts, A. Ibbotson, *Recl. Trav. Chim. Pays-Bas* **1994**, *113*, 465–474. (d) A. Baiker, T. Mallat, B. Minder, O. Schwalm, K. E. Simons, J. Weber in *Chiral Reactions in Heterogeneous Catalysis* (Eds: G. Jannes, V. Dubois), Plenum, New York, **1995**, pp 95–103. (e) R. L. Augustine, S. K. Tanielyan, *J. Mol. Catal. A: Chemical* **1996**, *112*, 93–104. (f) J. L. Margitfalvi, M. Hegedüs, E. Tfirst, *Stud. Surf. Sci. Catal.* **1996**, *101*, 241–250. (g) J. L. Margitfalvi, M. Hegedüs, E. Tfirst, *Tetrahedron: Asymmetry* **1996**, *7*, 571–580. (h) H.-U. Blaser, H.-P. Jalett, M. Garland, M. Studer, H. Thies, A. Wirth-Tilani, *J. Catal.* **1998**, *173*, 282–294.

184 (a) H. Bönnemann, G. A. Braun, *Angew. Chem. Int. Ed. Engl.* **1996**, 1992–1995. (b) H. Bönnemann, G. A. Braun, *Chem. Eur. J.* **1997**, *3*, 1200–1202.

185 (a) M. Schürch, N. Künzle, T. Mallat, A. Baiker, *J. Catal.* **1998**, *176*, 569–571. (b) N. Künzle, A. Szabo, M. Schürch, G. Wang, T. Mallat, A. Baiker, *Chem. Commun.* **1998**, 1377–1378.

186 (a) Y. Izumi, *Adv. Catal.* **1983**, *32*, 215–271. (b) A. Tai, T. Harada in *Tailored Metal Catalysts* (Ed: Y. Iwasawa), Reidel, Dordrecht, **1986**, pp 265–324. (c) T. Osawa, T. Harada, A. Tai, *J. Catal.* **1990**, *121*, 7–17. (d) A. Tai, T. Kikukawa, T. Sugimura, Y. Inoue, S. Abe, T. Osawa, T. Harada, *Bull. Chem. Soc. Jpn.* **1994**, *67*, 2473–2477. (e) T. Sugimura, T. Osawa, S. Nakagawa, T. Harada, A. Tai, *Stud. Surf. Sci. Catal.* **1996**, *101*, 231–240. (f) A. Tai, T. Sugimura in *Chiral Catalyst Immobilization and Recycling* (Eds: D.E. De Vos, I.F.J. Vankelecom, P.A. Jacobs), Wiley-VCH, Weinheim, **2000**, Chapter 8. (g) T. Sugimura, S. Nakagawa, A. Tai, *Bull. Chem. Soc. Jpn.* **2002**, *75*, 355–363.

187 (a) Y.I. Petrov, E.I. Klabunovskii, A.A. Balandin, *Kinet. Katal.* **1967**, *8*, 814–820. (b) Y. Nitta, T. Utsumi, T. Imanaka, S. Teranishi, *J. Catal.* **1986**, *101*, 376–388. (c) L. Fu, H.H. Kung, W.M.H. Sachtler, *J. Mol. Catal.* **1987**, *42*, 29–36. (d) G. Wittmann, G.B. Bartók, M. Bartók, G.V. Smith, *J. Mol. Catal.* **1990**, *60*, 1–10. (e) H. Brunner, M. Muschiol, T. Wischert, *Tetrahedron: Asymmetry* **1990**, *3*, 159–162. (f) G. Webb in *Chiral Reactions in Heterogeneous Catalysis* (Eds.: G. Jannes, V. Dubois), Plenum, New York, **1995**, pp 61–74.

188 (a) H. Schildknecht, K. Koob, *Angew. Chem.* **1971**, *83*, 110. (b) T. Shiba, S. Kusumoto, *J. Synth. Org. Chem. Jpn.* **1988**, *46*, 501–508. (c) M. Yoshikawa, T. Sugimura, A. Tai, *Agric. Biol. Chem.* **1989**, *53*, 37–40. (d) A. Tai, N. Morimoto, M. Yoshikawa, K. Uehara, T. Sugimura, T. Kikukawa, *Agric. Biol. Chem.* **1990**, *54*, 1753–1762. (e) T. Kikukawa, A. Tai, *Shokubai* **1992**, *34*, 182–190.

189 (a) H.U. Blaser, F. Spindler, M. Studer, *Appl. Catal. A: General* **2001**, *221*, 119–143. (b) H.-U. Blaser, M. Studer, A.G. Solvias in *Encyclopedia of Catalysis* (Ed: I.T. Horvás), Wiley-Interscience, New Jersey, **2003**, *1*, 481–516.

190 R. Schmid, M. Scalone in *Comprehensive Asymmetric Catalysis* (Eds: E. N. Jacobssen, A. Pfaltz, H. Yamamoto), Springer, Berlin, **1999**, Vol. 3, Chapter 41.2.

191 (a) K. Harada, T. Munegumi, S. Nomoto, *Tetrahedron Lett.* **1981**, *22*, 111–114. (b) I. Solodin, *Monatsh. Chem.* **1992**, *123*, 565–570.

192 (a) Y. Ohgo, Y. Natori, S. Takeuchi, J. Yoshimura, *Chem. Lett.* **1974**, 1327–1330. (b) Y. Ohgo, S. Takeuchi, Y. Natori, J. Yoshimura, *Bull. Chem. Soc. Jpn.* **1981**, *54*, 2124–2135.

193 R.W. Waldron, J.H. Weber, *Inorg. Chem.* **1977**, *16*, 1220–1225.

194 H. Kawano, Y. Ishii, M. Saburi, Y. Uchida, *J. Chem. Soc. Chem. Commun.* **1988**, 87–88.

195 A. Mezzetti, A. Tschumper, G. Consiglio, *J. Chem. Soc. Dalton Trans.* **1995**, 49–56.

196 H. Brunner, A. Terfort, *Tetrahedron: Asymmetry* **1995**, *6*, 919–922.

197 L. Shao, H. Kawano, M. Saburi, Y. Uchida, *Tetrahedron* **1993**, *49*, 1997–2010.

198 V. Blandin, J.-F. Carpentier, A. Mortreux, *Tetrahedron: Asymmetry* **1998**, *9*, 2765–2768.

199 D. Pini, A. Mandoli, A. Iuliano, P. Salvadori, *Tetrahedron: Asymmetry* **1995**, *6*, 1031–1034.

200 S.D. Rychnovsky, G. Griesgraber, S. Zeller, D.J. Skalitzky, *J. Org. Chem.* **1991**, *56*, 5161–5169.

201 (a) C.S. Poss, S.D. Rychnovsky, S.L. Schreiber, *J. Am. Chem. Soc.* **1993**, *115*, 3360–3361. (b) S.D. Rychnovsky, U.R. Khire, G. Yang, *J. Am. Chem. Soc.* **1997**, *119*, 2058–2059. (c) S.D. Rychnovsky, G. Yang, Y. Hu, U.R. Khire, *J. Org. Chem.* **1997**, *62*, 3022–3023.

202 Q. Fan, C. Yeung, A.S.C. Chan, *Tetrahedron: Asymmetry* **1997**, *8*, 4041–4045.

203 (a) A. Tai, T. Kikukawa, T. Sugimura, Y. Inoue, T. Osawa, S. Fujii, *J. Chem. Soc. Chem. Commun.* **1991**, 795–796. (b) H. Brunner, K. Amberger, J. Wiehl, *Bull. Soc. Chim. Belg.* **1991**, *100*, 571–583.

204 (a) T. Sugimura, T. Futagawa, A. Tai, *Chem. Lett.* **1990**, 2295–2298. (b) T. Sugi-

MURA, A. TAI, K. KOGURO, *Tetrahedron* **1994**, *50*, 11647–11658.

205 (a) M. KITAMURA, M. TOKUNAGA, R. NOYORI, *J. Am. Chem. Soc.* **1995**, *117*, 2931–2932. (b) I. GAUTIER, V. RATOVELOMANANA-VIDAL, P. SAVIGNAC, J.-P. GENET, *Tetrahedron Lett.* **1996**, *37*, 7721–7724.

206 M. KITAMURA, M. YOSHIMURA, N. KANDA, R. NOYORI, *Tetrahedron* **1999**, *55*, 8769–8785.

207 P. BERTUS, P. PHANSAVATH, V. RATOVELOMANANA-VIDAL, J.-P. GENÊT, A. R. TOUATI, T. HOMRI, B. B. HASSINE, *Tetrahedron: Asymmetry* **1999**, *10*, 1369–1380.

208 S. D. DE PAULE, L. PIOMBO, V. RATOVELOMANANA-VIDAL, C. GRECK, J.-P. GENÊT, *Eur. J. Org. Chem.* **2000**, 1535–1537.

209 Y. HIRAKI, K. ITO, T. HARADA, A. TAI, *Chem. Lett.* **1981**, 131–132.

210 (a) R. NOYORI, T. IKEDA, T. OHKUMA, M. WIDHALM, M. KITAMURA, H. TAKAYA, S. AKUTAGAWA, N. SAYO, T. SAITO, T. TAKETOMI, H. KUMOBAYASHI, *J. Am. Chem. Soc.* **1989**, *111*, 9134–9135. (b) M. KITAMURA, T. OHKUMA, M. TOKUNAGA, R. NOYORI, *Tetrahedron: Asymmetry* **1990**, *1*, 1–4.

211 J.-P. GENET, X. PFISTER, V. RATOVELOMANANA-VIDAL, C. PINEL, J.-A. LAFFITTE, *Tetrahedron Lett.* **1994**, *35*, 4559–4562.

212 (a) M. KITAMURA, M. TOKUNAGA, R. NOYORI, *J. Am. Chem. Soc.* **1993**, *115*, 144–152. (b) M. KITAMURA, M. TOKUNAGA, R. NOYORI, *Tetrahedron* **1993**, *49*, 1853–1860. (c) R. NOYORI, M. TOKUNAGA, M. KITAMURA, *Bull. Chem. Soc. Jpn.* **1995**, *68*, 36–56.

213 (a) J.-P. GENET, C. PINEL, S. MALLART, S. JUGE, S. THORIMBERT, J.-A. LAFFITTE, *Tetrahedron: Asymmetry* **1991**, *2*, 555–567. (b) J.-P. GENÊT, M. C. CANO DE ANDRADE, V. RATOVELOMANANA-VIDAL, *Tetrahedron Lett.* **1995**, *36*, 2063–2066.

214 M. KITAMURA, M. TOKUNAGA, T. PHAM, W. D. LUBELL, R. NOYORI, *Tetrahedron Lett.* **1995**, *36*, 5769–5772.

215 (a) N. FUKUDA, K. MASHIMA, Y. MATSUMURA, H. TAKAYA, *Tetrahedron Lett.* **1990**, *31*, 7185–7188. (b) K. INOGUCHI, K. ACHIWA, *Synlett* **1991**, 49–51. (c) U. SCHMIDT, V. LEITENBERGER, H. GRIESSER, J. SCHMIDT, R. MEYER, *Synthesis* **1992**, 1248–1254. (d) S. AKUTAGAWA in *Chirality in Industry* (Eds: A. N. COLLINS, G. N. SHELDRAKE, J. CROSBY), Wiley, Chichester, **1992**, Chapter 17. (e) P. M. WOVKULICH, K. SHANKARAN, J. KIEGIEL, M. R. USKOKOVIC, *J. Org. Chem.* **1993**, *58*, 832–839. (f) C. H. HEATHCOCK, J. C. KATH, R. B. RUGGERI, *J. Org. Chem.* **1995**, *60*, 1120–1130. (g) H. OHTAKE, S. YONISHI, H, TSUTSUMI, M. MURATA, *Abstracts of Papers, 69th National Meeting of the Chemical Society of Japan*, Kyoto, Chemical Society of Japan, Tokyo, **1995**, p 1030, 1H107. (h) J.-P. GENÊT, M. C. CAÑO DE ANDRADE, V. RATOVELOMANANA-VIDAL, *Tetrahedron Lett.* **1995**, *36*, 2063–2066. (i) M. NISHIZAWA, D. M. GARCÍA, R. MINAGAWA, Y. NOGUCHI, H. IMAGAWA, H. YAMADA, R. WATANABE, Y. C. YOO, I. AZUMA, *Synlett* **1996**, 452–454. (j) D. F. TABER, Y. WANG, *J. Am. Chem. Soc.* **1997**, *119*, 22–26. (k) E. COULON, M. CRISTINA, M. C. CAÑO DE ANDRADE, V. RATOVELOMANANA-VIDAL, J.-P. GENÊT, *Tetrahedron Lett.* **1998**, *39*, 6467–6470. (l) K. MAKINO, N. OKAMOTO, O. HARA, Y. HAMADA, *Tetrahedron: Asymmetry* **2001**, *12*, 1757–1762.

216 R. NOYORI, S. HASHIGUCHI, T. YAMANO in *Applied Homogeneous Catalysis with Organometallic Compounds 2nd edn.* (Eds: B. CORNILS, W. A. HERRMANN), Wiley-VCH, Weinheim, **2002**, Vol. 1, Chapter 2.9.

217 H. ADKINS, *Org. React.* **1954**, *8*, 1–27.

218 K. YOSHINO, Y. KAJIWARA, N. TAKAISHI, Y. INAMOTO, J. TSUJI, *J. Am. Oil Chem. Soc.* **1990**, *67*, 21–24.

219 D.-H. HE, N. WAKASA, T. FUCHIKAMI, *Tetrahedron Lett.* **1995**, *36*, 1059–1062.

220 L. FABRE, P. GALLEZOT, A. PERRARD, *J. Catal.* **2002**, *208*, 247–254.

221 (a) J. KONDO, N. DING, K. MARUYA, K. DOMEN, T. YOKOYAMA, N. FUJITA, T. MAKI, *Bull. Chem. Soc. Jpn.* **1993**, *66*, 3085–3090. (b) T. YOKOYAMA, T. SETOYAMA, N. FUJITA, T. MAKI, *Stud. Surf. Sci. Catal.* **1994**, *90*, 47–58.

222 (a) Y. SAKATA, C. A. VON TOL-KOUTSTAAL, V. PONEC, *J. Catal.* **1997**, *169*, 13–21. (b) Y. SAKATA, V. PONEC, *Appl. Catal. A: General* **1998**, *166*, 173–184.

223 K. NAGAYAMA, I. SHIMIZU, A. YAMAMOTO, *Bull. Chem. Soc. Jpn.* **2001**, *74*, 1803–1815.

224 (a) K.Y. Cheah, T.S. Tang, F. Mizukami, S. Niwa, M. Toba, Y.M. Choo, *J. Am. Oil Chem. Soc.* **1992**, *69*, 410–416. See also: (b) K. Tahara, E. Nagahara, Y. Itoi, S. Nishiyama, S. Tsuruya, M. Masai, *J. Mol. Catal. A: Chemical* **1996**, *110*, L5–L6.

225 M. Bianchi, G. Menchi, F. Francalanci, F. Piacenti, U. Matteoli, P. Frediani, C. Botteghi, *J. Organomet. Chem.* **1980**, *188*, 109–119.

226 P. Claus, M. Lucas, B. Lücke, T. Berndt, P. Birke, *Appl. Catal. A: General* **1991**, *79*, 1–18.

227 H.T. Teunissen, C.J. Elsevier, *Chem. Commun.* **1998**, 1367–1368.

228 K. Tahara, H. Tsuji, H. Kimura, T. Okazaki, Y. Itoi, S. Nishiyama, S. Tsuruya, M. Masai, *Catal. Today* **1996**, *28*, 267–272.

229 V.M. Deshpande, K. Ramnarayan, C.S. Narasimhan, *J. Catal.* **1990**, *121*, 174–182.

230 O.A. Ferretti, J.P. Bournonville, G. Mabilon, G. Martino, J.P. Candy, J.-M. Basset, *J. Mol. Catal.* **1991**, *67*, 283–294.

231 F. Th. van de Scheur, D.S. Brands, B. van der Linden, C.O. Luttikhuis, E.K. Poels, L.H. Staal, *Appl. Catal. A: General* **1994**, *116*, 237–257.

232 M. Studer, S. Burkhardt, H.-U. Blaser, *Adv. Synth. Catal.* **2001**, *343*, 802–808.

233 H.T. Teunissen, C.J. Elsevier, *Chem. Commun.* **1997**, 667–668.

234 U. Matteoli, G. Menchi, M. Bianchi, F. Piacenti, *J. Organomet. Chem.* **1986**, *299*, 233–238.

235 M.A. Kohler, M.S. Wainwright, D.L. Trimm, N.W. Cant, *Ind. Eng. Chem. Res.* **1987**, *26*, 652–656.

236 Y. Hara, H. Inagaki, S. Nishimura, K. Wada, *Chem. Lett.* **1992**, 1983–1986.

237 K. Nagayama, F. Kawataka, M. Sakamoto, I. Shimizu, A. Yamamoto, *Bull. Chem. Soc. Jpn.* **1999**, *72*, 573–580.

238 J.E. Lyons, *J. Chem. Soc. Chem. Commun.* **1975**, 412–413.

239 (a) Y. Hara, K. Wada, *Chem. Lett.* **1991**, 553–554. (b) Y. Hara, H. Kusaka, H. Inagaki, K. Takahashi, K. Wada, *J. Catal.* **2000**, *194*, 188–197.

240 G.L. Castiglioni, A. Vaccari, G. Fierro, M. Inversi, M. Lo Jacono, G. Minelli, I. Pettiti, P. Porta, M. Gazzano, *Appl. Catal. A: General* **1995**, *123*, 123–144.

241 G.L. Castiglioni, M. Ferrari, A. Guercio, A. Vaccari, R. Lancia, C. Fumagalli, *Catal. Today* **1996**, *27*, 181–186.

242 U.R. Pillai, E. Sahle-Demessie, *Chem. Commun.* **2002**, 422–423.

243 A. Baiker, *Chem. Rev.* **1999**, *99*, 453–473.

244 T. Ikariya, K. Osakada, Y. Ishii, S. Osawa, M. Saburi, S. Yoshikawa, *Bull. Chem. Soc. Jpn.* **1984**, *57*, 897–898.

245 P. Morand, M. Kayser, *J. Chem. Soc. Chem. Commun.* **1976**, 314–315.

246 (a) K. Osakada, M. Obana, T. Ikariya, M. Saburi, S. Yoshikawa, *Tetrahedron Lett.* **1981**, *22*, 4297–4300. (b) Y. Ishii, *Kagaku Kogyo* **1987**, *40*, 132–135.

247 M.W. Farlow, H. Adkind, *J. Am. Chem. Soc.* **1935**, *57*, 2222–2223.

248 Y. Inoue, H. Izumida, Y. Sasaki, H. Hashimoto, *Chem. Lett.* **1976**, 863–864.

249 (a) P.G. Jessop, T. Ikariya, R. Noyori, *Chem. Rev.* **1995**, *95*, 259–272. (b) W. Leitner, *Angew. Chem. Int. Ed. Engl.* **1995**, *34*, 2207–2221.

250 K. Kudo, N. Sugita, Y. Takezaki, *Nippon Kagaku Kaishi* **1977**, 302–309.

251 W. Leitner, E. Dinjus, F. Gassner, *J. Organomet. Chem.* **1994**, *457*, 257–266.

252 E. Graf, W. Leitner, *J. Chem. Soc. Chem. Commun.* **1992**, 623–624.

253 F. Gassner, W. Leitner, *J. Chem. Soc. Chem. Commun.* **1993**, 1465–1466.

254 P.G. Jessop, T. Ikariya, R. Noyori, *Nature* **1994**, *368*, 231–233.

255 D.J. Drury, J.E. Hamilton, Eur. Patent Appl. 0095321, **1983**.

256 C. Yin, Z. Xu, S.-Y. Yang, S.M. Ng, K.Y. Wong, Z. Lin, C.P. Lau, *Organometallics* **2001**, *20*, 1216–1222.

257 J.-C. Tsai, K.M. Nicholas, *J. Am. Chem. Soc.* **1992**, *114*, 5117–5124.

258 P.G. Jessop, T. Ikariya, R. Noyori, *Chem. Rev.* **1999**, *99*, 475–493.

259 (a) *Chemical Reviews: Supercritical Fluids (Special Thematic Issue)* (Ed: R. Noyori), American Chemical Society, Washington, DC, **1999**, Vol. 99, No. 2. (b) *Chemical Synthesis Using Supercritical Fluids* (Eds: P.G. Jessop, W. Leitner), Wiley-VCH, Weinheim, **1999**.

260 P. Munshi, A. D. Main, J. C. Linehan, C.-C. Tai, P. G. Jessop, *J. Am. Chem. Soc.* **2002**, *124*, 7963–7971.

261 I. S. Kolomnikov, T. S. Lobeeva, M. E. Vol'pin, *Izv. Akad. Nauk SSSR, Ser. Khim.* **1972**, 2329–2330.

262 H. Phala, K. Kudo, S. Mori, N. Sugita, *Bull. Inst. Chem. Res. Kyoto Univ.* **1985**, *63*, 63–71.

263 P. G. Lodge, D. J. H. Smith, Eur. Patent Appl. 0094785, **1983**.

264 P. G. Jessop, Y. Hsiao, T. Ikariya, R. Noyori, *J. Chem. Soc. Chem. Commun.* **1995**, 707–708.

265 O. Kröcher, R. A. Köppel, A. Baiker, *J. Chem. Soc. Chem. Commun.* **1997**, 453–454.

266 Z. Hong, Y. Cao, J. Deng, K. Fan, *Catal. Lett.* **2002**, *82*, 37–44.

267 P. Haynes, L. H. Slaugh, J. F. Kohnle, *Tetrahedron Lett.* **1970**, 365–368.

268 L. Vaska, S. Schreiner, R. A. Felty, J. Y. Yu, *J. Mol. Catal.* **1989**, *52*, L11–L16.

269 S. Schreiner, J. Y. Yu, L. Vaska, *Inorg. Chim. Acta* **1988**, *147*, 139–141.

270 Y. Kiso, K. Saeki, *Japan Kokai Tokkyo Koho* 36617, **1977**.

271 P. G. Jessop, Y. Hsiao, T. Ikariya, R. Noyori, *J. Am. Chem. Soc.* **1994**, *116*, 8851–8852.

272 O. Kröcher, R. A. Köppel, A. Baiker, *J. Chem. Soc. Chem. Commun.* **1996**, 1497–1498.

273 Y. Kayaki, Y. Shimokawatoko, T. Ikariya, *Adv. Synth. Catal.* **2003**, *345*, 175–179.

1.1.4
Enantioselective Reduction of C=N Bonds and Enamines with Hydrogen

Felix Spindler and Hans-Ulrich Blaser

1.1.4.1
Introduction

Despite some significant recent progress, the enantioselective hydrogenation of prochiral C=N groups (imines, oximes, hydrazones, etc.) and enamines to obtain the corresponding chiral amines still represents a major challenge. Whereas many highly enantioselective chiral catalysts have been developed for the asymmetric hydrogenation of alkenes and ketones bearing various functional groups, much fewer catalysts are effective for the hydrogenation of substrates with a C=N function (for pertinent recent reviews see [1–3]). There are several reasons that might explain this situation. On the one hand, the enantioselective hydrogenation of enamides and other C=C groups and later also of C=O compounds was so successful that most attention was directed to these substrates. On the other hand, C=N compounds have some chemical peculiarities that make their stereoselective reduction more complex than that of C=O and C=C compounds. Even though the preparation starting from the corresponding amine derivative and carbonyl compound is relatively simple, complete conversion is not always possible, and formation of trimers or oligomers can occur. In addition, the resulting C=N compounds are often sensitive to hydrolysis, and the presence of syn/anti as well as enamine isomers can be a problem for selective hydrogenation.

The nature of the substituent directly attached to the N-atom influences the properties (basicity, reduction potential etc.) of the C=N function more than the substituents at the carbon atom. For example, it was found that Ir–diphosphine catalysts that are very active for *N*-aryl imines are deactivated rapidly when applied for aliphatic imines [4] and that titanocene-based catalysts are active only for *N*-alkyl imines but not for *N*-aryl imines [5–7]. Oximes and other C=N–X compounds show even more pronounced differences in reactivity. The following sections give a short summary of the state of the art for the enantioselective hydrogenation of different classes of C=N groups and a critical assessment of the presently known catalytic systems. Only very selective or otherwise interesting catalysts have been included in Tabs. 1–4. Structures of chiral ligands are depicted in Fig. 1, and those of the substrates in Figs. 2-6.

1.1.4.2
Enantioselective Reduction of *N*-aryl Imines

N-Alkyl-2,6-disubstituted anilines with a stereogenic C-atom in the *α*-position are intermediates for a number of important acyl anilide pesticides, the most important example being the herbicide Metolachlor® [8, 9]. Because not all stereoisomers are biologically active, the stereoselective synthesis of the most effective ones is of industrial interest. This is the reason that the enantioselective hydrogenation of the imines **1–3** depicted in Fig. 2 will be discussed in somewhat more detail.

The hydrogenation of the imines **1a,b** has been extensively investigated by several research groups. While the first useful results were obtained with chiral Rh diphosphine catalysts [10], the first step toward a technically feasible catalyst was made with newly developed Ir diphosphine complexes [4]. Despite a significant tendency for deactivation, substrate/catalyst mole ratio (s/c) values of ≥10 000 and reasonable reaction rates were obtained for the hydrogenation of MEA-imine with an Ir–diop complex in presence of iodide ions (see entry 1.1 in Tab. 1). The hydrogenation of other *N*-aryl imines with similar structural elements showed that both the 2,6-alkyl substituents of the *N*-phenyl group and the methoxy substituent contribute to the high enantioselectivity. Replacing the methoxy group of the DMA-imine by an ethyl group led to a decrease in *ee* from 69% to 52%, and further replacement of the 2,6-dimethyl phenyl by a phenyl group led to a decrease in *ee* to 18% [4]. It is noteworthy that the phenyl group could be replaced by a 2,4-disubstituted thien-3-yl group (imine **2**) without loss in catalyst activity (entry 1.4). Despite these good results, both catalyst activity and productivity were insufficient for a technical application for a high volume product.

The final breakthrough on the way to a production process for the Metolachlor herbicide came in 1993 (Fig. 3) [9, 11]. A new class of Ir ferrocenyl diphosphine complexes turned out to be stable and in the presence of both acetic acid and iodide gave extraordinarily active and productive catalysts. An extensive ligand optimization led to the choice of $[Ir(COD)Cl]_2$–PPF-$PXyl_2$ (xyliphos) as optimal catalyst. At a hydrogen pressure of 80 bar and 50 °C using an S/C of >1 000 000, complete conversion can be reached within 3–4 h with an enantiomeric excess of

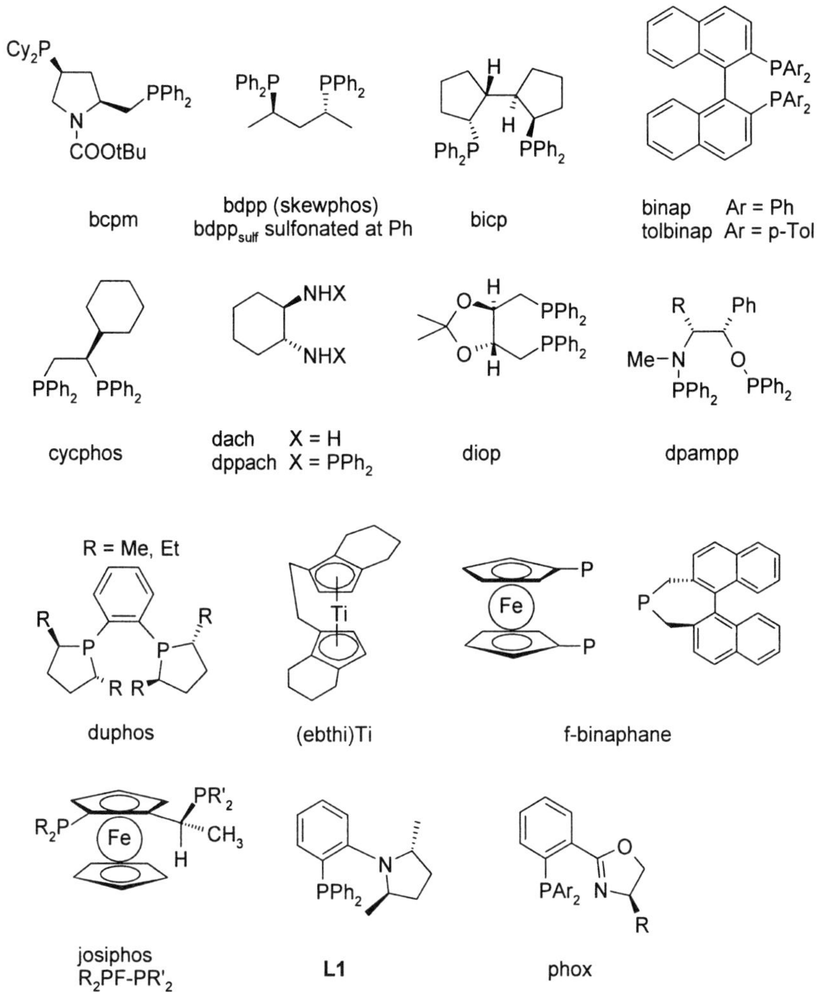

bcpm
bdpp (skewphos)
bdpp$_{sulf}$ sulfonated at Ph
bicp
binap Ar = Ph
tolbinap Ar = p-Tol
cycphos
dach X = H
dppach X = PPh$_2$
diop
dpampp
R = Me, Et
duphos
(ebthi)Ti
f-binaphane
josiphos
R$_2$PF-PR'$_2$
L1
phox

Fig. 1 Structures and abbreviations for chiral ligands.

1a R = Et
1b R = Me
2
3a R = Me Ar = Ph
3b R = H Ar = Ph
3c var. R Ar = o-Xyl

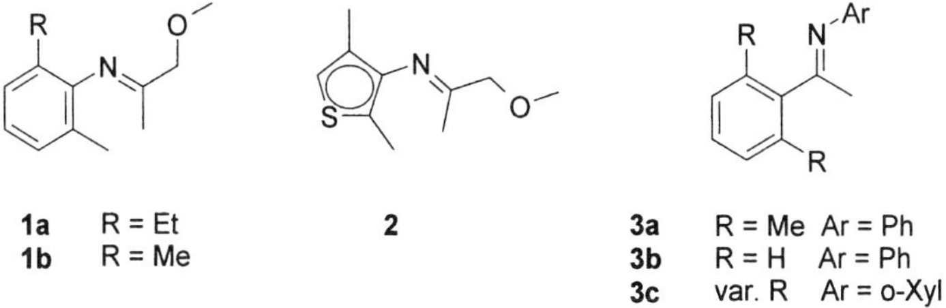

Fig. 2 Structures of N-aryl imines.

Tab. 1 Selected results for the enantioselective hydrogenation of *N*-aryl imines (structures in Fig. 2): Catalytic system, reaction conditions, enantioselectivity, productivity and activity

Entry	*Imine*	*Catalyst*	*p (bar)*	*ee (%)*	*S/C*	*TOF* (h^{-1})	*Ref.*
1.1	**1a**	Ir–diop/I^-	100	62	10000	200	4
1.2	**1a**	Ir–PPF-$PXyl_2/I^-/H^+$	80	78	1000000	350000	12
1.3	**1a**	Ir–PPF-PAr_2 [a]$/I^-/H^+$	80	87	5000	31	12
1.4	**2**	Ir–PPF-$PXyl_2/I^-/H^+$	60	80	100	200	12
1.5	**3a**	Ir–PPF-P(4-$CF_3Ph)_2/I^-/H^+$	80	96	200	n.a.	12
1.6	**1a** [b]	Ir–PPF-$PXyl_2/I^-/H^+$	80	10000	78	>600	13
1.7	**3b**	Ir–phox (sc-CO_2)	30	78	6800	2800	14
1.8	**3b**	Ir–phox	100	86	1400	1200	14
1.9	**3b**	Ru–duphos–dach	20	94	1000	50	15
1.10	**1b**	[Ir(diop)($OCOCF_3)_3$]	40	90	500	3	16
1.11	**3c**	Ir–f-binaphane	70	>99	100	2	17

a) Ar = 3,5-Me_2-4-NPr_2-Ph;
b) *in situ* formed from 2-methyl-6-ethyl-aniline + methoxyacetone

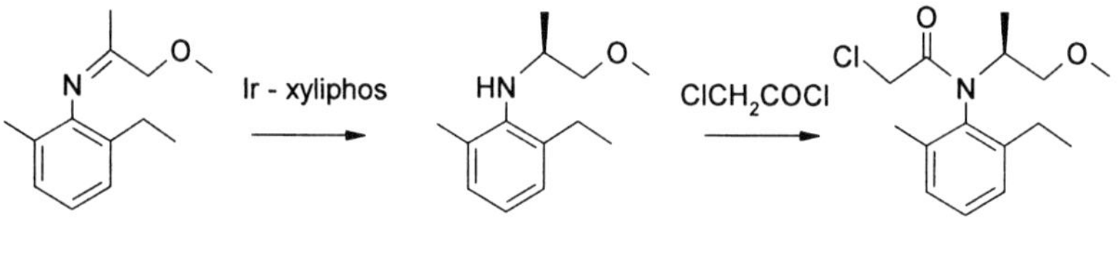

Fig. 3 Synthesis of *S-metolachlor.*

around 80% (entry 1.2). The best enantioselectivities of 87% were obtained with *N*-substituted xyliphos ligands, albeit with much lower activity (for an example, see entry 1.3). Scale-up presented no major problems, and the production plant was opened officially in November 1996. At the moment there is no convincing explanation for the remarkable effect of iodide and acid. The ferrocenyl diphosphine catalysts only exhibit the high enantioselectivity and especially the extraordinarily high activity and productivity when both additives are present. Even though the scope of this new catalytic system has not yet been fully determined, it was successfully applied to the hydrogenation of imines **2** (entry 1.4), **3a** (entry 1.5) and **9** (see below). In addition, it was shown for the first time that reductive alkylation of an amine via *in situ* formation of the corresponding imine is possible with a reasonable catalytic performance (entry 1.6).

Some further results are noteworthy for *N*-aryl imines. For the model substrate **3b**, the Ir–phox catalyst developed by Pfaltz achieved TON values of up to 6800 in supercritical CO_2, a considerable improvement over the catalytic performance in dichloromethane (entries 1.7, 1.8), and an Ru–duphos–dach complex gave up to 94% *ee* with acceptable productivity (entry 1.9). Osborn and Sablong [11] reported that completely halogen-free catalysts can also give very good enantioselectivities (e.g., 90% *ee* with imine **1b**) (entry 1.10), and an Ir–f-binaphane catalyst achieved *ees* >99% with several imines of the type **3c** (entry 1.11).

1.1.4.3
Enantioselective Reduction of *N*-alkyl Imines and Enamines

Up to now, few acyclic *N*-alkyl imines or the corresponding amines have been of practical industrial importance. Most studies reported herein were carried out with model substrates, especially with the *N*-benzyl imine of acetophenone and some analogs thereof. One reason for this choice could be the easy preparation of a pure crystalline starting material, another being that the chiral primary amines can be obtained by hydrogenolysis of the benzyl group. As can be seen in Tab. 2, there are several catalyst systems with fair to good *ees* and activities.

Enantioselectivities of >90% were reported for a Ti–ebthi catalyst (entry 2.1 of Tab. 2) and for some Rh diphosphine complexes (entries 2.2–2.4). Interestingly, the highest *ees* were obtained using sulfonated diphosphines ($bdpp_{sulf}$) in an aqueous biphasic medium (entry 2.3). The degree of sulfonation strongly affected the enantioselectivity: the Rh–monosulfonated bdpp gave 94% *ee* compared to

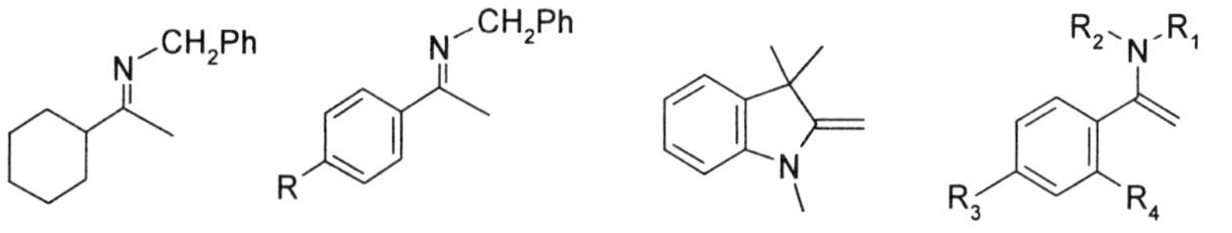

$N{-}CH_2Ph$ $N{-}CH_2Ph$ $R_2{-}N{-}R_1$ R N R_3 R_4

4 **5a** R = H **5b** R = OMe **6** **7**

Fig. 4 Structures of *N*-alkyl imines and enamines.

Tab. 2 Selected results for the enantioselective hydrogenation of *N*-alkyl imines and enamines (structures in Fig. 4): Catalytic system, reaction conditions, enantioselectivity, productivity and activity.

Entry	*Imine*	*Catalyst*	*p (bar)*	*ee (%)*	*S/C*	*TOF* (h^{-1})	*Ref.*
2.1	4	Ti–ebthi	5	92	20	4	7
2.2	5a	Rh–cycphos	100	91	100	0.7	18
2.3	5a	Rh–$bdpp_{sulf}$	70	96	100	16	19
2.4	5b	Rh–bdpp/AOT micelles	70	92	100	4.6	20
2.5	5a	Ir–L1	100	46	100	>36000	21
2.6	5a	Ru–dppach/dach	3	92	1500	23	22
2.7	6	Ir–diop/iodide	20	60–64	50	n.a.	25
2.8	7	Ti–ebthi	1–5	89–98	20	<1	26

65% *ee* with Rh–bdpp in MeOH, and the activity of the monosulfonated catalyst was higher by a factor of 5 [19]. A similar positive effect for the presence of a sulfo group was described by Buriak and Osborn [20] (entry 2.4). Unfortunately, catalytic activities and productivities for all these catalysts ranged from very low to modest. For this reason, the high TOF values claimed for the Ir–**L1** system are of special interest, but the *ee* is unfortunately very low (entry 2.5). Imine **5a** is also hydrogenated with good *ee* and TON using an Ru diphosphine diamine complex originally developed by Noyori (entry 2.6).

In addition to these results, we registered with interest the first example of a Pd–binap-catalyzed hydrogenation of a fluorinated α-imino ester in the presence of trifluoroacetic acid in fluorinated alcohols (*ee*s up to 91%, but very low TON and TOF) [23] and the claim by Magee and Norton [24] that cpW diphosphine complexes are able to hydrogenate imines via a novel ionic mechanism albeit with low *ee* and TOF.

The hydrogenation of enamines was even less investigated. Two results are of interest: **6** was hydrogenated in the presence of Ir/diop with *ee*s of 60–64% [25], and enamines of the type **7** were reduced in presence of Ti–(ebthi)Ti with enantioselectivities of 89–98% [26] (entries 2.7, 2.8). In both cases, catalyst activities and productivities are too low for practical purposes.

1.1.4.4 Enantioselective Reduction of Cyclic Imines

Cyclic imines do not have the problem of syn/anti isomerism, and therefore, in principle, higher enantioselectivities can be expected. This was indeed the case for the hydrogenation of several cyclic model substrates **8** using the Ti–ebthi with *ee*s up to 99% (entry 3.1 in Tab. 3), whereas enantioselectivities for acyclic imines were ≤90% [5, 7]. Unfortunately, these very selective catalysts operate at low S/C ratios and exhibit TOFs <3 h^{-1}. Iridium diphosphine catalysts in the presence of various additives look more promising because they show higher activities. With these catalyst systems, model compound **9** was reduced with *ee*s of 80–94% (entries 3.2–3.4), but the function of these additives is not known, and most were found empirically. Cyclic imines **10** are intermediates or models of biologically active compounds and can be reduced with *ee*s of 88–96% using Ti–ebthi or Ir–bcpm (entries 3.5 and 3.6).

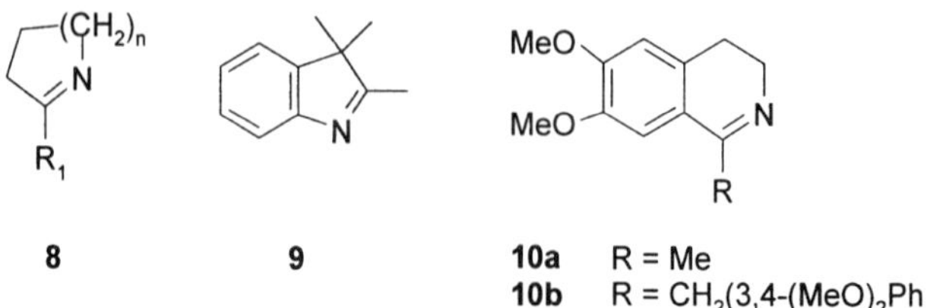

Fig. 5 Structures of cyclic imines.

Tab. 3 Selected results for the enantioselective hydrogenation of cyclic imines (structures in Fig. 5): Catalytic system, reaction conditions, enantioselectivity, productivity and activity

Entry	*Imine*	*Catalyst*	*p (bar)*	*ee(%)*	*S/C*	*TOF* (h^{-1})	*Ref.*
3.1	8	Ti–ebthi	5	99	100	0.6–2	6
3.2	9	Ir–bicp/phthalimide	70	95	200	2	27
3.3	9	Ir–Xyl_2PF-$PXyl_2/I^-/H^+$	40	93	250	56	12
3.4	9	Ir–Tol-binap/$PhCH_2NH_2$	60	90	100	5.5	28
3.5	**10a**	Ti–ebthi	5	96	20	0.4	6
3.6	**10b**	Ir–bcpm/F_4-phtalimide	100	88	100	4.5	29

1.1.4.5
Enantioselective Reduction of Miscellaneous C=N–X Systems

Less common types of C=N derivatives can be reduced enantioselectively as well. An interesting example is the hydrogenation of the *N*-acyl hydrazones **11** with the Rh/duphos catalyst (entry 4.1 in Tab. 4). This reaction was developed by Burk et al. [30] in analogy to the well-known hydrogenation of enamides. Indeed, the results are very impressive and confirm that the presence of a second group in the substrate molecule that is able to bind to the metal is beneficial for achieving high optical yields. The resulting *N*-acyl hydrazines can be reduced to the primary amine using SmI_2, but an effective technical solution for the cleavage of the N–N bond to obtain the primary amine without racemization is still lacking. Cyclic *N*-

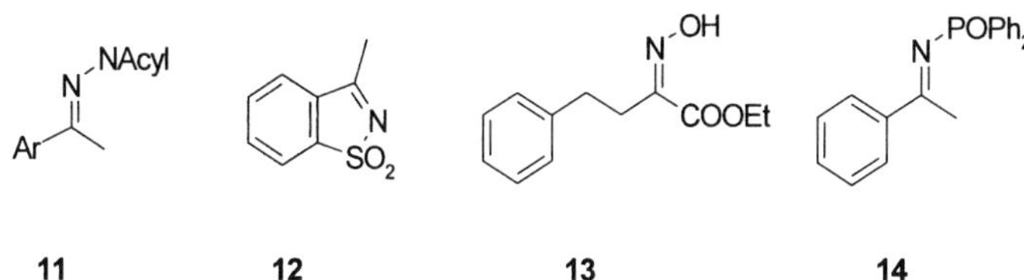

Fig. 6 Structures of miscellaneous C=N–X compounds..

Tab. 4 Selected results for the enantioselective hydrogenation miscellaneous C=N–X compounds (structures in Fig. 6): Catalytic system, reaction conditions, enantioselectivity, productivity and activity.

Entry	*Imine*	*Catalyst*	*p (bar)*	*ee(%)*	*S/C*	*TOF* (h^{-1})	*Ref.*
4.1	11	Rh–duphos	4	88–96	500	14–42	30
4.2	12	Ru–binap	4	99	90	6	31
4.3	13	Ir–dpampp/I^-	48	93	100 a)	0.5	33
4.4	14	Rh–Cy_2PF-PCy_2	70	99	500	500	36

a) conversion 22%.

sulfonyl-imines **12** can be hydrogenated with Ru/binap in good to very good optical yields (entry 4.2), whereas acyclic analogs are reduced with lower *ees* [32]. Oximes can be hydrogenated with very good enantioselectivity but low activity with the novel Ir–dpampp complex (entry 4.3), while Ru–binap [34] or Rh–binap [35] were also active but with modest *ees*. Phosphinyl imines **14** are highly suitable substrates for the Rh–josiphos-catalyzed hydrogenation, with *ees* up to 99% (entry 4.4).

1.1.4.6
Assessment of Catalysts

In this section, a critical assessment of the most efficient catalysts is given. Important factors with respect to preparative applications such as availability of precursors and ligands, ease of use, scope and specificity, catalyst activity, and productivity will be discussed.

Iridium complexes

Among the various catalyst types investigated in recent years for the hydrogenation of imines, Ir diphosphine complexes proved to be most versatile catalysts. The first catalyst of this type generated *in situ* from $[Ir(COD)Cl]_2$, a chiral diphosphine and iodide, was developed by the Ciba catalysis group in 1985 [4]. Ir ferrocenyl diphosphines (josiphos) complexes in the presence of iodide and acid are the most active and productive enantioselective catalysts for the hydrogenation of the *N*-aryl-imines. The josiphos ligands are quite stable and easy to tune to the special needs of the imine structure [12, 37], and a large variety are commercially available in technical quantities. Most other Ir complexes have not been studied in much detail, and many of these require the presence of additives such as phthalimide for good performance. Ligands such the f-binaphane or bicp by Zhang [17, 27] or the tol-binap are patent protected but are available commercially. The Ir–phox catalysts developed by Pfaltz [14] have not yet been explored enough to assess their potential for C=N hydrogenation.

Rhodium complexes

Rhodium diphosphine catalysts can be easily prepared from $[Rh(NBD)Cl]_2$ and a chiral diphosphine and are active for the hydrogenation of imines and *N*-acyl hydrazones. However, with most imine substrates they exhibit lower activities than the analogous Ir catalysts. The most selective diphosphine ligand is $bdpp_{sulf}$, and this is not easily available [19]. The Rh–duphos catalyst is very selective for the hydrogenation of *N*-acyl hydrazones, and, with a TOF value of up to 1000 h^{-1}, would be active enough for a technical application [30]. Rh josiphos complexes are the catalysts of choice for the hydrogenation of phosphinylimines [36]. Recently developed (pentamethylcyclopentyl)Rh–tosylated diamine or amino alcohol complexes are active for transfer hydrogenation for a variety of C=N functions and can be an attractive alternative for specific applications [38].

Ruthenium complexes

In contrast to the wide scope of the Ru–binap catalysts for the hydrogenation of substituted alkenes and ketones, their use in the hydrogenation of C=N groups is limited because of the tendency to deactivate in the presence of bases. Of more interest but not yet fully explored are Ru complexes containing both a diphosphine and a diamine [22]. (Arene)Ru complexes in the presence of tosylated diamines are also able to reduce imines under transfer hydrogenation conditions with high activities and high to very high optical yields [39].

Titanium complexes

Despite the remarkable enantioselectivities observed with the (ebthi)Ti catalyst for imine and enamine hydrogenation, we consider its technical potential rather small for the following reasons: the ligand is difficult to prepare, the activation of the catalyst precursor is difficult, for the moment the catalytic activity is far too low for preparative purposes, and, last but not least, its tolerance for other functional groups is low.

1.1.4.7 Summary

Tab. 5 gives a summary of the present state of the art for enantioselective hydrogenation organized according to effective catalytic systems. While more and more cases with high *ee*s are reported, progress concerning TON and TOF is much slower. Compared to the situation for the analogous C=C and C=O hydrogenation, there are still many areas where little systematic information is available. The most visible success stories, such as the (*S*)-metolachlor case, are rather “anecdotal” in nature. Nevertheless, they are proof that it is possible to hydrogenate C=N

Tab. 5 Typical ranges of reaction conditions, optical yields, turnover frequencies and substrate-to-catalyst ratios for the hydrogenation of C=N functions using various chiral catalytic systems.

Catalyst	*Substrate type*	*p (bar)*	*ee (%)*	*S/C*	*TOF (1/h)*
Ir–P^P	*N*-aryl imines	20–80	70–99	100–>1 000 000	2–>350 000
	Cyclic imines	40–70	80–97	100–1000	5–50
Rh–P^P	Imines	60–100	80–96	40–1000	0.1–50
	N-acyl hydrazones	4	70–96	500–1000	10–1000
	Phosphinylimines	70	90–99	100–500	100–500
Ru–P^P-N^N	Acyclic imines	3–20	60–94	500–1500	20–50
Ti–ebthi	Cyclic imines	5–33	98–99	20–100	0.4–2.4
	N-alkyl imines	130	53–85	20	<2

functions with not only adequate enantioselectivity but also high activity and productivity.

References

1 T. Ohkuma, M. Kitamura, R. Noyori in I. Ojima (Ed.), *Catalytic Asymmetric Synthesis* (2nd edn.), Wiley-VCH, Weinheim, **2000**, p. 1.

2 H. U. Blaser, F. Spindler in *Comprehensive Asymmetric Catalysis* (Eds.: E. N. Jacobsen, A. Pfaltz, H. Yamamoto), Springer, Berlin, **1999**, p 247.

3 H. U. Blaser, B. Pugin, F. Spindler in *Applied Homogeneous Catalysis by Organometallic Complexes* (2nd edn.) (Eds.: B. Cornils, W. A. Herrmann), Wiley-VCH, Weinheim, **2002**, p 1131.

4 F. Spindler, B. Pugin, H. U. Blaser, *Angew. Chem. Int. Ed. Engl.* **1990**, *29*, 558 and F. Spindler, B. Pugin, unpublished results.

5 C. A. Willoughby, S. L. Buchwald, *J. Am. Chem. Soc.* **1992**, *114*, 7562.

6 C. A. Willoughby, S. L. Buchwald, *J. Org. Chem.* **1993**, *58*, 7627.

7 C. A. Willoughby, S. L. Buchwald, *J. Am. Chem. Soc.* **1994**, *116*, 8952.

8 H. U. Blaser, F. Spindler, *Top. Catal.* **1997**, *4*, 275.

9 H. U. Blaser, H. P. Buser, K. Coers, R. Hanreich, H. P. Jalett, E. Jelsch, B. Pugin, H. D. Schneider, F. Spindler, A. Wegmann, *Chimia* **1999**, *53*, 275.

10 W. R. Cullen, M. D. Fryzuk, B. R. James, J. P. Kutney, G.-J. Kang, G. Herb, I. S. Thorburn, R. Spogliarich, *J. Mol. Catal.* **1990**, *62*, 243.

11 H. U. Blaser, *Adv. Synth. Catal.* **2002**, *344*, 17 and references therein.

12 H. U. Blaser, H. P. Buser, R. Häusel, H. P. Jalett, F. Spindler, *J. Organomet. Chem.* **2001**, *621*, 34.

13 H. U. Blaser, H. P. Buser, H. P. Jalett, B. Pugin, F. Spindler, *Synlett* **1999**, 867.

14 S. Kainz, A. Brinkmann, W. Leitner, A. Pfaltz, *J. Am. Chem. Soc.* **1999**, *121*, 6421.

15 C. Cobley, J. Henschke, J. Ramschen, WO 02/8169, **2001** (assigned to Chiro-Tech).

16 R. Sablong, J. A. Osborn, *Tetrahedron: Asymmetry* **1996**, *7*, 3059.

17 D. Xiao, X. Zhang, *Angew. Chem.* **2001**, *113*, 3533.

18 A. G. Becalski, W. R. Cullen, M. D. Fryzuk, B. R. James, G.-J. Kang, S. J. Rettig, *Inorg. Chem.* **1991**, *30*, 5002.

19 C. Lensink, E. Rijnberg, J. de Vries, *J. Mol. Cat. A: Chemical* **1997**, *116*, 199 and references cited.

20 J. Buriak, J. A. Osborn, *Organometallics* **1996**, *15*, 3161.

21 J. P. Cahill, A. P. Lightfoot, R. Goddard, J. Dust, P. J. Guiry, *Tetrahedron: Asymmetry* **1998**, *9*, 4307.

22 K. Abdur-Rashid, A. J. Lough, R. H. Morris, *Organometallics* **2001**, *20*, 1047.

23 H. Abe, H. Amii, K. Uneyama, *Org. Lett.* **2001**, *3*, 313.

24 M. P. Magee, J. R. Norton, *J. Am. Chem. Soc.* **2001**, *123*, 1779.

25 B. Pugin, Novartis Services AG, unpublished results.

26 N. Lee, S. L. Buchwald, *J. Am. Chem. Soc.* **1994**, *116*, 5985.

27 G. Zhu, X. Zhang, *Tetrahedron: Asymmetry* **1998**, *9*, 2415.

28 K. Tani, J. Onouchi, T. Yamagata, Y. Kataoka, *Chem. Lett.* **1995**, 955.

29 T. Morimoto, N. Suzuki, K. Achiwa, *Heterocycles* **1996**, *43*, 2557.

30 M. J. Burk, J. E. Feaster, *J. Am. Chem. Soc.* **1992**, *114*, 6266. M. J. Burk, J. Martinez, J. E. Feaster, N. Cosford, *Tetrahedron* **1994**, *50*, 4399.

31 W. Oppolzer, M. Wills, C. Starkemann, G. Bernardinelli, *Tetrahedron Lett.* **1990**, *31*, 4117.

32 A. Charette, A. Giroux, *Tetrahedron Lett.* **1996**, *37*, 6669.

33 Y. Xie, A. Mi, Y. Jiang, H. Liu, *Synth. Commun.* **2001**, *31*, 2767.

34 P. Krasik, H. Alper, *Tetrahedron: Asymmetry* **1992**, *3*, 1283.

35 A. S. C. Chan, C. Chen, C. Lin, M. Cheng, S. Peng, *J. Chem. Soc., Chem. Commun.* **1995**, 1767.

36 F. Spindler, H. U. Blaser, *Adv. Synth. Catal.* **2001**, *343*, 68.

37 For an overview see H. U. Blaser, W. Brieden, B. Pugin, F. Spindler, M. Studer, A. Togni, *Top. Catal.* **2002**, *19*, 3 and information on commercial ligands in www.solvias.com/E/main/services/solutionpackages/liganden.html

38 A. J. Blacker, WO9842643B1, **1997** (assigned to Avecia), see also information on www.avecia.com/pharms/ntventures/chiral_solutions/chemocatalysis.html

39 R. Noyori, S. Hashiguchi, *Acc. Chem. Res.* **1997**, *30*, 97.

1.2 Heterogeneous Hydrogenation: a Valuable Tool for the Synthetic Chemist

Hans-Ulrich Blaser, Heinz Steiner, and Martin Studer

1.2.1 Introduction

There is no doubt that hydrogen is the cleanest reducing agent and that hydrogenation is the most important catalytic method in synthetic organic chemistry. In his books on "Catalytic Hydrogenation in Organic Syntheses", P.N. Rylander [1] gives several reasons for this opinion: the scope of the reaction type is very broad and many functional groups can be hydrogenated with high selectivity; high conversions are usually obtained under relatively mild conditions in the liquid phase; because of the large pool of experience (documented in a number of books such as Rylander's [1–4]), the predictability for solving a given problem is high; and, last but not least, the process technology is well established and scale-up is usually not problematic. On the other hand, Loewenthal and Zass [5] call hydrogenation the "Cinderella of the Organic Experimentalist". As the main reason they mention the problem of using an explodable gas which is frequently under rather high pressure and in specialized equipment. This deters many synthetic chemists from applying the method.

In this chapter, we do not attempt to give a comprehensive review on catalytic hydrogenation and its application to organic synthesis, since many books [1–4] and reviews [6–8] do this already very competently. Our purpose is to familiarize the practicing synthetic chemist, both at university and in industry, with the opportunities available from heterogeneous hydrogenation technology. The most important catalysts (and their suppliers) are described, and the method of selecting a suitable catalyst is explained. It also is shown how the reaction medium and the reaction conditions can affect catalyst activity and selectivity and which equipment can be used. Some new developments for the hydrogenation of nitro arenes (see Section 1.2.5.1) and the topic of catalytic debenzylation (see Section 1.2.5.5) are described in more detail.

Transition Metals for Organic Synthesis, Vol. 2, 2nd Edition.
Edited by M. Beller and C. Bolm

ISBN: 3-527-30613-7

1.2.2
Some Special Features of Heterogeneous Catalysts

The application of solid catalysts has a number of practical consequences for the synthetic chemist [6]. While there is no doubt that the use of heterogeneous catalysts is often simple and very practical, working with solids also has some pitfalls and some of these are discussed in this chapter.

Separation, handling and work-up

Usually, solid catalysts can be separated from the reaction mixture by simple filtration. This allows convenient work-up and isolation of the desired product and is the most obvious advantage of a heterogeneous catalyst. Most heterogeneous hydrogenation catalysts are very easy and safe to handle; they do not require inert an atmosphere as many homogeneous catalysts do. Nevertheless, dry catalysts may catalyze the combustion of hydrogen or organic vapors (danger of ignition). Therefore, the catalyst should be wetted with the solvent under an inert atmosphere (Ar or CO_2) before addition to the reaction mixture, and it should not be sucked dry after filtration.

Accessibility of the active sites

Whereas 100% of a homogeneous catalyst is available for reaction, only the surface of solid catalysts can interact with the substrates. Therefore, it is important to create active phases with a high *dispersion,* i.e., where as many atoms as possible are exposed. Large surfaces areas are produced either by making very small crystallites, by generating thin films, or by the creation of a porous structure. However, larger particles are generally more stable exactly because they have a smaller surface area. The preparation of stable high surface area materials is therefore quite a challenge.

Diffusion problems

Another difference between homogeneous and heterogeneous reactions lies in the importance of diffusion (or transport) phenomena. Neither the removal of the heat of reaction (heat transport) nor the diffusion of reagents to the active site (mass transport) is of much concern in stirred homogeneous solutions. However, heterogeneous reactions on the surface of solids can be seriously affected by such effects. Several different diffusion steps can be distinguished (see Fig. 1) that can affect the rate and sometimes the selectivity of the reaction. Good stirring and the use of materials with adequate pore size are important. For reactions in the liquid phase, our experience has shown that with proper choice of the catalyst loading (i.e., percentage of catalyst present) the problems are minimal.

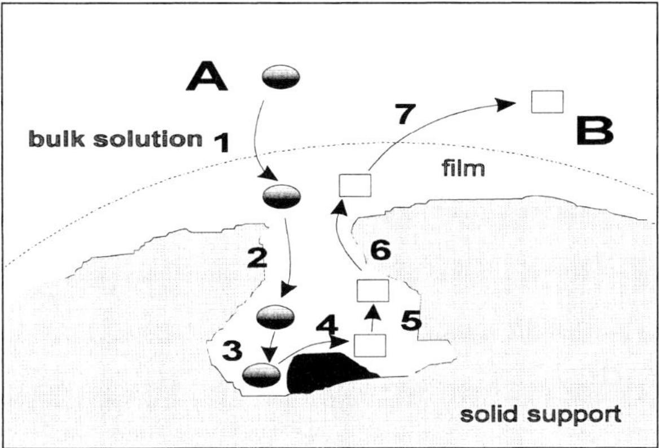

Fig. 1 Schematic representation of the different diffusion steps of substrate **A** and product **B**: (1) film diffusion of **A**, (2) pore diffusion of **A**, (3) adsorption of **A** on active site, (4) reaction, (5) desorption of **B**, (6) pore diffusion of **B**, (7) film diffusion of **B**.

Reproducibility and availability of catalysts

Heterogeneous catalysts cannot be characterized on a molecular level. The catalytic properties of two catalysts having the same description, e.g., 5% Pt/C, can vary considerably. It is well known that even small variations in the preparation procedure or in the impurities can alter the structural and chemical properties of a heterogeneous catalyst. This might or might not affect its catalytic or chemical activity. It is precisely this "might or might not" that leads to problems of reproducibility and frustration. For routine hydrogenations the problem is usually not significant, and more and more heterogeneous hydrogenation catalysts are being sold today with reasonably precise specifications (see below).

1.2.3 Hydrogenation Catalysts

The classical and most used hydrogenation catalysts are the noble metals Pt, Pd, Rh, and Ru supported on active carbon, Raney nickel, and some supported Ni and Cu catalysts. Since the active metal is present as very small particles on a support or as skeletal material, the specific surface area is generally very high. As a rule, catalysts are still chosen on an empirical basis by trial and error, and it is rarely understood why a given catalyst is superior to another one. Many factors influence the catalytic properties of such catalysts, and it is important to realize that even today it is not possible to adequately characterize a heterogeneous catalyst on a molecular level.

Of the numerous parameters of a heterogeneous hydrogenation catalyst that affect its catalytic performance, the following are the most important. *Type of metal*: As already mentioned, Pd, Pt, Rh, Ru, Ni, and Cu are used most often. Each metal has its own activity and selectivity profile (see Tab. 1). *Type of catalyst*: Noble metals are usually supported on a carrier; sometimes they are used as fine powders (Pd black and Pt black, PtO_2); Ni is most often applied as skeletal Raney nickel or supported on silica, Cu as Cu chromite. *Metal loading*: For noble metal catalysts 5% loading is standard. For Ni/SiO_2 the loading is 20–50%. The concentration of the metal is usually given in the description of the catalyst, e.g., 5 wt% of palladium metal on an active carbon support is designated as 5% Pd/C (for the calculation, the dry weight of the catalyst is used). *Type of support*: Charcoal (also called active carbon) is most common. Charcoals can adsorb large amounts of water. For safety reasons, many catalysts are sold with a water content of 50%. Aluminas and silicas as well as $CaCO_3$, $BaSO_4$ are also used for special applications.

Two commercial catalysts with the same designation (e.g., 5% Pt/C) can still differ significantly because of different supports or different preparation methods, leading to different catalyst parameters that sometimes correlate with the catalyst performance. Parameters for the active metal are the surface area, the dispersion (typically only 10–60% of the metal atoms are exposed), the size of the crystallites (typically in the range 20 to >200 Å), the location in the pores of the support, and the oxidation state (reduced or unreduced). Important support parameters are the particle size (for slurry catalysts typically 1–100 μm), the surface area (typically in the range of 100–1500 m^2/g), the pore structure (pore volume, pore size distribution) and acid-base properties.

1.2.3.1 Catalyst Suppliers

The following manufacturers supply a full range of hydrogenation catalysts (only European suppliers are listed): Degussa [9a], Engelhard [9b], Heraeus [9c], and Johnson Matthey [9d]. In addition, they have a great deal of know-how concerning which catalyst type is most suitable for a given problem. Our experience has shown that it is of advantage to find or optimize a suitable catalyst in close collaboration with the catalyst suppliers. This is especially true for the development of technical processes and when there is little hydrogenation experience or when a particular problem is to be solved. Screening and development should always be carried out with specified catalysts that can be supplied in technical quantities when needed. For laboratory use, Fluka and Aldrich Inorganics offer a wide variety of hydrogenation catalysts that are well suited for preparative purposes. With some exceptions, the catalyst manufacturer and the exact catalyst type is not specified.

The 2000/2001 catalogue of Aldrich [10a] lists 4 Ni catalysts; 16 Pd/C, 6 Pd/Al_2O_3, 2 Pd/$BaSO_4$, 1 Pd/$CaCO_3$, 1 Pd/$BaCO_3$, 1 Pd/$SrCO_3$, 7 Pd black/oxides/oxide hydrates, 1 Pd-Ba/$CaCO_3$ (Lindlar); 8 Pt/C, 6 Pt/Al_2O_3, 10 Pt black/PtO_2 (Adams)/Pt oxide hydrates; 2 Rh/C, 3 Rh/Al_2O_3, Rh black/oxide/oxide hydrates; 1 Ru/C, 4 Ru/Al_2O_3, and 4 Ru black/oxide/oxide hydrates.

The catalogue 2001/2002 of Fluka [10b] lists 1 Raney-Nickel catalyst, 5 Pd/C, 2 Pd/Al_2O_3, 2 Pd/$BaSO_4$, 2 Pd/$CaCO_3$ and Pd/$SrCO_3$; 4 Pt/C and 2 Pt black/oxide hydrate; 2 Rh/C, Rh/Al_2O_3; and 2 Rh oxide/oxide hydrates; Ru/Al_2O_3, Ru/C and 2 Ru oxide/oxide hydrate.

1.2.3.2
Choice of the Catalyst

As already mentioned, there are many variables that have an influence on the outcome of a catalytic reaction. For hydrogenation reactions the hierarchy of the variables is generally: metal > reaction medium > reaction conditions > catalyst support and type. This means that the choice of the active metal is the most important step when considering a catalytic hydrogenation.

Catalyst activity

Obviously, the catalyst has to be active for a desired transformation, and Tab. 1 lists the best metals for a number of frequently used reactions, together with some recommendations concerning the solvent. Except where otherwise noted, the reactions can be carried out at hydrogen pressures of 1–4 bar (1×10^5 to 4×10^5 Pa) and at temperatures of 20 to 40 °C. A very useful compilation, "The Catalytic Reaction Guide", that lists the optimal heterogeneous catalyst for 52 different transformations, is available from Johnson Matthey [9d].

Tab. 1 Preferred metal and solvent type for hydrogenations of important functional groups

Substrate	*Reaction*	*Catalyst*	*Solvent*
Azides	$RN_3 \rightarrow RNH_2$	Pd	Polar
Aromatic nitro groups	$ArNO_2 \rightarrow ArNH_2$	Ni, Pd, Pt	Various
Debenzylation	$ArCH_2X \rightarrow ArCH_3 + HX$ X=OH, OR, OCOR, NHR, X=Hal	Pd	Protic, acidic or basic
Alkenes	$R_2C{=}CR_2 \rightarrow R_2HC{-}CHR_2$	Pd, Pt, Rh	Various
Alkynes	$RC{\equiv}CR \rightarrow RHC{=}CHR$	Pd/Pb	Low polarity
Aliphatic ketones, aldehydes	$R_2CO \rightarrow R_2CHOH$	Ni, Ru, Pt, Rh	Polar [a)]
Aromatic, ketones, aldehydes	$ArCOR \rightarrow ArCH(OH)R$	Pd, Pt	Polar
Nitriles	$RCN \rightarrow RCH_2NH_2$	Ni, Rh	Basic [a)]
		Pd, Pt	Acidic
	3.2.1.3 $RCN \rightarrow (RCH_2)_2NH$	Rh	Acidic
Aryl halides	$ArX \rightarrow ArH$ X=Cl, Br, I	Pd	Basic
Acid chlorides	$RCOCl \rightarrow RCHO$	Pd	(Basic)
Oximes	$R_2C{=}NOR \rightarrow R_2CHNH_2$	Ni	Basic [a)]
		Pt, Pd	Acidic
Imines	$R_2C{=}NR \rightarrow R_2CHNHR$	Pd, Pt	Various
(Hetero)aromatic rings		Rh, Ru, Pt	Various [a)]
Furanes		Pd, Rh	
Pyrroles		Pt, Rh	Acidic [a)]

Reaction conditions: r.t., p 1–4 bar (higher p for [a)]), 20–150 °C

Catalyst selectivity

For the hydrogenation of multifunctional molecules it is usually not the catalyst activity that poses the most difficult problem but rather the selectivity of the catalyst. The functions to be converted and the functions to be preserved determine which metal has the best chance of having high selectivity. There exist a number of specialized books and reviews that address this central problem. Here, we would like to recommend the inexpensive bench top edition of P. N. Rylander's book *Hydrogenation Methods* [1] and Volume IV/1c of Houben-Weyl's *Methoden der organischen Chemie* [3], which is probably available in most libraries.

Besides catalyst activity and selectivity, there are further criteria to assess catalysts for a technical application such as catalyst productivity, chemical and thermal stability, sensitivity toward deviation of process parameters (e.g., temperature, impurities, etc.), and finally catalyst costs. The catalyst costs for noble metal catalysts consist of the following cost elements: *preparation* (including costs for the support etc.); *metal losses* (process and handling losses in the range of 1–10% are considered normal; recovery losses for Pd and Pt are 1–2%, for Rh and Ru ca. 10%); *metal recovery fees* of the catalyst manufacturer; *interest costs* for the noble metal inventory (usually treated as investment). The relative size of these cost elements varies depending on the specific situation. In our experience, total catalyst costs for 1 kg 5% Pd or Ru/C catalyst are $100–200, for 1 kg 5% Pt or Rh/C catalyst in the range of $200–600.

1.2.4 Hydrogenation Reactions

1.2.4.1 Reaction Medium and Process Modifiers

Catalytic hydrogenations on the laboratory scale are usually carried out in solution. The *choice of the solvent* affects not only the solubility of the reactants and products but can also very strongly influence the activity and selectivity of a catalyst. Solvents should not be hydrogenated under the particular reaction conditions. In the laboratory, only high-purity solvents should be used in order to minimize poisoning of the catalyst. Most often used are alcohols (MeOH, EtOH, *i*PrOH, BuOH), ethyl acetate, aromatic and aliphatic hydrocarbons, ethers such as *t*BuOMe, THF, dioxane (care has to be taken with Raney nickel), water, ketones, and acetic acid. In special cases, amides such as DMF, dimethylacetamide or *N*-methylpyrrolidone, and methylene chloride are used as well.

The application of organic *modifiers* is an important strategy to influence the properties, mainly the selectivity, of heterogeneous catalysts. This approach is especially attractive for the organic chemist since there is no need to prepare a new catalyst (which requires special know-how). Process modifiers are defined as (organic) additives that are added directly to the reaction mixture. Freifelder [2] gives a good overview on the effect of a wide variety of additives used in hydrogenation reactions. Well-known examples are the use of sulfur or nitrogen compounds, e.g., for the selective hydrogenation of acid chlorides to give aldehydes (Rosenmund system) or the selective hydrogenation of halogenated aromatic nitro groups. Interestingly, a metal surface can also be made chiral. For example, Pt catalysts modified by cinchona alkaloids are used for the enantioselective hydrogenation of α-ketoacid derivatives [8]. Other possibilities to tailor the properties of a metallic catalyst are the addition of a second metal (bimetallic catalysts), the surface modification by organometallic compounds, or the use of special supports.

1.2.4.2 Reaction Conditions

Especially for the production scale, it is important to carefully optimize all parameters of the catalytic system: catalyst, reaction medium, and reaction conditions. The quality of the optimization will strongly affect the costs of the hydrogenation step! The following parameters, which affect the process performance, can be influenced: *hydrogen pressure* (has a strong effect on the rate of reaction, sometimes also on selectivity); *temperature* (strongly affects rate and selectivity); *substrate concentration* (determines volume yield); *catalyst/substrate ratio* (depends on the catalyst activity and determines reaction time and costs); *agitation* (affects gas-liquid diffusion); *catalyst pre-treatment* (e.g., pre-reduction is sometimes necessary). In some cases, the continuous addition of unstable or dangerous substrate(s) should be considered.

1.2.4.3 Apparatus and Procedures

As already mentioned, hydrogenation reactions require special equipment. We can distinguish several levels of sophistication concerning pressure range, pressure control, temperature control, agitation, and measurement of hydrogen consumption.

Preparative reactions at normal pressure can be carried out using two-necked round bottom flasks with a magnetic stirrer. The hydrogen can be provided either from a hydrogen-filled balloon or a gas burette that allows the hydrogen consumption to be measured (for details see Loewenthal and Zass [5]). Pressures up to 4 bar and measurement of H_2 uptake can be handled with the well-known and reliable Parr Shaker, supplied by Labeq [11a]. Temperature control is not very good. Prices are in the order of $3000.

For higher pressures we would recommend the construction of special hydrogenation equipment with the necessary safety installations (rupture disc, expansion vessel, maybe reinforced cubicle, etc.). Depending on the size and construction material of the autoclave, the safety installations, and the accuracy of the measurement of hydrogen consumption, the price for such a system is between $20000 and $100000. Suppliers are Autoclave Engineers [11b], Büchi [11c], and others. We would also strongly recommend consulting colleagues who have practical experience with the set-up and running of a hydrogenation laboratory.

1.2.5 Selected Transformations

1.2.5.1 Hydrogenation of Aromatic Nitro Groups

Hydrogenation with heterogeneous catalysts is in most cases the method of choice for the conversion of aromatic nitro compounds to the corresponding anilines. Whereas the hydrogenation of simple nitroarenes poses little selectivity problems and is indeed carried out on very large scale, the situation is different if other reducible functional groups are present in the molecule. For details, we would like to refer the reader to Tab. 2 and to chapters in hydrogenation monographs [1a, 2a, 3a, 12a, 13a, 14a, 15a]. Several aspects are of importance and are discussed in detail. *Catalyst*: most metals are active, choice depends on the desired selectivity; *solvent*: alcohols often preferred, hydrocarbons and many others possible; *reactor type*: good agitation and effective cooling are essential. Other issues are the effect of pressure, impurities, or modifiers on rate and selectivity; the formation of desired or undesired intermediates (hydroxylamines, azo, azoxy, and hydrazo derivatives); and the mechanism of the desired hydrogenation and of side reactions. Safety and handling considerations are especially critical since nitro com-

Tab. 2 Selectivity profiles for the hydrogenation of aromatic nitro groups

Metal	*Function to be retained*					
	Ar-Hal [e]	*C≡C*	*C=C* [6]	*C=O*	*C≡N*	*Y* [g]*-Benzyl*
Pd	+**B** p520 –**B** p520	–**A** p109	±**C** p193 **B** p519	+**B** p528 **C** p194	+**B** p531 **C** p198	±**C** p200
Pt	+**A** p108 +[a), b)] **B** p520	+[a), b)] [16a]	+**C** p193 +[a), b)] [16a]	+**B** p528 +[a), b)] [16a]	±**A** p110 +[a), b)] [16a]	+**C** p198
Ru	±[b)] **B** p521 ±[d)] **B** p522	+**A** p109				+**B** p531
Rh	+**B** p520					
Ni	+**B** p522 +[b)] [21]		+**C** p193 ±**B** p518	+**B** p528 **C** p194	+**B** p531 **C** p196	+**C** p199 **B** p531

+ = selective; ± = partially selective; – = unselective.

a) Modified with second metal.
b) Nonmetallic modifier.
c) Metal sulfides or sulfided metal on support.
c) Hydrogen transfer process.
d) Rate of dehalogenation I > Br > Cl > F.
e) Rate of double bond hydrogenation: Mono > di > tri > tetra substituted.
f) Y = N, O.
g) No examples found.

References: A: Rylander [1], B: Houben-Weyl [3], C: Freifelder [2].

pounds are high energy content starting materials and because some intermediates and products are carcinogenic.

Here we will discuss in some more detail recent progress concerning chemoselectivity and hydroxylamine accumulation [17].

1.2.5.1.1 **Chemoselectivity**

Two novel catalyst systems were found to be selective for the selective hydrogenation of aromatic nitro groups tolerating functional groups such as C ≡ C, C=C, C ≡ N, C=N or C–Hal [16a, 17]:

- Pt/C catalysts, modified by H_3PO_2 and other low valent phosphorous additives and promoted by vanadium compounds highly effective in apolar solvents.
- Pt-Pb/$CaCO_3$ catalysts in the presence of small amounts of $FeCl_2$ and tetramethylammonium chloride were shown to be suitable for polar solvents.

Both catalyst systems work with commercially available components, have a wide substrate scope as depicted in Fig. 2, and are applied on a technical scale for several medium- to large-scale products.

1.2.5.1.2 **Hydroxylamine Accumulation**

Accumulation of hydroxylamines is problematic because of their potential for exothermic decomposition, their toxicity and their ability to form colored condensation products leading to quality problems. The suppression of hydroxylamine ac-

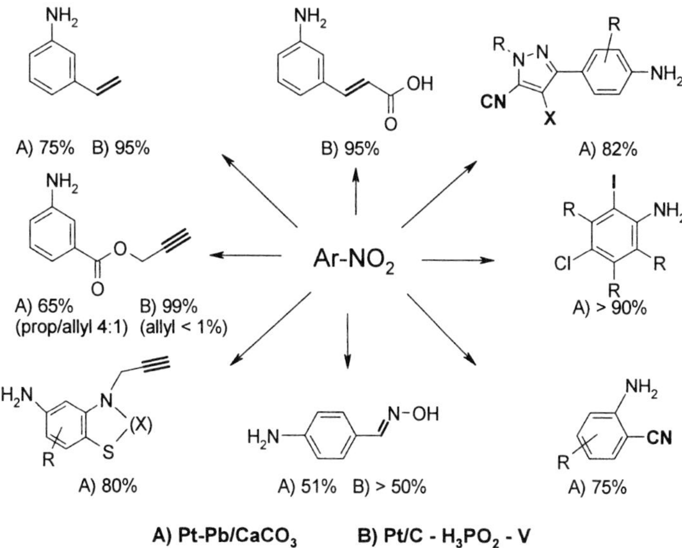

Fig. 2 Scope of the modified Pt catalysts (yields not optimized).

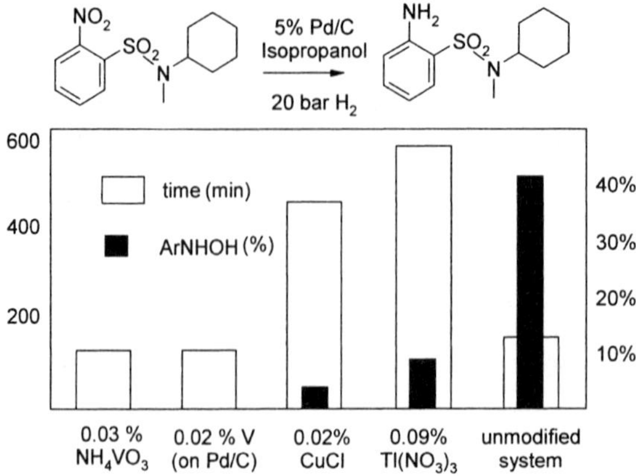

Fig. 3 Effect of the addition of promoters to Pd/C on hydrogenation time and maximum hydroxylamine accumulation.

cumulation is therefore a topic of industrial importance. Two recent publications described the addition of small amounts of metal, especially vanadium compounds to commercial Pt, Pd, and Ni catalysts [18, 19] leading to a dramatic decrease in the hydroxylamine accumulation, often to below 1% (for an example see Fig. 3). In addition, for Pd and Pt catalysts, the overall reaction with vanadium promoter was usually faster [18]. For Ni catalysts, the choice of the promoter is more difficult, and in some cases lower rates were observed [19]. The reaction products obtained with efficient promoters were whiter (cleaner) than the ones without. A mechanism called "catalytic by-pass" was proposed to explain the observed effects, whereby the vanadium promoters catalyze the disproportionation of the aryl hydroxylamine to give aniline and the nitroso intermediate that re-enters the catalytic cycle. As a consequence, the hydroxylamine accumulation is avoided and the aniline formation is accelerated.

1.2.5.2
Hydrogenation of Ketones

In the organic synthesis laboratory, ketones are usually reduced to alcohols by metal hydride reagents. Nevertheless, catalytic hydrogenation is the method of choice for diastereoselective reductions where H_2-addition occurs from the less hindered side; for reducing ketones on a larger scale; for the reduction of aromatic ketones to the corresponding methylene group; and for enantioselective reduction of α- and β-ketoesters using cinchona-modified Pt/Al_2O_3 or tartrate-modified Raney nickel catalysts, respectively [8]. Details of the hydrogenation of carbonyl groups can be found in Tab. 3 and in the hydrogenation monographs [1b, 2b, 3b, 12b, 13b, 14b, 15b]. Preferred *catalysts* are Pd, Pt and Ni; the structure of the ketone has a strong effect on rate and selectivity; and the chemo-, regio- and stereoselectivity can be controlled by catalyst, solvent, pH, modifiers, and the reaction conditions.

Tab. 3 Selectivity profiles for the hydrogenation of aldehydes and ketones

Metal	*Function to be retained*					
	Ar-Hal [e]	$C\equiv C$ [h]	$C=C$ [f]	$C\equiv N$	$ArNO_2$ [h]	Y[g]-*Benzyl*
Pd	±**B** p210 ±[b] **C** p307		±[a] **B** p224	±**C** p305		±**C** p306
Pt	+[a] **B** p210 +[a] **C** p307		+**B** p218 +[a] **B** p218			+**C** p306 +**B** p213
Ni	+**B** p210 +**C** p307		+[a] [16b] ±**B** p219	±**C** p305		±**C** p306 −**B** p213

Remarks and references see Tab. 2.

1.2.5.3
Hydrogenation of Alkenes

Usually, olefins are hydrogenated very easily with a wide variety of heterogeneous catalysts. However, for chemo- and especially for enantioselective hydrogenation, homogeneous catalysts are usually preferred. For details about hydrogenation of olefins with heterogeneous catalysts, we would refer to Tab. 4 and the monographs [1c, 2c, 3c, 12c, 13c, 14c, 15c, 20a]. Described in detail are the choice of catalysts; the mechanism of double bond hydrogenation (Horiuti and Polanyi); the problem of double bond migration and isomerization (effect of catalyst, substrate, hydrogen availability and reaction conditions); ways to influence the chemo-, regio- and stereoselectivity (catalysts, pressure, modifiers, solvent); and the effect of olefin structure on rate and selectivity.

Tab. 4 Selectivity profiles for the hydrogenation of alkenes

Metal	*Function to be retained*					
	Ar-Hal [e]	$C\equiv C$	$C=O$	$C\equiv N$	$ArNO_2$ [h]	Y[g]-*Benzyl*
Pd	+[b] [22] ±**C** p160		+[b] **A** p40 +**B** p161	+**B** p168 +**C** p157		+ [24] + [25]
Pt	+**C** p159	+ [23]	±**A** p161	+**B** p168 +**C** p157		+ [26] + [27]
Ni	+**C** p160		+**B** p161	−**B** p168		+**C** p158 + [28]

Remarks and references see Tab. 2.

1.2.5.4
Hydrogenation of Aromatic Rings

Heterogeneous catalytic hydrogenation is the method of choice for the reduction of carbocyclic and heterocyclic aromatic rings. However, depending on the type of aromatic ring system, the ease of reduction varies considerably. Most functional groups except carboxy functions are usually hydrogenated prior to the aromatic rings. Details can be found in [1 d, 2 d, 3 d, 12 d, 13 d, 14 d, 15 d, 20 b], where the following aspects are emphasized: type of catalyst (usually Rh, Ru, or Pt); the mechanism of the ring hydrogenation and side reactions like hydrogenolysis of substituents (halogen, hydroxy, alkoxy, amino); the considerable effect of ring type and substituents on rate and selectivity; ways to influence the chemo-, regio- and stereoselectivity (catalysts, solvents, pH, reaction conditions); and methods to obtain partially hydrogenated rings.

The asymmetric hydrogenation of (hetero) aromatic rings is an attractive way to chiral (hetero) cyclohexanes. While there are no successful examples of enantioselective reactions, the diastereoselective hydrogenation of carbocyclic or heterocyclic systems coupled to chiral auxiliaries such as proline or related compounds gave *de* values up to 95% [29–33] (for selected examples see Tab. 5). Usually, supported Rh catalysts show better performance than Ru catalysts, but in all cases laborious optimization and sometimes additives were necessary for good results. The issue of *cis* selectivity in the hydrogenation of disubstituted heterocyclic [34] and carbocyclic [35–37] rings was addressed by several groups. Usually after extensive process optimization, classical catalysts such as Rh and Raney Nickel were able to

Tab. 5 Diastereoselective hydrogenation of aromatic rings

Reaction	*Diastereoselectivity*	*Ref.*
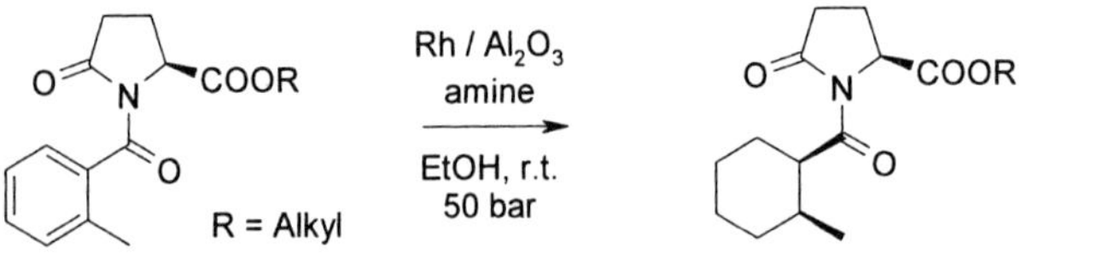 COOR, R = Alkyl; Rh / Al_2O_3, amine, EtOH, r.t., 50 bar → COOR	de up to 96% yield > 90%	28
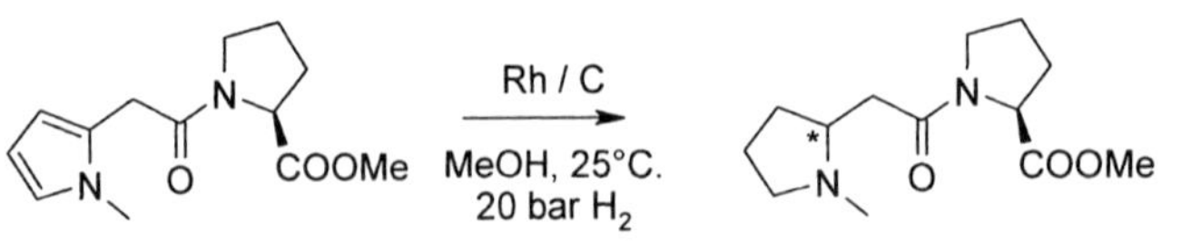 COOMe; Rh / C, MeOH, 25°C, 20 bar H_2 → COOMe	de up to 95%	31
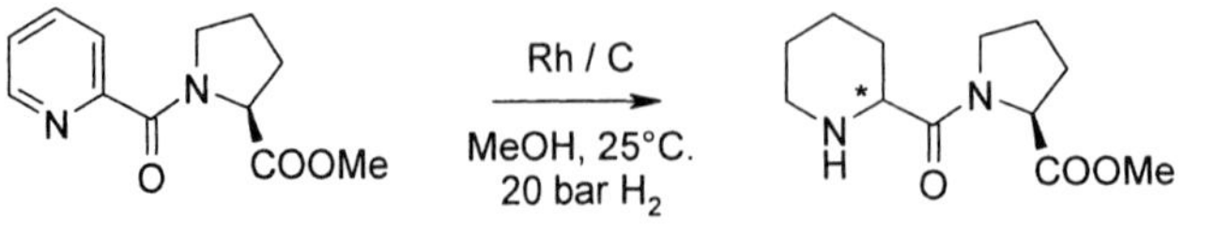 COOMe; Rh / C, MeOH, 25°C, 20 bar H_2 → COOMe	de 27%	32, 33

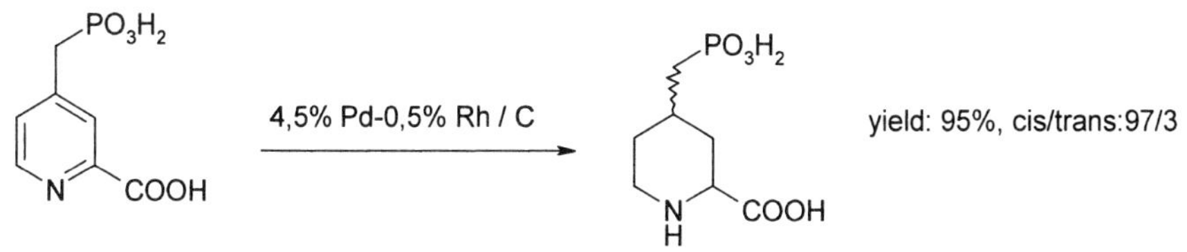

$$\text{4-(CH}_2\text{PO}_3\text{H}_2\text{)-pyridine-2-COOH} \xrightarrow{\text{4,5\% Pd-0,5\% Rh / C}} \text{4-(CH}_2\text{PO}_3\text{H}_2\text{)-piperidine-2-COOH}$$

yield: 95%, cis/trans:97/3

Fig. 4 Stereoselective hydrogenation of a substituted pyridine.

give satisfactory *cis* selectivities, but in some cases bimetallic systems [34, 35] proved to be superior.

A remarkable example of the synergism of bimetallic catalysts is the hydrogenation of pyridine-2-carboxylic acid derivatives as shown in Fig. 4. Surprisingly, a 4.5% Pd-0.5% Rh/C catalyst is twice as active as a 5% Rh/C catalyst and, in addition, shows better *cis* selectivity [34].

1.2.5.5
Catalytic Debenzylation

N- and *O*-benzyl groups are among the most useful protective groups in synthetic organic chemistry, and the method of choice for the removal of benzylic protecting groups is catalytic hydrogenolysis [1e, 2e, 3e, 38, 39]. Greene et al. [38], for example, list more than 20 different benzyl-type groups used for the protection of alcohols, phenols, esters, amines and amides. Usually, the hydrogenolysis is carried out under mild conditions and is quite selective. However, in multifunctional molecules selectivity and activity problems can be encountered. Even though there are many reports of selective debenzylations, generalization is not easy. From the rather empirical knowledge available, we have tried to extract the useful concepts and methods for obtaining high selectivity and activity.

1.2.5.5.1 Catalysts and Reaction Parameters

Many factors influence rate and selectivity of a debenzylation reaction: the nature of the benzyl group, the character of the protected group, the basicity of the substrate or product, steric and electronic effects, the type and amount of catalyst, solvents, modifiers, and the reducing agent. In the following paragraphs, we will describe the effects of these factors and discuss some mechanistic ideas. Most of the discussion will be restricted to the removal of *O*- and *N*-benzyl groups.

Catalysts

In most cases the catalysts of choice for both *N*-benzyl and *O*-benzyl groups are supported Pd catalysts that combine high activity for hydrogenolysis with a low tendency for the reduction of aromatic rings. The best catalysts seem to be 5–20% Pd/C catalysts with unreduced or oxidic metal present. High Pd concentrations are often beneficial even though the dispersion of the Pd becomes lower. Good results have been obtained with the original [40] and a carefully washed [41] Pearlman catalyst (20% $Pd(OH)_2/C$) even when other methods have failed. For the stepwise removal of different *O*-benzyl groups in a carbohydrate derivative, Pd/Al_2O_3 was more selective than Pd/C [42]. If dehalogenation is to be avoided, platinum or rhodium catalysts may be useful, but there is always the risk of ring saturation with these catalysts.

Reducing agent
Molecular hydrogen is the favorite hydrogen source for catalytic debenzylation. Most of the reactions are carried out at 1–3 bar hydrogen pressure. However, there are numerous reports describing hydrogen transfer reactions with donors like cyclohexene, cyclohexadiene, ammonium formate, or 2-propanol, often with good selectivity [7].

Solvents, modifiers, and promoters
Debenzylation reactions are very often carried out in alcohols and acetic acid. Non-protic solvents like THF [43] or toluene are also suitable, but the catalyst activity is sometimes lower. Mixtures of toluene and *o*-chlorotoluene seem to improve the selectivity for *N*-debenzylation versus hydrodechlorination. Sulfuric, nitric, and weak carboxylic acids like acetic acid promote debenzylation. Chemoselectivity can mainly be influenced by modifying the classical Pd/C catalysts. Amines can both promote and impede hydrogenolysis. We have found that the water content of the solvent frequently affects the activity of the catalyst. Recently, modification with ethylene diamine was shown to allow the selective removal of benzyl ethers while the *N*-Cbz (*N*-COOBn) group survived the hydrogenation of a variety of functional groups [44]. The addition of 2,2′-dipyridyl permitted the selective deprotection of *N*-Cbz and benzyl ethers in the presence of ArO–Bn groups [45], and with a Pd/C pyridine an ArO–Bn bond was cleaved in presence of an ArO-pOMeBn group [46].

1.2.5.5.2 Selective Removal of *O*-Benzyl Groups

Different *O*-benzyl groups
Most X-O-benzyl groups are removed very readily in neutral or acidic solutions. The rate of debenzylation increases in the order [1e] X=OH<*O*-alkyl<*O*-aryl<OH^+-alkyl<OH_2^+<OAc<$OCOCF_3$, i.e., with increasing electronegativity of the leaving group. The number of substituents on the benzylic carbon can influence the relative rate of debenzylation (Ph-CH_2-OH<Ph-CHR-OH>Ph-CR_2-OH (R=Aryl) [3e]. In monosaccharides, the reactivity strongly depends on the position of the benzyl group in the sugar.

O-benzyl in the presence of N-benzyl groups

O-benzyl groups are generally found to be somewhat easier to cleave than N-benzyl groups. There are exceptions, however, and modifiers – especially acids and organic bases – can reverse the selectivity. Seif et al. [47] described the influence of HCl or *n*-butyl amine on the debenzylation of *N,O*-dibenzyl-*p*-aminophenol in methanol. HCl strongly promoted N-debenzylation, whereas with *n*-butyl amine the O-benzyl was removed much faster than the N-benzyl. Bernotas and Cube [48] found rapid N- and no O-debenzylation for *N,O*-dibenzyl-1,2-aminoethanol with Pearlman's catalyst.

O-benzyl in the presence of other reducible functions

Other functional groups that are not easily hydrogenated by Pd usually survive debenzylation. Examples are the selective cleavage of benzyl esters in the presence of C=O bonds in aliphatic ketones and aldehydes, and nitriles [2e]. Selective debenzylation in the presence of halogens is no problem with aliphatic halogens. Aryl chlorides can be preserved if the substrate is only a weak base or neutral, or if the reaction is carried out in acidic medium [39]. Selective O-debenzylation in the presence of C=C, NO_2, C≡C, Ar-Br and Ar-I is difficult. Because of the low debenzylation activity of other metals, it is possible to hydrogenate many types of functional groups in the presence of O-benzyl groups, e.g., using Ni, Pt or Rh catalysts (see Tab. 1).

1.2.5.5.3 Selective Removal of N-Benzyl Groups

Different N-benzyl groups

Generally, the rate of N-debenzylation increases in the order quaternary ammonium salt > tertiary > secondary > primary N-benzyl group. If two or more benzyl groups are attached to a single nitrogen, stepwise removal is often possible [39, 49, 50]. This allows the synthesis of mixed secondary and tertiary amines by a debenzylation/alkylation sequence [49]. As above, differentiation between two benzyl groups attached to the same amid nitrogen is possible. N-benzyl amines can be selectively cleaved in the presence of N-benzyl amides.

In the presence of other reducible functional groups

Like for O-benzyl groups, the removal of N-benzyl groups is possible in the presence of aromatic halides (especially Cl and F), C=O (aliphatic and aromatic) and C≡N bonds. By adding acidic modifiers like HCl or acetic acid, the selectivity for N-debenzylation can be improved in the presence of halogens. On the one hand the reaction is accelerated by protonation of the nitrogen ("quaternization"), and on the other hand the removal of the halogen X is slowed down by the lack of an acceptor for H-X [39].

Since N-debenzylations are usually more difficult than O-debenzylations, the selective deprotection in the presence of C≡C or NO_2 groups is even more difficult, and no examples were found in an extensive literature search. C=C bonds are only known to survive an N-debenzylation if they are highly substituted and conjugated. Bornmann and Kuehne [51] for instance described the deprotection of a molecule also containing an α,β-unsaturated ester function that remained intact.

1.2.5.5.4 New Protecting Groups

There is strong interest in protective groups which can be removed selectively and easily. Fig. 5 depicts a series of recently published structures with interesting properties. 1-NAP and 4-QUI esters [52] have been cleaved with a homogeneous Pd complex and a formate donor. Benzyl esters, olefins, Ar-Br and other functional groups are tolerated. The highly selective removal of 1-NAP from *N*- and *O*-functions without affecting Bn and CFTB groups was reported for Pd/C–H_2 [53]. MPM-OAr groups survived the deprotection of Bn-OAr and CbzNH with Pd/C modified with pyridine but could easily be removed in absence of the pyridine modifier [54]. Tagging with fluoros benzyl groups allowed a clever combination of protection and fluoros phase chemistry with easy subsequent removal of the auxiliary group [55]. BOB-protected hydroxy groups were deprotected via hydrogenolysis/lactonization compatible with a number of fatty acid esters [56].

1.2.5.6 Chemoselective Hydrogenation of Nitriles

The hydrogenation of nitriles is one of the basic methods of obtaining primary amines, and diamines in particular are of high industrial importance. Unfortunately, the literature is rather scattered, the most up-to-date review having been written in 1994 [57]. We focus our summary of recent results on selectivity in favor of primary amines, catalyst deactivation, and functional group tolerance.

As well as primary amines, secondary and tertiary amines can be formed via condensation of reaction intermediates, and control of this chemoselectivity problem is one of the main issues of nitrile hydrogenation. Addition of ammonia is most widely used to improve the selectivity in favor of primary amines [57], but recently it was reported that less toxic strong bases such as NaOH [58, 59] and LiOH [60] are also effective for Raney Ni and Co catalysts. The OH^- ions not only prevent catalyst deactivation by inhibiting polyamine formation on the catalyst

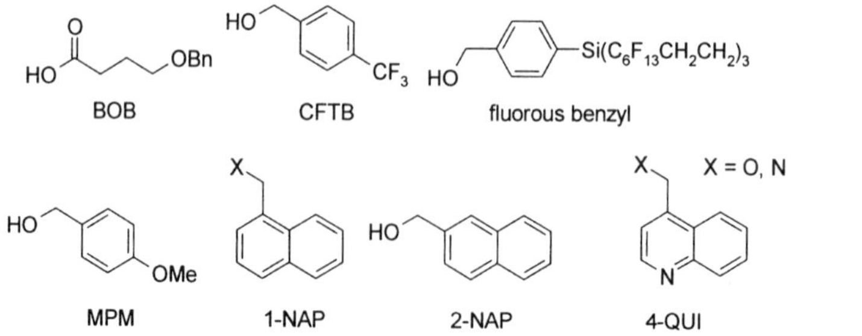

Fig. 5 New protective groups which can be removed via selective hydrogenolysis.

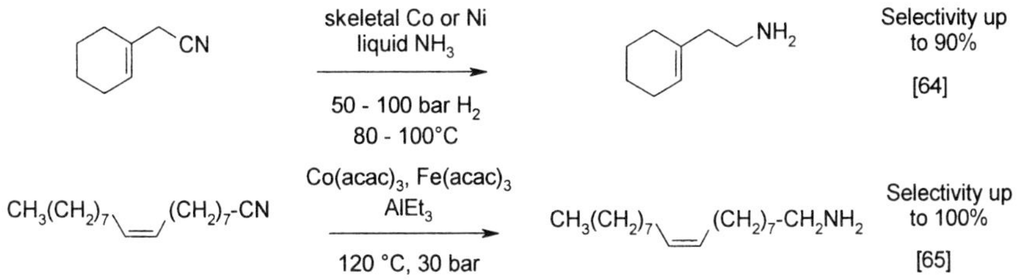

CN — skeletal Co or Ni, liquid NH_3, 50 - 100 bar H_2, 80 - 100°C → NH_2 — Selectivity up to 90% [64]

$CH_3(CH_2)_7$–CH=CH–$(CH_2)_7$-CN — $Co(acac)_3$, $Fe(acac)_3$, $AlEt_3$, 120 °C, 30 bar → $CH_3(CH_2)_7$–CH=CH–$(CH_2)_7$-CH_2NH_2 — Selectivity up to 100% [65]

Fig. 6 Chemoselective hydrogenation of CN bonds in the presence of a C=C bond.

surface for dinitrile hydrogenation [58], but also seem to block active sites responsible for by-product formation [59]. Pre-treatment of Ni and Co catalysts with CO, CO_2, aldehydes, or ketones also gave significantly less secondary amines [61].

For fine chemicals applications, functional group tolerance is an important issue. Substituents like aryl groups, benzylic functions or C-Hal are usually not reduced with skeletal Ni or Co catalysts. More difficult to conserve are heteroaromatic or heteroaryl-halogen functions, ketones, aldehydes, or a second CN group, but with the proper catalyst, solvent, and additives, success is often possible [62]. In contrast, the selective hydrogenation of CN groups in the presence of C=C bonds has long been an unsolved problem, particularly if they are conjugated or in close proximity in the molecule [62, 63]. If the C=C bond is sterically hindered [64] then high selectivity is possible in liquid ammonia, which not only inhibits the formation of secondary amines, but also improves the selectivity to the unsaturated amine, probably by forcing its desorption. Another case of chemoselective nitrile hydrogenation has been described for a fatty acid nitriles [65], where the selective hydrogenation of remote CN functions is possible with high selectivity applying a Ziegler-type Co-Fe catalyst even in the absence of NH_3 (see Fig. 6).

1.2.6 Conclusions and Outlook

Heterogeneous hydrogenation has developed to a rather mature methodology for both laboratory and industrial applications. Today, many commercial catalysts are available for a broad variety of different hydrogenation reactions. While research in heterogeneous hydrogenation was very active in the 1970s and 1980s, culminating in the well-known monographs by Rylander, Augustine, and Smith, only a few selected topics have received significant attention in recent years, as described in Section 1.2.5. This is in contrast to homogeneous hydrogenation, where especially enantioselective catalysts are a very hot topic, but where only a few industrial applications are known up to now. We are of the opinion that this situation will not change very quickly and that selective heterogeneous hydrogenation will continue to be a reliable method for synthesis planning and will play an even more important role in the manufacture of fine chemicals.

References

1 P.N. Rylander, *Hydrogenation Methods*, Academic Press, Bench top Edition, New York, **1990**. (a) p. 104; (b) p. 66; (c) p. 29; (d) p. 117; (e) p. 158.

2 M. Freifelder, *Practical Catalytic Hydrogenation*, Wiley-Interscience, New York, **1971**. (a) p. 168; (b) p. 282; (c) p. 127; (d) p. 168, 254; (e) p. 398.

3 Houben-Weyl, *Methoden der Organischen Chemie, Reduktionen I, Vierte Auflage, Band IV/1c*, Georg Thieme, Stuttgart, **1980**. (a) p. 490; (b) p. 189; (c) p. 145; (d) p. 254, 543; (e) p. 379.

4 G.V. Smith, F. Notheisz, *Heterogeneous Catalysis in Organic Chemistry*, Academic Press, San Diego, **1999**; S. Nishimura, *Handbook of Heterogeneous Catalytic Hydrogenation for Organic Synthesis*, Wiley, New York, **2001**.

5 H.J.R. Loewenthal, E. Zass, *A Guide for the Perplexed Organic Experimentalist*, John Wiley & Sons, Salzburg, **1990**.

6 H.U. Blaser, "Reactions at Surfaces: Opportunities and Pitfalls for the Organic Chemist" in *Modern Synthetic Methods* (Eds. B. Ernst, Ch. Leumann), Verlag Helvetica Chimica Acta, Basel, **1995**, p. 179.

7 R.A. Johnstone, A.H. Wilby, *Chem. Rev.* **1985**, *85*, 129.

8 H.U. Blaser, H.P. Jalett, M. Müller, M. Studer, *Catal. Today* **1997**, *37*, 441; T. Osawa, T. Harada, O. Takayasu, *Top. Catal.* **2000**, *13*, 155.

9 (a) Degussa AG, Geschäftsbereich Anorganische Chemieprodukte, Postfach, 63450 Hanau, Germany; (b) Engelhard de Meern B.V., Catalysts and Chemical Division, PO Box 19, 3454 ZG De Meern, The Netherlands; (c) Heraeus, Chemical Catalysts, Postfach 1553, 63450 Hanau 1, Germany; (d) Johnson Matthey, Process Catalysts, Orchard Road, Royston, Hertfordshire SG8 5HE, England.

10 (a) Aldrich Handbook of Fine Chemicals and Laboratory Equipment, Sigma-Aldrich Co.; (b) Fluka Laboratory Chemicals, Fluka Chemie AG, CH-9470 Buchs, Switzerland.

11 (a) Labeq, Laboratory Equipment AG, 8006 Zürich, Switzerland; (b) Autoclave Engineers Europe, F-Nogent sur Olle, Cedex, France; (c) Büchi AG, 8610 Uster, Switzerland.

12 P.N. Rylander, *Catalytic Hydrogenation in Organic Syntheses*, Academic Press, London, **1979**. (a) p. 258; (b) p. 81; (c) p. 309.

13 P.N. Rylander, *Catalytic Hydrogenation over Platinum Metals*, Academic Press, London, **1967**. (a) p. 169; (b) p. 189;(c) p. 145; (d) p. 254, 543.

14 R.L. Augustine, *Heterogeneous Catalysis for the Synthetic Chemist*, Marcel Dekker, Inc., New York Basel Hong Kong, **1996**. (a) p. 104; (b) p. 66; (c) p. 29; (d) p. 117.

15 F. Zymalkowski, *Katalytische Hydrierung*, Ferdinand Enke, Stuttgart, **1965**. (a) p. 234; (b) p. 91; (c) p. 40; (d) p. 178.

16 (A) P. Baumeister, H.U. Blaser, U. Siegrist, M. Studer, *Chem. Ind. (Dekker)* **1998**, *75*, 207. (b) P.S. Gradeff, G. Formica, *Tetrahedron Lett.* **1976**, *51*, 4681.

17 For more details see H.U. Blaser, U. Siegrist, H. Steiner, M. Studer in *Fine Chemicals through Heterogeneous Catalysis* (Eds.: R.A. Sheldon, H. van Bekkum), Wiley-VCH, Weinheim, **2001**, p. 389.

18 P. Baumeister, H.-U. Blaser, M. Studer, *Catal. Lett.* **1997**, *49*, 219.

19 M. Studer, S. Neto, H.U. Blaser, *Top. Catal.* **2000**, *13*, 205.

20 M. Bartók, *Stereochemistry of Heterogeneous Metal Catalysis*, John Wiley & Sons, Chichester, **1985**. (a) p. 53; (b) p. 251, 435, 469.

21 P. Baumeister, H.U. Blaser, W. Scherrer, *Stud. Surf. Sci. Catal.* **1991**, *59*, 321.

22 K. Kindler, H. Oelschläger, P. Heinrich, *Chem. Ber.* **1953**, *86*, 167.

23 C.J. Palmer, *Tetrahedron Lett.* **1990**, *31*, 2857.

24 L.A.M. Bastiaansen, P.M. van Lier, E.F. Godefroi, *Org. Synth., Coll. Vol.* **1990**, *7*, 287.

25 L. Jurd, G.D. Manners, *Synth. Commun.* **1980**, 618.

26 P.D. Leeson, B.J. Williams, R. Baker, T. Ladduwahetty, K.W. Moore, M. Row-

LEY, *J. Chem. Soc., Chem. Commun.* **1990**, 1578.

27 F. DININNO, *J. Am. Chem. Soc.* **1978**, *100*, 3251.

28 T. HANAYA, K. YASUDA, H. YAMAMOTO, H. YAMAMOTO, *Bull. Chem. Soc. Jpn.* **1993**, *66*, 2315.

29 M. BESSON, F. DELBECQ, P. GALLEZOT, S. NETO, C. PINEL , *Chem. Eur. J.* **2000**, *6*, 949.

30 V.S. RANADE, R. PRINS, *J. Catal.* **1999**, *185*, 479; V.S. RANADE, G. CONSIGLIO, R. PRINS, *Catal. Lett.* **1999**, *58*, 71; P. KUKULA, R. PRINS, *J. Catal.* **2002**, *208*, 404.

31 V. HADA, A. TUNGLER, L. SZEPESY, *Appl. Catal. A.* **2001**, *210*, 165.

32 L. HEGEDUS, V. HÁDA, A. TUNGLER, T. MÁTHÉ, L. SZEPESY, *Appl. Catal. A* **2000**, *201*, 107.

33 The *de* value of 98% given in ref. 78b had to be corrected; A. TUNGLER, PERSONAL COMMUNICATION.

34 H. STEINER, P. GIANNOUSIS, A. PISCHE-JACQUES, H.U. BLASER, *Top. Catal.* **2000**, *13*, 191.

35 R. BURMEISTER, A. FREUND, P. PANSTER, T. TACKE, S. WIELAND, *Stud. Surf. Sci. Catal.* **1995**, *92*, 343.

36 F. ROESSLER, H. HILPERT, *Proceedings 12th International Congress on Catalysis*, **2000**, CD-ROM, R124.

37 T.Q. HU, B.R. JAMES, J.S. RETTIG, C.-L. LEE, *Can. J. Chem.* **1997**, *75*, 1234.

38 T.W. GREENE, P.G.M. WUTS, *Protective Groups in Organic Synthesis, Second Edition*, John Wiley & Sons, Inc., New York, **1991**.

39 M. STUDER, H.U. BLASER, *J. Mol. Catal.* **1996**, *112*, 437.

40 W.M. PEARLMAN, *Tetrahedron Lett.* **1967**, *17*, 1663.

41 K. YOSHIDA, S. NAKAJIMA, T. WAKAMATSU, Y. BAN, M. SHIBASAKI, *Heterocycles* **1988**, *27*, 1167.

42 V.S. RAO, A.S. PERLIN, *Can. J. Chem.* **1983**, *61*, 652.

43 K.G. GRIFFIN, S. HAWKER, M.H. BHATTI, *Chem. Ind.* **1996**, *68*, 325.

44 K. HATTORI, H. SAJIKI, K. HIROTA, *Tetrahedron* **2000**, *56*, 8433 and references cited. H. SAJIKI, K. HATTORI, K. HIROTA, *Chem. Eur. J.* **2000**, *6*, 2200.

45 H. SAJIKI, K. HIROTA, *Tetrahedron* **1998**, *54*, 13981.

46 H. SAJIKI, H. KUNO, K. HIROTA, *Tetrahedron Lett.* **1997**, *38*, 399.

47 L.S. SEIF, M. PARTYAKA, J.E. HENGEVELD, *Catalysis of Organic Reactions* (Ed.: D.E. BLACKBURN) *Chem. Ind.* **1991**, *40*, 197.

48 R.C. BERNOTAS, R.V. CUBE, *Synth. Commun.* **1990**, *20*, 1209.

49 P.W. ERHARDT, *Synth. Commun.* **1983**, *13*, 103.

50 L. VELLUZ, G. AMIARD, R. HEYMES, *Bull. Soc. Chim. Fr.* **1954**, 1012.

51 W.G. BORNMANN, M.E. KUEHNE, *J. Org. Chem.* **1992**, *57*, 1752.

52 A. BOUTROS, J.-Y. LEGROS, J.-C. FIAUD, *Tetrahedron* **2000**, *56*, 2239.

53 E.A. PAPAGEORGIOU, M.J. GAUNT, J. YU, J.B. SPENCER, *Org. Lett.* **2000**, *2*, 1049 and references cited.

54 H. SAJIKI, H. KUNO, K. HIROTA, *Tetrahedron Lett.* **1997**, *38*, 399.

55 D.P. CURRAN, R. FERRITTO, H. YE, *Tetrahedron Lett.* **1998**, *39*, 4937.

56 M.A. CLARK, B. GANEM, *Tetrahedron Lett.* **2000**, *41*, 9523.

57 C. DE BELLEFON, P. FOUILLOUX, *Catal. Rev.* **1994**, *36*, 459.

58 A.M. ALLGEIER, M.W. DUCH, *Chem. Ind. (Dekker)* **2001**, *82*, 229.

59 S.N. THOMAS-PRYOR, T.A. MANZ, Z. LIU, T.A. KOCH, S.K. SENGUPTA, W.N. DELGASS, *Chem. Ind. (Dekker)* **1998**, *75*, 195.

60 T.A. JOHNSON, D.P. FREYBERGER, *Chem. Ind. (Dekker)* **2001**, *82*, 201.

61 O.G. DEGISCHER, F. ROESSLER, *EP 1108469* (**2001**) assigned to F. Hoffmann La Roche AG.

62 P. TINAPP in *Methoden der organischen Chemie* (HOUBEN-WEYL, Reduktionen Teil 1) **1980**, 111.

63 G.D. YADAV, M.R. KHARKARA, *Appl. Catal. A: General* **1995**, *123*, 115.

64 W. POEPEL, J. GAUBE, *DECHEMA Monogr.* **1991**, *122*, 189.

65 B. FELL, J. SOJKA, *Fett Wiss. Technol.* **1991**, *93*, 79.

1.3
Transferhydrogenations

Serafino Gladiali and Elisabetta Alberico

1.3.1
Introduction

This chapter is intended to update the previous version which appeared in the first edition of this book and which covered the literature up to the end of 1997 [1]. Since that time, the importance of transferhydrogenation as a methodology for the reduction of unsaturated compounds has increased further. The number of papers and reports dealing with this subject which appeared in the period 1998–2002, the end date of the literature coverage of the present survey, is much greater than that in the previous five-year term, 1993–1997 [2]. The same holds true for the number of research groups which have entered this area of catalysis for the first time. Apparently, chemists have become more conscious of the potential of this technique and are more comfortable with its application in the reduction of organic compounds.

The research efforts during this period have led to significant advances in the development of new catalysts of higher activity/selectivity, in the understanding of the reaction mechanisms (particularly of the Ru-catalyzed reactions), in exploring unconventional approaches driven by green chemistry principles, and in exploiting the potential of enzyme-metal-coupled catalysis in kinetic dynamic resolution processes. These subjects are addressed in some detail in dedicated sections of this chapter.

As a natural consequence of the increased familiarity developed by chemists toward this synthetic tool, an asymmetric H-transfer reaction has been for the first time scaled up to the multi-kilo range. The time seems ripe for setting up practical applications of this reaction to the industrial synthesis of fine chemicals.

1.3.2
General Background

Hydrogen transfer reactions (H-transfer or transferhydrogenation) are those processes where hydrogen is moved from a hydrogen donor, DH_2, to a hydrogen acceptor, A or A-X, by the action of a suitable metal catalyst. The net result is the

Transition Metals for Organic Synthesis, Vol. 2, 2nd Edition.
Edited by M. Beller and C. Bolm

ISBN: 3-527-30613-7

addition of hydrogen to a multiple bond of an unsaturated substrate A to give the hydrogenated product AH_2 or, less frequently, the reductive cleavage of the A-X single bond of a substrate prone to hydrogenolysis to afford A-H and H-X. At the same time the hydrogen donor is converted into its oxidized counterpart D.

$$\text{H-D-H} + \text{A} \underset{}{\overset{\text{Metal}}{\rightleftharpoons}} \text{D} + \text{H-A-H}$$

$$\text{H-D-H} + \text{A-X} \xrightarrow{\text{Metal}} \text{D} + \text{A-H} + \text{H-X} \quad (1)$$

Obvious advantages of transferhydrogenation over catalytic hydrogenation by molecular hydrogen are the elimination of safety restrictions associated with the use of pressure vessels, and the reduction of risks.

Formation of a co-product is the main drawback of H-transfer processes compared with alternative methodologies for the reduction of organic compounds. As the co-product is produced in a stoichiometric amount, its presence in the reaction mixture can be the cause of undesired complications in the isolation of the reaction product and in the separation of the catalyst. Additionally, the co-product D can itself be a hydrogen acceptor, and if it is not removed it can compete with the substrate for the reduction until eventually an equilibrium is attained.

These limitations can be circumvented by selecting, where possible, a metal catalyst which can tolerate and activate formic acid, formates, or hydrazine as H-donors. In this case, the co-product (CO_2 or N_2) is removed from the catalyst as soon as it is formed, and the reaction is completely irreversible. When these conditions are fulfilled, H-transfer reduction can be hardly rivaled by alternative methodologies of hydrogenation.

1.3.2.1 Mechanism

The most significant advances in this area have been achieved in the case of the Ru-catalyzed transferhydrogenation of aryl alkyl ketones and imines, where independent investigations by different research groups have produced a body of results which, taken together, provide an additional mechanistic path for these processes.

In the case of Ru-mediated transferhydrogenations, it has been demonstrated that, depending on the structure of the non-transferable ligands coordinated to the metal, the reaction can proceed through different routes which involve a ruthenium monohydride or a ruthenium dihydride as the active catalytic species.

Ruthenium monohydride complexes such as **3** are the species delivering the hydrogen to the substrate when aminoalcohols or *N*-monoarylsulfonylated diamines are used as *N,N*- or *N,O*-bidentate chelating ligands in conjunction with arene ruthenium precatalysts such as $[RuCl_2(\eta^6\text{-arene})]_2$. A key intermediate in this process is the coordinatively unsaturated 16-electron complex **2** which originates

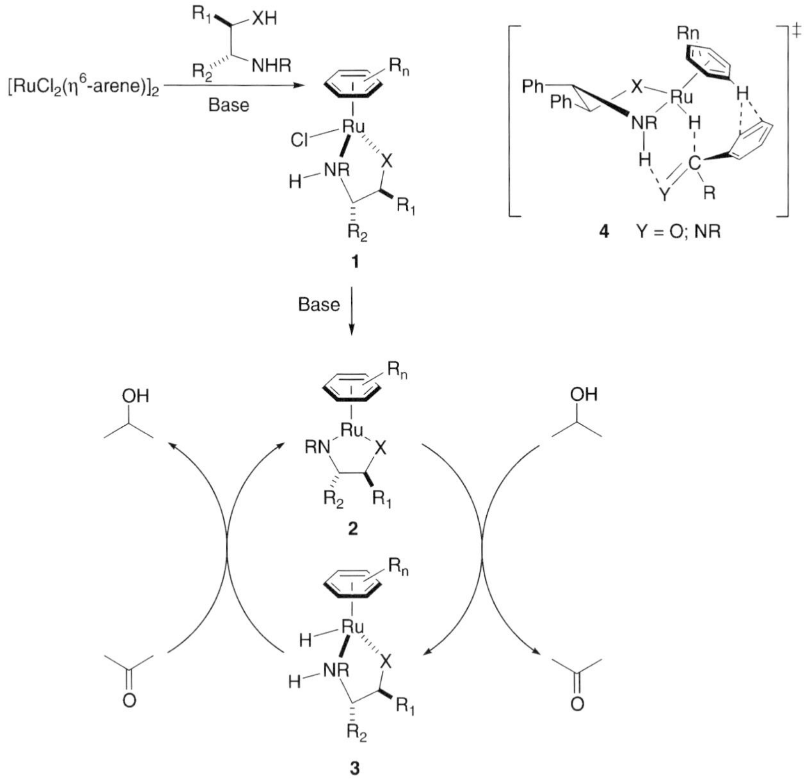

Scheme 1 Catalytic cycle of Ru-promoted H-transfer via metal-ligand bifunctional catalysis.

from the base-promoted elimination of hydrogen chloride from the 18-electron Ru complex **1**. Compound **2** reacts with the H-donor to afford the monohydride **3**, from which it is regenerated by dehydrogenation at every catalytic cycle. The basic steps of this mechanism are illustrated in Scheme 1.

The transfer of hydrogen takes place in a concerted process involving a six-membered pericyclic transition state **4**, where the two hydrogens, one from the metal and the other from the nitrogen, simultaneously move to the unsaturated functional group of the substrate. Hydrogen bond and dipolar interactions between the carbonyl group and the hydrogens on Ru and on N provide the conditions for the periplanarity of the transition state and for the appropriate docking of the substrate to the catalyst.

The most innovative aspect of this mechanism is that the reaction takes place in an outer coordination sphere of the metal and does not imply prior activation of the substrate by coordination to the metal center. Thus, formation of the C-H bond, the decisive step of the reduction, does not occur via a migratory insertion of the substrate into a metal-hydride bond, as envisaged in the classical "hydridic route" mechanism [3], but via "metal-ligand bifunctional catalysis", as originally proposed by

Noyori in the first papers on this subject [4]. This conclusion, originally drawn from experiments, was later corroborated by theoretical calculations performed at various levels of accuracy on catalysts of this type containing different model ligands [5]. Further convincing evidence for a concerted transfer of hydrogen occurring outside the coordination sphere of the metal from OH to N and from CH to Ru has recently been provided on the basis of kinetic isotopic effect [5 d].

The second main difference in the two mechanisms is that all the species involved in "metal-ligand bifunctional catalysis" feature anionic instead of neutral ligands in the coordination sphere of the metal. Thus, bidentate chelating ligands with a protonated donor center –XH of appropriate acidity (X=O or NY, where Y is an electron-withdrawing group) adjacent to a basic sp^3-nitrogen donor are necessary for this pathway to be enabled in transferhydrogenation. *N*-monosulfonylated diamines and β-aminoalcohols both meet this prerequisite. Treatment with a stoichiometric amount of base, KOH or similar, is enough to activate them to catalysis.

As a general trend, aminoalcohols provide catalysts of higher activity than diamine derivatives, and monosubstitution at the vicinal basic nitrogen exerts a positive effect on the asymmetric bias of the reaction. In contrast, *N,N*-disubstituted ligands are poorly suited for this chemistry and produce complexes which are practically devoid of any catalytic activity.

This metal-ligand bifunctional catalysis is quite efficient, and alkyl aryl ketones can be smoothly reduced to the relevant carbinols in high chemical and optical yields upon stirring at room temperature in 2-propanol in the presence of a catalytic amount of an Ru(II)-complex **1** and one molar equivalent of a base. Some of these catalysts can also tolerate formic acid/triethylamine as a reducing agent, and this allows asymmetric H-transfer reductions to be run at high substrate concentration with high conversions and with no erosion of optical purity.

The activity of these catalysts closely reflects the propensity of the substrates to accept hydrogen. This correlates well with the polarity of the double bond, carbonyl derivatives being reduced faster than imines and these much faster than alkenes.

The enantioselectivity of this reaction is basically addressed by the chiral geometry adopted by the five-membered chelate ring of the anionic ligand. This is dictated by the configuration of the stereocenters of the carbon backbone [6]. For phenyl-substituted 2-aminoalcohols or 1,2-diamines, the handedness of the 1-phenylethanol obtained from the reduction of acetophenone derives from the C(1) configuration, 1*R*-stereoisomers consistently giving the (*R*)-alcohol. An *R*-configuration at C(1) in phenyl-substituted ligands results in a λ conformation of the penta-atomic ring and in an *S* configuration at the Ru stereocenter. The presence of an additional substituent such as methyl at C(2) exerts a favorable effect on the stereoselectivity of the reaction irrespective of the configuration of the second stereocenter, either *S* or *R*.

The transfer of hydrogen from the relevant Ru-catalyst **3** to an aryl alkyl ketone or ketimine occurs preferentially to the sterically congested *Si*-enantioface as depicted in **4** (Scheme 1) rather than to the less encumbered *Re*-face. Most important for the asymmetric bias of the reaction is the CH/π attraction developed in the transition state between the arene ligand and the aryl group of the substrate.

This contributes to improvement of the stereoselectivity by stabilizing the more congested transition state in preference to the less crowded transition state and provides the driving force necessary to surmount the more sterically demanding reaction path [7].

Metal-ligand bifunctional catalysis is also operative in the H-transfer reductions promoted by the dinuclear ruthenium hydride **5**, generally referred to as the Shvo catalyst (Scheme 2) [8]. This complex is notable because it does not require any base for activation to catalysis and because it displays an intriguing catalytic profile. For these reasons, in addition to H-transfer reductions it has been successfully exploited in a range of different transferhydrogenation processes including Tischenko disproportionation [9], Oppenauer-type oxidation [10], and lipase-assisted dynamic kinetic resolution of secondary alcohols [11].

Partial dissociation of the dimer **5** in solution releases the two fragments **6** and **7**, which both play an active role in the catalytic cycle of transferhydrogenation (Scheme 2). The former one, **6**, is the reducing agent, and, upon delivering hydrogen to the unsaturated substrate, is converted into the 16-electron complex **7**, which, on its part, regenerates **6** upon extracting hydrogen from the hydrogen donor. The dimer **5**, which should be regarded as the resting state of the real catalyst, regulates the relative concentration of the active species **6** and **7** through the equilibrium constant.

In complex **6**, the hydroxy cyclopentadienyl fragment acts as an anionic ligand toward the metal, providing at the same time the acidic site necessary for assembling the substrate and for the bifunctional catalysis to operate.

The Shvo catalyst reacts with benzaldehyde 30 times faster than with acetophenone and reduces benzaldimines 25 times faster than it reduces benzaldehyde

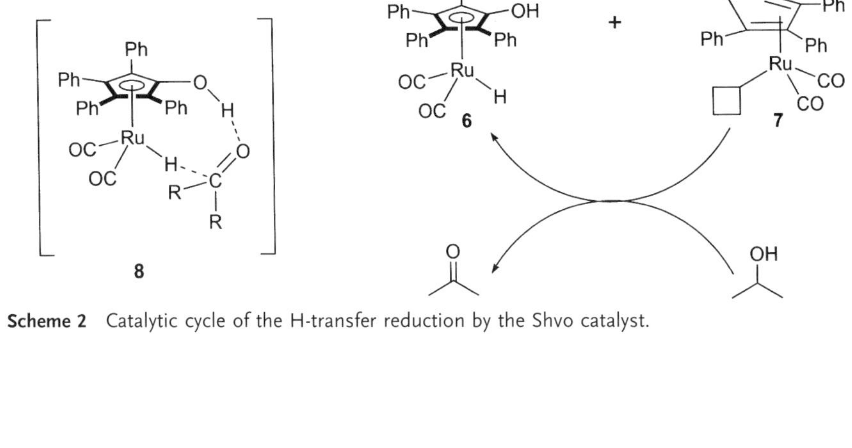

Scheme 2 Catalytic cycle of the H-transfer reduction by the Shvo catalyst.

[12]. This apparent dichotomy is accounted for by the higher electrophilicity of the formyl carbon and by the presence of a fairly acidic OH group, which provides a higher concentration of protonated substrate in the case of the imine. Detailed kinetic and isotopic labeling studies on H-transfer reductions by this complex are in keeping with the transition state depicted in **8** involving a concerted transfer of a proton from OH and of a hydride from Ru to the substrate [12].

When the ancillary non-transferable ligands in the coordination sphere of the metal are neutral and not anionic, then Ru-catalyzed hydrogen transfer proceeds through a different mechanistic path which is reliant on a Ru-dihydride instead of on a Ru-monohydride intermediate.

The case of the triphenylphosphino dihydride **11** is the most illustrative of this chemistry and has been investigated in detail (Scheme 3). The dihydride **11** is formed when the corresponding dichloride **9** is treated with base in isopropanol [13]. The Ru-dihydride readily reduces ketones to alcohols, whereas the corresponding hydrochloride **10**, which is produced as an intermediate compound, is unreactive. This observation, taken together with kinetic evidences [13], supports the view that the dihydride **11** is the real catalyst in the transferhydrogenation of ketones catalyzed by Ru-complexes featuring neutral ligands like triphenylphosphine in the coordination sphere.

Isotopic labeling studies of the $RuCl_2(PPh_3)_3$-catalyzed transferhydrogenation of ketones by secondary carbinols have shown that in this process the hydrogens transferred from the H-donor to the substrate do not retain their identity, but that they are scrambled between the two sites of the carbonyl group. This means that the migration of the hydrogens of the H-donor does not occur selectively, i.e. from the O-H and from the C-H of the donor, respectively, to the oxygen and to the carbon of the carbonyl group, as happens in the case of the monohydride mechanism.

For this scrambling to be accounted for, a mixed hydride-deuteride **16** must be involved in the step of the catalytic cycle where the hydrogens are transferred to the carbonyl group. The reaction sequence leading to the formation of **16** from the relevant dichloride and the mechanism of the H-transfer reduction mediated by Ru-dihydrides is illustrated in Scheme 3.

The dichloride **9** reacts with the deuterated alcohol **13** in the presence of base to afford selectively the dideuteride **12**. Addition of this deuteride to acetophenone gives an alkoxy ruthenium complex **14**, which undergoes reductive elimination to provide the highly reactive 14-electron Ru(0)-complex **15**. Oxidative addition of the O-H bond of C-deuterated alcohol **13** to **15** affords the alkoxy derivative **16a**, which, upon β-hydride elimination, provides the mixed hydride-deuteride **17**. Addition of this species to acetophenone leads to a mixture of alkoxy complexes **16a** and **16b**, which are the immediate precursors of the scrambled alcohol.

The question whether the Ru-catalyzed hydrogen transfer relies on a mono- or on a di-hydride intermediate can be addressed by monitoring the extent of the deuterium maintained in the *rac*-alcohol after the racemization of enantiopure α-deuterio-α-phenylethanol (**13**) by a suitable Ru-complex has been completed [14]. From the relevant mechanisms, it follows that all the deuterium should be re-

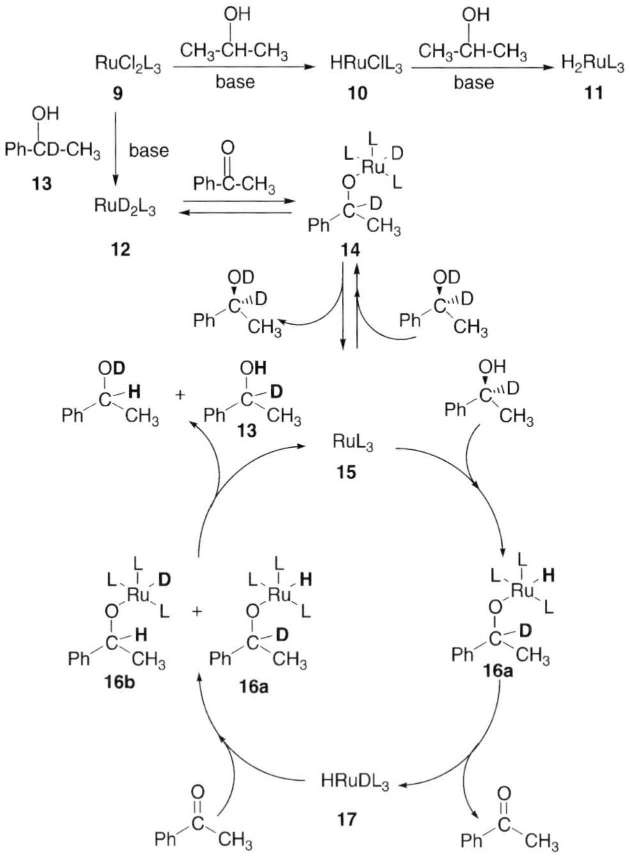

Scheme 3 Catalytic cycle of Ru-dihydride mediated H-transfer.

tained in the case of a monohydride-based reaction path, whereas reduction to about half should be observed with a dihydride-based mechanism. The results obtained on a wide range of complexes featuring different metals and different ligands are substantially consistent with this expectation and indicate that the monohydride mechanism is operating in the case of rhodium and iridium catalysts, while both mechanisms can operate with ruthenium, the choice being determined by the nature of the ancillary ligands [15].

Mechanistic investigations on metals other than ruthenium have produced less conclusive results. While the reduction of ketones by rhodium(III) and iridium(III) complexes with chiral *N*-monosulfonylated diamines **19** apparently depends on the same metal-ligand bifunctional catalysis operating with ruthenium [16], the conclusions are less clear-cut when other *N*-protonated ligands are used around d^8-metal ions. Computational studies on rhodium(I) complexes with diamine ligands support the intermediacy of a rhodium monohydride with one diamine and one ancillary cyclooctadiene ligand coordinated [17]. For the migration of the hydride from this intermediate to the substrate, both a two-step [18] and a concerted mechanism [19] have been validated by calculations.

In the case of Ir(I) complexes, a pronounced dependence of the stereoselectivity on the nature of the H-donor seems better accommodated with the coexistence of different competitive mechanisms [20], including the so-called "direct H-transfer" [1].

1.3.2.2 Hydrogen Donors and Promoters

Isopropanol and formic acid/triethylamine are the sources of hydrogen by far most used in transferhydrogenation. Isopropanol is a good solvent for most substrates, and most complexes can be dissolved in it without extensive decomposition. The lifetime of the relevant catalysts in this solution is usually long enough, even at reflux temperature, to allow the reaction to be completed.

When isopropanol is the H-donor, a base is usually required for the activation of the starting complex to catalysis. Sodium or potassium carbonates, hydroxides, or alkoxides at various concentrations have been employed for this purpose. Quite a few catalytic precursors do not require any base (Shvo catalyst) or need just two equivalents per metal atom (Noyori's and similar catalysts).

Formic acid and its derivatives have the advantage that, unlike isopropanol, their dehydrogenation is irreversible, but there are some restrictions to their use. Many complexes undergo fast decomposition on attempted dissolution in formic acid and other completely lose their catalytic activity, probably because the acid inhibits one of the steps of the activation process promoted by the base.

1.3.2.3 Catalysts

1.3.2.3.1 Metals

The most efficient catalysts devised so far and the most extensively used are centered on Ru, Rh and Ir complexes in d^6 and d^8 electronic configuration. These should be regarded as the metals of choice, while, with the notable exception of Os, other second or third row elements seem to be far less suited for this catalysis. This trend is even more apparent in the case of asymmetric reductions. The peculiar behavior of these three privileged metals can be better appreciated by comparing the results obtained in the asymmetric transferhydrogenation of acetophenone by the relevant complexes containing the same chiral ligand. Scheme 4 shows the most significant cases where this comparison can be made.

Making allowance for the fact that the reaction conditions are not strictly identical in all the cases, in general Ru-catalysts are the most efficient with the majority of the reported ligands.

Enantioselectivities of practical significance (*ee*>90%) are achieved with the Ts-diamine **19** (Ru), with the aminophosphines **26** and **28** (Ru), and with the Ts-diamine **20** (Rh). Among the other *d*-block elements, only osmium in combination

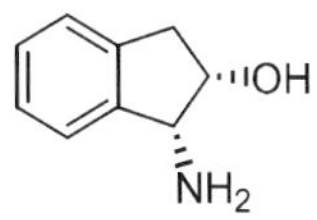

18 (21)		
Ru	70	91
Os	84	97

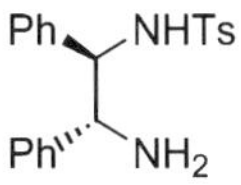

19 (16, 22)		
Ru	95	97
Rh	14	90
Ir	87	92
Co	1	0

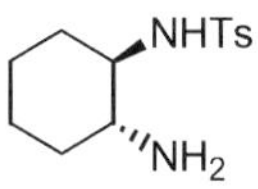

20 (16, 23)		
Ru	97	89
Rh	85	97
Ir	36	96

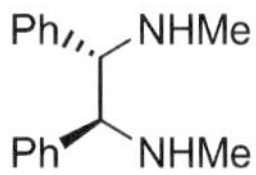

21 (22b, 24)		
Rh	100	67
Co	8	58

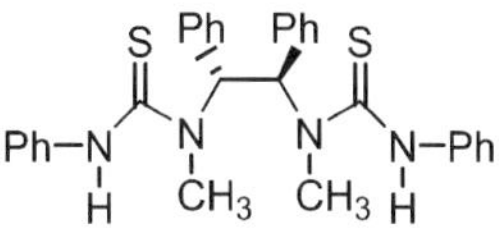

22 (25)		
Ru	94	89
Rh	97	63
Ir	94	36
Co	9	63

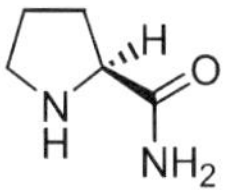

23 (26a)		
Ru	90	79
Rh	66	56

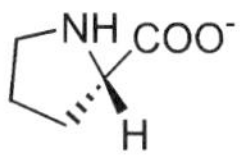

24 (2f, 27)		
Ru	97	75
Os	70	72
Rh	9	60
Ir	16	64

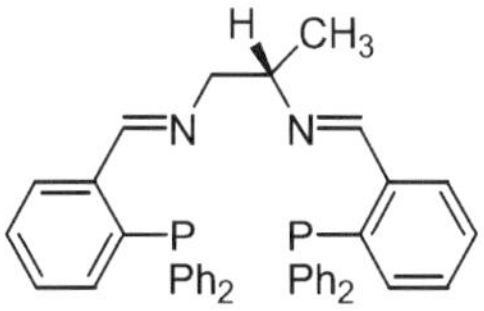

25 (28)		
Ru	63	26
Rh	98	15

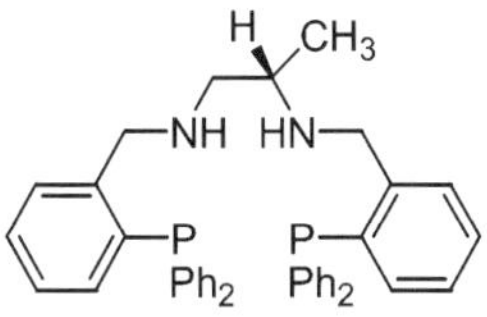

26 (28)		
Ru	90	91
Rh	95	74

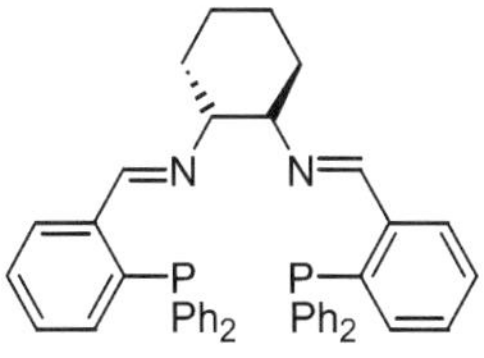

27 (4, 29a)		
Ru	3	18
Rh	40	40

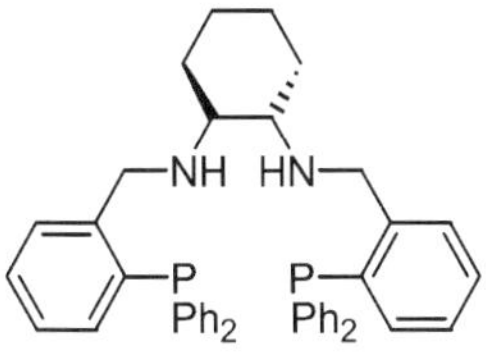

28 (4, 29)		
Ru	92	95
Rh	97	91

Scheme 4 Yield (%) and *ee* (%) obtained in the asymmetric transferhydrogenation of acetophenone in *i*PrOH by different metal complexes with the same chiral ligand **X** (the relevant reference is reported in parentheses).

with the aminoindanol **18** is able to provide *ees* in the range of excellence (98%) [21 c].

Most important, the handedness of the product does not change upon changing the metal and depends only on the configuration of the ligand.

The application of first row transition metals in transferhydrogenation is still in its infancy. Thus far, either the catalytic activity (Co [22 b, 25 a, 25 b]) and/or the enantioselectivity (Cr [30]) are low or the chiral version of the catalyst has not yet been developed (Ni [31]), even if it looks promising in terms of activity and substrate scope.

1.3.2.3.2 **Ligands**

A selection of the ligands of recent introduction (1998–2002) in transferhydrogenation is given in Scheme 5. The ligands are classified as anionic or neutral, depending on whether or not they possess a protonated donor center –XH of appropriate acidity, as this has an important bearing on the mechanism of transferhydrogenation. The *ees* obtained in the reduction of acetophenone are also quoted for chiral ligands.

Aminoalcohols feature the highest ligand acceleration effect, and TOF_{50} values as high as 8500 with 96% *ee* are observed with the catalyst prepared *in situ* from $[RuCl_2(\eta^6\text{-}p\text{-cymene})]_2$ and **30**. Noyori's complex **1** (Scheme 1, X= –NHTos; ligand **19**) seems the catalyst with the broadest scope, as it provides significant *ees* with a large variety of substrates. When $[RuCl_2(\eta^6\text{-arene})]_2$ is employed as precatalyst with protic ligands, "metal-ligand bifunctional catalysis" is expected to operate (Section 1.3.2.1). Albeit less stringent than the anionic ligand, the η^6-arene fragment contributes significantly to the performance of these catalysts, and polyalkylated arenes of increasing steric demand generally provide higher *ees* at the expense of reactivity.

This has been ascribed to the contribution of polyalkylated arenes to the stabilization of the CH/π interaction developed in the transition state (Scheme 1) through an improved electron donation and/or attractive secondary interaction [5 c, 33, 58]. Among neutral ligands, the most successful are the oxazolinylferrocenylphosphines (**53**), whose Ru-catalysts are able to reduce with extremely high enantioselectivity not only alkyl aryl ketones but also some dialkyl ketones, and the polydentate ligands **26** and **28** (Scheme 4), which ensure a deeper chiral concave pocket around the metal. Complexes of carbenes and aryl pincer ligands are emerging catalysts of outstanding activity for transferhydrogenation: TOF_{50} values as high as 27 000 have been achieved in the reduction of cyclohexanone with Ru(II)-complexes containing **45 a** as a tridentate anionic ligand [43].

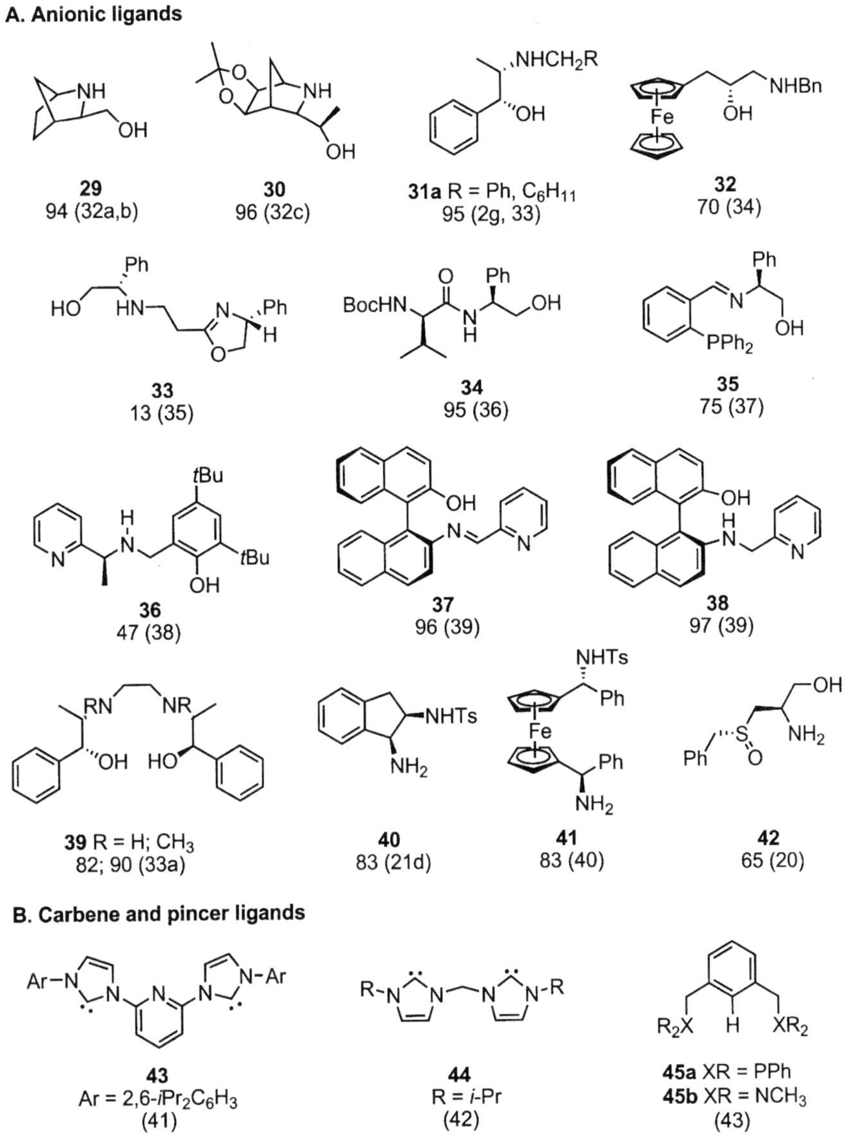

A. Anionic ligands

29 94 (32a,b)

30 96 (32c)

31a R = Ph, C_6H_{11} 95 (2g, 33)

32 70 (34)

33 13 (35)

34 95 (36)

35 75 (37)

36 47 (38)

37 96 (39)

38 97 (39)

39 R = H; CH_3 82; 90 (33a)

40 83 (21d)

41 83 (40)

42 65 (20)

B. Carbene and pincer ligands

43 Ar = 2,6-*i*$Pr_2C_6H_3$ (41)

44 R = *i*-Pr (42)

45a XR = PPh
45b XR = NCH_3
(43)

Scheme 5 Ligands for transferhydrogenation of acetophenone: %*ee*, relevant reference is reported in parentheses.

1.3.3 Substrates

Transferhydrogenation is one of the methodologies best suited for the reduction of C=O, C=N, activated C=C, and N=X groups to form the saturated counter-

C. Neutral ligands

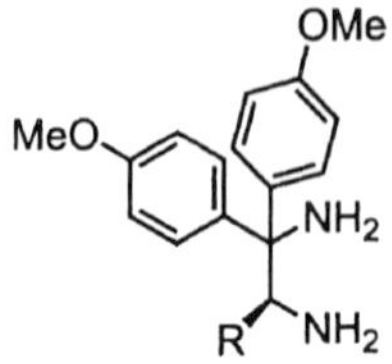

46 R = CH_2Ph
78 (44)

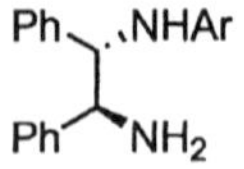

47
Ar = *p*-($PhCH_2N(Ph)SO_2)C_6H_4$
60 (45)

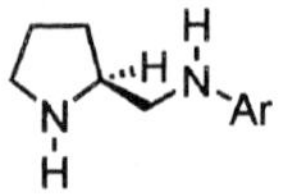

48
Ar = 2-naphtyl
91 (46)

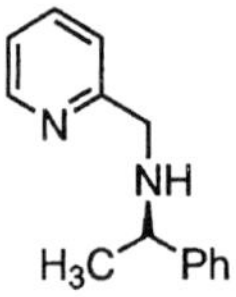

49
59(47)

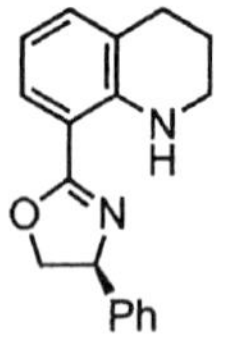

50
83 (48)

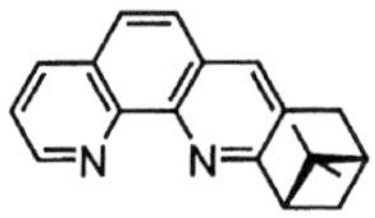

51
14 (49)

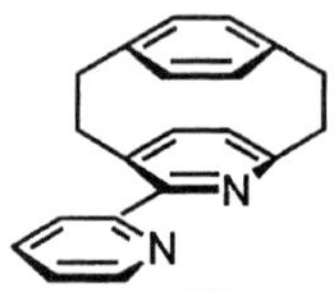

52
31 (50)

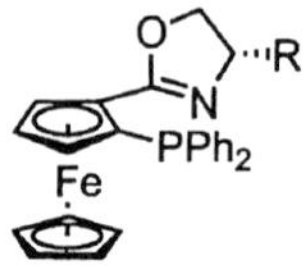

53 R = *i*-Pr, Ph
>99.6 (51)

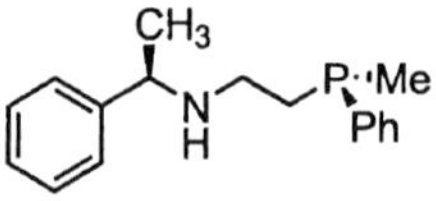

54
57 (52a)

55
80 (20)

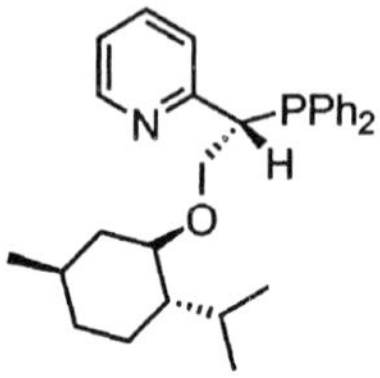

56
60 (53)

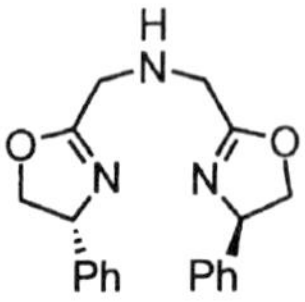

57
97 (54a)

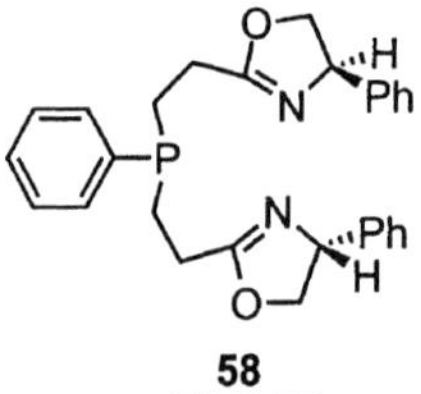

58
79 (54b)

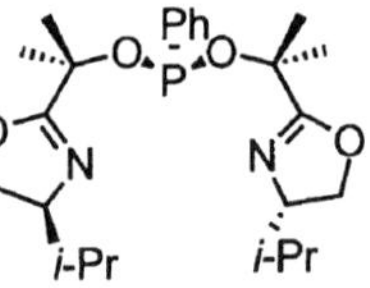

59
45 (55)

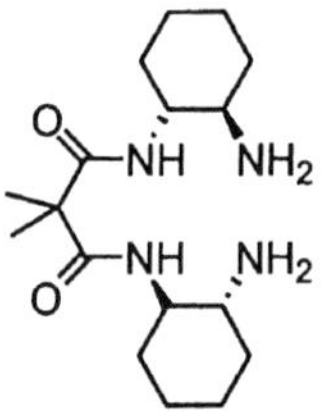

60
62 (56)

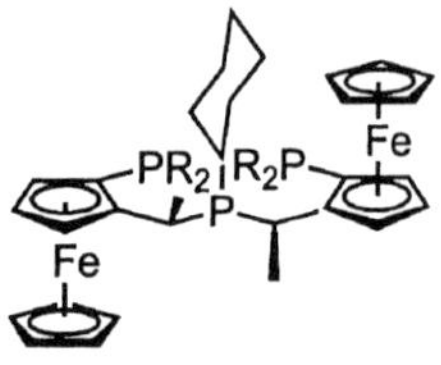

61 R = Ph
72 (57)

Scheme 5 (cont.) Rh catalyst precursors were used with ligands **44** and **51**.
Ir catalyst precursors were used with ligands **46**, **52** and **55**.
HCOOH was the hydrogen source when ligands **40**, **41**, **42** and **49** were used.

parts. Other unsaturated compounds such as simple alkenes, alkynes, nitriles, and epoxides [1] have been reduced as well, but the scope of these reactions is not as general.

1.3.3.1
Ketones and Aldehydes

Aryl alkyl ketones are the substrates of choice for transferhydrogenation, and the assessment of most of the catalysts has been done on these. A selection of the most significant results obtained with acetophenone are collected in Scheme 5 and a survey of the results obtained on the H-transfer reduction of simple ketones, quoting the best *ees*, is given in Tab. 1.

A wide range of ring-substituted acetophenones have been reduced by transferhydrogenation from isopropanol in the presence of different metal catalysts. As for phenyl alkyl ketones, increasing the branching of the alkyl group results in a reduction in the reaction rate and a modest decrease in the enantioselectivity.

Regardless of the nature of the substituent in almost all these cases, *ees* higher than 95% can be obtained. The reduction of *m*-trifluoromethyl-acetophenone, a key step in the preparation of a commercial fungicide, can be performed on a scale of up to 100 kg batches using Noyori's catalyst (**1**) and formic acid/triethylamine as the reductant at a substrate/metal ratio as high as 5000:1 [60].

In spite of an isolated excellent result recorded on derivatives with a *t*-alkyl group, dialkyl ketones have so far failed to provide *ees* in the range of excellence and should be considered even now to be poorly suited substrates for H-transfer

Tab. 1 Asymmetric transferhydrogenation of ketones

Substrate	***ee (%)***	***Ref.***	***Substrate***	***ee (%)***	***Ref.***
A. Ring-substituted acetophenone			**B. Phenyl alkyl ketones**		
o-Me	>99.9	51b	Ph-CO-Et	>99.7	51b
m-	>99.9	51b	Ph-CO-*i*Pr	94	51b
p-	>99.3	51b	Ph-CO-*t*Bu	93	51c
o-OMe [a]	95	26b			
m-	98	32c			
p-	97	59	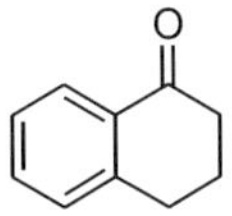 (1-tetralone)	99	59
m-NH_2	99	32c			
o-Br	99	26b			
m-	>99.7	51b			
p-	>99.3	51b	**C. Dialkyl ketones**		
o-Cl	97	54a			
m-	>99.7	51b	R=*n*-hexyl	36	51c
p-	99	51b	R=*c*-hexyl	63	54b
o-CF_3	96	16	R=*t*-Bu	>99	51b
m-	97	16			
p-	88	32b			
m-NO_2	91	32c	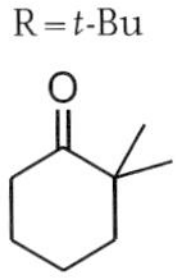 (2,2-dimethylcyclohexanone)	98	51b
p- [a]	89	32b			
p-CN	94	54b			

a) Slighltly higher e.e. have been obtained by using a Sm-based catalyst, see Ref. 1

Tab. 2 Asymmetric transferhydrogenation of functionalized ketones

Substrate	*ee (%)*	*Ref.*	*Substrate*		*ee (%)*	*Ref.*
Ph-CO-CH$_2$X			**R^1-CO-CH$_2$-CO-R^2**			
X=Cl	97	62	R^1=Me	R^2=Et	56	3b
OH	94	63	Me	*t*-Bu	68	3b
CN	98	64	Ph	Et	94	68a
N_3	92	64	Ph	*t*-Bu	89	68a
NO_2	98	64	MeCH(OH)CH$_2$	*t*-Bu	71 (*syn*)	68b
NHCOO*t*-Bu	99	65				
			Ph-CO-CHR-CO-Ph			
Me-CO-X			R=H		99.8	69
X=CH$_2$OMe	66	2g	Me		94.5	69
o-C$_5$H$_5$N	95	66				
			Ph-CO-CO-R			
Ph-CO-X			R=Ar		>99	70a
X=SiMe$_3$	98	67	Me		99	70b
			Et		95	70b

reduction [51b]. Notably, the carbinol arising from *t*-butyl ketones has a comparable *ee*, but the opposite configuration compared to the products obtained from other less branched alkyl phenyl ketones [25b].

The asymmetric deuterohydrogenation of benzaldehydes has been successfully accomplished in 98% *ee* with deuteroformic acid as the D-donor [61]. Even if the synthetic scope of this reaction is modest, this is a real novelty, because until recently H-transfer catalysts were unsuitable for the reduction of the formyl group. Stereoselectivities are lower with conjugated aldehydes and disappointingly modest with aliphatic substrates.

Tab. 2 reports a selection of the most significant results obtained with bifunctional ketones.

Reduction of these substrates usually proceeds with excellent *ees*, but sometimes it is affected by inhibition of the catalyst either by the reduction product [71] or, in the case of *β*-diketones, by the substrate [68c]. This is a real risk for Ru-complexes with chiral aminoalcohols, where displacement of the ligand with deactivation of the catalyst can occur [2g]. Thanks to the stepwise nature of the process, the reduction of 1,2-diketones by the Noyori's catalyst (**1**) (X= –NHTos; ligand **19**) can be used for the selective production of *α*-hydroxyketones [70a] or *anti*-1,2-diols [70b].

1.3.3.2
Conjugated C–C Double Bond

Although H-transfer hydrogenation of C–C double bond is a thermodynamically favored process even when alcohols are used as H-donors, only conjugated C–C double bonds are reduced easily, simple alkenes and dienes being poorly reactive.

Conjugated acid derivatives are selectively hydrogenated at the C–C double bond by reaction with formic acid in the presence of Rh-catalysts with chelating diphosphines [1]. Values of *ee* higher than 90% are not unusual, as in the case of the chiral diphosphine **62** (Scheme 6) or (*R*,*R*)-2,4-bis(diphenylphosphino)pentane [72].

In the H-transfer reduction of α,β-unsaturated carbonyl derivatives, competition between vinyl and carbonyl group hydrogenation is expected. In general, the reduction proceeds preferentially at the carbonyl group, producing the corresponding unsaturated carbinol, as in the case of aminoprolinate complexes of Ru [27 b] and Rh [27 a]. The regioselective reduction of the oxo group of diketone **63** (Scheme 6) proceeds with high stereoselectivity in the presence of an Ru-catalyst with chiral aminoalcohol ligands to give the isophorone derivative **64** in over 95% *ee* [73].

Carbon-carbon triple bonds are resistant to reduction, and chiral propargylic alcohols **66** (Scheme 6) are accessible in over 95% *ee* by transferhydrogenation from isopropanol with Noyori's catalyst [74 a]. This reaction has been exploited in the stereocontrolled synthesis of a β-ionol glycoside [74 b].

Quite a few catalysts show the opposite selectivity and reduce preferentially the C–C double bond instead of the carbonyl group of α,β-unsaturated carbonyl derivatives. Chlorobenzylidene ketones are selectively converted into the saturated ketones in the presence of $RuCl_2(PPh_3)_3$ and ethylene glycol as the H-donor [75]. A range of α,β-unsaturated carbonyl derivatives have been hydrogenated at the vinyl group with isopropanol in the presence of Ir complexes with 1,3-bis(diphenylphosphino)propane and cesium carbonate [76]. This process is intriguing because no significant over-reduction of the substrate is observed notwithstanding the fact that in separate experiments the complex has been shown to efficiently reduce ketones to alcohols.

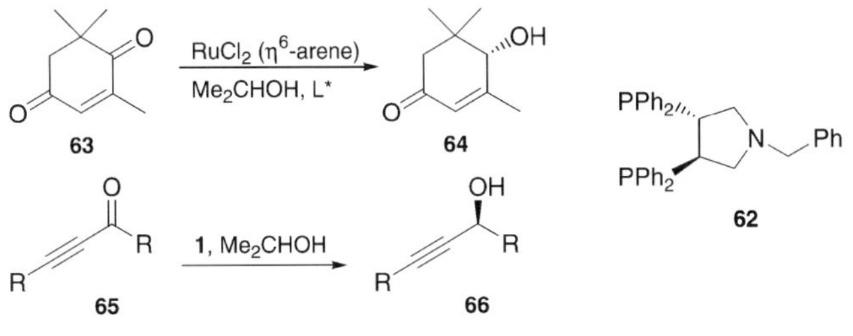

63 → 64: $RuCl_2(\eta^6\text{-arene})$, Me_2CHOH, L*

65 → 66: **1**, Me_2CHOH

62: PPh_2, PPh_2, N, Ph

Scheme 6 H-Transfer reduction of conjugated carbonyl compounds.

1.3.3.3
Imines and Other Nitrogen Compounds

H-transfer hydrogenation of imines deserves particular attention because it provides an expedient route to (chiral) amines, a class of compounds of remarkable biological interest. In general, nitrogen-containing functional groups are best reduced using formic acid or a derivative as the H-donor. For instance, with formic acid/triethylamine, chiral tetrahydroquinolines and chiral sultams have been obtained from the corresponding imines **67**, **69** and **70** with remarkably high *ees* using Noyori's mono-tosyldiamine ligand **19** either with $[Cp^*RhCl_2]_2$ [77] or with $[RuCl_2(\eta^6\text{-arene})]_2$ [4, 78] as catalysts (Scheme 7, Tab. 3). The asymmetric H-transfer reduction of suitable dihydroquinoline intermediates has been successfully exploited in the key step of the new total synthesis of morphine [79] and in the preparation of an isoquinoline-based pharma [80]. Primary amines and α-amino acids are accessible in similar manner via reductive amination by ammonium formate of ketones and α-keto acids in the presence $[Cp^*RhCl_2]_2$ [81].

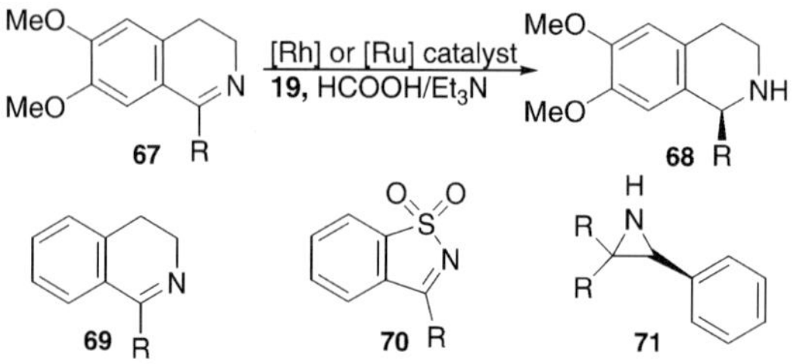

Scheme 7 H-Transfer reduction of imines.

Tab. 3 H-Transfer reduction of imines

Substrates	*R*	*Ru*		*Rh*	
		ee (%)	*Ref.*	*ee (%)*	*Ref.*
67	Me	95	4	90	77
	Ph	84	4	4	77
	3,4-$(CH_3O)_2C_6H_3$	84	4	3	77
	o-Br-phenyl	99	78b	–	–
69	*o*-NH_2-phenyl	85	78b	–	–
	o-Br-phenyl	94	78b	–	–
70	Me	–	–	68	77
	Butyl	–	–	67	77
	t-Butyl	91	78a	–	–
	Benzyl	93	78a	68	77
	m-Cl-phenyl	69	77	81	77

Even if less frequently used with these substrates, isopropanol is the H-donor of choice when the reduction of imines is performed with the Shvo catalyst. An inversion in the normal scale of reactivity of the substrates is observed with this catalyst, and apparently the reduction proceeds faster on the substrates featuring the less electrophilic sp^2-carbon. Thus, imines react faster than the parent oxo derivative [12] and ketimines faster than aldimines, the rate of reduction increasing further in the presence of electron-donating groups on the sp^2-carbon [82]. The conceivable formation of a protonated species, made possible by the absence of a basic promoter, may account for this otherwise puzzling behavior. In the presence of a suitable oxidant, the Shvo catalyst is able to promote the dehydrogenation of amines to imines [83].

In isopropanol and in the presence of a base, Ru-catalysts derived from the chiral aminoalcohol **30** efficiently convert azirines into chiral aziridines **71** in good yield and stereoselectivity [84]. This is the first successful case ever reported of asymmetric reduction of this type of substrate.

Nitro compounds can be reduced to amines by different heterogeneous catalysts such as Ni-Raney, Pd/C or Pt/C. Under similar conditions, primary amines can be obtained as well by reduction of other unsaturated functional groups containing nitrogen, such as azobenzenes, oximes, azides, and hydrazones. In all these reactions the H-donors of most general use are formic acid or hydrazine and their derivatives. A combination of these two H-donors, hydrazinium monoformate, is more effective than the isolated parent compounds, and even substrates inert to H-transfer reduction, such as nitriles, are converted into amines using this reagent in the presence of Raney nickel [85].

1.3.3.4
Other Substrates

Transfer hydrogenolysis has been successfully exploited for the cleavage of C-heteroatom bonds at the benzylic carbon. This techique adds to the traditional protocols for deblocking some of the protective groups of most frequent use in peptide synthesis. Heterogeneous Pd-derivatives in combination with a variety of H-donors (formates, hydrazine, cyclohexadienes, etc.) are the catalysts of choice for this purpose. Under microwave irradiation in ethylene glycol, hydrogenolysis and hydrogenation can occur simultaneously, and even an isolated C-C double bond can be saturated using ammonium formate as the H-donor [85].

1.3.4
Miscellaneous H-Transfer Processes

1.3.4.1
Kinetic Resolution and Dynamic Kinetic Resolution

As the H-transfer reduction of ketones with secondary carbinols is reversible, the same catalysts used for the reduction of the carbonyl group can be exploited in the oxidation of alcohols, and even primary alcohols can be converted into aldehydes in this way [86]. This provides the rationale for accomplishing the kinetic resolution of racemic secondary carbinols. In this process, chiral Ru-complexes with the aminoindanol **18** provide products with over 90% *ee* from a wide range of secondary carbinols [87].

Simultaneous introduction of two stereogenic centers in high stereoselectivity is achievable in the H-transfer reduction of 2-alkyl-1,3-dicarbonyl compounds with formic acid/triethylamine by Noyori's catalyst (1) [88]. This is an example of transition metal-catalyzed kinetic dynamic resolution, which is possible because of the presence of a configurationally labile stereocenter in 2-alkyl-3-hydroxy ketones.

In recent years, dynamic kinetic resolution has gained increased consideration as a suitable technology for asymmetric synthesis, and its applications have been expanded to include even substrates devoid of configurationally labile stereocenters [89]. This second-order asymmetric transformation is rendered possible by a tandem process which combines an enzymatic resolution of a racemate with a suitable transition metal-catalyzed reaction which provides for the racemization of the unreactive enantiomer.

The reversibility of H-transfer reduction of ketones with secondary carbinols provides the means for a dynamic kinetic resolution of racemic carbinols to be accomplished by coupling an enzyme, which converts just one of the enantiomers, with a H-transfer catalyst, which takes on the task of racemizing the unreactive antipode. The mild conditions required for this tandem oxidation-reduction to be accomplished preserve the activity of the enzyme and make the overall process practically feasible in a one-pot procedure. For this purpose, Ru-catalysts with different enzymes have been applied with remarkable success to a range of secondary alcohols [90]. More recently, this technique has been successfully extended to primary amines [91].

1.3.4.2
Green H-Transfer Processes

Since catalysis is considered a "foundation pillar" of green chemistry [92], the increased attention paid to sustainability has prompted the introduction of green chemistry concepts in transferhydrogenation. Among the topics taken up, catalyst recycling and the use of enviromentally benign solvents or solventless systems have received particular attention.

Since the cleanest processes use no catalyst, the high-temperature uncatalyzed H-transfer reduction of aldehydes and ketones by alcohols deserves the first mention here [93]. This intriguing reaction proceeds smoothly at 225 °C, producing the expected alcohol in good yield and selectivity.

Several water-soluble H-transfer catalysts have been developed for use in aqueous or biphasic or liquid-supported H-transfer catalysis. Iridium(III) catalysts have been employed in transferhydrogenation, reductive amination, and dehalogenation of water-soluble carbonyl compounds with formates at room temperature [94]. Ru-catalysts with sulfonated Noyori-type ligands (**19**) promote the transferhydrogenation of aryl alkyl ketones in aqueous solvents in over 95% *ee*, albeit at a lower rate than that in the original systems [95]. In the same reaction, Ru-complexes with proline amides afford similar *ee*s in aqueous biphasic system, a performance comparable with that in the homogeneous phase. Catalytic activity and recyclability are improved by the addition of surfactants [96].

Transferhydrogenation of ketones can be performed in fluorous biphasic systems (FBS) using perfluorinated aldimine ligands to induce the solubilization of the catalyst. Reactants and products can then be separated by freezing out the fluorous phase at 0 °C. The *ee*s obtained with iridium(I) complexes in FBS conditions are encouraging (56% with acetophenone) and are higher than those obtained with the corresponding non-fluorinated ligands [97]. Ionic liquids have been introduced as solvents for the reduction of a variety of functional groups with Pd/C and formate salts under microwave irradiation [98].

Improved catalyst performances have sometimes been obtained using polymer-supported complexes. Imprinting techniques have been exploited in the preparation of the polymeric supports with interesting results [99]. The subject has been reviewed recently [100].

References

1 S. Gladiali, G. Mestroni, in M. Beller, C. Bolm (Eds), *Transition Metals for Organic Synthesis*, Wiley-VCH, **1998**, 97.

2 Reviews: (a) T. Naota, H. Takaya, S.-I., Murahashi, *Chem. Rev.* **1998**, *98*, 2599; (b) V. Fehring, R. Selke, *Angew. Chem. Int. Ed.* **1998**, *37*, 1827; (c) M.J. Palmer, M. Wills, *Tetrahedron: Asymmetry* **1999**, *10*, 2045; (d) M. Wills, M. Palmer, A. Smith, J. Kenny, T. Walsgrove, *Molecules* **2000**, *5*, 4; (e) J.-E. Bäckvall, *J. Organomet. Chem.* **2002**, *652*, 105. (f) D. Carmona, M.P. Lamata, L. Oro, *Eur. J. Inorg. Chem.* **2002**, 2239. (g) K. Everaere, A. Mortreux, J.-F. Carpentier, *Adv. Synth. Catal.* **2003**, *345*, 67.

3 (a) S. Gladiali, L. Pinna, G. Delogu, S. De Martin, G. Zassinovich, G. Mestroni, *Tetrahedron: Asymmetry* **1990**, *1*, 635; (b) G. Zassinovich, G. Mestroni, S. Gladiali, *Chem. Rev.* , **1992**, *92*, 1051.

4 R. Noyori, S. Hashiguchi, *Acc. Chem. Res.* **1997**, *30*, 97.

5 (a) D.A. Alonso, P. Brandt, S.J.M. Nordin, P.G. Andersson, *J. Am. Chem. Soc.* **1999**, *121*, 9580. (b) M. Yamakawa, H. Ito, R. Noyori, *J. Am. Chem. Soc.* **2000**, *122*, 1466. (c) D.G.I. Petra, J.N.H. Reek, J.-W. Handgraaf, E.J. Meijer, P. Dierkes, P.C. J. Kamer, J. Brussee, H.E. Schoemaker, P.W.N.M. van Leeuwen, *Chem. Eur. J.* **2000**, *6*, 2818. (d) C.P. Casey, J.B. Johnson, *J. Org. Chem.* **2003**, *68*, 1998.

6 R. Noyori, M. Yamakawa, S. Hashiguchi, *J. Org. Chem.* **2001**, *66*, 7931.

7 M. Yamakawa, I. Yamada, R. Noyori, *Angew. Chem. Int. Ed.* **2001**, *40*, 2818.

8 Y. Shvo, D. Czarkie, Y. Rahamin, *J. Am. Chem. Soc.* **1986**, *108*, 7400.

9 N. Menashe, Y. Shvo, *Organometallics* **1991**, *10*, 3885.

10 M.L. Almeida, M. Beller, G.-Z. Wang, J.-E. Bäckvall, *Chem. Eur. J.* **1996**, *2*, 1533.

11 (a) A.L.E. Larsson, B.A. Persson, J.-E. Bäckvall, *Angew. Chem. Int. Ed.* **1997**, *36*, 121. (b) H.M. Jung, J.H. Koh, M.-J. Kim, J. Park, *Organometallics* **2001**, *20*, 3370 and references cited therein.

12 C.P. Casey, S.W. Singer, D.R. Powell, R.K. Hayashi, M. Kavana, *J. Am. Chem. Soc.* **2001**, *123*, 1090.

13 A. Aranyos, G. Csjernyik, K.J. Szabò, J.-E. Bäckvall, *Chem. Commun.* **1999**, 351.

14 Y.R.S. Laxmi, J.-E. Bäckvall, *Chem. Commun.* **2000**, 611.

15 O. Pàmies, J.-E. Bäckvall, *Chem. Eur. J.* **2001**, *7*, 5052.

16 K. Murata, T. Ikariya, R. Noyori, *J. Org. Chem.* **1999**, *64*, 2186.

17 M. Bernard, V. Guiral, F. Delbecq, F. Fache, P. Sautet, M. Lemaire, *J. Am. Chem. Soc.* **1998**, *120*, 1441.

18 V. Guiral, F. Delbecq, P. Sautet, *Organometallics* **2000**, *19*, 1589.

19 M. Bernard, F. Delbecq, P. Sautet, F. Fache, M. Lemaire, *Organometallics* **2000**,19, 5715.

20 D.G.I. Petra, P.C.J. Kamer, A.L. Speck, H.E. Schoemaker, P.W.N.M. van Leeuwen, *J. Org. Chem.* **2000**, *65*, 3010.

21 (a) M. Palmer, T. Walsgrove, M. Wills, *J. Org. Chem.* **1997**, *62*, 5226. (b) M. Wills, M. Gamble, M. Palmer, A. Smith, J. Studley, J. Kenny, *J. Mol. Catal., A: Chem.* **1999**, *146*, 139. (c) M. Palmer, J. Kenny, T. Walsgrove, A.M. Kawamoto, M. Wills, *J. Chem. Soc., Perkin Trans. 1* **2002**, 416.

22 (a) S. Hashiguchi, A. Fujii, J. Takehara, T. Ikariya, R. Noyori, *J. Am. Chem. Soc.* **1995**, *117*, 7562. (b) R. Halle, A. Bréhéret, E. Schulz, C. Pinel, M. Lemaire, *Tetrahedron: Asymmetry* **1997**, *8*, 2101.

23 K. Püntener, L. Schwink, P. Knochel, *Tetrahedron Lett.* **1996**, *37*, 8165.

24 F. Touchard, M. Bernard, F. Fache, F. Delbecq, V. Guiral, P. Sautet, M. Lemaire, *J. Organomet. Chem.* **1998**, *567*, 133.

25 (a) F. Touchard, P. Gamez, F. Fache, M. Lemaire, *Tetrahedron Lett.* **1997**, *38*, 2275. (b) F. Touchard, F. Fache, M. Lemaire, *Tetrahedron: Asymmetry* **1997**, *8*, 3319.

26 (a) J.W. Faller, A.R. Lavoi, *Organometallics* **2001**, *20*, 5245. (b) H.Y. Rhyoo, Y.-A. Yoon, H.-J. Park, Y.K. Chung, *Tetrahedron Lett.* **2001**, *42*, 5045.

27 (a) D. Carmona, F.J. Lahoz, R. Atencio, L.A. Oro, M.P. Lamata, F. Viguri, E. San José, C. Vega, J. Reyes, F. Joó, A. Kathó, *Chem. Eur. J.* **1999**, *5*, 1544. (b) A. Kathó, D. Carmona, F. Viguri, C.D. Remacha, J. Kovács, F. Joó, L.A. Oro, *J. Organomet. Chem.* **2000**, *593-594*, 299. (c) T. Ohta, S. Nakahara, Y. Shigemura, K. Hattori, I. Furukawa, *Chem. Lett.* **1998**, 491. (d) T. Ohta, S. Nakahara, Y. Shigemura, K. Hattori, I. Furukawa, *Appl. Organometal. Chem.* **2001**, *15*, 699.

28 (a) J.-X. Gao, P.-P. Xu, X.-D. Yi, C. Yang, H. Zhang, S. Cheng, H.-L. Wan, K. Tsai, T. Ikariya, *J. Mol. Catal., A: Chem.* **1999**, *147*, 105. (b) J.-X. Gao, X.-D. Yi, P.-P. Xu, C.-L. Tang, H. Zhang, H.-L. Wan, T. Ikariya, *J. Mol. Catal., A: Chem.* **2000**, *159*, 3.

29 (a) J.-X. Gao, X.-D. Yi, P.-P. Xu, C.-L. Tang, H.-L. Wan, T. Ikariya, *J. Organomet. Chem.* **1999**, *592*, 290. (b) J.-X. Gao, X.D. Yi, C.-L. Tang, P.-P. Xu, H.-L. Wan, *Polym. Adv. Technol.* **2001**, *12*, 716.

30 J.-J. Brunet, R. Chauvin, P. Leglaye, *Eur. J. Inorg. Chem.* **1999**, 713.

31 (a) M.D. La Page, B.R. James, *Chem. Commun.* **2000**, 1647. (b) S. Iyer, A.K. Sattar, *Synth. Commun.* **1998**, *28*, 1721. (c) P. Phukan, S. Sudalai, *Synth. Commun.* **2000**, *30*, 2401.

32 (a) D.A. Alonso, D. Guijarro, P. Pinho, O. Temme, P.G. Anderson, *J. Org. Chem.* **1998**, *63*, 2749. (b) D.A. Alonso, S.J.M. Nordin, P. Roth, T. Tarnai, P. G. Anderson, *J. Org. Chem.* **2000**, *65*, 3116. (c) S.J.M. Nordin, P. Roth, T. Tarnai, D.A. Alonso, P. Brandt, P.G. Anderson, *Chem. Eur. J.* **2001**, *7*, 1431.

33 (a) D.G.I. Petra, P.C.J. Kamer, P.W.N. M. van Leeuwen, K. Goubitz, A.M. van Loon, J.G. de Vries, H.E. Schoemaker, *Eur. J. Inorg. Chem.* **1999**, 2335. (b) K. Everaere, J.F. Carpentier, A. Mortreux, M. Builliard, *Tetrahedron: Asymmetry* **1999**, *10*, 4083. (c) C.G. Frost, P. Mendonça, *Tetrahedron: Asymmetry* **2000**, *11*, 1845.

34 A. Patti, S. Pedotti, *Tetrahedron: Asymmetry* **2003**, *14*, 597.

35 Y. Jiang, Q. Jiang, G. Zhu, X. Zhang, *Tetrahedron Lett.* **1997**, *38*, 6565.

36 I.M. Pastor, P. Västilä, H. Adolfsson, *Chem. Commun.* **2002**, 2046.

37 H-L. Kwong, W-S. Lee, T-S. Lai, W-T. Wong, *Inorg. Chem. Commun.* **1999**, *2*, 66.

38 H. Brunner, M. Niemetz, *Monatsh. Chem.* **2002**, *133*, 115.

39 H. Brunner, F. Henning, M. Weber, *Tetrahedron: Asymmetry* **2002**, *13*, 37.

40 L. Schwink, T. Ireland, K. Püntener, P. Knochel, *Tetrahedron: Asymmetry* **1998**, *9*, 1143.

41 A. A. Danopoulos, S. Winston, W. B. Motherwell, *Chem. Commun.* **2002**, 1376.

42 M. Albrecht, R. H. Crabtree, J. Mata, E. Peris, *Chem. Commun.* **2002**, 32.

43 P. Dani, T. Karlen, R. A. Gossage, S. Gladiali, G. van Koten, *Angew. Chem. Int. Ed.* **2000**, *39*, 743.

44 S. Inoue, K. Nomura, S. Hashiguchi, R. Noyori, Y. Izawa, *Chem. Lett.* **1997**, 957.

45 C. G. Frost, P. Mendonça, *Tetrahedron: Asymmetry* **1999**, *10*, 1831.

46 M. Aitali, S. Allaoud, A. Karim, C. Meliet, A. Mortreux, *Tetrahedron: Asymmetry* **2000**, *11*, 1367.

47 E. Mizushima, H. Ohi, M. Yamaguchi, T. Yamagishi, *J. Mol. Catal., A: Chem.* **1999**, *149*, 43.

48 Y.-B. Zhou, F.-Y. Tang, H.-D. Xu, X.-Y. Wu, J.-A. Ma, Q.-l. Zhou, *Tetrahedron: Asymmetry* **2002**, *13*, 469.

49 J.-C. Moutet, L. Y. Cho, C. Duboc-Toia, S. Ménage, E. C. Riesgo, R. P. Thummel, *New J. Chem.* **1999**, *23*, 939.

50 U. Wörsdörfer, F. Vögtle, M. Nieger, M. Waletzke, S. Grimme, F. Glorius, A. Pfaltz, *Synthesis* **1999**, *4*, 597.

51 (a) T. Sammakia, E. L. Stangeland, *J. Org. Chem.* **1997**, *62*, 6104. (b) Y. Nishibayashi, I. Takei, S. Uemura, M. Hidai, *Organometallics* **1999**, *18*, 2291. (c) Y. Arikawa, M. Ueoka, K. Matoba, Y. Nishibayashi, M. Hidai, S. Uemura, *J. Organomet. Chem.* **1999**, *572*, 163.

52 (a) A. M. Maj, K. P. Pietrusiewicz, I. Suisse, F. Agbossou, A. Mortreux, *J. Organomet. Chem.* **2001**, *626*, 157. (b) A. M. Maj, K. P. Pietrusiewicz, I. Suisse, F. Agbossou, A. Mortreux, *Tetrahedron: Asymmetry* **1999**, *10*, 831.

53 H. Yang, M. Alvarez-Gressier, N. Lugan, R. Mathieu, *Organometallics* **1997**, *16*, 1401.

54 (a) Y. Jiang, Q. Jiang, X. Zhang, *J. Am. Chem. Soc.* **1998**, *120*, 3817. (b) Y. Jiang, Q. Jiang, G. Zhu, X. Zhang, *Tetrahedron Lett.* **1997**, *38*, 215.

55 P. Braunstein, F. Naud, A. Pfaltz, S. J. Rettig, *Organometallics* **2000**, *19*, 2676.

56 C. M. Marson, I. Schwarz, *Tetrahedron Lett.* **2000**, *41*, 8999.

57 P. Barbaro, C. Bianchini, A. Togni, *Organometallics* **1997**, *16*, 3004.

58 J. Takehara, S. Hashiguchi, A. Fujii, S.-I. Inoue, T. Ikariya, R. Noyori, *Chem. Commun.* **1996**, 233.

59 A. Fujii, S. Hashiguchi, N. Uematsu, T. Ikariya, R. Noyori, *J. Am. Chem. Soc.* **1996**, *118*, 2521.

60 (a) K. Tanaka, M. Katsurada, F. Ohno, Y. Shiga, M. Oda, M. Miyagi, J. Takehara, K. Okano, *J. Org. Chem.*, **2000**, *65*, 432. (b) M. Miyagi, J. Takehara, S. Collet, K. Okano, *Org. Process Res. Dev.*, **2000**, *4*, 346.

61 I. Yamada, R. Noyori, *Org. Lett.* **2000**, *2*, 3425.

62 T. Hamada, T. Torii, K. Izawa, R. Noyori, T. Ikariya, *Org. Lett.* **2002**, *4*, 4373.

63 D. J. Cross, J. A. Kenny, I. Houson, L. Campbell, T. Walsgrove, M. Wills, *Tetrahedron: Asymmetry* **2001**, *12*, 1801.

64 M. Watanabe, K. Murata, T. Ikariya, *J. Org. Chem.* **2002**, *67*, 1712.

65 A. Kawamoto, M. Wills, *J. Chem. Soc., Perkin Trans. I* **2001**, 1916.

66 K. Okano, K. Murata, T. Ikariya, *Tetrahedron Lett.* **2000**, *41*, 9277.

67 J. Cossrow, S. D. Rychnovsky, *Org. Lett.* **2002**, *4*, 147.

68 (a) K. Everaere, J.-F. Carpentier, A. Mortreux, M. Bulliard, *Tetrahedron: Asymmetry* **1999**, *10*, 4663. (b) K. Everaere, N. Franceschini, A. Mortreux, J.-F. Carpentier, *Tetrahedron Lett.* **2002**, *43*, 2569. (c) K. Everaere, A. Mortreux, M. Bulliard, J. Brussee, A. van der Gen, G. Nowogrocki, J.-F. Carpentier, *Eur. J. Org. Chem.* **2001**, 275.

69 J. Cossy, F. Eustache, P. I. Dalko, *Tetrahedron Lett.* **2001**, *42*, 5005.

70 (a) K. Murata, K. Okano, M. Miyagi, H. Iwane, R. Noyori, T. Ikariya, *Org. Lett.*

1999, *1*, 1119. (b) T. Koike, K. Murata, T. Ikariya, *Org. Lett.* **2000**, *2*, 3833.
71 J.A. Kenny, M.J. Palmer, A.R.C. Smith, T. Walsgrove, M. Wills, *Synlett* **1999**, 1615.
72 A.M. d'A. Rocha Gonsalves, J.C. Bayòn, M.M. Pereira, M.E.S. Serra, J.P.R. Pereira, *J. Organomet. Chem.* **1998**, *553*, 199.
73 M. Henning, K. Püntener, M. Scalone, *Tetrahedron: Asymmetry* **2000**, *11*, 1849.
74 (a) K. Matsumura, S. Hashiguchi, T. Ikariya, R. Noyori, *J. Am. Chem. Soc.* **1997**, *119*, 8738. (b) Y. Yamano, Y. Watanabe, N. Watanabe, M. Ito, *Chem. Pharm. Bull.* **2000**, *48*, 2017.
75 S. Mukhopadhyay, A. Yaghmur, Y. Sasson, *Org. Process Res. Dev.* **2000**, *4*, 571.
76 S. Sakaguchi, T. Yamaga, Y. Ishii, *J. Org. Chem.* **2001**, *66*, 4710.
77 J. Mao, D.C. Baker, *Org. Lett.* **1999**, *1*, 841.
78 (a) K.H. Ahn, C. Ham, S.-K. Kim, C.-W. Cho, *J. Org. Chem.* **1997**, *62*, 7047. (b) E. Vedejs, P. Trapencieris, E. Suna, *J. Org. Chem.* **1999**, *64*, 6724.
79 G.J. Meuzelaar, M.C.A. van Vliet, L. Maat, R.A. Sheldon, *Eur. J. Org. Chem.* **1999**, 2315.
80 V. Samano, J.A. Ray, J.B. Thompson, R.A. Mook Jr., D.K. Jung, C.S. Koble, M.T. Martin, E.C. Bigham, C.S. Regitz, P.L. Feldman, E.C. Boros, *Org. Lett.* **1999**, *1*, 1993.
81 M. Kitamura, D. Lee, S. Hayashi, S. Tanaka, M. Yoshimura, *J. Org. Chem.* **2002**, *67*, 8685.
82 J.S.M. Samec, J.-E. Bäckvall, *Chem. Eur. J.* **2002**, *8*, 2955.
83 A.H. Éll, J.S.M. Samec, C. Brasse, J.-E. Bäckvall, *Chem. Commun.* **2002**, 1144.
84 P. Roth, P.G. Andersson, P. Somfai, *Chem. Commun.* **2002**, 1752.
85 S. Gowda, D.C. Gowda, *Tetrahedron* **2002**, *58*, 2211.
86 T. Suzuki, K. Morita, M. Tsuchida, K. Hiroi, *J. Org. Chem.* **2003**, *68*, 1601.
87 J.W. Faller, A.R. Lavoie, *Org. Lett.* **2001**, *3*, 3703.
88 F. Eustache, P.I. Dalko, J. Cossy, *Org. Lett.* **2002**, *4*, 1263.
89 F.F. Huerta, A.B.E. Minidis, J.-E. Bäckvall, *Chem. Soc. Rev.* **2001**, *30*, 321.
90 (a) O. Pamiès, J.-E. Bäckvall, *J. Org. Chem.* **2002**, *67*, 9006 and references cited therein. (b) D. Lee, E.A. Huh, M.-J Kim, H.M. Jung, J.H. Koh, J. Park, *Org. Lett.* **2000**, *2*, 2377 and references cited therein.
91 O. Pamiès, A.H. Éll, J.S.M. Samec, N. Hermanns, J.-E. Bäckvall, *Tetrahedron Lett.* **2002**, *43*, 4699.
92 P.T. Anastas, M.M. Kirchhof, *Acc. Chem. Res.* **2002**, *35*, 686.
93 L. Bagnell, C.R. Strauss, *Chem. Commun.* **1999**, 287.
94 (a) S. Ogo, N. Makihara, Y. Watanabe, *Organometallics* **1999**, *18*, 5470. (b) S. Ogo, N. Makihara, Y. Kaneko, Y. Watanabe, *Organometallics* **2001**, *20*, 4903.
95 (a) C. Bubert, J. Blacker, S.M. Brown, J. Crosby, S. Fitzjohn, J.P. Muxworthy, T. Thorpe, J.M.J. Williams, *Tetrahedron Lett.* **2001**, *42*, 4037. (b) T. Thorpe, J. Blacker, S.M. Brown, C. Bubert, J. Crosby, S. Fitzjohn, J.P. Muxworthy, J.M.J. Williams, *Tetrahedron Lett.* **2001**, *42*, 4041.
96 (a) H.Y. Rhyoo, H.-J. Park, W.H. Suh, Y.K. Chung, *Tetrahedron Lett.* **2002**, *43*, 269. (b) H.Y. Rhyoo, H.-J. Park, Y.K. Chung, *Chem. Commun.* **2001**, 2064.
97 D. Maillard, C. Nguefack, G. Pozzi, S. Quici, B. Valadé, D. Sinou, *Tetrahedron: Asymmetry* **2000**, *11*, 2881.
98 H. Berthold, T. Schotten, H. Hönig, *Synthesis* **2002**, 1607.
99 K. Polborn, K. Severin, *Chem. Eur. J.* **2000**, *6*, 4604 and references cited therein.
100 C. Saluzzo, M. Lemaire, *Adv. Synth. Catal.* **2002**, *344*, 915.

1.4
Hydrosilylations

1.4.1
Hydrosilylation of Olefins

K. Yamamoto and T. Hayashi

1.4.1.1
Introduction

The enormous progress in the field of transition metal-catalyzed reactions directed toward organic synthesis is continuing. Ample research activities in the catalytic hydrosilylation of alkenes and alkynes are still devoted to obtaining an insight into the mechanisms of hydrosilylation, including chemo-, regio- or stereoselective hydrosilylation. Catalytic hydrosilylation is a versatile synthetic method of obtaining organosilicon compounds. The value of this hydrosilylation has been further augmented by protocols for converting the silyl group to other functional groups

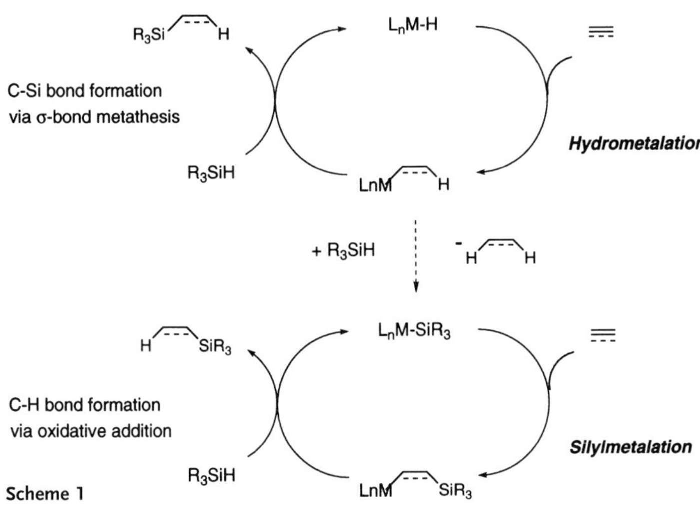

Scheme 1

Transition Metals for Organic Synthesis, Vol. 2, 2nd Edition.
Edited by M. Beller and C. Bolm

ISBN: 3-527-30613-7

[1]. For example, certain silyl groups can be converted to hydroxyl groups by the Tamao-Fleming oxidation [2].

This catalysis is generally thought to proceed by either *hydrometalation* (Chalk-Harrod mechanism) or *silylmetalation* (so-called modified Chalk-Harrod mechanism) as one of the key steps, and the essential features of the catalytic cycles are depicted in Scheme 1 [3].

In the present chapter, we attempt to give an account of the significant reports which have appeared since 1998, when a brief review was published in the first edition of this book [4].

1.4.1.2 Hydrosilylation of Alkenes

1.4.1.2.1 Mechanistic Studies of Hydrosilylation Catalyzed by Groups 9 and 10 Metal Complexes

A theoretical study of the platinum(0)-catalyzed hydrosilylation of ethylene has been reported, in which a higher bond strength for the Pt-Si bond and a weaker *trans*-influence of the corresponding silyl group with electronegative substituents is predicted to favor the Chalk-Harrod mechanism [5]. The first alkene-platinum-silyl complexes were presented, and the facile *hydrometalation* rather than *silylmetalation* of the coordinated alkene provides the experimental support for the sequence of insertion steps as predicted above [6]. Detailed study of *in situ* determination of the active catalyst was reported using Karstedt's catalyst, Pt_2(1,3-divinyl-1,1,3,3-tetramethyldisiloxane)$_3$, as a precursor [7].

Dehydrogenative silylation of 1-alkenes to afford 1-alkenylsilanes, if selective at all, seems to be very useful in synthetic purposes. The reaction is akin to hydrosilylation, but catalyzed often by rhodium or ruthenium cluster complexes with the necessary use of trialkylsilanes. It is argued that the *silylmetalation* must precede the *hydrometalation* in a key step of the catalytic loop [4].

In this connection, a theoretical study of the rhodium(I)-catalyzed hydrosilylation of ethylene clearly shows that the reaction takes place through the *silylmetalation* pathway unlike the platinum-catalyzed one. This is because ethylene undergoes insertion into the Rh-$SiMe_3$ bond with a moderate activation barrier, but insertion into the Pt-SiR_3 bond with a much higher activation barrier [8].

Although a thorough understanding of the effects of silicon substituents on the catalytic aspects of transition metal behavior is lacking, there are studies addressing the subject of structure-property relationships in reactions of catalytically active metal complexes with hydrosilanes [9, 10].

Several reports have previously appeared on the hydrosilylation of alkenes, where the catalytic cycle is proposed to involve *silylmetalation*. It is also demonstrated that $Co_2(CO)_8$-catalyzed hydrosilylation of allyl acetate does proceed via a *silylmetalation* pathway on the basis of an elaborate crossover experiment [11].

1.4.1.2.2 **Hydrosilylations of Alkenes of Synthetic Value**

Almost all transition metal-catalyzed hydrosilylations of alkenes end up with a *cis*-addition of a hydrosilane across the carbon-carbon double bond of the substrate. Lewis acid-assisted methodologies providing alternative *trans* stereochemistry have been known for some years [12a], and now a highly efficient method for *trans*-selective hydrosilylation of alkenes catalyzed by the Lewis acid $B(C_6F_5)_3$ has been developed [12b].

The heterogeneous or heterogenized homogeneous version is revisited. Thus, PtO_2 is a versatile and powerful hydrosilylation catalyst for a wide variety of functionalized alkenes, especially for aminated alkenes, to produce silane coupling agents. Highly reproducible results and easy separation of the catalyst were secured (Scheme 2) [13]. With regard to the production of γ-aminopropylsilanes, a $CoCl_2$-catalyzed hydrosilylation of acrylonitrile may be of industrial merit [14]. Remarkable activity, selectivity, and stability of a polymer-supported platinum catalyst were found in room temperature, solvent-less alkene hydrosilylation [15].

The concept of fluorous biphasic separation has been applied in the rhodium(I)-based catalysts for hydrosilylation of 1-alkenes and fluorinated alkenes, the fluorous phase containing the catalyst that is to be recycled [16]. The hydrosilylation of polyfluoroalkene in dense carbon dioxide has been reported [17].

Certain functionalized alkenes are hydrosilylated in their own right. A hydroxorhodium complex was found to be a highly efficient catalyst for 1,4-hydrosilylation of α,β-unsaturated carbonyl compounds to give selectively enoxysilanes (precursor of the Mukaiyama aldol reactions), diastereoselectivity being only moderate [18]. Similarly, generation of (*E*)-silylketene acetals from the rhodium-catalyzed hydrosilylation of methyl acrylate with Cl_2MeSiH was applied to a two-step reductive aldol reaction [19]. Tandem cyclization/hydrosilylation of functionalized 1,6-heptadienes and 1,7-octadienes has been developed using a cationic palladium complex as catalyst (Brookhart's catalyst) (Scheme 3) [20]. Although reversibility of *silylpalladation* in palladium-catalyzed cyclization/hydrosilylation has not been established, the fact that an exclusive formation of *trans,cis* diastereomer of **B** from **A** was observed provides evidence for reversible *silylpalladation* under the conditions.

Rhodium(I)- and platinum(0)-catalyzed hydrosilylation of alkenes (and alkynes) using dimethyl(2-pyridyl)silane ($2\text{-}PyMe_2SiH$) exhibited a marked difference in re-

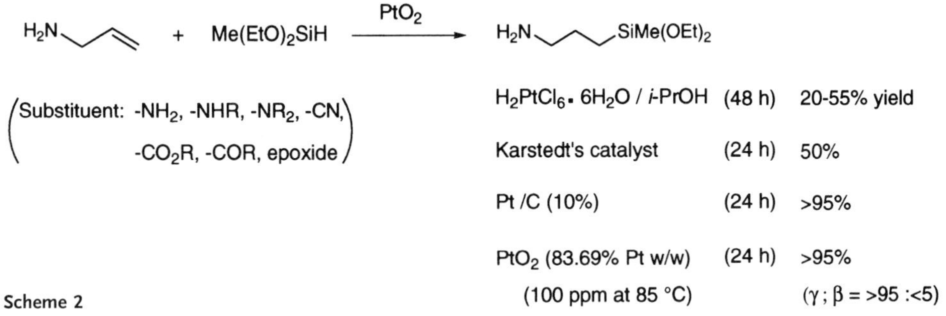

H_2N–allyl + $Me(EtO)_2SiH$ $\xrightarrow{PtO_2}$ $H_2N(CH_2)_3SiMe(OEt)_2$

(Substituent: $-NH_2$, $-NHR$, $-NR_2$, $-CN$, $-CO_2R$, $-COR$, epoxide)

Catalyst	Time	Yield
$H_2PtCl_6 \cdot 6H_2O$ / *i*-PrOH	(48 h)	20-55% yield
Karstedt's catalyst	(24 h)	50%
Pt /C (10%)	(24 h)	>95%
PtO_2 (83.69% Pt w/w) (100 ppm at 85 °C)	(24 h)	>95% (γ ; β = >95 :<5)

Scheme 2

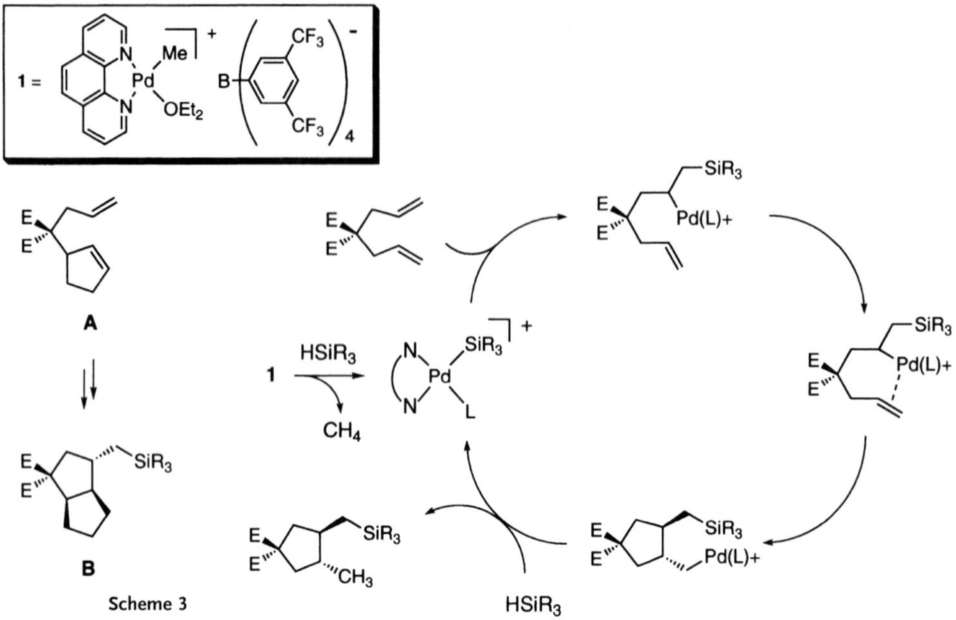

Scheme 3

activity between both cases. Salient features of exceptionally high reactivity only in the rhodium-catalyzed hydrosilylation of various alkenes with this silane are discussed in terms of the coordination-induced *silylmetalation*. Besides the mechanistic arguments, it is of value to apply a “phase tag” technique using a 2-PyMe_2Si group, which enables easy separation of the product and reuse of the catalyst (Scheme 4) [21].

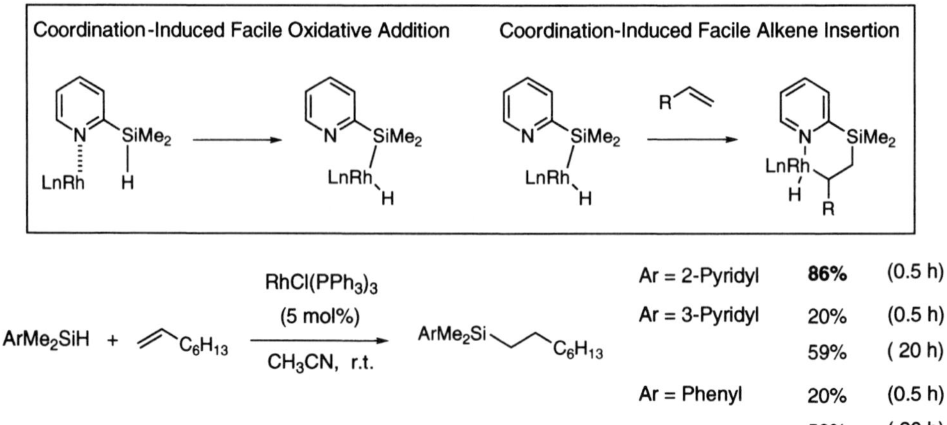

Scheme 4

Finally, lanthanocene catalysts in selective organic synthesis that includes hydrosilylation have been reviewed [22]. There are recent studies of the hydrosilylation of styrene, which relies on early transition metal complexes as the catalyst, regioselectivity being only moderate [23].

1.4.1.3 Hydrosilylation of Alkynes

1.4.1.3.1 Mechanistic Aspects

Alkenylsilanes, which are widely used intermediates for organic synthesis, could be efficiently prepared by the transition metal-catalyzed addition of hydrosilanes to alkynes [1]. The major concern in this conversion is selectivity. Hydrosilylation of 1-alkynes may give a primary mixture of three isomeric alkenylsilanes (Scheme 5). For example, Pt-catalyzed hydrosilylation of 1-pentyne with triethylsilane was reported to give an 89:11 mixture of β-(*E*) and α-isomers. Therefore, complete control of the selectivity is generally difficult, and considerable effort has been devoted to the improvement of selectivity. The selectivity depends on various reaction factors, e.g., substituents on both the alkyne and hydrosilane, the catalyst metal species, solvent, and even reaction temperature. While a cationic Rh(I) complex-catalyzed hydrosilylation of 1-alkynes with triethylsilane gives exclusively the β-(*E*) isomer [24], neutral Rh(I)- or Ir(I)-catalyzed hydrosilylation of 1-alkynes has been known to form more or less selectively the β-(*Z*) isomer, indicative of formal *trans* addition of the hydrosilane across the carbon-carbon triple bond, which is understood in terms of the Ojima-Crabtree postulation [24].

1.4.1.3.2 Stereo- and Regioselective Hydrosilylations of 1-Alkynes: Products of Particular Value

Among the alkenyl-metal species, alkenylsilanes are of particular value because of their low toxicity, ease of handling, and simplicity of by-product removal. Particularly significant is the potential of alkenylsilanes as nucleophilic partners in Pd-catalyzed cross-coupling reactions. Alkenylsilanes are also useful as acceptors in conjugate addition reactions, as masked ketones through the Tamao-Fleming oxidation, and as terminators for cation cyclization (Scheme 6) [25].

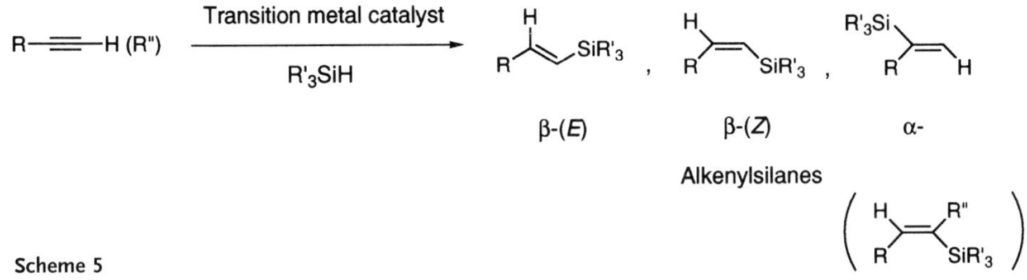

R—≡—H (R")
Transition metal catalyst
R'$_3$SiH
β-(*E*)
β-(*Z*)
α-
Alkenylsilanes

Scheme 5

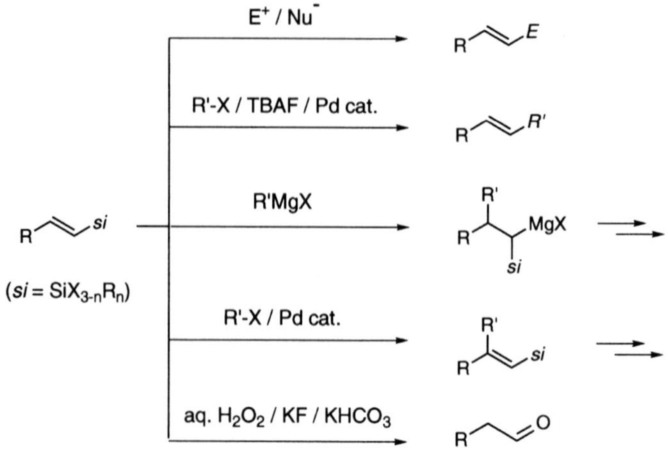

E^+ / Nu^-

R'-X / TBAF / Pd cat.

R'MgX

(si = $SiX_{3-n}R_n$)

R'-X / Pd cat.

aq. H_2O_2 / KF / $KHCO_3$

Scheme 6

Stereodivergent synthesis of either (*E*)- or (*Z*)-2-phenylethenylsilanes was achieved in the hydrosilylation of phenylacetylene, catalyzed by a few rhodium or ruthenium complexes, respectively: $RhI(PPh_3)_3$ vs $RhCl(PPh_3)_3$/NaI [26], $[Cp^*Rh(binap)][SbF_6]_2$ vs $[Cp^*RhCl_2]_2$ [27], and $RuHCl(CO)(PPh_3)_3$ vs $Ru(SiMe_2Ph)Cl(CO)(PPr^i_3)_2$ [28].

With $[RuCl_2(p\text{-cymene})_2]_2$ as a catalyst, extremely high stereoselectivity was observed in the hydrosilylation of certain functionalized 1-alkynes under mild conditions to afford β-(*Z*)-alkenylsilanes, the origin of the high stereoselectivity being unclear. In addition, a strong directing effect was observed in the hydrosilylation of alkynes having a hydroxyl group at the β position to the triple bond, and the reaction proceeded to give α-isomers with an excellent regioselectivity, despite rather low yields due mainly to the competitive *O*-silylation [29].

Although significant progress toward providing stereodefined 1,2-substituted alkenylsilanes has been achieved, there is no reported general access to 1,1-disubstituted alkenylsilanes, and very little is known about selectivity in accessing trisubstituted ones, e.g., $RCH{=}C(R'')SiR'_3$ (see Scheme 5). This subject of research, which involves an essentially novel regioselectivity affording 1-silyl-1-alkenes (α-isomers), has recently been reported by two groups [25, 30]. In this Ru-catalyzed hydrosilylation of 1-alkynes, the presence of a bulky Cp* (Cp* = pentamethylcyclopentadienyl) ligand in the ruthenium(II) complex appears to be indispensable for obtaining α-isomers selectively.

With the sterically demanding $[Cp^*Ru(MeCN)_3]^+PF_6^-$ as a catalyst of choice, a variety of terminal alkynes was found to be amenable to the reaction with either triethylsilane or triethoxysilane, and good yields and good regioselectivity (branched : linear from 9 : 1 to 20 : 1) are maintained through a wide range of substrates (Scheme 7) [25]. The reaction could even be extended to internal alkynes, e.g., treatment of 4-octyne under the standard conditions afforded clean conversion to a single alkenylsilane, (*Z*)-1-propyl-1-(triethoxysilyl)-1-pentene, in quantita-

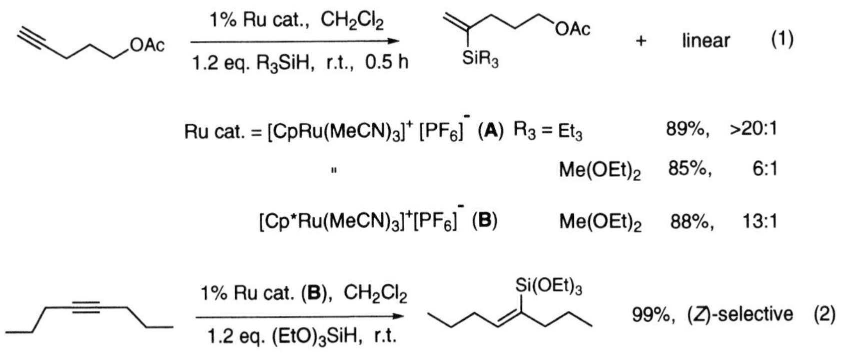

1% Ru cat., CH_2Cl_2; 1.2 eq. R_3SiH, r.t., 0.5 h

OAc → SiR_3 / OAc + linear (1)

Ru cat. = $[CpRu(MeCN)_3]^+ [PF_6]^-$ (**A**) $R_3 = Et_3$ 89%, >20:1

" $Me(OEt)_2$ 85%, 6:1

$[Cp^*Ru(MeCN)_3]^+[PF_6]^-$ (**B**) $Me(OEt)_2$ 88%, 13:1

1% Ru cat. (**B**), CH_2Cl_2; 1.2 eq. $(EtO)_3SiH$, r.t.

→ $Si(OEt)_3$ 99%, (*Z*)-selective (2)

Scheme 7

tive yield. The fact that *trans* hydrosilylation takes place exclusively raises questions whether any equilibration rationale (the Ojima-Crabtree postulation [24]) is viable. In this connection, it should be mentioned that *in situ* formed polynuclear aggregates of ruthenium complexes play an important role in the *trans* addition of hydrosilanes to 1-alkynes to afford the (Z)-isomers [31].

A ruthenium catalyst precursor bearing a bulky and electron-donating Cp* ligand, $Cp^*RuH_3(PPh_3)$, was also found to mediate hydrosilylation of simple or certain functionalized 1-alkynes with specifically Cl_2MeSiH to give 2-silyl-1-alkenes highly selectively [30]. It is postulated that the *silylmetalation* of 1-alkynes with *in situ* formed $Cp^*Ru(SiMeCl_2)(PPh_3)$ is responsible for the selective formation of 2-silylated-1-alkenes (branched : linear from 8 : 1 to 33 : 1).

Although several cyclization/addition protocols employing α,ω-diynes are well known, the cyclization/hydrosilylation of 1,6-heptadiynes, which has been carried out using Ni(0) or Rh(I) complex [32], remained rather problematic, despite an easy access to cationic palladium-catalyzed reactions of 1,6-heptadienes [20]. Now a versatile route for the synthesis of a 1,2-dialkylidenecycloalkane skeleton has been developed by a cationic platinum(II)-catalyzed cyclization/hydrosilylation of either 1,6- or 1,7-diynes [33]. The silylated 1,2-dialkylidenecyclopentanes and a 1,2-dialkylidenecyclohexane with high *Z*-selectivity were subjected to a range of transformations including protodesilylation, *Z/E* isomerization, and [4+2] cycloaddition with dienophiles.

1.4.1.4
Catalytic Asymmetric Hydrosilylation of Alkenes

Catalytic asymmetric hydrosilylation of alkenes has been attracting our continuous attention in recent years because of its synthetic utility as well as its mechanistic interest [34]. In this section, we describe some of the new results reported since 1998, when the previous review appeared in this treatise. Considerable progress has been made in the palladium-catalyzed asymmetric hydrosilylation of styrenes and 1,3-dienes. New types of asymmetric hydrosilylation, i.e. cyclization-hydrosilylation of 1,6-dienes in the presence of a cationic palladium catalyst and some yttrium-catalyzed reactions have appeared (see below).

1.4.1.4.1 Palladium-catalyzed Asymmetric Hydrosilylation of Styrenes with Trichlorosilane

The asymmetric hydrosilylation that has been most extensively studied recently is the palladium-catalyzed hydrosilylation of styrene derivatives with trichlorosilane in the presence of palladium catalysts coordinated with chiral monodentate phosphine ligands. The MOP ligands whose chirality is due to the 1,1′-binaphthyl axial chirality [35] were modified for higher enantioselectivity in the catalytic asymmetric hydrosilylation of styrenes [36]. It turned out that the introduction of two trifluoromethyl groups onto the phenyl rings of the diphenylphosphino group on the H-

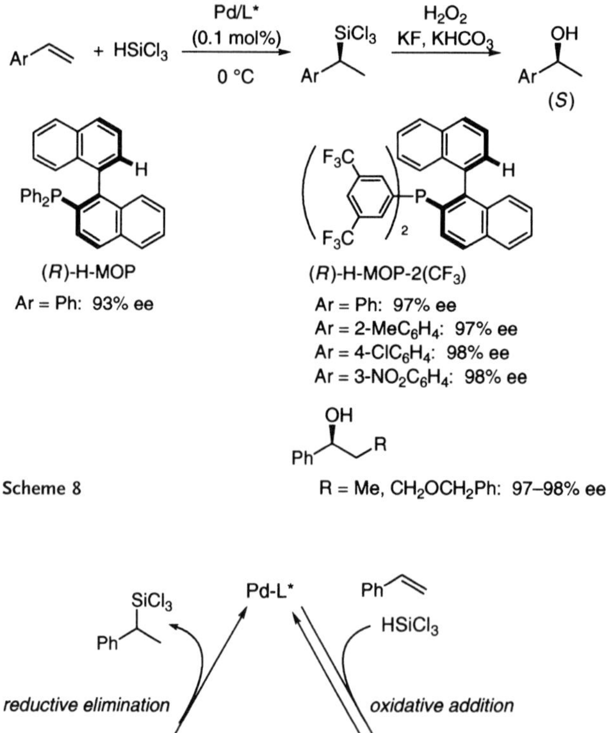

Ar + $HSiCl_3$ → (Pd/L* (0.1 mol%), 0 °C) → Ar–CH($SiCl_3$)–CH_3 → (H_2O_2, KF, $KHCO_3$) → Ar–CH(OH)–CH_3 (*S*)

(*R*)-H-MOP
Ar = Ph: 93% ee

(*R*)-H-MOP-2(CF_3)
Ar = Ph: 97% ee
Ar = 2-MeC_6H_4: 97% ee
Ar = 4-ClC_6H_4: 98% ee
Ar = 3-$NO_2C_6H_4$: 98% ee

R = Me, CH_2OCH_2Ph: 97–98% ee

Scheme 8

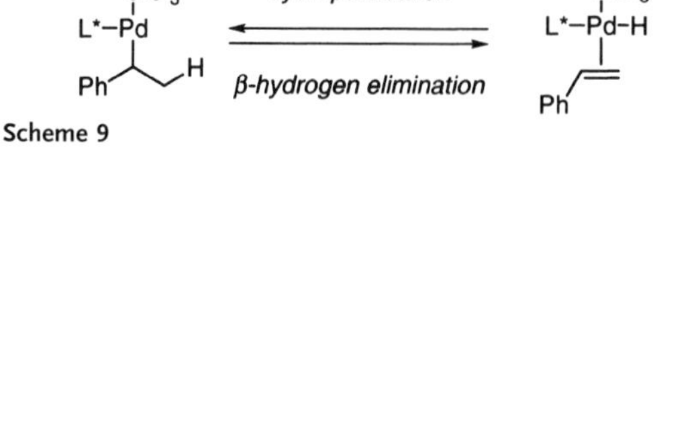

Pd-L*
oxidative addition
hydropalladation
β-hydrogen elimination
reductive elimination

Scheme 9

MOP ligand greatly enhances the enantioselectivity and catalytic activity of its palladium complex (Scheme 8). Thus, the hydrosilylation of styrene with trichlorosilane in the presence of 0.1 mol% of the palladium catalyst coordinated with (*R*)-H-MOP-2(CF_3) was completed within 1 h at 0 °C to give a quantitative yield of 1-phenyl-1-(trichlorosilyl)ethane, whose enantiomeric purity was determined to be 97% *ee* by oxidation to (*S*)-1-phenylethanol. Under the same conditions, the standard H-MOP ligand gave (*S*)-1-phenylethanol of 93% *ee*. The palladium complex of (*R*)-H-MOP-2(CF_3) also catalyzed the hydrosilylation of substituted styrenes on the phenyl ring or at the β-position to give the corresponding chiral benzylic alcohols in over 96% *ee*. Deuterium-labeling studies on the hydrosilylation of regiospecifically deuterated styrene revealed that β-hydrogen elimination from 1-phenylethyl(silyl)palladium intermediate is very fast compared with reductive elimination giving the hydrosilylation product when ligand H-MOP-2(CF_3) is used (Scheme 9). The catalytic cycle involving a hydropalladation step is supported by the formation of an indane derivative in the hydrosilylation of *o*-allylstyrene [37]. The palladium-catalyzed asymmetric hydrosilylation of (*E*)-1-aryl-2-(trichlorosilyl)ethenes, which are readily generated by platinum-catalyzed hydrosilylation of arylacetylenes, opened up a new method of preparing optically active 1,2-diols from arylacetylenes (Scheme 10) [38].

The asymmetric hydrosilylation of styrenes has also been studied by the use of several types of other chiral monophosphine ligands (Scheme 11). Moderate to high enantioselectivity has been reported with axially chiral biaryl-based monophosphine ligands [39, 40] and monophosphine ligands on planar chiral ferrocenes [41] and η^6-arene(tricarbonyl)chromium [42]. Recently, Johannsen reported that very high enantioselectivity is realized by use of one of the chiral phosphoramidite ligands which include the axially chiral (*S*)-1,1′-binaphthol [43]. The most enantioselective is that substituted with bis((*R*)-1-phenylethyl)amino group on the phosphorus atom, which gave (*R*)-1-phenylethanol of 99% *ee*. The high enantioselectivity was also observed for the styrenes substituted with electron-withdrawing groups on the phenyl.

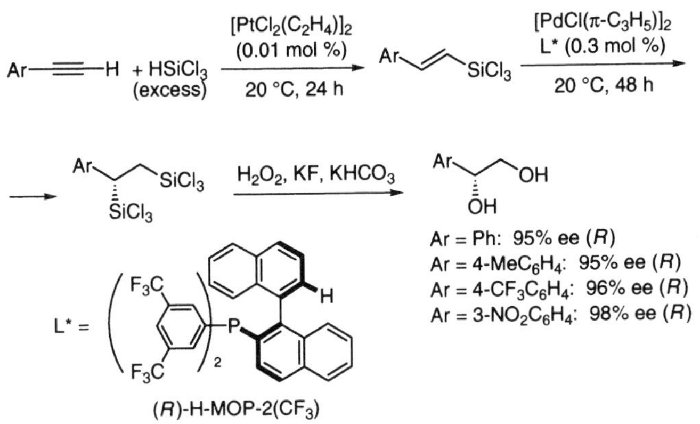

Ar—≡—H + $HSiCl_3$ (excess) → [$[PtCl_2(C_2H_4)]_2$ (0.01 mol %), 20 °C, 24 h] → Ar–CH=CH–$SiCl_3$ → [$[PdCl(\pi\text{-}C_3H_5)]_2$, L* (0.3 mol %), 20 °C, 48 h] → Ar–CH($SiCl_3$)–CH_2SiCl_3 → [H_2O_2, KF, $KHCO_3$] → Ar–CH(OH)–CH_2OH

Ar = Ph: 95% ee (*R*)
Ar = 4-MeC_6H_4: 95% ee (*R*)
Ar = 4-$CF_3C_6H_4$: 96% ee (*R*)
Ar = 3-$NO_2C_6H_4$: 98% ee (*R*)

L* = (*R*)-H-MOP-2(CF_3)

Scheme 10

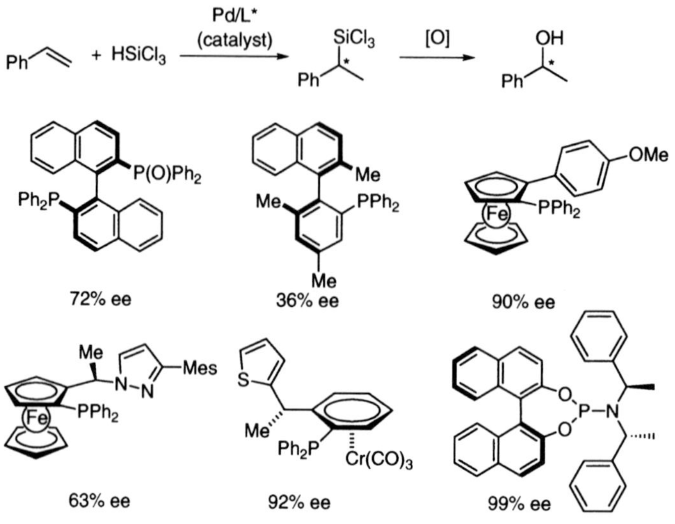

Scheme 11

1.4.1.4.2 **Palladium-catalyzed Asymmetric Hydrosilylation of 1,3-Dienes with Trichlorosilane**

Palladium-catalyzed asymmetric hydrosilylation of 1,3-dienes with trichlorosilane is another synthetically useful asymmetric reaction, because the reaction produces enantiomerically enriched allylsilanes which are chiral reagents giving, for example, homoallyl alcohols on reaction with aldehydes. Similarly to the palladium-catalyzed hydrosilylation of styrenes, monodentate phosphine ligands are used because the palladium complexes coordinated with chelating bisphosphine ligands are much less active than those of monophosphine ligands for 1,3-dienes. In the hydrosilylation of cyclic 1,3-dienes, Ar-MOP ligands, which are substituted with aryl groups at the 2′ position of the MOP ligands, were found to be more enantioselective than MeO-MOP or H-MOP (Scheme 12) [44]. Of the aryl groups at the 2′ position examined, 3,5-dimethyl-4-methoxyphenyl was most enantioselective, giving allylsilanes of 90% *ee* and 79% *ee* in the reaction of 1,3-cyclopentadiene and 1,3-cyclohexadiene, respectively. The Ar-MOP ligand containing the *n*-octyl group at 6 and 6′ positions showed higher enantioselectivity than that lacking the long-chain alkyl group [45]. The higher solubility of the dioctylated ligand in the reaction system realized high catalytic activity at a low reaction temperature.

For linear 1,3-dienes, the MOP ligands are not so effective as for cyclic 1,3-dienes. The highest enantioselectivity for 1,3-decadiene was 77% *ee*, which was reported by use of the dioctylated Ar-MOP ligand [35]. One of the bis(ferrocenyl)monophosphine ligands, which have two planar chiral ferrocenyl groups on the phosphorus atom, is more effective than the MOP ligands (Scheme 13) [46]. The ferrocenylphosphine (*S*)-(*R*)-bisPPFOMe-Ar, where the aryl group is 3,5-$(CF_3)_2C_6H_3$ gave the corresponding allylic silanes of highest enantioselectivity in

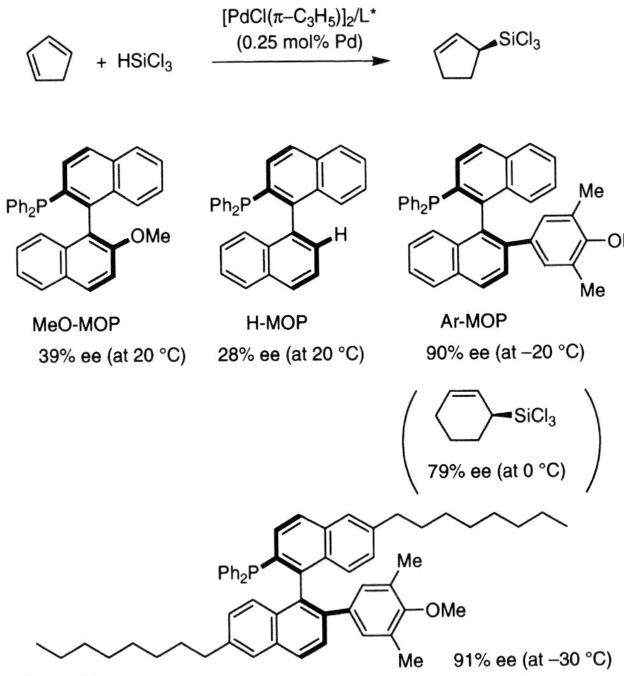

$[PdCl(\pi\text{-}C_3H_5)]_2/L^*$ (0.25 mol% Pd)

+ $HSiCl_3$ → $SiCl_3$

MeO-MOP 39% ee (at 20 °C)

H-MOP 28% ee (at 20 °C)

Ar-MOP 90% ee (at −20 °C)

($SiCl_3$ 79% ee (at 0 °C))

91% ee (at −30 °C)

Scheme 12

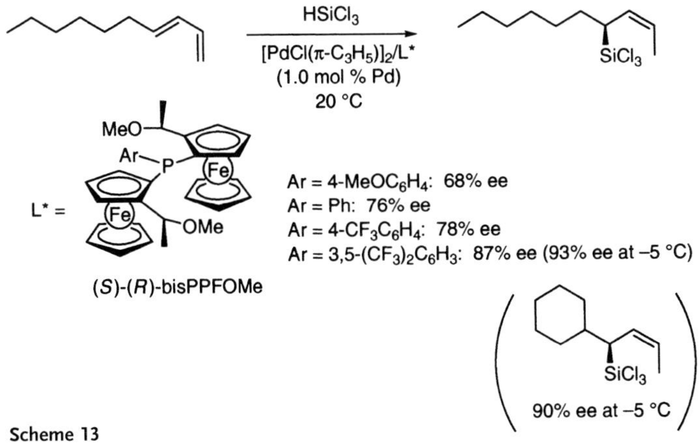

$HSiCl_3$, $[PdCl(\pi\text{-}C_3H_5)]_2/L^*$ (1.0 mol % Pd), 20 °C

$SiCl_3$

L* = (S)-(R)-bisPPFOMe

Ar = 4-$MeOC_6H_4$: 68% ee
Ar = Ph: 76% ee
Ar = 4-$CF_3C_6H_4$: 78% ee
Ar = 3,5-$(CF_3)_2C_6H_3$: 87% ee (93% ee at −5 °C)

($SiCl_3$ 90% ee at −5 °C)

Scheme 13

the palladium-catalyzed hydrosilylation of 1,3-decadienes (93% *ee*) and 1-cyclohexyl-1,3-butadiene (90% *ee*).

A new type of asymmetric hydrosilylation which produces axially chiral allenylsilanes by use of a palladium catalyst coordinated with the bisPPFOMe ligand has been reported recently [47]. The hydrosilylation of 1-buten-3-ynes substituted with

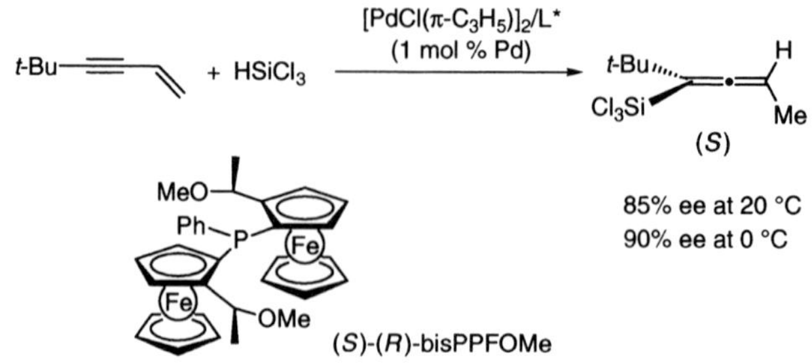

$$t\text{-Bu}-\equiv-\text{CH}{=}\text{CH}_2 + \text{HSiCl}_3 \xrightarrow[\text{(1 mol \% Pd)}]{[\text{PdCl}(\pi\text{-C}_3\text{H}_5)]_2/\text{L}^*} (S)\text{-allene}$$

(S)-(R)-bisPPFOMe

85% ee at 20 °C
90% ee at 0 °C

Scheme 14

sterically bulky groups such as *tert*-butyl at the acetylene terminus took place in a 1,4-fashion to give allenyl(trichloro)silanes with high selectivity. The highest enantioselectivity (90% *ee*) was observed in the reaction of 5,5-dimethyl-1-hexen-3-yne with trichlorosilane catalyzed by the bisPPFOMe-palladium complex (Scheme 14).

1.4.1.4.3 **Palladium-catalyzed Asymmetric Cyclization-Hydrosilylation**

The palladium-catalyzed cyclization-hydrosilylation of 1,6-dienes (Scheme 3) [20] has been extended to asymmetric synthesis using 4-substituted 2-(2-pyridinyl)-2-oxazoline ligands in place of phenanthroline. The oxazoline substituted with an isopropyl group was most enantioselective, giving the cyclization-hydrosilylation product in 87% *ee* in the reaction of dimethyl diallylmalonate with triethylsilane (Scheme 15) [48]. A little higher enantioselectivity was observed in the reaction with $HSiMe_2OSiPh_2Bu\text{-}t$ or $HSiMe_2CHPh_2$ [49]. The carbon-silicon bond in the silylated carbocycles was oxidized with hydrogen peroxide in the presence of fluoride into the carbon-oxygen bond to give the corresponding enantiomerically enriched alcohol.

Very recently, 1,6-enynes were reported to undergo cyclization-hydrosilylation in the presence of a cationic rhodium coordinated with biphemp as a chiral ligand (Scheme 16) [50]. For example, the reaction of 4,4-dicarbomethoxy-1-octene-6-yne with triethylsilane at 70 °C gave the cyclic alkenylsilane in 92% *ee*.

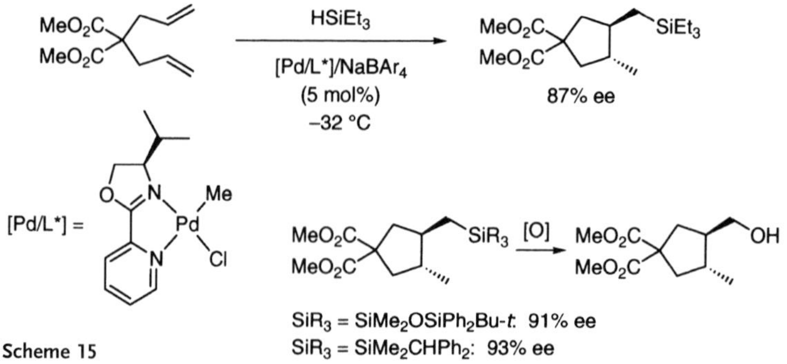

$HSiEt_3$, $[Pd/L^*]/NaBAr_4$ (5 mol%), −32 °C

87% ee

$[Pd/L^*]$ =

SiR_3 → [O]

$SiR_3 = SiMe_2OSiPh_2Bu\text{-}t$: 91% ee
$SiR_3 = SiMe_2CHPh_2$: 93% ee

Scheme 15

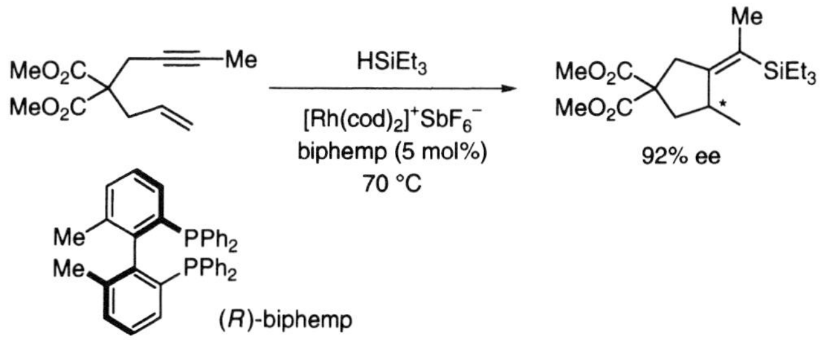

Scheme 16

1.4.1.4.4 Asymmetric Hydrosilylation with Yttrium as a Catalyst

The yttrium hydride {[2,2′-bis(*tert*-butyldimethylsilylamido)-6,6′-dimethylbiphenyl]-YH(THF)}$_2$, conveniently generated *in situ* from [2,2′-bis(*tert*-butyldimethylsilylamido)-6,6′-dimethylbiphenyl]YMe(THF)$_2$ demonstrated its high catalytic activity in olefin hydrosilylation. This system represents the first use of a d^0 metal complex with non-Cp ligands for the catalytic hydrosilylation of olefins. Hydrosilylation of norbornene with $PhSiH_3$ gave the corresponding product in 90% *ee* (Scheme 17) [51].

The yttrocene hydride which has a planar chiral cyclopentadienyl ring is an effective catalyst for the asymmetric cyclization-hydrosilylation of α,ω-dienes [52]. As

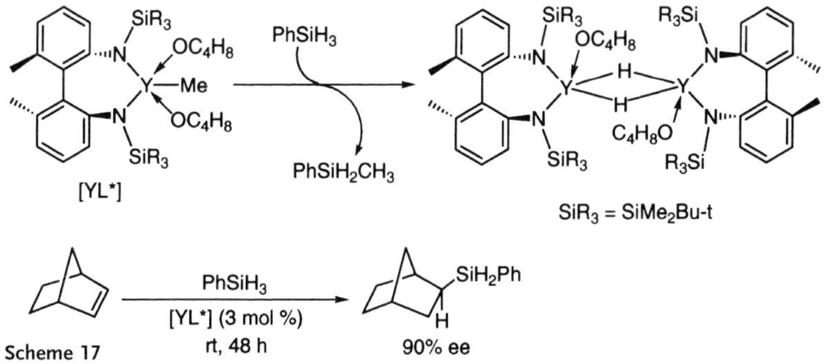

Scheme 17

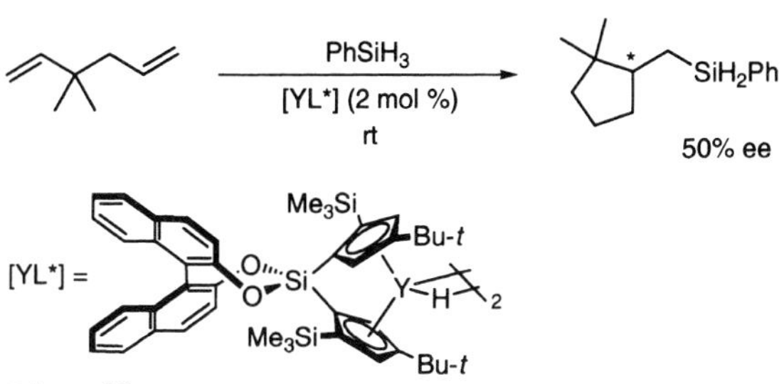

Scheme 18

a very good example, the reaction of a 1,5-hexadiene with phenylsilane proceeded with 50% enantioselectivity to give the chiral cyclopentylmethylsilane in high yield (Scheme 18).

References

1 *Comprehensive Handbook on Hydrosilylation* (Ed.: B. Marciniec), Pergamon Press, Oxford, **1992**; M.A. Brook, *Silicon in Organic, Organometallic, and Polymer Chemistry*, John Wiley & Sons, New York, **2000**.

2 K. Tamao in *Advances in Silicon Chemistry* (Ed.: G.L. Larson), JAI Press, London, **1996**, *3*, 1; I. Fleming, *ChemTracts: Org. Chem.* **1996**, 1.

3 Y. Maruyama, K. Yamamura, T. Sagawa, H. Katayama, and F. Ozawa, *Organometallics* **2002**, *19*, 1308; for a pertinent review, see F. Ozawa, *J. Organomet. Chem.* **2000**, *611*, 332.

4 K. Yamamoto and T. Hayashi in *Transition Metals for Fine Chemicals and Organic Synthesis* (Eds.: M. Beller, C. Bolm), Wiley-VCH, Weinheim, **1998**, Vol. 2, pp 120–140.

5 S. Sakaki, N. Mizoe, and M. Sugimoto, *Organometallics* **1998**, *17*, 2510.

6 A.K. Roy and R.B. Taylor, *J. Am. Chem. Soc.* **2002**, *124*, 9510.

7 J. Stein, L.N. Lewis, Y. Gao, and R.A. Scott, *J. Am. Chem. Soc.* **1999**, *121*, 3693.

8 S. Sasaki, M. Sugimoto, M. Fukuhara, M. Sugimoto, H. Fujimoto, and S. Matsuzaki, *Organometallics* **2002**, *21*, 3788.

9 Y. Nishihara, M. Takemura, and K. Osakada, *Organometallics* **2002**, *21*, 825.

10 F.R. Lemke, K.J. Galat, and W.J. Youngs, *Organometallics* **1999**, *18*, 1419.

11 N. Chatani, T. Kodama, Y. Kajikawa, H. Murakami, F. Kakiuchi, S. Ikeda, and S. Murai, *Chem. Lett.* **2000**, 14.

12 (a) Y.-S. Song, B.R. Yoo, G.-H. Lee, and I.N. Jung, *Organometallics* **1999**, *18*, 3109. (b) M. Rubin, T. Schwier, and V. Gevorgyan, *J. Org. Chem.* **2002**, *67*, 1936, and references therein.

13 N. Sabourault, G. Mignani, A. Wagner, and C. Mioskowski, *Org. Lett.* **2002**, *4*, 2117.

14 M. Chauhan, B.P.S. Chauhan, and P. Boudjouk, *Tetrahedron Lett.* **1999**, *40*, 4127.

15 R. Drake, R. Dunn, C. Sherrington, and S.J. Thomson, *Chem. Commun.* **2000**, 1931.

16 E. de Wolf, E.B.-J. Deelman, and G. van Koten, *Organometallics* **2001**, *20*, 3686.

17 L.-N. He, J.-C. Choi, and T. Sakakura, *Tetrahedron Lett.* **2001**, *42*, 2169.

18 A. Mori and T. Kato, *Synlett* **2002**, 1167.

19 C.-X. Zhao, J. Bass, and J.P. Morken, *Org. Lett.* **2001**, *3*, 2839.

20 X. Wang, H. Chakrapani, C.N. Stengone, and R.A. Widenhoefer, *J. Org. Chem.* **2001**, *66*, 1755, and references therein.

21 K. Itami, K. Mitsudo, A. Nishino, and J. Yoshida, *J. Org. Chem.* **2002**, *67*, 2645.

22 G.A. Molander and J.C.A. Romero, *Chem. Rev.* **2002**, *102*, 2161.

23 T.I. Gountchev and T. Don Tilley, *Organometallics* **1999**, *18*, 5661; A.A. Trifonov, T.P. Spaniol, and J. Okuda, *Organometallics* **2001**, *20*, 4869.

24 R. Takeuchi, S. Nitta, and D. Watanabe, *J. Org. Chem.* **1995**, *60*, 3045; For the Ojima-Crabtree postulation, see I. Ojima, N. Clos, R.J. Donovan, and P. Ingallina, *Organometallics* **1991**, *9*, 3127; R.S. Tanke and R.H. Crabtree, *J. Am. Chem. Soc.* **1990**, *112*, 7984.

25 B.M. Trost and Z.T. Ball, *J. Am. Chem. Soc.* **2001**, *123*, 12726.

26 A. Mori, E. Takehisa, H. Kajiro, K. Hirabayashi, Y. Nishihara, and T. Hiyama, *Chem. Lett.* **1998**, 443.

27 J.W. Faller and D.G. D'Alliessi, *Organometallics* **2002**, *21*, 1743.

28 H. Katayama, K. Taniguchi, M. Kobayashi, T. Sagawa, T. Minami, and F. Ozawa, *J. Organomet. Chem.* **2002**, *645*, 192.

29 V. Na and S. Chang, *Org. Lett.* **2000**, *2*, 1887.

30 Y. Kawanami, Y. Sonoda, T. Mori, and K. Yamamoto, *Org. Lett.* **2002**, *4*, 2825.

31 M. Martin, E. Sola, F. J. Lahoz, and L. A. Oro, *Organometallics* **2002**, *21*, 4027; see also S. M. Maddock, C. E. F. Rickard, W. R. Roper, and L. Wright, *Organometallics* **1996**, *15*, 1793.

32 T. Muraoka, I. Matsuda, and K. Itoh, *Organometallics* **2002**, *21*, 3650.

33 (a) X. Wang, H. Chakrapani, J. W. Madine, M. A. Keyerleber, and R. A. Widenhoefer, *J. Org. Chem.* **2002**, *67*, 2778. (b) T. Uno, S. Wakayanagi, Y. Sonoda, and K. Yamamoto, *Synlett*, **2003**, 1997.

34 For reviews: T. Hayashi in *Comprehensive Asymmetric Catalysis* (Eds.: E. N. Jacobsen, A. Pfaltz, and H. Yamamoto), Springer, Berlin, **1999**, Vol. 1, pp 319–333; J. Tang and T. Hayashi in *Catalytic Heterofunctionalization* (Eds.: A. Togni and H. Grützmacher), Wiley-VCH, Weinheim, **2001**, pp 73-90; H. Nishiyama and K. Itoh in *Catalytic Asymmetric Synthesis* (Ed.: I. Ojima), Wiley-VCH, New York, **2000**, pp 111–143.

35 For a pertinent review on the MOP ligands: T. Hayashi, *Acc. Chem. Res.* **2000**, *33*, 354.

36 (a) T. Hayashi, S. Hirate, K. Kitayama, H. Tsuji, A. Torii, and Y. Uozumi, *Chem. Lett.* **2000**, 1272. (b) T. Hayashi, S. Hirate, K. Kitayama, H. Tsuji, A. Torii, and Y. Uozumi, *J. Org. Chem.* **2001**, *66*, 1441.

37 Y. Uozumi, H. Tsuji, and T. Hayashi, *J. Org. Chem.* **1998**, *63*, 6137.

38 T. Shimada, K. Mukaide, A. Shinohara, J. W. Han, and T. Hayashi, *J. Am. Chem. Soc.* **2002**, *124*, 1584.

39 S. Gladiali, S. Pulacchini, D. Fabbri, M. Manassero, and M. Sansoni, *Tetrahedron Asymmetry* **1998**, *9*, 391.

40 G. Bringmann, A. Wuzik, M. Breuning, P. Henschel, K. Peters, and E.-M. Peters, *Tetrahedron Asymmetry* **1999**, *10*, 3025.

41 (a) H. L. Pedersen and M. Johannsen, *Chem. Commun.* **1999**, 2517. (b) H. L. Pedersen and M. Johannsen, *J. Org. Chem.* **2002**, *67*, 7982. (c) G. Pioda and A. Togni, *Tetrahedron Asymmetry* **1998**, *9*, 3903.

42 I. Weber and G. B. Jones, *Tetrahedron Lett.* **2001**, *42*, 6983.

43 J. F. Jensen, B. Y. Svendsen, T. V. la Cour, H. L. Pedersen, and M. Johannsen, *J. Am. Chem. Soc.* **2002**, *124*, 4558.

44 T. Hayashi, J. W. Han, A. Takeda, J. Tang, K. Nohmi, K. Mukaide, H. Tsuji, and Y. Uozumi, *Adv. Synth. Catal.* **2001**, *343*, 279.

45 (a) J. W. Han and T. Hayashi, *Chem. Lett.* **2001**, 976. (b) J. W. Han and T. Hayashi, *Tetrahedron Asymmetry* **2002**, *13*, 325.

46 J. W. Han, N. Tokunaga, and T. Hayashi, *Helvetica Chimica Acta* **2002**, *85*, 3848.

47 J. W. Han, N. Tokunaga, and T. Hayashi, *J. Am. Chem. Soc.* **2001**, *123*, 12915.

48 (a) N. S. Perch and R. A. Widenhoefer, *J. Am. Chem. Soc.* **1999**, *121*, 6960. (b) N. S. Perch, T. Pei, and R. A. Widenhoefer, *J. Org. Chem.* **2000**, *65*, 3836.

49 (a) T. Pei and R. A. Widenhoefer, *Tetrahedron Lett.* **2000**, *41*, 7597. (b) T. Pei and R. A. Widenhoefer, *J. Org. Chem.* **2001**, *66*, 7639.

50 H. Chakrapani, C. Liu, and R. A. Widenhoefer, *Org. Lett.* **2003**, *5*, 157.

51 T. I. Gountchev and T. D. Tilley, *Organometallics* **1999**, *18*, 5661.

52 A. R. Muci and J. E. Bercaw, *Tetrahedron Lett.* **2000**, *41*, 7609.

1.4.2
Hydrosilylations of Carbonyl and Imine Compounds

Hisao Nishiyama

1.4.2.1
Hydrosilylation of Carbonyl Compounds

Metal-catalyzed hydrosilylations of carbonyl compounds have been investigated for a long time as one of the important and convenient reduction methods to obtain secondary alcohols [1]. The reaction with organohydrosilane reagents can easily be manipulated by the usual Schlenk technique without the need for handling the hazardous gas, hydrogen, in hydrogenation reactions. After a simple hydrolysis work-up of the silyl ether adducts, the desired secondary alcohols are isolated by distillation or chromatography. A small amount of side reaction giving silylenol ethers can occur, depending on the reaction conditions. The adoption of optically active ligands facilitates asymmetric reduction of prochiral ketones such as acetophenone (Scheme 1). The first enantioselective hydrosilylation of prochiral ketones was reported in 1972 with the platinum complex of an optically active monophosphine ligand, the enantioselectivity being below 20% [2]. In the search for more efficient catalysts, investigations were shifted to optically active rhodium catalysts, after the Wilkinson type of rhodium catalysts had been successfully applied to the asymmetric hydrogenation of olefins.

1.4.2.1.1 Rhodium Catalysts

In 1973, three research groups reported the enantioselective reductions of acetophenone with rhodium catalysts of chiral monodentate or bidentate phosphine ligands, obtaining 29–43% *ee*, the selectivity varying depending on the hydrosilane used, e.g. Me_2PhSiH, $MePh_2SiH$, or Ph_2SiH_2 [3]. In 1981, the rhodium catalyst Glucophinite, derived from glucose, attained the maximum 65% *ee* for the reduction of acetophenone (Scheme 2) [4]. In 1977, for reduction of ketoesters such as

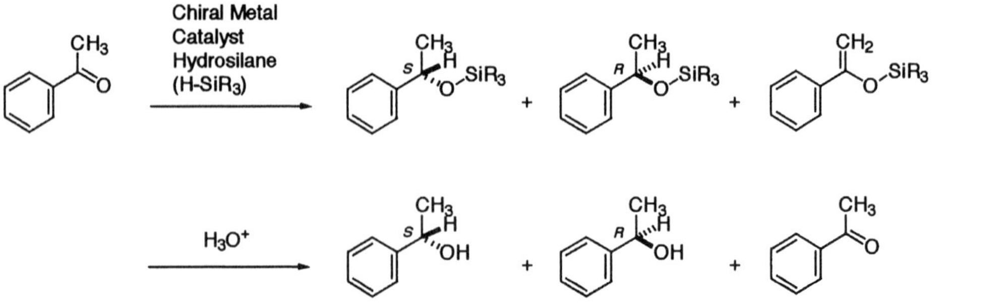

Chiral Metal Catalyst Hydrosilane (H-SiR_3)

H_3O^+

Scheme 1

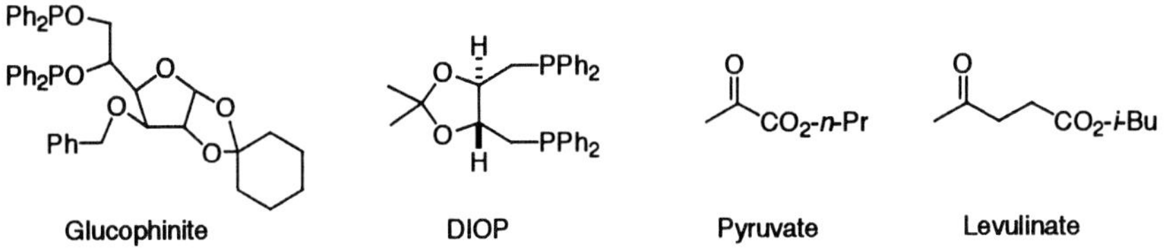

Glucophinite DIOP Pyruvate Levulinate

Scheme 2

pyruvate and levulinate, the combination of Rh-DIOP catalyst and *α*-NpPhSiH was reported to reach 84–86% *ee* [5].

Nitrogen-based ligands, pyridine-imines, e.g., Ppei, which were derived from 2-pyridincarboxaldehyde by condensation with optically active amines such as (*S*)-phenylethylamine or (–)-3-aminomethylpinane, emerged in 1982 for hydrosilylation of ketones, reaching 79% *ee* for acetophenone with $[Rh(COD)Cl]_2$ (1 mol% of Rh) (Scheme 3) [6]. Further improvements of *ee* were attained, reaching 89% *ee* with the bidentate pyridine-monooxazoline [7–10], e.g., *tert*-Bu-Pymox, derived from (*S*)-*tert*-leucinol [9], and up to 97.6% *ee* with bidentate pyridine-thiazoline derived from L-cysteine ethyl ester, named Pythia [6, 8]. With Pythia, 3-methoxyphenyl methyl ketone was reduced in 93% *ee* [8]. Although these nitrogen-based ligands have to be used in most cases in large excess with respect to rhodium metal (ca. 0.5–1.0 mol%) to attain higher enantioselectivity, they are readily accessible from optically active natural precursors [10]. If ketones of substrates are liquid, the reaction can be carried out under solvent-free conditions. THF, ether, benzene, toluene, CH_2Cl_2, and CCl_4 can be used as solvents. Most of the reaction proceeds below room temperature (down to –78 °C).

In 1989, 2,6-bis(oxazolinyl)pyridine (Pybox) was introduced as a C_2 chiral tridentate ligand to obtain high efficiency for asymmetric hydrosilylation of ketones with diphenylsilane (Scheme 4) [9]. By using $RhCl_3$(Pybox-*ip*) complex, which was fully characterized by X-ray analysis, aromatic ketones were reduced to secondary alcohols in over 90% yields, in 95% *ee* for acetophenone, 99% *ee* for 1-tetralone, 94% *ee* for 1-acetonaphthone, and 95% *ee* for ethyl levulinate with the assistance of excess Pybox and silver ions [9, 12]. The remote electron-withdrawing substituent (ex. $X=CO_2Me$) on the 4-position of the pyridine skeleton of Pybox enhanced the reaction rate and increased the enantioselectivity [13].

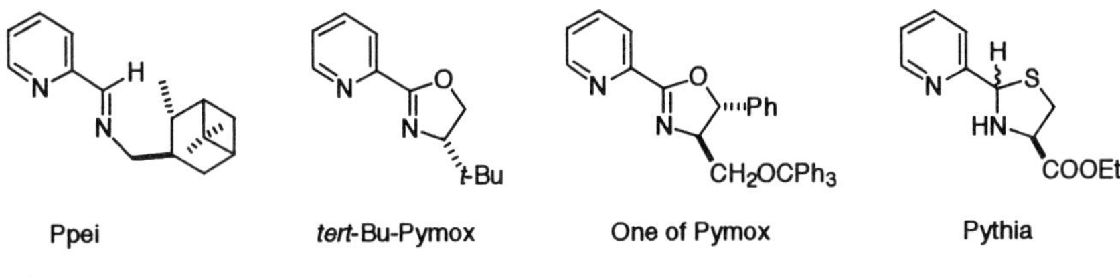

Scheme 3

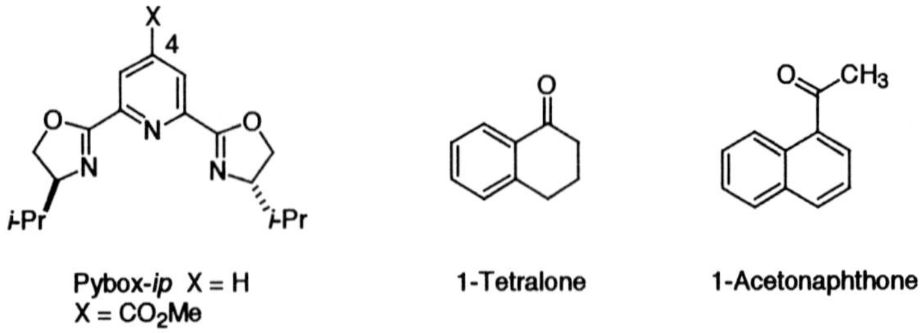

Pybox-*ip* X = H
X = CO_2Me

1-Tetralone

1-Acetonaphthone

Scheme 4

Many nitrogen-based ligands including oxazoline ligands and sparteine were reported in 1990–1998 by different groups, but the enantioselectivity was in the middle range, with 90% *ee* as a maximum (Scheme 5) [14–21].

Into these new types of nitrogen-based ligands, a diphosphine ligand was introduced in 1994. A bis-phosphinoferrocene named Trap, having the special feature of the wide bite angle 164° (almost *trans*-chelating), achieved a high enantioselectivity of over 90% for the first time as chiral phosphines (Scheme 6) [22]. Acetophenone was reduced in 92% *ee* and 88% yield with the combination of Bu-Trap, $Rh(COD)Cl/_2$, and diphenylsilane at –40 °C, and in 94% *ee* and 89% yield by the use of Et-Trap-H, a planar-chirality ligand [23]. In particular, acetylferrocene and 1-acetylcyclohexene were reduced with 95–97% *ee* with Bu-Trap, $[Rh(COD)_2]BF_4$ (1 mol% of Rh) [24]. Et-Trap proved to be efficient for bulky ketones such as acetyladamantane (96% *ee*, 78% yield) and ethyl 2,2-dimethyl-3-oxo-butyrate (98% *ee*, 80% yield) [24]. The bidentate diphosphine ligand DuPhos was applied to intramolecular hydrosilylation [25]. A new P-chiral bisphosphine-ferrocene ligand was used to attempt the reduction of several aromatic ketones with 1-naphthylphenylsilane, giving up to 92% *ee* for acetophenone [26].

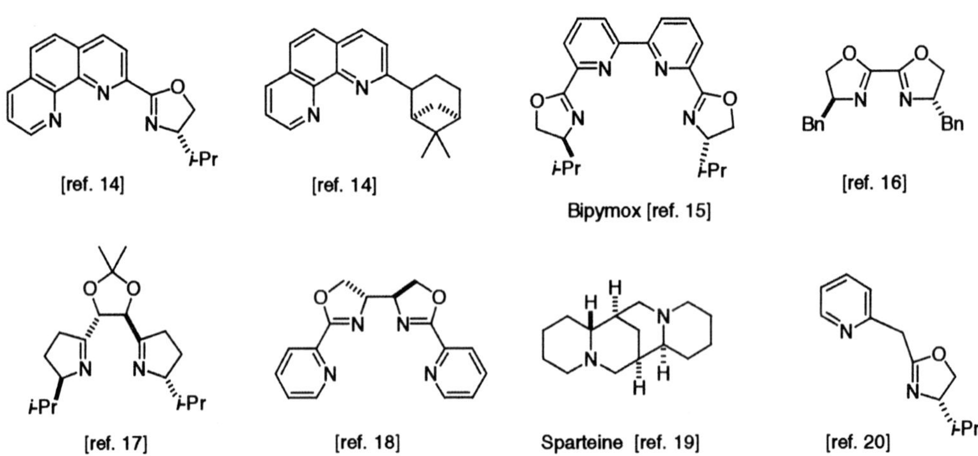

Scheme 5

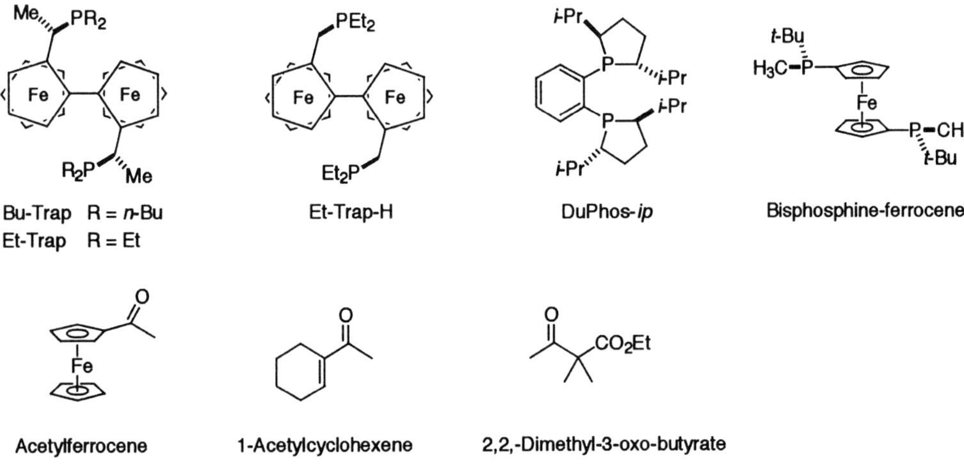

Scheme 6

In the later 1990s, a variety of chiral multi-dentate ligands with mixed hetero atoms, such as P/N [27, 28, 30, 31], P/Se [29], P/S [32, 33] and phosphite-based ligands appeared and were examined for hydrosilylation of ketones (Scheme 7) [27–36]. Acetophenone was reduced in 94% *ee* with the indane type of phosphine-oxazo-

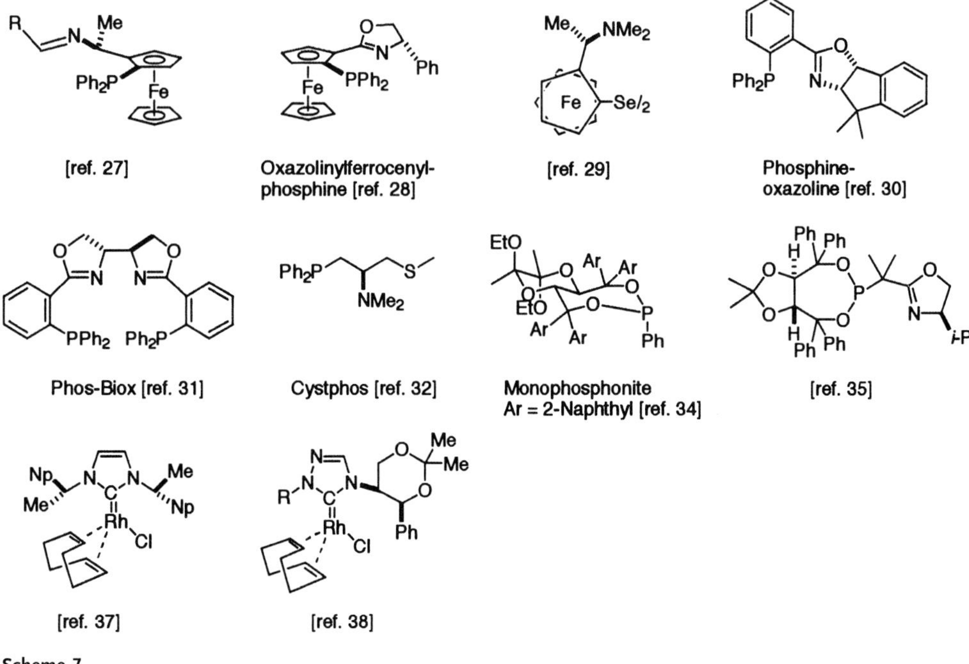

Scheme 7

line [30]. Phos-Biox, which is a tetradentate P-N-N-P ligand, exhibited the high efficiency of 97% *ee* for acetophenone with 0.25 mol% of Rh catalyst [31]. A unique chiral cyclic monophosphonite ligand was synthesized and evaluated in the hydrosilylation to give 86% *ee* for acetophenone [34]. A phosphite ligand containing a chiral tetraaryl-dioxolanedimethanols (TADDOL) skeleton, and chiral oxazoline gave 88% *ee* for acetophenone and notably 95% *ee* for *t*-butyl methyl ketone [35].

Interestingly, heterocyclic carbene complexes of rhodium proved to be active catalysts for the asymmetric hydrosilylation of ketones, resulting in moderate enantioselectivity [37, 38].

1.4.2.1.2 **Iridium Catalysts**

Using iridium catalysts, the hydrosilylations of ketones proceed smoothly. Highly enantioselective reaction giving >90% *ee* was first reported in 1995 with chiral diphenyloxazolinyl-ferrocenylphosphine, named DIPOF (Scheme 8) [29]. The hydrosilylation of acetophenone with $Ir(COD)Cl/_2$ (1 mol% of Ir to ketone) and diphenylsilane at 0 °C for 20 h gave 1-phenylethanol in almost quantitative yield and 96% *ee* (*S*), while the same reaction with $Rh(COD)Cl/_2$ in place of $Ir(COD)Cl/_2$ resulted in 91% *ee* with the opposite absolute configuration (*R*), interestingly. Several ketones were also subjected to this reaction, giving higher *ee*s of over 90%.

1.4.2.1.3 **Ruthenium Catalysts**

Chiral ruthenium catalyst derived from $RuCl_2(PPh_3)$(oxazolinylferrocenyl-phosphine) with the aid of $Cu(OTf)_2$ was reported in 1998 to attain high enantioselectivities, >95% for acetophenone, 97% for propiophenone, and 95% for butyrophenone [39]. Chiral tridentate P,P,N-ligand containing two phosphines and one pyridine was employed for ruthenium-catalyzed hydrosilylation, giving a middle range of enantioselectivity [40]. In comparison with BINAP and Pybox, the ruthenium-catalyzed hydrosilylation of ketones was carried out using tol-BINAP, giving an improved result for the reduction of acetophenone (Scheme 8) [41].

1.4.2.1.4 **Copper Catalysts**

In 2001, an extremely high level of efficiency and enantioselectivity was attained by the use of a chiral diphosphine-copper catalyst, after screening a number of chiral

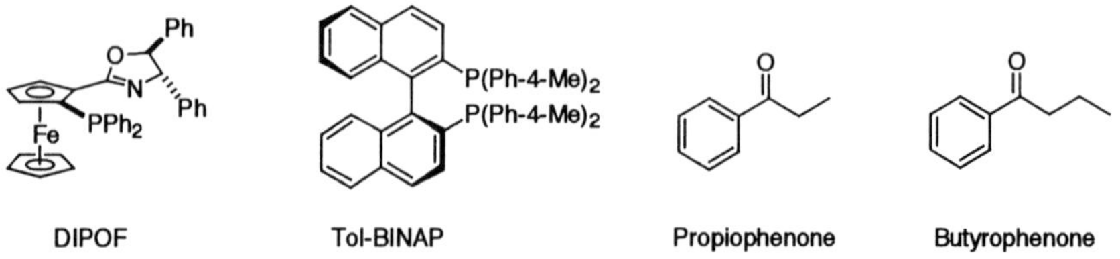

Scheme 8

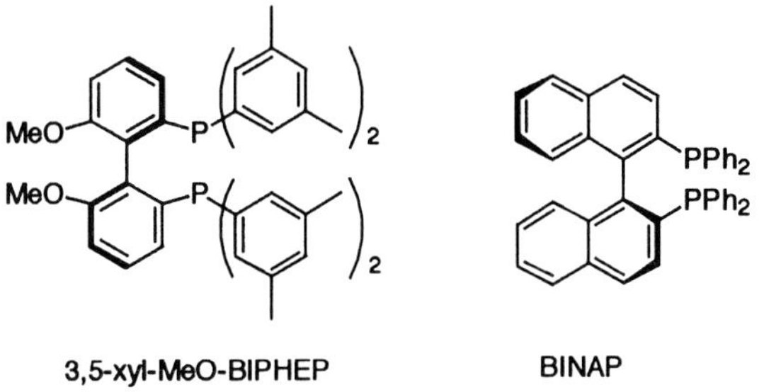

3,5-xyl-MeO-BIPHEP BINAP

Scheme 9

phosphine ligands in the combination of Stryker's reagent $[Cu(PPh_3)H]_6$ and polymethylhydrosilane (PMHS) as an inexpensive hydrogen source resulting in 75–86% *ee* for several aromatic ketones (Scheme 9) [42]. Eventually, the combination catalyst of CuCl (3 mol%), NaO-*t*-Bu (3 mol%), and 3,5-xyl-MeO-BIPHEP significantly improved the enantioselectivity, to 94% *ee* for acetophenone, 97% *ee* for propiophenone, 95% *ee* for 2-acetonaphthone, and 92% *ee* for 1-tetralone in toluene with PMHS at −78 °C. It was found that the mol% of the ligand could be lowered to 0.005 mol%, which corresponds to a substrate-to-ligand ratio of 20 000 : 1 with 0.5 mol% of copper(I), without any decrease of enantiomeric excess.

The catalysts derived from copper fluoride and BINAP was investigated with Ph_3SiH, giving 92% *ee* for butyrophenone [43].

1.4.2.1.5 Titanium Catalysts

In 1988, a titanium complex, $TiCp_2Ph_2$, was reported to act as a catalyst for the hydrosilylation of ketones with hydrosilanes, giving secondary alcohols [44, 45].

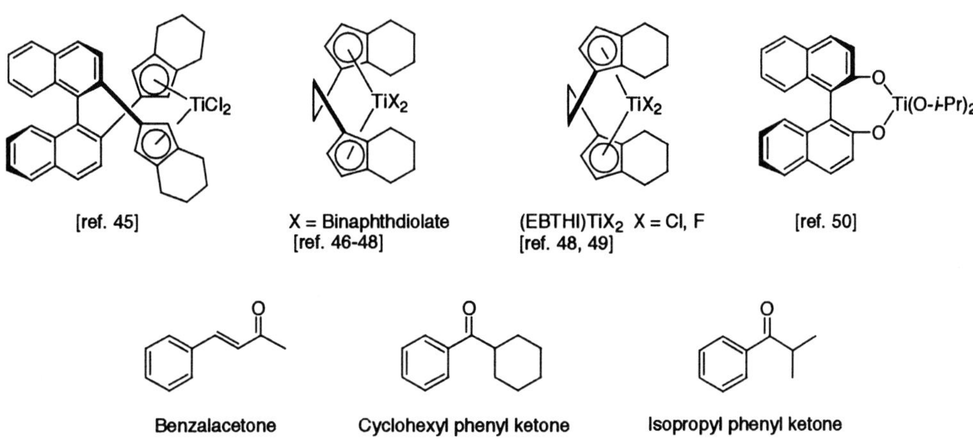

Scheme 10

Asymmetric variations first appeared in 1994 by using binaphthyl-biscyclopentadienyl titanium chloride and biscyclopentadienyl titanium binaphthdiolate (Scheme 10) [45, 46–48]. The latter catalyst exhibited high activity for the hydrosilylation of aromatic ketones with PMHS (5-fold excess to ketone) with 97% *ee* for acetophenone and 95% *ee* for 2-acetonaphthone [47]. However, 1-acetylcyclohexene and benzalacetone gave reduced enantioselectivity [47]. In these catalysts, initial treatment with alkyllithium is essential to produce the active titanium hydride species. Ethylenebis(tetrahydroindenyl)titanium chloride (EBTHI)$TiCl_2$ [48], (EBTHI)TiF_2 [49], and Ti-Binolate [50] exhibited activity for this reaction with diphenylsilane. In the case of the titanium fluoride catalyst, alcoholic additives such as methanol and ethanol improved the turnover number of the catalyst and the enantioselectivity with PMHS [49]. Isopropyl phenyl ketone was reduced after activation of the titanium fluoride catalyst (1 mol%) to give the secondary alcohol with PMHS in 87% yield and with 98% *ee*. 1-Acetylcyclohexene and cyclohexyl phenyl ketone were also reduced with 96–98% *ee*. Moreover, no alkyllithium reagents were necessary for activation.

1.4.2.2 Hydrosilylation of Imine Compounds

1.4.2.2.1 Rhodium Catalysts

It is synthetically of importance that the asymmetric reduction of imines and their derivatives ($R_2C{=}NR$) with hydrosilanes and transition metal catalysts can provide optically active primary or secondary amines [1]. In 1985, it was reported that, in the presence of chiral rhodium catalysts, the reduction of ketimines proceeds very smoothly, giving the middle range of enantioselectivity in high yields. For example, with rhodium-DIOP catalyst (2 mol%) and diphenylsilane (Ph_2SiH_2), the ketimine derived from acetophenone and benzylamine could be reduced to the *N*-benzyl-phenylethylamine with 65% *ee* (Scheme 11) [51]. However, the enantioselectivity with rhodium catalysts has not so far been improved.

1.4.2.2.2 Titanium Catalysts

In 1996, an extremely high enantioselectivity was obtained by the use of (tetrahydroindenyl)titanium(IV) fluoride (EBTHI-TiF_2) (1 mol%) and phenylsilane ($PhSiH_3$) (1.5 eq. to imine), which also exhibited high efficiency for the asymmetric hydrosilylation of ketones (Scheme 12) [52]. The *N*-methylimine and the cyclic imine were efficiently reduced at room temperature for 12 h to give the cor-

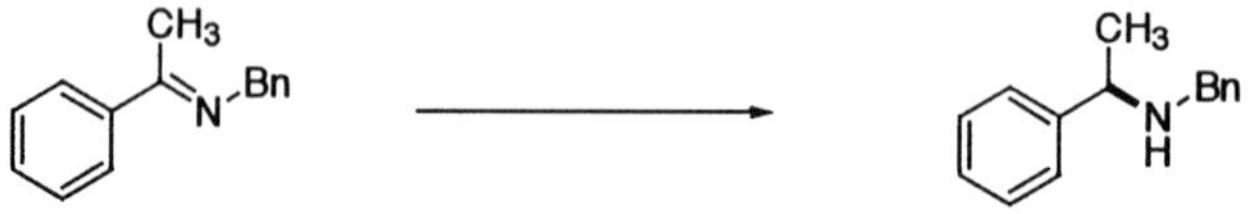

CH_3 N Bn → CH_3 N H Bn

Scheme 11

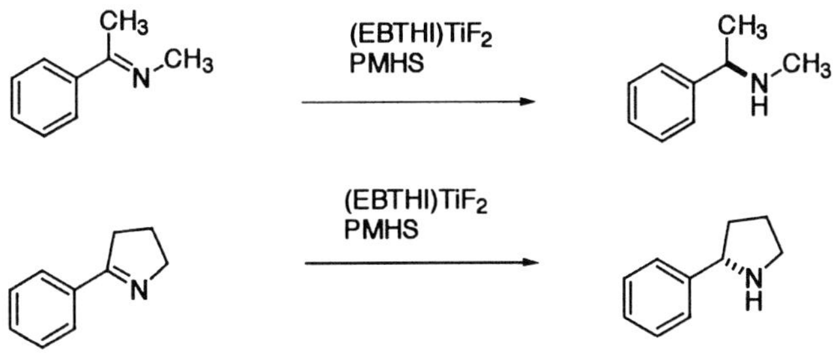

Scheme 12

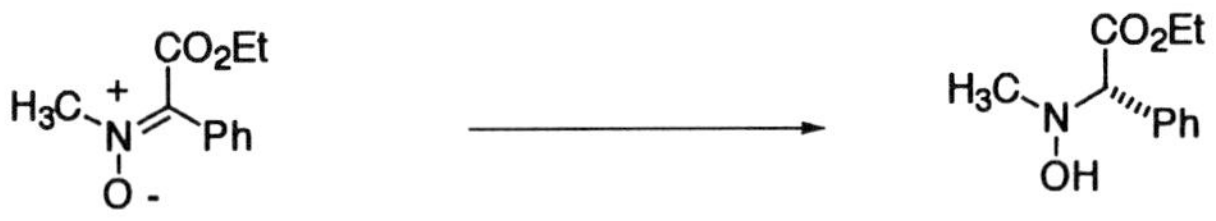

Scheme 13

responding amines, phenylethylamine and pyrrolidine, in 94% and 97% yield and with 97% *ee* and 99% *ee*, respectively. In this catalysis, polymethylhydrosiloxane (PMHS) can also be used as a hydride source in the combination of primary alkylamines such as *tert*-butylamine as an additive [53]. This convenient process could also be applied to the reduction of *N*-aryl-substituted imines [54].

1.4.2.2.3 **Ruthenium Catalysts**

The ruthenium complex, $RuCl_2(PPh_3)$(oxazolinylferrocenyl-phosphine), was applied to a catalyst for the reduction of the cyclic imine in toluene at 0 °C with diphenylsilane to give 88% *ee* in 60% yield [39]. In addition, the combination of DIPOF and iridium complex $[Ir(COD)Cl]_2$ (1 mol% of Ir to imine) also provides a new catalytic system to reduce the imine, with 89% *ee*, in ether at 0 °C [55].

The C=N bond of nitrones was reduced with Ru_2Cl_2(Tol-BINAP)$_2$(Et_3N) (1 mol%) and diphenylsilane at 0 °C to give the hydroxylamine derivatives [56] (Scheme 13). The hydroxylamine was obtained in 91% *ee*. It is important that the hydroxylamines could be converted to the corresponding optically active amines.

References

1 I. Ojima, K. Hirai in *Asymmetric Synthesis* (Ed.: J. D. Morrison), Academic Press, Orlando, **1985**, *5*, 103. H. Brunner, *Methoden Org. Chem. (Houben Weyl) 4th edn.*, **1995**, *E 21 d*, 4074. H. Brunner, H. Nishiyama, K. Itoh in *Catalytic Asymmetric Synthesis* (Ed.: I. Ojima), VCH, New York, **1993**, 303. H. Nishiyama, K. Itoh in *Catalytic Asymmetric Synthesis, 2nd edn.* (Ed.: I. Ojima), VCH, New York, **2000**, 111. H. Nishiyama in *Comprehensive Asymmetric Catalysis I* (Eds.: E. N. Jacobsen, A. Pfaltz, H. Yamamoto), Springer, Berlin, **1999**, 267.

2 K. Yamamoto, T. Hayashi, M. Kumada, *J. Organomet. Chem.* **1972**, *46*, C65.

3 J.-C. Poulin, W. Dumont, T.-P. Dang, H.B. Kagan, *C.R. Acad, Sci., Ser. C* **1973**, *277*, 41. K. Yamamoto, T. Hayashi, M. Kumada, *J. Organomet. Chem.* **1973**, *54*, C45. I. Ojima, T. Kogure, Y. Nagai, *Chem. Lett.* **1973**, 541.

4 T.H. Johnson, K.C. Klein, S. Tomen, *J. Mol. Catal.* **1981**, *12*, 37.

5 I. Ojima, T. Kogure, Y. Nagai, *Tetrahedron Lett.* **1974**, 1889. I. Ojima, T. Kogure, M. Kumagai, *J. Org. Chem.* **1977**, *42*, 1671. I. Ojima, T. Tanaka, T. Kogure, *Chem. Lett.* **1981**, 823.

6 H. Brunner, G. Riepl, *Angew. Chem., Int. Ed. Engl.* **1982**, *21*, 377. H. Brunner, G. Riepl, H. Weitzer, *Angew. Chem., Int. Ed. Engl.* **1983**, *22*, 331.

7 H. Brunner, U. Obermann, P. Wimmer, *J. Organomet. Chem.* **1986**, *316*, C1. H. Brunner, U. Obermann, *Chem. Ber.* **1989**, *122*, 499.

8 H. Brunner, A. Kürzinger, *J. Organomet. Chem.* **1988**, *346*, 413.

9 H. Nishiyama, H. Sakaguchi, T. Nakamura, M. Horihata, M. Kondo, K. Itoh, *Organometallics* **1989**, *8*, 846.

10 G. Balavoine, J.C. Client, I. Lellouche, *Tetrahedron Lett.* **1989**, *39*, 5141.

11 H. Brunner, P. Brandl, *J. Organomet. Chem.* **1990**, *390*, C81.

12 H. Nishiyama, M. Kondo, T. Nakamura, K. Itoh, *Organometallics* **1991**, *10*, 500.

13 H. Nishiyama, S. Yamaguchi, M. Kondo, K. Itoh, *J. Org. Chem.* **1992**, *57*, 4306. Cf. S.B. Park, K. Murata, H. Matsumoto, H. Nishiyama, *Tetrahedron: Asymmetry* **1995**, *6*, 2487.

14 S. Gladiali, L. Pinna, G. Delogu, E. Graf, H. Brunner, *Tetrahedron: Asymmetry* **1990**, *1*, 937.

15 H. Nishiyama, S. Yamaguchi, S.B. Park, K. Itoh, *Tetrahedron: Asymmetry* **1993**, *4*, 143.

16 G. Helmchen, A. Krotz, K.T. Ganz, D. Hansen, *Synlett.* **1991**, 257.

17 Y. Imai, W. Zang, T. Kida, Y. Nakatsuji, I. Ikeda, *Tetrahedron: Asymmetry* **1996**, *7*, 2453.

18 S.-G. Lee, C.W. Lim, C.E. Song, I.O. Kim, C.-H. Jun, *Tetrahedron: Asymmetry* **1997**, *8*, 2027.

19 H. Alper, Y. Goldberg, *Tetrahedron: Asymmetry* **1992**, *3*, 1055.

20 M.D. Fryzuk, L. Jafarpour, S.J. Rettig, *Tetrahedron: Asymmetry* **1998**, *9*, 3191.

21 H. Brunner, R. Störiko, *Eur. J. Org. Chem.* **1998**, 783. H. Brunner, R. Störiko, N. Bernhard, *Tetrahedron: Asymmetry* **1998**, *9*, 407.

22 M. Sawamura, R. Kuwano, Y. Ito, *Angew. Chem. Int. Ed. Engl.* **1994**, *33*, 111. M. Sawamura, R. Kuwano, J. Shirai, Y. Ito, *Synlett.* **1995**, 347. R. Kuwano, M. Sawamura, J. Shirai, M. Takahashi, Y. Ito, *Tetrahedron Lett.* **1995**, *36*, 5239.

23 R. Kuwano, T. Uemura, M. Saitoh, Y. Ito, *Tetrahedron Lett.* **1999**, *40*, 1327.

24 R. Kuwano, M. Sawamura, J. Shirai, M. Takahashi, Y. Ito, *Bull. Chem. Soc. Jpn.* **2000**, *73*, 485.

25 M.J. Burk, J.E. Feaster, *Tetrahedron Lett.* **1992**, *33*, 2099.

26 H. Tsuruta, T. Imamoto, *Tetrahedron Lett.* **1999**, *10*, 877.

27 T. Hayashi, C. Hayashi, Y. Uozumi, *Tetrahedron: Asymmetry* **1995**, *6*, 2503.

28 Y. Nishibayashi, K. Segawa, K. Ohe, S. Uemura, *Organometallics* **1995**, *14*, 5486.

29 Y. Nishibayashi, J.D. Singh, K. Segawa, S. Fukuzawa, K. Ohe, S. Uemura, *J. Chem. Soc., Chem. Commun.* **1994**, 1375. Y. Nishibayashi, K. Segawa, J.D. Singh, S. Fukuzawa, K. Ohe, S. Uemura, *Organometallics* **1996**, *15*, 370.

30 A. Sudo, H. Yoshida, K. Saigo, *Tetrahedron: Asymmetry* **1997**, *8*, 3205.

31 S.-G. Lee, C.W. Lim, C.E. Song, I.O. Kim, *Tetrahedron: Asymmetry* **1997**, *8*, 4027.

32 J.W. Faller, K.J. Chase, *Organometallics* **1994**, *13*, 989.

33 M. Hiraoka, A. Nishikawa, T. Morimoto, K. Achiwa, *Chem. Pharm. Bull.* **1998**, *46*, 704.

34 D. Haag, J. Runsink, H.-D. Scharf, *Organometallics* **1998**, *17*, 398.

35 D. Heldmann, D. Seebach, *Helv. Chim. Acta* **1999**, *82*, 1096.

36 S.D. Pastor, S.P. Shum, *Tetrahedron: Asymmetry* **1998**, *9*, 543.

37 W.A. Herrmann, L.J. Goosse, C. Köcher, G.R. Artus, *Angew. Chem. Int. Ed.* **1996**, *35*, 2805.
38 D. Enders, H. Gielen, K. Breuer, *Tetrahedron: Asymmetry* **1997**, *8*, 4027.
39 Y. Nishibayashi, I. Takei, S. Uemura, M. Hidai, *Organometallics* **1998**, *17*, 3420.
40 G. Zhu, M. Terry, X. Zhang, *J. Organomet. Chem.* **1997**, *547*, 97.
41 C. Moreau, C. Frost, B. Murrer, *Tetrahedron Lett.* **1999**, *40*, 5617.
42 B.H. Lipshutz, K. Noson, W. Chrisman, *J. Am. Chem. Soc.* **2001**, *123*, 12917.
43 S. Sirol, J. Courmarcel, N. Mostefai, O. Riant, *Org. Lett.* **2001**, *3*, 4111.
44 T. Nakano, Y. Nagai, *Chem. Lett.* **1988**, 481.
45 S.C. Berk, K.A. Kreutzer, S.L. Buchwald, *J. Am. Chem. Soc.* **1991**, *113*, 5093. K.J. Barr, S.C. Berk, S.L. Buchwald, *J. Org. Chem.* **1994**, *59*, 4323.
46 R.L. Halterman, T.M. Ramsey, Z. Chen, *J. Org. Chem.* **1994**, *59*, 2642.
47 M.B. Carter, B. Schiøtt, A. Gutiérrez, S.L. Buchwald, *J. Am. Chem. Soc.* **1994**, *116*, 11667.
48 S. Xin, J.F. Harrod, *Can. J. Chem.* **1995**, *73*, 999.
49 J. Yun, S.L. Buchwald, *J. Am. Chem. Soc.* **1999**, *121*, 5640.
50 H. Imma, M. Mori, T. Nakai, *Synlett.* **1996**, 1229.
51 N. Langlois, T.-P. Dang, H.B. Kagan, *Tetrahedron Lett.* **1973**, 4865. H.B. Kagan, N. Langlois, T.-P. Dang, *J. Organomet. Chem.* **1975**, *90*, 353. R. Becker, H. Brunner, S. Mahboobi, W. Wiegrebe, *Angew. Chem. Int. Ed. Engl.* **1985**, *24*, 995.
52 X. Verdaguer, U.E.W. Lange, M.T. Reding, S.L. Buchwald, *J. Am. Chem. Soc.* **1996**, *118*, 6784. J. Yun, S.L. Buchwald, *J. Org. Chem.* **2000**, *65*, 767. M.T. Reding, S.L. Buchwald, *J. Org. Chem.* **1998**, *63*, 6344.
53 X. Verdaguer, U.E.W Lange, S.L. Buchwald, *Angew. Chem. Int. Ed.* **1998**, *37*, 1103.
54 M.C. Hanse, S.L. Buchwald, *Org. Lett.* **2000**, *2*, 713.
55 I. Takei, Y. Nishibayashi, Y. Arikawa, S. Uemura, M. Hidai, *Organometallics* **1999**, *18*, 2271.
56 S. Murahashi, S. Watanabe, T. Shiota, *J. Chem. Soc., Chem. Commun.* **1994**, 725.

1.5
Transition Metal-Catalyzed Hydroboration of Olefins

Gregory C. Fu

1.5.1
Introduction

The transition metal-catalyzed hydroboration of olefins was first reported in 1985 by Nöth, who employed rhodium and ruthenium complexes for this transformation (e.g., Eq. 1) [1].

catalytic $Rh(PPh_3)_3Cl$, 20 °C; 83% (1)

A reasonable mechanism for the rhodium-catalyzed pathway is: oxidative addition of the boron hydride to Rh(I), olefin complexation, β-migratory insertion, and then reductive elimination (Fig. 1). Since Nöth's pioneering discovery, a range of other transition metals have been shown to accelerate the hydroboration of olefins, some via pathways that differ from the rhodium-catalyzed process [2].

In this review, we will highlight certain developments in metal-catalyzed hydroborations of olefins that have been reported during the period 1999–2002. By reason of space limitations, the discussion will focus on just a few of the many interesting aspects of this fascinating field.

1.5.2
Catalytic Asymmetric Hydroboration of Olefins

The catalytic asymmetric hydroboration of olefins has been reviewed in an article published in 1999 [3]. Since that time, several new catalyst systems have been described that furnish excellent selectivity. For example, Guiry has reported that a quinazoline-derived ligand provides good regio- and enantioselection in rhodium-catalyzed hydroborations of β-substituted styrene derivatives (Fig. 2) [4].

Transition Metals for Organic Synthesis, Vol. 2, 2nd Edition.
Edited by M. Beller and C. Bolm

ISBN: 3-527-30613-7

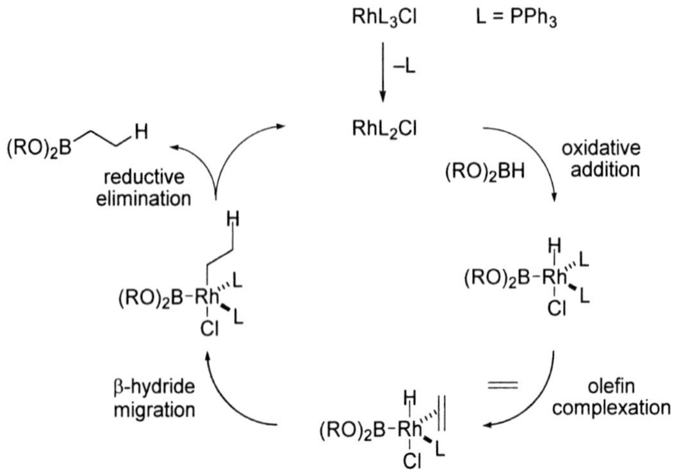

Fig. 1 A possible mechanism for rhodium-catalyzed hydroborations of olefins.

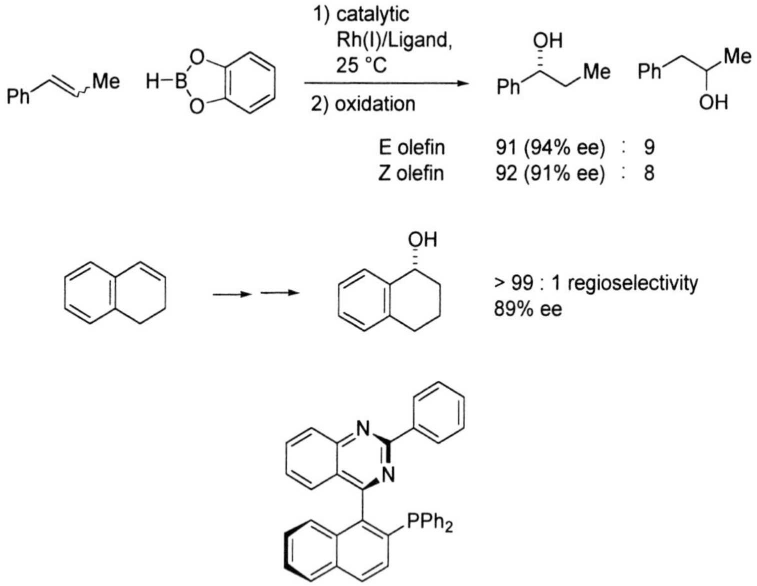

Fig. 2 Rhodium-catalyzed enantioselective hydroboration with a quinazoline-derived ligand.

Chan has also applied a new *P,N*-ligand to rhodium-catalyzed asymmetric hydroborations of olefins (Eq. 2). For para-substituted styrenes, the *ee* of the alcohol correlates with the donating/withdrawing nature of the substituent [5].

1) catalytic Rh(I)/Ligand, 0 °C
2) oxidation
>99 : 1 regioselectivity

Ar =	
4-ClC$_6$H$_4$	79% ee
C$_6$H$_5$	90% ee
4-MeOC$_6$H$_4$	94% ee

(2)

Knochel has developed a new family of C_2-symmetric bisphosphines and established their utility in enantioselective rhodium-catalyzed hydroborations of styrene and styrene derivatives (Eq. 3) [6]. In addition, through a screening process, Schmalz discovered a bidentate phosphine-phosphite that is effective for the regio- and stereoselective hydroboration of styrene (Eq. 3) [7].

1) catalytic Rh(I)/Ligand
2) oxidation

Knochel: –35 °C (85% yield) >99 (92% ee) : 1

Schmalz: –78 °C (63% yield) 95 (91% ee) : 5

(3)

Building on earlier work on desymmetrizations of norbornenes via rhodium-catalyzed hydroboration, in 2002 Bonin and Micouin described a Rh(I)/BDPP-catalyzed addition to a bicyclic hydrazine that generates an exo alcohol in good *ee* (Eq. 4) [8]. Interestingly, when Ir(I)/BDPP is used as the catalyst, the opposite enantiomer of the alcohol is produced preferentially (~ 35% *ee*) [9].

N–CO₂Bn
N–CO₂Bn
H–B
1) catalytic Rh(I)/Ligand, –50 °C
2) oxidation
HO
N–CO₂Bn
N–CO₂Bn
exclusively exo
84% ee
91% yield
Me
PPh₂
PPh₂
Me

(4)

Finally, Brown has described interesting kinetic resolutions of 1,2-dihydronaphthalenes in the presence of Rh(I)/QUINAP (Eq. 5) [10].

1) catalytic Rh(I)/Ligand, r.t.
H–B
0.40 equiv
2) oxidation
kinetic resolution
Me
OH
Me
Me

k (fast-reacting enantiomer) / k (slow-reacting enantiomer) ~20

PPh₂
N

(5)

1.5.3
Applications of Transition Metal-Catalyzed Hydroboration in Synthesis

There have been several reports that metal-catalyzed hydroboration furnished a solution to a challenge that could not be addressed satisfactorily by uncatalyzed hydroboration methods. For example, during the course of a synthesis of a dipeptide isostere, Rich needed to achieve a selective hydroboration of a terminal olefin in the presence of a γ-lactone (Eq. 6) [11]. Unfortunately, a variety of conventional hydroborating agents (e.g., disiamylborane, 9-BBN, and dicyclohexylborane) provide a low yield of the desired alcohol, because of reduction of the lactone. In contrast, rhodium-catalyzed hydroboration proceeds smoothly without any detectable formation of the lactol.

1) catalytic Rh(I)/Ligand, r.t.; catecholborane (H–B); 2) oxidation

Ph, BocHN → Ph, BocHN, OH; 70% (6)

As part of a medicinal chemistry program focused on the central nervous system, Bunch required a selective route to the endo alcohol illustrated in Eq. (7) [12]. Borane and dialkylboranes were, however, inadequate for the task. Fortunately, $Rh(PPh_3)_3Cl$ catalyzes the hydroboration with high diastereoselectivity and in good yield.

Boc, N, EtO_2C — 1) hydroboration; 2) oxidation → Boc, N, EtO_2C, CH_2OH (7)

BH_3	89% endo, 50% yield
Cy_2BH	50% endo, yield not determined
9-BBN	77% endo, yield not determined
catalytic $Rh(PPh_3)_3Cl$, catecholborane, 25 °C	**95% endo, 83% yield**

1.5.4 Transition Metal-Catalyzed Hydroboration in Supercritical CO_2

In 2000, Baker and Tumas reported an intriguing study of rhodium-catalyzed hydroboration reactions in supercritical CO_2 ($scCO_2$) [13]. Mixing fluorinated phosphines with $Rh(hfacac)(cyclooctene)_2$ provides homogeneous $scCO_2$ solutions that catalyze the hydroboration of 4-vinylanisole by catecholborane. Interestingly, the course of this reaction can be markedly solvent dependent. Thus, using the partially fluorinated trialkylphosphine illustrated in Eq. (8), in THF or in a fluorinated solvent an unfortunate mixture of hydroboration and side products is obtained; in contrast, in $scCO_2$, the hydroborated olefin is generated cleanly as a single regioisomer.

catalytic Rh(hfacac)(cyclooctene)$_2$

C_6F_{13}CH$_2$CH$_2$PCy$_2$

40 °C

Ar = 4-MeOC$_6$H$_4$
hfacac = hexafluoroacetylacetonate

(8)

solvent:	A	B	side products
THF	32	34	34
$CF_3C_6F_{11}$	25	41	34
$scCO_2$	100	0	0

1.5.5 Summary

The report by Nöth in 1985 that transition metals can catalyze the hydroboration of olefins added an exciting new dimension to this powerful transformation. During the past few years, a variety of significant developments have been described, including the discovery of new chiral catalysts, fresh applications in target-oriented synthesis, and the observation of intriguing reactivity patterns in environmentally benign solvents.

References

1 Mannig, D., Nöth, H. *Angew. Chem., Int. Ed. Engl.* **1985**, *24*, 878–879.

2 For earlier reviews, see: (a) Fu, G.C., Evans, D.A., Muci A.R. In *Advances in Catalytic Processes* (Ed.: M.P. Doyle), JAI, Greenwich, CT, 1995, *1*, 95–121. (b) Beletskaya, I., Pelter, A. *Tetrahedron* **1997**, *53*, 4957–5026.

3 Hayashi, T. in *Comprehensive Asymmetric Catalysis*; Jacobsen, E.N., Pfaltz, A., Yamamoto, H., Eds., Springer, New York, **1999**; Chapter 9.

4 (a) McCarthy, M., Guiry, P.J. *Polyhedron* **2000**, *19*, 541–543. (b) McCarthy, M., Hooper, M.W., Guiry, P.J. *Chem. Commun.* **2000**, 1333–1334.

5 Kwong, F.Y., Yang, Q., Mak, T.C.W., Chan, A.S.C., Chan, K.S. *J. Org. Chem.* **2002**, *67*, 2769–2777.

6 Demay, S., Volant, F., Knochel, P. *Angew. Chem. Int. Ed.* **2001**, *40*, 1235–1238.

7 Blume, F., Zemolka, S., Fey, T., Kranich, R., Schmalz, H.-G. *Adv. Synth. Catal.* **2002**, *344*, 868–883.

8 Perez Luna, A., Ceschi, M.-A., Bonin, M., Micouin, L., Husson, H.-P., Gougeon, S., Estenne-Bouhtou, G., Marabout, B., Sevrin, M., George, P. *J. Org. Chem.* **2002**, *67*, 3522–3524.

9 Perez Luna, A., Bonin, M., Micouin, L., Husson, H.-P. *J. Am. Chem. Soc.* **2002**, *124*, 12098–12099.

10 Maeda, K., Brown, J. M. *Chem. Commun.* **2002**, 310–311.

11 Brewer, M., Rich, D.H. *Org. Lett.* **2001**, *3*, 945–948.

12 Bunch, L., Liljefors, T., Greenwood, J.R., Frydenvang, K., Bräuner-Osborne, H., Krogsgaard-Larsen, P., Madsen, U. *J. Org. Chem.* **2003**, *67*, 1489–1495.

13 Carter, C.A.G., Baker, R.T., Nolan, S. P., Tumas, W. *Chem. Commun.* **2000**, 347–348.

2

Oxidations

Transition Metals for Organic Synthesis, Vol. 2, 2nd Edition.
Edited by M. Beller and C. Bolm

ISBN: 3-527-30613-7

2.1
Basics of Oxidations

Roger A. Sheldon and Isabel W. C. E. Arends

2.1.1
Introduction

Oxidation and reduction are pivotal reactions in organic synthesis. On the one hand, catalytic hydrogenation has broad scope and is widely applied on an industrial scale. Catalytic oxidation with dioxygen, in contrast, is an important technology in bulk chemicals manufacture but has a much narrower scope in organic synthesis in general. There are two underlying reasons for the lack of selectivity/scope of oxidations with dioxygen compared with reduction with hydrogen. First, owing to the triplet nature of its ground state, dioxygen undergoes free-radical reactions with organic molecules, in the absence and in the presence of (metal) catalysts. In some cases this is the desired reaction but more often than not it leads to the formation of undesirable side-products. Second, the thermodynamically stable product of the reaction of organic molecules with dioxygen is carbon dioxide, and hence it is difficult to achieve high selectivities to the desired partial oxidation products. In contrast, hydrogen does not react with organic molecules in the absence of a catalyst, and the desired product is generally the thermodynamically stable one.

Consequently, the great challenge in oxidation catalysis is to promote the desired pathway at the expense of the ubiquitous free-radical autoxidation pathway. Alternatively, the problem can be circumvented by employing an oxygen transfer reaction (analogous to hydrogen transfer instead of hydrogenation) in which a reduced form of oxygen, e.g., hydrogen peroxide, is the oxidant (see later). Catalytic oxidations are also important in the context of Green Chemistry. Traditionally, oxidations in the fine chemicals industry have been generally performed with stoichiometric amounts of inorganic oxidants, such as chromium(VI) reagents, permanganate, and manganese dioxide, resulting in the formation of copious amounts of (often toxic) inorganic waste. Increasingly stringent environmental regulation has rendered such methods prohibitive and created an urgent need for greener, catalytic alternatives that employ dioxygen or hydrogen peroxide as the stoichiometric oxidant.

Transition Metals for Organic Synthesis, Vol. 2, 2nd Edition.
Edited by M. Beller and C. Bolm

ISBN: 3-527-30613-7

Catalytic oxidations can be conveniently divided into three groups:

1. Free radical autoxidations
2. Direct oxidation of the substrate by the (metal) oxidant followed by re-oxidation of its reduced form by dioxygen
3. Oxygen transfer processes.

In the following discussion, the fundamental steps involved in the different categories of oxidation mechanisms will be delineated.

2.1.2
Free-Radical Autoxidations

As noted above, dioxygen reacts with organic molecules via a free radical chain process involving initiation, propagation, and termination steps (Reactions 1–5).

$$\text{Initiation}\ \ In_2 \xrightarrow{R_i} 2\,In^{\bullet} \tag{1}$$

$$In^{\bullet} + RH \longrightarrow R^{\bullet} + InH \tag{2}$$

$$\text{Propagation}\ \ R^{\bullet} + O_2 \xrightarrow{\text{very fast}} RO_2^{\bullet} \tag{3}$$

$$RO_2^{\bullet} + RH \xrightarrow{k_p} RO_2H + R^{\bullet} \tag{4}$$

$$\text{Termination}\ \ 2RO_2^{\bullet} \xrightarrow{k_t} RO_4R \longrightarrow \text{non-radical products} \tag{5}$$

The susceptibility of a particular molecule to autoxidation is determined by the ratio $k_p/[2k_t]^{1/2}$, which is usually referred to as its oxidizability [1]. The reaction can be started by adding an initiator which undergoes homolytic thermolysis at the reaction temperature to produce chain-initiating radicals. The initiator could be the alkyl hydroperoxide product, although relatively high temperatures (>100 °C) are generally required for the thermolysis of hydroperoxides. Alternatively, chain-initiating radicals can be generated at lower temperatures by reaction of trace amounts of alkyl hydroperoxides with variable valence metals, e.g. cobalt, manganese, iron, cerium, etc. (Reactions 6–8).

$$RO_2H + M^n \longrightarrow RO^{\bullet} + M^{n+1}OH \tag{6}$$

$$RO_2H + M^{n+1} \longrightarrow RO_2^{\bullet} + M^n + H^+ \tag{7}$$

$$\text{Net reaction:}\ 2\,RO_2H \xrightarrow{M^n/M^{n+1}} RO^{\bullet} + RO_2^{\bullet} + H_2O \tag{8}$$

In such processes the metal ion acts (in combination with ROOH) as an initiator rather than a catalyst. Herein lies the basic problem in interpreting metal-catalyzed oxidation processes. The catalyst is almost always capable of undergoing valence changes, which makes it difficult to distinguish (desirable) heterolytic processes from the ubiquitous free-radical autoxidation initiated via Reactions 6 and 7. Even when alkyl hydroperoxides or hydrogen peroxide are used as oxygen transfer agents (see later) homolytic decomposition of the hydroperoxide via one-electron transfer processes can lead to the formation of dioxygen via subsequent chain decomposition of the hydroperoxide (Reactions 9 and 10), resulting in free radical autoxidation[1)]. This possibility has not been recognized by many authors and leads, inevitably, to misinterpretation of results. It is therefore recommended that reactions should be performed in the presence of a free-radical scavenger, e.g., a phenol, to eliminate any free-radical chain process. Another simple test, which should always be performed in oxidations with alkyl hydroperoxides or hydrogen peroxide, is to purge the reaction mixture with a constant stream of an inert gas, thus removing oxygen and preventing autoxidation.

$$RO^\bullet + ROOH \longrightarrow ROH + ROO^\bullet \quad (9)$$

$$ROO^\bullet + ROO^\bullet \longrightarrow 2\,RO^\bullet + O_2 \quad (10)$$

Bromide ion has a synergistic effect on (metal-catalyzed) autoxidations [2] by changing the propagation steps to the energetically more favorable steps shown in Reactions 11–13. The bromide atoms can be generated by one-electron oxidation of bromide ions by, e.g., cobalt(III) or manganese(III) (Reaction 14). This forms the basis for several commercial processes for the oxidation of methyl-substituted aromatics to the corresponding carboxylic acids, e.g., *p*-xylene to terephthalic acid using cobalt and/or manganese in combination with bromide ion as the catalyst [2].

$$Br^\bullet + RH \longrightarrow + R^\bullet \quad (11)$$

$$R^\bullet + O_2 \longrightarrow RO_2^\bullet \quad (12)$$

$$RO_2^\bullet + HBr \longrightarrow RO_2H + Br^\bullet \quad (13)$$

$$M^{III} + Br^- \longrightarrow M^{II} + Br^\bullet \quad (14)$$

However, HBr is not a suitable additive in some autoxidations as it would catalyze the rearrangement of intermediate hydroperoxides to unwanted by-products, which may even be inhibitors, e.g., phenol from cumene hydroperoxide. An interesting recent development in this context is the discovery by Ishii and coworkers [3] that *N*-hydroxyphthalimide (NHPI) can function in the same way as HBr. This

1) We note that reaction of two secondary or primary alkylperoxy radicals with each other generally leads to non-radical products (equimolar amounts of alcohol, ketone or aldehyde and oxygen via the Russell mechanism [1]) as in Reaction 5. On the other hand for tertiary alkylperoxy radicals this pathway is not available and Reaction 10 prevails. This explains why the rate of termination for primary and secondary alkylperoxy radicals is orders of magnitude higher than that of tertiary alkylperoxy radicals [1].

leads to an alternative autoxidation scheme in which the radical derived from NHPI, and referred to as PINO, is responsible for chain propagation (see Reactions 15–17). NHPI in turn efficiently traps the intermediate alkylperoxy radicals, increasing the ratio of propagation to termination and, hence, both the selectivity and rate of the autoxidation.

$$\text{N–O}^\bullet + \text{RH} \longrightarrow \text{N–OH} + \text{R}^\bullet \qquad (15)$$

$$\text{R}^\bullet + \text{O}_2 \longrightarrow \text{ROO}^\bullet \qquad (16)$$

$$\text{ROO}^\bullet + \text{N–OH} \longrightarrow \text{ROOH} + \text{N–O}^\bullet \qquad (17)$$

By analogy with the bromide-based systems (see above), the PINO radical can be generated via oxidation of NHPI with, e.g., cobalt(III) or manganese(III). Ishii and co-workers [3, 4] have described the selective aerobic oxidation of a wide variety of substrates using the combination of NHPI and a metal salt, mainly cobalt, under remarkably mild conditions. For example, toluene was oxidized to benzoic acid at ambient temperature, a reaction which is usually performed at temperatures in excess of 100 °C. Interestingly, the metal salt acts as an initiator for the autoxidation but also catalyzes the decomposition of intermediate hydroperoxides. Since the metal salt acts as an initiator, we envisaged that it could be replaced by organic initiators, leading to an NHPI-catalyzed oxidation that would afford the alkyl hydroperoxide in high selectivity (since there is no metal present to decompose it). This was confirmed in the autoxidation of cumene, ethylbenzene, and cyclohexylbenzene as relevant hydrocarbon substrates [5]. Cyclohexylbenzene, for example, afforded the corresponding tertiary hydroperoxide (Reaction 18) in ca. 98% selectivity at 32% conversion using as little as 0.5 mol% NHPI together with the product hydroperoxide (2 mol%) as the initiator [5, 6]. This result is quite remarkable when one considers that cyclohexylbenzene contains ten secondary C-H bonds in addition to the single tertiary C-H bond. This highly selective autoxidation forms the basis for a co-product-free route from benzene to phenol [6]. The starting material is prepared from benzene, via selective hydrogenation to cyclohexene and subsequent Friedel-Crafts alkylation, and the cyclohexanone co-product can be dehydrogenated to phenol (Reaction 19).

Similarly, the Ishii group has exploited NHPI-catalyzed autoxidations to perform a number of interesting oxidative transformations [3, 4], and it is clear that the full potential of this chemistry still has to be realized.

$$\text{PhC}_6\text{H}_{11} \xrightarrow[\text{0.5 mol\% NHPI, 2 mol\% CHBHP}]{O_2} \text{PhC}_6\text{H}_{10}\text{OOH} \quad (18)$$

$$\text{PhC}_6\text{H}_{10}\text{OOH} \xrightarrow{H^+} \text{PhOH} + \text{cyclohexanone} \quad (\text{cyclohexanone} \xrightarrow{-2\,H_2} \text{phenol}) \quad (19)$$

2.1.3
Direct Oxidation of the Substrate by the (Metal) Oxidant

Many catalytic oxidations employing dioxygen as the stoichiometric oxidant proceed via a pathway in which the substrate undergoes direct oxidation by the (metal) catalyst. This is followed by re-oxidation of the reduced form of the catalyst by dioxygen (Reactions 20 and 21). Many gas phase oxidations involve such a pathway, in which a surface oxometal species (usually a metal oxide) oxidizes the substrate, i.e. lattice oxygen is incorporated, and the reduced form is re-oxidized by dioxygen. This is generally referred to as the Mars-van Krevelen mechanism [7]. In the liquid phase, where there are relatively high concentrations of substrate, autoxidation chain lengths are often long, and it is difficult to compete with the ubiquitous free-radical chain autoxidation. In the gas phase, in contrast, substrate concentrations are much lower, and a Mars-van Krevelen pathway is more likely.

$$M = O + S \longrightarrow M + SO \quad (20)$$

$$M + 1/2\ O_2 \longrightarrow M = O \quad (21)$$

A wide variety of aerobic oxidations mediated by monooxygenase enzymes are similarly thought [8] to involve oxygen transfer from a high-valent oxoiron intermediate to the substrate (M=O in Reaction 25 is, e.g., $Fe^{V}=O$). However, in this case a stoichiometric cofactor is needed for the overall process, in which one atom of dioxygen is incorporated in the substrate and the other oxygen atom is reduced to water (Reaction 22).

$$RH + O_2 + DH_2 \xrightarrow{\text{monooxygenase}} ROH + D + H_2O$$
$$D/DH_2 = \text{cofactor} \quad (22)$$

Since monooxygenases are able to catalyze a wide variety of industrially relevant oxidations, e.g., hydroxylation of relatively unreactive C-H bonds and olefin epoxidation, extensive studies of biomimetic systems have been aimed at circumventing the need for a cofactor [9]. Indeed, the Holy Grail of catalytic oxidations is to design a suprabiotic system that is able to catalyze direct oxidation of relevant hydrocarbon substrates via a Mars-van Krevelen mechanism in the liquid phase. However, an effective system has not been forthcoming. Most biomimetic approaches involve the use of a reduced form of dioxygen, e.g., hydrogen peroxide, or employ a sacrificial reductant.

An example of the latter is the Mukaiyama method [10], which employs an aldehyde as the sacrificial reductant (Reaction 23). This method produces the corresponding carboxylic acid as the coproduct, which, in the context of commodity chemicals manufacture, is not a viable proposition.

$$RH + R'HCO + O_2 \xrightarrow{\text{catalyst}} ROH + R'CO_2H \tag{23}$$

Mars-van Krevelen type oxidations in the liquid phase have been, in principle, demonstrated with ruthenium complexes of sterically hindered porphyrins [11] or phenanthrolines [12] and with ruthenium polyoxometalates [13]. However, turnover numbers were generally low and have not yet resulted in the design of a truly effective catalyst for the direct oxidation of relevant hydrocarbons with dioxygen. As noted earlier, a major problem is to design a Mars-van Krevelen system that can effectively compete with the ubiquitous free radical autoxidation.

Direct oxidation of a substrate by a metal oxidant can involve either a homolytic or a heterolytic mechanism. An example of the former is the autoxidation of alkylaromatics in the presence of relatively high concentrations (>0.1 M) of cobalt(III) acetate in acetic acid, which involves rate-limiting one-electron oxidation of the substrate, affording the corresponding cation radical (Reaction 24). Subsequent elimination of a proton affords the benzylic radical (Reaction 25), which reacts with oxygen to form the corresponding peroxy radical (Reaction 26). The primary product, the corresponding aldehyde, is formed by reaction of the benzylperoxy radical with cobalt(II), with concomitant formation of cobalt(III) (Reaction 27) to complete the catalytic cycle.

$$ArCH_3 + Co^{III} \longrightarrow [ArCH_3]^{+\bullet} + Co^{II} \tag{24}$$

$$[ArCH_3]^{+\bullet} \longrightarrow ArCH_2^{\bullet} + H^+ \tag{25}$$

$$ArCH_2^{\bullet} + O_2 \longrightarrow ArCH_2OO^{\bullet} \tag{26}$$

$$ArCH_2OO^{\bullet} + Co^{II} \longrightarrow ArCH(H)\text{-}O\text{-}O\text{-}Co^{III} \longrightarrow ArCHO + HOCo^{III} \tag{27}$$

Similarly, a Mars-van Krevelen pathway involving an oxometal species as the oxidant (see earlier) can involve either a homolytic or a heterolytic pathway. In the former case free radical autoxidation is circumvented if the radical, produced at the oxide surface, does not diffuse away from it but undergoes further reaction resulting in incorporation of lattice oxygen. A similar situation pertains to aerobic oxidations catalyzed by iron-dependent monooxygenases. Reaction of the putative oxoiron intermediate with a hydrocarbon could involve either a homolytic, stepwise or a heterolytic, concerted insertion of an oxygen atom. In the former (Reaction 28), an alkyl radical intermediate reacts with the iron center via the so-called oxygen rebound mechanism [14] to afford the product and iron(III). This process can presumably compete effectively with free-radical chain autoxidation because the radical is produced in the active site of the enzyme and not in the bulk solution, reminiscent of the situation on the surface of a metal oxide.

$$Fe^{V} = O + RH \longrightarrow Fe^{IV} - OH + R^{\bullet} \longrightarrow Fe^{III} + ROH \quad (28)$$

Heterolytic mechanisms for the direct oxidation of substrates generally involve a two-electron oxidation of a coordinated substrate molecule. Typical examples are the palladium(II)-catalyzed oxidations of alkenes (Wacker process, Reactions 29 and 30) [15] and the oxidative dehydrogenation of alcohols (Reaction 31) catalyzed by palladium and other noble metals [16].

$$RCH = CH_2 + Pd^{II} + H_2O \longrightarrow RCOCH_3 + Pd^0 + 2\,H^+ \quad (29)$$

$$Pd^0 + 2\,H^+ + 1/2\,O_2 \longrightarrow Pd^{II} + H_2O \quad (30)$$

$$R_2CHOH + Pd^{II} \longrightarrow R_2C = O + Pd^0 + 2\,H^+ \quad (31)$$

2.1.4 Catalytic Oxygen Transfer

One way of avoiding the selectivity problems associated with the use of dioxygen as the stoichiometric oxidant is to use a reduced form of dioxygen, e.g., H_2O_2 or RO_2H as a single oxygen donor in a catalytic oxygen transfer process (Reaction 32).

$$S + XOY \xrightarrow{\text{catalyst}} + XY \quad (32)$$

S = substrate; SO = oxidized substrate
XOY = H_2O_2, RO_2H, R_3NO, NaOCl, $KHSO_5$, etc.

Catalytic oxygen transfer processes are widely applied in organic synthesis, e.g., in olefin epoxidations, dihydroxylations, aminohydroxylations, alcohol oxidations, heteroatom oxidations, Baeyer-Villiger oxidations, etc. [17].

Virtually all of the transition elements and several main group elements are known to catalyze oxygen transfer processes [17]. A variety of single oxygen donors can be used (Tab. 1). In addition to price and ease of handling, two important considerations influencing the choice of oxygen donor are the weight percentage of available oxygen and the nature of the co-product. The former has a direct bearing on the volumetric productivity (kg product per unit reactor volume per

Tab. 1 Oxygen donors

Donor	*% Active Oxygen*	*Coproduct*
H_2O_2	47.0 (14.1) [a]	H_2O
N_2O	36.4	N_2
O_3	33.3	O_2
CH_3CO_3H	21.1	CH_3CO_2H
tert-BuO_2H	17.8	*tert*-BuOH
HNO_3	25.4	NO_x
NaOCl	21.6	NaCl
NaO_2Cl	35.6	NaCl
NaOBr	13.4	NaBr
$C_5H_{11}NO_2$ [b]	13.7	$C_5H_{11}NO$
$KHSO_5$	10.5	$KHSO_4$
$NaIO_4$	7.5	$NaIO_3$
PhIO	7.3	Phl

a) Figure in parentheses refers to 30% aq. H_2O_2.
b) *N*-Methylmorpholine-*N*-Oxide (NMO).

unit time) and the latter is important in the context of environmental acceptability. With these criteria in mind, it is clear that hydrogen peroxide is preferred, from both an economic and an environmental viewpoint. Generally speaking, organic co-products are more easily recycled than inorganic ones, e.g., the co-products from RO_2H and amine oxides can be recycled via reaction with H_2O_2. The overall process produces water as the co-product but requires one extra step compared with the corresponding reaction with H_2O_2. With inorganic oxygen donors environmental considerations are relative. Sodium chloride and potassium bisulfate are obviously preferable to the heavy metal salts (Cr, Mn, etc.) produced in classical stoichiometric oxidations. The choice of oxidant may be governed by the ease of recycling, e.g., NaOBr may be preferred over NaOCl, as NaBr can in principle be reoxidized with H_2O_2. A disadvantage of peroxides as oxygen donors is possible competition from metal-catalyzed homolytic decomposition pathways (see earlier) leading to nonselective free radical autoxidation.

Heterolytic oxygen transfer processes can be divided into two categories based on the nature of the active oxidant: an oxometal or a peroxometal species (Fig. 1).

Catalysis by early transition elements (Mo, W, Re, V, Ti, Zr, etc.) generally involves high-valent peroxometal complexes, whereas later transition elements (Ru, Os) and first row elements (Cr, Mn, Fe) mediate oxygen transfer via oxometal species. Some elements, e.g., vanadium, occupy an intermediate position and can operate via either mechanism, depending on the substrate.

Reactions that typically involve peroxometal pathways are alkene epoxidations, alcohol oxidations, Baeyer-Villiger oxidations of ketones, and heteroatom (N and S) oxidations. Oxometal species tend to be stronger oxidants capable of oxidizing a wider variety of substrate types, e.g., hydroxylation of C-H bonds and dihydroxy-

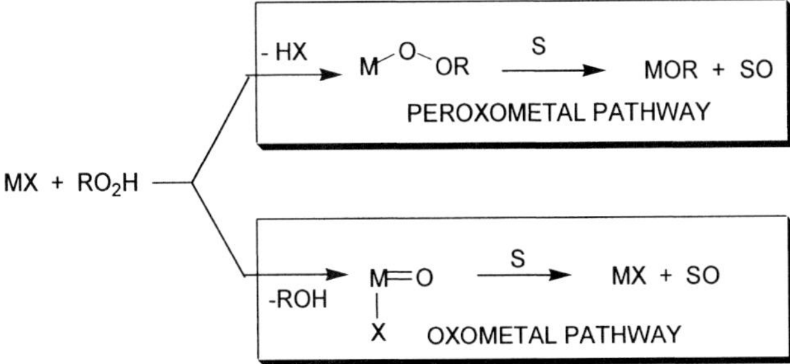

Fig. 1 Peroxo versus oxometal pathways.

lation and oxidative cleavage of olefinic bonds, in addition to the above-mentioned transformations.

Oxygen transfer processes are also catalyzed by certain organic molecules [18], which can be categorized on the same basis as metal catalysts. For example, ketones catalyze a variety of oxidations with monoperoxysulfate ($KHSO_5$) [19]. The active oxidant is the corresponding dioxirane, and hence the reaction can be construed as involving a "peroxometal" pathway (Reactions 33 and 34).

$$R_2C{=}O + HSO_5^- \longrightarrow R_2C\langle{}^{O}_{O}| + HSO_4^- \tag{33}$$

$$R_2C\langle{}^{O}_{O}| + S \longrightarrow R_2C{=}O + SO \tag{34}$$

Similarly, the TEMPO (2,2,6,6-tetramethylpiperidinyloxyl)-catalyzed oxidations of alcohols with hypochlorite [20] involve the corresponding oxoammonium cation as the active oxidant (Reaction 35) and can be viewed as an "oxometal" pathway.

$$R^1R^2CH(OH) + \text{oxoammonium } (N^+{=}O) \longrightarrow R^1R^2C{=}O + \text{N–OH}; \quad \text{N–OH} \xrightarrow{NaOCl} N^+{=}O \tag{35}$$

Although many of these oxygen transfer agents are often economically viable in the context of the production of high-value-added fine chemicals, there is a trend toward replacing them, where possible, with "cleaner" dioxygen or hydrogen peroxide. Sodium hypochlorite, for example, suffers from the disadvantage of forming chlorinated by-products, and transport and storage of peracetic acid has been severely curtailed for safety reasons.

It has been shown that hypochlorite can be replaced, in TEMPO-mediated oxidations, by a combination of a metal catalyst (Cu or Ru) and dioxygen [21, 22]. In the case of ruthenium it was shown that the role of TEMPO is to promote the re-oxidation of a ruthenium hydride species formed in the initial dehydrogenation of the alcohol substrate [22]. We note that this now becomes an example of a catalytic oxidation of the second category: direct oxidation by metal oxidant followed by re-oxidation with dioxygen. Other effective methods for alcohol oxidations, involving direct oxidation of the substrate by the metal oxidant followed by re-oxidation of the reduced form by dioxygen, include the use of [n-Pr_4N][RuO_4] [23] or water-soluble palladium(II) complexes in an aqueous biphasic system [24]. Alternatively, hydrogen peroxide can be used as an oxygen transfer oxidant in the oxidation of alcohols catalyzed by early transition elements, such as tungsten [25].

Similarly, it has been shown [26] that persulfate ($KHSO_5$) can be replaced by CH_3CN/H_2O_2 in asymmetric epoxidations involving a chiral dioxirane as the active oxidant [27]. This presumably involves the intermediate formation of the peroxyimidate ($CH_3C(OOH)=NH$, the Payne reagent).

In Baeyer-Villiger oxidations of ketones, which are widely applied in organic synthesis, there is a marked trend toward replacing the traditional percarboxylic acid oxidant with aqueous hydrogen peroxide as an oxygen transfer agent, in conjunction with a metal catalyst. For example, according to recent reports [28] heterogeneous Sn-containing catalysts are effective with hydrogen peroxide as the oxidant. Arylselenenic acids are also effective catalysts for Baeyer-Villiger oxidations with hydrogen peroxide, via the intermediate formation of perselenenic acids [29].

2.1.5
Ligand Design in Oxidation Catalysis

Many of the major challenges in oxidation chemistry involve very demanding transformations, such as the selective oxidation of unactivated C-H bonds, which require powerful oxidants. This presents a dilemma: if an oxidant is powerful enough to oxidize an unactivated C-H bond then, by the same token, it will readily oxidize most ligands, which may contain C-H bonds that are more active than the targeted bond in the substrate. The low operational stability of, for example, heme-dependent monooxygenases and peroxidases is a direct consequence of the facile oxidative destruction of the porphyrin ligand. Nature's solution to this problem is simple: *in vivo* the organism synthesizes fresh enzyme to replace that destroyed. *In vitro* this is not a viable option. In this context it is worth noting that many metal complexes that are routinely used as oxidation catalysts contain ligands, e.g., acetylacetonate, Schiff's bases, that are rapidly destroyed under oxidizing conditions. This fact is often not appreciated by authors of publications on catalytic oxidations.

Collins [30] has addressed the problem of ligand design in oxidation catalysis in some detail and developed guidelines for the rational design of oxidatively robust ligands. It essentially involves replacing all reactive C-H bonds in the ligand with other, more stable, bonds and ensuring that there are no hydrolytically labile moieties in the molecule. It is also worth emphasizing, in this context, that an additional requirement has to be fulfilled: the desired catalytic pathway should compete effectively with the ubiquitous free-radical autoxidation.

2.1.6
Enantioselective Oxidations

One category of oxidations in which ligand design is quintessential is enantioselective oxidations. It is difficult to imagine enantioselective oxidation without a requirement for chiral organic ligands. Hence, the task is to design ligands that not only endow the (metal) catalyst with the desired activity and enantioselectivity but also are stable and recyclable.

Much progress has been achieved in enantioselective oxidations over the last two decades. Because of the relatively low volumes and high added value of the products, enantioselective oxidations allow for the use of more expensive and/or environmentally less attractive oxidants, such as hypochlorite, *N*-methylmorpholine-*N*-oxide, and even potassium ferricyanide. It goes beyond the scope of this chapter to discuss enantioselective oxidations. Suffice it to say that they predominantly employ the use of single oxygen donors as primary oxidants and involve the very same oxometal and peroxometal pathways observed in the absence of chiral ligands. For example, Sharpless epoxidations with ROOH/Ti(IV) [31] involve a peroxometal pathway, while Jacobsen epoxidation with NaOCl/Mn(III) [32] involves an oxometal pathway. Similarly, other enantioselective oxidations can be rationalized on the basis of the standard oxometal/peroxometal pathways in conjunction with chiral recognition mediated by an appropriate chiral ligand.

2.1.7
Concluding Remarks

This chapter is concerned with the basics of catalytic oxidations. Most other catalytic processes are child's play compared with the complications encountered in oxidation processes, largely owing to the competing free radical pathways occurring even in the absence of the catalyst. A rudimentary understanding of the fundamental processes arising when organic molecules are subjected to dioxygen or peroxides, in the presence of (metal) catalysts, is a *conditio sine qua non* for the design of selective oxidation procedures. One could go so far as to say that researchers should be required to demonstrate competence in these basics before embarking on the development of a selective and sustainable catalytic oxidation.

References

1 R.A. Sheldon, J.K. Kochi, *Metal-Catalyzed Oxidations of Organic Compounds*, Academic Press, New York, **1981**.

2 W. Partenheimer, *Catal. Today* **1995**, *23*, 69–158.

3 Y. Ishii, S. Sakaguchi, T. Iwahama, *Adv. Synth. Catal.* **2001**, *343*, 393–427.

4 T. Iwahama, S. Sakaguchi, Y. Ishii, *Tetrahedron Lett.* **1998**, *39*, 9059–9062; T. Iwahama, S. Sakaguchi, Y. Ishii, *Chem. Commun.* **1999**, 727–728; Y. Ishii, T. Iwa-

hama, S. Sakaguchi, K. Nakyama, Y. Nishiyama, *J. Org. Chem.* **1996**, *61*, 4520–4526; Y. Tashiro, T. Iwahama, S. Sakaguchi, Y. Ishii, *Adv. Synth. Catal.* **2001**, *343*, 220–225; T. Iwahama, S. Sakaguchi, Y. Nishiyama, Y. Ishii, *Tetrahedron Lett.* **1999**, *36*, 6923–6926.

5 I. W. C. E. Arends, M. Sasidharan, S. Chatel, R. A. Sheldon, C. Jost, M. Duda, A. Kühnle, in *Catalysis of Organic Reactions*, D. G. Morrell, Ed., Marcel Dekker, New York, **2002**, pp. 143–156; see also O. Fukuda, S. Sakaguchi, Y. Ishii, *Adv. Synth. Catal.* **2001**, *343*, 809–813.

6 I. W. C. E. Arends, M. Sasidharan, A. Kühnle, M. Duda, C. Jost, R. A. Sheldon, *Tetrahedron* **2002**, *58*, 9055–9061.

7 P. Mars, D. W. van Krevelen, *Chem. Eng. Sci. Spec. Suppl.* **1954**, *3*, 41–59.

8 P. R. Ortiz de Montellano, J. D. Voss, *Nat. Prod. Rep.* **2002**, *19*, 477–493; D. A. Kopp, S. J. Lippard, *Curr. Opin. Chem. Biol.* **2002**, *6*, 568–576; M. Merkx, D. A. Kopp, M. H. Sazinsky, J. L. Blazyk, J. Muller, S. J. Lippard, *Angew. Chem. Int. Ed.* **2001**, *40*, 2782–2807.

9 *Biomimetic Oxidations Catalyzed by Transition Metal Complexes*, B. Meunier, Ed., Imperial College Press, London, **1999**.

10 T. Mukaiyama, in *The Activation of Dioxygen and Homogeneous Catalytic Oxidation*, D. H. R. Barton, A. E. Martell, D. T. Sawyer, Eds., Plenum, New York, **1993**, pp. 133–146; T. Yamada, K. Imagawa, T. Nagata, T. Mukaiyama, *Chem. Lett.* **1992**, 2231–2234.

11 J. T. Groves, R. Quinn, *J. Am. Chem. Soc.* **1985**, *107*, 5790–5792; B. Scharbert, E. Zeisberger, E. Paulus, *J. Organomet. Chem.* **1995**, *493*, 143–147.

12 A. S. Goldstein, R. H. Beer, R. S. Drago, *J. Am. Chem. Soc.* **1994**, *116*, 2424–2429.

13 R. Neumann, M. Dahan, *Nature* **1997**, *388*, 353–355; R. Neumann, A. M. Khenkin, M. Dahan, *Angew. Chem. Int. Ed. Engl.* **1995**, *34*, 1587–1589.

14 J. T. Groves, G. A. McClusky, R. E. White, M. J. Coon, *Biochem. Biophys. Res. Commun.* **1978**, *81*, 154–160.

15 E. Monflier, A. Mortreux, in *Aqueous Phase Organometallic Catalysis*, B. Cornils, W. A. Herrmann, Eds., VCH, Weinheim, **1997**, pp. 513–518.

16 R. A. Sheldon, I. W. C. E. Arends, A. Dijksman, *Catal. Today* **2000**, *57*, 157–166.

17 R. A. Sheldon, *Top. Curr. Chem.* **1993**, *164*, 21–43; R. A. Sheldon, *Bull. Soc. Chim. Belg.* **1985**, *94*, 651–670.

18 W. Adam, C. R. Saha-Möller, P. A. Ganeshpure, *Chem. Rev.* **2001**, *101*, 3499–3548.

19 W. Adam, R. Curci, J. O. Edwards, *Acc. Chem. Res.* **1989**, *22*, 205–211; W. Adam, A. K. Smerz, *Bull. Soc. Chim. Belg.* **1996**, *105*, 581–599.

20 J. M. Bobbitt, M. C. L. Flores, *Heterocycles* **1988**, *27*, 509–533; A. E. J. de Nooy, A. C. Besemer, H. van Bekkum, *Synthesis* **1996**, *10*, 1153–1174.

21 M. F. Semmelhack, C. R. Schmid, D. A. Cortes, C. S. Chou, *J. Am. Chem. Soc.* **1984**, *106*, 3374–3376.

22 A. Dijksman, I. W. C. E. Arends, R. A. Sheldon, *Chem. Commun.* **1999**, 1591–1593; A. Dijksman, A. Marino-Gonzalez, A. Mairati, I. Payeras, I. W. C. E. Arends, R. A. Sheldon, *J. Am. Chem. Soc.* **2001**, *123*, 6826–6833.

23 I. E. Marko, P. R. Giles, M. Tsukazaki, I. Chellé-Regnaut, C. J. Urch, S. M. Brown, *J. Am. Chem. Soc.* **1997**, *119*, 12661–12662; R. Lenz, S. V. Ley, *J. Chem. Soc. Perkin Trans.* **1997**, 3291–3292.

24 G. J. ten Brink, I. W. C. E. Arends, R. A. Sheldon, *Science* **2000**, *287*, 1636–1639.

25 K. Sato, M. Aoki, J. Takagi, R. Noyori, *J. Am. Chem. Soc.* **1997**, *119*, 12386–12387; K. Sato, J. Takagi, M. Aoki, R. Noyori, *Tetrahedron Lett.* **1998**, *39*, 7549–7552; O. Bortolini, V. Conte, F. Di Furia, G. Modena, *J. Org. Chem.* **1986**, *51*, 2661–2663.

26 L. Shu, Y. Shi, *Tetrahedron Lett.* **1999**, *40*, 8721–8724.

27 Y. Tu, Z. X. Wang, Y. Shi, *J. Am. Chem. Soc.* **1996**, *118*, 9806–9807; Z. X. Wang, Y. Tu, M. Frohn, J. R. Zhang, Y. Shi, *J. Am. Chem. Soc.* **1997**, *119*, 11224–11235.

28 A. Corma, L. T. Nemeth, M. Renz, S. Valencia, *Nature* **2001**, *412*, 423–425; A. Corma, M. T. Navarro, L. Nemeth, M. Renz, *Chem. Commun.* **2001**, 2190–2191;

U. R. Pillai, E. Sahle-Demessie, *J. Mol. Catal. A: Chem.* **2003**, *191*, 93–100.
29 G. J. ten Brink, J. M. Vis, I. W. C. E. Arends, R. A. Sheldon, *J. Org. Chem.* **2001**, *66*, 2429–2433.
30 T. J. Collins, *Acc. Chem. Res.* **1994**, *27*, 279–285.
31 R. A. Johnson, K. B. Sharpless, in *Catalytic Asymmetric Synthesis*, I. Ojima, Ed., VCH, Berlin, **1993**, pp. 103–158.
32 E. N. Jacobsen, in *Catalytic Asymmetic Synthesis*, I. Ojima, Ed., VCH, Berlin, **1993**, pp. 159–202.

2.2
Oxidations of C–H Compounds Catalyzed by Metal Complexes

Georgiy B. Shul'pin

2.2.1
Introduction

Selective and efficient oxidative functionalization of aliphatic C–H bonds is one of the very important goals of organic chemistry. However, a practical realization of this task meets serious difficulties, especially in the case of saturated hydrocarbons, because of to the extremely high inertness of alkanes (which are the "noble gases of organic chemistry"). Alkanes do not usually react with "normal" reagents that easily oxidize much more reactive olefins, alcohols, amines etc. The usual solvents for organic synthesis, such as alcohols or ketones, are often not appropriate for reactions with saturated hydrocarbons, since a solvent and not a substrate is oxidized in this case. Moreover, alkanes are oxygenated to give products which are more reactive than the starting substrates, and naturally, if an excess of an oxidant is used, substantial over-oxidation to give undesirable compounds will take place.

Fortunately, during the last few decades, new systems based on metal complexes have been discovered which allow us to oxidize saturated hydrocarbons under relatively mild conditions, and these reactions are relatively efficient [1]. It is necessary to emphasize, however, that efficiencies of alkane oxygenations are usually noticeably lower in comparison with, for example, olefin epoxidations (which are also often catalyzed with transition metal complexes) or oxidation of alcohols to ketones. An over-oxidation can be avoided if an excess of an alkane over an oxidizing reagent is employed, but in this case the yield of products based on the starting hydrocarbon will be much less than quantitative. Typically, yields of 10–30% based on either a starting alkane or an oxidant can be considered as good. Certain groups connected with methylene fragments of molecules can dramatically enhance the reactivity of C–H bonds. For example, oxygenation of benzylic or allylic positions (activated by aryl or olefin fragments, respectively) or reactions of ethers (activated by alkoxy groups) proceed much more easily than the corresponding oxidations of cyclohexane and especially normal hexane. On the other hand, electron-deficient substituents (–CN, $-NO_2$, –COOH) make the neighboring CH_2 groups less reactive, and such liquids as acetonitrile, nitromethane, or acetic acid are appropriate solvents for oxidations of alkanes including even methane and ethane.

Transition Metals for Organic Synthesis, Vol. 2, 2nd Edition.
Edited by M. Beller and C. Bolm

ISBN: 3-527-30613-7

An important parameter for metal-catalyzed alkane oxidations is turnover number (TON), which is given by the total moles of products produced per mole of a catalyst. In some cases, parameters such as turnover number per hour or minute (turnover frequency, TOF) are used. The TON parameter is more preferable from the "synthetic" point of view because in some cases a very rapid initial reaction (with high TOF in min^{-1} or even in s^{-1}) can soon stop, and the final TON will be quite low. The range of possible solvents for alkane oxidations is very narrow. As mentioned above, usually liquids containing C–H bonds "deactivated" by electron-withdrawing substituents are used as solvents; these are acetic acid, acetonitrile, nitromethane, and methylene chloride. Pyridine (Gif-oxidations) or perfluorinated liquids have been employed in some "exotic" cases. Water – which is resistant to the action of normal oxidants – is a very attractive solvent, but it was used only in the case of lower alkanes (methane and ethane), because they are relatively soluble (under pressure) in aqueous solutions. The same can be said about concentrated sulfuric acid.

This chapter deals with oxidative activation of C–H bonds in saturated and aromatic hydrocarbons as well as in some other C–H-containing compounds (e.g., in olefins) by metal complexes in solutions under mild conditions (that is at temperatures lower than 100–150 °C) with predominant emphasis on synthetic aspects of described reactions. Oxygenations (i.e., insertion of an oxygen atom into the C–H bond) published during last few years are mainly considered; earlier work has been described in books and reviews [1].

From the mechanistic point of view, C–H activation processes can be divided into three types. The first group includes reactions involving "true", "organometallic" activation of the C–H bond, i.e., reactions where organometallic derivatives (i.e., compounds containing a metal–carbon σ-bond) are formed as an intermediate or as the final product. In the second group, we include reactions in which the contact between the complex and the C–H bond is only via a complex ligand during the process of the C–H bond cleavage. The σ-C–M bond is not generated directly at any stage. In these reactions the function of the metal complex usually consists in abstracting an electron or a hydrogen atom from the hydrocarbon. Finally, in the processes that belong to the third type, a complex activates initially not the hydrocarbon but another reactant (for example, hydrogen peroxide or molecular oxygen). The reactive species formed (for example, hydroxyl radical) then attacks the hydrocarbon molecule without any participation of the metal complex in the latter process. The metal catalyst does not take part in the direct "activation" of the C–H bond by the radical. The hydrocarbon oxidations in living cells under the action of certain metal-containing enzymes proceed as reactions of the second or third type [1]. Although these oxidations occur via the formation of reactive radicals, they are selective and give the products and energy necessary for microorganisms. Biodegradation of hydrocarbons also requires metal-containing enzymes [1]. It is very interesting that microorganisms are known to degrade hydrocarbons under strictly anoxic conditions [2].

Three types of oxidative activation of C–H bonds

First ("organometallic"):

$$\gt C{-}H + M \longrightarrow \gt C{-}M \xrightarrow{\text{Oxidant X}} \gt C{-}X$$

Second ("via a ligand at M"):

$$\gt C{-}H + O{=}M \longrightarrow \gt C\bullet + HO{-}M \longrightarrow \gt C{-}OH$$

Third ("activation of other reactant"):

$$M + H_2O_2 \longrightarrow HO\bullet \xrightarrow{\gt C{-}H} \gt C\bullet \rightarrow\rightarrow\rightarrow \gt C{-}X$$

Organometallic stoichiometric and catalytic activation of C–H bonds in alkanes and arenes gives rise to hydrocarbon functionalization (numerous examples can be found in [1a–d]; see also certain recent publications [3]). Although the mechanisms of the reaction with C–H bonds are in many cases unknown, we can state that metal-catalyzed oxygenations (i.e., processes of oxygen atom insertion) of saturated hydrocarbons rarely begin from the formation of the σ-C–M bond (the first type of activation). An unambiguous example of the organometallic (first type) activation is the Shilov reaction [1, 4], which enables the oxidation of alkanes in aqueous solutions under catalysis by platinum(II) complexes. The first step of the reaction is the formation of a σ-alkyl platinum(II) derivative, which is then oxidized by platinum(IV) present in the solution to give alkanol (and also alkyl chloride):

$$\text{Alk–H} + \text{Cl}-\text{Pt}^{\text{II}}-\text{Cl} \rightarrow \text{Alk–Pt}^{\text{II}}\text{–Cl} + \text{HCl} \tag{1.1}$$

$$\text{Alk–Pt}^{\text{II}}\text{–Cl} + \text{Pt(IV)} + \text{H}_2\text{O} \rightarrow \text{Alk}-\text{OH} + \text{Pt}^{\text{II}} + \text{Pt(II)} + \text{HCl} \tag{1.2}$$

Since the first step proceeds via a direct contact of the C–H bond with a voluminous Pt^{II}-containing species, the reaction exhibits an "unusual" bond selectivity, i.e. the stronger C–H bonds of methyl groups react faster than the weaker secondary and tertiary C–H bonds: $1^\circ > 2^\circ > 3^\circ$. Hexachloroplatinate, used originally as the stoichiometric oxidant, is obviously a very inconvenient reagent because it is too expensive to be used in the synthesis. In recent years, attempts to employ other cheap oxidants have been made. Sames and co-workers found that salts $CuCl_2$ and $CuBr_2$ can regenerate the active platinum species [5a]. In the L-valine oxidation by the system K_2PtCl_4–$CuCl_2$, TONs attained 20 and isolated yields of lactones were up to 35%.

$$\text{2-amino-3-methylbutanoic acid} \xrightarrow[\text{H}_2\text{O, 160 °C, 10 h}]{\text{catalyst: K}_2\text{PtCl}_4\text{ oxidant: CuCl}_2} \text{lactone } (\text{NH}_3^+) \quad (1.3)$$

Interestingly, the C–H bond functionalization occurred with regio- and stereoselectivity; anti and syn lactones were produced in a 3:1 ratio.

Thorn and co-workers used the Pt(II)–Pt(IV)–H_2O_2 system to hydroxylate *n*-propanol selectively to 1,3-propanediol [5 b], but the efficiency was very low: the amount of 1,3-propanediol corresponded to 1.3 turnovers of the entire platinum content (about 0.09 h^{-1}). Periana et al. [5 c] found that platinum complexes derived from the bidiazine ligand family catalyze the oxidation of methane by sulfuric acid to a methanol derivative, giving a one-pass yield of >70% based on methane. These complexes are very stable in concentrated sulfuric acid at relatively high temperature and are among the most effective catalysts for methane conversion. Under the action of platinum, arenes give biaryl compounds with good yields [1, 5 d].

Unlike the reactions mentioned above, almost all processes described in this chapter can be considered to belong to the second and third types of C–H bond activation. From the mechanistic point of view, all hydrocarbon oxidations occurring in living organisms [1, 6 a–c] are also of the second or third types (i.e. they do not involve organometallic activation). However, it has been shown in a recent publication [6d] that aryl C–H activation occurs in copper complexes with triazamacrocyclic ligands, which is a model of the hydroxylation performed by a binuclear copper enzyme tyrosinase. The authors noted that "while the generally accepted enzymatic mechanism does not involve direct aryl C–H activation by a Cu^{II} center, no current data precludes it". Metal complexes that are models of certain metal-containing enzyme centers often efficiently oxygenate saturated and aromatic hydrocarbons. Metal derivatives of porphyrins play an important role in hydrocarbon functionalization [6 e, f], and not only in oxygenation processes [6 g]. Another field which gives models of metal-catalyzed and enzymatic processes (and consequently helps us to understand their mechanisms) is activation of C–H bonds in the gas phase [7]. Finally, metal-catalyzed oxidations of hydrocarbons are processes of great importance both for laboratory and industrial practice [8]. In this chapter we will consider only recent publications devoted to metal-catalyzed liquid-phase reactions. Different sections of the chapter are devoted to functionalization by certain oxidative reagents.

2.2.2 Oxidation with Molecular Oxygen

Doubtless, molecular oxygen (and especially air) is the most cheap, convenient, and green [9] oxidation agent in organic chemistry. Thermodynamically, the formation of oxygen-containing products from saturated hydrocarbons and molecular oxygen is always favorable because oxidation reactions are highly exothermic. The complete oxidation of alkanes by air (burning) to produce water and carbon dioxide is a very important source of energy. There can also be partial oxidation (autoxidation) of saturated hydrocarbons producing various valuable organic substances, e.g., alkyl hydroperoxides, alcohols, and ketones or aldehydes.

Non-catalyzed autoxidation [1 b] of saturated hydrocarbons in the liquid phase is usually a branched-chain process. Hydroperoxides are the intermediates in liquid phase oxidation. Let us consider first the mechanism of non-catalyzed oxidation. The following classical scheme represents the typical mechanism of liquid-phase hydrocarbon oxidation.

Chain initiation: $$\mathrm{RH + O \rightarrow R^{\bullet} + HOO^{\bullet}} \qquad (A)$$

Chain propagation: $$\mathrm{R^{\bullet} + O_2 \rightarrow ROO^{\bullet}} \qquad (B)$$

$$\mathrm{ROO^{\bullet} + RH \rightarrow ROOH + R^{\bullet}} \qquad (C)$$

Chain branching: $$\mathrm{ROOH \rightarrow RO^{\bullet} + HO^{\bullet}} \qquad (D)$$

or $$\mathrm{2\,ROOH \rightarrow RO^{\bullet} + ROO^{\bullet} + H_2O} \qquad (D')$$

Chain termination: $$\mathrm{R^{\bullet} + R^{\bullet} \rightarrow R\text{–}R} \qquad (E)$$

$$\mathrm{ROO^{\bullet} + R^{\bullet} \rightarrow ROOR} \qquad (F)$$

$$\mathrm{ROO^{\bullet} + ROO^{\bullet} \rightarrow ROH + R'COR'' + O_2} \qquad (G)$$

Highly reactive radicals $\mathrm{RO^{\bullet}}$ and $\mathrm{HO^{\bullet}}$ can take part in the following fast steps:

$$\mathrm{HO^{\bullet} + RH \rightarrow H_2O + R^{\bullet}} \qquad (C')$$

$$\mathrm{RO^{\bullet} + RH \rightarrow ROH + R^{\bullet}} \qquad (C'')$$

Relatively unstable chemical initiators can be used to initiate the reaction generating alkyl radicals, $\mathrm{R^{\bullet}}$. For example, in the case of azobis(isobutyronitrile) (AIBN, In–N=N–In) [10 a], the chain initiation step

$$\mathrm{In\text{–}N{=}N\text{–}In \rightarrow 2\,In^{\bullet} + N_2} \qquad (A')$$

$$\mathrm{In^{\bullet} + RH \rightarrow InH + R^{\bullet}} \qquad (A'')$$

is much more efficient in comparison with stage (*A*). An alkane oxidation can be also initiated by any other free radicals, $X^{\bullet}$, which are capable of abstracting the hydrogen atom from an alkane:

$$X^{\bullet} + RH \rightarrow XH + R^{\bullet} \qquad (A''')$$

Very reactive hydroxyl and alkoxyl radicals are among potential initiators of alkane oxidations. Often, at low temperature and at least at the beginning of the reaction, the catalyzed oxidation of an alkane, RH, initiated with H_2O_2 or *tert*-BuOOH (see below), gives rise almost exclusively to the corresponding alkyl hydroperoxide, ROOH. This supports the view that chain-branching steps *D* and *D'*, which can give the alcohol in the propagation step *C'''*, are not involved in the alkane hydroperoxidation mechanism.

An important question arises whether the radical-*chain* oxidation of the alkane with molecular oxygen is possible at low temperature or the oxidation reaction occurs as a simple radical-initiated process with the rate less than the initiation rate. The simplified mechanism of an initiated non-branched radical-chain liquid-phase oxidation of a hydrocarbon, RH, can be described by the following equations:

$$RH \rightarrow R^{\bullet} \qquad \text{(chain initiation with rate } W_i\text{)} \qquad (2.1)$$

$$R^{\bullet} + O_2 \rightarrow ROO^{\bullet} \qquad \text{(chain propagation)} \qquad (2.2)$$

$$ROO^{\bullet} + RH \rightarrow ROOH + R^{\bullet} \qquad \text{(chain propagation)} \qquad (2.3)$$

$$ROO^{\bullet} + ROO^{\bullet} \rightarrow \text{non-radical products} \qquad \text{(chain termination)} \qquad (2.4)$$

In this scheme, $R^{\bullet}$ are alkyl radicals, $ROO^{\bullet}$ are peroxyl radicals, Eq. (2.1) corresponds to the stage of radical generation with the rate W_i , Eqs. (2.2) and (2.3) represent the cycle of chain propagation, and Reaction 2.4 is the chain termination step. Reaction (2.3) is the crucial step for the classical radical chain route. Let us assume that at low temperature the sole terminal product of the oxidation is alkyl hydroperoxide, ROOH. The results of the kinetic analysis of the scheme 2.1–2.4 are summarized in Tab. 1 [10b]. It follows from the data of this table that for a hydrocarbon such as cyclohexane, and especially for the much more inert ethane and methane at 30 °C, we have in principle no reason to consider the possibility of the chain process according to scheme 2.1–2.4. Even at 100 °C for the 10% transformation of ethane and methane via mechanism 2.1–2.4, the reactions will take 6.5 and 30 days, respectively, and the highest possible rate of the chain process is extremely low for these hydrocarbons. For hydrocarbons with weak C–H bonds, such as tetralin and cyclohexene (allylic methylenes), as well as cumene, corresponding transformations via route 2.3 will take less than one hour and consequently are quite possible.

Ions of transition metals are often used in catalytic low-temperature alkane oxidations. While the classical radical-chain mechanism of the alkane oxidation (steps 2.2, 2.3 and 2.4) remains unchanged, catalysts take part in the initiation

Tab. 1 Highest possible rates of the hydrocarbon *radical-chain* low-temperature oxidation and the minimum possible times $\tau^0_{0.1}$ for the transformation of these hydrocarbons to the extent of 10% [a)]

No.	Hydrocarbon	Rate W_i^{max} (mol dm^{-3} s^{-1})		Time $\tau^0_{0.1}$	
		30 °C	100 °C	30 °C	100 °C
1	Tetralin	9.1×10^{-6}	1.0×10^{-4}	1.7 hours	0.2 hours
2	Cyclohexene	6.3×10^{-6}	5.1×10^{-5}	2.5 hours	0.3 hours
3	Ethylbenzene	8.3×10^{-8}	7.3×10^{-6}	7.5 days	2.1 hours
4	Toluene	2.3×10^{-10}	1.0×10^{-7}	7.5 years	6.5 days
5	Cumene	2.0×10^{-6}	1.2×10^{-4}	7.5 hours	0.2 hours
6	Cyclohexane	5.5×10^{-11}	1.0×10^{-7}	32.5 years	6.5 days
7	Ethane	3.2×10^{-12}	1.0×10^{-7}	550 years	6.5 days
8	Methane	7.9×10^{-14}	2.1×10^{-8}	22000 years	30 days

a) The parameters have been calculated assuming a hydrocarbon concentration of 1.0 mol dm^{-3} (although in many cases it is lower).

stage, inducing the initiator decomposition to produce free radicals. This case is not different from the initiated alkane oxidation considered above, and here all parameters estimated previously can also be used. On the other hand, complex ions of transition metals could effectively interact with the alkyl hydroperoxide formed in the oxidation process even at relatively low temperature. This interaction results in the peroxide decomposition, generating free radicals. In this case we have to add to the scheme 2.1–2.4 the following stages (M is an ion of transition metal in oxidized or reduced form):

$$\mathrm{ROOH} + \mathrm{M}(ox) \rightarrow \mathrm{ROO}^{\bullet} + \mathrm{H}^{+} + \mathrm{M}(red) \tag{2.5}$$

$$\mathrm{ROOH} + \mathrm{M}(red) \rightarrow \mathrm{RO}^{\bullet} + \mathrm{HO}^{-} + \mathrm{M}(ox) \tag{2.6}$$

$$\mathrm{RO}^{\bullet} + \mathrm{RH} \rightarrow \mathrm{ROH} + \mathrm{R}^{\bullet} \tag{2.7}$$

It should be noted that if in the case of mechanism 2.1–2.4 the final product of the reaction is alkyl hydroperoxide, ROOH, the mechanism 2.5–2.7, which is a branching one, gives rise to the alcohol as a main product. The analysis in steady-state approximation of the kinetic scheme taking into account 2.5–2.7 leads us to the conclusion that the rate of the ROOH decomposition with participation of a catalyst is only 1.5 times higher than the rate of hydrocarbon consumption in the chain termination step. In the latter case the composition of the products can be dramatically changed. Since in some metal-catalyzed oxidations, at least at low (<50 °C) temperatures and at least at the beginning of the reaction, cyclohexane and normal alkanes are transformed only into alkyl hydroperoxides, we can disregard a mechanism involving steps 2.5–2.7. Taking this into account, we conclude that, in accordance with data summarized in Tab. 1, the classical radical-chain mechanism 2.1–2.4 should be neglected for this case. However, the increment of this pathway might be expected for easily oxidizable hydrocarbons such as tetralin, cyclohexene, etc.

It should be emphasized that the analysis described above does not exclude the possibility of the oxidation of cyclohexane, methane, and other alkanes having strong C–H bonds at a high rate via a radical *non-chain* mechanism, for example, according to the third type of C–H bond activation:

$$H_2O_2 + M \rightarrow HO^{\bullet} \tag{2.8}$$

$$HO^{\bullet} + RH \rightarrow R^{\bullet} \tag{2.9}$$

$$R^{\bullet} + O_2 \rightarrow ROO^{\bullet} \tag{2.10}$$

$$ROO^{\bullet} + H^+ + e^- \rightarrow ROOH \tag{2.11}$$

$$ROOH \rightarrow \text{more stable products (ketones, aldehydes, alcohols)} \tag{2.12}$$

It can be seen, however, that in this case an oxidizing system requires a source of stoichiometric amounts of radicals initiating the oxidation.

In practice, autoxidation of C–H compounds is usually carried out in the presence of various metal complexes, and its mechanism involves reaction 2.6 as a crucial step (Haber-Weiss decomposition of hydroperoxides). Radicals $RO^{\bullet}$ are more reactive in comparison with $ROO^{\bullet}$, which makes possible the chain mechanism via step 2.7. The reactions occur typically at temperatures around 100 °C without solvents, in the presence of surfactants or in inert solvents (e.g., acetic acid). Only compounds containing relatively weak C–H bonds can be oxidized: alkylbenzenes, olefins into allylic position, high branched and normal alkanes. Some examples of these oxidations are presented in Tab. 2. Alkyl hydroperoxides and/or alcohols and ketones are the products of these reactions. In some cases alkyl chain dehydrogenation occurs. The oxidation of (–)-α-pinene catalyzed by the cobalt derivative gave verbenone with good yield [11 n]:

(2.13)

The system consisting of cobalt or manganese acetate and sodium bromide catalyzes a very efficient autoxidation of methylarenes to corresponding arenecarboxylic acids in acetic acid (cobalt-bromide catalysis, which is the basis for the industrial MC/Amoco process) [12 a]. Saturated hydrocarbons cannot be oxidized by this method. Branched hydrocarbons (isobutane) and even cycloalkanes (cyclooc-

Tab. 2 Autoxidation of hydrocarbons with relatively weak C–H bonds in the presence of metal complexes

No.	Hydrocarbons	Catalysts	Ref.
1	Ethylbenzene	Nickel bis(acetylacetonate) and nickel bis(enaminoacetonate)	11a
2	Cumene	Transition metal salts supported on polymer	11b
3	*p*-Xylene [a)]	$MnBr_2$	11c
4	Alkylaromatics [b)]	$[H_2F_6NaV^VW_{17}O_{56}]^{8-}$	11d
5	Adamantane and 1,3-dimethyladamantane [b)]	$K_5FeSi(OH_2)W_{11}O_{39} \cdot 3H_2O$ and $Na_6MnSi(OH_2)W_{11}O_{39}$ supported on Al_2O_3	11e
6	Ethylbenzene	Cobalt bis(acetylacetonate), cationic surfactant	11f
7	1-Hexene [c)]	$[Ir(CH_3CN)_4NO_2]^{2+}$	11g
8	Cyclohexene	Vanadyl Schiff base complexes	11h, i
9	Cyclohexene, tetralin	$VO(acac)_2$	11j
10	Cyclohexene	Bimetallic Pd(II) complex	11k
11	Tetrahydrofuran [d)]	Mo/Ru complexes	11l
12	Isochroman [e)]	Multi-Cu oxidase laccase	11m

a) Selective oxidation to terephthalic acid in supercritical H_2O at ca. 400 °C.
b) Simultaneous oxygenation and oxydehydrogenation.
c) Conversion 79%, products: 1,2-epoxyhexane (24), 1-hexen-3-one (26), 2-hexenal (20), 1-hexen-3-ol (5), 2-hexen-1-ol (3).
d) The oxidation catalyzed with $[Ru(CO)_2Cp]_2$ gave γ-butyrolactone (TON = 290) as well as propylformate (TON = 48).
e) Corresponding lactone was obtained with yield 10% in the oxidation in the presence by TEMPO (2,2′,6,6′-tetramethylpiperidine-*N*-oxide).

tane) can be efficiently oxygenated with molecular oxygen under catalysis with certain metal porphyrins and some other complexes (Lyons system) [12b–d].

Ishii and coworkers described the oxidation of organic compounds including alkanes by molecular oxygen catalyzed by *N*-hydroxyphthalimide (NHPI) combined with $Co(acac)_n$ ($n = 2, 3$) or transition metal salts [12e]. The analogous system "NHPI–ammonium hexanitratocerate(IV)" enables C–H bonds to be functionalized under argon atmosphere [12f]:

$$PhCH_2CH_3 + EtCN \rightarrow PhCH(CH_3)NHCOEt \tag{2.14}$$

Metal complexes also catalyze the hydroxylation of aromatics with molecular oxygen. Thus, heteropolyacid $H_6PMo_9V_3O_{40}$ encapsulated in mesoporous MCM-41 and microporous VPI-5 molecular sieves catalyzes the transformation of benzene to phenol with a TON of 800 [12g].

Catalytic splitting of C–C bonds in alcohols and ketones occurs with simultaneous cleavage of C–H bonds. For example, treatment of a mixture of cyclohexanone and cyclohexanone (KA-oil, a very important intermediate for the production of nylon) by the Ishii oxidation system gives rise to the BaeyerVilliger products [12h]:

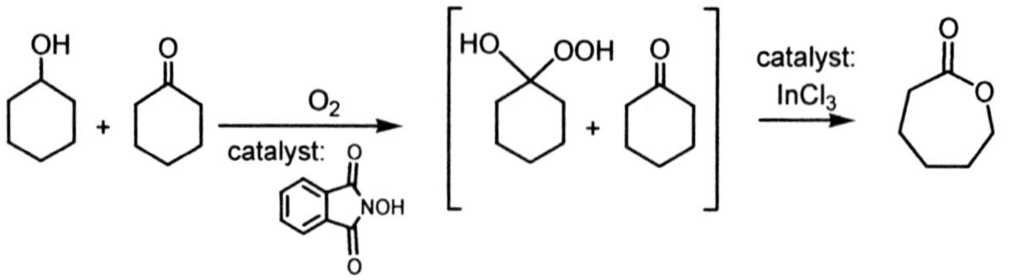

(2.15)

It has recently been shown by Sheldon and co-workers that NHPI, in the absence of any metal complex, catalyzes the selective oxidation of cyclohexylbenzene to cyclohexylbenzene-1-hydroperoxide [12 i]. This reaction provides the basis for a new coproduct-free route to phenol.

Deep catalytic oxidation of cyclohexanone derivatives (as well as other ketones) [12 j–l] affords corresponding acids, for example [12 l]:

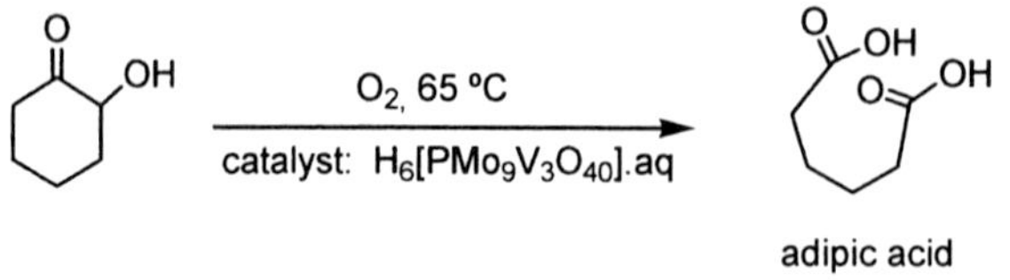

(2.16)

This reaction is reminiscent of biological oxidations catalyzed by catechol dioxygenases [12 k, m].

2.2.3
Combination of Molecular Oxygen with a Reducing Agent

Unlike the case of dioxygenases, which insert both atoms from the O_2 molecule into the substrate, the biological oxidation of hydrocarbons catalyzed by monooxygenases is coupled with the oxidation of electron donors, such as NADH or NADPH. The donor in biological oxidation is believed to transfer its electrons initially to the metal ion, which is subsequently oxidized by an oxygen molecule. It should be noted that hydrocarbons could play the role of reductants, although it is very difficult to abstract proton or hydrogen from these compounds. Some chemical systems based on metal complexes and involving molecular oxygen as the oxidant require a reducing agent which can easily provide the system either with electron or with hydrogen atom. Tab. 3 summarizes examples of the aerobic oxidations with participation of a reductant.

Copper complex **3.1** in the presence of pivaldehyde catalyzes aerobic oxidation of racemic 2-arylcyclohexanones to afford the corresponding lactones with enantioselectivities of up to 69% *ee* [13 o]:

(3.1)

Tab. 3 Autoxidation of hydrocarbons in the presence of reducing agents

No.	Hydrocarbons	Catalysts	Reductant	Efficiency	Ref.
1	Cyclohexane[a)]	$FeCl_3$-picolinic acid	H_2S	Conversion 36%	13a
2	Alkanes	Copper salts	Aldehydes	Yield 4.3% (on converted cyclohexane)	13b
3	Adamantane and alkylaromatics	Metal acetylaceto-nates	3-Methylpropanal	Conversion of adamantane 81%	13c
4	Cyclohexane, *n*-hexane	Immobilized Fe carboxylate complex	Mercaptane, PPh_3	Yield TON=119	13d
5	Methane[c)]	V complexes	Zn/CF_3CO_2H	TON=11	13e
6	Indane copper derivatives		Isobutyraldehyde	Conversion 49%	13f
7	Cyclohexane	Porphyrinato-iron(III)	Ascorbic acid	TON=17	13g
8	Adamantane	Fe oxo/peroxo pivalate	Zn/Pivalic acid	Low product yield	13h
9	Methane	$Pd/C+Cu(MeCO_2)_2$	H_2	TON=13	13i
10	Cyclooctane	$NaAuCl_4$	Zn/CH_3CO_2H	TON=10	13j
11	Cyclooctane[d)]	No catalyst	Acetaldehyde	Yield 22%	13k
12	Benzene cyclohexane	Pd/Al_2O_3+V or Fe	H_2	TOF up to 59 h^{-1}	13l
13	Benzene	Pt/SiO_2 + $V(acac)_3$	H_2		13m
14	Benzene[e)] cyclohexane	VO_3^-	Ascorbic acid, Zn/CH_3CO_2H	TON up to 78	13n

a) Under Gif conditions, i.e. in MeCN in the presence of 4-*tert*-butylpyridine.
b) In the presence of *t*-butyl hydroperoxide.
c) In CF_3CO_2H.
d) In the presence of compressed carbon dioxide at 42–90 °C. Predominant formation of cyclooctanone.
e) The oxidation occurs in acetonitrile only in the presence of pyridine, pyrazinic acid, and acetic acid; no reaction if ascorbic acid is dissolved in the reaction medium.

In concluding the two first sections, we can state that in the aerobic oxidation two mechanistic pathways lead to the C–H activations. These can be conventionally called the "dioxygenase" route (insertion of both oxygen atoms from the O_2 molecule) and the "monooxygenase" route (insertion of only one oxygen from the O_2 while the second oxygen is reduced to water by a reductant). Obviously, the dioxygenase type is more profitable from the practical point of view, because the monooxygenase type requires "non-productive" use of a reducing agent (and also "non-productive" use of half of the oxygen, which is not so important because air is a very cheap reagent). The dioxygenase type can be successfully used for the oxidation of compounds with C–H and C–C bonds activated by neighboring oxo or hydroxy groups. An example is shown in Eq. (2.16). As stated above, the radical-chain oxidation of saturated hydrocarbons under mild conditions is possible only in the case of compounds containing relatively weak C–H bonds. Usually this process is non-selective and gives many products. The monooxygenase route is much more selective: hydroxylation of even lower alkanes can be carried out at room temperature like biological oxidations (methane monooxygenase [1, 6b], cytochrome P450 [1, 13p]). A very important variant of this route is the use of oxidants containing oxygen in a "reduced form". These are called "oxygen atom donors". The next sections are devoted to oxidations by oxygen atom donors.

2.2.4
Hydrogen Peroxide as a Green Oxidant

Hydrogen peroxide is a very convenient oxidant and also the cheapest (after molecular oxygen and air). Moreover, like dioxygen, it is a "green" reagent because water is the only by-product in these oxidations [14a]. It can be used in laboratory practice and also in industrial production of relatively expensive products (for the large-scale production of simple alcohols from alkanes, hydrogen peroxide would appear to be too expensive).

In the absence of catalysts under mild conditions (low temperatures, usual solvents) hydrogen peroxide does not react even with compounds containing weak C–H bonds. Certain metal complexes catalyze not only the "non-productive" decomposition of H_2O_2 to H_2O and O_2 but also alkane oxygenation. Corresponding alkyl hydroperoxides are usually formed, at least at the beginning of the reaction. Alkyl hydroperoxides formed in H_2O_2 oxidations can be determined quantitatively if the solution samples are injected into the GC equipment before and after treatment with triphenylphosphine [1b, d, 14b].

One of the most efficient systems for alkane oxidation, proposed recently, is based on the dinuclear manganese(IV) derivative $[L_2Mn_2O_3](PF_6)_2$ (**4.1**) (L=1,4,7-trimethyl-1,4,7-triazacyclononane). Complex **4.1** catalyzes very efficient oxygenation of various organic compounds in acetonitrile or nitromethane only if a carboxylic acid is present in small concentration in the reaction mixture [15a, b]. Light (methane, ethane, propane, normal butane, and isobutane) and higher (*n*-hexane and *n*-heptane, decalin, cyclohexane, methylcyclohexane, etc.) alkanes can easily be oxidized by the "H_2O_2–**4.1**–CH_3CO_2H" system at room temperature, at 0 °C, and even at –22 °C. Turnover numbers of 3300 have been attained after 1–2 h, and the yield of oxygenated products was 46% based on the alkane. The oxidation initially affords the corresponding alkyl hydroperoxide as the predominant product. However, this compound decomposes in the course of the reaction to produce the corresponding ketone and alcohol.

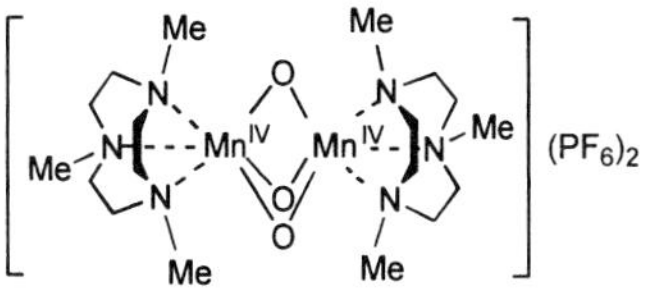

4.1 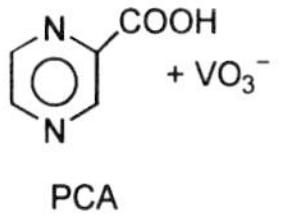 **4.2** 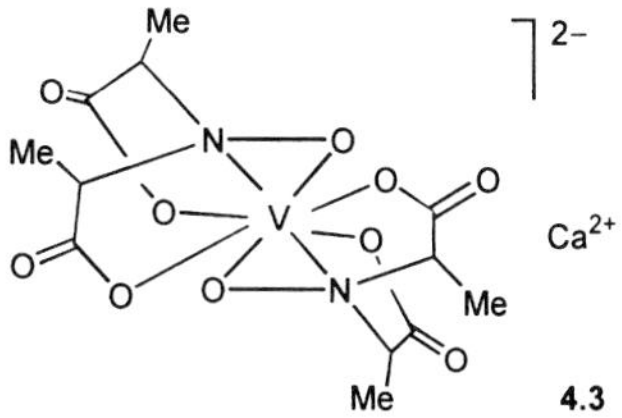 **4.3**

Regio and bond selectivities of the reaction are high: C(1):C(2):C(3):C(4) ≈ 1:40:35:35 and 1°:2°:3° is 1:(15–40):(180–300). The reaction with *cis*- or *trans*-isomers of decalin gives (after treatment with PPh_3) alcohols hydroxylated in the tertiary position with a *cis/trans* ratio of ~2 in the case of *cis*-decalin and a *trans/cis* ratio of ~30 in the case of *trans*-decalin. It has been proposed [15b] that catalytically active species containing an $Mn^{III}Mn^{IV}$ fragment is formed in the solution. The alkane oxidation begins with hydrogen atom abstraction from the alkane by oxygen-centered radical or radical-like species. The active oxidant is probably a dinuclear manganese complex (HOO–)MnMn(=O), and the reaction occurs via an "oxygen-rebound mechanism" between radical $R^\bullet$ and the HOO– group to produce ROOH with retention of stereochemistry. Alkyl radicals ($R^\bullet$) can also partially escape from the solvent cage and react with dioxygen to generate $ROO^\bullet$ and subsequently ROOH with some loss of stereochemistry.

The soluble manganese(IV) complex containing as ligands 1,4,7-triazacyclononane moieties bound to a polymeric chain also catalyzes oxidation of alkanes, and the presence of relatively small amount of acetic acid is obligatory for this reaction [15c]. It is interesting that the oxidation of alkanes and olefins exhibits some features (kinetic isotope effect, bond selectivities) that distinguish this system from an analogous system based on dinuclear Mn(IV) complex **4.1**. A combination of $MnSO_4$ and 1,4,7-trimethyl-1,4,7-triazacyclononane in the presence of oxalate, ascorbate, or citrate buffers catalyzes the oxidation of arylalkanes with hydrogen peroxide [15d]. Ethylbenzene was oxidized at 40 °C with TOF = 188 h^{-1}.

Any soluble vanadium derivative, for example $^nBu_4NVO_3$, $VOSO_4$, $VO(acac)_2$, can be used as a catalyst in combination with pyrazine-2-carboxylic acid (PCA) as co-catalyst (combination **4.2**) for the oxidations with hydrogen peroxide in acetonitrile solution [16]. At low temperatures, the predominant product of alkane oxidation is the corresponding alkyl hydroperoxide, while alcohols and ketones or aldehydes are formed simultaneously in smaller amounts. This alkyl hydroperoxide then slowly decomposes to produce the corresponding ketone and alcohol. Atmospheric oxygen takes part in this reaction; in the absence of air the oxygenation reaction does not proceed. Thus, in alkane oxidation, hydrogen peroxide plays the role of a promoter while atmospheric oxygen is the true oxidant. The oxidation of *n*-heptane by the reagent under consideration exhibits low selectivity: C(1) : C(2) : C(3) : C(4) $\approx$ 1 : 4 : 4 : 4. Methane, ethane, propane, *n*-butane, and isobutane can also be readily oxidized in acetonitrile by the same reagent. In addition to the primary oxidation products (alkyl hydroperoxides), alcohols, aldehydes or ketones, and carboxylic acids are obtained with high total turnover numbers (at 75 °C after 4 h: 420 for methane and 2130 for ethane) and H_2O_2 efficiency. Methane can also be oxidized in aqueous solution, giving in this case methanol as the product (after 20 h at 20 °C the turnover number is 250). The reagent also oxygenates arenes to phenols and alcohols to ketones, and hydroperoxidizes the allylic position in olefins. The crucial step of the oxidation by the reagent "O_2–H_2O_2–VO_3^- – pyrazine-2-carboxylic acid" is the very efficient generation of $HO^\bullet$ radicals. These radicals abstract a hydrogen atom from the alkane, RH, to generate the alkyl radical, $R^\bullet$. The latter reacts rapidly with an O_2 molecule affording the peroxo radical, $ROO^\bullet$, which is then transformed simultaneously into three products: alkyl hydroperoxide, ketone, and alcohol. The proposed mechanism of $HO^\bullet$ generation involves the reduction of V(V) species by the first molecule of H_2O_2 to give a V(IV) derivative. No oxidation occurs in the absence of pyrazine-2-carboxylic acid. The possible role of pyrazine-2-carboxylic acid is its participation (in the form of a ligand at the vanadium center) in the proton transfer, which gives the hydroperoxy derivative of vanadium.

Zeolite-encapsulated vanadium complexes with picolinic acid are also efficient (although less so) in hydrocarbon oxidations [17a, b]. Synthetic amavadine (present in *Amanita* fungi) models, for example, complex **4.3**, exhibits haloperoxidase activity and catalyzes [17c] in the presence of HNO_3 oxo-functionalization of alkanes and aromatics with TONs up to 10. Alkanes can be oxidized by hydrogen peroxide using vanadium-containing polyphosphomolybdate $[PMo_{11}VO_{40}]^{4-}$ as catalyst in acetonitrile [17d] or trifluoroacetic anhydride [17e, f]. Complex $K_{0.5}(NH_4)_{5.5}[MnMo_9O_{32}]$ is a catalyst for phenol hydroxylation with 30% H_2O_2 in methanol [17g].

Mono- and dinuclear iron complexes with various *N*-containing ligands are good catalysts for the H_2O_2 oxidations of hydrocarbons. These complexes, for example **4.4** [18a–c], **4.5** [18d], **4.6** [18e], and **4.7** [18f] mimic non-heme enzymes (see also certain recent publications [18g, h]). In alkane oxidations, the TONs vary from 2–5 to 100–150. In some cases (complexes like **4.4**, chiral complex **4.5**), the reaction proceeds stereospecifically; the hydroxylation with complex **4.5** is partially enantioselective. The reactions in acetonitrile catalyzed by compounds **4.6** and **4.7** can be dramatically accelerated by adding picolinic acid or PCA.

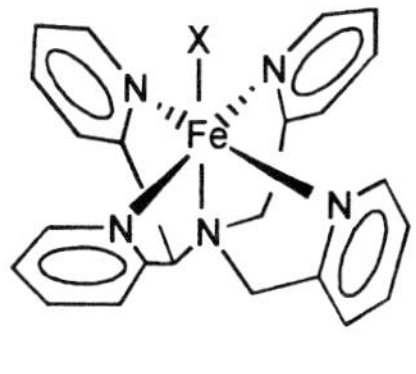

4.4

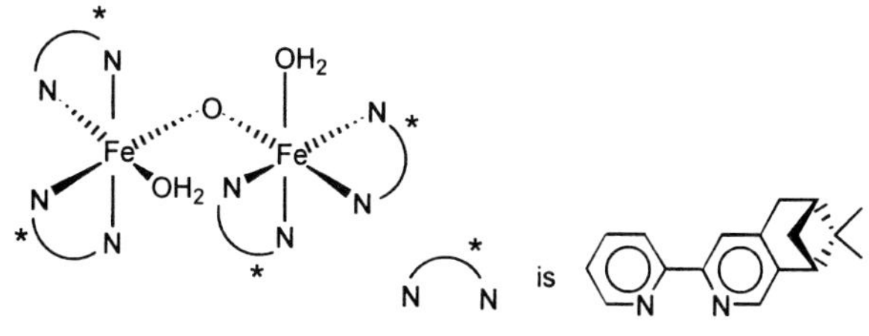

4.5

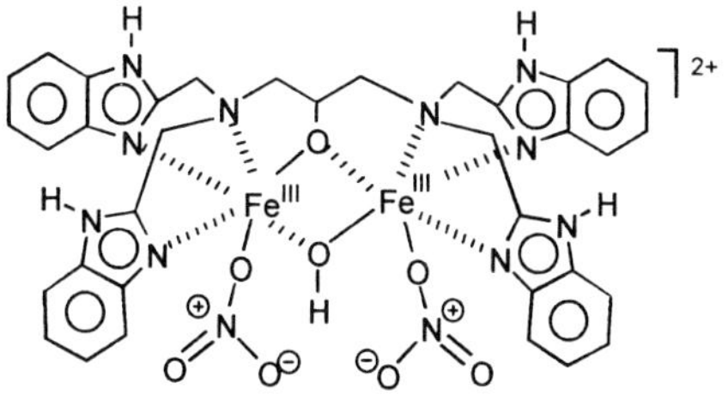

cation of compound **4.6**

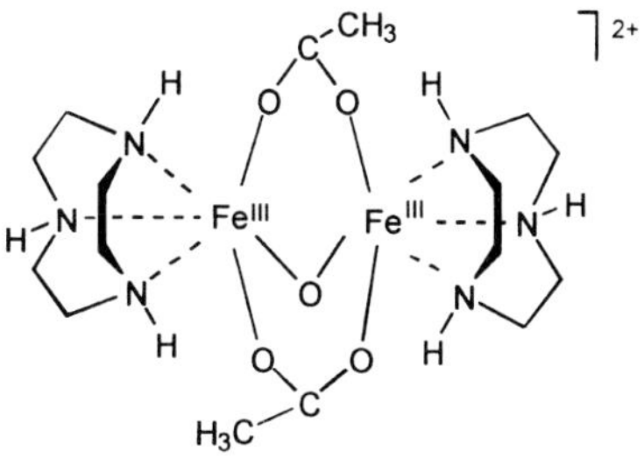

cation of compound **4.7**

Oxidations catalyzed by metalloporphyrins [19] can be considered as models of biological processes occurring under the action of cytochrome P450 and some other heme enzymes [1]. Some other hydrogen peroxide oxidations that are catalyzed by synthetically prepared soluble metal complexes, solid compounds, and even enzymes are summarized in Tab. 4.

Tab. 4 Examples of catalytic H_2O_2 oxidations of hydrocarbons

No.	Hydrocarbons	Catalysts	Solvent	Efficiency	Ref.
1	Methylbenzenes	$MoO(O_2)(QO)_2$ [a)]	Acetonitrile	Yields up to 95% based on substrate	20a
2	Alkanes	Fe derivatives	py/AcOH [b)]	TONs 2–30	20b
3	Alkanes	$NaAuCl_4$, $ClAuPPh_3$	Acetonitrile	TON = 520	13j
4	Ethane and other alkanes	CrO_3	Acetonitrile	TON = 620 for the case of ethane	20c
5	Methane, ethane and other alkanes	$OsCl_3$	Acetonitrile	TON = 102 for ethane, 150 for propane	20d
6	Alkanes	$Ni(ClO_4)_2$–TMTACN	Acetonitrile	TON = 66	20e
7	Alkanes	H_2PtCl_6	Acetonitrile	TON = 44	20e
8	Toluene	Peroxidase	Water		20f

a) QOH is 8-quinolinol.
b) Gif systems. For Gif chemistry see [1, 20g].

Metal complexes also catalyze the oxidation of arenes to phenols or quinones [21]. Methyltrioxorhenium [22a] and Ti- and Fe-containing zeolites [22b] are catalysts for the practically important oxidation of methylnaphthalene to menadione (vitamin K_3):

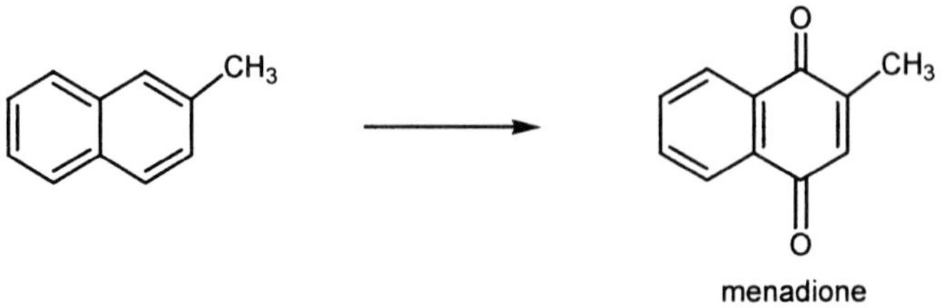

menadione

(4.1)

It has recently been shown that the oxidation in acetic acid occurs without any catalyst [22c].

2.2.5
Organic Peroxy Acids

Peracetic acid oxidizes hydrocarbons if Ru/C [23a], Ti-containing zeolite [23b], and manganese-porphyrins [23c] are used as catalysts. Copper salts, for example, $Cu(ClO_4)_2$ and some complexes, particularly $Cu(CH_3CN)_4BF_4$, taken in small concentrations (for example, 10^{-5} mol dm^{-3}) are also efficient in alkane oxidations with peroxyacetic acid in acetonitrile solution at 60 °C [24a]. The reaction gives rise to the formation of alkyl hydroperoxides as main products and occurs with low bond selectivity. Total turnover number attains 1900. Various vanadium complexes (particularly, n-Bu_4NVO_3) catalyze alkane oxidations by peroxyacetic acid in acetonitrile at 60 °C [24b]. The reaction gives a mixture of corresponding ketones, alcohols, and alkylacetates; formation of alkyl hydroperoxides can be detected (by reduction with triphenylphosphine) only at the beginning of the reaction. Bond selectivities of the oxidation are not high, which testifies to the formation of free radicals. Analogous "modeling" reactions with H_2O_2 in acetonitrile in the presence of acetic acid or in pure acetic acid gave alkyl hydroperoxides as main products.

Copper(I) complexes catalyze allylic oxidations by *tert*-butylperbenzoate [24c–f]. Metal-porphyrins [25a, b] and metal (Mn, Fe, Co) perchlorates [25c] are good catalysts for C–H oxidations with *meta*-chloroperbenzoic acid, for example [25a]:

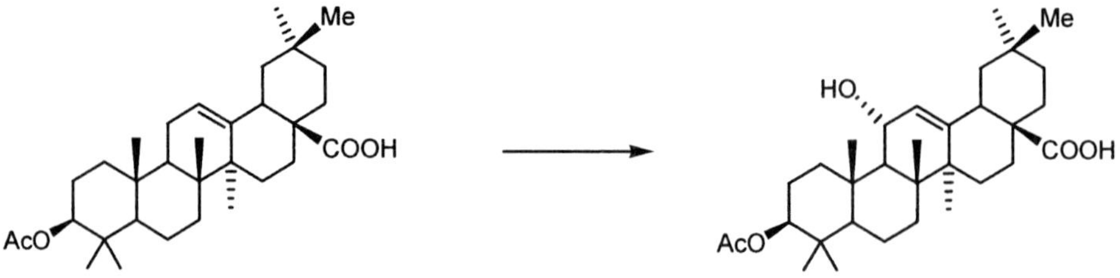

(5.1)

It is interesting that peroxy acids can oxidize alkanes even in the absence of metal catalysts [25d, e]. Finally, manganese derivative **4.1** catalyzes efficient alkane oxidation with peroxyacetic and *meta*-chloroperbenzoic acids [25f].

2.2.6 Alkyl Hydroperoxides as Oxidants

Recent examples of hydrocarbon oxidations with alkyl hydroperoxide (usually, *tert*-butylhydroperoxide) are listed in Tab. 5. As Meunier wrote [26v], "many hydroxylation reactions with alkyl hydroperoxides in the presence of transition-metal complexes are not due to a metal-centered active species, but to a free-radical process initiated by $RO^{\bullet}$". Alkyl hydroperoxide can act as a radical initiator and as a source of molecular oxygen [26t]. For example, the oxidation of cyclohexane, CyH, in the presence of cobalt compounds includes the following stages:

$$t\text{-BuOOH} + \text{Co}^{\text{III}} \rightarrow t\text{-BuOO}^{\bullet} + \text{Co}^{\text{II}} + \text{H}^{+} \qquad (A)$$

$$t\text{-BuOOH} + \text{Co}^{\text{II}} \rightarrow t\text{-BuO}^{\bullet} + \text{Co}^{\text{III}} + \text{H}^{+} \qquad (B)$$

$$t\text{-BuO}^{\bullet} + \text{CyH} \rightarrow t\text{-BuOH} + \text{Cy}^{\bullet} \qquad (C)$$

$$\text{Cy}^{\bullet} + \text{O}_2 \rightarrow \text{CyOO}^{\bullet} \qquad (D)$$

$$\text{CyOO}^{\bullet} + \text{Co}^{\text{II}} + \text{H}^{+} \rightarrow \text{CyOOH} + \text{Co}^{\text{III}} \qquad (E)$$

$$\text{CyOOH} + \text{Co}^{\text{III}} \rightarrow \text{CyO}^{\bullet} + \text{Co}^{\text{II}} + \text{H}^{+} \qquad (F)$$

$$\text{CyOOH} + \text{Co}^{\text{II}} \rightarrow \text{CyO}^{\bullet} + \text{Co}^{\text{III}} + \text{H}^{+} \qquad (G)$$

2.2.7 Oxidation with Sulfur-containing Peroxides

Bagrii and co-workers [27a] described oxidation of 1,3-dimethyladamantane and cyclooctane with potassium permonosulfate. Manganese and iron complexes of alkylated tetrapyridylporphyrin were used as catalysts. The latter was either dissolved in a reaction medium or adsorbed on a layered aluminosilicate [27a]. Tab. 6 shows some other recent examples of metal-catalyzed hydrocarbon oxidations with permonosulfate, HSO_5^-. This anion can be used either as Oxone® ($KHSO_4$/K_2SO_4/$2KHSO_5$) in a biphasic solvent containing water and an organic liquid or as an organic-soluble salt, for example, Ph_4PHSO_5.

Fujiwara and co-workers used $K_2S_2O_8$ as oxidant in various Pd-catalyzed C–H activation processes [3b, c, 28]. For example, benzene, toluene and other aromatic hydrocarbons can be carboxylated by Pd(II) acetate catalyst with CO in trifluoroacetic acid at room temperature to give the aromatic carboxylic acids [28b].

Tab. 5 Hydrocarbon oxidation with *tert*-butyl hydroperoxide

No.	Hydrocarbons	Catalysts	Solvent	Efficiency	Ref.
1	Benzylic and allylic C–H	$CuCl_2$	CH_2Cl_2, phase-transfer catalyst	Conversion up to 96%, selectivity 100%	26a
2	Cyclohexane	Fe(III) and Cu(II) complexes	None	Conversion 4–5%, TON=70–90	26b
3	*p*-Xylene	Zeolite-encapsulated Co and Mn complexes	None	Conversion up to 60%	26c
4	Alkanes	$[Fe_2O(\eta^1\text{-}H_2O)(\eta^1\text{-}OAc)(TPA)_2]^{3+}$ a)	Water	TON=238 for cyclohexane	26d
5	Cyclohexane	μ-Hydroxo diiron(II) with L b)	Acetonitrile	Yield up to 46% based on oxidant	26e
6	Alkylaromatics	Silicate xerogels containing Co	None	Acetophenone from ethyl-benzene: conversion 65%, selectivity >99%	26f
7	Ethylbenzene	Cu(I) complexes	MeCN/py	TON up to 34	26g
8	Alkanes	Ru oxo complexes	Acetone	Yield up to 89% (ethylbenzene)	26h
9	Pinane	Encaged metal phthalocyanines in Y zeolites	Acetone/*t*-BuOH	Conversion 80%, selectivity 90%	26i
10	Cyclohexane	Ru(III) complex	CH_2Cl_2	Yield up to 38% based on oxidant consumed	26j
11	Steroids	Co acetate	Acetonitrile	Yield up to 86%	26k
12	Alkanes, alkylaromatics	Dimanganese(III) complex	CH_2Cl_2	Yield 4.1% (cyclohexane), 4.4% (toluene)	26l
13	Benzene	Ru(III) complex	CH_2Cl_2	TON=890	26m
14	Alkanes	Compound **4.1**	Acetonitrile	TON up to 2000	15a, 26n
15	Isopropyl arenes	Cu salt-crown ether	None	Yield up to 82%	26o
16	Toluene, propylbenzene etc.	Mn(II) complexes	CH_2Cl_2	Isolated yield >85%	26p
17	Unsaturated steroids	Immobilized Co(II), Cu(II), Mn(II), V(II) complexes			26q
18	Toluene, cyclohexane	Mn_4O_{46}+cubane complexes	None	TON=7 (cyclohexane), 101 (toluene)	26r
19	Cyclohexane	Phthalocyanine Fe(II)	Water/methanol	Yield 8.6%	26s
20	Cyclohexane	Immobilized Co acetate oligomers	None	Yield up to 3%	26t
21	Alkanes	Vanadium complexes	Acetonitrile	TONs up to 250 (cyclohexane)	26u
22	Alkanes	Cu(I) and Cu(II) complexes	Acetonitrile	TONs up to 2000 (cyclohexane)	24a

a) TPA is tris[(2-pyridil)methyl]amine.
b) L is 1,4,10,13-tetrakis(2-pyridyl)methyl-1,4,10,13-tetraaza-7,16-dioxacyclooctadecane.

Tab. 6 Examples of catalytic oxidations of hydrocarbons with permonosulfate

No.	Hydrocarbons	Catalysts	Solvent	Efficiency	Ref.
1	Ethylbenzene	Mn(III) porphyrins	Dichloroethane	Yields up to 86%	27b
2	Cycloalkanes	Mn(III) porphyrins	Two-phase [a)]	Yield 43% (cyclohexane)	27c
3	Cyclohexane [b)]	Metal sulfo-phthalocyanines	Two-phase	Yields up to 100%	27d
4	Cyclooctane	Mn tetraphenyl-porphyrin	CH_2Cl_2	Yield 12% (cyclooctane)	27e

a) Solid Oxone®/dichloroethane in the presence of a phase transfer reagent.
b) Oxidation to adipic acid.

2.2.8 Iodosobenzene as an Oxidant

Iodosobenzene, PhIO, is widely used in metal-catalyzed oxidations of various hydrocarbons (Tab. 7). Jitsukawa et al. [30a] found that the catalytic activities of the ruthenium complexes **8.1** containing different substituents at the pyridine 6-position can be fine-tuned. Complexes containing electron-withdrawing groups (for example, R = *t*-BuCONH) promote the epoxidation of cyclohexene, whereas those containing electron-releasing groups (for example, R = *t*-BuCH$_2$NH) promote mainly the adamantane hydroxylation.

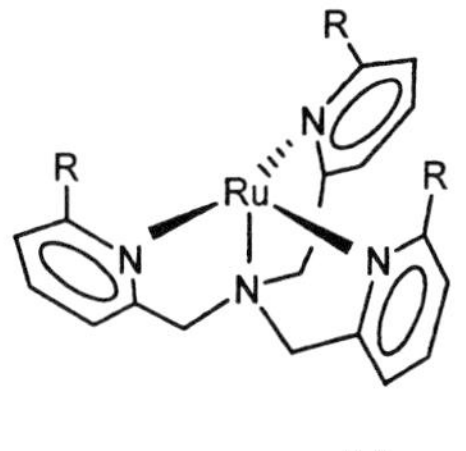

8.1

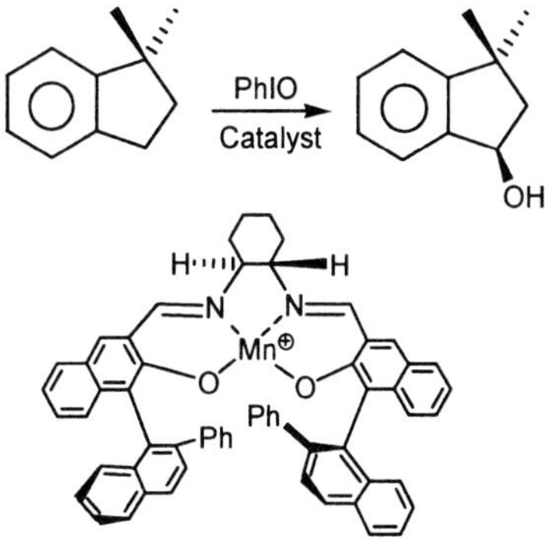

PhIO / Catalyst

Catalyst **8.2** (8.1)

Hydroxylations with iodosobenzene often proceed selectively. Thus, oxidation (Reaction 8.1) of 1,1-dimethylindane catalyzed by optically active manganese complex **8.2** gives the corresponding alcohol with *ee* up to 60% and yield 10% [30b]. Sames and co-workers were able to ketonize exclusively one benzylic position of the 5,6,7,8-tetrahydro-2-naphthol covalently bound to a metal catalyst center **8.3** [30c].

8.3

1) PhIO
2) NaOH

(8.2)

Selective hydroxylations of steroids with artificial cytochrome P450 enzymes have been carried out by Breslow and co-workers [30d, e], e.g., hydroxylation of ester derivative **8.4** of androstan-3,17-diol to alcohol **8.5** catalyzed by the Mn(III) complex of porphyrin **8.6** [30e].

PhIO
Catalyst: 8.6-Mn^{III}

8.4 8.5

β-cyclodextrin

8.6 (8.3)

In the synthesis of bromopyrrole alkaloids, When and Du Bois [30f] employed as one of the steps oxidative cyclization of compound **8.7**, which, under the action of $PhI(OAc)_2$ in the presence of rhodium catalyst, gave smoothly and stereospecifically the oxathiazinane product **8.8**.

Tab. 7 Examples of oxidations of hydrocarbons with iodosobenzene catalyzed by metalloporphyrins

No.	Hydrocarbons	Catalysts	Solvent	Efficiency	Ref.
1	Ethylbenzene	Fe(III) porphyrins	Benzene cyclohexane	Yields up to 73% (cyclohexane, based on PhIO)	29a
2	Cyclohexane	Acetylglycosylated Fe-, Mn-porphyrins		TONs up to 11	29b
3	Cyclohexane, adamantane	Homogeneous and supported Mn(III) porphyrins	CH_2Cl_2 [a)]	Yields up to 44% (cyclohexane, based on PhIO)	29c
4	Cyclohexane	Sterically hindered Fe(III) porphyrins	None	Yields up to 72%	29d
5	Cyclohexane	Mn(III) porphyrins	Dichloroethane	Yields up to 92%	29e
6	2-Methylbutane	μ-Oxo-bismetallo-porphyrins	Chlorobenzene	Yields up to 7%, based on PhIO	29f

a) In the presence of co-catalysts (pyridine, imidazole).

PhI(OAc)$_2$

cat: Rh$_2$(OAc)$_4$

8.7 → **8.8** (8.4)

2.2.9
Oxidations with Other Reagents

In recent years, various oxidants have been employed that are less common in comparison with hydrogen peroxide and alkyl hydroperoxides. Some of them cannot be considered as "green" reagents, for example, hypochlorite. Selected examples of such oxidations are summarized in Tab. 8. The oxidation of ethylbenzenes **9.1** with 2,6-dichloropyridine *N*-oxide (proposed earlier by Higuchi and co-workers [32a]), catalyzed by porphyrin **9.2**, gave corresponding alcohols **9.3** with *ee* up to 75% [32b].

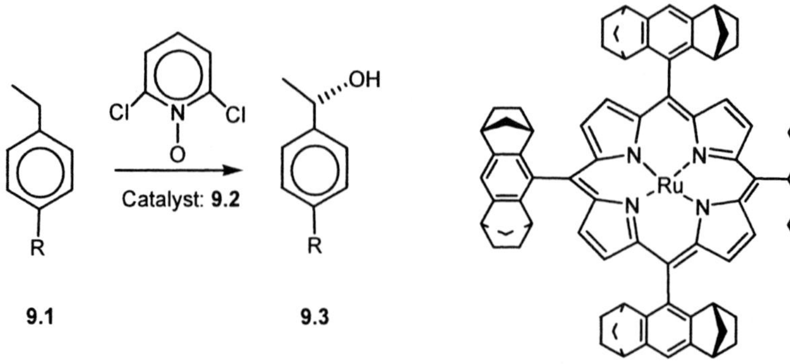

Lee and Fuchs [32c] described very recently an unprecedented Cr-catalyzed chemospecific oxidation by H_5IO_6 of compound **9.4** to the corresponding hemiacetal **9.5**. The reaction proceeds at very low temperature (–40 °C) and gives the product in 69% yield.

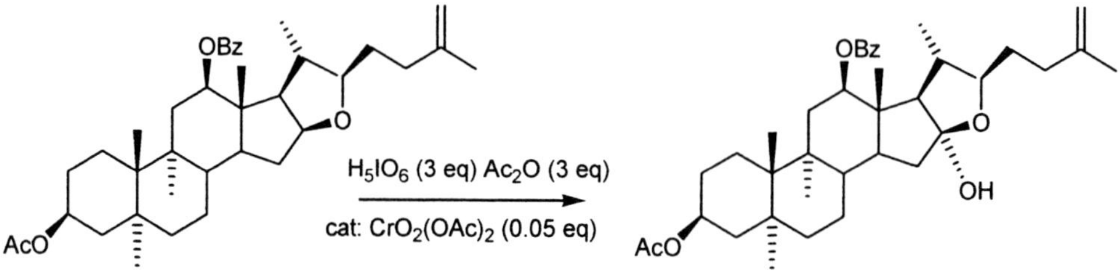

(9.2)

Tab. 8 Examples of oxidations of hydrocarbons with various oxidants

No.	*Hydrocarbons*	*Oxidant*	*Catalyst*	*Efficiency*	*Ref.*
1	Xylenes	Hypochlorite	$RuCl_3$	Yield 98% (4-chloro-2-methyl-benzoic acid)	31 a
2	1,4-Dimethyl cyclohexane	Perchloric acid	Polyphenyl-ferrosiloxane		31 b
3	Polycondensed aromatics	$NaBrO_3$	Nafion–Ce(IV) and Nafion–Cr(III)	Yields up to 95%	31 c
4	Cycloalkanes, arylalkanes	$NaIO_4$	Supported Mn(III) complexes	Yields up to 60%	31 d
5	Arenes	H_5IO_6	CrO_3	Yields up to 90%	31 e
6	Limonene	$CuCl_2$	$PdCl_2$	Conversion up to 92%	31 f
7	Alkylarenes	PMSO [a)]	$[PMo_{12}O_{40}]^{3-}$	TON = 300 (anthracene)	31 g
8	NBMA	TMAO [b)]	Cu(II) complexes	Yields up to 98%	31 h

a) Phenylmethylsulfoxide.

b) *N*-Benzoyl-2-methylalanine (NBMA) is *ortho*-hydroxylated stereoselectively by trimethylamine *N*-oxide (TMAO).

References

1 (a) G.B. Shul'pin, *Organic Reactions Catalyzed by Metal Complexes*, Nauka, Moscow, **1988** (in Russian); (b) A.E. Shilov, G.B. Shul'pin, *Activation and Catalytic Reactions of Saturated Hydrocarbons in the Presence of Metal Complexes*, Kluwer Academic Publishers, Dordrecht Boston London, **2000**, see for example: Chapter IX (Homogeneous catalytic oxidation of hydrocarbons by molecular oxygen), Chapter X (Homogeneous catalytic oxidation of hydrocarbons by peroxides and other oxygen atom donors), Chapter XI (Oxidation in living cells and its chemical models); (c) A.E. Shilov, G.B. Shul'pin, *Chem. Rev.* **1997**, *97*, 2879–2932; (d) G.B. Shul'pin, *J. Mol. Catal. A: Chem.* **2002**, *189*, 39–66.

2 F. Widdel, R. Rabus, *Curr. Opin. Biotechnol.* **2001**, *12*, 259–276.

3 (a) J. Halpern, *Pure Appl. Chem.* **2001**, *73*, 209–220; (b) C. Jia, D. Piao, J. Oyamada, W. Lu, T. Kitamura, Y. Fujiwara, *Science* **2000**, *287*, 1992–1995; (c) C. Jia, T. Kitamura, Y. Fujiwara, *Acc. Chem. Res.* **2001**, *34*, 633–639; (d) V.V. Grushin, W.J. Marshall, D.L. Thorn, *Adv. Synth. Catal.* **2001**, *343*, 161–165; (e) T. Ishiyama, J. Takagi, J.F. Hartwig, N. Miyaura, *Angew. Chem. Int. Ed.* **2002**, *41*, 3056–3058; (f) S.R. Klei, J.T. Golden, P. Burger, R.G. Bergman, *J. Mol. Catal. A: Chem.* **2002**, *189*, 79–94; (g) K. Krogh-Jespersen, M. Czerw, A.S. Goldman, *J. Mol. Catal. A: Chem.* **2002**, *189*, 95–110; (h) H.M.L. Davies, *J. Mol. Catal. A: Chem.* **2002**, *189*, 125–135.

4 (a) B.S. Williams, K.I. Goldberg, *J. Am. Chem. Soc.* **2001**, *123*, 2576–2587; (b) J. Procelewska, A. Zahl, D. van Eldik, H.A. Zhong, J.A. Labinger, J.E. Bercaw, *Inorg. Chem.* **2002**, *41*, 2808–2810; (c) V.V. Rostovtsev, L.M. Henling, J.A. Labinger, J.E. Bercaw, *Inorg. Chem.* **2002**, *41*, 3608–3619; (d) A.G. Wong-Foy, L.M. Henling, M. Day, J.A. Labinger, J.E. Bercaw, *J. Mol. Catal. A: Chem.* **2002**, *189*, 3–16.

5 (a) B.D. Dangel, J.A. Johnson, D. Sames, *J. Am. Chem. Soc.* **2001**, *123*, 8149–8150; (b) N. DeVries, D.C. Roe, D.L. Thorn, *J. Mol. Catal. A: Chem.* **2002**, *189*, 17–22; (c) R.A. Periana, D.J. Taube, S. Gamble, H. Taube, T. Satoh, H. Fujii, *Science* **1998**, *280*, 560–564; (d) W. V. Konze, B.L. Scott, G.J. Kubas, *J. Am. Chem. Soc.* **2002**, *124*, 12550–12556.

6 (a) E.I. Solomon, *Inorg. Chem.* **2001**, *40*, 3656–3669; (b) D. Lee, S.J. Lippard, *Inorg. Chem.* **2002**, *41*, 827–837; (c) V.P. Bui, T. Hudlicky, T.V. Hansen, Y. Stenstrom, *Tetrahedron Lett.* **2002**, *43*, 2839–2841; (d) X. Ribas, D.A. Jackson, B. Donnadieu, J. Mahía, T. Parella, R. Xifra, B. Hedman, K.O. Hodgson, A. Llobet, T.D.P. Stack, *Angew. Chem. Int. Ed.* **2002**, *41*, 2991–2994; (e) G.B. Maravin, M.V. Avdeev, E.I. Bagriy, *Neftekhimiya* **2000**, *40*, 3–21 (in Russian); (f) J.T. Groves, *J. Porphyrins Phthalocyanines* **2000**, *4*, 350–352; (g) A.P. Nelson, S.G. DiMagno, *J. Am. Chem. Soc.* **2000**, *122*, 8569–8570.

7 (a) M. Brönstrup, C. Trage, D. Schröder, H. Schwarz, *J. Am. Chem. Soc.* **2000**, *122*, 699–704; (b) M. Brönstrup, D. Schröder, H. Schwarz, *Chem. Eur. J.* **2000**, *6*, 91–103; (c) Y. Shiota, K. Yoshizawa, *J. Am. Chem. Soc.* **2000**, *122*, 1217–1232.

8 U. Schuchardt, D. Cardoso, R. Sercheli, R. Pereira, R.S. da Cruz, M.C. Guerreiro, D. Mandelli, E.V. Spinacé, E.L. Pires, *Appl. Catal. A: General* **2001**, *211*, 1–17.

9 (a) P.T. Anastas, J.C. Warner, *Green Chemistry: Theory and Practice*, Oxford University Press, New York, **1998**; (b) P.T. Anastas, M.M. Kirchhoff, T.C. Williamson, *Appl. Catal. A: General* **2001**, *221*, 3–13; (c) P.T. Anastas, M.M. Kirchhoff, *Acc. Chem. Res.* **2002**, *35*, 686–694.

10 (a) A. Goosen, C.W. McCleland, D.H. Morgan, J.S. O'Connell, A. Ramplin, *J. Chem. Soc. Perkin Trans. 2* **1993**, 401–404; (b) Yu.N. Kozlov, G.B. Shul'pin, "Can methane and other alkanes be oxidized in solutions at low temperature via a classical radical-chain mechanism?", *The Chemistry Preprint Server*, http://preprint.chemweb.com/physchem/0106002, **2001**, pp. 1–9.

11 (a) L. I. Matienko, L. A. Mosolova, *Russ. Chem. Bull.* **1999**, *48*, 55–60; (b) Y. F. Hsu, C. P. Cheng, *J. Mol. Catal. A: Chem.* **1998**, *136*, 1–11; (c) P. A. Hamley, T. Ilkenhans, J. M. Webster, E. Garcia-Verdugo, E. Venardou, M. J. Clarke, R. Auerbach, W. B. Thomas, K. Whiston, M. Poliakoff, *Green Chem.* **2002**, *4*, 235–238; (d) A. M. Khenkin, R. Neumann, *Inorg. Chem.* **2000**, *39*, 3455–3462; (e) A. I. Nekhaev, R. S. Borisov, V. G. Zaikin, E. I. Bagrii, *Petrol. Chem.* **2002**, *42*, 238–245; (f) T. V. Maksimova, T. V. Sirota, E. V. Koverzanova, A. M. Kashkai, O. T. Kasaikina, *Petrol. Chem.* **2002**, *42*, 46–50; (g) P. J. Baricelli, V. J. Sánchez, A. J. Pardey, S. A. Moya, *J. Mol. Catal. A: Chem.* **2000**, *164*, 77–84; (h) D.aM. Boghaei, S. Mohebi, *J. Mol. Catal. A: Chem.* **2002**, *179*, 41–51; (i) D. M. Boghaei, S. Mohebi, *Tetrahedron* **2002**, *58*, 5357–5366; (j) L. J. Csányi, K. Jáky, G. Galbács, *J. Mol. Catal. A: Chem.* **2002**, *179*, 65–72; (k) A. K. El-Qisiari, H. A. Qaseer, P. M. Henry, *Tetrahedron Lett.* **2002**, *43*, 4229–4231; (l) T. Straub, A. M. P. Koskinen, *Inorg. Chem. Commun.* **2002**, *5*, 1052–1055; (m) F. d'Acunzo, P. Baiocco, M. Fabbrini, C. Galli, P. Gentili, *Eur. J. Org. Chem.* **2002**, 4195–4201; (n) M. K. Lajunen, T. Maunula, A. M. P. Koskinen, *Tetrahedron* **2000**, *56*, 8167–8171.

12 (a) P. D. Metelski, V. A. Adamian, J. H. Espenson, *Inorg. Chem.* **2000**, *39*, 2434–2439; (b) J. E. Lyons, P. E. Ellis, Jr., S. N. Shaikh, *Inorg. Chim. Acta* **1998**, *270*, 162–168; (c) K. T. Moore, I. T. Horváth, M. J. Therien, *Inorg. Chem.* **2000**, *39*, 3125–3139; (d) J. Haber, L. Matachowski, K. Pamin, J. Poltowicz, *J. Mol. Catal. A: Chem.* **2000**, *162*, 105–109; (e) for the Ishii oxidation reaction, see a book [1b], a review: Y. Ishii, S. Sakaguchi, T. Iwahama, *Adv. Synth. Catal.* **2001**, *343*, 393–427; and recent articles: Y. Tashiro, T. Iwahama, S. Sakaguchi, Y. Ishii, *Adv. Synth. Catal.* **2001**, *343*, 220–225; O. Fukuda, S. Sakaguchi, Y. Ishii, *Adv. Synth. Catal.* **2001**, *343*, 809–813; A. Shibamoto, S. Sakaguchi, Y. Ishii, *Tetrahedron Lett.* **2002**, *43*, 8859–8861; (f) S. Sakaguchi, T. Hirabayashi, Y. Ishii, *Chem. Commun.* **2002**, 516–517; (g) L. C. Passoni, F. J. Luna, M. Wallau, R. Buffon, U. Schuchardt, *J. Mol. Catal. A: Chem.* **1998**, *134*, 229–235; (h) O. Fukuda, S. Sakaguchi, Y. Ishii, *Tetrahedron Lett.* **2001**, *42*, 3479–3481; (i) I. W. C. E. Arends, M. Sasidharan, A. Kühnle, M. Duda, C. Jost, R. A. Sheldon, *Tetrahedron* **2002**, *58*, 9055–9061; (j) J.-M. Brégeault, F. Launay, A. Atlamsani, *C. R. Acad. Sci Paris, Ser. IIc, Chemistry* **2001**, *4*, 11–26; (k) M. Vennat, P. Herson, J.-M. Brégeault, G. B. Shul'pin, *Eur. J. Inorg. Chem.* **2003**, 908–917; (l) L. El Aakel, F. Launay, A. Atlamsani, J.-M. Brégeault, *Chem. Commun.* **2001**, 2218–2219; (m) R. Yamahara, S. Ogo, H. Masuda, Y. Watanabe, *J. Inorg. Biochem.* **2002**, *88*, 284–294.

13 (a) D. H. R. Barton, T. Li, J. MacKinnon, *Chem. Commun.* **1997**, 557–558; (b) N. Komiya, T. Naota, Y. Oda, S.-I. Murahashi, *J. Mol. Catal. A: Chem.* **1997**, *117*, 21–37; (c) M. M. Dell'Anna, P. Mastrorilli, C. F. Nobile, *J. Mol. Catal. A: Chem.* **1998**, *130*, 65–71; (d) K. Miki, T. Furuya, *Chem. Commun.* **1998**, 97–98; (e) I. Yamanaka, K. Morimoto, M. Soma, K. Otsuka, *J. Mol. Catal. A: Chem.* **1998**, *133*, 251–254; (f) H. Rudler, B. Denise, *J. Mol. Catal. A: Chem.* **2000**, *154*, 277–279; (g) J.-W. Huang, W.-Z. Huang, W.-J. Mei, J. Liu, S.-G. Hu, L.-N. Ji, *J. Mol. Catal. A: Chem.* **2000**, *156*, 275–278; (h) R. Çelenligil-Çetin, R. J. Staples, P. Stavropoulos, *Inorg. Chem.* **2000**, *39*, 5838–5846; (i) E. D. Park, Y.-S. Hwang, J. S. Lee, *Catal. Commun.* **2001**, *2*, 187–190; (j) G. B. Shul'pin, A. E. Shilov, G. Süss-Fink, *Tetrahedron Lett.* **2001**, *42*, 7253–7256; (k) N. Theyssen, W. Leitner, *Chem. Commun.* **2002**, 410–411; (l) J. E. Remias, A. Sen, *J. Mol. Catal. A: Chem.* **2002**, *189*, 33–38; (m) T. Miyake, M. Hamada, H. Niwa, M. Nishizuka, M. Oguri, *J. Mol. Catal. A: Chem.* **2002**, *178*, 199–204; (n) G. B. Shul'pin, E. R. Lachter "Aerobic hydroxylation of hydrocarbons catalysed by vanadate ion", *The Chemistry Preprint Server,* http://preprint.chemweb.com/biochem/0204001, **2002**, pp. 1–5; *J. Mol. Catal. A: Chem.* **2003**, *197*, 65–71; (o) C. Bolm, G. Schlingloff, F. Bienewald, *J. Mol. Cat-*

al. A: Chem. **1997**, *117*, 347–350; (p) E.T. Farinas, U. Schwaneberg, A. Glieder, F.H. Arnold, *Adv. Synth. Catal.* **2001**, *343*, 601–606.

14 (a) *Catalytic Oxidations with Hydrogen Peroxide as Oxidant* (Ed.: G. Strukul), Kluwer Academic Publishers, Dordrecht, **1992**; C.W. Jones, *Applications of Hydrogen Peroxide and Derivatives*, The Royal Society of Chemistry, Cambridge, **1999**; T.J. Collins, *Acc. Chem. Res.* **2002**, *35*, 782–790; (b) G.B. Shul'pin, "Alkane oxidation: estimation of alkyl hydroperoxide content by GC analysis of the reaction solution samples before and after reduction with triphenylphosphine", *The Chemistry Preprint Server*, http://preprint.chemweb.com/orgchem/0106001, **2001**, pp. 1–6.

15 (a) G.B. Shul'pin, G. Süss-Fink, L.S. Shul'pina, *J. Mol. Catal. A: Chem.* **2001**, *170*, 17–34; (b) G.B. Shul'pin, G.V. Nizova, Yu. N. Kozlov, I.G. Pechenkina, *New J. Chem.* **2002**, *26*, 1238–1245; (c) G.V. Nizova, C. Bolm, S. Ceccarelli, C. Pavan, G.B. Shul'pin, *Adv. Synth. Catal.* **2002**, *344*, 899–905; (d) T.H. Bennur, S. Sabne, S.S. Deshpande, D. Srinivas, S. Sivasanker, *J. Mol. Catal. A: Chem.* **2002**, *185*, 71–80.

16 (a) Yu.N. Kozlov, G.V. Nizova, G.B. Shul'pin, *Russ. J. Phys. Chem.* **2001**, *75*, 770–774; (b) G.B. Shul'pin, Yu.N. Kozlov, G.V. Nizova, G. Süss-Fink, S. Stanislas, A. Kitaygorodskiy, V.S. Kulikova, *J. Chem. Soc., Perkin Trans. 2* **2001**, 1351–1371; (c) M.H.C. de la Cruz, Yu. N. Kozlov, E.R. Lachter, G.B. Shul'pin, "Kinetics and mechanism of the benzene hydroxylation by the 'O_2–H_2O_2–vanadium derivative–pyrazine-2-carboxylic acid' reagent", *The Chemistry Preprint Server*, http://preprint.chemweb.com/orgchem/0205002, **2002**, pp. 1–9; *New J. Chem.* **2003**, *27*, 634–638.

17 (a) A. Kozlov, K. Asakura, Y. Iwasawa, *J. Chem. Soc., Faraday Trans.* **1998**, *94*, 809–816; (b) A. Kozlov, A. Kozlova, K. Asakura, Y. Iwasawa, *J. Mol. Catal. A: Chem.* **1999**, *137*, 223–237; (c) P.M. Reis, J.A.L. Silva, J.J.R. Fraùsto da Silva, A.J.L. Pombeiro, *Chem. Commun.* **2000**, 1845–1856; (d) G. Süss-Fink, L. Gonzalez, G.B. Shul'pin, *Appl. Catal., A: General* **2001**, *217*, 111–117; (e) Y. Seki, N. Mizuno, M. Misono, *Appl. Catal. A: General* **2000**, *194/195*, 13–20; (f) Y. Seki, J.S. Min, M. Misono, N. Mizuno, *J. Phys. Chem. B* **2000**, *104*, 5940–5944; (g) S. Lin, Y. Zhen, S.-M. Wang, Y.-M. Dai, *J. Mol. Catal. A: Chem.* **2000**, *156*, 113–120.

18 (a) G. Roelfes, M. Lubben, R. Hage, L. Que, Jr., B. L. Feringa, *Chem. Eur. J.* **2000**, *6*, 2152–2159; (b) K. Chen, M. Costas, L. Que, Jr., *J. Chem. Soc., Dalton Trans.* **2002**, 672–679; (c) M. Costas, L. Que, Jr., *Angew. Chem. Int. Ed.* **2002**, *41*, 2179–2181; (d) Y. Mekmouche, C. Duboc-Toia, S. Ménage, C. Lambeaux, M. Fontecave, *J. Mol. Catal. A: Chem.* **2000**, *156*, 85–89; (e) G.V. Nizova, B. Krebs, G. Süss-Fink, S. Schindler, L. Westerheide, L. Gonzalez Cuervo, G.B. Shul'pin, *Tetrahedron* **2002**, *58*, 9231–9237; (f) G.B. Shul'pin, C.V. Nizova, Yu. N. Kozlov, L. Gonzalez-Cuervo, G. Süss-Fink, *Adv. Synth. Catal.* **2004**, *346*, 317–332; (g) S. Nishino, H. Hosomi, S. Ohba, H. Matsushima, T. Tokii, Y. Nishida, *J. Chem. Soc., Dalton Trans.* **1999**, 1509–1513; (h) K. Chen, L. Que, Jr., *J. Am. Chem. Soc.* **2001**, *123*, 6327–6337.

19 (a) E. Baciocchi, T. Boschi, L. Cassioli, C. Galli, A. Lapi, P. Tagliatesta, *Tetrahedron Lett.* **1997**, *38*, 7283–7286; (b) A. M. d'A. Rocha Gonsalves, A.C. Serra, *J. Porphyrins Phthalocyanines* **2000**, *4*, 599–604; (c) J.-F. Bartoli, K. Le Barch, M. Palacio, P. Battioni, D. Mansuy, *Chem. Commun.* **2001**, 1718–1719.

20 (a) R. Bandyopadhyay, S. Biswas, S. Guha, A.K. Mukherjee, R. Bhattacharyya, *Chem. Commun.* **1999**, 1627–1628; (b) P. Stavropoulos, R. Çelenigil-Çetin, A.E. Tapper, *Acc. Chem. Res.* **2001**, *34*, 745–752; U. Schuchardt, M.J.D.M. Jannini, D.T. Richens, M.C. Guerreiro, E.V. Spinacé, Tetrahedron **2001**, *57*, 2685–2688; (c) G.B. Shul'pin, G. Süss-Fink, L.S. Shul'pina, *J. Chem. Res. (S)* **2000**, 576–577; (d) G.B. Shul'pin, G. Süss-Fink, L.S. Shul'pina, *Chem. Commun.*, **2000**, 1131–1132; G.B. Shul'pin, G. Süss-Fink, *Petrol. Chem.* **2002**, *42*, 233–237; (e) G.B. Shul'pin, *J. Chem. Res. (S)*

2002, 351–353; (f) R. Russ, T. Zelinski, T. Anke, *Tetrahedron Lett.* **2002**, *43*, 791–793; (g) H. B. Dunford, *Coord. Chem. Rev.* **2002**, *233/234*, 311–318.

21 (a) J. Jacob, J. H. Espenson, *Inorg. Chim. Acta* **1998**, *270*, 55–59; (b) J.-F. Bartoli, V. Mouries-Mansuy, K. Le Barch-Ozette, M. Palacio, P. Battioni, D. Mansuy, *Chem. Commun.* **2000**, 827–828; (c) N. A. Alekar, V. Indira, S. B. Halligudi, D. Srinavas, S. Gopinathan, C. Gopinathan, *J. Mol. Catal. A: Chem.* **2000**, *164*, 181–189; (d) D. Bianchi, R. Bortolo, R. Tassinari, M. Ricci, R. Vignola, *Angew. Chem. Int. Ed.* **2000**, *39*, 4321–4323; (e) G. L. Elizarova, L. G. Matvienko, A. O. Kuzmin, E. R. Savinova, V. N. Parmon, *Mendeleev Commun.* **2001**, *11*, 15–17.

22 (a) W. A. Herrmann, J. J. Haider, R. W. Fischer, *J. Mol. Catal. A: Chem.* **1999**, *138*, 115–121; (b) O. A. Anunziata, L. B. Pierella, A. R. Beltramone, *J. Mol. Catal. A: Chem.* **1999**, *149*, 255–261; (c) S. Narayanan, K. V. V. S. B. S. R. Murthy, K. M. Reddy, N. Premchander, *Appl. Catal. A: General* **2002**, *228*, 161–165.

23 (a) S.-I. Murahashi, N. Komiya, Y. Hayashi, T. Kumano, *Pure Appl. Chem.* **2001**, *73*, 311–314; (b) T. Sooknoi, J. Limtrakul, *Appl. Catal. A: General* **2002**, *233*, 227–237; (c) S. Banfi, M. Cavazzini, G. Pozzi, S. V. Barkanova, O. K. Kaliya, *J. Chem. Soc., Perkin Trans 2* **2000**, 871–877; S. V. Barkanova, E. A. Makarova, *J. Mol. Catal. A: Chem.* **2001**, *174*, 89–105; S. V. Barkanova, O. K. Kaliya, E. A. Luk'yanets, *Mendeleev Commun.* **2001**, *11*, 116–118.

24 (a) G. B. Shul'pin, J. Gradinaru, Yu. N. Kozlov, *Org. Biomol. Chem.* **2003**, *1*, 3611–3617; (b) L. Gonzalez Cuervo, Yu. N. Kozlov, G. Süss-Fink, G. B. Shul'pin, *J. Mol. Catal. A: Chem.* **2004**, in press; (c) G. Schlingloff, C. Bolm, in *Transition Metals for Organic Synthesis* (Eds.: M. Beller, C. Bolm), Wiley-VCH, **1998**, *2*, 193–199. (d) J. Le Bras, J. Muzart, *J. Mol. Catal. A: Chem.* **2002**, *185*, 113–117; (e) G. Chelucci, G. Loriga, G. Murineddu, G. A. Pinna, *Tetrahedron Lett.* **2002**, *43*, 3601–3604; (f) G. Chelucci, A. Iuliano, D. Muroni, A. Saba, *J. Mol. Catal. A: Chem.* **2003**, *191*, 29–33.

25 (a) T. Konoike, Y. Araki, Y. Kanda, *Tetrahedron Lett.* **1999**, *40*, 6971–6974; (b) H. R. Khavasi, S. S. H. Davarani, N. Safari, *J. Mol. Catal. A: Chem.* **2002**, *188*, 115–122; (c) W. Nam, J. Y. Ryu, I. Kim, C. Kim, *Tetrahedron Lett.* **2002**, *43*, 5487–5490; (d) C. J. Moody, J. L. O'Connell, *Chem. Commun.* **2000**, 1311–1312; (e) N. Komiya, S. Noji, S.-I- Murahashi, *Chem. Commun.* **2001**, 65–66; (f) J. R. Lindsay Smith, G. B. Shul'pin, *Tetrahedron Lett.* **1998**, *39*, 4909–4912; G. B. Shul'pin, J. R. Lindsay-Smith, *Russ. Chem. Bull.* **1998**, *47*, 2379–2386.

26 (a) G. Rothenberg, L. Feldberg, H. Wiener, Y. Sasson, *J. Chem. Soc., Perkin Trans. 2* **1998**, 2429–2434; (b) U. Schuchardt, R. Pereira, M. Rufo, *J. Mol. Catal. A: Chem.* **1998**, *135*, 257–262; (c) C. R. Jacob, S. P. Varkey, P. Ratnasamy, *Appl. Catal. A: General* **1999**, *182*, 91–96; (d) K. Neimann, R. Neumann, A. Rabion, R. M. Buchanan, R. H. Fish, *Inorg. Chem.* **1999**, *38*, 3575–3580; (e) J.-M. Vincent, S. Béarnais-Barbry, C. Pierre, J.-B. Verlhac, *J. Chem. Soc., Dalton Trans.* *1999*, 1913–1914; (f) M. Rogovina, R. Neumann, *J. Mol. Catal. A: Chem.* **1999**, *138*, 315–318; (g) M. Costas, A. Llobet, *J. Mol. Catal. A: Chem.* **1999**, *142*, 113–124; (h) C.-M. Che, K.-W. Cheng, M. C. W. Chan, T.-C. Lau, C.-K. Mak, *J. Org. Chem.* **2000**, *65*, 7996–8000; (i) A. A. Valente, J. Vital, *J. Mol. Catal. A: Chem.* **2000**, *156*, 163–172; (j) D. Chatterjee, A. Mitra, *Inorg. Chem. Commun.* **2000**, *3*, 640–644; (k) J. A. R. Salvador, J. H. Clark, *Chem. Commun.* **2001**, 33–34; (l) G. Blay, I. Fernández, T. Giménez, J. R. Pedro, R. Ruiz, E. Pardo, F. Lloret, M. C. Muoz, *Chem. Commun.* **2001**, 2102–2103; (m) D. Chatterjee, A. Mitra, S. Mukherjee, *J. Mol. Catal. A: Chem.* **2001**, *165*, 295–298; (n) G. B. Shul'pin, *Petrol. Chem.* **2001**, *41*, 405–412; (o) J. Zawadiak, D. Gilner, R. Mazurkiewicz, B. Orlinska, *Appl. Catal. A: General* **2001**, *205*, 239–243; (p) J.-F. Pan, K. Chen, *J. Mol. Catal. A: Chem.* **2001**, *176*, 19–22; (q) J. A. R. Salvador, J. H. Clark, *Green*

Chem. **2002**, *4*, 352–356; (r) T. G. Carrell, S. Cohen, G. C. Dismukes, *J. Mol. Catal. A: Chem.* **2002**, *187*, 3–15; (s) N. Grootboom, T. Nyokong, *J. Mol. Catal. A: Chem.* **2002**, *179*, 113–123; (t) M. Nowotny, L. N. Pedersen, U. Hanefeld, T. Maschmeyer, *Chem. Eur. J.* **2002**, *8*, 3724–3731; (u) G. B. Shul'pin, J. Gradinaru, Yu. N. Kozlov, *Org. Biomol. Chem.* **2003**, *1*, 2303–2306; (v) B. Meunier, *Chem. Rev.* **1992**, *92*, 1411–1456.

27 (a) M. V. Avdeev, E. I. Bagrii, G. B. Maravin, Yu. M. Korolev, R. S. Borisov, *Petrol. Chem.* **2000**, *40*, 391–398; (b) A. Cagnina, S. Campestrini, F. Di Furia, P. Ghiotti, *J. Mol. Catal. A: Chem.* **1998**, *130*, 221–231; (c) L. Cammarota, S. Campestrini, M. Carrieri, F. Di Furia, P. Ghiotti, *J. Mol. Catal. A: Chem.* **1999**, *137*, 155–160; (d) N. d'Alessandro, L. Liberatore, L. Tonucci, A. Morvillo, M. Bressan, *New J. Chem.* **2001**, *25*, 1319–1324; (e) D. Mohajer, A. Rezaeifard, *Tetrahedron Lett.* **2002**, *43*, 1881–1884.

28 (a) Y. Fujiwara, C. Jia, *Pure Appl. Chem.* **2001**, *73*, 319–324; (b) W. Lu, Y. Yamaoka, Y. Taniguchi, T. Kitamura, K. Takaki, Y. Fujiwara, *J. Organomet. Chem.* **1999**, *580*, 290–294; (c) P. M. Reis, J. A. L. Silva, A. F. Palavra, J. J. R. Fraùsto da Silva, T. Kitamura, Y. Fujiwara, A. J. L. Pombeiro, *Angew. Chem. Int. Ed.* **2003**, *42*, 821–823.

29 (a) Z. Gross, L. Simkovich, *Tetrahedron Lett.* **1998**, *39*, 8171–8174; (b) X.-B. Zhang, C.-C. Guo, J.-B. Xu, R.-Q. Yu, *J. Mol. Catal. A: Chem.* **2000**, *154*, 31–38; (c) F. G. Doro, J. R. Lindsay Smith, A. G. Ferreira, M. D. Assis, *J. Mol. Catal. A: Chem.* **2000**, *164*, 97–108; (d) J. Duxiao, S. Lingying, Z. Shenjie, G. Mingde, *J. Chem. Res. (S)* **2001**, 24–25; (e) F. S. Vinhado, C. M. C. Prado-Manso, H. C. Sacco, Y. Imamoto, *J. Mol. Catal. A: Chem.* **2001**, *174*, 279–288; (f) C.-C. Guo, X.-Q. Liu, Z.-P. Li, D.-C. Guo, *Appl. Catal. A: General* **2002**, *230*, 53–60.

30 (a) K. Jitsukawa, Y. Oka, H. Einaga, H. Masuda, *Tetrahedron Lett.* **2001**, *42*, 3467–3469; (b) T. Hamada, R. Irie, J. Mihara, K. Hamachi, T. Katsuki, *Tetrahedron* **1998**, *54*, 10017–10028; (c) R. F. Moreira, P. M. When, D. Sames, *Angew. Chem. Int. Ed.* **2000**, *39*, 1618–1621; (d) J. Yang, R. Breslow, *Angew. Chem. Int. Ed.* **2000**, *39*, 2692–2694; R. Breslow, J. Yan, S. Belvedere, *Tetrahedron Lett.* **2002**, *43*, 363–365; R. Breslow, Z. Fang, *Tetrahedron Lett.* **2002**, *43*, 5197–5200; (e) R. Breslow, J. Yang, J. Yan, *Tetrahedron* **2002**, *58*, 653–658; (f) P. M. When, J. Du Bois, *J. Am. Chem. Soc.* **2002**, *124*, 12950–12951.

31 (a) Y. Sasson, A. E.-A. A. Quntar, A. Zoran, *Chem. Commun.* **1998**, 73–74; (b) V. S. Kulikova, M. M. Levitsky, A. F. Shestakov, A. E. Shilov, *Russ. Chem. Bull.* **1998**, *47*, 435–437; (c) T. Yamato, N. Shinoda, T. Kanakogi, *J. Chem. Res. (S)* **2000**, 522–523; (d) V. Mirkhani, S. Tangestaninejad, M. Moghadam, *J. Chem. Res. (S)* **1999**, 722–723; (e) S. Yamazaki, *Tetrahedron Lett.* **2001**, *42*, 3355–3357; (f) L. E. Firdoussi, A. Baqqa, S. Allaoud, B. A. Allal, A. Karim, Y. Castanet, A. Mortreux, *J. Mol. Catal. A: Chem.* **1998**, *135*, 11–22; (g) A. M. Khenkin, R. Neumann, *J. Am. Chem. Soc.* **2002**, *124*, 4198–4199; (h) W. Buijs, P. Comba, D. Corneli, H. Pritzkow, *J. Organometal. Chem.* **2002**, *641*, 71–80.

32 (a) T. Shingaki, K. Miura, T. Higuchi, M. Hirobe, T. Nagano, *Chem. Commun.* **1997**, 861–862; (b) R. Zhang, W.-Y. Yu, T.-S. Lai, C.-C. Che, *Chem. Commun.* **1999**, 1791–1792; (c) S. M. Lee, P. L. Fuchs, *J. Am. Chem. Soc.* **2002**, *124*, 13978–13979.

2.3 Allylic Oxidations

2.3.1 Palladium-Catalyzed Allylic Oxidation of Olefins

Helena Grennberg and Jan-E. Bäckvall

2.3.1.1 Introduction

2.3.1.1.1 General

Allylic acetates are important intermediates in organic synthesis, their particular usefulness deriving from the facile and efficient metal-catalyzed replacement of the acetoxy leaving group by a wide range of nucleophiles [1]. Allylic acetates are often prepared from the corresponding allylic alcohol, which in turn can be obtained by fairly expensive hydride reduction of carbonyl compounds [2]. Procedures for direct allylic functionalization of easily available olefins with introduction of an oxygen functionality are thus of synthetic interest [3]. Apart from radical-initiated reactions [4], selenium-based [5] and transition metal-based [6] reactions have attracted considerable interest. So far, the palladium-catalyzed oxidations of olefins are among the most practical and useful procedures for the preparation of allylic acetates, and thus of allylic alcohol derivatives.

2.3.1.1.2 Oxidation Reactions with Pd(II)

Pd(II)-olefin and -allyl complexes

Palladium(II) salts that are soluble in organic media participate in several reaction types, many of which involve the formation of Pd(II)-olefin complexes. Such complexes readily (reversibly) react with nucleophiles such as water, alcohols, carboxylates, stabilized carbanions, and amines (Fig. 1a), predominantly from the face opposite to that of the metal (*trans* attack), thus forming a new carbon-nucleophile bond and a carbon-metal σ-bond. The σ-complex obtained is usually quite reactive and unstable, and can undergo a number of synthetically useful transformations [7]. The σ-complexes obtained from conjugated dienes rapidly rearrange to form π-allyl complexes (Fig. 1a) that are often stable enough to be isolated [8]. π-Allyl complexes can also be obtained from alkenes possessing allylic hydrogens in a process known as allylic C-H bond activation (Fig. 1b) [9].

Transition Metals for Organic Synthesis, Vol. 2, 2nd Edition.
Edited by M. Beller and C. Bolm

ISBN: 3-527-30613-7

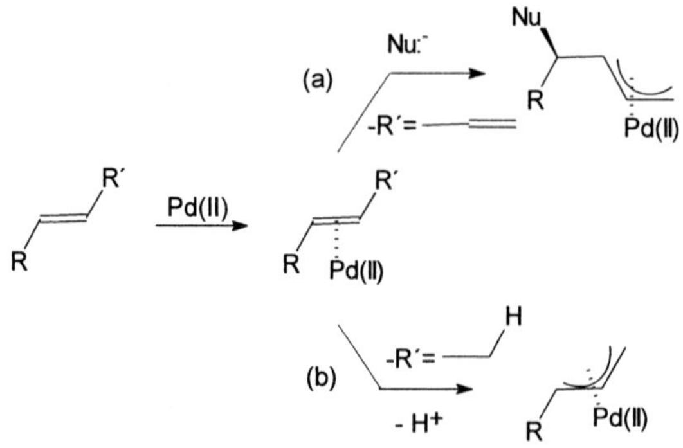

Fig. 1 Formation of (π-allyl)palladium complexes from conjugated dienes (1 a) and by cleavage of an allylic C-H bond (1 b).

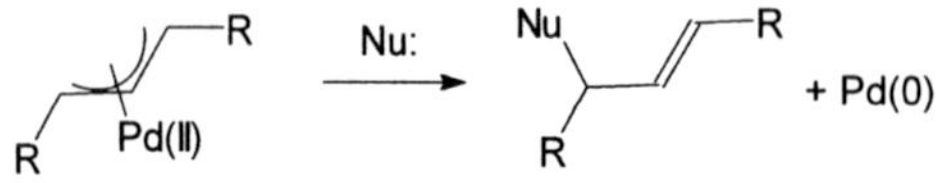

Fig. 2 Nucleophilic attack on a (π-allyl)palladium complex.

The allyl moiety of (π-allyl)palladium complexes can react with nucleophiles, giving an allylically substituted olefin (Fig. 2). This is a key reaction step in many palladium-catalyzed or -mediated reactions [1, 7, 10]. Thus, nucleophilic attack by acetate on a (π-allyl)palladium complex yields an allylic acetate, but halides, alcohols, stabilized carbanions, and amines can also be used.

Regeneration of Pd(II)

The transformation according to Figs. 1 and 2 ultimately produces palladium(0), while palladium(II) is required to activate the substrates. Thus, if such a process is to be run with catalytic amounts of the noble metal, a way to rapidly regenerate palladium(II) in the presence of both substrate and product is required. This requirement may cause problems, and reaction conditions often have to be tailored to fit a particular type of transformation. For palladium-catalyzed oxidation of olefins to allylic acetates, the processes employing $Pd(OAc)_2$ with *p*-benzoquinone (BQ) as reoxidant or electron transfer mediator [11, 12, 13] has proven to be selective, robust, and applicable to a range of substrates and nucleophiles, in contrast to earlier processes employing, for example, $PdCl_2$-$CuCl_2$ or Pd(II)-HNO_3 oxidation systems [14, 15]. More recently, procedures have been developed that also employ the quinone in catalytic amounts with peroxides [16, 17] or activated molecular oxygen as the stoichiometric oxidant. In the latter approach (Scheme 1), the hydroquinone is reoxidized to BQ by molecular oxygen in a process catalyzed by a metal-macrocycle [18, 19], a heteropolyacid [20], or a metal salt [19, 21]. In these biomimetic oxidations the only products formed are the organic oxidation product

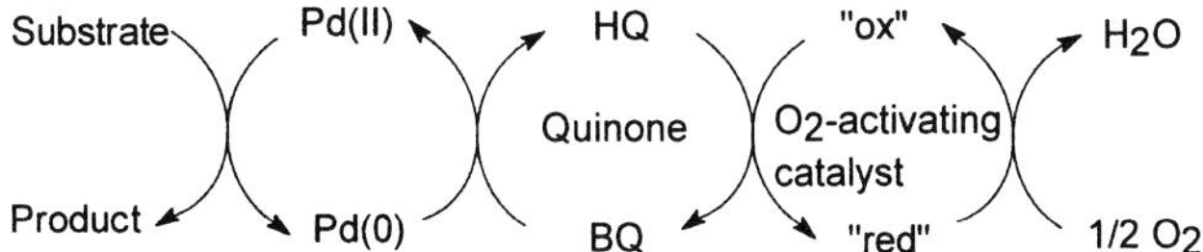

Scheme 1 Aerobic biomimetic three-component catalytic system.

and water. This is the case also in processes employing palladium dimers [22] or clusters [23] with air or molecular oxygen as stoichiometric oxidant.

2.3.1.2 Palladium-Catalyzed Oxidation of Alkenes: Allylic Products

2.3.1.2.1 Intermolecular Reactions

The palladium-quinone-based allylic acetoxylation of olefins is a synthetically useful method for the preparation of intermediates for organic synthesis (Fig. 3). In particular, five-, six-, and seven-membered cycloolefins are oxidized to their corresponding allylic carboxylates [11, 12]. Cyclohexene is quantitatively oxidized to 1-acetoxy-2-cyclohexene in only 2 h in acetic acid at 50–60 °C with molecular oxygen as stoichiometric oxidant, employing an aerobic three-component catalytic system (Scheme 1) with Co(Salophen) as oxygen-activating catalyst [18]. Other metal-macrocyclic oxygen-activating catalysts [18, 19] or a heteropolyacid [20] can also be employed.

Substituents and ring size have a large influence on the outcome of the reactions. Larger rings often require longer reaction times, whereas substituents, with some exceptions, do not affect the reaction rates [12]. On the other hand, the substituted cycloalkenes generally form several isomeric products, with total yields in the range of 50–85%. From 1-phenylcyclohexene, two isomeric allylic acetates are obtained, whereas 1-methylcyclohexene gives two major and two minor products (Fig. 4a). For 3- and 4-methyl-substituted cyclohexenes, an even larger number of products are observed (Fig. 4b), whereas 4-carbomethoxycyclohexene yielded only two major regioisomers, each as the anti stereoisomer (Fig. 4c).

$(CH_2)_n$ —[cat $Pd(OAc)_2$, oxidant, HOAc; 50 °C, 16-90 h]→ OAc, $(CH_2)_n$

n	isolated (conversion)
1	66 (95)
2	quantitative
3	73 (98)
4	35 (60)

Fig. 3 Allylic acetoxylation of simple cyclic olefins.

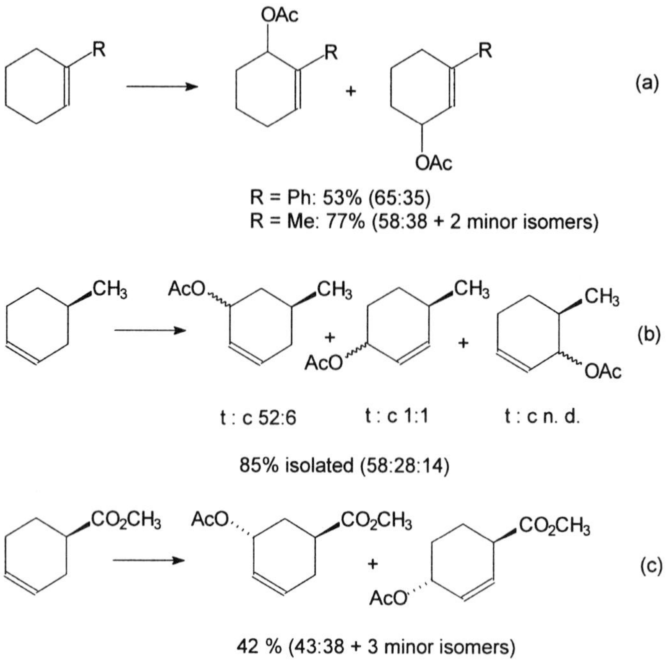

OAc R R R + (a)
OAc

R = Ph: 53% (65:35)
R = Me: 77% (58:38 + 2 minor isomers)

CH_3 AcO CH_3 CH_3 CH_3 + + (b) AcO OAc

t : c 52:6 t : c 1:1 t : c n. d.

85% isolated (58:28:14)

CO_2CH_3 AcO CO_2CH_3 CO_2CH_3 + (c) AcO

42 % (43:38 + 3 minor isomers)

Fig. 4 Allylic acetoxylation of 1-substituted cyclohexenes (a), 4-methylcyclohexene (b) and 4-carbomethoxycyclohexene (c).

The allylic carboxylates that are the primary product of an allylic acetoxylation of olefins often possess allylic hydrogens, and thus a second oxidation may occur. If 1-acetoxy-2-cyclohexene is treated with $Pd(OAc)_2$ and BQ under forced allylic acetoxylation conditions [11, 12], 1,4-diacetoxy-2-cyclohexene can be isolated in small amounts as a 1:1 mixture of *cis* and *trans* isomers [25 b, 28], together with unreacted starting material. Compared to other substituted cycloolefins, 1-acetoxy-2-cyclohexene is thus less reactive, but it is more selective since only one regioisomer of the possible (π-allyl)palladium intermediates is formed. 1,4-Diacetoxy-2-cyclohexene can be obtained directly from cyclohexene in a reaction where the main isolated product is the monoallylic acetate. A more powerful method for the preparation of 1,4-diallylically substituted alkenes is presented in the next section.

Allylic acetoxylation is normally carried out in acetic acid at moderate temperatures (50–60 °C), although there are some reports of reactions at room temperature [11 a, 12, 22, 34]. Since the acetic acid solvent is nucleophilic, the nucleophile is present in large excess, and the predominant product has been the acetoxy-substituted one. Changing the solvent to CH_2Cl_2 containing the desired carboxylate nucleophile as its carboxylic acid is a useful extension [16].

A very interesting development is an asymmetric reaction with a chiral bimetallic palladium catalyst which oxidizes cyclohexene to its allylic acetate with an *ee* of 52% [22]. A more general alternative is, however, the transformation of racemic allylic acetates into enantiomerically enriched allylic alcohols by palladium-catalyzed deracemization [24].

2.3.1.2.2 Mechanistic Considerations

The mechanism of the intermolecular quinone-based allylic acetoxylation has been studied using a 1,2-dideuterated cyclohexene (Scheme 2) [25]. The olefin is activated by coordination to the metal (step i). Then (step ii), cleavage of an allylic carbon-hydrogen bond leads to a (π-allyl)palladium intermediate, which, after activation by coordination of a benzoquinone [26], subsequently is attacked (step iii) by the acetate nucleophile at either allyl terminus to give allylic acetate and Pd(0).

The observation that allylic acetoxylation of substituted and linear olefins gives several isomeric products is understood if the selectivity of each step is taken into account: (i) the olefin, for example 4-methylcyclohexene in Fig. 4b, may coordinate palladium from both faces, and (ii) the two resulting stereoisomeric π-olefin-palladium complexes can each form two regioisomeric (π-allyl)palladium complexes. These are, in turn (iii), attacked by the nucleophile at either allyl terminus. The situation is further complicated by (iv) the unique property of the carboxylate nucleophiles in that they may add to the (π-allyl) ligand via two different path-

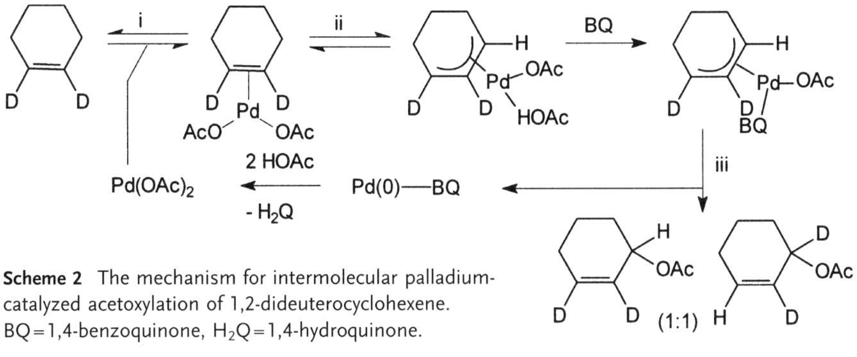

Scheme 2 The mechanism for intermolecular palladium-catalyzed acetoxylation of 1,2-dideuterocyclohexene. BQ=1,4-benzoquinone, H_2Q=1,4-hydroquinone.

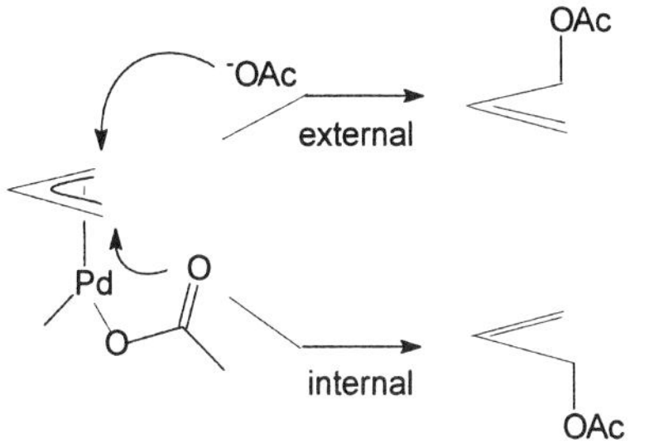

Fig. 5 Internal vs external attack by acetate on a (π-allyl)palladium complex.

ways (Fig. 5), either via the external *trans* pathway or, by first being coordinated to the metal, via internal *cis* migration [27–29].

Olefin rearrangement [9 b, 25 b, 30] and 1,3-allylic transposition [31] of the acetoxy substituent could further complicate the outcome of the reaction. The latter process is, however, too slow under the conditions of the quinone-based allylic acetoxylation to account for the observed product distributions [25 b]. In the presence of acids stronger than acetic acid, e.g., methanesulphonic acid or trifluoroacetic acid, the products obtained are the corresponding homoallylic acetates [25 b, 32].

Taken together, factors (i)–(iv) can account for (theoretically) $2\times2\times2\times2=16$ reaction pathways, which for unsubstituted alkenes lead to a racemic product and may give as many as eight isomeric products for substituted cycloalkenes. The observed unequal isomer distribution indicates that one or more reaction steps are selective.

The first step is considered to be unselective [12], whereas in the (π-allyl)-formation step, there seems to be some degree of substrate-dependent preference for the formation of one regioisomeric complex over the other. As the allyl termini of substituted (π-allyl) ligands have unequal electron density [33] and thus different reactivity towards nucleophiles, step (iii) may proceed with regioselectivity. The most easily controllable factor is (iv), as the mode of attack of acetate depends on the ligands on the palladium and the concentration and identity of the nucleophile [27, 28].

In the ideal case, only one stereo- and regioisomer of the four possible (π-allyl)-palladium complexes is formed, which is then attacked at only one of the allyl termini by one of the two possible pathways. No such case has, however, been reported for an unsymmetrical olefin, and although studies with the important objective of improving selectivity by the addition of ligands [34], strong acids [32] or different palladium-oxidant combinations [35] have been carried out, the problem of regiocontrol still remains.

2.3.1.2.3 **Intramolecular Reactions**

Cycloalkenes with a nucleophilic substituent give rise to bicyclic allylic oxidation products. The nucleophilic atom could be oxygen or nitrogen, leading to bicyclic ethers and amines, respectively [23 b], or carbon in the form of a tethered allene (Fig. 6) [36].

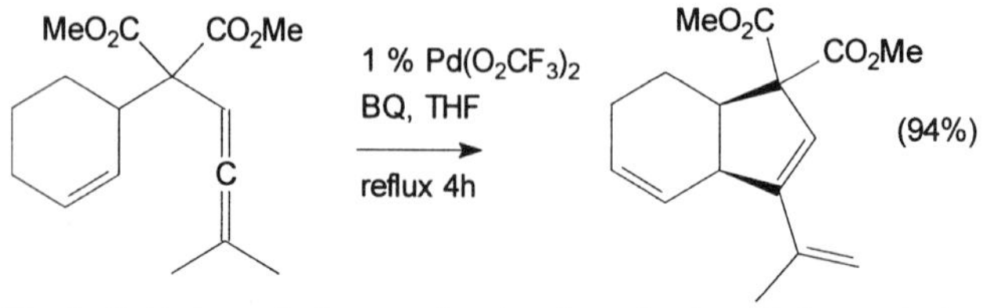

Fig. 6 Oxidative palladium-catalyzed carbocyclization of allene-substituted cyclohexene.

2.3.1.3 Palladium-Catalyzed Oxidation of Conjugated Dienes: Diallylic Products

2.3.1.3.1 1,4-Oxidation of 1,3-Dienes

Palladium-catalyzed 1,4-oxidations of conjugated dienes constitute a group of synthetically useful regio- and stereoselective transformations where a wide range of nucleophiles can be employed (Fig. 7). The reaction proceeds smoothly at room temperature, and the conditions are much milder than those required for the related allylic acetoxylation of monoalkenes discussed in the previous section.

The mechanism of the 1,4-oxidation strongly resembles that of the allylic oxidation of alkenes discussed in the previous section (Scheme 2). The first step in the reaction sequence (Scheme 3 and Fig. 1) is a coordination of the metal to the diene (i), thus activating it toward a reversible [37] regio- and stereo-selective *trans*-acetoxypalladation (ii) of one of the double bonds. This step produces a (σ-alkyl)-palladium species which rearranges to a (π-allyl)palladium complex. This is then, after coordination of the activating ligand *p*-benzoquinone (BQ) (step iii) [26], attacked by the second nucleophile either in a bimolecular reaction leading to a *cis* product (meso) or intramolecularly to give the *trans* product (racemate) [28]. In contrast to the allylic oxidation of alkenes, the reaction steps leading to the formation of the (π-allyl)palladium complex are regio- and stereoselective, although not enantioselective under standard conditions. The high selectivity for the 1,4-substituted product, due to differences in electron density at the two allyl termini of the

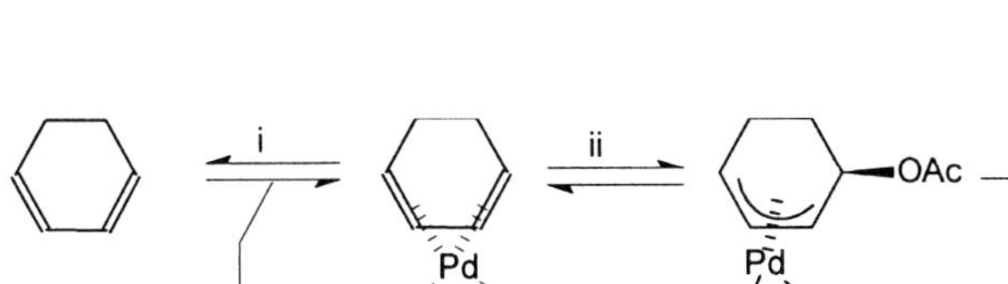

Fig. 7 Palladium-catalyzed 1,4-oxidation of 1,3-dienes.

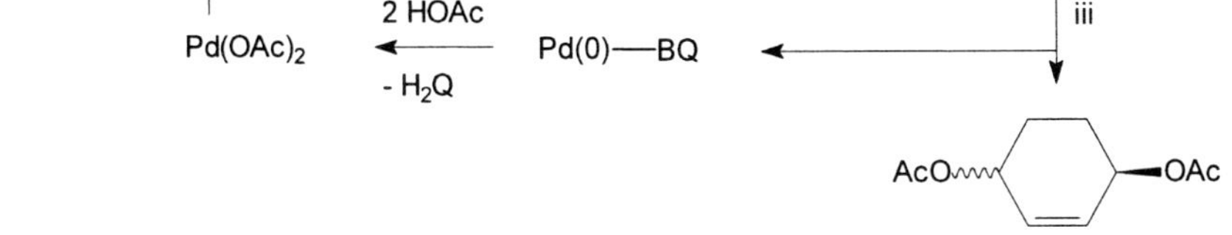

Scheme 3 General mechanism for palladium-catalyzed 1,4-diacetoxylation of 1,3-cyclohexadiene. BQ = 1,4-benzoquinone, H_2Q = 1,4-hydroquinone.

intermediate(π-allyl)palladium complex [33], was observed also in the allylic oxidation of 1-acetoxy-2-cyclohexene [25 b].

As was the case in the allylic acetoxylation of olefins, BQ has been used as either the stoichiometric oxidant or as a catalytic electron transfer mediator in combination with stoichiometric oxidants such as MnO_2 [13, 28] or molecular oxygen (Scheme 1) [18, 20 b]. It is also possible to electrochemically reoxidize the BQ catalyst [38].

2.3.1.3.2 **Intermolecular 1,4-Oxidation Reactions**

Diacetoxylation

A selective catalytic reaction that gives high yields of 1,4-diacetoxy-2-alkenes is obtained in acetic acid in the presence of a lithium carboxylate and benzoquinone [28]. The reaction (Fig. 7 and 8) proceeds in high yields and high selectivity for cyclic as well as acyclic dienes. An interesting observation is that it is possible to control the relative stereochemistry at distant carbons in an acyclic system. Since the reaction generally is slower than that observed for cyclic dienes, the competing Diels-Alder reaction becomes more important. The best results are obtained using a two-phase system of acetic acid and pentane or hexane, thus keeping the diene concentration low in the quinone-containing acetic acid phase.

An important feature of the 1,4-diacetoxylation reaction is the ease by which the relative stereochemistry of the two acetoxy substituents can be controlled (Fig. 8).

This is achieved by utilizing the ability of carboxylate nucleophiles to attack either externally or internally in a predictable fashion (Fig. 5) [28]. By variation of the concentration of chloride ions, a selectivity for either the *trans*-diacetate (Fig. 8 a) or the *cis*-diacetate (Fig. 8 b) is obtained. The selectivity for the *trans* product at chloride-free conditions is further enhanced if the reaction is carried out in the presence of a sulfoxide co-catalyst [39]. When a chiral sulfoxide-substituted quinone catalyst was used, the *trans* diacetate of 2-phenyl-1,3-cyclohexadiene was obtained with an *ee* of 45% [40].

In contrast to 1,3-cyclohexadienes, 1,3-cycloheptadienes give mainly *cis*-1,4-diacetate at standard conditions. At slightly elevated temperature [20 b, 28, 39] and no

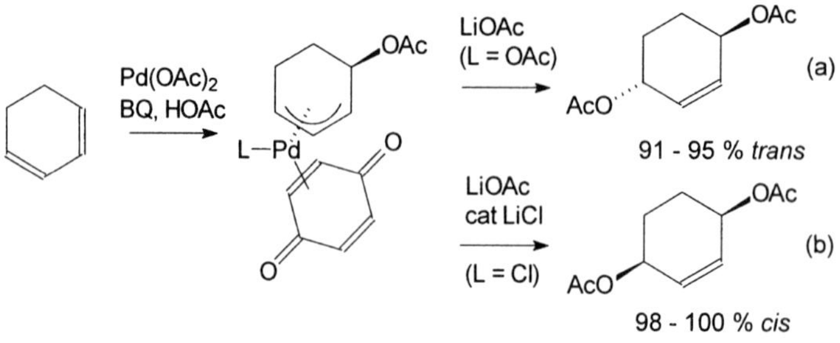

Fig. 8 The ligand-dependent relative stereochemistry in the 1,4-oxidation exemplified by the oxidation of 1,3-cyclohexadiene.

added acetate salt, a 2:1 *trans* to *cis* ratio is obtained. The internal migration is thus less favored in seven-membered ring systems, probably because of a steric crowding in the intermediate (σ-allyl)palladium complex [28].

Several substituted cyclic 1,3-dienes have been studied to determine the scope of the reaction [28]. For 1,3-cyclohexadienes carrying a methyl substituent in one of the olefinic positions, only one 1,4-oxidation product was observed (Fig. 9 a, b). 5-Methyl-1,3-cyclohexadiene reacted both in the presence and in the absence of chloride salt to give diastereoisomeric mixtures of diacetates, differing in the orientation of the acetoxy groups relative to the methyl substituent (Fig. 9 c). This indicates a poor facial selectivity in the coordination of the diene to the metal for these substrates, which is better (3:1) in the absence than in the presence (1.4:1) of chloride. On the other hand, 5-carbomethoxy-1,3-cyclohexadiene gave a diastereomeric ratio of 9:1 under these conditions (Fig. 9 d). Apparently, a carbomethoxy substituent results in a more diastereoselective reaction than a methyl substituent in both allylic acetoxylation of cyclic olefins (Fig. 4 c) and 1,4-diacetoxylation of conjugated dienes.

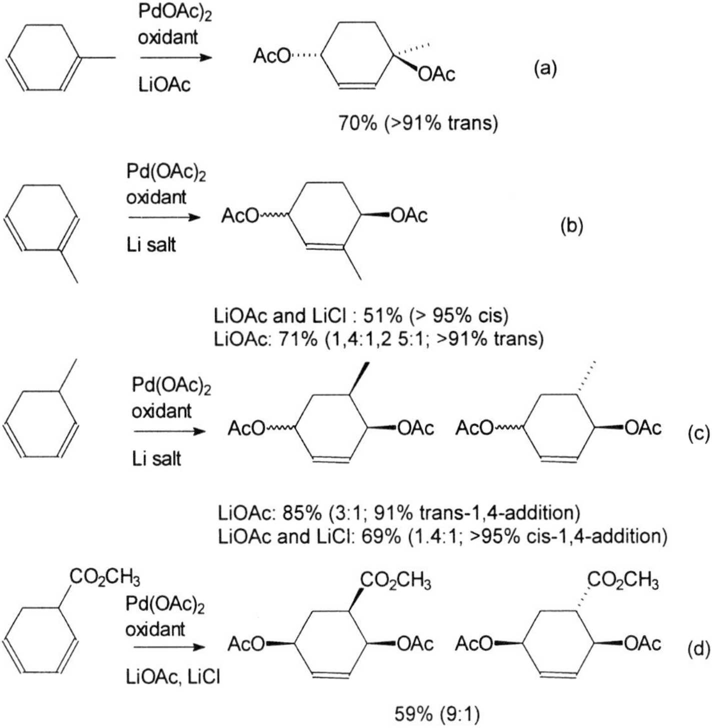

Fig. 9 1,4-Diacetoxylation of 1-methyl-1,3-cyclohexadiene (a), 2-methyl-1,3-cyclohexadiene (b), 5-methyl-1,3-cyclohexadiene (c), and 5-carbomethoxy-1,3-cyclohexadiene (d).

Enzymatic hydrolysis of *cis-meso*-1,4-diacetoxy-2-cyclohexene is a useful alternative to the enantioselective oxidation [40], which yields *cis*-1-acetoxy-4-hydroxy-2-cyclohexene in more than 98% *ee* [41], thus giving access to a useful starting material for enantioselective synthesis [42].

Dialkoxylation

If the reaction is carried out in an alcoholic solvent, *cis*-1,4-dialkoxides can be obtained [43]. The reaction is highly regio- and stereoselective in cyclic systems, and internal acyclic dienes gave a 1,4-dialkoxylation with the double bond of E configuration. It was found that the presence of a catalytic amount of non-nucleophilic acid was necessary in order to get a reaction catalytic in palladium. Acidic conditions seem to be a requirement for the electron transfer from palladium to coordinated quinone [18, 44]. Also, this reaction can be enantioselective by use of a chiral benzoquinone catalyst (Fig. 10) [45].

Incorporation of Two Different Nucleophiles

The use of two different oxygen nucleophiles can lead to unsymmetrical dicarboxylates [46] or alkoxy-carboxylates [47]. Since the reactivities of the two allylic C-O bonds are different, further transformations can be carried out at one allylic position without affecting the other. Of higher synthetic value is, however, a procedure run in the presence of a stoichiometric amount of LiCl (Fig. 11). In this process, it is possible to obtain *cis*-1-acetoxy-4-chloro-2-alkenes in high 1,4-selectivity and high chemical yield [48].

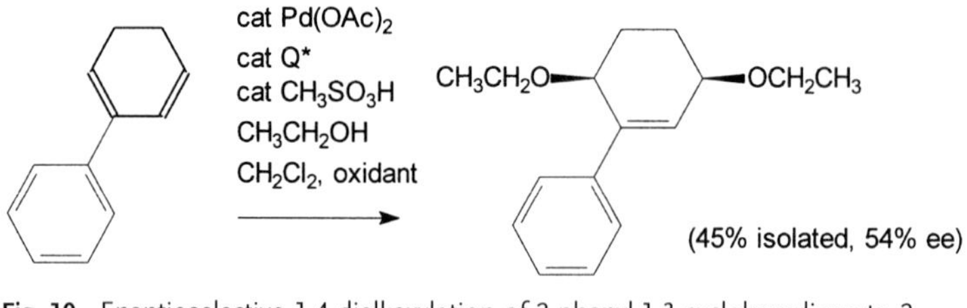

Fig. 10 Enantioselective 1,4-dialkoxylation of 2-phenyl-1,3-cyclohexadiene to 2-phenyl-(1S*,4R*)-diethoxycyclo-2-ene.

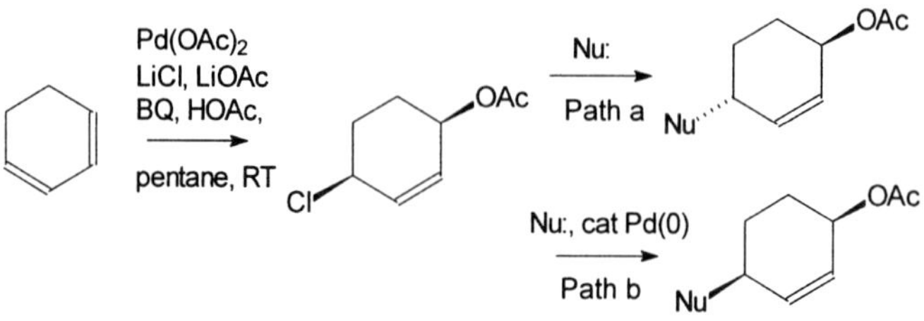

Fig. 11 1,4-Chloroacetoxylation of 1,3-cyclohexadiene and subsequent manipulation of the chloro substituent.

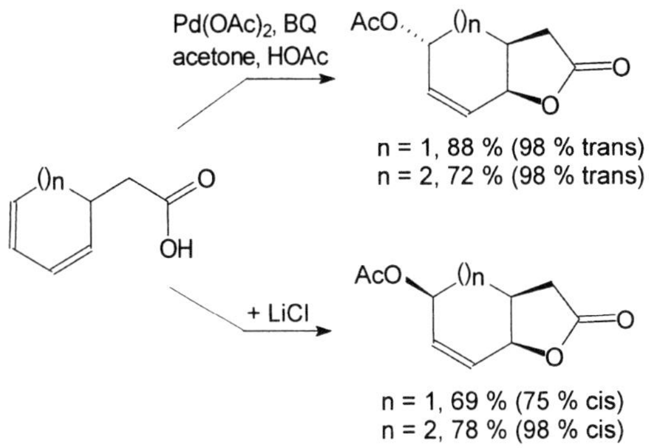

Pd(OAc)$_2$, BQ acetone, HOAc

n = 1, 88 % (98 % trans)
n = 2, 72 % (98 % trans)

+ LiCl

n = 1, 69 % (75 % cis)
n = 2, 78 % (98 % cis)

Fig. 12 Intramolecular 1,4-oxidation of carboxylato-tethered conjugated dienes to give allylic lactones.

The halogen substituent is the more reactive group of the chloroacetate. The halogen atom can be replaced by other nucleophiles both in classical nucleophilic substitution (Fig. 11, Path a) and in palladium-catalyzed substitution (Fig. 11, Path b) [49]. The methodology has been applied to the synthesis of some natural products [50].

2.3.1.3.3 Intramolecular 1,4-Oxidation Reactions

Intramolecular versions of the 1,4-oxidations have been developed. The internal nucleophile can be a carboxylate [51], an alkoxide [52], a nitrogen functionality [23c, 53], or a stabilized or masked carbon anion [54], which adds to the palladium-activated diene (Scheme 3) to form a *cis*-fused hetero- or carbocycle. In analogy with the intermolecular reaction (Fig. 8), the stereochemical outcome of the second attack can be controlled to yield either an overall *trans*- or *cis*-diallylically functionalized product.

With internal nucleophiles linked to the 1-position of the 1,3-diene, spirocyclization occurs [52, 53]. The synthetic utility of the method has been demonstrated in the total syntheses of heterocyclic natural products [55].

References

1 (a) J. Tsuji *Organic Synthesis with Palladium Compounds*, Springer, Heidelberg, **1980**. (b) R. F. Heck *Palladium Reagents in Organic Synthesis*, Academic Press, New York **1985**. (c) B. M. Trost, T. R. Verhoeven, in *Comprehensive Organometallic Chemistry* G. Wilkinson, Ed. *Pergamon* Oxford, **1982**, *8*, 799–838.

2 Very selective reactions are observed in the presence of lanthanide salts: (a) J.-L. Luche, *J. Am. Chem. Soc.* **1978**, *100*, 2226. (b) J.-L. Luche, L. Rodriguez-Hahn, P. Crabbé, *J. Chem. Soc. Chem. Commun.* **1978**, 601. (c) A. P. Marchand, W. D. LaRoe, G. V. S. Sharma, S. Chan-

der Suri, D. S. Reddy, *J. Org. Chem.* **1986**, *51*, 1622.
3 Encyclopedia of Reagents for Organic Synthesis, L. A. Paquette, Ed., John Wiley & Sons, New York **1995**.
4 For example: L. M. Stephenson, M. R. Grdina, M. Orfanopoulos, *Acc. Chem. Res.* **1980**, *13*, 419.
5 (a) M. A. Umbreit, K. B. Sharpless, *J. Am. Chem. Soc.* **1977**, *99*, 5526. (b) L. M. Stevenson, D. R. Speth, *J. Org. Chem.* **1979**, *44*, 4683. (c) K. B. Sharpless, R. F. Lauer, *J. Am. Chem. Soc.* **1972**, *94*, 7154. (d) K. B. Sharpless, R. F. Lauer, *J. Org. Chem.* **1974**, *39*, 429. (e) H. J. Reich, *J. Org. Chem.* **1974**, *39*, 428. (f) H. J. Reich, S. Wollowitz, J. E. Trend, F. Chow, D. F. Wendelborn, *J. Org. Chem.* **1978**, *43*, 1697. (g) T. Hori, K. B. Sharpless, *J. Org. Chem.* **1978**, *43*, 1689.
6 J. Muzart, *Bull. Soc. Chim. Fr.* **1986**, 65.
7 For example: (a) L. S. Hegedus *Transition Metals in the Synthesis of Complex Organic Molecules*, University Science Books, Mill Valley **1994**. (b) J. Tsuji, *Palladium Reagents and Catalysts, Innovations in Organic Synthesis*, John Wiley & ons, Chichester **1997**
8 For example: F. Bökman, A. Gogoll, O. Bohman, L. G. M. Pettersson, H. O. G. Siegbahn, *Organometallics* **1992**, *11*, 1784.
9 For example: (a) R. G. Brown, R. V. Chaudhar, J. M. Davidsson, *J. Chem. Soc. Dalton Trans.* **1977**, 176. (b) B. M. Trost, P. M. Metzner, *J. Am. Chem. Soc.* **1980**, *102*, 3572. (c) J. E. Bäckvall, K. Zetterberg, B. Åkermark, in *Inorganic Reactions and Methods*, A. P. Hagen, Ed., VCH, **1991**, *12A*, 123.
10 J. E. Bäckvall, in *Advances in Metal-Organic Chemistry*, JAI Press Inc., **1989**, pp 135–175.
11 (a) A. Heumann, B. Åkermark, *Angew. Chem. Int. Ed. Engl.* **1984**, *23*, 453. (b) A. Heumann, B. Åkermark, S. Hansson, T. Rein, *Organic Synthesis*, *68*, 109.
12 S. Hansson, A. Heumann, T. Rein, B. Åkermark *J. Org. Chem.* **1990**, *55*, 975.
13 J. E. Bäckvall, R. E. Nordberg, E. Björkman, C. Moberg, *J. Chem. Soc. Chem. Commun.* **1980**, 943.
14 (a) P. M. Henry in *Palladium-Catalyzed Oxidation of Hydrocarbons*, Reidel Publishing Co, Dordrecht, **1980**, pp 103. (b) P. M. Henry, G. A. Ward, *J. Am. Chem. Soc.* **1971**, *93*, 1494.
15 (a) S. Wolfe, P. G. C. Campbell, *J. Am. Chem. Soc.* **1971**, *93*, 1497. (b) S. Wolfe, P. G. C. Campbell, *J. Am. Chem. Soc.* **1971**, *93*, 1499. (c) E. N. Frankel, W. K. Rohwedder, W. E. Neff, D. Weisleder, *J. Org. Chem.* **1975**, *40*, 3272.
16 B. Åkermark, E. M. Larsson, J. D. Oslob, *J. Org. Chem.* **1994**, *59*, 5729.
17 C. Jia, P. Müller, H. Mimoun, *J. Mol. Catal. A* **1995**, *101*, 127.
18 (a) J. E. Bäckvall, R. B. Hopkins, H. Grennberg, M. M. Mader, A. K. Awasthi, *J. Am. Chem. Soc.* **1990**, *112*, 5160. (b) J. Wöltinger, J. E. Bäckvall, A. Zsigmond, *Chem. Eur. J.* **1999**, *5*, 1460.
19 S. E. Byström, E. M. Larsson, B. Åkermark, *J. Org. Chem.* **1990**, *55*, 5674.
20 (a) H. Grennberg, K. Bergstad, J. E. Bäckvall, *J. Mol. Catal. A* **1996**, *113*, 355. (b) K. Bergstad, H. Grennberg, J. E. Bäckvall, *Organometallics* **1998**, *17*, 45.
21 E. M. Larsson, B. Åkermark, *Tetrahedron Lett.* **1993**, *34*, 2523.
22 A. K. El-Quisiari, H. A. Quaseer, P. M. Henry, *Tetrahedron Lett.* **2002**, 4229.
23 (a) R. C. Larock, T. R. Hightower, *J. Org. Chem.* **1993**, *58*, 5298. (b) M. Rönn, J. E. Bäckvall, P. G. Andersson, *Tetrahedron Lett.* **1995**, *36*, 7749. (c) M. Rönn, P. G. Andersson, J. E. Bäckvall, *Acta Chem. Scand.* **1997**, *51*, 773.
24 B. J. Lüssem, H.-J. Gais, *J. Am. Chem. Soc.* **2003**, *125*, 6066.
25 (a) H. Grennberg, V. Simon, J. E. Bäckvall, *J. Chem. Soc. Chem. Commun.* **1994**, 265. (b) H. Grennberg, J. E. Bäckvall, *Chem. Eur. J.* **1998**, *4*, 1083.
26 J. E. Bäckvall, A. Gogoll, *Tetrahedron Lett.* **1988**, *29*, 2243.
27 J. E. Bäckvall, R. E. Nordberg, D. Wilhelm, *J. Am. Chem. Soc.* **1985**, *107*, 6892.
28 J. E. Bäckvall, S. E. Byström, R. E. Nordberg, *J. Org. Chem.* **1984**, *49*, 4619.
29 H. Grennberg, V. Langer, J. E. Bäckvall, *J. Chem. Soc. Chem. Commun.* **1991**, 1190.

30 (a) G.W. Parshall, S.D. Ittel, in *Homogenous Catalysis*, John Wiley & Sons, New York, **1992**. (b) R. Cramer, R.V. Lindsey Jr, *J. Am. Chem. Soc.* **1966**, *88*, 3534.
31 For example: (a) L.E. Overman, F.M. Knoll, *Tetrahedron Lett.* **1979**, 321. (b) J. Clayden, E.W. Collington, S. Warren, *Tetrahedron Lett.* **1992**, *33*, 7039. (c) P.M. Henry, *J. Am. Chem. Soc.* **1972**, *94*, 5200.
32 B. Åkermark, S. Hansson, T. Rein, J. Vågberg, A. Heumann, J.E. Bäckvall, *J. Organomet. Chem.* **1989**, *369*, 433.
33 (a) K.J. Szabó, *J. Am. Chem. Soc.* **1996**, *118*, 7818. (b) K.J. Szabó, *Chem. Eur. J.* **1997**, *3*, 592. (c) C. Jonasson, M. Kritikos, J.E. Bäckvall, K.J. Szabó, *Chem. Eur. J.* **2000**, *6*, 432.
34 J.E. McMurry, P. Kocovsky, *Tetrahedron Lett.* **1984**, *25*, 4187.
35 (a) Cf. Refs. 11, 12 and 16–21. (b) H. Grennberg: unpublished results.
36 J. Franzén, J.E. Bäckvall, *J. Am. Chem. Soc.* **2003**, *125*, 6056.
37 A. Thorarensen, A. Palmgren, J.E. Bäckvall, unpublished.
38 J.E. Bäckvall, A. Gogoll, *J. Chem. Soc. Chem. Commun.* **1987**, 1236.
39 H. Grennberg, A. Gogoll, J.E. Bäckvall, *J. Org. Chem.* **1991**, *56*, 5808.
40 A. Thorarensen, A. Palmgren, K. Itami, J.E. Bäckvall, *Tetrahedron Lett.* **1997**, *38*, 8541.
41 R.J. Kazlaukas, A.N.E. Weissfloch, A.T. Rappaport, L.A. Cuccia, *J. Org. Chem.* **1991**, *56*, 2656.
42 (a) H.E. Schink, J.E. Bäckvall, *J. Org. Chem.* **1992**, *57*, 1588. (b) J.E. Bäckvall, R. Gatti, H.E. Schink, *Synthesis* **1993**, 343. (c) R.G.P. Gatti, A.L.E. Larsson, J.E. Bäckvall, *J. Chem. Soc. Perkin 1* **1997**, 577. (d) A.L.E. Larsson, R.G.P. Gatti, J.E. Bäckvall, *J. Chem. Soc. Perkin 1* **1997**, 2873. (e) C.R. Johnson, S.J. Bis, *J. Org. Chem.*, **1995**, *60*, 615. (f) C. Jonasson, M. Rönn, J.E. Bäckvall, *J. Org. Chem.* **2000**, *65*, 2122.
43 (a) J.E. Bäckvall, J.O. Vågberg, *J. Org. Chem.* **1988**, *53*, 5695. (b) E. Hupe, K. Itami, A. Aranyos, K.J. Szabó, J.E. Bäckvall, *Tetrahedron* **1998**, *54*, 5375.
44 H. Grennberg, A. Gogoll, J.E. Bäckvall, *Organometallics* **1993**, *12*, 1790.
45 K. Itami, A. Palmgren, A. Thorarensen, J.E. Bäckvall, *J. Org. Chem.* **1998**, *63*, 6466.
46 J.E. Bäckvall, J.O. Vågberg, R.E. Nordberg, *Tetrahedron Lett.* **1984**, *25*, 2717.
47 E. Hupe, K. Itami, A. Aranyos, K.J. Szabó, J.E. Bäckvall, *Tetrahedron* **1998**, *54*, 5375.
48 J.E. Bäckvall, J.E. Nyström, R.E. Nordberg, *J Am. Chem. Soc.* **1985**, *107*, 3676.
49 (a) J.E. Nyström, T. Rein, J.E. Bäckvall, *Organic Synthesis* **1989**, *67*, 105. (b) J.E. Bäckvall, J.O. Vågberg, *Organic Synthesis* **1992**, *69*, 38.
50 For example: (a) J.E. Bäckvall, H.E. Schink, Z.D. Renko, *J. Org. Chem.* **1990**, *55*, 826. (b) H.E. Schink, H. Pettersson, J.E. Bäckvall, *J. Org. Chem.* **1991**, *56*, 2769. (c) D. Tanner, M. Sellén, J.E. Bäckvall, *J. Org. Chem.* **1989**, *54*, 3374. (d) A. Palmgren, A.L.E. Larsson, J.E. Bäckvall, P. Helquist, *J. Org. Chem.* **1999**, *64*, 836.
51 (a) J.E. Bäckvall, K.L. Granberg, P.G. Andersson, R. Gatti, A. Gogoll, *J. Org. Chem.* **1993**, *58*, 5445. (b) J.E. Bäckvall, *Pure Appl. Chem.* **1992**, *64*, 429. (c) J.E. Bäckvall, P.G. Andersson, *J. Am. Chem. Soc.* **1992**, *114*, 6374.
52 K. Itami, A. Palmgren, J.E. Bäckvall, *Tetrahedron Lett.* **1998**, *39*, 1223.
53 (a) P.G. Andersson, J.E. Bäckvall, *J. Am. Chem. Soc.* **1992**, *114*, 8696. (b) A. Palmgren, K. Itami, J.E. Bäckvall, Manuscript.
54 (a) J.E. Bäckvall, Y.I.M Nilsson, P.G. Andersson, R.G.P. Gatti, J. Wu, *Tetrahedron Lett.* **1994**, *35*, 5713. (b) Y.I.M. Nilsson. R.G.P. Gatti, P.G. Andersson, J.E. Bäckvall, *Tetrahedron* **1996**, *52*, 7511. (c) J.E. Bäckvall, Y.I.M. Nilsson, R.G.P. Gatti, *Organometallics* **1995**, *14*, 4242. (d) A.M. Castao, J.E. Bäckvall, *J. Am. Chem. Soc.* **1995**, *117*, 560. (e) A.M. Castao, B.A. Persson, J.E. Bäckvall, *Chem. Eur. J.* **1997**, *3*, 482. (f) M. Rönn, P.G. Andersson, J.E. Bäckvall, *Tetrahedron Lett.* **1997**, *38*, 3603.
55 (a) J.E. Bäckvall, P.G. Andersson, G.B. Stone, A. Gogoll, *J. Org. Chem.* **1991**, *56*, 2988. (b) Y.M.I. Nilsson, A. Aranyos, P.G. Andersson, J.E. Bäckvall, *J. Org. Chem.*, **1996**, *61*, 1825.

2.3.2
Kharasch-Sosnovsky Type Allylic Oxidations

Jacques Le Paih, Gunther Schlingloff, and Carsten Bolm

2.3.2.1
Introduction

Among the various oxidative functionalizations of olefins, *allylic* oxidation reactions are among the most attractive for organic synthesis. They allow the introduction of a functional group at the *allylic* position of an alkene without reacting at the double bond, and the functional group can then be elaborated further. In the last decade, several procedures have been developed for this purpose [1–3]. Most of them involve the use of stoichiometric amounts of metals. However, since reagent efficiency has gained increasing attention, the demand for catalytic methods for synthetic purposes has steadily increased. This account will therefore focus on the preparation of allylic alcohols (and derivatives thereof) from olefins using *catalytic quantities* of metals. Particular emphasis will be given to *asymmetric* acyloxylation reactions.

2.3.2.2
Background

Besides palladium-catalyzed reactions [4], selenium(IV)-mediated allylic oxidations of alkenes are of great synthetic value [5, 6], in particular, since Sharpless introduced *tert*-butyl hydroperoxide (TBHP) as a re-oxidant for selenium dioxide [7]. Several modifications of this procedure have been reported [8–11]. Reactions with stoichiometric amounts of metal salts, e.g., the acetates of lead(IV), mercury(II), and manganese(III) have been reviewed earlier [12]. Catalytic versions are also known. For example, the use of catalytic quantities of cobalt complexes and molecular oxygen as oxidant either with [13] or without co-reducing agent [14] is an attractive alternative. Furthermore, several Gif-type catalysts [15], cytochrome P-450 [16] and related systems [17] have also been described to be effective in allylic oxidations of alkenes. Other systems are based on ruthenium [18], titanium [19], and vanadium [20] catalysts. Finally, enzymes can be used for this process [21]. The product yields, however, are generally low because of the formation of significant amounts of by-products.

2.3.2.3
Copper-Catalyzed Allylic Acyloxylation

In general, reactions of olefins with organic peroxides [22] (and particularly with peresters) result in complex product mixtures. Transformations of this type are therefore considered to be of low synthetic value. The addition of catalytic amounts of copper salts, however, considerably increases the selectivity of the reaction [23]. Thus, a clean substitutive acyloxylation at the allylic position of the olefinic substrate occurs (Kharasch-Sosnovsky reaction) (Eq. 1) [24, 25].

$$R^1\text{-}CH{=}CH\text{-}CH_2\text{-}R^2 + R^3C(O)\text{-}O\text{-}O\text{-}R^4 \xrightarrow[-\,R^4OH]{\text{cat. } Cu^+/Cu^{++}} R^1\text{-}CH(OCOR^3)\text{-}CH{=}CH\text{-}R^2 + R^1\text{-}CH{=}CH\text{-}CH(OCOR^3)\text{-}R^2 \quad (1)$$

The reaction works particularly well with cyclic olefins. For example, the reaction of *tert*-butyl perbenzoate and cyclohexene in the presence of cuprous bromide at 80 °C yields 70% of cyclohex-1-en-3-yl benzoate (cf. Tab. 1, entry 1) [24]. Either cuprous or cupric halides and carboxylates can be used. In some cases isomerization of the starting olefin is observed when bromide ions are present [26]. Typical solvents are benzene, acetone, or acetonitrile. Under ambient conditions the oxidation is slow [2], and elevated reaction temperatures or relatively long reaction time are common. With cheap olefins a large excess with respect to the peroxide is often employed. Mechanistic investigations revealed the decisive role of the copper catalyst [3, 27, 28]. The reaction is initiated by reductive cleavage of the perester by a cuprous salt. A copper(II) carboxylate and an alkoxy radical are generated (Eq. 2), and the latter then abstracts a hydrogen atom of the substrate (Eq. 3). Product formation stems from trapping of the resulting allylic radical by the copper(II) carboxylate followed by ligand coupling and exclusion of the metal (Eq. 4). By this process copper(I) is regenerated which then reenters the catalytic cycle. Because of the large number of reagent combinations, the scope of the reaction is significantly broadened when a mixture of hydroperoxide and acid is used instead of the perester [29]. In these cases, carboxylate/hydroxide ligand exchange occurs after activation of the peroxide by the metal ion (Eqs. 5 and 6).

$$R^3C(O)\text{-}O\text{-}O\text{-}R^4 + Cu^+ \longrightarrow Cu(II)OCOR^3 + R^4O^\bullet \quad (2)$$

$$R^4O^\bullet + RH \longrightarrow R^\bullet + R^4OH \quad (3)$$

$$R^\bullet + Cu(II)OCOR^3 \longrightarrow ROCOR^3 + Cu(I) \quad (4)$$

$$t\text{BuOOH} + Cu(I) \longrightarrow t\text{BuO}^\bullet + Cu(II)OH \quad (5)$$

$$Cu(II)OH + R^3CO_2H \longrightarrow Cu(II)OCOR^3 + H_2O \quad (6)$$

In the absence of acids, peroxides are obtained [23]. The assumption that in these reactions the alkoxy radical is the only species responsible for substrate activation has been questioned by Minisci et al. [15, 30]. Accordingly, when the selectivity of TBHP oxidations is discussed, the presence of peroxy radicals formed by hydrogen abstraction from the peroxide by *tert*-butoxy radicals ($k=2.5\times10^8$ $M^{-1}s^{-1}$ [31])

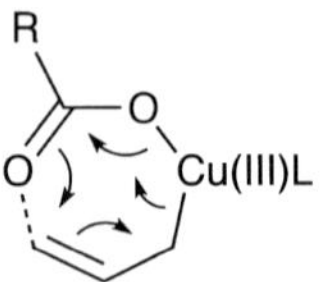

Scheme 1 Aerobic biomimetic three-component catalytic system.

must be taken into account [27], because those radicals may also contribute to hydrogen abstraction at the olefin [32].

As suggested by Beckwith and Zavitsas, a copper(III) species mediates the delivery of the carboxylate to the allylic radical in a seven-membered transition state (Scheme 1) [33].

Early studies of the acyloxylation of all three isomeric butenes (*cis*, *trans*, and terminal) suggested the presence of the same intermediate in all three reactions [28]. In general, the thermodynamically *less favorable* isomer is the preferred product when terminal olefins are used. For example, 1-hexene and *tert*-butyl peracetate gives the 3-acetoxy derivative predominantly with only small amounts of the isomeric 1-acetoxy being formed (Tab. 1, entry 2) [27]. This finding was rational-

Tab. 1 Examples of the copper-catalyzed allylic acyloxylation

Entry	*Substrate*	*Product(s)*	*Yield (%)*	*Ref.*
1		OBz	70	[24]
2		OAc + OAc > 90 : <10	90	[27]
3	optically active	OH + OH rac.	57 (after hydrolysis)	[35]
4	Ph	Ph OAc + Ph OAc 71 : 29	45–50	[27]
5	Ph	Ph OAc + Ph OAc 100 : 0	40–45	[27]
6		OBz	74	[36]
7		OBz	47	[36]

ized by assuming a lower energy barrier for a transition state in which the metal center is attached to the least substituted carbon atom [33]. Aromatic hydrocarbons, although prone to oxidation at the benzylic position, react sluggishly, since an allylic radical can only be generated with concomitant loss of aromaticity [33]. Recently, theoretical studies using perturbation interactions between Frontier Molecular Orbitals have been utilized to explain the regioselectivity of allylic oxidation reactions [34].

When optically active bicyclo[3.2.1]octene-2 was used as substrate, the product was racemic, which hints at the formation of a (symmetrical) allylic radical (Tab. 1, entry 3) [35]. In contrast to this, allyl benzene and β-methyl styrene gave markedly different results upon reaction with peresters, indicating that the degree of rearrangement is substrate and reagent dependent (entries 4, 5) [27]. Alkynes such as 3-hexyne can also be used as substrates (cat. CuCl, 100 °C), giving rise to protected propargylic alcohols (entry 6) [36]. Tetramethyl allene was also oxidized under copper catalysis (entry 7).

Singh and co-workers reported on a remarkable acceleration of allylic acyloxylations when bases such as DBN or DBU were added [37]. In this case, good conversion and high selectivity within a few hours at room temperature were achieved. Recently, excellent yields of cyclohexenyl benzoate were obtained with $Cu(CH_3CN)_4BF_4$ as catalyst and a 1:1 ratio of cycloalkene and *tert*-butylperbenzoate using benzotrifluoride as solvent [38].

Recycling procedures have also been studied, using water- [39] and fluorous-soluble [40] catalysts as well as Cu-exchanged zeolites [41]. In all cases, good conversions were observed even after several cycles.

2.3.2.3.1 Asymmetric Acyloxylation with Chiral Amino Acids

The use of optically active acids [42] in the Kharasch-Sosnovsky reaction with cyclic olefins was reported as early as 1965, giving products of low diastereomeric excess [43]. Asymmetric inductions by using chiral ligands such as salicylidenes or amino acids were later reported in a patent [44]; however, the enantioselectivities remained low (Eq. 7).

cat. Cu^{n+}, cat. L*

RCO_3tBu or $TBHP/RCO_2H$

OCOR

L* = e.g. (proline, CO_2H, NH) or (bicyclic amino acid, CO_2H, NH)

refs. [44], [46-51] ref. [53]

(7)

Despite these (and a few other [45]) early results on asymmetric allylic acyloxylations, further progress in this area remained rather limited until 1995. Muzart and co-workers carefully optimized [46–49] the reaction conditions introduced by

Araki and Nagase [44] and achieved a maximum enantioselectivity of 59% *ee* (67% yield) in the oxidation of cyclopentene in the presence of proline (at 40 °C) [46]. Several other chiral amino acids gave less satisfactory results. On the basis of results by UV spectroscopy, the copper complex CuL_2 (where LH = proline) was suggested to play a dominant role. This proposal was supported by the fact that after the reaction most of this complex could be reisolated [48].

As reported by Feringa and co-workers, additional metals such as copper bronze significantly improved catalyst activity and enantioselectivity [50]. Thus, with proline as chiral modifier and *tert*-butyl peracetate, cyclohexenyl acetate with 57% *ee* was isolated at 70% conversion. The asymmetric induction was found to be almost independent of the nature of the oxidant, *tert*-butyl peroxyacetate being best in terms of reactivity. A higher enantioselection as well as a faster reaction was observed when several equivalents of anthraquinone were added [51]. Asymmetric amplification studies revealed opposite non-linear effects [52] in the anthraquinone and the anthraquinone-free reaction, indicating the complex nature of the reaction system. It was also found that when optically active cyclohexenyl propionate (59% *ee*) was added in the acyloxylation of cyclopentene, the enantiomeric excess of the re-isolated propionate had dropped to 51% *ee* [50]. A Claisen-type rearrangement was proposed to explain this result.

Södergren and Andersson prepared unnatural proline-like *α*-amino acids and tested them under standard reaction conditions (cf. Eq. 7) [53]. Enantioselectivities of 60–65% *ee* were found for products derived from cyclopentene and cyclohexene.

2.3.2.3.2 **Asymmetric Acyloxylation with Chiral Oxazolines**

The use of bisoxazolines [54] (Scheme 2) in copper-catalyzed allylic oxidations was reported by Pfaltz and co-workers in 1995 [45, 55–57]. In the presence of 6–8 mol% of bisoxazoline **1** (R^1 = Me, R^2 = H, R^3 = *t*-Bu), a remarkable *ee* of 84% (61% yield at 68% conversion, –20 °C) was achieved in the transformation of cyclopentene. Moreover, cycloheptene, a notoriously difficult case, gave 82% *ee* (44% yield, **1** with R^1 = Me, R^2 = H, R^3 = *i*-Pr), although the reaction was slow. With 1-methyl cyclohexene a mixture of isomeric olefinic benzoates with different enantioselectivities (13–90% *ee*) was obtained. The authors suggested that these high levels of enantiocontrol were a result of the interaction between the allylic radical and the chiral copper complex followed by internal carboxylate transfer, in accord with the model of Beckwith and Zavitsas [33].

The same ligand type was independently studied by Andrus et al. [58]. In order to avoid a possible oxidative degradation of the oxazoline core during the oxidation, modified bisoxazolines bearing additional substituents at the heterocycles were synthesized. A maximum *ee* of 81% for cyclopentenyl benzoate was stated (ligand **1**, R^1 = R^2 = Me, R^3 = Ph). The oxidation of terminal olefins was also investigated, the results being moderate in terms of yield and optical activity (13–50% yield, 0–36% *ee* for allyl benzene and 1-octene). Further studies focused on a variation of the perester moiety. With the idea of weakening the O-O bond, peroxybenzoates

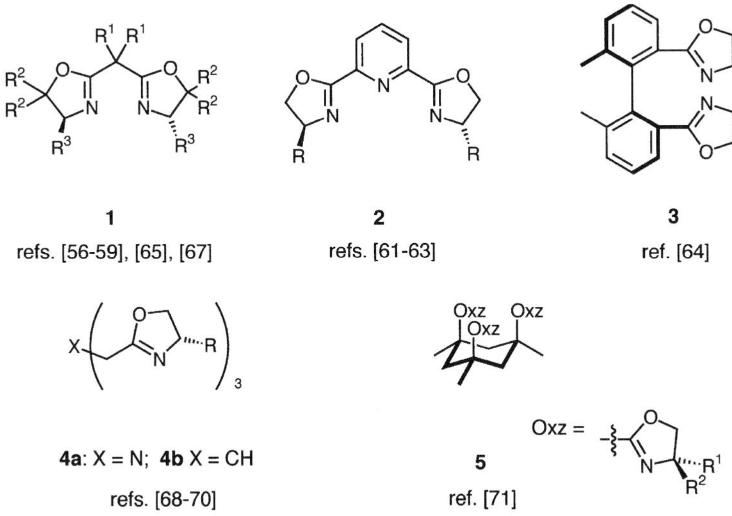

Scheme 2 Chiral oxazolines ligands for the asymmetric Kharasch-Sosnovsky reaction.

bearing withdrawing substituents on the arene were applied [59]. As predicted, reactivity and the stereoselectivity depended on the substitution pattern, and for the *para*-nitro derivative 99% *ee* was achieved in the transformation of cyclopentene using a Cu(I) complex bearing a ligand of type **1**. Unfortunately, however, the conversion remain low even after long reaction times (41% after 8 d) [60].

DarraGupta and Singh used bis(oxazolinyl)pyridines (**2**) as chiral ligands for the acyloxylation of cyclic olefins [61]. Activation of Cu(II)-triflate to give the active Cu(I) species was accomplished with phenylhydrazine. The addition of 4 Å molecular sieves improved the catalyst efficiency. Maximum enantioselectivities of 59 and 81% *ee* for cyclopentene and cyclohexene, respectively, were thus obtained. Generally, working with TBHP and acid instead of peresters gave products with lower enantiomeric excesses. In the case of cycloheptene, however, the use of this combination proved beneficial (39% yield, 25% *ee*) [61]. With a perester and another Cu(I) source, cycloheptene and cyclooctene were converted to the corresponding allylic esters in low yields but with good enantioselectivities (72% and 81% *ee*, respectively) [62]. The addition of phenylhydrazine resulted in shorter reaction times. Oxidations of 1-substituted 1-cyclohexenes catalyzed by copper complexes bearing bis(oxazolinyl)pyridines **2** led to mixtures of regioisomeric peroxides with different regio- and enantioselectivities, depending on the ligand and the 1-substituent [63]. Bisoxazolines with additional axial chirality, such as bis-*o*-tolyl-Box (**3**), gave the oxidation product of cyclohexene in 76% yield having 73% *ee* [64]. Good yields have also been achieved in catalyses with other bisoxazolines, but the enantioselectivities remain low [65, 66]. Asymmetric propargylic acyloxylations were performed using ligand **2**, and with 1-phenyl hexyne as substrate a product with 51% *ee* was obtained in 92% yield after several days at 40 °C [67].

Katsuki and co-workers synthesized optically active trisoxazolines **4a** for use as ligands in models for non-heme oxygenases [68–70]. Whereas several Fe(II) and Fe(III) complexes were catalytically inactive, the corresponding copper complexes effected allylic oxidation of cyclic olefins with *tert*-butyl peroxybenzoate as the oxidant. The best results were achieved when working in acetone, and a maximum *ee* of 93% (30% yield) was obtained at –20 °C with cyclopentene as the substrate (83% *ee*, 81% yield at 0 °C). Carbon analog **4b** gave similar results in terms of reactivity and selectivity [69]. The application of trisoxazolines **4** in the oxidative asymmetrization of racemic alkenes was also studied [69, 70]. In particular, transformations of dicyclopentadiene derivatives such as **6** (Eq. 8) were investigated. According to the mechanism discussed above, the oxidation proceeds via *meso*-radical **10** and leads to optically active allyl benzoates **7–9** with multiple asymmetric centers.

6 (4 eq.) (racemate) → [$PhCO_3tBu$ (1 eq.), cat. $Cu(OTf)_2$, **4a**, $(CH_2Cl)_2$, –20 °C, 200 h; 38%; via **10**] → **7** + **8** + **9**

7 = 81% ee; **8** = 58% ee; **9** = 85% ee

7 : **8** : **9** = 2.9 : 1.2 : 1 (8)

In this particular example (Eq. 8), the three isomers **7**, **8** and **9** were formed in a 2.9:1.2:1 ratio, respectively [69]. After a reaction time of 200 h the combined yield was 38%. Isomer **7** stems directly from *meso*-radical **10**, whereas the other two, **8** and **9**, result from intermediates formed by double-bond migrations and hydrogen shifts. The enantiomeric excesses of all three products were different, ranging from 58% *ee* for **8** to 85% *ee* for **9**.

Finally, C_3-symmetric trisoxazolines **5** were introduced by Bolm and co-workers [71]. They are derived from Kemp's triacid, and their application in copper-catalyzed asymmetric allylic oxidations leads to moderate enantioselectivities (49% *ee*, 29% yield).

2.3.2.3.3 Asymmetric Acyloxylation with Chiral Bipyridines and Phenanthrolines

Recently, chiral bipyridines and phenanthrolines [72] were applied as ligands in Kharasch-Sosnovsky reactions (Scheme 3). For example, after 12 h at 0 °C the combination of bipyridine **11** and $Cu(OTf)_2$ in the presence of phenyl hydrazine gave 80% yield of a cyclopentenol ester having 59% *ee* [73]. Performing the reaction at ambient temperature led to a higher reactivity, but decreased the enantioselectivity (85% yield and 48% *ee* in 30 min).

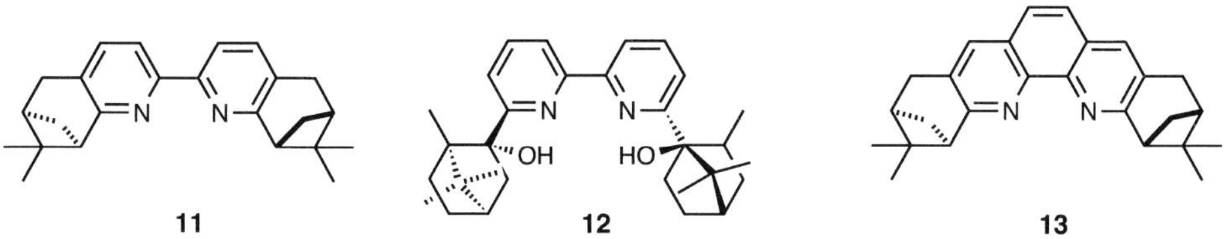

11
ref. [73]

12
ref. [74]

13
ref. [75]

Scheme 3 Chiral bipyridines and phenantholines used as ligands for the asymmetric Kharasch-Sosnovsky reaction.

Bipyridines bearing hydroxyl groups can also be used as ligands in copper-catalyzed allylic oxidations. Thus, catalytic quantities of CuBr and bipyridine **12** gave the oxidation product of cyclopentene in 88% yield having 61% *ee* (after 48 h at room temperature) [74]. Finally, chiral phenanthroline **13** has been applied in this transformation. After a relatively short reaction time (30 min), cyclopentene was converted into the corresponding cyclopentenol ester having 57% *ee* (86% yield) [75, 76].

2.3.2.4
Perspectives

As illustrated above, the Kharasch-Sosnovsky reaction is a practical method for the synthesis of allylic alcohol derivatives [77]. Compared to most procedures for allylic oxidations, the chemoselectivities are generally excellent. *Asymmetric* allylic acyloxylations of olefins [25] proceed with promising enantioselectivities, but they still suffer from the limited substrate scope [78]. In addition, to become a valuable method for organic synthesis, control of the stoichiometry (olefin as the limiting agent) is yet to be generalized, as well as the extension of the reaction from simple symmetrical to more complex substrates.

References

1 (a) P.C. Bulman Page, T.J. McCarthy, in *Comprehensive Organic Synthesis* (Eds.: B.M. Trost, I. Fleming), Pergamon, Oxford, **1991**, *7*, 83; (b) *Comprehensive Organic Transformations*, 2nd edn (Ed.: R.C. Larock), Wiley, New York, **1999**, section 3, p. 978; (c) *Oxidations in Organic Chemistry*, M. Hudlicky, Series: Monograph series (ACS), **1990**, 84.

2 D.J. Rawlinson, G. Sosnovsky, *Synthesis* **1972**, 1.

3 G. Sosnovsky, D.J. Rawlinson, in *Organic Peroxides* (Ed.: D. Swern), Wiley, New York, **1970**, *1*, 585.

4 J.E. Bäckvall, H. Grennberg, cf. Chapter 2.3.1.

5 J. Drabowicz, M. Mikolajczyk, *Top. Curr. Chem.* **2000**, *208*, 143.

6 (a) J.A. Marshall, R.C. Andrews, *J. Org. Chem.* **1985**, *50*, 1602; (b) N.R. Schmuff, B.M. Trost, *J. Org. Chem.* **1983**, *48*, 1404.

7 M.A. Umbreit, K.B. Sharpless, *J. Am. Chem. Soc.* **1977**, *99*, 5526.

8 (a) T. Hori, K.B. Sharpless, *J. Org. Chem.* **1978**, *43*, 1689; (b) Y. Nishibayashi, S. Uemura, *Top. Curr. Chem.* **2000**, *208*, 235; (c) A.K. Jones, T.E. Wilson in *Oxidizing and Reducing Agents* (Eds.: S.D. Burke, R.L. Danheiser), Wiley, New York, **2000**, p. 61.

9 K. Uneyama, H. Matsuda, S. Torii, *J. Org. Chem.* **1984**, *49*, 4315.

10 B. Chhabra, K. Hayano, T. Ohtsuka, H. Shirahama, T. Matsumoto, *Chem. Lett.* **1981**, 1703.

11 D.R. Andrews, D.H.R. Barton, K.P. Cheng, J.-P. Finet, R.H. Hesse, G. Johnson, M.M. Pechet, *J. Org. Chem.* **1986**, *51*, 1635.

12 D.J. Rawlinson, G. Sosnovsky, *Synthesis* **1973**, 567.

13 (a) M.M. Reddy, T. Punniyamurthy, J. Iqbal, *Tetrahedron Lett.* **1995**, *36*, 159; (b) T. Punniyamurthy, J. Iqbal, *Tetrahedron Lett.* **1994**, *35*, 4003.

14 H. Alper, M. Harustiak, *J. Mol. Cat.* **1993**, *84*, 87.

15 F. Minisci, F. Fontana, S. Araneo, F. Recupero, L. Zhao, *Synlett* **1996**, 119.

16 H. Fretz, W.-D. Woggon, R. Voges, *Helv. Chim. Acta* **1989**, *72*, 391.

17 (a) A.J. Appleton, S. Evans, J.R. Lindsay Smith, *J. Chem. Soc., Perkin Trans. 2* **1996**, 281; (b) T. Konoike, Y. Araki, Y. Kanda, *Tetrahedron Lett.* **1999**, *40*, 6971; (c) A. Böttcher, M.W. Grinstaff, J.A. Labinger, H.B. Gray, *J. Mol. Catal. A: Chem.* **1996**, *113*, 191; (d) J.E. Lyons, P.E. Ellis, EP 0527623 A2, **1993**.

18 (a) R. Neumann, C. Abu-Gnim, *J. Am. Chem. Soc.* **1990**, *112*, 6025; (b) L.K. Stultz, M.H.V. Huynh, R.A. Binstead, M. Curry, T.J. Meyer, *J. Am. Chem. Soc.* **2000**, *122*, 5984.

19 W. Adam, M. Braun, A. Griesbeck, V. Lucchini, E. Staab, B. Will, *J. Am. Chem. Soc.* **1989**, *111*, 203.

20 T. Hirao, S. Mikami, M. Mori, Y. Oshiro, *Tetrahedron Lett.* **1991**, *32*, 1741.

21 (a) S. Flitsch, G. Grogan, D. Ashcroft in *Enzyme Catalysis in Organic Synthesis* (Eds.: K. Drauz, H. Waldmann), Wiley-VCH, Weinheim, **2002**, *3*, 1065; (b) S.R. Sirimanne, S.W. May, *J. Am. Chem. Soc.* **1988**, *110*, 7560.

22 (a) R.A. Sheldon, *Top. Curr. Chem.* **1993**, *164*, 21; (b) Y. Sawaki in *Organic Peroxides* (Ed.: W. Ando), Wiley, Chichester, **1992**, p. 425; (c) J. Meijer, A.H. Hogt, B. Fischer, *Acros Organics: Chemistry Review Prints*, No. 6.

23 M.S. Kharasch, A. Fono, *J. Org. Chem.* **1958**, *23*, 324.

24 M.S. Kharasch, G. Sosnovsky, *J. Am. Chem. Soc.* **1958**, *80*, 756.

25 Overviews: (a) M.B. Andrus, J.C. Lashley, *Tetrahedron* **2002**, *58*, 845; (b) J. Eames, M. Watkinson, *Angew. Chem.* **2001**, *113*, 3679; *Angew. Chem. Int. Ed.* **2001**, *40*, 3567; (c) J.-M. Brunel, O. Legrand, G. Buono, *C.R. Acad. Sci. Paris-Série II* **1999**, 19; (d) T. Katsuki in *Comprehensive Asymmetric Catalysis* (Eds.: E. N. Jacobsen, A. Pfaltz, H. Yamamoto), Springer, Berlin Heidelberg, **1999**, *2*, 791.

26 J.K. Kochi, *J. Am. Chem. Soc.* **1962**, *84*, 774.

27 C. Walling, A.A. Zavitsas, *J. Am. Chem. Soc.* **1963**, *85*, 2084.

28 J.K. Kochi, *Science* **1967**, 415.

29 M.S. Kharasch, A. Fono, *J. Org. Chem.* **1958**, *23*, 325.

30 F. Minisci, F. Fontana, S. Araneo, F. Recupero, S. Banfi, S. Quici, *J. Am. Chem. Soc.* **1995**, *117*, 226.

31 H. Paul, R.D. Small, J.C. Scaiano, *J. Am. Chem. Soc.* **1978**, *100*, 4520.

32 J.M. Mayer, *Chemtracts: Org. Chem.* **1996**, *9*, 242.

33 A.L.J. Beckwith, A.A. Zavitsas, *J. Am. Chem. Soc.* **1986**, *108*, 8230.

34 G. Rothenberg, Y. Sasson, *Tetrahedron* **1999**, *55*, 561.

35 H.L. Goering, U. Mayer, *J. Am. Chem. Soc.* **1964**, *86*, 3753.

36 H. Kropf, R. Schröder, R. Fölsing, *Synthesis* **1977**, 894.

37 G. Sekar, A. DattaGupta, V.K. Singh, *Tetrahedron Lett.* **1996**, *37*, 8435.

38 J. Le Bras, J. Muzart, *J. Mol. Catal. A: Chem.* **2002**, *185*, 113.

39 J. Le Bras, J. Muzart, *Tetrahedron Lett.* **2002**, *43*, 431.

40 F. Fache, O. Piva, *Synlett* **2002**, 2035.

41 S. Carloni, B. Frullanti, R. Maggi, A. Mazzancani, F. Bigi, G. Sartori, *Tetrahedron Lett.* **2000**, *41*, 8947.
42 P.I. Dalko, L. Moisan, *Angew. Chem.* **2001**, *113*, 3840; *Angew. Chem. Int. Ed.* **2001**, *40*, 3726.
43 D.B. Denny, R. Napier, A. Cammarata, *J. Org. Chem.* **1965**, *30*, 3151.
44 M. Araki, T. Nagase, *Ger. Offen.* 2625030, **1976**; *Chem. Abstr.* **1977**, *86*, 120886r.
45 See also in: D. Zeller, Diploma thesis, University of Basel, **1992**.
46 A. Levina, J. Muzart, *Tetrahedron: Asymmetry* **1995**, *6*, 147.
47 A. Levina, J. Muzart, *Synth. Commun.* **1995**, *25*, 1789.
48 A. Levina, F. Hénin, J. Muzart, *J. Organomet. Chem.* **1995**, *494*, 165.
49 J. Muzart, *J. Mol. Cat.* **1991**, *64*, 381.
50 M.T. Rispens, C. Zondervan, B.L. Feringa, *Tetrahedron: Asymmetry* **1995**, *6*, 661.
51 C. Zondervan, B.L. Feringa, *Tetrahedron: Asymmetry* **1996**, *7*, 1895.
52 (a) C. Bolm, in *Advanced Asymmetric Synthesis* (Ed.: G.R. Stephenson), Chapman & Hall, London, **1996**, p. 9; (b) C. Girard, H.B. Kagan, *Angew. Chem.* **1998**, *110*, 3088; *Angew. Chem. Int. Ed.* **1998**, *37*, 2922; (c) H.B. Kagan, D. Fenwick, *Top. Stereochem.* **1999**, *22*, 257; (d) K. Soai, T. Shibata, in *Catalytic Asymmetric Synthesis*, 2nd ed. (Ed.: I. Ojima), Wiley-VCH, New York, **2000**, p. 699; (e) H.B. Kagan, *Synlett* **2001**, 888; (f) H.B. Kagan, *Adv. Synth. Catal.* **2001**, *343*, 227.
53 M.J. Södergren, P.G. Andersson, *Tetrahedron Lett.* **1996**, *37*, 7577.
54 (a) C. Bolm, *Angew. Chem.* **1991**, *103*, 556; *Angew. Chem. Int. Ed. Engl.* **1991**, *30*, 542; (b) A.K. Ghosh, P. Mathivanan, J. Cappiello, *Tetrahedron: Asymmetry* **1998**, *9*, 1.
55 A.S. Gokhale, A.B.E. Minidis, A. Pfaltz, *Tetrahedron Lett.* **1995**, *36*, 1831.
56 (a) See also ref. 43; (b) C. Bolm, D. Zeller, K. Weickhardt, unpublished results.
57 J. Clariana, J. Comelles, M. Moreno-Manas, A. Vallribera, *Tetrahedron: Asymmetry* **2002**, *13*, 1551.
58 M.B. Andrus, A.B. Argade, M.G. Pamment, *Tetrahedron Lett.* **1995**, *36*, 2945.
59 M.B. Andrus, X. Chen, *Tetrahedron* **1997**, *53*, 16229.
60 M.B. Andrus, Z. Zhou, *J. Am. Chem. Soc.* **2002**, *124*, 8806.
61 A. DattaGupta, V.K. Singh, *Tetrahedron Lett.* **1996**, *37*, 2633.
62 G. Sekar, A. DattaGupta, V.K. Singh, *J. Org. Chem.* **1998**, *63*, 2961.
63 M. Schulz, R. Kluge, F. Gadissa Gelalcha, *Tetrahedron: Asymmetry* **1998**, *9*, 4341.
64 (a) M.B. Andrus, D. Asgari, J.A. Sclafani, *J. Org. Chem.* **1997**, *62*, 9365; (b) M.B. Andrus, D. Asgari, *Tetrahedron* **2000**, *56*, 5775.
65 J.S. Clark, K.F. Tolhurst, M. Taylor, S. Swallow, *J. Chem. Soc., Perkin Trans. 1* **1998**, 1167.
66 C.J. Fahrni, *Tetrahedron* **1998**, *54*, 5465.
67 J.S. Clark, K.F. Tolhurst, M. Taylor, S. Swallow, *Tetrahedron Lett.* **1998**, *39*, 4913.
68 (a) K. Kawasaki, S. Tsumura, T. Katsuki, *Synlett* **1995**, 1245; (b) K. Kawasaki, T. Katsuki, *Tetrahedron* **1997**, *53*, 6337.
69 Y. Kohmura, T. Katsuki, *Tetrahedron Lett.* **2000**, *41*, 3941.
70 Y. Kohmura, T. Katsuki, *Synlett* **1999**, 1231.
71 T.-H. Chuang, J.-M. Fang, C. Bolm, *Synth. Commun.* **2000**, *30*, 1627.
72 G. Chelucci, R.P. Thummel, *Chem. Rev.* **2002**, *102*, 3129.
73 (a) A.V. Malkov, I.R. Baxendale, M. Bella, V. Langer, J. Fawcett, D.R. Russel, D.J. Mansfield, M. Valko, P. Kocovsky, *Organometallics* **2001**, *20*, 673; (b) A.V. Malkov, M. Bella, V. Langer, P. Kocovsky, *Org. Lett.* **2000**, *2*, 3047.
74 W.-S. Lee, H.-L. Kwong, H.-L. Chan, W.-W. Choi, L.-Y. Ng, *Tetrahedron: Asymmetry* **2001**, *12*, 1007.
75 G. Chelucci, G. Loriga, G. Murineddu, G.A. Pinna, *Tetrahedron Lett.* **2002**, *43*, 3601.
76 For a related system see: C. Bolm, J.-C. Frison, J. Le Paih, C. Moessner, *Tetrahedron Lett.*, in print.
77 J. Muzart, *Bull. Chem. Soc. Fr.* **1986**, 65.
78 For a summary of recent advances in asymmetric benzylic oxidations, see ref. 25d.

2.4
Metal-Catalyzed Baeyer-Villiger Reactions

Carsten Bolm, Chiara Palazzi, and Oliver Beckmann

2.4.1
Introduction

In 1899, Baeyer and Villiger investigated reactions of ketones with Caro's reagent, peroxymonosulfuric acid, and observed a previously unknown transformation: an oxygen atom was inserted into the C-C bond in position α to the carbonyl group, affording esters and lactones [1]. Later, the term "Baeyer-Villiger reaction" or "Baeyer-Villiger rearrangement" was coined for this new type of oxidation.

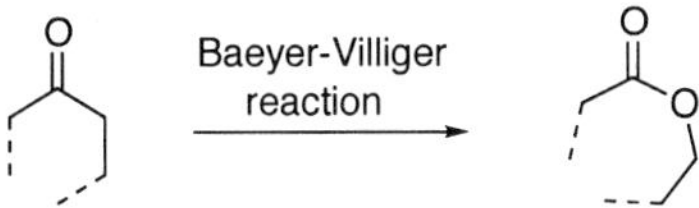

Various peroxy compounds such as peracids, hydrogen peroxide, and alkyl peroxides can be used as oxidizing agents for acyclic and cyclic ketones, respectively [2]. Acids, bases, enzymes [3, 4] and metal-containing reagents are known to catalyze Baeyer-Villiger reactions. The latter type of catalysis, i.e. the metal-promoted oxidation, is the topic of this chapter.

2.4.2
Metal Catalysis

The presence of metals can influence Baeyer-Villiger reactions in several respects. For instance, metals such as $SnCl_4$, $Bi(OTf)_3$, or $BF_3 \cdot Et_2O$ [5] can catalyze the addition of peroxy species to the carbonyl group of the substrate, or they may promote the rearrangement of the resulting intermediary perhemiketal. In 1978, a molybdenum catalyst for Baeyer-Villiger reactions was described [6]. In the presence of peroxomolybdenum complex **1**, the oxidation of cycloalkanones with 90% H_2O_2 was achieved, the corresponding lactones being obtained in 10–83% yield.

Transition Metals for Organic Synthesis, Vol. 2, 2nd Edition.
Edited by M. Beller and C. Bolm

ISBN: 3-527-30613-7

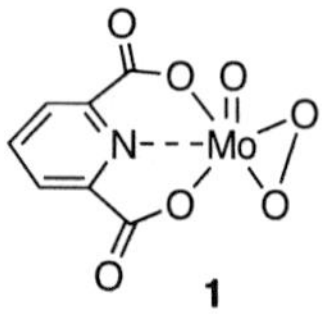

1

This work has been re-investigated, and it was proposed that complex **1** only serves as an acid catalyst, with hydrogen peroxide as the effective oxidant [7]. The bis(peroxo) complex of methyl trioxorhenium (MTO) [8] has been used by Herrmann as catalyst. In acetonitrile the oxidation of cyclobutanone **2** to lactone **3** was accelerated by a factor of ten, and no significant decomposition of hydrogen peroxide was observed [9].

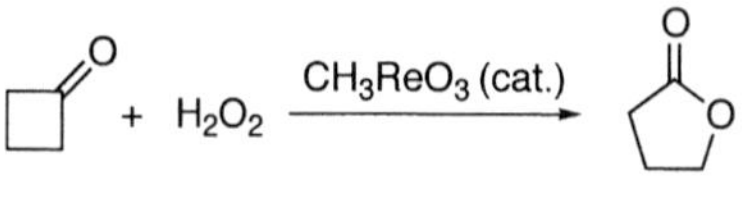

$$\mathbf{2} + H_2O_2 \xrightarrow{CH_3ReO_3\ (cat.)} \mathbf{3}$$

Another suitable catalyst for the activation of hydrogen peroxide was found through the use of platinum complexes. In 1991, Strukul reported that cationic platinum diphosphine complexes of the type $[(P\text{-}P)Pt(CF_3)(CH_2Cl_2)$ (P-P = diphosphine)$]^+$ such as **4** catalyzed Baeyer-Villiger oxidations of cyclic ketones with H_2O_2 [10a]. Detailed studies revealed that the transformation involved the coordination of the ketone to a vacant coordination site of the platinum complex, and this was followed by nucleophilic attack of free hydrogen peroxide on the attached carbonyl group [10b].

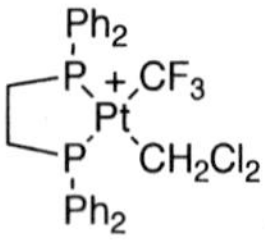

Ph2 P + CF3 Pt P CH2Cl2 Ph2 **4**

Another variant, which makes use of a heterogeneous catalyst based on an Sn-zeolite, was reported by Corma et al. [11]. Selective oxidation of cyclic ketones with hydrogen peroxide was made possible through incorporation of 1.6 wt% of tin into the framework of the zeolite.

Mukaiyama and co-workers found another catalyst system, which utilized nickel(II) complexes as catalysts and combinations of aliphatic aldehydes and dioxygen as oxidant [12, 13].

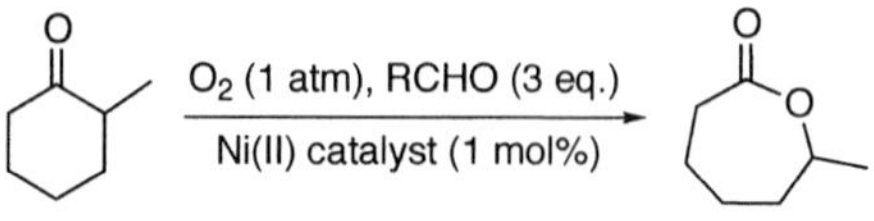

$$\mathbf{5} \xrightarrow[\text{Ni(II) catalyst (1 mol\%)}]{O_2\ (1\ atm),\ RCHO\ (3\ eq.)} \mathbf{6}$$

Mechanistic details of this catalysis remained largely unknown. It is, however, reasonable to assume that under these reaction conditions peroxo species were formed. In this scenario, the metal initiates a radical chain reaction and produces acyl radicals, which then participate in the autoxidation of the aldehyde, producing the peroxides required for the Baeyer-Villiger reaction [14–16].

In related studies, Murahashi et al. used Fe_2O_3 as the catalyst and applied the aforementioned aldehyde/dioxygen method in the synthesis of 4-benzyloxy β-lactam **8** [17].

OSiR3 … Ph, N H, 7 → O_2, C_6H_5CHO, Fe_2O_3 (1 mol%) → OSiR3 … O–C(=O)–Ph, N H, 8

A combination of iron(0) and iron(III) catalysts allowed the successive oxidation of cyclohexane (**9**) to the corresponding ε-caprolactone **11** via cyclohexanone (**10**) as intermediate [18].

9 → O_2, CH_3CHO, CH_3CO_2H (cat.), Fe(0) (cat.) → 10 → O_2, C_6H_5CHO, CH_3CO_2H (cat.), Fe(III) (cat.) → 11

A similar system for the conversion of cycloalkanones to lactones was found to work in compressed carbon dioxide as solvent. In this case no additional metal was required, because the stainless steel of the autoclave was most probably responsible for the initiation of the radical process which led to peroxy species [19].

The first metal-catalyzed *asymmetric* Baeyer-Villiger reactions were reported in 1994 [20, 21]. In the presence of 1 mol% of Bolm's chiral copper complex (*S*,*S*)-**14**, racemic 2-phenylcyclohexanone (**12**) was oxidized to give optically active lactone (*R*)-**13** in 41% yield with up to 69% *ee* [20].

O_2 (1 atm), *t*BuCHO, (*S,S*)-**14** (1 mol%)

rac-**12** → (*R*)-**13** + (*S*)-**12**

(*S,S*)-**14**

With respect to cyclohexanones, the scope of this reaction remained limited: only 2-aryl-substituted compounds were reactive enough to give the corresponding optically active lactones. Cyclobutanone derivatives, on the other hand, were readily oxidized by (*S,S*)-**14** (1 mol%). Prochiral cyclobutanones **15** gave optically active lactones **16**. However, the enantioselectivity in this process generally remained moderate (up to 47% *ee*) [22, 23]. Only concavely shaped tricyclic ketone **17** [24] afforded the corresponding lactone **18** with 91% *ee* [25].

O_2, *t*BuCHO, (*S,S*)-**14** (1 mol%)

15 → **16**

R = alkyl, aryl

17 **18**

In the same year, 1994, an asymmetric version of Strukul's catalyst was described [21]. With chiral platinum complexes, optically active lactones could be obtained. The best result (58% *ee*) was achieved in the oxidation of 2-pentylcyclopentanone (**19**) using the platinum cationic complex **21**, which is based on an optically active diphosphine ligand [(*R*)-BINAP]. Lactone **20** was formed regioselectively and resulted from a kinetic resolution of racemic ketone **19**.

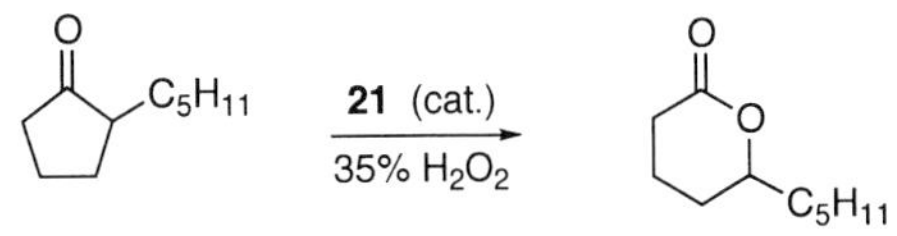

Other cationic metal complexes were found to catalyze asymmetric Baeyer-Villiger reactions with H_2O_2 (or its urea adduct) as terminal oxidant. In this context, Katsuki reported Co(III)(salen) complex **24** to be an efficient catalyst for the asymmetric Baeyer-Villiger oxidation of 3-substituted cyclobutanones such as **22** [26]. The efficiency of this cobalt catalyst was attributed to its *cis-β* structure, which had two vicinal coordination sites that became vacant during the catalysis. A chiral palladium complex with phosphino-pyridine **25** as ligand gave enantioselectivities up to 80% *ee* in oxidations of simple 4-substituted cyclobutanones [27]. The conversion of **17** afforded *ent*-**18** with >99% *ee*.

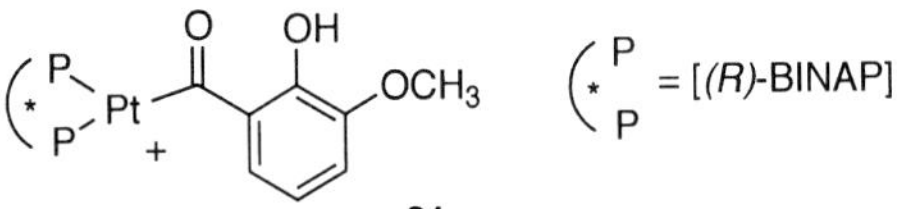

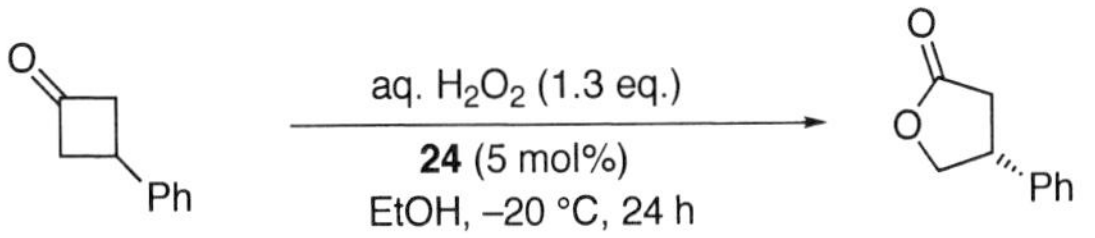

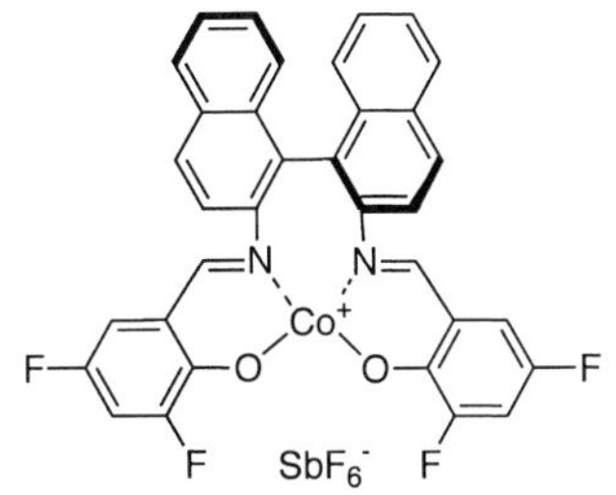

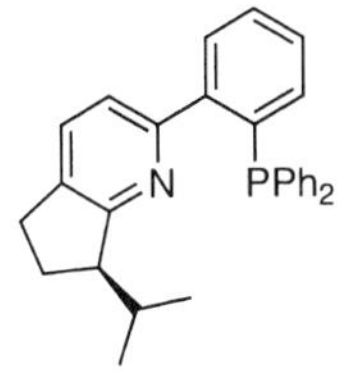

A system based on the combination of enantiopure 2,2′-dihydroxy-1,1′-binaphthyl (BINOL) and Me_2AlCl using a hydroperoxide as oxidant also proved to be capable of promoting the enantioselective Baeyer-Villiger oxidation of cyclobutanones to the corresponding γ-butyrolactones [28]. Modification of the substitution pattern of the binaphthyl ligand brought about a significant increase in the enantioselectivity. The introduction of electron-withdrawing groups, such as bromine and especially trimethylsilylacetylene, in the position 6 and 6′ of the binaphthol had a positive influence on the enantioselectivity, affording lactones with up to 81% *ee* at full conversion [29]. Use of an enantiopure hydroperoxide as oxidant had only a minor effect on the enantioselectivity [28b]. An asymmetric version of the Baeyer-Villiger reaction making use of magnesium as metal precursor was also devised [30]. Again, the combination of enantiopure BINOL and a properly chosen magnesium reagent gave rise to species that oxidized prochiral cyclobutanones in yields of up to 91 % and an enantioselectivity of up to 65% *ee*. A chiral oxazoline-based diselenide in combination with $Yb(OTf)_3$ and H_2O_2 as oxidant afforded lactones with up to 19% *ee* [31].

Besides the catalytic versions described above, a few methods were derived that used stoichiometric amounts of a metal and a chiral ligand. Thus, cyclobutanone derivatives have been oxidized by chiral titanium [32] and zirconium [33] reagents using hydroperoxides as oxidants. Furthermore, with the overstoichiometric use of $ZnEt_2$ and an amino alcohol as chiral ligand, Kotsuki and coworkers made 3-phenyl cyclobutanone **22** react to the corresponding lactone **23** with an *ee* value of up to 39%. In this system, dioxygen served as the oxidizing agent [34].

An approach different from the aforementioned ones was introduced by Seebach and Aoki, in that they synthesized a nonracemic oxidant first, and this which was then employed under base catalysis [35]. The readily accessible TADDOL-derived hydroperoxide **28** oxidized bicyclooctanone **26** exclusively to lactone **27**, which had 50% *ee*.

cat. **28**, 0.7 DBU, 0.7 eq. of LiCl

THF, –30 °C

rac-**26**

(+)-**27**

50% *ee*

70% conversion

28

Ph Ph H OOH OH H Ph Ph

0.6 eq.

2.4.3 Perspectives

More than a century after its discovery, the Baeyer-Villiger reaction has reached a remarkable level of synthetic value, making it an almost indispensable tool in organic synthesis. More recently, metal catalysts have been developed which allow asymmetric Baeyer-Villiger oxidations of racemic or prochiral ketones for the synthesis of optically active products. The scope of these new variants of the Baeyer-Villiger reaction is still rather limited, and enantioselectivities exceeding 95% *ee* have only been achieved in selected examples. However, the first steps have now been taken, and catalysts with higher selectivities would appear to be feasible in the near future.

References

1 A. Baeyer, V. Villiger, *Ber.* **1899**, *32*, 3625.

2 Reviews: (a) C. H. Hassal, *Org. React.* **1957**, *9*, 73. (b) G. R. Krow, *Org. React.* **1993**, *43*, 251. (c) C. Bolm, in *Advances in Catalytic Processes* (Ed.: M. P. Doyle), JAI Press, Greenwich, **1997**, *2*, 43. (d) G. Strukul, *Angew. Chem.* **1998**, *110*, 1256; *Angew. Chem. Int. Ed.* **1998**, *37*, 1198. (e) C. Bolm, in *Comprehensive Asymmetric Catalysis*, (Eds.: E. N. Jacobsen, A. Pfaltz, H. Yamamoto), Springer, Stuttgart, **1999**, *2*, 803. (f) M. Renz, B. Meunier, *Eur. J. Org. Chem.* **1999**, 737. (g) C. Bolm, in *Peroxide Chemistry* (Ed.: W. Adam), Wiley-VCH, Weinheim, **2000**, p. 494.

3 (a) V. Alphand, R. Furstoss, in *Enzyme Catalysis in Organic Synthesis* (Eds.: K. Drauz, H. Waldmann), VCH, Weinheim, **1995**, p. 745. (b) K. Faber, *Biotransformations in Organic Chemistry*, Springer, Berlin, **1995**, p. 203. (c) R. Azerad, *Bull. Chem. Soc. Fr.* **1995**, *132*, 17. (d) M. Kayser, G. Chen, J. Stewart, *Synlett* **1999**, 153. (e) V. Alphand, R. Furstoss, in *Asymmetric Oxidation Reactions: A Practical Approach* (Ed.: T. Katsuki), University Press, Oxford, **2001**, p. 214. (f) M. D. Mihovilovic, B. Müller, P. Stannetty, *Eur. J. Org. Chem.* **2002**, 3711.

4 For Baeyer-Villiger reactions in fluorinated solvents and flavine-catalyzed transformations of this type, see: (a) K. Neimann, R. Neumann, *Org. Lett.* **2000**, *2*, 2861. (b) A. Berkessel, M. R. M. Andreae, H. Schmickler, J. Lex, *Angew. Chem.* **2002**, *114*, 4661; *Angew. Chem. Int. Ed.* **2002**, *41*, 4481. (c) C. Mazzini, J. Lebreton, R. Furstoss, *J. Org. Chem.* **1996**, *61*, 8. (d) Asymmetric variant: S.-I. Murahashi, S. Ono, Y. Imada, *Angew. Chem.* **2002**, *124*, 2472; *Angew. Chem. Int. Ed.* **2002**, *41*, 2366.

5 (a) J. D. McClure, P. H. Williams, *J. Org. Chem.* **1962**, *27*, 24. (b) S. Matsubara, K. Takai, H. Nozaki, *Bull. Chem. Soc. Jpn.* **1983**, *56*, 2029. (c) M. Suzuki, H. Takada, R. Noyori, *J. Org. Chem.* **1982**, *47*, 902. (d) R. Göttlich, K. Yamakoshi, H. Sasai, M. Shibasaki, *Synlett* **1997**, 971. (e) X. Hao, O. Yamazaki, A. Yoshida, J. Nishikido, *Tetrahedron Lett.* **2003**, *44*, 4977. (f) M. M. Alam, R. Varala, S. R. Adapa, *Synth. Commun.* **2003**, *33*, 3055. (g) For a $SnCl_4$-mediated Baeyer-Villiger reaction with chiral acetals and mCPBA see: T. Sugimura, Y. Fujiwara, A. Tai, *Tetrahedron Lett.* **1997**, *38*, 6019.

6 S. E. Jacobson, R. Tang, F. Mares, *J. Chem. Soc. Chem. Commun.* **1978**, 888.

7 S. Campestrini, F. Di Furia, *J. Mol. Cat.* **1993**, *79*, 13.

8 W. A. Herrmann, *J. Organomet. Chem.* **1995**, *500*, 149.

9 (a) W. A. Herrmann, R. W. Fischer, J. D. G. Correia, *J. Mol. Cat.* **1994**, *94*, 213. (b) See also: A. M. F. Phillips, C. Romão, *Eur. J. Org. Chem.* **1999**, 1767.

10 (a) M. Del Todesco Frisone, F. Pinna, G. Strukul, *Stud. Surf. Sci. Catal.* **1991**, *66*, 405. (b) M. Del Todesco Frisone, F. Pinna, G. Strukul, *Organometallics* **1993**, *12*, 148.

11 (a) A. Corma, L. T. Nemeth, M. Renz, S. Valencia, *Nature* **2001**, *412*, 423. (b) A. Corma, M. T. Navarro, L. T. Nemeth, M. Renz, *Chem. Commun.* **2001**, 2190. (c) M. Renz, T. Blasco, A. Corma, V. Fornés, R. Jensen, L. Nemeth, *Chem. Eur. J.* **2002**, *8*, 4708. (d) A. Corma, M. T. Navarro, M. Renz, *J. Catal.* **2003**, *219*, 242. (e) DFT calculations on Sn-catalyzed Baeyer-Villiger reactions with H_2O_2: R. R. Seve, T. W. Root, *J. Phys. Chem. B* **2003**, *107*, 10848. (f) R. R. Severand, T. W. Root, *J. Phys. Chem. B* **2003**, *107*, 10521.

12 T. Yamada, K. Takahashi, K. Kato, T. Takai, S. Inoki, T. Mukaiyama, *Chem. Lett.* **1991**, 641.

13 Review: T. Mukaiyama, T. Yamada, *Bull. Chem. Soc. Jpn.* **1995**, *68*, 17.

14 (a) J. R. McNesby, C. A. Heller, *Chem. Rev.* **1954**, *54*, 325. (b) B. Phillips, F. C. Frostick, P. S. Starcher, *J. Am. Chem. Soc.* **1957**, *79*, 5982.

15 For a detailed study of metal-catalyzed autoxidations see: D. R. Larkin, *J. Org. Chem.* **1990**, *55*, 1563.

16 For mechanistic proposals in related nickel-catalyzed olefin epoxidations, see:

(a) Y. Nishida, T. Fujimoto, N. Tanaka, *Chem. Lett.* **1992**, 1291. (b) P. Laszlo, M. Levart, *Tetrahedron Lett.* **1993**, *34*, 1127. (c) S.C. Jarboe, P. Beak, *Org. Lett.* **2000**, *2*, 357 and references therein.

17 S.-I. Murahashi, Y. Oda, T. Naota, *Tetrahedron Lett.* **1992**, *33*, 7557.

18 S.-I. Murahashi, Y. Oda, T. Naota, *J. Am. Chem. Soc.* **1992**, *114*, 7913.

19 C. Bolm, C. Palazzi, G. Francio, W. Leitner, *Chem. Commun.* **2002**, *15*, 1588.

20 (a) C. Bolm, G. Schlingloff, K. Weickhardt, *Angew. Chem.* **1994**, *106*, 1944; *Angew. Chem. Int. Ed. Engl.* **1994**, *33*, 1848. (b) C. Bolm, G. Schlingloff, *J. Chem. Soc. Chem. Commun.* **1995**, 1247. (c) C. Bolm, T.K.K. Luong, O. Beckmann, in *Asymmetric Oxidation Reactions: A Practical Approach* (Ed.: T. Katsuki), University Press, Oxford, **2001**, p. 147. (d) See also: Y. Peng, X. Feng, K. Yu, Z. Li, Y. Jiang, C.-H. Yeung, *J. Organomet. Chem.* **2001**, *619*, 204.

21 (a) A. Gusso, C. Baccin, F. Pinna, G. Strukul, *Organometallics* **1994**, *13*, 3442. (b) G. Strukul, A. Varagnolo, F. Pinna, *J. Mol. Cat.* **1997**, *117*, 413. (c) For the conversion of *meso*-substrates, see: C. Paneghetti, R. Gavagnin, F. Pinna, G. Strukul, *Organometallics* **1999**, *18*, 5057.

22 G. Schlingloff, Ph.D. thesis, University of Marburg, **1995**.

23 C. Bolm, G. Schlingloff, T.K.K. Luong, *Synlett* **1997**, 1151.

24 (a) D.R. Kelly, C.J. Knowles, J.G. Mahdi, I.N. Taylor, M.A. Wright, *J. Chem. Soc. Chem. Commun.* **1995**, 729. (b) D.R. Kelly, C.J. Knowles, J.G. Mahdi, M.A. Wright, I.N. Taylor, D.E. Hibbs, M.B. Hursthouse, A.K. Mish'al, S.M. Roberts, P.W.H. Wan, G. Grogan, A.J. Willetts, *J. Chem. Soc. Perkin Trans 1* **1995**, 2057.

25 C. Bolm, G. Schlingloff, F. Bienewald, *J. Mol. Cat. A* **1997**, *117*, 347.

26 (a) T. Uchida, T. Katsuki, *Tetrahedron Lett.* **2001**, *42*, 6911. (b) A. Watanabe, T. Uchida, K. Ito, T. Katsuki, *Tetrahedron Lett.* **2002**, *43*, 4481. (c) T. Uchida, T. Katsuki, K. Ito, S. Akashi, A. Ishii, T. Kuroda, *Helv. Chim. Acta* **2002**, *85*, 3078.

27 K. Ito, A. Ishii, T. Kuroda, T. Kasuki, *Synlett* **2003**, 643.

28 (a) C. Bolm, O. Beckmann, C. Palazzi, *Can. J. Chem.* **2001**, *79*, 1593. (b) C. Bolm, O. Beckmann, T. Kühn, C. Palazzi, W. Adam, P. Bheema Rao, C.R. Saha-Möller, *Tetrahedron: Asymmetry* **2001**, *12*, 2441. (c) C. Bolm, J.-C. Frison, Y. Zhang, W.D. Wulff, *Synlett*, in press.

29 (a) C. Palazzi, Ph.D. thesis, RWTH University of Aachen, **2002**. (b) C. Bolm, C. Palazzi, J.-C. Frison, to be published.

30 C. Bolm, O. Beckmann, A. Cosp, C. Palazzi, *Synlett* **2001**, *9*, 1461.

31 (a) Y. Miyake, Y. Nishibayashi, S. Uemura, *Bull. Chem. Soc. Jpn.* **2002**, *75*, 2233. (b) For non-asymmetric Se-catalyzed Baeyer-Villiger reactions, see: G.J. ten Brink, J.M. Vis, W.C.E. Arends, R.A. Sheldon, *J. Org. Chem.* **2001**, *66*, 2429.

32 (a) M. Lopp, A. Paju, T. Kanger, T. Pehk, *Tetrahedron Lett.* **1996**, *37*, 7583. (b) T. Kanger, K. Kriis, A. Paju, T. Pehk, M. Lopp, *Tetrahedron: Asymmetry* **1998**, *9*, 4475.

33 C. Bolm, O. Beckmann, *Chirality* **2000**, *12*, 523.

34 T. Shinohara, S. Fujioka, H. Kotsuki, *Heterocycles* **2001**, *55*, 237.

35 M. Aoki, D. Seebach, *Helv. Chim. Acta* **2001**, *84*, 187.

2.5 Asymmetric Dihydroxylation

Hartmuth C. Kolb and K. Barry Sharpless

2.5.1 Introduction

The synthetic organic chemist has obtained a variety of powerful tools in recent years due to the development of many new asymmetric processes. Especially useful are the carbon–heteroatom bond-forming reactions, since the resulting functionality can be readily manipulated to produce many important classes of compounds. In addition, bonds to heteroatoms are chemically much easier to form than carbon–carbon bonds.

Simple olefins are the most fundamental synthetic intermediates, being inexpensive products of the petrochemical industry. More complex olefins are also readily available due to the existence of a set of predictable and powerful reactions for their construction. Olefins are inert to a wide range of conditions, which increases their utility as 'masked' 1,2-difunctionalized intermediates, whose functionality is dramatically revealed upon the oxidative addition of heteroatoms. Last but not least, the resulting '1,2-placement' of heteroatom groups is otherwise difficult to achieve.

It is not surprising, therefore, that the oxidative addition of heteroatoms to olefins has been a fruitful area in recent years (Scheme 1). A number of transition-metal-mediated methods for the epoxidation [1, 2], oxidative cyclization [8], aminohydroxylation [9], halohydrin formation [5], and dihydroxylation [3] have emerged.

A common feature of most of these processes is the phenomenon of *ligand acceleration* [10], wherein a metal-catalyzed process turns over faster in the presence of a coordinating ligand (Scheme 2). This causes the reaction to be funneled through the ligated pathway with the additional consequence that the ligand may leave its 'imprint' on the selectivity-determining step. Hence, the ligand can influence the chemo-, regio- and stereoselectivity of the reaction in a profound way, since *ligand acceleration* ensures that the unligated pathway moves into the background. The principle of *ligand acceleration* is proving to be a powerful tool for discovering new reactivity and new asymmetric processes [10].

One of the processes that greatly benefit from ligand acceleration is the asymmetric dihydroxylation of olefins by osmium(VIII) complexes. Criegee observed the accelerating influence of tertiary amines in the 1930s [11]. However, it was not

Transition Metals for Organic Synthesis, Vol. 2, 2nd Edition.
Edited by M. Beller and C. Bolm

ISBN: 3-527-30613-7

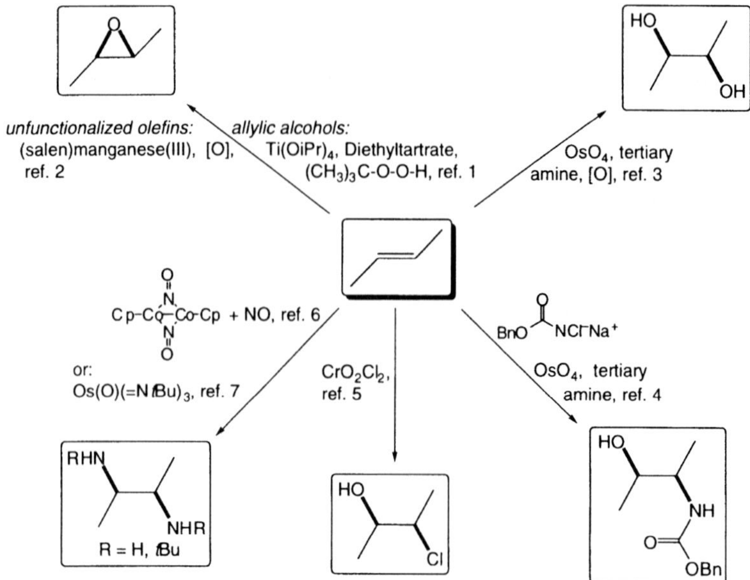

Scheme 1 Transition-metal mediated suprafacial 1,2-difunctionalization of olefins.

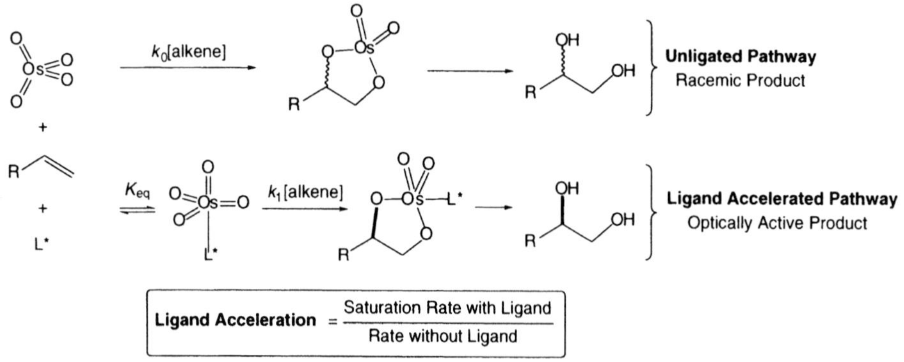

Unligated Pathway
Racemic Product

Ligand Accelerated Pathway
Optically Active Product

Ligand Acceleration = $\frac{\text{Saturation Rate with Ligand}}{\text{Rate without Ligand}}$

Scheme 2 Ligand-accelerated catalysis–dihydroxylation of olefins [10].

until 1979 that Hentges and Sharpless developed an asymmetric process based on this principle [12] (Scheme 3).

Cinchona alkaloid derivatives **1** and **2** (R–Ac) were chosen as chiral ligands in order to ensure adequate coordination to the metal center during the reaction. The authors were able to obtain moderate enantioselectivities using stoichiometric amounts of OsO_4, thereby demonstrating that it is possible to establish an asymmetric process based on the ligand acceleration phenomenon. Another reason for the choice of the cinchona system was the availability of two 'pseudoenantiomeric'

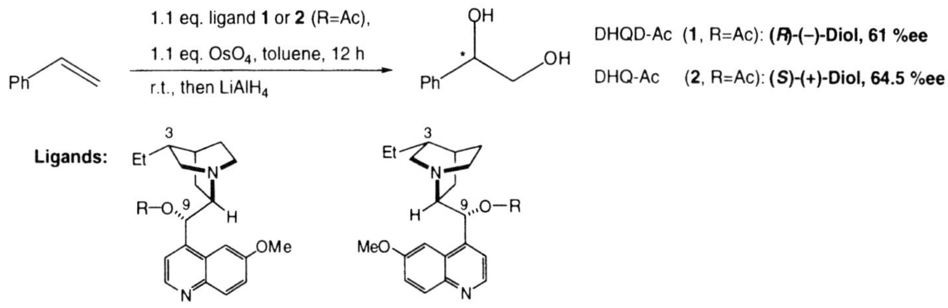

1.1 eq. ligand **1** or **2** (R=Ac),
1.1 eq. OsO_4, toluene, 12 h
r.t., then $LiAlH_4$

DHQD-Ac (**1**, R=Ac): **(*R*)-(−)-Diol, 61 %ee**
DHQ-Ac (**2**, R=Ac): **(*S*)-(+)-Diol, 64.5 %ee**

Ligands:

Dihydroquinidine **1** (DHQD) *Dihydroquinidine* **2** (DHQ)

Scheme 3 Stoichiometric asymmetric dihydroxylation using cinchona ligands [12].

alkaloids, dihydroquinidine **1** (R=H) and dihydroquinine **2** (R=H), which provide access to diols of opposite configuration, even though these two alkaloids are diastereomers, not enantiomers, due to the presence of the side chain at C-3. Not unexpectedly then, the enantioselectivities usually differ and quinine-derived ligands typically give slightly inferior results.

The first asymmetric dihydroxylation systems were stoichiometric in the expensive reagent OsO_4. Earlier work had shown, however, that a catalytic system can be established if the reaction product, an osmium(VI) glycolate, is recycled by oxidizing the metal back to osmium(VIII) [3]. A number of conditions for the reoxidation have been developed over the years and the most common protocols utilize H_2O_2 [13], alkyl hydroperoxides [14], tertiary amine *N*-oxides, e.g. *N*-methylmorpholine *N*-oxide [15], and inorganic salts, e.g. chlorates [16]. In recent years other reoxidation methods have emerged, among them the $K_3Fe(CN)_6/K_2CO_3$ system [17] and electrochemical procedures [18, 19].

A number of advantageous features have turned the osmium-catalyzed asymmetric dihydroxylation process into a powerful method for asymmetric synthesis:

1) the reaction is stereospecific leading to 1,2-*cis*-addition of two OH groups to the olefin;
2) it typically proceeds with high chemoselectivity and enantioselectivity;
3) the face selectivity is reliably predicted using a simple 'mnemonic device' and exceptions are very rare (Sect. 2.5.3.3);
4) the reaction conditions are simple and the reaction can be easily scaled up (Sect. 2.5.3.1);
5) it has broad scope, tolerating the presence of most organic functional groups – even some sulfur(II) containing functional groups [20, 21];
6) the reaction never makes mistakes, i.e. the product is always a diol derived from *cis*-addition and side products, such as epoxides or *trans*-diols, are never observed;
7) it usually exhibits a high catalytic turnover number, allowing low catalyst loading and good yields;

8) it makes use of inexpensive substrates;
9) it provides access to chemically very useful 1,2-difunctionalized intermediates which are set up for further manipulation.

2.5.2
The Mechanism of the Osmylation

Despite all the predictability of the AD reaction in a practical sense, the actual mechanism leading to the stereospecific transfer of two OH groups onto the olefin remains a mystery. Some mechanistic insight can be gleaned from studies of the reaction products of closely related d^0 transition metal species with olefins. Complexes of Mn(VII), Tc(VII), Re(VII), and Ru(VIII) are able to effect analogous transformations. It is interesting to note, however, that these d^0 complexes (especially MnO_4^- and RuO_4) typically form more side products than the osmium(VIII) system, giving rise to epoxides, and overoxidation products, such as ketols and C–C cleavage products. In addition, oxidative cyclization is observed in the case of 1,5-dienes [22].

In the 1970s, Sharpless et al. performed extensive studies on olefin oxidations by d^0 metal species, e.g. CrO_2Cl_2, OsO_4, and osmium(VIII) imido complexes [23]. Organometallic intermediates were invoked in the reactions of CrO_2Cl_2 with olefins, in order to rationalize why *all the primary products* (including the epoxides) derive from suprafacial addition of the heteroatoms to the olefin (Scheme 4). It was proposed that a chromaoxetane intermediate **4**, formed via insertion from an initial metal/olefin π-complex **3**, rearranges to the epoxide **5**. In analogy, formation of furans **7** by oxidative cyclization of 1,5-dienes **6** with permanganate may proceed via a metallaoxetane mechanism (Scheme 4) [22, 24]. The most striking feature of this transformation is the strictly suprafacial addition of all three oxygen atoms.

In analogy, the osmylation of olefins was suggested to proceed via osmaoxetanes **9** and **10** which rearrange to the product in the rate determining step [23] (Scheme 5). Even though this stepwise mechanism has been termed '[2+2]' mechanism, it should be noted that this designation is only a formalism describing the

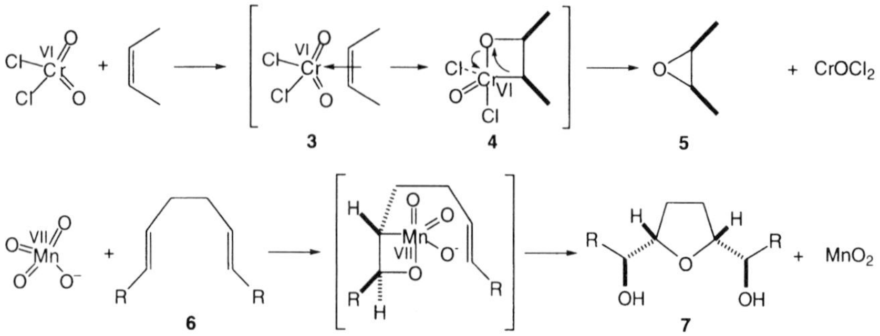

Scheme 4 Proposed metallaoxetane mechanisms for the formation of *cis*-epoxides and tetrahydrofurans in the CrO_2Cl_2 and MnO_4^- oxidation of olefins [22–24].

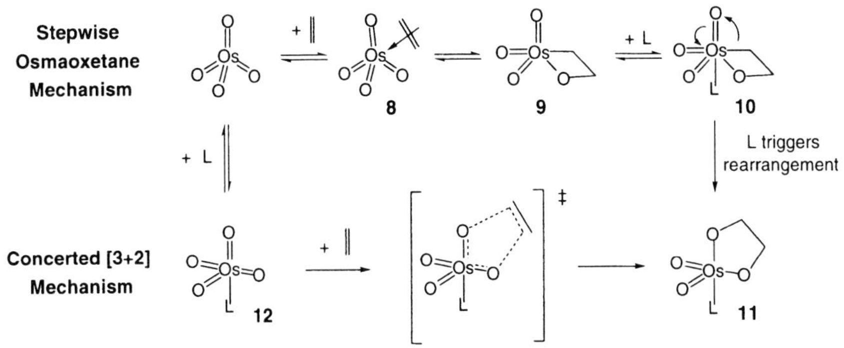

Stepwise Osmaoxetane Mechanism

Concerted [3+2] Mechanism

Scheme 5 Schematic presentation of the concerted [3+2] mechanism [11, 25, 29] and the stepwise osmaoxetane mechanism [23, 26–28].

overall formation of the four-membered osmaoxetane ring. Formation of the oxametallacycle, if it is involved, would almost certainly proceed by reversible insertion from a prior intermediate, an alkene-Os(VIII) π-complex **8**. The preferred term is, therefore, 'stepwise osmaoxetane' mechanism as opposed to '[2+2]' mechanism.

The popular alternative mechanism for the osmylation had already been proposed by Boseken in 1922 [25]. He noticed the similarity between the osmylation and permanganate oxidation of alkenes, leading to diols, and suggested a direct transfer of the oxo groups of OsO_4 to the olefin (Scheme 5). This [3+2] mechanism was also adopted by Criegee [11] and has since been favored by most organic chemists, probably due to its similarity to well-known organic cycloaddition reactions not involving metals.

Both mechanisms are currently under consideration, because they are kinetically indistinguishable [30, 36], and both are able to rationalize the characteristic features of the reaction, most notably the phenomenon of ligand acceleration [10, 37]. Ligand acceleration in the [3+2] mechanism could arise from the decrease of the O=Os=O angle upon formation of the trigonal bipyramidal $OsO_4^{\bullet}$ ligand complex **12** [31–33] (Scheme 5). This would reduce the strain of the five-membered ring transition state, thereby accelerating the reaction. However, electrochemical studies have shown that the $OsO_4^{\bullet}$ ligand complexes are much weaker oxidants than free OsO_4 [26]. The positive effect of the ligand on the transition state geometry would have to outweigh this strongly deleterious electronic effect.

Ligand acceleration in the stepwise osmaoxetane mechanism would be explained if coordination of the ligand to the oxetane **9** triggered its rearrangement to glycolate **11** (Scheme 5, **9** → **10** → **11**) [26, 30].

Since the metals are the most electron-deficient centers in OsO_4 and the other d^0 metal oxo complex oxidants (*vide supra*), the metallaoxetane mechanism fits their electrophilic behavior better in reactions with olefins. Their electrophilic nature [26] is inconsistent with the single step [3+2] mechanism, since that mechanism invokes attack by the olefin π-bond on two partially negatively charged oxygen atoms of the oxo groups [34]. The nucleophilic character of the oxygen end of the Os=O groups will of course become even more pronounced upon ligation of the amine ligand. A recent kinetic study established parabolic Hammett relationships for both substituted styrenes and stilbenes **13** (Scheme 6) [26]. Normally, electron-donating substituents on the aromatic olefin increase the reaction rate, i.e. the reaction is electrophilic (negative ρ). However, in the presence of both a strongly coordinating ligand (quinuclidine or DMAP) and a strongly electron withdrawing substituent on the olefin (e.g. a nitro group) the reaction becomes nucleophilic (positive ρ). This curvature of the free enthalpy plot is much less pronounced in the presence of DHQD-CLB, a typical AD ligand, and the negative slope indicates that under the usual AD conditions the reaction remains electrophilic. The nonlinear Hammett plots suggest the participation of at least two different mechanistic paths: (1) an electrophilic path (operating under the usual AD conditions), which is entirely consistent with the metallaoxetane mechanism; and (2) a nucleophilic path, which operates only under exceptional circumstances (i.e. highly electron-deficient olefins *and* unhindered, strongly basic ligands). At present, the latter situation best fits expectations based on a concerted [3+2] mechanism for the reaction of an electron-poor olefin with an $OsO_4 \bullet$ ligand complex.

Mechanistic studies of the osmylation reaction are complicated by the irreversible nature of the reaction, making it impossible to study its microscopic reverse and thereby gain more information on the energetic profile of this unique transformation. A major advance was made when Gable and co-workers [40–42] realized that the reverse process could be examined by studying the extrusion of alkenes from Re-diolate complexes **15** and **16** [43].

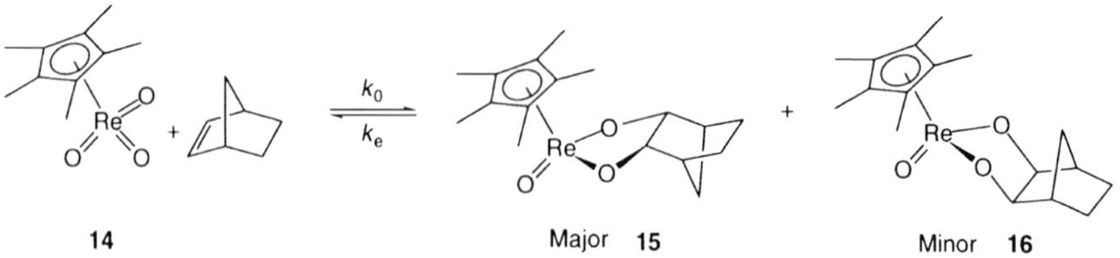

k_o, k_e

14 Major **15** Minor **16**

Cp* trioxorhenium(VII) **14** behaves just like tetroxoosmium(VIII), i.e. OsO_4, in its ability to oxidize strained olefins with formation of glycolate complexes, e.g. **15** and **16.** Gable and co-workers discovered that they could tune the position of the equilibrium by adjusting the olefin's strain energy. Norbornene was perfect, being just about at the thermodynamic balance point. Consequently, it is possible to examine the kinetics in both directions and thus obtain activation parameters for both alkene extrusion and alkene oxidation.

The strain of the olefin has a large effect on the activation enthalpy of diolate *formation,* due to the change from sp^2 to sp^3 hybridization in the transition state of the rate-determining step (Scheme 7) [40]. However, the degree of strain in the olefin was varied over a wide range (e.g. norbornene → ethylene) and found to have very little influence on the activation enthalpy of olefin *extrusion,* suggesting

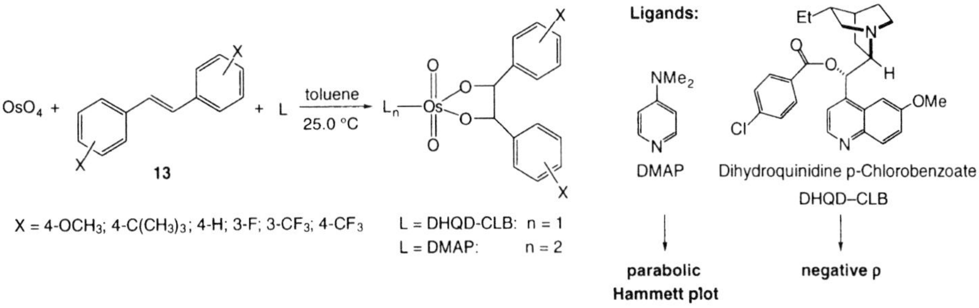

Scheme 6 Hammett studies for the DMAP- and DHQD-CLB-accelerated osmylations of *trans*-stilbenes in toluene at 25 °C [26].

that sp^3 hybridization of the reacting carbon centers is maintained in the transition state of the rate-determining step. These data are inconsistent with a concerted [3+2]-like mechanism, wherein the developing olefinic strain energy would be expected to have a substantial effect on the enthalpy of alkene extrusion (Scheme 7). In contrast, a stepwise process, proceeding via rhenaoxetane **18**, can readily explain the experimental data, since rehybridization of the reacting carbon center is minimal upon migration from oxygen to Re. The carbon atom bound to oxygen remains sp^3 hybridized en route to the oxametallacycle **18**, and it is worth noting that the two-carbon fragment in this four-membered intermediate experiences almost no ring strain due to the very long Re–O and Re–C bonds.

Studies of the influence of the ring puckering on the extrusion of olefin from diolate complexes **17** [41] as well as rate measurements [42] support a stepwise

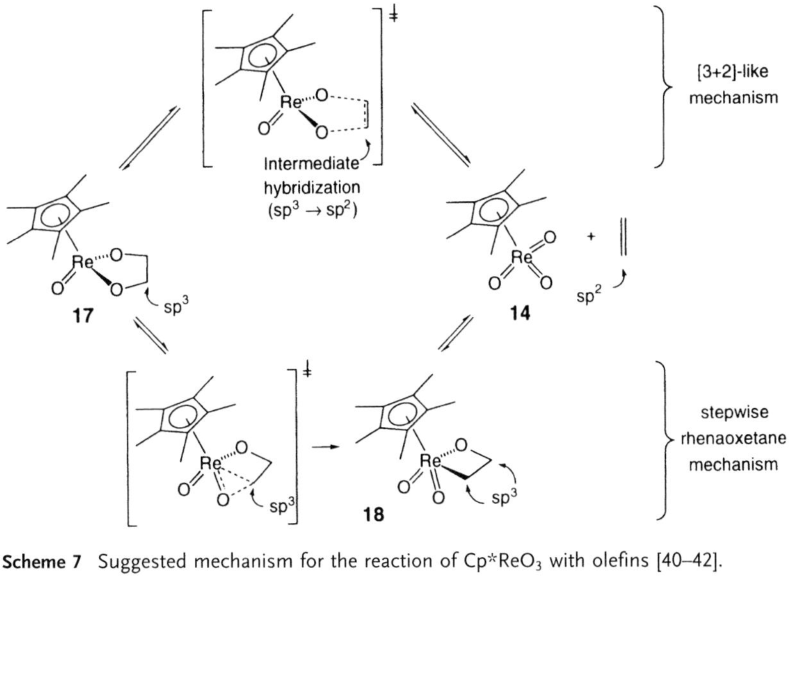

Scheme 7 Suggested mechanism for the reaction of $Cp^{*}ReO_3$ with olefins [40–42].

metallaoxetane mechanism for the olefin extrusion and its microscopic reverse, the diolate formation from $Cp^{*}ReO_3$ **14** and olefin (Scheme 7). A similar stepwise mechanism may, therefore, operate in the analogous osmylation reaction, considering the similarity between Cp* trioxorhenium(VII) and tetraoxoosmium(VIII) complexes.

Further evidence for a stepwise mechanism in the osmylation of olefins stems from variable temperature studies. It was shown that both the enantioselectivity as well as the chemoselectivity of stoichiometric ligand-assisted dihydroxylations exhibit nonlinear temperature relationships [44, 45]. Consequently, there have to be at least two selectivity-determining levels, requiring the presence of a reaction intermediate [30]. The break in the modified Eyring plots further requires the two transition states leading to and from this intermediate to have unequal temperature dependencies. These observations are inconsistent with the concerted [3+2] mechanism, while the stepwise osmaoxetane mechanism can easily rationalize this behavior.

To date, it has not been possible to detect an osmaoxetane intermediate in the osmylation reactions [46]. However, computational *ab initio* studies have shown that osmaoxetanes are minima on the energy surface of the system [27, 28, 47]. A thorough investigation of all the isomeric forms of such an intermediate [47] has suggested **19** to be the most favorable structure for the unligated pathway and **20** for the ligated pathway (Fig. 1).

Unfortunately, *ab initio* calculations cannot exclude either mechanism and both paths are feasible. Despite the apparent stability of metallaoxetanes, recent studies favor the concerted [3+2] mechanism over the stepwise pathway based on calculated transition state energies [29]. However, great care should be taken in the interpretation of *energetic* data, especially with respect to potential transition states, due to the approximations underlying the calculations (especially the basis sets and their application to oxo complexes of heavy metals) and the problem of finding the 'correct' transition state geometries [48]. These calculations grossly underestimate the contributions of π-bonding to the stability of osmium-oxo complexes [48] and energetic data have to be validated by checking them against experimental values of analogous systems, e.g. the Re-diolate system [40–42]. Thus, *ab initio* calculations have not been able to solve the mechanistic dichotomy yet and face selectivity models for the AD reaction have been developed for both mechanisms (cf. Sect. 2.5.3.4).

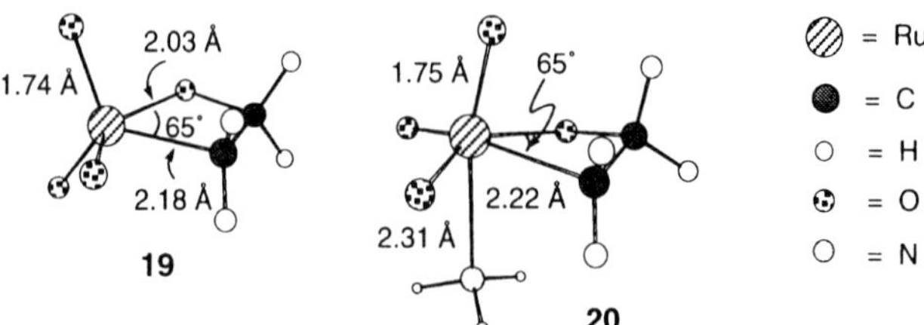

Fig. 1 *Ab initio* structures of ruthenaoxetanes [27, 47].

2.5.3
Development of the Asymmetric Dihydroxylation

2.5.3.1
Process Optimization

An important measure for the value of any catalytic process is its turnover rate. Mechanistic investigations are invaluable for the optimization of a catalytic process with respect to both catalytic turnover and enantioselectivity. The asymmetric dihydroxylation is one of the examples where this interplay between mechanistic investigation and optimization has led to a very successful process.

The first catalytic version of the asymmetric dihydroxylation was based on the Upjohn process, using *N*-morpholine-*N*-oxide (NMO, **21**) as the stoichiometric reoxidant [49]. It was found, however, that the enantioselectivities in the catalytic version were almost always inferior to those obtained under stoichiometric conditions. Mechanistic studies revealed that the culprit is a second catalytic dihydroxylation cycle (Scheme 8), which proceeds with poor-to-no face selectivity, since it does not involve the chiral ligand [50].

The primary cycle proceeds with high face selectivity, since it involves the chiral ligand in its selectivity-determining step, the formation of the osmium(VI) glycolate **22**. The latter is oxidized to the osmium(VIII) glycolate **23** by the co-oxidant (NMO) resulting in loss of the chiral ligand. Intermediate **23** plays a crucial role in determining the selectivity for it lies at the point of bifurcation of the 'good' and 'bad' catalytic cycles. The desired path involves hydrolysis of **23** to OsO_4 and the optically active 1,2-diol. Whereas the undesired, secondary cycle is entered when **23** reacts instead with a second molecule of olefin, yielding the osmium(VI)

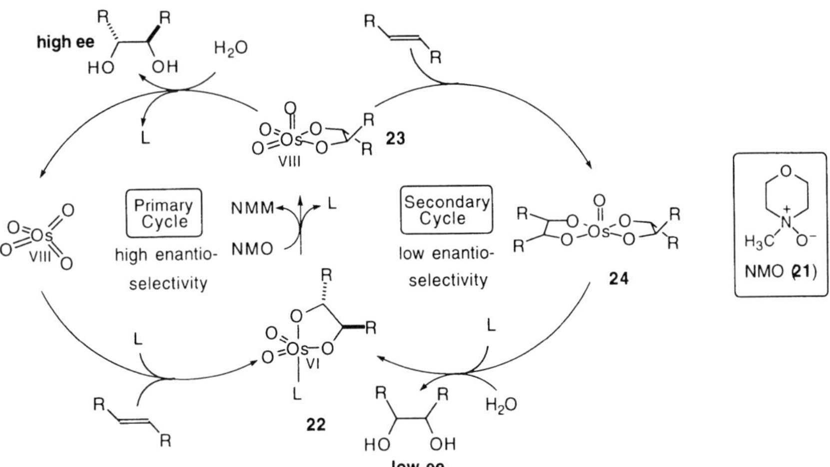

Scheme 8 The two catalytic cycles for the asymmetric dihydroxylation using NMO as co-oxidant [50].

bisglycolate **24** and thence 1,2-diol of low enantiopurity. This mechanistic insight enabled Wai and Sharpless to develop an optimized version of the asymmetric Upjohn process based on slow addition of the olefin [50]. The slow addition ensured a low olefin concentration in the reaction mixture, thereby favoring hydrolysis of the pivotal osmium(VIII) trioxoglycolate intermediate **23** over its alternative fate – entry into the non-selective secondary cycle.

Another, and technically simpler protocol was then developed [51]. The process is based on the use of $K_3Fe(CN)_6$ as the stoichiometric reoxidant [17] and it employs heterogeneous solvent systems, typically *tert*-butanol/water. The reason for the success of this system is that the olefin osmylation and osmium re-oxidation steps are uncoupled, since they occur in different phases (Scheme 9).

The actual osmylation takes place in the organic layer, giving rise to the osmium(VI) glycolate **22**. This osmium(VI) complex cannot be oxidized to an osmium(VIII) glycolate, because of the absence of the inorganic stoichiometric oxidant, $K_3Fe(CN)_6$, in the organic layer. Consequently, the second catalytic cycle cannot occur. Further reaction requires hydrolysis of the osmium(VI) glycolate **22** to the 1,2-diol and a water soluble inorganic osmium(VI) species **25**, which enters the basic aqueous layer ready to be oxidized by $K_3Fe(CN)_6$ to OsO_4. The latter returns to the organic phase, completing the catalytic cycle. The enantiomeric purities of diols obtained under these heterogeneous conditions are essentially identical to those obtained under stoichiometric conditions.

The above discussion has suggested how the catalytic variant of the dihydroxylation might be influenced by the events that take place *after* the olefin osmylation step. In actuality, for virtually all cases of catalytic dihydroxylation, hydrolysis of

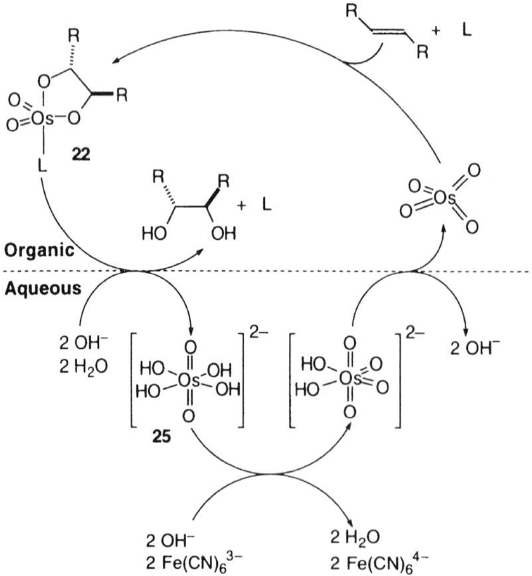

Scheme 9 Catalytic cycle of the AD reaction with $K_3Fe(CN)_6$ as the co-oxidant [51].

the osmium(VI) glycolate products **22** is the turnover limiting step. This is especially true for sterically hindered olefins, and a key goal for improving these catalytic processes has been, and remains facilitation of glycolate hydrolysis. Advances on this front translate directly to higher turnover rates. Amberg and Xu discovered that alkylsulfonamides, e.g. $MeSO_2NH_2$, considerably accelerate the hydrolysis of the osmium(VI) glycolate **22** under the heterogeneous conditions, and the reaction times can be up to 50 times shorter in the presence of this additive [52, 53]. The sulfonamide effect enables satisfactory turnover rates with most olefins, even with some tetrasubstituted olefins [54]. One equivalent of this auxiliary reagent should be added to every AD reaction except for terminal olefins.

Further development of the system led to the formulation of a reagent mixture, called AD-mix [52, 55], which contains all the ingredients for the asymmetric dihydroxylation under the heterogeneous conditions, including $K_2OsO_2(OH)_4$ as a nonvolatile osmium source. This commercially available formulation [56] makes the reaction very easy to perform. In a typical experiment, 1 mmol of olefin is added at 0 °C to the reaction mixture consisting of 1.4 g AD-mix [57], 1 equivalent of methanesulfonamide (except for terminal olefins) and 10 ml 1:1 *tert*-butanol/H_2O. The heterogeneous reaction mixture should be stirred vigorously until the reaction is complete.

AD-mix-β (1 mol % $(DHQD)_2PHAL$, 0.2 mol-% OsO_4, 3 eq. K_2CO_3), 1-2 eq. $MeSO_2NH_2$, 1:1 $^tBuOH/H_2O$, 0°C, 15.5 h

96 % yield; 97 %ee

$(DHQD)_2PHAL$

The sulfonamide effect ensures satisfactory turnover rates for most olefins. However, sterically hindered [54, 58] or electronically deactivated olefins [59–61] may require further rate enhancements. This can be achieved by performing the reaction at room temperature and increasing the amounts of OsO_4 and ligand from the typical AD-mix concentrations of 0.4 mol% OsO_4 and 1 mol% ligand to 1–2 mol% OsO_4 and 5 mol% ligand. In addition, up to 3 equivalents of methanesulfonamide may be employed for sterically very hindered olefins in order to facilitate glycolate hydrolysis. Thus, even tetrasubstituted [54] as well as electron deficient olefins [59–61] give useful results under these more powerful AD conditions.

2.5.3.2
Ligand Optimization

Since the initial discovery of the cinchona alkaloid system a large number of derivatives (>400) have been screened as chiral ligands for the asymmetric dihydroxylation. This systematic structure activity study has revealed that the cinchona molecule (Fig. 2) is ideally set-up for the asymmetric dihydroxylation [39], providing the basis both for high ligand acceleration and for high asymmetric induction.

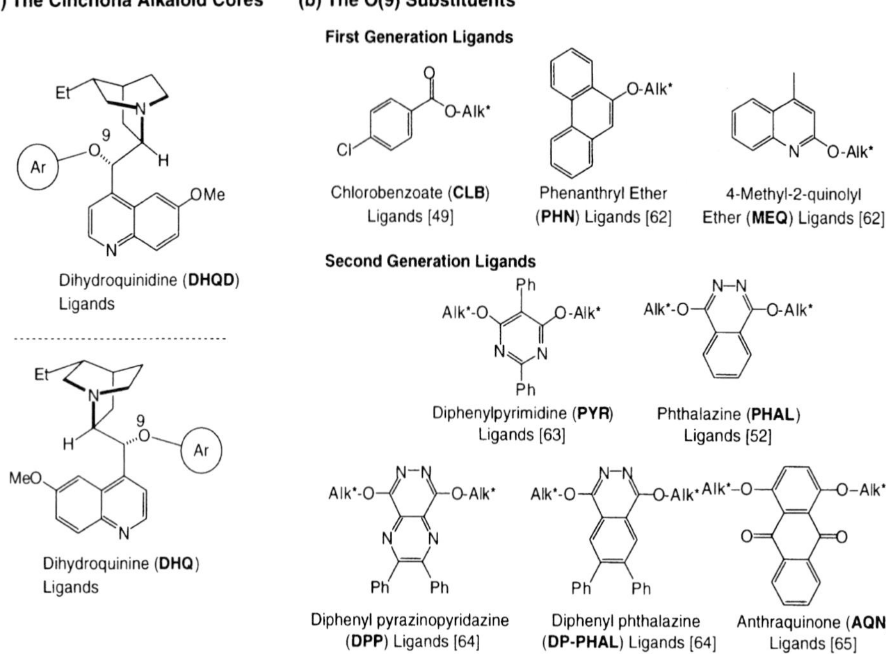

Fig. 2 Structural motif of AD ligands.

The most significant improvements in ligand performance were achieved by optimizing the O(9) substituent. In contrast, modifications to the cinchona core were rarely beneficial.

All of the most successful ligands have one structural feature in common – an *aromatic* group in the O(9) substituent (Fig. 2). The beneficial effect of an aromatic group at O(9) can be understood in terms of stacking interactions with the substituents of the substrate in the transition state of the selectivity-determining step (cf. Sect. 2.5.3.4).

Based on the historical development, the cinchona derivatives are classified as first and second generation ligands (Fig. 2b). These ligand generations have distinct structural features. All of the first generation ligands are 'monomeric' in a sense that they are formed by a formal 1:1 combination of a cinchona alkaloid molecule with an aromatic molecule. The second generation ligands are 'dimeric', since they combine *two* molecules of the alkaloid which are held apart by an aromatic spacer.

The recommended ligands for each substrate class will be discussed in Section 2.5.3.5.

2.5.3.3
Empirical Rules for Predicting the Face Selectivity

Despite the mechanistic uncertainties, the face selectivity of the dihydroxylation can reliably be predicted using an empirical 'mnemonic' device (Scheme 10) [39, 52, 66]. The plane of the olefin is divided into four quadrants and the substituents are placed into these quadrants according to a simple set of rules. The SE quadrant is sterically inaccessible and, with few exceptions, no substituent other than hydrogen can be placed here. The NW quadrant, lying diagonally across from the SE quadrant, is slightly more open and the NE quadrant appears to be quite spacious. The SW quadrant is special in that its preferences are ligand dependent. Even though this SW quadrant normally accepts the largest group, especially in the case of PYR ligands, it is especially attractive for *aromatic* groups in the case of PHAL ligands [66]. An olefin which is placed into this plane according to the above constraints receives the two OH groups from above, i.e. from the β-face, in the case of DHQD-derived ligands and from the bottom, i.e. from the α-face, in the case of DHQ derivatives.

2.5.3.3.1 The Mnemonic Device – Ligand-specific Preferences

In certain cases it may be difficult to judge which one of the olefin substituents should be placed into the SW quadrant. This especially applies to 1,1-disubstituted olefins [66–68] and to *cis*-1,2-disubstituted olefins [69–73] owing to the '*meso*-problem'. Studies with these olefin classes have shown that the pure steric size of a group is not by itself a measure for its propensity to be in the SW quadrant. Also the kind and the properties of the substituents have to be taken into account and compared with the ligand-specific preferences for the SW quadrant. The following rules for these ligand preferences were derived partially from face selectivity studies [66–68] and partially from the existing mechanistic models (cf. Sect. 2.5.3.4).

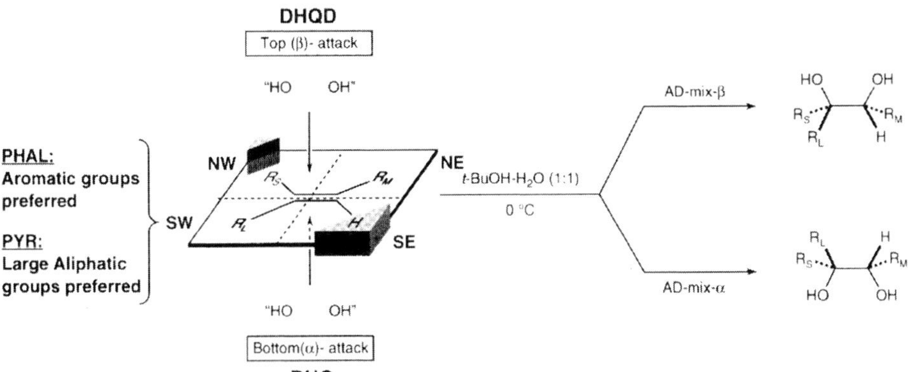

Scheme 10 The mnemonic device for predicting the face selectivity [66].

PHAL ligands show the following preferences for the SW quadrant [66–68]:

Aromatic groups ≫ *n*-alkyl > branched alkyl > oxygenated residues.

Recent studies have revealed that oxygenated residues, e.g. acyloxymethyl/alkoxymethyl [68] or phosphinoxides [74], have a very small preference for the ligand's binding pocket (i.e. the SW quadrant) and even the small methyl group can compete with these groups (Tab. 1). Studies with 1,1-disubstituted olefins have shown that pyrimidine (PYR) ligands have very different preferences for the SW quadrant [66, 67] and the steric size of a substituent is much more important than in the PHAL system (Tab. 2). Thus, the enantioselectivity correlates with the steric volume of a group [75], which translates to the following order of preference for the SW quadrant:

branched alkyl > long *n*-alkyl (length ≥ 3) > aromatic residues > short *n*-alkyl

These results demonstrate that the higher preference of the PYR ligand for aliphatic groups can actually lead to a reversal of face selectivity (Tab. 2): as the aliphatic chain is elongated, the preference for it being in the SW quadrant increases, resulting in the opposite face selectivity (compare entries 1–2 with entries 3–4). The same applies for branching in the olefin substituent (entry 5 vs. entry 6). A similar observation was made by Krysan with the sterically very hindered 3-methylidene-benzofurans [67].

Tab. 1 Application of the mnemonic device for PHAL ligands and 1,1-disubstituted olefins (allylic alcohol derivatives [68], phosphinoxides [74]).

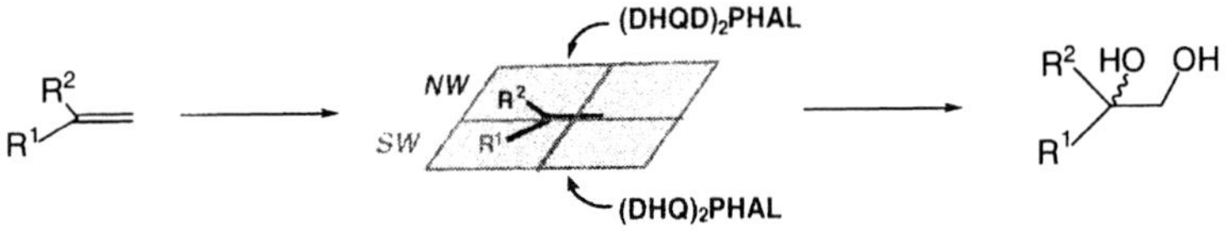

(DHQD)$_2$PHAL
NW SW R^2 R^1
R^2 R^1 → → R^2 HO OH R^1
(DHQ)$_2$PHAL

Quadrant	*Olefin*	*R*	*Ligand*	*% ee*	*Product*
NW –	RO	*t*BuPh$_2$Si		91	RO, OH OH
SW –	Me	Bn	(DHQD)$_2$PHAL	31	
		Piv	*β*	11	Me
NW –	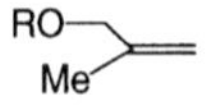 RO	*t*BuPh$_2$Si	(DHQ)$_2$PHAL	47	RO, OH OH
SW –	Me	Piv	*α*	15	Me
NW –	Ph$_2$P(=O)	Me	(DHQD)$_2$PHAL	55	Ph$_2$P(=O), OH OH
SW –	R	Ph	*β*	86	R

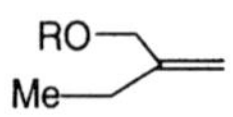

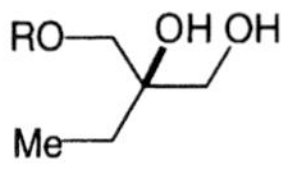

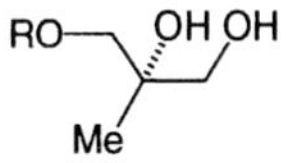

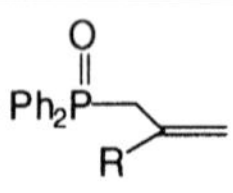

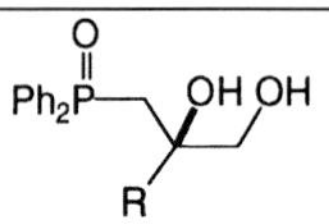

Tab. 2 Application of the mnemonic device for PYR ligands and 1,1-disubstituted olefins [66, 67]

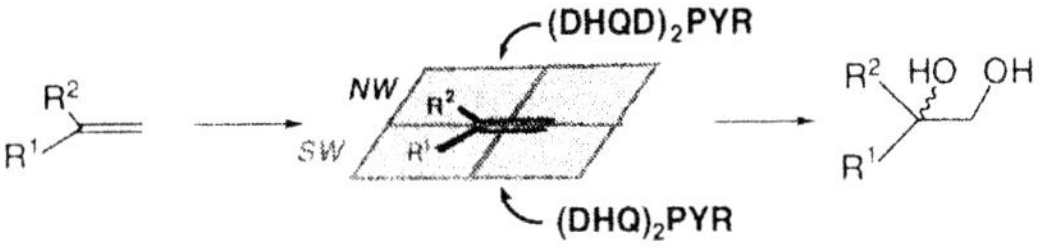

Entry	Quadrant	Olefin	% ee	Major enantiomer
1	NW – H_3C SW – Ph	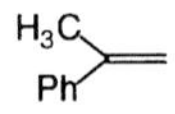	69	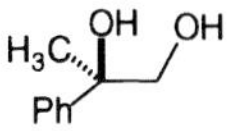
2	NW – H_5C_2 SW – Ph	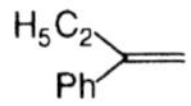	20	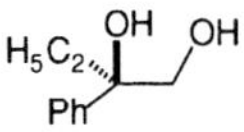
3	NW – Ph SW – H_7C_3	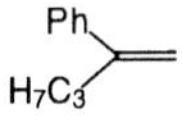	–16	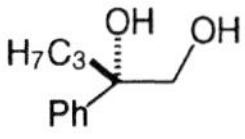
4	NW – Ph SW – $H_{13}C_6$	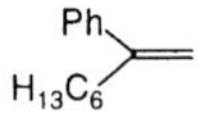	–35	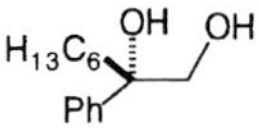
5	NW – SW –	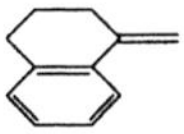	60	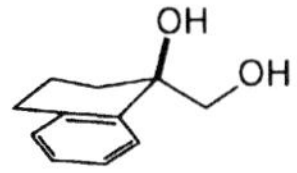
6	NW – SW –	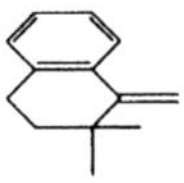	–59	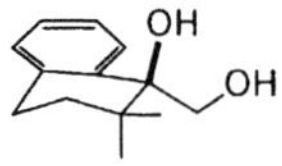

2.5.3.3.3 The Mnemonic Device – Exceptions

The empirical mnemonic device is very reliable in terms of predicting the sense of face selectivity. However, a few exceptions have appeared in recent years, mostly observed with terminal olefins. The asymmetric dihydroxylation of certain *ortho*-substituted allyl benzenes in the presence of phthalazine ligands have been shown to give facial selectivities opposite to those predicted by the mnemonic device (Tab. 3, entry 1) [76–78]. Interestingly, this exceptional behavior seems limited to the second ligand generation, because the first generation phenanthryl ether ligand gave the expected absolute stereochemistry (entry 2) [76]. Furthermore, *trans*-olefins in the same series react with the expected face selectivity even with the phthalazine ligands (entry 3), thereby demonstrating that exceptions are so far limited to the class of terminal olefins.

In summary, the mnemonic device is a simple tool for predicting the facial selectivity of the AD reaction. However, reliable predictions require the intrinsic preferences of each ligand to be taken into account. Thus, the SW quadrant is especially attractive for aromatic groups in the PHAL system, while aliphatic groups are preferred in the PYR system. PYR ligands are, therefore, the ligands

Tab. 3 Exceptions to mnemonic device predictions [76]

Entry	*Substrate*	*Ligand*	*% ee*	*Major enantiomer*	*Mnemonic device obeyed?*
1	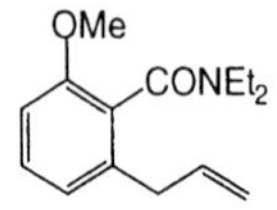	$(DHQD)_2PHAL$	16%	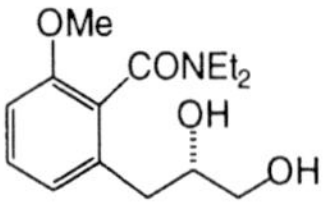	N
2		DHQD-PHN	40%	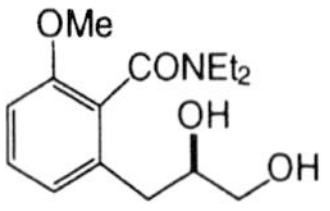	Y
3	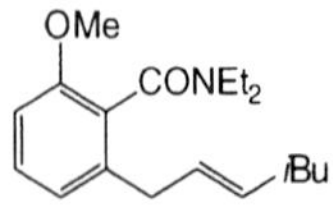	$(DHQ)_2PHAL$	81%	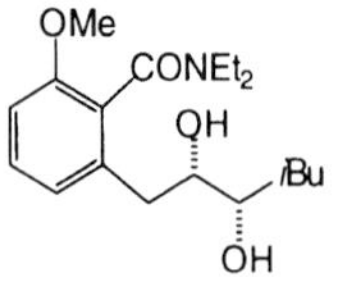	Y

of choice for aliphatic and/or sterically congested olefins, while PHAL ligands are better for aromatic substrates. These simple rules allow the prediction of the face selectivities even in difficult cases (1,1-disubstituted olefins) and very few exceptions are known. These mainly involve monosubstituted olefins.

2.5.3.4 Mechanistic Models for the Rationalization of the Face Selectivity

The development of mechanistic models for the origin of the high face selectivity in the AD reaction is hampered by the uncertainties regarding the mechanism of the osmylation step (cf. Sect. 2.5.2). Models based on both the [3+2] and the stepwise osmaoxetane mechanisms have been advanced and they have converged to the same basic principle: the face selectivity is thought to arise from a reaction of the olefin or a related organometallic derivative within a chiral binding pocket, which is set up by the ligand's aromatic groups. Both models are able to rationalize the especially good selectivities observed with olefins carrying aromatic substituents, since these aromatic groups allow a tight fit into the chiral binding pocket. Despite these superficial similarities, both models differ in the exact location and the shape of the hypothetical binding pocket and in the underlying mechanism of the osmylation reaction.

The model proposed by the Corey group is based on the [3+2] mechanism and features a U-shaped binding pocket, set up by the two parallel methoxyquinoline units (Fig. 3a) [35, 79]. Obviously, this model is limited to the second generation ligands, since ligands from the first generation lack the second methoxyquinoline system. OsO_4 is coordinated to one of the two quinuclidine groups and it is

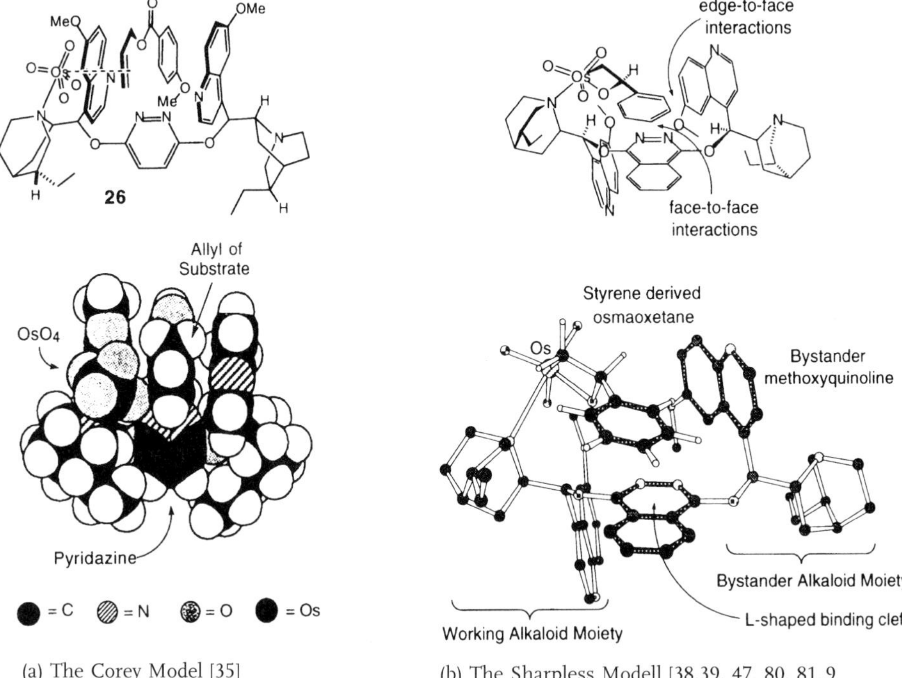

(a) The Corey Model [35]

(b) The Sharpless Modell [38,39, 47, 80, 81,,9

Fig. 3 Face selectivity models.

bound in a staggered conformation. The substrate is suggested to be pre-complexed to this ligand •OsO_4 complex **26** in a two-site binding mode, involving aryl–aryl interactions of the aromatic residue of the substrate with the ligand's two parallel methoxyquinoline units, in addition to contacts between the olefinic π-orbital and low-lying *d*-orbitals of Os(VIII) [35]. This complexation requires the equatorial oxygen atoms of the OsO_4 complex to be in an eclipsed conformation with the C–N bonds of the quinuclidine ((a) The Corey Model [35] (b) The Sharpless model [38, 39, 47, 80, 81]) system and it gives rise to a 20 electron complex – both highly unfavorable events. One axial and one equatorial oxo group of the ligand •OsO_4 complex **26** are suggested to be involved in the [3+2] cycloaddition to the olefin, leading to the glycolate. The face selectivity is thought to arise from selective rate acceleration for the 'correct' diastereomeric ensemble, which is ascribed to the favorable arrangement for the complex shown in Fig. 3a as well as a relief of eclipsing interactions due to rotation about the N–Os bond. However, the latter effect would be expected to be negligible, because of the long Os–N distance in the complex (2.48 Å). Apparently, dihydroxylation of the opposite olefin face is disfavored due to the lack of a simultaneous interaction of the olefin's substituent with the binding pocket and the double bond with both oxo groups.

The Sharpless model is based on the stepwise osmaoxetane mechanism and an L-shaped binding cleft is proposed (Fig. 3b) [38, 39, 47, 80, 81]. The latter is formed by the aromatic linker (typically phthalazine) as the floor and the methoxyquinoline unit as a perpendicular wall. This structure is one of the most stable conformations of the ligand [39]. One of the olefin's substituents, most favorably an aromatic group, snugly fits into this chiral binding pocket as shown for styrene in Fig. 3b. This model readily explains the observed match between aromatic groups in both the substrate and the PHAL ligand with respect to both enantioselectivity and rate acceleration [39], since these aromatic groups enable an especially good stabilization of the oxetane-like transition state due to both offset-parallel interactions between the aromatic substituent of the olefin with the phthalazine floor as well as favorable edge-to-face interactions with the 'bystander' methoxyquinoline ring. The metallaoxetane is expected to be energetically above the ground states [27, 47] so that the transition states flanking it should have considerable oxetane character (Hammond postulate). With this assumption, the relative stabilities of both diastereotopic transition states can be estimated by comparing the relative energies of both diastereomeric metallaoxetane/ligand intermediates **27A** and **27B** (Fig. 4). A molecular mechanics model has been developed based on the MM2* force field in MacroModel [80].

Enantioselectivity may arise chiefly by the interplay of two opposing factors: transition state *stabilizing* interactions between one of the oxetane substituents (R_c) and the binding pocket, and transition state *destabilizing* interactions between another oxetane substituent (H_a) and H(9) of the ligand (cf. Fig. 4, structures **27A**

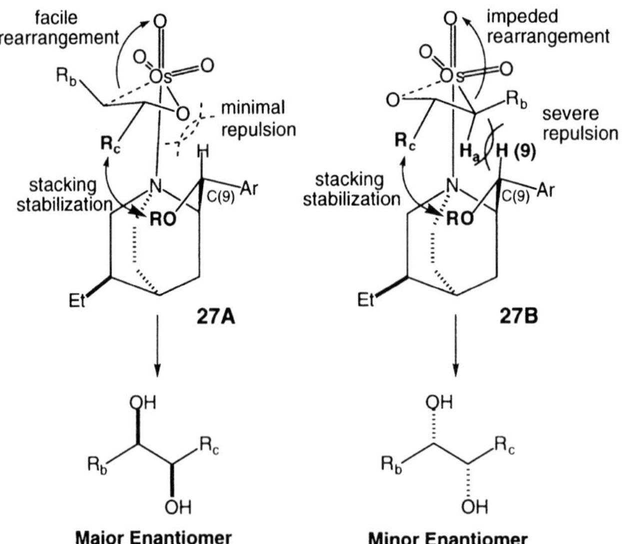

Fig. 4 Rationalization for enantiofacial selectivity in the AD reaction based on the interplay of attractive and repulsive interactions [47, 80].

and **27 B**). Structures **27 A** and **27 B** are diastereomers leading to the major and minor enantiomer. Both diastereomeric structures allow the favorable stacking interactions with the ligand leading to an overall acceleration of the reaction. This may be the origin of the high acceleration which is observed especially with aromatic substrates in the presence of the phthalazine ligand. However, structure **27 B**, leading to the minor enantiomer, is selectively *destabilized due* to greater repulsive interactions with H(9) of the ligand. Thus, the AD is primarily dependent on noncovalent interactions both with respect to face selectivity and ligand acceleration. Attractive interactions force the system into a transition state arrangement for the disfavored diastereomer **27 B**, wherein the net effect of the noncovalent interactions is nil – attraction and repulsion having off-set each other. A second level of selectivity may result from an impeded rearrangement of oxetane **27 B**, due to increased H_a–H(9) interactions in the course of the rearrangement. These two levels of selectivity may add up to the high overall selectivity typically observed in the AD reaction. More recent *ab initio* studies have led to a refinement of the model, suggesting that oxetane ring puckering and dipole–dipole interactions may play an additional role in the face selection process [47].

The Sharpless model can readily be extended to the first generation ligands, since the floor of the 'binding pocket' remains intact, and even the lower face selectivities can be rationalized. These arise from less tight binding in the transition state due to the lack of the bystander aromatic system and consequently the loss of edge-to-face interactions.

2.5.3.5 The Cinchona Alkaloid Ligands and their Substrate Preferences

The ligands with the broadest scope belong to the second generation (cf. Fig. 2). The phthalazine ligands (PHAL) are most widely used, due to their ready availability and their broad substrate scope [52]. This ligand class is used in the AD-mix formulation [55–57]. PHAL ligands react especially well when aromatic groups are present, and remarkably high enantioselectivities are observed when the aromatic substituents appear in certain optimal locations/patterns [39]. One such case is *trans*-stilbene for which the enantioselectivity is as high as 99.8% [82]. However, PHAL ligands give inferior results with aliphatic olefins, especially if they are branched near the double bond or if they have very small substituents.

Recent developments have provided ligands with even broader scope than that of the PHAL derivatives. The data in Tab. 4 show that the PHAL ligands have been superseded by DPP, DP-PHAL [64], and AQN ligands [65].

The substrate recommendations for each ligand class are summarized below.

Anthraquinone (AQN) ligands

The anthraquinone ligands are especially well-suited for almost all olefins having aliphatic substituents [65]. Even diols derived from allyl halides or allyl alcohols can now be obtained with satisfactory enantiomeric purity, thereby giving access to valuable chiral building blocks. *The AQN derivatives are the ligands of choice for*

Tab. 4 Comparison of the second generation ligands [52, 63–65]. The best result for each olefin is printed in bold

	(DHQD)₂-PHAL	*(DHQD)₂-PYR*	*(DHQD)₂-DPP*	*(DHQD)₂-AQN*	*Diol config.*
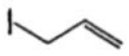	63	70	68	**83**	*S*
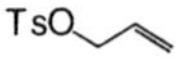 TsO	40			**83**	*S*
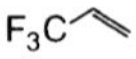 F_3C	63	64		**81**	*S*
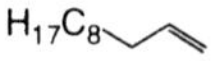 $H_{17}C_8$	84	89	89	**92**	*R*
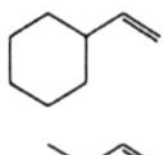	88	**96**	89	86	*R*
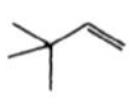	64	**92**	59		*R*
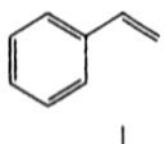	97	80	**99**	89	*R*
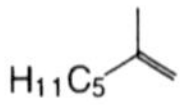 $H_{11}C_5$	78	76	78	**85**	*R*
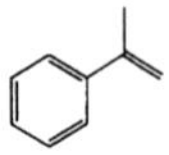	94	69	**96**	82	*R*
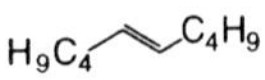 H_9C_4 C_4H_9	97	88	96	**98**	*R, R*
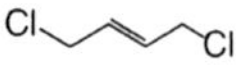 Cl Cl	94			**96**	*S, S*
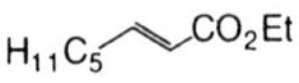 $H_{11}C_5$ CO_2Et	**99**			**99**	2 *S*, 3 *R*

the AD reaction, except for olefins with aromatic or sterically demanding substituents. (However, for reasons of availability, the PHAL derivatives are likely to remain the 'best' ligands for some time.)

Pyrimidine (PYR) ligands
The pyrimidine ligands are the ligands of choice for olefins with sterically demanding substituents [63].

Diphenyl pyrazinopyridazine (DPP) and diphenyl phthalazine (DP-PHAL) ligands
These ligands give improved enantioselectivities for almost all olefins except for terminal alkyl olefins which are better served by the AQN or PYR ligands [64]. Even *cis*-1,2-disubstituted olefins give improved face selectivities with these ligands. The DPP ligand is normally slightly superior to the DP-PHAL ligand. *The DPP derivatives are the optimal ligands for aromatic olefins and for certain cis-1,2-disubstituted olefins.*

Tab. 5 The recommended ligands for each olefin class [52, 54, 63–65, 69]

Olefin class	R-CH=CH₂	$R^1R^2C{=}CH_2$	cis-$R^1CH{=}CHR^2$	trans-$R^1CH{=}CHR^2$	$R^1R^2C{=}CHR^3$	$R^1R^2C{=}CR^3R^4$
Preferred ligand	*R = Aromatic* DPP, PHAL	*R^1,R^2 = Aromatic* DPP, PHAL	*Acyclic* IND	*R^1, R^2 = Aromatic* DPP, PHAL	PHAL, DPP, AQN	PYR, PHAL
	R = Aliphatic AQN	*R^1, R^2 = Aliphatic* AQN	*Cyclic* PYR, DPP, AQN	*R^1, R^2 = Aliphatic* AQN		
	R = Branched PYR	*R^1, R^2 = Branched* PYR				

Indoline (IND) ligands

Cis-1,2-disubstituted olefins generally are poor substrates for the AD reaction and *the IND derivatives are normally the ligands of choice* [69]. However, in certain cases better results are obtained with the new second generation ligands [64, 65, 70, 71, 73]. The recommended ligands for each olefin class are listed in Tab. 5.

References

1 (a) T. Katsuki, V.S. Martin, *Org. React.* **1996**, *48*, 1–299. (b) R.A. Johnson, K.B. Sharpless, *Catalytic Asymmetric Synthesis* (Ed.: I. Ojima), VCH, New York, **1993**, pp. 101–158.

2 (a) T. Katsuki, *J. Mol Catal. A: Chem.* **1996**, *113*, 87–107. (b) E.N. Jacobsen, *Catalytic Asymmetric Synthesis* (Ed.: I. Ojima), VCH, New York, **1993**, 159–202.

3 (a) H.C. Kolb, M.S. VanNieuwenhze, K.B. Sharpless, *Chem. Rev.* **1994**, *94*, 2483–2547. (b) M. Schröder, *Chem. Rev.* **1980**, *80*, 187–213.

4 (a) G. Li, H.-T. Chang, K.B. Sharpless, *Angew. Chem., Int. Ed. Engl.* **1996**, *35*, 451–453. (b) G. Li, K.B. Sharpless, *Acta Chem. Scand.* **1996**, *50*, 649–651.

5 K.B. Sharpless, A.Y. Teranishi, J.-E. Bäckvall, *J. Am. Chem. Soc.* **1977**, *99*, 3120–3128.

6 P.N. Becker, M.A. White, R.C. Bergman, *J. Am. Chem. Soc.* **1980**, *102*, 5676–5677.

7 A.O. Chong, K. Oshima, K.B. Sharpless, *J. Am. Chem. Soc.* **1977**, *99*, 3420–3426.

8 (a) F.E. McDonald, T.B. Towne, *J. Org. Chem.* **1995**, *60*, 5750–5751. (b) R.M. Kennedy, S. Tang, *Tetrahedron Lett.* **1992**, *33*, 3729–3732. (c) S. Tang, R.M. Kennedy, *Tetrahedron Lett.* **1992**, *33*, 5299–5302. (d) S. Tang, R.M. Kennedy, *Tetrahedron Lett.* **1992**, *33*, 5303–5306. (e) R.S. Boyce, R.M. Kennedy, *Tetrahedron Lett.* **1994**, *35*, 5133–5136.

9 (a) G. Li, H.-T. Chang, K.B. Sharpless, *Angew. Chem., Int. Ed. Engl.* **1996**, *35*, 451–454. (b) G. Li, K.B. Sharpless, *Acta Chem. Scand.* **1996**, *50*, 649–651. (c) J. Rudolph, P.C. Sennhenn, C.P. Vlaar, K.B. Sharpless, *Angew. Chem., Int. Ed. Engl.* **1996**, *35*, 2810–2813. (d) G. Li,

H. H. Angert, K. B. Sharpless, *Angew. Chem., Int. Ed. Engl.* **1996**, *35*, 2813–2817. (e) R. Angelaud, Y. Landais, K. Schenk, *Tetrahedron Lett.* **1997**, *38*, 1407–1410.

10 D. J. Berrisford, C. Bolm, K. B. Sharpless, *Angew. Chem., Int. Ed. Engl.* **1995**, *34*, 1059–1070.

11 (a) R. Criegee, *Justus Liebigs Ann. Chem.* **1936**, *522*, 75–93. (b) R. Criegee, *Angew. Chem.* **1937**, *50*, 153–155. (c) R. Criegee, *Angew. Chem.* **1938**, *51*, 519–520. (d) R. Criegee, B. Marchand, H. Wannowias, *Justus Liebigs Ann. Chem.* **1942**, *550*, 99–133.

12 S. G. Hentges, K. B. Sharpless, *J. Am. Chem. Soc.* **1980**, *102*, 4263–4265.

13 (a) N. A. Milas, S. Sussman, *J. Am. Chem. Soc.* **1936**, *58*, 1302–1304. (b) N. A. Milas, J. H. Trepagnier, J. T. Nolan Jr., M. I. Iliopulos, *J. Am. Chem. Soc.* **1959**, *81*, 4730–4733.

14 K. B. Sharpless, K. Akashi, *J. Am. Chem. Soc.* **1976**, *98*, 1986–1987.

15 (a) W. P. Schneider, A. V. McIntosh, US Patent 2,769,824, Nov. 6, **1956.** (b) V. VanRheenen, R. C. Kelly, D. Y. Cha, *Tetrahedron Lett.* **1976**, 1973–1976.

16 K. A. Hofmann, *Chem. Ber.* **1912**, *45*, 3329–3338.

17 M. Minato, K. Yamamoto, J. Tsuji, *J. Org. Chem.* **1990**, *55*, 766–768.

18 S. Torii, P. Liu, N. Bhuvaneswari, C. Amatore, A. Jutand, *J. Org. Chem.* **1996**, *61*, 3055–3060.

19 S. Torii, P. Liu, H. Tanaka, *Chem. Lett.* **1995**, 319–320.

20 P. J. Walsh, P. T. Ho, S. B. King, K. B. Sharpless, *Tetrahedron Lett.* **1994**, *55*, 5129–5132.

21 K. Ohmori, S. Nishiyama, S. Yamamura, *Tetrahedron Lett.* **1995**, *36*, 6519–6522.

22 (a) E. Klein, W. Rojahn, *Tetrahedron* **1965**, *21*, 2353–2358. (b) D. M. Walba, M. D. Wand, M. C. Wilkes, *J. Am. Chem. Soc.* **1979**, *101*, 4396–4397. (c) R. Amouroux, G. Folefoc, F. Chastrette, M. Chastrette, *Tetrahedron Lett.* **1981**, *22*, 2259–2262. (d) D. M. Walba, C. A. Przybyla, C. B. Walker Jr., *J. Am. Chem. Soc.* **1990**, *112*, 5624–5625.

23 K. B. Sharpless, A. Y. Teranishi, J.-E. Bäckvall, *J. Am. Chem. Soc.* **1977**, *99*, 3120–3128.

24 For a general review of metallaoxetanes, see: K. A. Jørgensen, B. Schiøtt, *Chem. Rev.* **1990**, *90*, 1483–1506.

25 J. Böseken, *Rec. Trav. Chim.* **1922**, *41*, 199.

26 D. W. Nelson, A. Gypser, P. T. Ho, H. C. Kolb, T. Kondo, H.-L. Kwong, D. McGrath, A. E. Rubin, P.-O. Norrby, K.aP. Gable, K. B. Sharpless, *J. Am. Chem. Soc.* **1997**, *119*, 1840–1858.

27 P.-O. Norrby, H. C. Kolb, K. B. Sharpless, *Organometallics* **1994**, *13*, 344–347.

28 A. Veldkamp, G. Frenking, *J. Am. Chem. Soc.* **1994**, *116*, 4937–4946.

29 (a) S. Dapprich, G. Ujaque, F. Maseras, A. Lledós, D. G. Musaev, K. Morokuma, *J. Am. Chem. Soc.* **1996**, *118*, 11660–11661. (b) M. Torrent, L. Deng, M. Sola, T. Ziegler, *Organometallics* **1997**, *16*, 13–19. (c) U. Pidun, C. Boehme, G. Frenking, *Angew. Chem., Int. Ed. Engl.* **1996**, *35*, 2817–2820.

30 P.-O. Norrby, K. P. Gable, ***J.*** *Chem. Soc., Perkin Trans. 2* **1996**, 171–178.

31 A wide variety of ligands, including acetate, halides, and azides have been found to accelerate stoichiometric osmylation reactions: K. B. Sharpless, P. J. Walsh, unpublished results.

32 J. S. Svendsen, I. Marko, E. N. Jacobsen, C. P. Rao, S. Bott, K. B. Sharpless, *J. Org. Chem.* **1989**, *54*, 2263–2264.

33 R. M. Pearlstein, B. K. Blackburn, W. M. Davis, K. B. Sharpless, *Angew. Chem., Int. Ed. Engl.* **1990**, *29*, 639–641.

34 (a) J. C. Green, M. F. Guest, I. H. Hillier, S. A. Jarret-Sprague, N. Kaltosyannis, M. A. MacDonald, K. H. Sze, *Inorg. Chem.* **1992**, *31*, 1588–1594. (b) P. Pykko, J. Li, T. Bastug, B. Fricke, D. Kolb, *Inorg. Chem.* **1993**, *32*, 1525–1526.

35 E. J. Corey, M. C. Noe, *J. Am. Chem. Soc.* **1996**, *118*, 319–329.

36 D. W. Nelson, W. Derek, K. B. Sharpless, K. Barry, *Reevaluation of the kinetics of the catalytic asymmetric dihydroxylation of alkenes.* Book of Abstracts, 213th ACS National Meeting, San Francisco, April 13–17 (1997), ORGN-616. CODEN: 64AOAA AN 1997:162878.

37 E. N. Jacobsen, I. Marko, M. B. France, J. S. Svendsen, K. B. Sharpless, *J. Am. Chem. Soc.* **1989**, *111*, 737–739.

38 H. C. Kolb, P. G. Andersson, Y. L. Bennani, G. A. Crispino, K.-S. Jeong, H.-L. Kwong, K. B. Sharpless, *J. Am. Chem. Soc.* **1993**, *115*, 12226–12227.

39 H. C. Kolb, P. G. Andersson, K. B. Sharpless, *J. Am. Chem. Soc.* **1994**, *116*, 1278–1291.

40 K. P. Gable, T. N. Phan, *J. Am. Chem. Soc.* **1994**, *116*, 833–839.

41 K. P. Gable, J. J. J. Juliette, *J. Am. Chem. Soc.* **1995**, *117*, 955–962.

42 K. P. Gable, J. J. J. Juliette, *J. Am. Chem. Soc.* **1996**, *118*, 2625–2633.

43 (a) W. A. Herrmann, D. Marz, E. Herdtweck, A. Schaefer, W. Wagner, H.-J. Kneuper, *Angew. Chem.* **1987**, *99*, 462–464. (b) W. A. Herrmann, M. Floel, J. Kulpe, J. K. Felixberger, E. Herdtweck, *J. Organomet. Chem.* **1988**, *355*, 297–313. (c) W. A. Herrmann, D. W. Marz, E. Herdtweck, *J. Organomet. Chem.* **1990**, *394*, 285–303.

44 T. Göbel, K. B. Sharpless, *Angew. Chem., Int. Ed. Engl.* **1993**, *32*, 1329–1331.

45 D. W. Nelson, A. Gypser, P. T. Ho, H. C. Kolb, T. Kondo, H.-L. Kwong, D. V. McGrath, A. E. Rubin, P.-O. Norrby, K. P. Gable, K. B. Sharpless, *J. Am. Chem. Soc.* **1997**, *119*, 1840–1858.

46 D. V. McGrath, G. D. Brabson, L. Andrews, K. B. Sharpless, unpublished results.

47 P.-O. Norrby, H. Becker, K. B. Sharpless, *J. Am. Chem. Soc.* **1996**, *118*, 35–42.

48 A. K. Rappe, unpublished results.

49 E. N. Jacobsen, I. Marko, W. S. Mungall, G. Schröder, K. B. Sharpless, *J. Am. Chem. Soc.* **1988**, *110*, 1968–1970.

50 J. S. M. Wai, I. Markó, J. S. Svendsen, M. G. Finn, E. N. Jacobsen, K. B. Sharpless, *J. Am. Chem. Soc.* **1989**, *111*, 1123–1125.

51 H.-L. Kwong, C. Sorato, Y. Ogino, H. Chen, K. B. Sharpless, *Tetrahedron Lett.* **1990**, *31*, 2999–3002.

52 K. B. Sharpless, W. Amberg, Y. L. Bennani, G. A. Crispino, J. Hartung, K.-S. Jeong, H.-L. Kwong, K. Morikawa, Z.-M. Wang, D. Xu, X.-L. Zhang, *J. Org. Chem.* **1992**, *57*, 2768–2771.

53 For example, in the absence of $MeSO_2NH_2$, *trans*-5-decene was only partially (70%) converted to the corresponding diol after 3 days at 0 °C, whereas the diol was isolated in 97% yield after only 10 h at 0 °C in the presence of this additive.

54 K. Morikawa, J. Park, P. G. Andersson, T. Hashiyama, K. B. Sharpless, *J. Am. Chem. Soc.* **1993**, *115*, 8463–8464.

55 Recipe for the preparation of 1 kg of AD-mix-α or AD-mix-β: potassium osmate $[K_2OsO_2(OH)_4$ (1.04 g) and $(DHQ)_2PHAL$ (for AD-mix-α) or $(DHQD)_2PHAL$ for AD-mix-β) (5.52 g) were ground together to give a fine powder, then added to powdered $K_3Fe(CN)_6$ (699.6 g) and powdered K2CO3 (293.9 g), and finally mixed thoroughly in a blender for *c*. 30 min.

56 The PHAL-, PYR-, and AQN-based ligands, the AD-mixes, and the parent cinchona alkaloids are all available from Aldrich Chemical Co.

57 1.4 g of AD-mix, needed for the AD of 1 mmol of olefin, contain the following amounts of reagents: 1.46 mg (0.004 mmol) of $K_2OsO_2(OH)_4$, 7.73 mg (0.01 mmol) of $(DHQ)_2PHAL$ or $(DHQD)_2PHAL$, 980 mg (3 mmol) of $K_3Fe(CN)_6$, and 411 mg (3 mmol) of K_2CO_3.

58 M. A. Brimble, D. D. Rowan, J. A. Spicer, *Synthesis* **1995**, 1263–1266.

59 Y. L. Bennani, K. B. Sharpless, *Tetrahedron Lett.* **1993**, *34*, 2079–2082.

60 P. J. Walsh, K. B. Sharpless, *Synlett* **1993**, 605–606.

61 K. C. Nicolaou, E. W. Yue, S. La Greca, A. Nadin, Z. Yang, J. E. Leresche, T. Tsuri, Y. Naniwa, F. De Riccardis, *Chem. Eur. J.* **1995**, *7*, 467–494.

62 K. B. Sharpless, W. Amberg, M. Beller, H. Chen, J. Hartung, Y. Kawanami, D. Lübben, E. Manoury, Y. Ogino, T. Shibata, T. Ukita, *J. Org. Chem.* **1991**, *56*, 4585.

63 G. A. Crispino, K.-S. Jeong, H. C. Kolb, Z.-M. Wang, D. Xu, K. B. Sharpless, *J. Org. Chem.* **1993**, *58*, 3785–3786.

64 H. Becker, S. B. King, M. Taniguchi, K. P. M. VanHessche, K. B. Sharpless, *J. Org. Chem.* **1995**, *60*, 3940–3941.

65 H. Becker, K. B. Sharpless, *Angew. Chem., Int. Ed. Engl.* **1996**, *35*, 448–450. The published procedure for the synthesis of $(DHQD)_2AQN$ is performed in THF using *n*-BuLi as the base. However, NaH in DMF gives better results: Sharpless et al, unpublished results.
66 K. P. M. VanHessche, K. B. Sharpless, *J. Org. Chem.* **1996**, *61*, 7978–7979.
67 D. J. Krysan, *Tetrahedron Lett.* **1996**, *37*, 1375–1376.
68 K. J. Hale, S. Manaviazar, S. A. Peak, *Tetrahedron Lett.* **1994**, *35*, 425–428.
69 L. Wang, K. B. Sharpless, *J. Am. Chem. Soc.* **1992**, *114*, 7568–7570.
70 (a) T. Yoshimitsu, K. Ogasawara, *Synlett* **1995**, 257–259. (b) S. Takano, T. Yoshimitsu, K. Ogasawara, *J. Org. Chem.* **1994**, *59*, 54–57.
71 L. Xie, M.T. Crimmins, K.-H. Lee, *Tetrahedron Lett.* **1995**, *36*, 4529–4532.
72 W.-S. Zhou, W.-G. Xie, Z.-H. Lu, X.-F. Pan, *Tetrahedron Lett.* **1995**, *36*, 1291–1294.
73 Z.-M. Wang, K. Kakiuchi, K. S. Sharpless, J. *Org. Chem.* **1994**, *59*, 6895–6897.
74 P. O'Brien, S. Warren, *J. Chem. Soc., Perkin Trans. 1* **1996**, 2129–2138.
75 K. P. M. VanHessche, K. B. Sharpless, submitted.
76 P. Salvadori, S. Superchi, F. Minutolo, *J. Org. Chem.* **1996**, *61*, 4190–4191.
77 D. L. Boger, J. A. McKie, T. Nishi, T. Ogiku, *J. Am. Chem. Soc.* **1996**, *118*, 2301–2302.
78 D. L. Boger, J. A. McKie, T. Nishi, T. Ogiku, *J. Am. Chem. Soc.* **1997**, *119*, 311–325.
79 (a) E. J. Corey, M. C. Noe, A. Y. Ting, *Tetrahedron Lett.* **1996**, *37*, 1735–1738. (b) M. C. Noe, E. J. Corey, *Tetrahedron Lett.* **1996**, *37*, 1739–1742. (c) E. J. Corey, M. C. Noe, M. J. Grogan, *Tetrahedron Lett.* **1996**, *37*, 4899–4902. (d) E. J. Corey, M. C. Noe, A. Guzman-Perez, *J. Am. Chem. Soc.* **1995**, *117*, 10817–10824. (e) E. J. Corey, A. Guzman-Perez, M. C. Noe, *J. Am. Chem. Soc.* **1995**, *117*, 10805–10816. (f) E. J. Corey, A. Guzman-Perez, M. C. Noe, *J. Am. Chem. Soc.* **1994**, *116*, 12109–12110. (g) E. J. Corey, M. C. Noe, S. Sarshar, *Tetrahedron Lett.* **1994**, *35*, 2861–2864. (h) E. J. Corey, M. C. Noe, M. J. Grogan, *Tetrahedron Lett.* **1994**, *35*, 6427–6430.
80 P. O. Norrby, H. C. Kolb, K. B. Sharpless, *J. Am. Chem. Soc.* **1994**, *116*, 8470–8478.
81 H. Becker, P. T. Ho, H. C. Kolb, S. Loren, P.-O. Norrby, K. B. Sharpless, *Tetrahedron Lett.* **1994**, *35*, 7315–7318.
82 G. A. Crispino, P. T. Ho, K. B. Sharpless, *Science* **1993**, *259*, 64–66.

2.5.4
Asymmetric Dihydroxylation – Recent Developments

Kilian Muñiz

2.5.4.1
Introduction

The asymmetric, osmium-catalyzed conversion of unfunctionalized olefins into diols is nowadays recognized as one of the most versatile and efficient asymmetric catalytic reactions [2]. It is regarded as a universally applicable reaction, and its general importance in all areas of asymmetric synthesis has gained its principal inventor, K. B. Sharpless, the 2001 Nobel Prize in Chemistry [3]. In this chapter, recent developments in this area will be discussed, paying special attention to modification of reaction conditions and the development of novel asymmetric dihydroxylation (AD) processes.

2.5.4.2
Homogeneous Dihydroxylation

2.5.4.2.1 Experimental Modifications

It is nowadays widely believed that the development of the original Sharpless system has come to an end. The recommended AD reaction conditions for a broad variety of substrates have already been given in Chapter 2.5.

However, the number of additional variations, for both achiral and asymmetric dihydroxylations, has recently grown rapidly. Sharpless reported the use of phenylboronic acid as the hydrolyzing reagent in the presence of NMO as terminal oxidant. When carried out in anhydrous dichloromethane, this procedure provides access to boronic esters of high purity and diminishes overoxidation, a sometimes serious side reaction of unprotected diols. Since boronic esters remain in solution under the conditions employed, the present protocol is optimal for multi-step dihydroxylation of polyenes. For example, single-step perhydroxylation of squalene and of cyclic triols has been achieved by this method, the latter having been formed as unusual isomers. Deprotection of the boronic esters with aqueous hydrogen peroxide liberates the free diols and polyols in high purity and high yields [4].

Other work has been aimed at replacing the common terminal oxidants (iron hexacyanoferrate, NMO) by more economically and ecologically benign reoxidants. Bäckvall has reported on triple catalytic systems that use hydrogen peroxide as the terminal oxidant [5, 6]. The main aspect of this procedure is a biomimetic selective electron-transfer reaction in which a flavin hydroperoxide (**1**) generates the common oxidant NMO (**4**) from *N*-methyl morpholine (NMM, **3**). The reduced flavin **2** is then reoxidized by hydrogen peroxide. Thereby, the established osmium catalysis with NMO as the reoxidant for OsO_4 is left unchanged (Scheme 2). As expected, the asymmetric version remained uneffected by the additional oxidative processes, and enantioselectivities reached values up to 99%. The superiority of NMO as an oxidant was proven for other tertiary amines which yielded lower conversion [7]. A useful extension of this concept was developed when the chiral cinchona alkaloid ligand itself acted as the reoxidant. It was found that in the H_2O_2/flavin system, $(DHQD)_2PHAL$ could be converted to its *N*-oxide, which then promoted regeneration of Os(VIII), while its original role in stereoinduction remained uneffected [8]. Enantioselectivities of up to 99% were obtained, which in some cases were slightly higher than those obtained with the original Sharpless system.

To a lesser extent, the use of *m*CPBA [9] and vanadyl acetylacetoate [10] instead of hydrogen peroxide also proved successful for various classes of olefins, although an enantioselective reaction has not been reported.

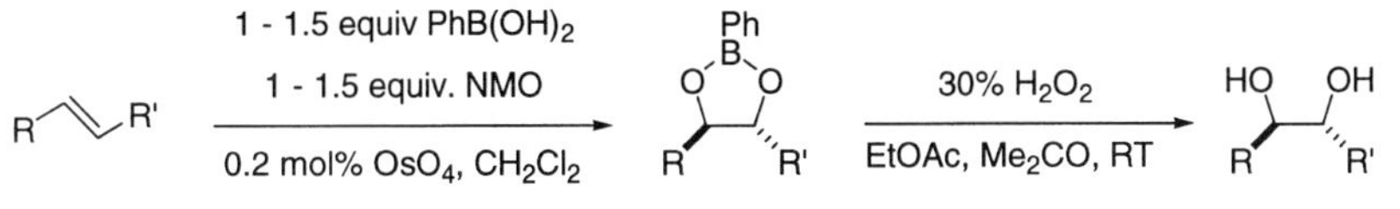

R–CH=CH–R' → (1 - 1.5 equiv $PhB(OH)_2$; 1 - 1.5 equiv. NMO; 0.2 mol% OsO_4, CH_2Cl_2) → phenylboronic ester (Ph, B, O, O, R, R') → (30% H_2O_2; EtOAc, Me_2CO, RT) → HO, OH, R, R' diol

Scheme 1 Catalytic dihydroxylation with phenylboronic acid as hydrolyzing agent.

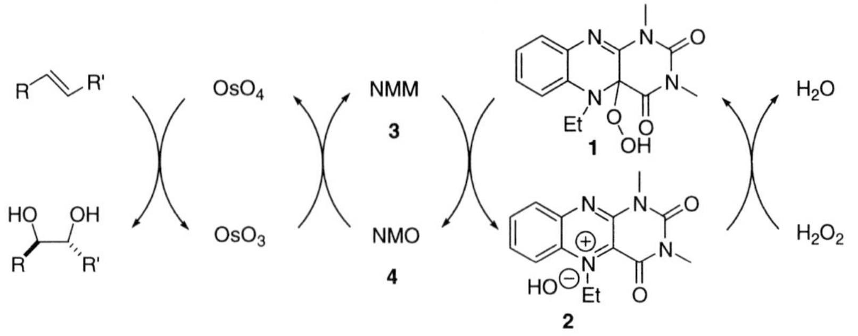

Scheme 2 Multicomponent reoxidation system for AD with hydrogen peroxide (achiral cycle shown). NMO = *N*-methyl morpholine *N*-oxide.

A major breakthrough was achieved by Beller, who reported AD reaction with molecular oxygen as the terminal oxidant [11–13]. Importantly, both oxygen atoms could be transferred, making this process one of the most atom-economic protocols known to date, since it is free of by-products from the terminal oxidant. An oxygen atmosphere at ambient pressure is sufficient, and the only modification with regard to the Sharpless conditions consists of an increase in pH, for which an optimum value of 10.4 was determined [14]. Furthermore, the reaction displays broad functional group tolerance and is compatible with the enantioselective version employing cinchona alkaloid ligands. However, these AD reactions lead to enantioselectivities that are lower than the ones from the classical Sharpless system. On the other hand, *trans*-stilbene, which is a superb substrate for Sharpless AD, gives very much poorer results under aerobic oxidation conditions and suffers overoxidative cleavage of the internal C-C bond, which results in a very selective formation of benzaldehyde [15, 16]. However, this problem could be overcome by changing the solvent system to water/isobutyl methylketone. Other interesting reoxidation systems for osmiumtrioxide include selenoxides, which result in equilibrium with selenides and Os(VIII). Enantioselectivity can be induced in these reactions when the usual cinchona alkaloids are employed [17].

2.5.4.2.2 **Kinetic Resolutions**

Because of its inherent high selectivity, AD has continously been investigated for its potential in kinetic resolution procedures [18]. While there have been various attempts to develop these reactions [19], the success rate still remains very low. This might in part be a result of the high stereochemical dominance of the cinchona alkaloid ligands that override stereochemical information in the substrate. Still, the most efficient kinetic resolution is the one of C-76, a chiral carbon allotrope, which had been achieved by stoichiometric asymmetric dihydroxylation [20]. Recently, a stoichiometric AD kinetic resolution has been reported for a complex of OsO_4 and a chiral diamine [21, 22]. Regarding catalytic conditions, the most successful example to date consists of an AD-derived kinetic resolution of atropisomeric amides, which has been claimed to proceed with selectivity factors of up to 26 [23].

2.5.4.2.3 Mechanistic Discussion

The fundamental question concerning the course of asymmetric dihydroxylation has remained unanswered. At present, neither the [2+2] nor the [3+2] mechanism (see Chapter 2.5.2 in the first edition) can be ruled out completely. However, data in favor of the latter mechanism was obtained from experimental kinetic isotope effects (KIE) and was in agreement with transition structure/KIE calculations [24]. These results predict a highly symmetrical transition state and a [3+2] cycloaddition as the rate-determining step. Additional experimental results all favor such a single-step concerted mechanism [25, 26]. Within the mechanistic context, work on the mechanistic elucidation of stereoselective AD reactions has been extended. Both experimental [27, 28] and theoretical [29] investigations into substrate binding have appeared, and Corey has employed his AD transition-state model for the design of a novel cinchona alkaloid ligand that recognizes the terminal olefin in polyisoprenoid substrates [30].

2.5.4.2.4 Directed Dihydroxylation Reactions

The application of suitable coordination sites within a given substrate in order to direct the incoming reagent in a regio- or stereoselective manner is a widely known concept in preparative organic synthesis [31]. However, it had only been applied to AD reactions to a lesser extent. Regarding the dihydroxylation of cyclic allylic alcohols, Kishi had reported that the reaction proceeds with high anti-selectivity [27, 32]. Such a stereochemical outcome can easily be achieved from dihydroxylation under the common Upjohn conditions. Reaction sequences toward the opposite all-*syn* stereochemistry were investigated by Donohoe and take advantage of hydrogen bridges between the osmium tetroxide reagent and an allylic heteroatom [33, 34]. To this end, modification of the OsO_4 reagent was necessary, and tmeda was found to be the most efficient additive. It is presumed that coordination of this bidentate donor to osmium drastically enhances electron density and thereby renders the oxo groups more prone to a hydrogen-bonding scenario that exercises stereochemical control in favor of the desired *syn*-addition (Model B, Scheme 3). Hydroxyl groups of allylic alcohols led to a significant preference for *syn*- over anti-dihydroxylation, and the more elaborate hydrogen donor trichloroacetamide was found to be the functional group of choice. Thus, treatment of an allylic trichloroacetamide such as **5** with an equimolar amount of osmium tetroxide/tmeda gave rise to a product **7a** with more than 25:1 diastereomeric ratio, and the structure of the resulting chelated osmate ester **6** was proven by X-ray analysis. Because of the chelating stability of the tmeda ligand, these adducts do not undergo simple hydrolytic cleavage, and osmium removal had to be carried out with either HCl in methanol, aqueous Na_2SO_3, or ethylenediamine.

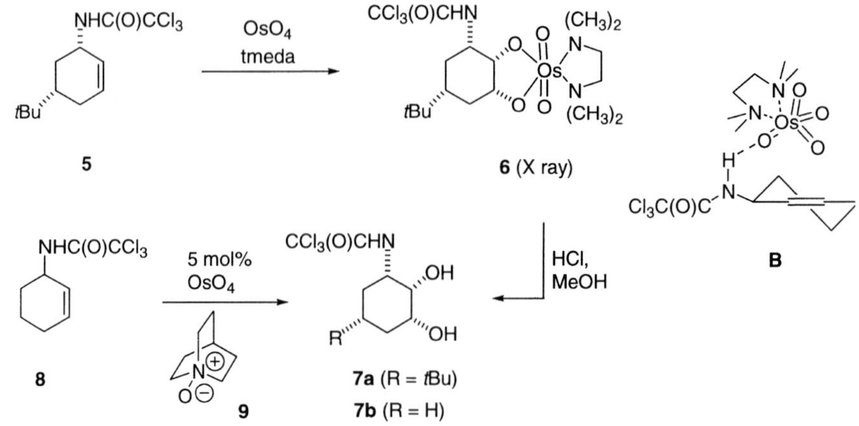

Scheme 3 Intramolecular dihydroxylation through hydrogen bonding. tmeda = *N,N,N′,N′*-tetramethyl ethylenediamine.

Since the Os/tmeda moiety is removed under conditions that are incompatible with a catalytic reaction, a mono-amine was necessary to render the process catalytic. Here, the well-known quinine moiety worked best when its *N*-oxide monohydrate **9** was employed as both the terminal oxidant and as precursor to the actual ligand for ligation to osmium. Moderate to high diastereomeric ratios could be obtained for these reactions. For example, **7b** together with its *trans*-isomer are formed from dihydroxylation of **8** in a ratio of 82:18.

2.5.4.2.5 Secondary-Cycle Catalysis

It was in the early days of Os-mediated dihydroxylation that Criegee isolated both mono- and bisglycolate complexes of Os(VI), thereby indicating that the synthesis of two product molecules from one molecule of OsO_4 upon reoxidation is the thermodynamically preferred reaction [35]. In the area of catalytic dihydroxylation, Sharpless coined the term *secondary cycle* for this reaction sequence [36]. In the original AD reaction, which required the phenomenon of chiral ligand acceleration [37], this was an unwanted reaction path, since it was shown that the chiral ligand does not participate in this catalytic diol formation, thus leading to products with very low or no enantiomeric excess. This is the direct consequence of slow diol hydrolysis of the initially formed osma(VI) glycol ester **10**, which under catalytic conditions undergoes fast reoxidation to **11** and thereby enables a second dihydroxylation to furnish the bisglycolate **12**, which represents the resting catalyst form.

However, recent results from the area of catalytic AA reaction (see Chapter 2.6.3.2.5) suggested that a certain class of olefins bearing polar functionalities such as amides and carboxylates represent privileged substrates in that their oxidation proceeds almost exclusively within the second cycle [38]. Apparently, the rate-limiting step, the hydrolysis of the bisglycoxylate **12**, is dramatically enhanced by

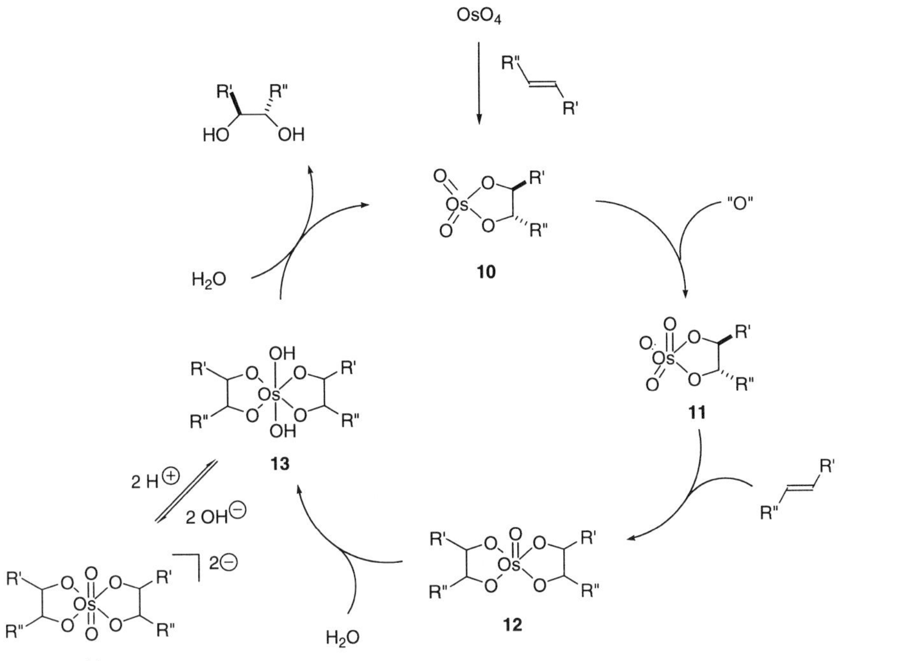

Scheme 4 Second-cycle dihydroxylation.

the presence of these polar functional groups, and, of these, carboxylic acids have been the most successful ones. The process described so far is initiated by a first reaction of osmium tetroxide and the substrate itself to give **10**. Nevertheless, since the reaction of preformed diols with osmium(VI) salts had been reported to form monoglycolates as well [35], initiation of the catalytic cycle by addition of external ligands to the common potassium osmate salt should also be possible. A recent screening by Sharpless revealed that a variety of acids had a beneficial impact on the catalysis and that the optimum pH range is 4–6 [39]. Several additional advantages are believed to result at this pH range: 4-methyl morpholine formed from the terminal oxidant NMO is neutralized, and the formation of a catalytically inert dioxosmate dianion (**14**, formed from hydration of **12** and deprotonation of the resulting compound **13**) is prevented. Among the many acids that were screened, citric acid gave an exceptionally stable catalyst, most probably because of the formation of a chelated osmium(VIII) species **C** (Scheme 5), which prevents catalyst decomposition from disproportionation. Moreover, contamination of the products with residual Os is essentially avoided because of this chelation.

Chiral non-racemic reaction sequences were developed for replacing the achiral diol or hydroxyl carboxylate with a chiral ligand such as tartaric acid [40]. While this chiral pool derivative proved suitable, albeit at amounts of about 25 mol%, related *N*-tosylated α,β-hydroxy amino acids were determined to be the ligands of

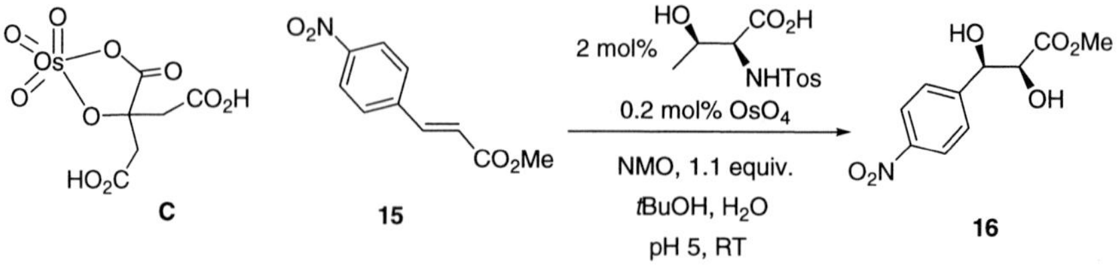

Scheme 5 Second-cycle AD reaction.

choice. Not incidentally, the corresponding ester precursors are the products from first cycle AA reactions. Thus, dihydroxylation of 4-nitro-cinnamic ester **15** in the presence of only 0.2 mol% osmium tetroxide gives the diol **16** with 70% *ee* [40].

2.5.4.2.6 **Polymer Support**

In view of the high cost of both osmium compounds and chiral ligands, extensive work has been undertaken to replace them by reusable derivatives. Within this approach, a variety of soluble and insoluble ligands on polymer support were developed [41]. However, these reaction modifications could not overcome the drawback of significant osmium leaching. This is the consequence of the original homogeneous procedure that had been developed for monomeric unbound cinchona alkaloid ligands and makes use of a significant rate enhancement for the chiral ligand-complexed osmium tetroxide compared with the uncomplexed one (*ligand accelerated catalysis*) [37]. Because of this inherently reversible complexation, osmium recovery by complexation to the polymer-supported ligand must be virtually impossible. A catalytic asymmetric dihydroxylation with fully reusable catalyst has been devised by Kobayashi [42]. His approach relied on microencapsulated osmium tetroxide that could be recovered by filtration techniques, while the chiral ligand was reisolated by acid/base extraction. This system can be used for several runs without loss in yield or *ee*. For the AD of (*E*)-methylstyrene, it was possible to scale up this procedure to a 100 mmol reaction to give 91% yield and 89% *ee* at 1 mol% Os loading [42a].

In an alternative approach, osmium tetroxide was immobilized on ion exchangers, which allows for continuous dihydroxylation reactions, and the strong binding of the Os to the resin ensures that equimolar amounts of chiral ligand are sufficient to obtain the maximum enantioselectivities. However, the amount of osmium was still 1 mol%, a much higher amount than in homogeneous reactions [43]. Finally, efficient recyclability and reuse of Os has been achieved by changing the solvent to an ionic liquid [44, 45]. In this way, the volatility of osmiumtetroxide is suppressed and recovery does not constitute any problem. The yields have been proven to vary only slightly within several consecutive runs, and addition of DMAP was found to greatly enhance the catalyst stability for one of the systems [44].

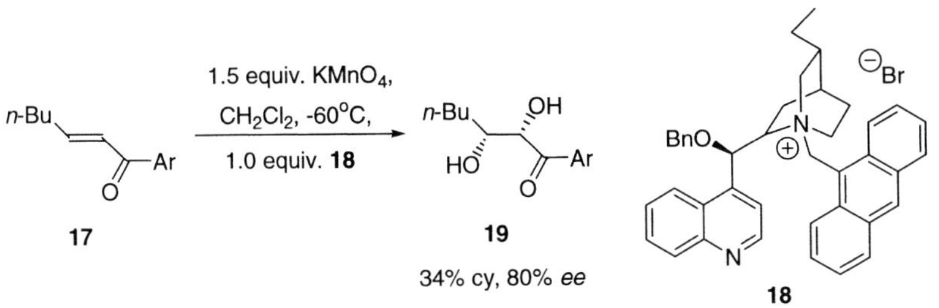

Scheme 6 Asymmetric stoichiometric dihydroxylation with $KMnO_4$. Ar=*p*-$(CH_3O)C_6H_4$.

2.5.4.3
Alternative Oxidation Systems

Finally, in view of the still high cost of Os metal, the search for alternative metals continues. For example, the interesting ruthenium tetroxide-catalyzed dihydroxylation with $NaIO_4$ as terminal oxidant [46] has been converted into a stereoselective diol synthesis employing *α,β*-unsaturated carboxamides containing Oppolzer sultams as chiral auxiliaries leading to diastereomeric excesses of up to 80% [47, 48]. Also, iron complexes have emerged as promising catalyst systems for the dihydroxylation of unfunctionalized olefins in the presence of hydrogen peroxide as oxidant [49].

An interesting dihydroxylation of enones such as **17** in the presence of equimolar amounts of a chiral phase transfer reagent **18** and permanganate as oxidant has been reported to proceed with moderate enantioselectivity (Scheme 6). At present, the substrate scope appears rather limited since neutral olefins give inferior results. Clearly, despite all attempts to develop other systems, the Sharpless catalytic AD reaction in homogeneous phase represents the method of choice for enantioselective catalytic diol synthesis.

Acknowledgement
The continuous financial support provided by the Fonds der Chemischen Industrie is gratefully acknowledged.

References

1 For an in-depth discussion of this system, see the preceding chapter.
2 (a) H.C. Kolb, M.S. VanNieuwenhze, K.B. Sharpless, *Chem. Rev.* **1994**, *94*, 2483; (b) C. Bolm, J.P. Hildebrand, K. Muiz, in *Catalytic Asymmetric Synthesis* (Ed.: I. Ojima), Wiley-VCH, Weinheim **2000**, p. 299; (c) I.E. Marko, J.S. Svendsen in *Comprehensive Asymmetric Catalysis II* (Eds: E.N. Jacobsen, A. Pfaltz, H. Yamamoto), Springer, Berlin **1999**, p. 713; (d) H. Becker, K.B. Sharpless in *Asymmetric Oxidation Reactions: A Practical Approach* (Ed.: T. Katsuki), Oxford University Press, London **2001**, p. 81; (e) M. Beller, K.B. Sharpless in *Applied Homogeneous Catalysis* (Eds.: B. Cornils,

W. A. Herrmann), VCH, Weinheim **1996**, p. 1009.

3 www.nobel.se/chemistry/laureates/2001/index.html

4 (a) A. Gypser, D. Michel, D. S. Nirschl, K. B. Sharpless, *J. Org. Chem.* **1998**, *63*, 7322; (b) earlier work: H. Sakurai, N. Iwasawa, K. Narasaka, *Bull. Chem. Soc. Jpn.* **1996**, *69*, 2585.

5 K. Bergstad, S. Y. Jonsson, J.-E. Bäckvall, *J. Am. Chem. Soc.* **1999**, *121*, 10424.

6 S. Y. Jonsson, K. Färnegårdh, J.-E. Bäckvall, *J. Am. Chem. Soc.* **2001**, *123*, 1365.

7 For a discussion: (a) K. Bergstad, J.-E. Bäckvall, *J. Org. Chem.* **1999**, *63*, 6650. (b) A. B. E. Minidis, J.-E. Bäckvall, *Chem. Eur. J.* **2001**, *7*, 297.

8 S. Y. Jonsson, H. Adolfsson, J.-E. Bäckvall, *Org. Lett.* **2001**, *3*, 3463.

9 K. Bergstad, J. J. N. Piet, J.-E. Bäckvall, *J. Org. Chem.* **1999**, *64*, 2545.

10 A. H. Ell, S. Y. Jonsson, A. Borje, H. Adolfsson, J.-E. Bäckvall, *Tetrahedron Lett.* **2001**, *42*, 2569.

11 Short review: T. Wirth, *Angew. Chem. Int. Engl.* **2000**, *39*, 334.

12 C. Döbler, G. Mehltretter, M. Beller, *Angew. Chem. Int. Ed.* **1999**, *38*, 3026.

13 C. Döbler, G. Mehltretter, U. Sundermeier, M. Beller, *J. Am. Chem. Soc.* **2000**, *122*, 10289.

14 There is evidence for pH dependence in Sharpless AD reactions: G. Mehltretter, C. Döbler, U. Sundermeier, M. Beller, *Tetrahedron Lett.* **2000**, *41*, 8083.

15 C. Döbler, G. Mehltretter, U. Sundermeier, M. Beller, *J. Organomet. Chem.* **2001**, *621*, 70.

16 For a recent osmium-catalyzed ozonolysis: B. R. Travis, R. S. Narayan, B. Borhan, *J. Am. Chem. Soc.* **2002**, *124*, 3824.

17 (a) A. Krief, C. Colaux-Castillo, *Pure Appl. Chem.* **2002**, *74*, 107; (b) A. Krief, A. Destree, V. Durisotti, N. Moreau, C. Smal, C. Colaux-Castillo, *Chem. Commun.* **2001**, 558; (c) A. Krief, C. Castillo-Colaux, *Tetrahedron Lett.* **1999**, *40*, 4189; (d) A. Krief, C. Castillo-Colaux, *Synlett* **2001**, 501.

18 J. M. Keith, J. F. Larrow, E. N. Jacobsen, *Adv. Synth. Catal.* **2001**, *343*, 5.

19 For example: (a) H. S. Christie, D. P. G. Hamon, K. L. Tuck, *Chem. Commun.* **1999**, 1989; (b) D. P. G. Hamon, K. L. Tuck, H. S. Christie, *Tetrahedron* **2001**, *57*, 9499; (c) T. Yokomatsu, T. Yamagishi, T. Sada, K. Suemune, S. Shibuya, *Tetrahedron* **1998**, *54*, 781.

20 J. M. Hawkins, A. Meyer, *Science* **1993**, *260*, 1918.

21 R. Hodgson, T. Majid, A. Nelson, *J. Chem. Soc., Perkin Trans. 1* **2002**, 1631.

22 For related complexes in asymmetric dihydroxylation, see: (a) E. J. Corey, S. Sarshar, M. D. Azimioara, R. C. Newbold, M. C. Noe, *J. Am. Chem. Soc.* **1996**, *118*, 7851; (b) K. Tomioka, M. Nakajima, K. Koga, *J. Am. Chem. Soc.* **1987**, *109*, 6213; (c) E. J. Corey, P. DaSilva Jardine, S. Virgil, P.-W. Yuen, R. D. Connell, *J. Am. Chem. Soc.* **1989**, *111*, 9243; (d) S. Hanessian, P. Meffre, M. Girard, S. Beaudoin, J.-Y. Sancéau, Y. Bennani, *J. Org. Chem.* **1993**, *58*, 1991.

23 R. Rios, C. Jimeno, P. J. Carroll, P. J. Walsh, *J. Am. Chem. Soc.* **2002**, *124*, 10272.

24 A. J. DelMonte, J. Haller, K. N. Houk, K. B. Sharpless, D. A. Dingleton, T. Strassner, A. A. Thomas, *J. Am. Chem. Soc.* **1997**, *119*, 9907.

25 M. Torrent, M. Sola, G. Frenking, *Chem. Rev.* **2000**, *100*, 439.

26 Selected work: (a) P. O. Norrby, T. Rasmussen, J. Haller, T. Strassner, K. N. Houk, *J. Am. Chem. Soc.* **1999**, *121*, 10186; (b) G. Ujaque, F. Maseras, A. Lledos, *J. Am. Chem. Soc.* **1999**, *121*, 1317; (c) P. Gisdakis, N. Rosch, *J. Am. Chem. Soc.* **2001**, *123*, 697.

27 Review: J. K. Cha, N.-S. Kim, *Chem. Rev.* **1995**, *95*, 1761.

28 A. Bayer, J. S. Svendsen, *Eur. J. Org. Chem.* **2001**, 1769.

29 N. Moitessier, C. Henry, C. Len, Y. Chapleur, *J. Org. Chem.* **2002**, *67*, 7275.

30 E. J. Corey, J. H. Zhang, *Org. Lett.* **2001**, *3*, 3211.

31 A. H. Hoveyda, D. A. Evans, G. C. Fu, *Chem. Rev.* **1993**, *93*, 1307.

32 (a) J. K. Cha, W. J. Christ, Y. Kishi, *Tetrahedron* **1984**, *40*, 2247 and literature cited; (b) see also [27] and [34].

33 T.J. Donohoe, K. Blades, P.R. Moore, M.J. Waring, J.J.G. Winter, M. Helliwell, N.J. Newcombe, G. Stemp, *J. Org. Chem.* **2002**, *67*, 7946 and literature cited.

34 T.J. Donohoe, *Synlett* **2002**, 1223.

35 (a) R. Criegee, *Liebigs Ann. Chem.* **1936**, *522*, 75; (b) R. Criegee, B. Marchand, H. Wannowius, *Liebigs Ann. Chem.* **1936**, *550*, 99.

36 J.S.M. Wai, I. Markó, J.S. Svendsen, M.G. Finn, E.N. Jacobsen, K.B. Sharpless, *J. Am. Chem. Soc.* **1989**, *111*, 1123.

37 D. J. Berrisford, C. Bolm, K.B. Sharpless, *Angew. Chem., Int. Ed. Engl.* **1995**, *34*, 1059.

38 (a) A.E. Rubin, K.B. Sharpless, *Angew. Chem. Int. Ed. Engl.* **1997**, *36*, 2637; (b) W. Pringle, K.B. Sharpless, *Tetrahedron Lett.* **1999**, *40*, 5150; (c) V.V. Fokin, K.B. Sharpless, *Angew. Chem. Int. Ed.* **2001**, *40*, 3455; see also: (d) H.C. Kolb, M.G. Finn, K.B. Sharpless, *Angew. Chem. Int. Ed.* **2001**, *40*, 2004.

39 P. Dupau, R. Epple, A.A. Thomas, V.V. Fokin, K.B. Sharpless, *Adv. Synth. Catal.* **2002**, *344*, 421.

40 M.A. Andersson, R. Epple, V.V. Fokin, K.B. Sharpless, *Angew. Chem. Int. Ed.* **2002**, *41*, 2004.

41 Reviews: (a) C. Bolm, A. Gerlach, *Eur. J. Org. Chem.* **1998**, 21; (b) C.E. Song, S.-g. Lee, *Chem. Rev.* **2002**, *102*, 3495; (c) P. Salvadori, D. Pini, A. Petri, *Synlett* **1999**, 1181; (d) D.J. Gravert, K.D. Janda, *Chem. Rev.* **1997**, *97*, 489.

42 (a) S. Kobayashi, M. Endo, S. Nagayama, *J. Am. Chem. Soc.* **1999**, *121*, 11229; (b) S. Nagayama, M. Endo, S. Kobayashi, *J. Org. Chem.* **1998**, *63*, 6094; (c) S. Kobayashi, T. Ishida, R. Akiyama, *Org. Lett.* **2001**, *3*, 2649; (d) see also: S.V. Ley, C. Ramarao, A.-L. Lee, N. Østergaard, S.C. Smith, I.M. Shirley, *Org. Lett.* **2003**, *5*, 185.

43 (a) B.M. Choundary, N.S. Chowdari, M.L. Kantam, K.V. Raghavan, *J. Am. Chem. Soc.* **2001**, *123*, 9220. (b) B.M. Choudary, N.S. Chowdari, K. Jyothi, M.L. Kantam, *J. Am. Chem. Soc.* **2002**, *124*, 5341; (c) see also: J.W. Yang, H. Han, E.J. Roh, S.-g. Lee, C.E. Song, *Org. Lett.* **2002**, *4*, 4685.

44 Q. Yao, *Org. Lett.* **2002**, *4*, 2197.

45 R. Yanada, Y. Takemoto, *Tetrahedron Lett.* **2002**, *43*, 6849.

46 (a) T.K.M. Shing, V.W.-F. Tai, E.K.W. Tam, *Angew. Chem. Int. Ed. Engl.* **1994**, *33*, 2312; (b) T.K.M. Shing, E.K.W. Tam, W.-F. Chung, I.H.F. Chung, Q. Jiang, *Chem. Eur. J.* **1996**, *2*, 50; (c) T.K.M. Sing, E.K.W. Tam, *Tetrahedron Lett.* **1999**, *40*, 2179.

47 A.W.M. Lee, W.H. Chan, W.H. Yuen, P.F. Xia, W.Y. Wong, *Tetrahedron Asymmetry* **1999**, *10*, 1421.

48 For the corresponding substrates in osmium-catalyzed reactions, see: (a) W. Oppolzer, J.P. Barras, *Helv. Chim. Acta* **1987**, *70*, 1666; (b) L. Colombo, C. Gennari, G. Poli, C. Scolastico, *Tetrahedron Lett.* **1985**, *26*, 5459; (c) S. Hatakeyama, Y. Matsui, M. Suzuki, K. Sakurai, S. Takano, *Tetrahedron Lett.* **1985**, *26*, 6485.

49 (a) M. Costas, A.K. Tipton, K. Chen, D.-H. Jo, L. Que, Jr., *J. Am. Chem. Soc.* **2001**, *123*, 6722; (b) K. Chen, L. Que, Jr., *Angew. Chem. Int. Ed.* **1999**, *38*, 2227; (c) K. Chen, M. Costas, J. Kim, A.K. Tipton, L. Que, Jr. *J. Am. Chem. Soc.* **2002**, *123*, 3026; (d) J.Y. Ryu, J. Kim, M. Costas, K. Chen, W. Nam, L. Que, Jr., *Chem. Commun.* **2002**, 1288.

50 (a) R.A. Bhunnoo, Y. Hu, D.I. Lainé, R.C.D. Brown, *Angew. Chem. Int. Ed.* **2002**, *41*, 3479; (b) for related oxidative cyclization: R.C.D. Brown, J.F. Kelly, *Angew. Chem. Int. Ed.* **2001**, *40*, 4496.

2.6 Asymmetric Aminohydroxylation

Hartmuth C. Kolb and K. Barry Sharpless

2.6.1 Introduction

A wealth of biomolecules and biologically active compounds formally derive from 1,2-hydroxyamines. The great abundance of the 1,2-hydroxyamine substructure calls for good methods to construct it. Certainly, one of the most efficient ways to achieve this goal is to utilize the masked 1,2-functional group relationship in olefins. The latter are arguably the most useful starting materials for the synthetic chemist, since they are readily available and the double bond is set up for 1,2-functionalization by face-selective oxidation [1, 2]. While powerful methods for the enantioselective addition of *identical* heteroatoms to double bonds exist, the development of methods for the delivery of *two different* heteroatoms, an oxygen atom and a nitrogen atom, is hampered by the presence of another challenging problem – that of *regioselectivity* (Scheme 1).

Recent advances in the field of d^0 transition-metal-catalyzed olefin oxidation have led to a considerable improvement of the well-known racemic variant of the osmium-catalyzed aminohydroxylation reaction [3, 4], a close relative of the osmium-catalyzed dihydroxylation reaction [1]. The metal catalyzes the suprafacial addition

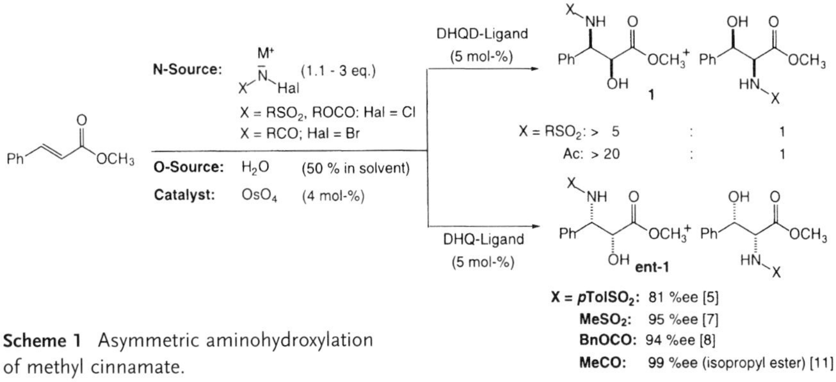

N-Source: M^+ $X\bar{N}Hal$ (1.1 - 3 eq.)
X = RSO_2, ROCO: Hal = Cl
X = RCO; Hal = Br
O-Source: H_2O (50 % in solvent)
Catalyst: OsO_4 (4 mol-%)

DHQD-Ligand (5 mol-%) → **1** + regioisomer
DHQ-Ligand (5 mol-%) → **ent-1** + regioisomer

X = RSO_2: > 5 : 1
Ac: > 20 : 1

X = *p*TolSO₂: 81 %ee [5]
MeSO₂: 95 %ee [7]
BnOCO: 94 %ee [8]
MeCO: 99 %ee (isopropyl ester) [11]

Scheme 1 Asymmetric aminohydroxylation of methyl cinnamate.

Transition Metals for Organic Synthesis, Vol. 2, 2nd Edition.
Edited by M. Beller and C. Bolm

ISBN: 3-527-30613-7

of a nitrogen atom, coming from an N-acyl or N-sulfonyl chloramine salt, and an oxygen atom, coming from water, to the double bond [5–11] (Scheme 1).

Three different selectivity issues have to be addressed in the development of the asymmetric aminohydroxylation reaction (AA): enantioselectivity, regioselectivity, and chemoselectivity. The latter concerns the formation of diol as the main side product of the AA reaction, since both paths are catalyzed by d^0 osmium complexes (cf. AD reaction [1]). Enantioselectivity can be induced using the cinchona alkaloid ligands known from the asymmetric dihydroxylation (AD) system [1]. Interestingly, these ligands give the same sense of facial selectivity in both asymmetric processes (Scheme 2), suggesting that the factors governing the selectivity are very similar [12]. Thus, the enantiofacial selectivity can be predicted using the mnemonic device from the AD system.

The cinchona ligands are not only responsible for enantioselectivity, they also improve both chemoselectivity and regioselectivity. Thus, the selectivity for the benzylic amine **1** in the Chloramine-T based AA of methyl cinnamate (Scheme 1) increases from 2:1 to >5:1 when the cinchona ligand is employed [5].

The catalytically active species in the reaction most likely is an imidotrioxo osmium(VIII) complex **2** which is formed *in situ* from the osmium reagent and the stoichiometric nitrogen source, i.e. the chloramine (Scheme 3). Experiments under stoichiometric conditions have shown that imidotrioxo osmium(VIII) com-

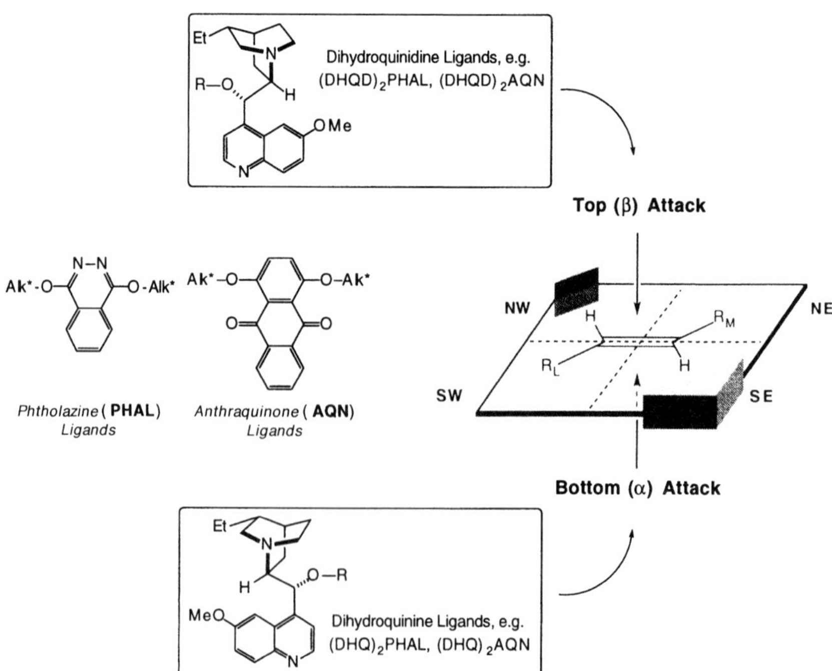

Scheme 2 Mnemonic device for the prediction of the face selectivity.

plexes transfer both the nitrogen atom and one of the oxygen atoms onto the substrate [13]. The major regioisomer normally has the nitrogen atom placed distal to the most electron withdrawing group of the substrate. A stepwise mechanism [5, 14], proceeding via the osmaazetidine **3**, can readily explain this observation, since the osmium atom is the most electrophilic center of the reagent. This mechanism is analogous to that proposed for the AD reaction [1, 14].

Even though the racemic reaction has been known since the 1970s [3, 4], only recent advances and mechanistic insights have made the asymmetric version possible. Interestingly, the success of this new system crucially depends on one inconspicuous reaction parameter – water [5, 8]. Earlier aminohydroxylation protocols had utilized only a few equivalents of water, leading to poor catalytic turnover. Heavy metal salts, e.g. silver(I) or mercury(II) salts, were added to enhance the reactivity of the chloramine in order to establish a catalytic process [3, 4]. Chang and Li discovered that the catalytic turnover increases considerably upon increasing the amount of water in the system and best results are obtained in solvent systems containing 50% water. The new conditions obviate the need for heavy metals. Water probably accelerates the turnover limiting step, the hydrolysis of the osmium(VI) hydroxyamine complex, and this example again demonstrates that *all the steps* of the catalytic cycle have to be considered when optimizing a catalytic process. A more general conclusion is that catalytic processes are exceedingly sensitive to the reaction parameters and many potentially powerful processes may have been overlooked, just because of one missing step which in turn was inoperable due to one unoptimized reaction parameter.

In analogy to the AD reaction, two catalytic cycles may be operating in the aminohydroxylation reaction [7] (Scheme 4). The primary cycle involves the chiral ligand, allowing it to exert its beneficial influence on enantio-, regio- and chemoselectivity. The competing secondary cycle is independent of the ligand and this cycle should, therefore, be avoided by careful optimization of the conditions. The reaction of trioxoimidoosmium(VIII) complex **2** with the olefin leads to the osmium(VI) azaglycolate complex **4** (step a^1), which is oxidized by the chloramine to the dioxoimidoosmium(VIII) azaglycolate complex **5** (step o). As in the AD reaction, this osmium(VIII) complex has two options: the desired path involves its hydrolysis (h^1) and thus re-entry into the primary cycle. The undesirable secondary cycle is entered by reaction of **5** with a second molecule of olefin (a^2), leading to the bisazaglycolate **6**. How may the system be influenced in favor of the desired primary cycle? First, the amount of water present in the system influences the rate of hydrolysis – the turnover limiting step. A large water content not only

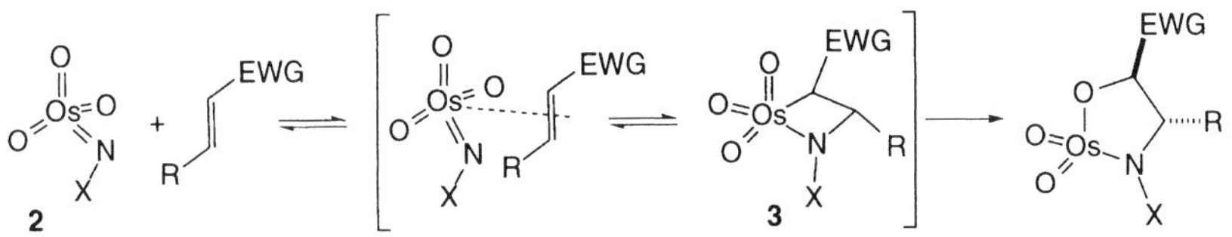

Scheme 3 Proposed stepwise osmaazetidine mechanism [14].

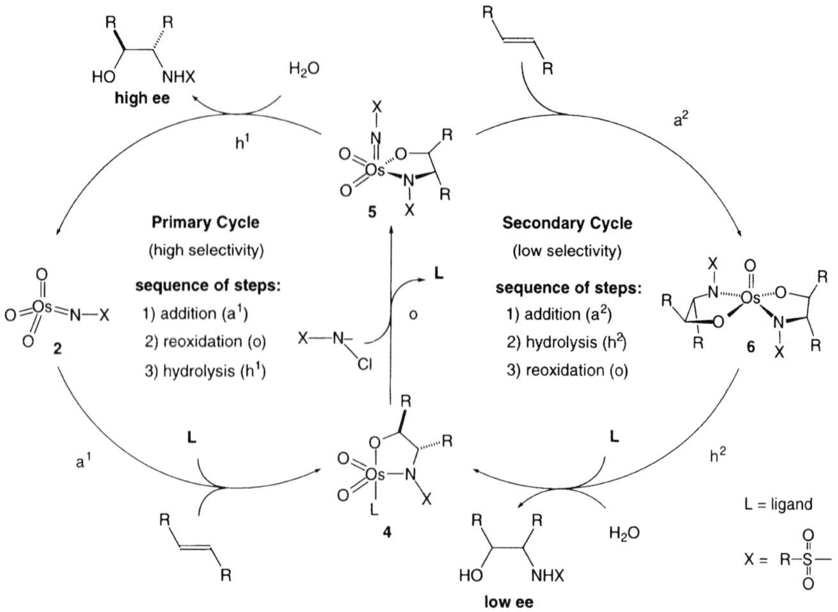

Scheme 4 The two catalytic cycles proposed for the AA reaction [7].

increases the overall turnover, but it also favors the primary over the secondary cycle, since the osmium(VIII) azaglycolate **5** is hydrolyzed (step h^1) at a rate which is fast enough to prevent its reaction with a second molecule of olefin (step a^2). Another factor is the nature of group X of the stoichiometric nitrogen source. Big and hydrophobic groups retard hydrolysis and thus have a deleterious effect on the reaction. This is in accord with the experimental observation that smaller groups lead to better enantio-, regio- and chemoselectivity [7].

2.6.2
Process Optimization of the Asymmetric Aminohydroxylation Reaction

2.6.2.1
General Observations – Comparison of the Three Variants of the AA Reaction

The nitrogen atom transferred in the AA reaction always carries a substituent X. The earlier racemic procedures provided hydroxysulfonamides [3] and hydroxycarbamates [4]. It proved possible to further extend the scope to hydroxyacetamides [11] and to develop asymmetric variants of all three systems. Thus, three types of enantiomerically enriched *N*-protected hydroxyamines can be prepared (Tab. 1), depending on the choice of the stoichiometric *N*-source: sulfonamides [5–7, 9], carbamates [8–10] and carboxamides [11].

Tab. 1 The three variants of the asymmetric aminohydroxylation reaction using cinnamates as substrates and the PHAL class of ligands

	Reagent	***Major product***		
Sulfonamide variant [5–7, 9]	O O S R NCl Na	RSO_2NH O Ph O*i*Pr OH	R = Me: 65% yield 94% *ee*	R = *p*-Tol: 51–66% yield 81–89% *ee*
Carbamate variant [8–10]	O R O NCl Na	O R O NH O Ph OCH_3 OH	R = Bu: 65% yield 94% *ee*	R = Et: 78% yield 99% *ee*
Amide variant [11]	O R NLi Br	O R NH O Ph O*i*Pr OH	81% yield 99% *ee*	

The outcome of the asymmetric aminohydroxylation process, with respect to yield, enantio- and regioselectivity, is greatly influenced by a number of reaction parameters, e.g. the type of starting material, the ligand, the solvent, the nature of the stoichiometric nitrogen source as well as the size of its substituent.

The type of nitrogen source

Even though the sulfonamide variant was the first to be developed into an asymmetric reaction [5] it has since been superseded by the carbamate and amide versions in terms of substrate scope, yield, and selectivity. The latter two protocols show the desired phenomenon of ligand acceleration, while the Chloramine-T procedure, leading to toluenesulfonamides, is actually inhibited by the cinchona alkaloid ligand in some cases (ligand deceleration). As a general rule, the smaller the nitrogen substituent the better the results (cf. Tab. 1). This holds true especially for the sulfonamide variant, which gives much better turnover numbers, yields, enantio- and regioselectivities with Chloramine-M (leading to *β*-hydroxy methanesulfonamides) than with Chloramine-T [7].

Availability of the stoichiometric nitrogen source

Some stoichiometric nitrogen sources for the AA reaction, e.g. Chloramine-T (*p*-tolSO_2NClNa) or *N*-bromoacetamide, are commercially available. However, the chloramine reagent can also be readily prepared and used *in situ* by treating the appropriate sulfonamide or urethane with *tert*-butylhypochlorite and sodium hydroxide in water (cf. Sect. 2.6.2.3 for a representative procedure) [8–10]. The resultant aqueous solution of the chloramine salt is then diluted with the organic co-solvent (*n*-propanol, *tert*-butanol or MeCN) and used directly for the aminohydroxylation reaction.

The solvent

Solvent systems containing 50% water are employed in the AA reaction to ensure a high catalytic turnover by enhancing the rate of hydrolysis. Typically, alcoholic co-solvents, such as *tert*-butanol or *n*-propanol, give superior results with respect to enantio- and regioselectivity compared to acetonitrile [7, 9, 10]. However, the latter solvent leads to slightly higher turnover numbers [7] and sometimes to higher chemical yields, due to the formation of less diol side product [10]. The acetonitrile/water (1:1) system is, therefore, well suited for the Chloramine-T variant of the reaction [5, 9], even though the greater solubility of sulfonamide byproducts may complicate product isolation. Alcoholic solvent systems offer advantages in terms of work-up, since the products are quite often insoluble, allowing them to be isolated simply by filtration of the reaction mixture (solution-to-solid AA) [5, 6, 8, 10]. While the best solvent system for the carbamate-based AA reaction is a 1:1 mixture of *n*-propanol/H_2O [8, 9], the *tert*-butyl carbamate-based AA reaction should be performed in a solvent containing larger amounts of the alcohol (2:1 *n*-propanol/H_2O) in order to suppress diol formation [10].

Reagent amounts

Good turnovers are normally obtained using catalytic amounts of osmium (4 mol% $K_2OsO_2(OH)_4$) and cinchona alkaloid ligand (5 mol%) and an excess of the nitrogen source (3 equivalents in the sulfonamide and carbamate variants, 1.1 equivalents in the amide variant). However, these amounts may be reduced for reactive substrates and cinnamates may be successfully aminohydroxylated with as little as 1.5 mol% osmate and 1 mol% ligand [11].

Scope

The scope of the AA reaction depends considerably on the type of stoichiometric nitrogen source. The carbamate and acetamide variants have a much broader substrate scope than the sulfonamide version. The latter gives good results mainly with disubstituted olefins, e.g. cinnamates [5–7], while monosubstituted olefins, e.g. styrene, lead to poor chemical yields as well as low regioselectivity (2:1 mixture) and enantioselectivity (50–70% *ee*) [8]. In contrast, styrenes are among the best substrates for the carbamate [8, 10] and amide versions [11], allowing the products to be isolated in good yield (>60%) and enantioselectivity (>90% *ee*). In addition, the regioselectivity of the carbamate- and acetamide-based AA of styrenes can be controlled by the choice of reaction conditions (*vide infra*).

The nitrogen atom is added preferentially to the center distal to the most electron withdrawing group [5] (cf. Schemes 1 and 3). With styrenes, the preferred regioisomer normally is the benzylic amine **8** [8, 10]. However, a most welcome feature of the reaction with these aromatic terminal olefins is the control over the regioselectivity which one can exert by choosing the appropriate solvent, ligand and nitrogen source (Tab. 2). The following rules for controlling the regioselectivity in the AA of styrene-like substrates apply for both the carbamate [10] and amide [11] variants of the reaction:

Solvent influence: *n*-Propanol/H_2O (1:1) favors the benzylic amine **8**, MeCN/H_2O (1:1) favors the benzylic alcohol **7**.
Ligand influence: PHAL ligands favor the benzylic amine **8**, AQN ligands favor the benzylic alcohol **7**.
Nitrogen source: The carbamate variant favors the benzylic amine **8**, the amide version favors the benzylic alcohol **7**.

Depending on the type of product desired, the following reaction conditions should be chosen:
Desired product: Benzylic alcohol **7**: Use MeCN/H_2O and AQN ligands. Benzylic amine **8**: Use *n*-PrOH/H_2O and PHAL ligands.

Enantioselectivity
The asymmetric aminohydroxylation reaction gives the highest enantioselectivities when chloramines with small substituents are employed [7, 8] (Tab. 3). Large substituents most likely inhibit the hydrolysis of the azaglycolate intermediate **5** (Scheme 4), thereby favoring the nonselective second cycle. In addition, large residues may compete with the olefin's substituents for the binding pocket of the ligand leading to a further deterioration of the enantioselectivity. In general, the carbamate and amide versions give superior selectivities compared to the sulfonamide variant. The solvent system also influences the selectivity and *n*-PrOH/H_2O quite often gives the best results.

2.6.2.2
The Sulfonamide Variant [5–7, 9]

The AA reaction was discovered based on Chloramine-T as the nitrogen source [5]. Subsequent studies have revealed that the size of the sulfonamide group has a tremendous influence on the outcome of the reaction – the smaller the residue the better the results [7] (cf. Tab. 1 and 3). Thus, the methanesulfonamide-based Chloramine-M reagent generally gives superior results in terms of enantio- and regioselectivity, catalytic turnover, and yield, compared to Chloramine-T. Additionally, the Chloramine-M system shows ligand acceleration, while the toluenesulfonamide based system is ligand decelerated. Also the product isolation is simpler, since excess sulfonamide can be readily removed by aqueous base extraction or by vacuum sublimation.

Many hydroxysulfonamides are poorly soluble in the alcohol/water solvents employed in the AA reaction, causing them to crystallize from the reaction mixture. This greatly facilitates product isolation, allowing it to be collected simply by filtration of the reaction mixture: *solution-to-solid* AA reaction [6] (Eq. 1) and *solid-to-solid* AA reaction [5] (Eq. 2).

Tab. 2 Controlling the regioselectivity in the AA of styrene derivates [10, 11]

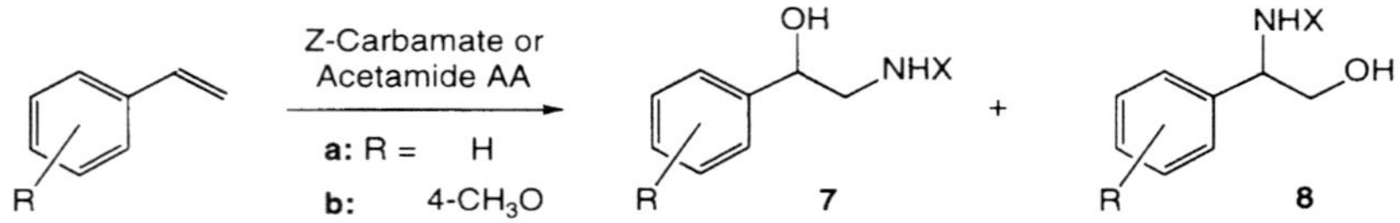

Z-Carbamate or Acetamide AA

a: R = H

b: 4-CH_3O

7 (OH / NHX) + 8 (NHX / OH)

Substrate	*Solvent*	*Ligand*	*Z-Carbamate variant* [a]			*Acetamide variant* [a]				*Comments*
			Ratio 7:8	***%ee 7***	***values 8***	***Ratio 7:8***	***%ee 7***	***values 8***		
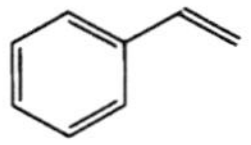 (styrene)	*n*-PrOH	PHAL	1:3	–	93	1:1.1	83	91	⟵	best conditions for **benzylic amine**
	MeCN	PHAL				6.1:1	88	–		
	MeCN	AQN				13:1	88	–	⟵	**best conditions for benzylic alcohol**
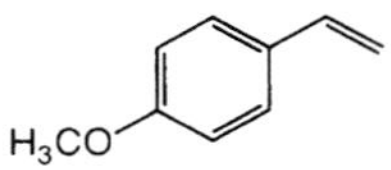 H_3CO (4-methoxystyrene)	*n*-PrOH	PHAL	1:3	–	93	1:2.5	62	96	⟵	best conditions for **benzylic amine**
	MeCN	PHAL				2.4:1	84	85		
	MeCN	AQN				9:1	86	–	⟵	**best conditions for benzylic alcohol**
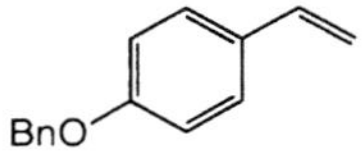 BnO (4-benzyloxystyrene)	*n*-PrOH	PHAL	1:7	–	93				⟵	best conditions for **benzylic amine**
	MeCN	PHAL	1:3							
	MeCN	AQN	3:1						⟵	**best conditions for benzylic alcohol**

a) Ref. [10]; the DHQ-derived ligands were used, leading to *S*-configured amino alcohols; reaction conditions: 3 equivalents NaOH, 3 equivalents *N*-chloro benzyl carbamate, 4 mol% $K_2OsO_2(OH)_4$, 5 mol% ligand, r.t.

b) Ref. [11]; the DHQD-derived ligands were used, leading to *R*-configured amino alcohols; reaction conditions: 1.0 equivalent KOH, 1.1 equivalents *N*-bromoacetamide, 4 mol% $K_2OsO_2(OH)_4$, 5 mol% ligand, 4 °C.

Ph–CH=CH–CO–OCH3 (54.5 g)
2.5 mol-% $(DHQ)_2PHAL$, 2.0 mol-% $K_2OsO_2(OH)_4$, 3.5 eq. $TsNClNa{\bullet}3H_2O$, r.t., 1:1 *t*-BuOH/H_2O
→ TsNH, OH, Ph, OCH_3
81.1 g (69 % yield, 82 %ee)
Product isolation by filtration (1)

Ph–CH=CH–Ph (10.5 g)
5 mol-% $(DHQ)_2PHAL$, 4 mol-% $K_2OsO_2(OH)_4$, 3 eq. $TsNClNa{\bullet}3H_2O$, r.t., 1:1 *t*-BuOH/H_2O
→ TsNH, Ph, Ph, OH
16.1 g (78 % yield, 64 %ee)
Product isolation by filtration (2)

Most methanesulfonamides crystallize more readily than toluenesulfonamides, making it possible to further enhance the enantiomeric excess by recrystallization from ethyl acetate/diethyl ether or ethyl acetate/hexane systems [7] (Eq. 3).

5 mol-% $(DHQ)_2PHAL$, 4 mol-% $K_2OsO_2(OH)_4$, 3 eq. $MeSO_2NClNa$, 1:1 *n*PrOH/H_2O, r.t., 18 h
→ CH_3SO_2NH, OH
49 % yield
before recrystallization: 66 %ee
after recrystallization: 99 %ee
(ethyl acetate/hexane) (3)

Sulfonamides have unique synthetic value, since the sulfonyl group sufficiently acidifies the N–H bond to allow facile *N*-alkylation under basic conditions [9, 15, 16a] (Eq. 4).

TsNH, O, P–OEt, OEt, OH, MeO
76 %ee
i. MsCl, TEA, CH_2Cl_2, -10 °C
ii. K_2CO_3, DMF, 25°C
→ Ts, N, O, P–OEt, OEt, MeO
86 % yield (4)

The synthetic utility of sulfonamides is limited only by their high stability, requiring forcing conditions for their removal. Recently, a very mild method for the cleavage of nosyl amides, based on the nucleophilic aromatic substitution with thiolate anion, has been developed by Fukuyama et al. [15]. Unfortunately, the nosyl amide-based AA system gives inferior results to the toluene- or methanesulfonamide systems and this class of sulfonamides is, therefore, not readily accessible by the AA reaction [17]. Other methods involve the reductive cleavage of sulfonamides under Birch conditions [3a, 16] or with Red-Al [18]. In addition, 33% HBr in acetic acid has been used to cleave toluenesulfonamides (Eq. 5) [6]. The -amino acid **9** is a precursor for the Taxol C13 side chain.

Tab. 3 Enantioselectivities obtained with $(DHQ)_2PHAL$

Entry	***Substrate***	***Products***	***p-TlSO$_2$NClNa***[a]		
			Regio-select.	***% ee***	***% yield***
1	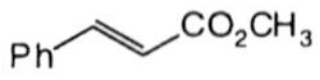 Ph, CO_2CH_3	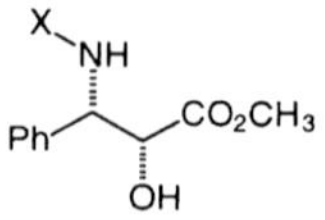 X, NH, Ph, CO_2CH_3, OH	≥5:1	81	64
2	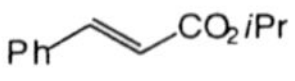 Ph, CO_2iPr	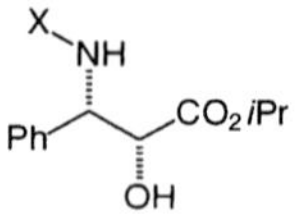 X, NH, Ph, CO_2iPr, OH			
3	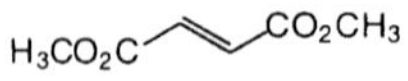 H_3CO_2C, CO_2CH_3 ref. (f)	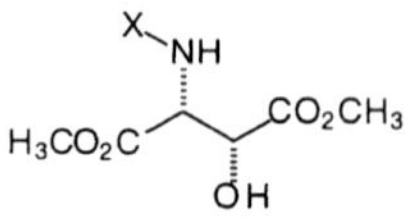 X, NH, H_3CO_2C, CO_2CH_3, OH		77	65
4	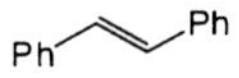 Ph, Ph	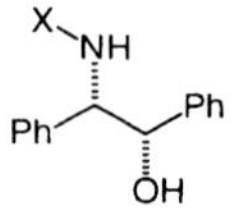 X, NH, Ph, Ph, OH		62	52
5	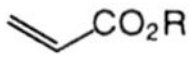 CO_2R ref. (f)	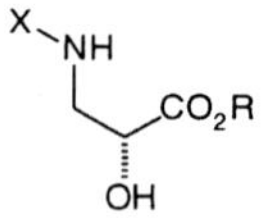 X, NH, CO_2R, OH			
6	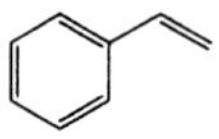	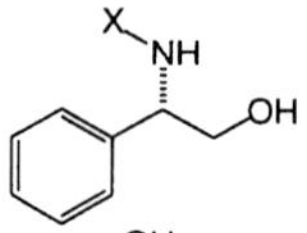 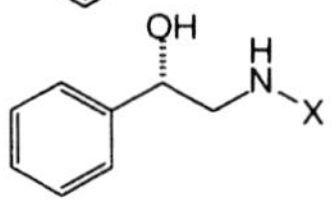 X, NH, OH; OH, H, N, X			
7	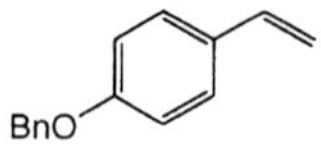 BnO	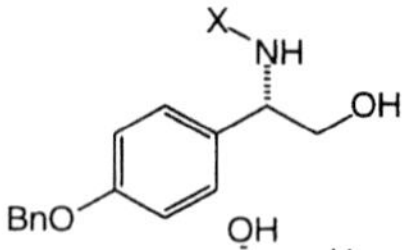 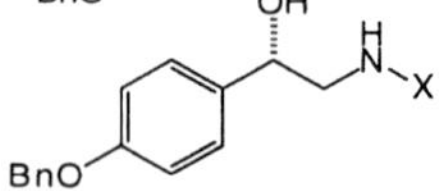 X, NH, OH, BnO; OH, H, N, X, BnO			

$MeSO_2NClNa$ [b)]			*BnOCONClNa* [c)]			*t-BuOCONClNa* [d)]			*$H_3CCONHBr/LiOH$* [e)]		
Regio-select.	*% ee*	*% yield*	*Regio-select.*	*% ee*	*% yield*	*Regio-select.*	*% ee*	*% yield*	*Regio-select.*	*% ee*	*% yield*
91:9	95	65		94	65						
95:5	94	65							>20:1	99	81
	95	76		84	55						
	75	71								94	50
				84	89				>20:1	89	46
				($R=C_2H_5$)					($R=CH_3$)		
			3:1	93	60						
			7:1	93	76	5:1	99	68			

Tab. 3 (cont.)

Entry	*Substrate*	*Products*	*p-TlSO₂NClNa* [a)]		
			Regio-select.	*% ee*	*% yield*
8		X–NH / OH (BnO-aryl); OH / NH–X (BnO-aryl)			

a) Ref. 5; 5 mol% $(DHQ)_2PHAL$, 4 mol% $K_2OsO_2(OH)_4$; 3 equivalents Chloramine-T; 1:1 MeCN/H_2O, r.t.
b) Ref. 7; 5 mol% $(DHQ)_2PHAL$, 4 mol% $K_2OsO_2(OH)_4$; 3 equivalents Chloramine-M; 1:1 *n*-PrOH/H_2O, r.t.
c) Ref. 8; 5 mol% $(DHQ)_2PHAL$, 4 mol% $K_2OsO_2(OH)_4$; 3 equivalents benzyl carbamate/*t*-BuOCl/NaOH; 1:1 *n*-PrOH/H_2O, r.t.

TsNH, O, Ph, OCH_3, OH — 33 % HBr/HOAc, 75°C, 10 h, then Amberlite IR-120 → NH_2, O, Ph, OH, OH **9** (5)

Recent work has shown that the AA based on 2-trimethylsilylethanesulfonamide gives comparable results to the Chloramine-M variant (Eq. 6) [17]. The resulting *β*-hydroxy-2-trimethylethanesulfonamides **10** can be cleaved by treatment with fluoride, following Weinreb et al.'s method [19].

Ph, O, O*i*Pr — 5 mol-% $(DHQ)_2PHAL$, 4 mol-% $K_2OsO_2(OH)_4$, 1:1 MeCN/H_2O, 3.1 eq. Me_3Si~SO_2NH_2, 3.05 eq. NaOH, 3 eq. *t*BuOCl → Me_3Si, O_2S, NH, O, Ph, O*i*Pr, OH **10**
48 % yield, 83:17 regioselectivity, 70 %ee (6)

2.6.2.3
The Carbamate Variant [8–10]

The carbamate variant of the AA reaction has a much broader scope than the sulfonamide-based versions and even some terminal olefins are good substrates (cf. Tab. 3) [8, 10]. Additionally, carbamates are of considerable synthetic value, since the protecting group is cleavable under very mild conditions. The carbamate-based AA shows ligand acceleration for all substrates in contrast to the sulfonamide sys-

MeSO₂NClNa [b]			*BnOCONClNa* [c]			*t-BuOCONClNa* [d]			*H₃CCONHBr/LiOH* [e]		
Regio-select.	**% *ee***	**% *yield***	***Regio-select.***	**% *ee***	**% *yield***	***Regio-select.***	**% *ee***	**% *yield***	***Regio-select.***	**% *ee***	**% *yield***
			10 : 1	99	68	7 : 1	98	70			

d) Ref. 10; 6 mol% $(DHQ)_2$PHAL, 4 mol% $K_2OsO_2(OH)_4$; *tert*-butyl carbamate/*t*/BuOCl/NaOH; 2:1 *n*-PrOH/H_2O, 0 °C.
e) Ref. 11; 5 mol% $(DHQ)_2$PHAL, 4 mol% $K_2OsO_2(OH)_4$; 1.1 equivalents AcNBrH/LiOH; 1:1 *t*-BuOH/H_2O, 4 °C.
f) The reaction was performed in 1:1 MeCN/H_2O.

tem, which is inhibited by the ligand in certain instances. Depending on the stoichiometric nitrogen source, ethyl, benzyl, or *tert*-butyl carbamates, **11**, **12**, **13**, are formed (Scheme 5). The selectivity trends parallel those of the sulfonamide reaction in that smaller groups typically give better results.

Best results are obtained with 1:1 *n*-propanol/water as the solvent. The *tert*-butyl carbamate version requires a 2:1 *n*-propanol/water ratio to suppress diol for-

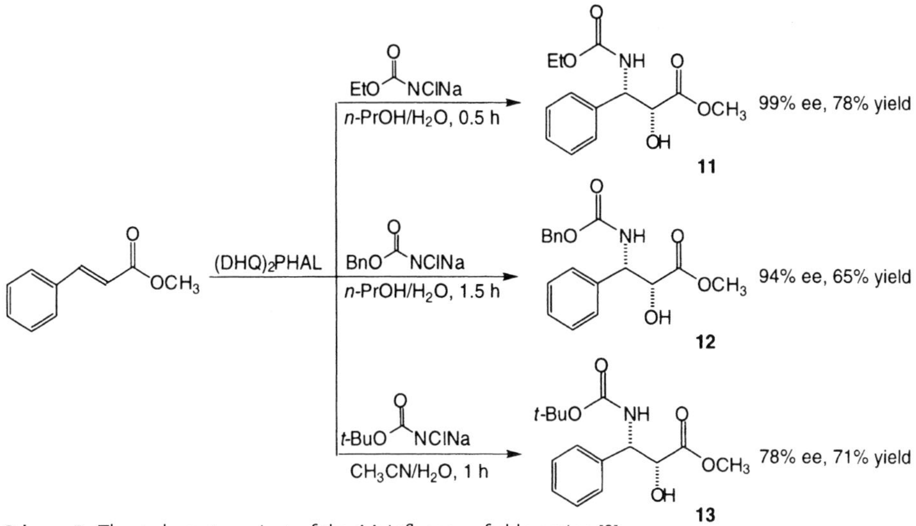

Scheme 5 The carbamate variant of the AA influence of chloramine [8].

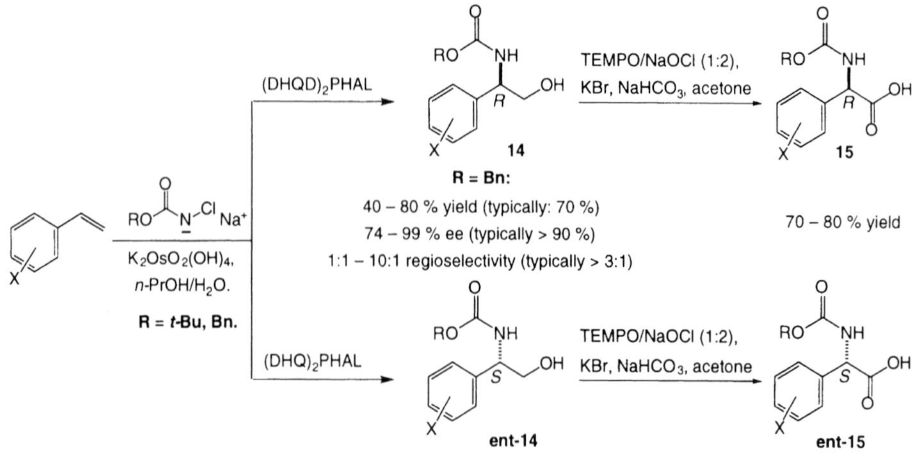

Starting Material	Amino alcohol 14	Arylglycin 15
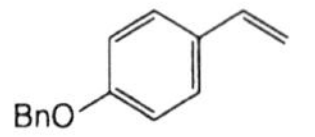 BnO	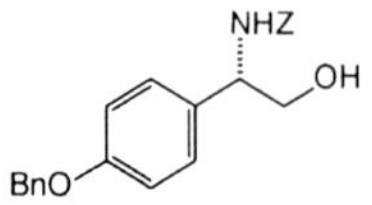 BnO, NHZ, OH 9:1 regioselectivity 76 % yield 98 % ee	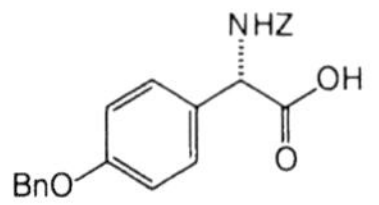 BnO, NHZ, OH 75 – 80 % yield
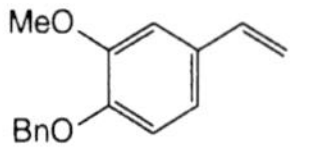 MeO, BnO	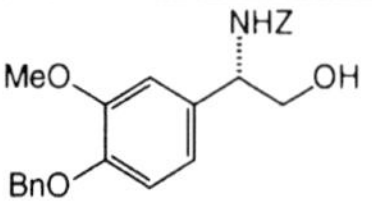 MeO, BnO, NHZ, OH 3:1 regioselectivity 71 % yield 99 % ee	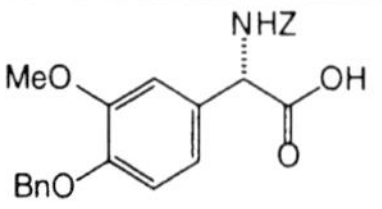 MeO, BnO, NHZ, OH 71 % yield
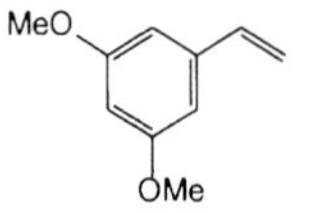 MeO, OMe	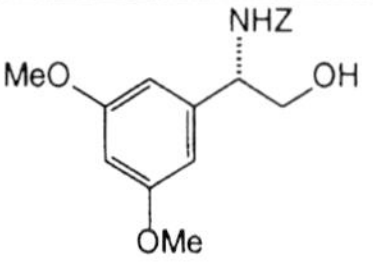 MeO, OMe, NHZ, OH 3:1 regioselectivity 71 % yield 90 % ee	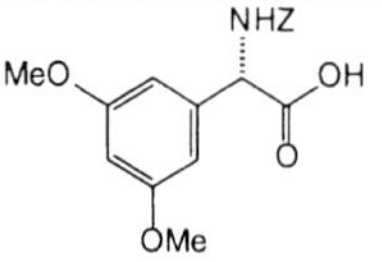 MeO, OMe, NHZ, OH 73 % yield
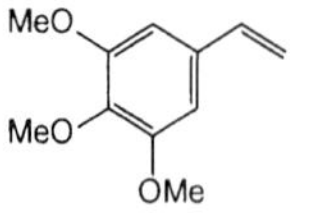 MeO, MeO, OMe	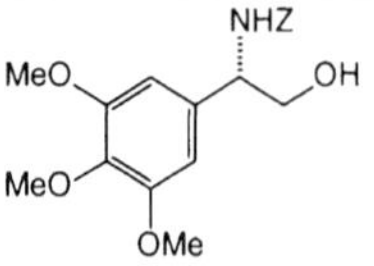 MeO, MeO, OMe, NHZ, OH 3:1 regioselectivity 65 % yield 95 % ee	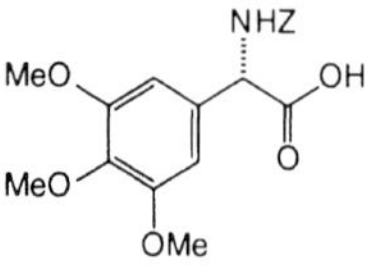 MeO, MeO, OMe, NHZ, OH 75 % yield

Scheme 6 Synthesis of enantiomerically enriched arylglycines [10].

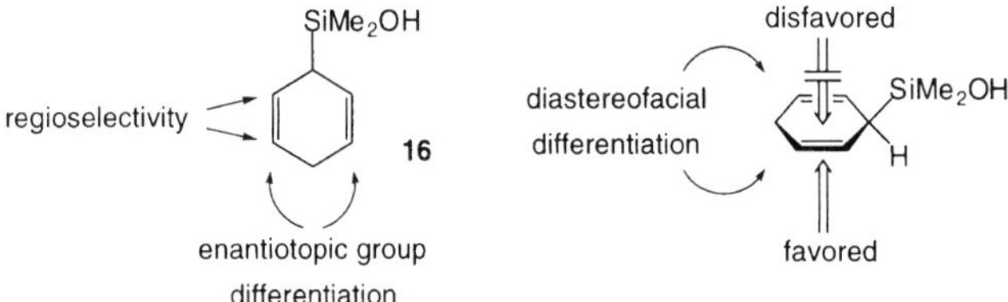

Scheme 7 Selectivity issues in the AA of silyl-2,5-cyclohexadiene (**16**).

mation [10]. An added advantage is that the products are often insoluble in the reaction mixture, allowing them to be isolated by filtration. The chloramines are prepared *in situ* and used without purification [8].

Enantiomerically enriched arylglycins **15** are readily accessible using an AA/oxidation sequence (Scheme 6) [10]. The oxidation of the *N*-protected aminoalcohol intermediate **14** to the carboxylic acid may be performed using the ruthenium-catalyzed periodic acid protocol [20]. However, best results are obtained with TEMPO/NaOCl [21], allowing the amino acid **15** to be isolated in good yield even in the presence of electron-rich aromatic systems. This oxidation step works equally well on the crude mixture of the two AA regioisomers, since the benzylic alcohol isomer is converted into the nonpolar ketocarbamate which is removed from the desired aminoacid derivative by simple trituration [10].

The desymmetrization of silyl-2,5-cyclohexadiene (**16**) by asymmetric aminohydroxylation has recently been investigated by Landais and co-workers [22]. This system provides a challenging test for the AA reaction, since three selectivity issues have to be addressed: (1) enantiotopic group differentiation, (2) diastereofacial differentiation, and (3) regioselectivity.

The reaction was found to proceed with complete anti-diastereoselectivity as well as >98% regioselectivity in favor of the hydroxy carbamate **17**. The excellent selectivity for the sterically more encumbered regioisomer **17** is probably due to the electronic directing influence of the silyl group and it is in full accord with the osmaazetidine mechanism involving electrophilic attack by the d^0 metal center (cf. Scheme 3). Even though the $(DHQ)_2PYR$ ligand provided only moderate enantioselectivity (68% *ee*), the optical purity could be raised to >99% *ee* by a single recrystallization of the allylic alcohol intermediate **18**. The latter is a key intermediate for the synthesis of amino cyclitols, e.g. **19**.

2.6.2.4
The Amide Variant [11]

The amide version of the AA reaction is comparable in scope to the carbamate-based system. Terminal olefins, e.g. styrenes, belong to the best substrates for this reaction [11]. Even ethyl acrylate reacts with good regio- (>20:1) and enantioselectivity (89% *ee*) to give ethyl *N*-acetyl isoserine (cf. Tab. 3).

The regioselectivity in the amide-based AA reaction of styrenes is highly solvent and ligand dependent (cf. Tab. 2) and the benzylic alcohol 7 is intrinsically fa-

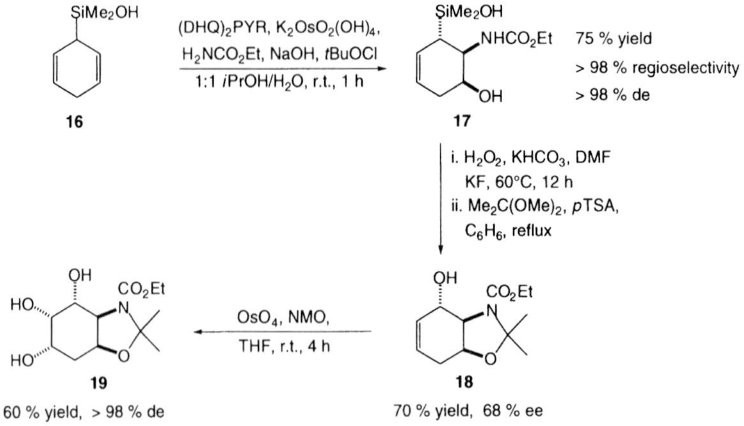

Scheme 8 Synthesis of amino cyclitols.

vored over the benzylic amine **8**. Thus, the regioselectivity is reversed compared to the carbamate version of the reaction.

Decomposition of the anionic *N*-halo amide reagent ($RCONX^-$) by Hoffmann rearrangement can be prevented by using the *N*-bromo, in place of the less stable *N*-chloro analog, and by keeping the temperature near 4 °C. A major advantage compared to the other versions of the reaction (i.e. sulfonamide or carbamate AA) is the fact that just 1.1 equivalents of *N*-bromoacetamide are needed, instead of 3 equivalents. This greatly simplifies product isolation especially on a large scale. Thus, 3-phenylisoserine (**21**), a precursor for the Taxol C13 side chain, was synthesized on a 120 g scale (Scheme 9). In this example, just 1.5 mol% of $K_2OsO_2(OH)_4$ and 1 mol% of $(DHQ)_2$PHAL are sufficient to achieve excellent yields and enantiomeric purities.

As before, reversal of regioselectivity is observed when the AQN class of ligands is used [11a] (Eq. 7). Thus, the AA of ethyl *m*-nitrocinnamate (**22**) with *N*-bromobenzamide in the presence of $(DHQ)_2$AQN using 1:1 chlorobenzene/H_2O as the solvent system provided the α-benzamido-β-hydroxyester **23** with excellent regioselectivity.

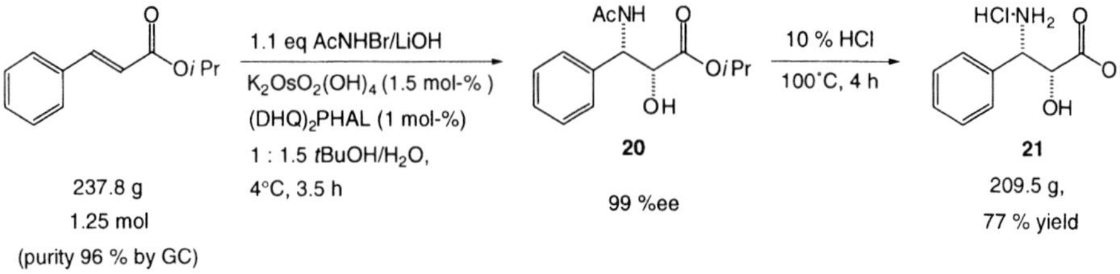

Scheme 9 Large-scale synthesis of 3-phenylisoserine [11a].

22 → **23** (7)

5% $(DHQ)_2AQN$
4% $K_2OsO_2(OH)_4$
1.1 eq PhCONHBr
1.02 eq n-Bu_4NOH
1:1 $PhCl/H_2O$, 4 °C

53 % yield, 91 % ee
15: 1 regioselectivity

In summary, the asymmetric aminohydroxylation reaction has evolved into a reliable and predictable process in just two years after the initial reports [5]. The reaction provides synthetically very useful *N*-protected 1,2-aminoalcohol derivatives starting from readily available olefinic precursors. In addition the reaction is easy to scale-up, since the products are often crystalline and insoluble in the reaction mixture, allowing them to be isolated by filtration. Both the enantioselectivity and the regioselectivity may be controlled by carefully adjusting the reaction parameters, i.e. the ligand, the solvent and the stoichiometric nitrogen source.

References

1 (a) Cf. the chapter on 'Catalytic Asymmetric Dihydroxylation'. (b) H. C. Kolb, M. S. VanNieuwenhze, K. B. Sharpless, *Chem. Rev.* **1994**, *94*, 2483–2547.

2 (a) T. Katsuki, *J. Mol. Catal. A: Chem.* **1996**, *113*, 87–107. (b) E. N. Jacobsen, Asymmetric catalytic epoxidation of unfunctionalized olefins in *Catalytic Asymmetric Synthesis* (Ed.: I. Ojima), VCH, New York, **1993**, pp. 159–202. (c) T. Katsuki, V. S. Martin, *Org. React.* **1996**, *48*, 1–299. (d) R. A. Johnson, K. B. Sharpless, Catalytic asymmetric epoxidation of allylic alcohols in *Catalytic Asymmetric Synthesis* (Ed.: I. Ojima), VCH, New York, **1993**, pp. 101–158.

3 (a) K. B. Sharpless, A. O. Chong, K. Oshima, *J. Org. Chem.* **1976**, *41*, 177–179. (b) E. Herranz, K. B. Sharpless, *J. Org. Chem.* **1978**, *43*, 2544–2548. (c) E. Herranz, K. B. Sharpless, *Org. Synth.* **1981**, *61*, 85–93.

4 (a) E. Herranz, S. A. Biller, K. B. Sharpless, *J. Am. Chem. Soc.* **1978**, *100*, 3596–3598. (b) E. Herranz, K. B. Sharpless, *J. Org. Chem.* **1980**, *45*, 2710–2713. (c) E. Herranz, K. B. Sharpless, *Org. Synth.* **1981**, *61*, 93–97.

5 G. Li, H.-T. Chang, K. B. Sharpless, *Angew. Chem., Int. Ed. Engl.* **1996**, *35*, 451–454.

6 G. Li, K. B. Sharpless, *Acta Chem. Scand.* **1996**, *50*, 649–651.

7 J. Rudolph, P. C. Sennhenn, C. P. Vlaar, K. B. Sharpless, *Angew. Chem., Int. Ed. Engl.* **1996**, *35*, 2810–2813.

8 G. Li, H. H. Angert, K. B. Sharpless, *Angew. Chem., Int. Ed. Engl.* **1996**, *35*, 2813–2817.

9 A. A. Thomas, K. B. Sharpless, *J. Org. Chem.* **1999**, *64*, 8379.

10 K. L. Reddy, K. B. Sharpless, *J. Am. Chem.* **1998**, *120*, 1207.

11 (a) M. Bruncko, G. Schlingloff, K. B. Sharpless, unpublished results. (b) M. Bruncko, G. Schlingloff, K. B. Sharpless, *Angew. Chem., Int. Ed. Engl.* **1997**, *36*, 1483.

12 For the Sharpless model, see: (a) H. C. Kolb, P. G. Andersson, Y. L. Bennani, G. A. Crispino, K.-S. Jeong, H.-L. Kwong, K. B. Sharpless, *J. Am. Chem. Soc.* **1993**, *115*, 12226. (b) H. C. Kolb,

P. G. ANDERSSON, K. B. SHARPLESS, *J. Am. Chem. Soc.* **1994**, *116*, 1278–1291. (c) P.-O. NORRBY, H. BECKER, K. B. SHARPLESS, *J. Am. Chem. Soc.* **1996**, *118*, 35–42. (d) P.-O. NORRBY, H. C. KOLB, K. B. SHARPLESS, *J. Am. Chem. Soc.* **1994**, *116*, 8470–8478. For the Corey model, see: (e) E. J. COREY, M. C. NOE, *J. Am. Chem. Soc.* **1996**, *118*, 319–329. (f) E. J. COREY, M. C. NOE, A. Y. TING, *Tetrahedron Lett.* **1996**, *37*, 1735–1738. (g) M. C. NOE, E. J. COREY, *Tetrahedron Lett.* **1996**, *37*, 1739–1742. (h) E. J. COREY, M. C. NOE, M. J. GROGAN, *Tetrahedron Lett.* **1996**, *37*, 4899–4902. (i) E. J. COREY, M. C. NOE, A. GUZMAN-PEREZ, *J. Am. Chem. Soc.* **1995**, *117*, 10817–10824. (j) E. J. COREY, A. GUZMAN-PEREZ, M. C. NOE, *J. Am. Chem. Soc.* **1995**, *117*, 10805–10816. (k) E. J. COREY, A. GUZMAN-PEREZ, M. C. NOE, *J. Am. Chem. Soc.* **1994**, *116*, 12109–12110. (l) E. J. COREY, M. C. NOE, S. SARSHAR, *Tetrahedron Lett.* **1994**, *35*, 2861–2864. (m) E. J. COREY, M. C. NOE, M. J. GROGAN, *Tetrahedron Lett.* **1994**, *35*, 6427–6430.

13 (a) K. B. SHARPLESS, D. W. PATRICK, L. K. TRUESDALE, S. A. BILLER, *J. Am. Chem. Soc.* **1975**, *97*, 2305–2307. (b) A. O. CHONG, K. OSHIMA, K. B. SHARPLESS, *J. Am. Chem. Soc.* **1977**, *99*, 3420–3426. (c) D. W. PATRICK, L. K. TRUESDALE, S. A. BILLER, K. B. SHARPLESS, *J. Org. Chem.* **1978**, *43*, 2628–2638. (d) S. G. HENTGES, K. B. SHARPLESS, *J. Org. Chem.* **1980**, *45*, 2257–2259.

14 K. B. SHARPLESS, A. Y. TERANISHI, J.-E. BÄCKVALL, *J. Am. Chem. Soc.* **1977**, *99*, 3120–3128.

15 T. FUKUYAMA, C.-K. JOW, M. CHEUNG, *Tetrahedron Lett.* **1995**, *36*, 6373–6374.

16 (a) J.-E. BÄCKVALL, K. OSHIMA, R. E. PALERMO, K. B. SHARPLESS, *J. Org. Chem.* **1979**, *44*, 1953–1957; sodium naphthalide in glyme has also been used for the reductive cleavage of sulfonamides, see: (b) S. JI, L. B. GANTLER, A. WARING, A. BATTISTI, S. BANK, W. D. CLOSSON, *J. Am. Chem. Soc.* **1967**, *89*, 5311–5312.

17 K. B. SHARPLESS ET AL., unpublished results.

18 E. H. GOLD, E. BABAD, *J. Org. Chem.* **1972**, *37*, 2208–2210.

19 S. M. WEINREB, D. M. DEMKO, T. A. LESSEN, *Tetrahedron Lett.* **1986**, *27*, 2099–2102.

20 (a) P. H. J. CARLSEN, T. KATSUKI, V. S. MARTIN, K. B. SHARPLESS, *J. Org. Chem.* **1981**, *46*, 3936–3938. (b) J. M. CHONG, K. B. SHARPLESS, *J. Org. Chem.* **1985**, *50*, 1560–1563.

21 (a) P. L. ANELLI, C. BIFFI, F. MONTANARI, S. QUID, *J. Org. Chem.* **1987**, *52*, 2559–2562. (b) T. INOKUCHI, S. MATSUMOTO, T. NISHIYAMA, S. TORII, *J. Org. Chem.* **1990**, *55*, 462–466. (c) T. MIYAZAWA, T. ENDO, S. SHIIHASHI, M. OKAWARA, *J. Org. Chem.* **1985**, *50*, 1332–1334.

22 R. ANGELAUD, Y. LANDAIS, K. SCHENK, *Tetrahedron Lett.* **1997**, *38*, 1407–1410.

2.6.3
Asymmetric Aminohydroxylation – Recent Developments

Kilian Muñiz

2.6.3.1
Introduction

The catalytic conversion of unfunctionalized olefins into aminoalcohols has been recognized as the second fundamental osmium-based oxidative olefin functionalization after asymmetric catalytic dihydroxylation (Chapters 2.5 and 2.5.1), and it has already been reviewed several times [1]. While there have been various advances, and the asymmetric aminohydroxylation (AA) reaction in the presence of chiral cinchona alkaloid ligands nowadays represents an established asymmetric catalytical process, it still suffers from a lack in substrate scope (electron-rich olefins), catalyst activity, and, most importantly, chemoselectivity leading to significant amounts of diol side products. In this chapter, important developments in the area of asymmetric aminohydroxylation that appeared after the contribution by Kolb and Sharpless (Chapters 2.6.1 and 2.6.2) are discussed.

2.6.3.2
Recent Developments

2.6.3.2.1 Nitrogen Sources and Substrates

Initial work dealt with the introduction of suitable nitrene precursors for *in situ* generation of the reactive imido trioxoosmium(VIII) species [2, 3].

To this end, a set of orthogonal protecting groups at the nitrenoid nitrogen were introduced, resulting in AA reactions that employed the respective haloamine salts of tosylamides, carbamates and amides. Over the past years, these substrate classes have been joined by further nitrene precursors that include *N*-halo salts of *tert*-butylsulfonamide (**1**) [4], of primary amides such as **2** [5], and of 2-TMS ethyl carbamate (**3**) [6]. The latter is especially interesting since it promotes AA under standard conditions (4 mol% osmium salt and 5 mol% ligand) to give products with exceptionally high enantiomeric excesses and high regioselectivities. Moreover, deprotection of the nitrogen can be conveniently achieved with TBAF under very mild conditions.

An observation by Jerina that the *N*-chloro salt of an adenosine derivative underwent aminohydroxylation at a racemic dihydrodiol substrate [7] led to the introduction of this class of compounds as nitrene precursors [8]. Although the reactions proceeded in good yield and with excellent regioselectivities, the solvent system appeared to be limited to alcohol-water mixtures. Apparently, chiral non-racemic substituents in the 9-position of adenosine had no influence, since equal mixtures of diastereomers were obtained from unsymmetrical substrates. Use of a cinchona alkaloid ligand had no beneficial stereochemical influence, although a rate enhancement could be detected. However, when nitrenoids of related amino-substituted heterocycles such as 2-amino pyrimidine (**5**) were investigated, a highly efficient AA was discovered (Scheme 1) [9].

This time, the reaction course was influenced positively by the chiral ligand, and aminohydroxylation of stilbene with $(DHQ)_2PHAL$ yielded a product **6** with 97% *ee*. Related heterocycles gave similar results presuming the *N*-chlorination was carried out in absolute alcohol to avoid ring halogenation, and again the solvent for the AA had to be an alcohol-water mixture. Since stilbene had been a rather problematic substrate in former AA examples, these reactions are especially noteworthy. The high enantiomeric excesses could also be achieved for other substrates ranging from hydrocarbons to cinnamates.

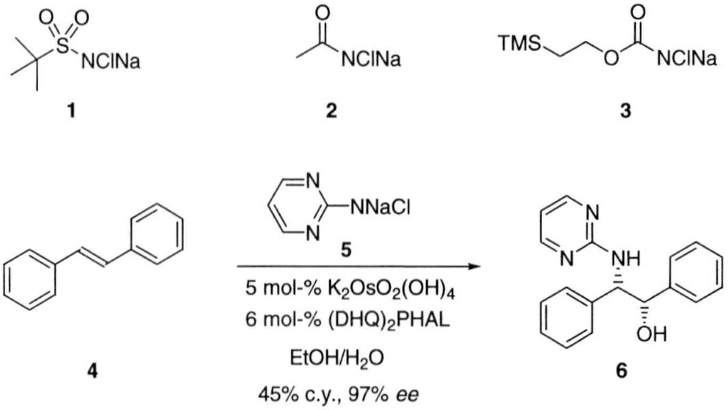

NClNa NClNa TMS NClNa

1 2 3

NNaCl

5

5 mol-% $K_2OsO_2(OH)_4$

6 mol-% $(DHQ)_2PHAL$

$EtOH/H_2O$

45% c.y., 97% *ee*

4 6

Scheme 1 New nitrene precursors and AA reaction with amino-substituted heterocycles.

In view of the concerted stereodefined introduction of a vicinal aminoalcohol moiety, asymmetric aminohydroxylation has undergone various applications in synthesis. Among the many examples, there are several in the synthesis of natural products [10] and compounds of general biological interest [11]. It has also been extensively used in the synthesis of natural and non-natural α-amino acids [12–14] and in the search for new aryl serin derivatives. This structural motif has been of particular interest because of the natural occurrence of an isoserine derivative as the side chain in the powerful antitumor agent TAXOL® (paclitaxel). While the synthesis of the original side chain was the very first application of catalytic AA and was described by Sharpless himself [15, 16], the interest in a potential relationship between structural variation and biological activity [17] has led to various examples of asymmetric aminohydroxylations on aryl acrylic esters [16, 18], including heteroaromatic moieties [19].

Another application concerning biorelevant molecules has been the AA reaction of unsaturated phosphonates with chloramine-T or the *N*-chloro ethoxycarbamate salt [20]. Unfortunately, this example is restricted to vinyl phosphonate and its β-aryl derivatives. While the yields remained rather low, enantioselectivities were low only in the case of chloramine-T but high for carbamate-based AA. In these reactions, the products could all be crystallized to enantiopurity in a single step. Interestingly, all reactions occurred with complete regioselectivity, introducing the nitrogen into the benzylic position of the product.

2.6.3.2.2 **Regioselectivity**

In contrast to the related asymmetric dihydroxylation (AD) reaction, in which two identical heteroatoms are introduced, the simultaneous introduction of different heteroatoms raises the immediate question of regioselectivity with respect to unsymmetrical substrates. Apart from some examples with regioselectivities of 20 : 1 and higher [4, 9, 21], this feature still requires a general solution. The reversal of regioselectivity has been described for the use of $(DHQ)_2AQN$ and $(DHQD)_2AQN$ as ligands, although these reactions still produce mixtures of regioisomers [13]. Based on transition state model structures for the active imidoosmium-alkaloid ligand complex, Janda developed AA reactions that were controlled by a combination of steric and electronic effects and by the incorporation of suitable protecting groups into the substrates [22]. While each of these factors exercises only a limited influence, their combination gave an excellent tool for controlling regioselectivity. The model could be applied further to certain cases with a reversal in regioselectivity that was induced by precise complementarity of substrate and catalyst shape.

Two elegant solutions to the problem of regioselectivity employ chemical modifications after the AA reaction itself. In their AA reaction of styrene, Sharpless and Reddy submitted the regioisomeric product mixture to oxidation, thereby producing an achiral amino ketone (**11**), and the desired phenyl glycine (**10**) with 93% *ee* could be isolated by simple acid-base extraction [12]. In a related case, Barta, Reider, and co-workers carried out AA reactions on substituted *β*-methyl styrenes. With the regioselectivities not exceeding a ratio of 5:1, the aminoalcohols were converted into the corresponding oxazolidinones, which could be separated. However, it was discovered that the base-mediated cyclization occurred at rates

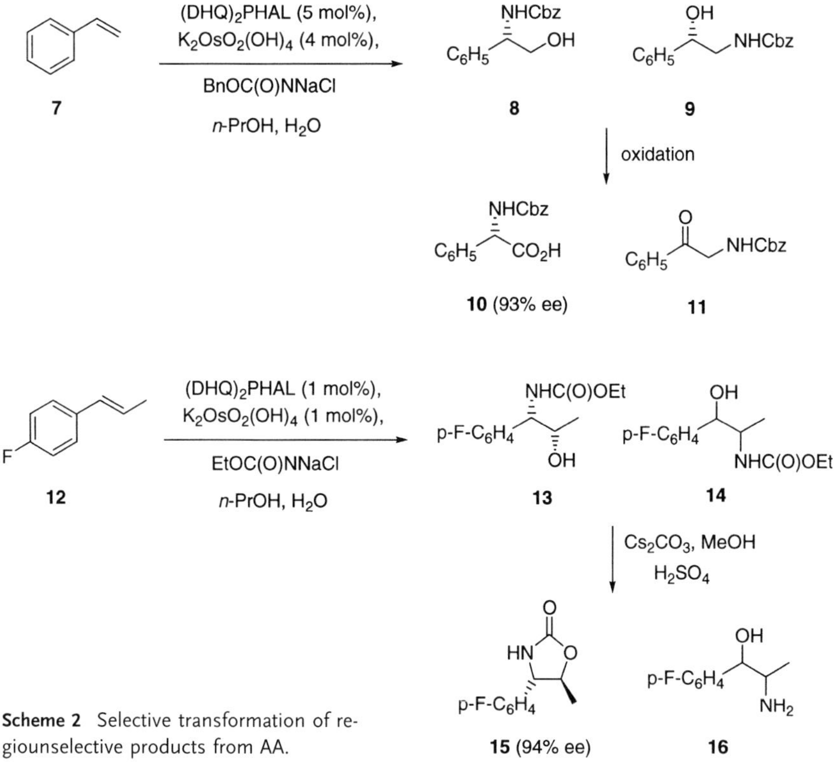

Scheme 2 Selective transformation of regiounselective products from AA.

sufficiently different to distinguish the major regioisomer from the minor one. Thus, the desired oxazolidinone (**15**) could be obtained in up to 94% *ee* as a single isomer, while the minor compound was converted to the free aminoalcohol **16** under the conditions of acidic quench [23].

2.6.3.2.4 Intramolecular Aminohydroxylation

A promising path to overcome the inherent regioselectivity problems of aminohydroxylation consists of an intramolecular reaction sequence. Here, it was chosen to start from allylic alcohols and thus to incorporate the respective amino group into a carbamate functionality. The intramolecular aminohydroxylation reaction then furnished the desired hydroxyl imidazolidinones with complete regioselectivity and stereospecificity. The best chemical yield was accomplished in the presence of $(DHQ)_2PHAL$ as ligand, although the reaction gave only racemic product [24]. A subsequent investigation of carbamates derived from cyclic allylic alcohols such as **17** gave the desired all-*syn* products such as **19** in yields between 50 and 65%. The reaction is initiated upon treatment of **17** with basic *tert*-butyl hypochlorite to give the *N*-chloro carbamate salt **18** and, upon exposure to a catalytic amount of osmium(VI), yields the hydroxyl imidazolidinone **19** in a diastereo- and regioselective manner (Scheme 3). Again, a superiority of Sharpless cinchona alkaloids over other achiral ligands was observed, but no kinetic resolution could be achieved. Moreover, when an enantiopure carbamate derived from dehydromenthol was submitted to this aminohydroxylation, there was no rate difference for the two pseudo-enantiomeric ligands $(DHQ)_2PHAL$ and $(DHQD)_2PHAL$, respectively. For the given transformation of **17** into **19**, a control experiment was carried out in the absence of water, and the intermediary azaglycol osmate could be trapped upon addition of tmeda to furnish complex **20**. Its structure was confirmed by X-ray analysis and permitted a first insight into a modified intermediate of the catalytic aminohydroxylation. Treatment with aqueous sodium sulfite trans-

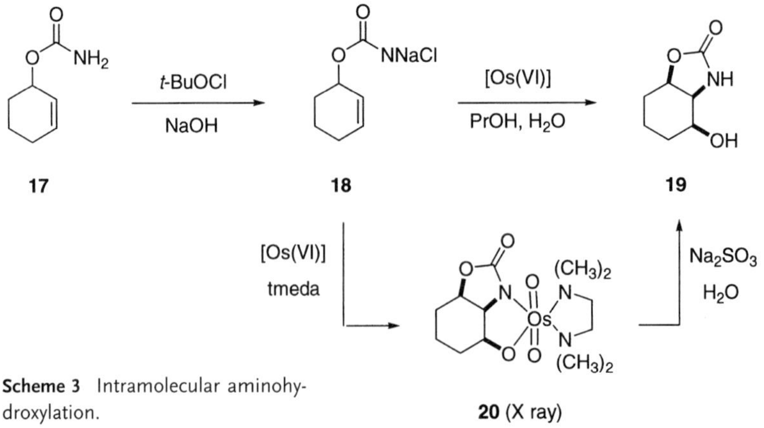

Scheme 3 Intramolecular aminohydroxylation.

formed **20** into the final product **19**, thereby mimicking the hydrolysis of the catalytic cycle [25].

2.6.3.2.5 "Secondary-Cycle" Aminohydroxylations

Two results from earlier investigations on the substrate scope of the AA deserve special attention.

In 1997 and 1999, Sharpless and his co-worker observed that certain substrates such as cinnamic amides [26] and Baylis-Hillman adducts [27] yielded regioisomeric products **22**, **23** and **25**, **26** that were essentially racemic, even in cases in which a large amount of chiral cinchona alkaloid ligand was employed. The same behavior was later uncovered for free carboxylic acids such as acrylic and fumaric acid (**27**) and their derivatives (Scheme 4) [28]. Regarding the aminohydroxylation of Baylis-Hillman adducts, both the complete regioselectivity and the high stereoselectivity in favor of the all-*syn* isomer (98:2 for **22**:**23**) is remarkable [27].

In all these cases, the reaction proceeds exclusively within the secondary cycle. This behavior has already been discussed in depth for the related dihydroxylation (see Chapter 2.5.1). Apparently, the polar moieties of free carboxylic acids, amides and related groups introduce a lipophilic scenario which promotes direct cleavage of the intermediary bis(azaglycolate) (**A**). This is a unique result, since the related bisglycolate complexes omitting these functional groups are found to be extremely stable [29]. As a direct result, this ligand-independent process is unselective regarding enantioselectivity and not very selective regarding regioselectivity (it is 5:1 for the given example of aminohydroxylation on amide **24** and in the range of 1.6:1 to 3:1 for cinnamic acids). Nevertheless, this new reaction variant displays two particularly interesting features, in that it requires only a low catalyst loading (0.1–2 mol% of Os) and slightly more than the stoichiometric amount of nitre-

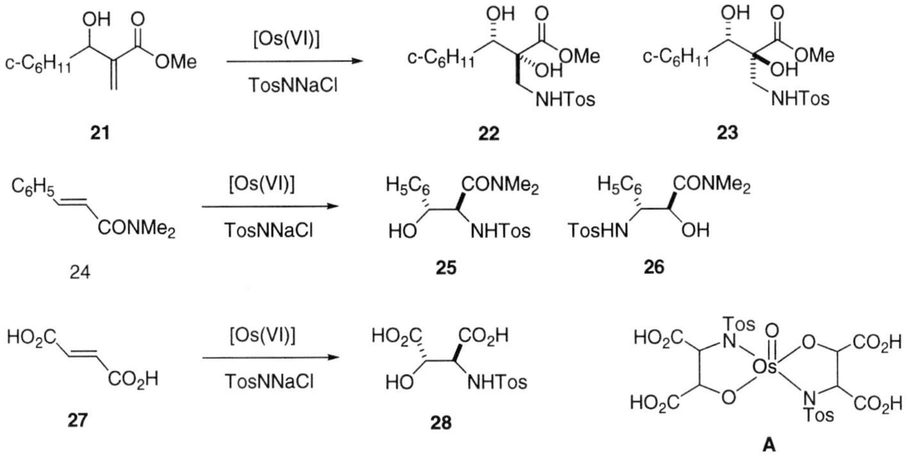

Scheme 4 Privileged substrates for second-cycle aminohydroxylation.

noid. The latter feature is remarkable, since the first-cycle AA requires at least three equivalents of nitrenoid in order to obtain high yields. Moreover, the reaction can be run at higher concentration (up to 0.8 molar) and proceeds in alcohol-water mixtures or even in water itself. For all substrates bearing free-acid functionality, a neutralization of the substrate is required prior to aminohydroxylation. In some cases, the two regioisomers derived from an aminohydroxylation of free carboxylic acids display dramatically different solubilities. For example, in the aminohydroxylation of cinnamic acid in water, one regioisomer precipitates readily, while the other remains in solution [28, 30]. So far, chloramine-T and its *tert*-butyl-substituted chloramine counterpart [4] have been the only nitrene precursors in these reactions [31]. In principle, other known haloamine salts should be applicable as well. It is further assumed that the above-mentioned unselective aminohydroxylations employing *N*-chloro salts of adenine derivatives belong to this class of reactions as well [8]. In this case, the size of the adenine moiety and its precise heteroatom arrangement are believed to be the reason that the intermediary bis-(azaglycolester) undergoes hydrolysis at a sufficiently high rate to render the whole process efficient.

Moreover, the development of an efficient AA within the secondary cycle has been possible thanks to the use of tosylated amino alcohols, as they are readily provided by the standard AA reaction employing cinchona alkaloids. Thus, chiral non-racemic compounds such as **29** serve as ligands since they promote formation of a ligated osmium catalyst precursor **30** that is oxidized to the actual cata-

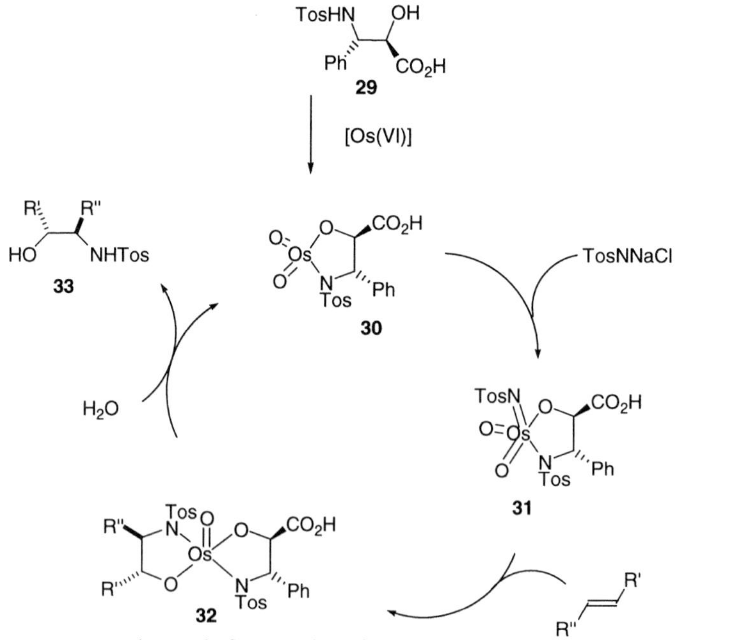

TosHN OH Ph CO_2H **29**

[Os(VI)]

R' R'' HO NHTos **33**

O O Os O CO_2H N Tos Ph **30**

TosNNaCl

H_2O

TosN O=Os O O CO_2H N Tos Ph **31**

R'' Tos N O Os O CO_2H R''' O N Tos Ph **32**

R' R''

Scheme 5 Catalytic cycle for AA with preformed enantiopure aminoalcohol ligands.

lyst **31**. Subsequent AA and hydrolysis of the bis(azaglycolate) **32** furnishes the desired enantioenriched aminoalcohol **33** and regenerated catalyst precursor **30** (Scheme 5). In all cases, regioselectivities of 2:1 together with high yields of 75% or more were obtained for catalyses in the presence of 0.2 mol% Os. The reaction was found to require an optimum amount of about 2 mol% of the chiral ligand. Higher amounts of ligand had no effect on rate, selectivity, or enantioselectivity. Under these conditions, enantiomeric excesses were as high as 59% [32].

2.6.3.3
Vicinal Diamines

The conversion of the vicinal amino alcohol functionality into 1,2-diamines has been described. A first example by Janda [33] was followed by a more detailed study aimed at threo- and erythro-selectivity [34]. In addition, the synthesis of diamines starting from styrenes has been reported [35, 36]. However, in most of these cases, the transformation of vicinal aminoalcohols into diamines has been rather tedious, and this has inspired the quest for a more efficient synthesis.

2.6.3.4
Asymmetric Diamination of Olefins

Asymmetric catalytic dihydroxylation and aminohydroxylation both being at a highly sophisticated stage, one might wonder about the remaining reaction of a concerted transfer of two amino moieties from a bisimido osmium complex onto olefinic C-C bonds [37–39]. This reaction sequence, which would result in an asymmetric diamination (ADA), remains elusive [40]. To date, several drawbacks are known. For example, an *in situ* generation of bisimido complexes of osmium has yet not been developed, and the known osmium complexes are not prone to undergo coordination to the standard cinchona alkaloid ligands [41]. In accord with these findings, only an achiral reaction course has been achieved for the known imido osmium reagents [38]. As in the case of the related aminohydroxylation, electron-poor olefins are more prone to diamination than their neutral or electron-rich counterparts, and the respective products of these reactions are isolated as stable osmaimidazolidine adducts. A stereoselective diamination has been accomplished by use of acrylic esters containing a chiral non-racemic alcohol component. Thus, for the reaction of (–)-8-phenyl menthyl-substituted acrylic esters **33**, a discrimination of the two diastereotopic faces of the C-C double bond becomes possible, and the products were isolated as chiral osmaimidazolidines **35a–c** and **36a–c** with ratios of up to 90:10 (Scheme 6).

When the chiral olefin was exchanged for commercially available bis[(–)-menthyl] fumarate (**38**), the corresponding osmaimidazolidines **39a** and **40a** were obtained with 82:18 d.r. Upon use of the trisimido reagent **37**, the diamination resulted in an even higher ratio of 95:5 for **39b** and **40b**. However, because of the basic lone pairs of the nitrogen moieties, the resulting osmaimidazolidines are extremely stable. This renders all modification and especially removal of the os-

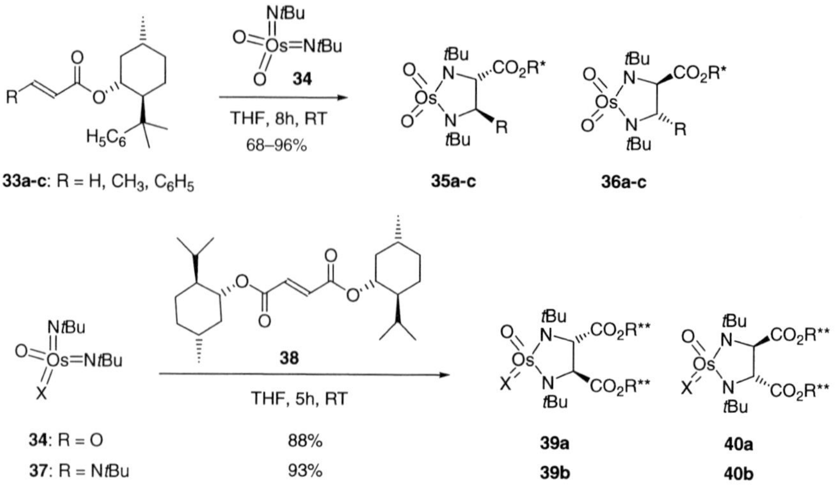

Scheme 6 Asymmetric diamination (ADA) of olefins.

mium moiety extremely difficult [42b]. Certainly, a catalytic ADA reaction will have to search for new reactivity in the area of osmium imido complexes [43].

Acknowledgement

Support from the Fonds der Chemischen Industrie is gratefully acknowledged.

References

1 (a) C. Bolm, J. P. Hildebrand, K. Muñiz, in *Catalytic Asymmetric Synthesis* (Ed.: I. Ojima), Wiley-VCH, Weinheim **2000**, p. 299; (b) G. Schlingloff, K. B. Sharpless in *Asymmetric Oxidation Reactions: A Practical Approach* (Ed.: T. Katsuki), Oxford University Press, London **2001**, p. 104; (c) D. Nilov, O. Reiser, *Adv. Synth. Catal.* **2002**, *344*, 1169; (d) P. O'Brien, *Angew. Chem. Int., Ed. Engl.* **1999**, *38*, 326; (e) G. Casiraghi, G. Rassu, F. Zanardi, *Chemtracts-Organic Chemistry* **1997**, *10*, 318; (f) O. Reiser, *Angew. Chem., Int. Ed. Engl.* **1996**, *35*, 1308.

2 For original work by Sharpless on isolated *tert*-butylimido trioxoosmium(VIII) reagent or catalytic achiral aminohydroxylation employing chloramine-T as terminal oxidant: (a) K. B. Sharpless, A. O. Chong, K. Oshima, *J. Org. Chem.* **1976**, *41*, 177; (b) K. B. Sharpless, E. Herranz, *J. Org. Chem.* **1978**, *43*, 2544; (c) E. Herranz, S. A. Biller, K. B. Sharpless, *J. Am. Chem. Soc.* **1978**, *100*, 3596; (d) E. Herranz, K. B. Sharpless, *J. Org. Chem.* **1980**, *45*, 2710; (e) E. Herranz, K. B. Sharpless, *Org. Synth.* **1981**, *61*, 85; (f) E. Herranz, K. B. Sharpless, *Org. Synth.* **1981**, *61*, 93.

3 For stereoselective reactions employing the *tert*-butylimido trioxoosmium(VIII) reagent, see: (a) H. Rubenstein, J. S. Svendsen, *Acta Chem. Scand.* **1994**, *48*, 439; (b) S. Pinheiro, S. F. Pedraza, F. M. C. Farias, A. S. Conçalves, P. R. R. Costa, *Tetrahedron Asymmetry* **2000**, *11*, 3845. The exclusive formation of aminoalcohols in the AA of pinene derivatives is noteworthy since other investigations

on aminohydroxylation with *tert*-butylimido trioxoosmium(VIII) report diol formation as well.

4 A.V. Gontcharov, H. Liu, K.B. Sharpless, *Org. Lett.* **1999**, *1*, 1949.

5 Z.P. Demko, M. Bartsch, K.B. Sharpless, *Org. Lett.* **2000**, *2*, 2221.

6 K.L. Reddy, K.R. Dress, K.B. Sharpless, *Tetrahedron Lett.* **1998**, *39*, 3667.

7 A.S. Pilcher, H. Yagi, D.M. Jerina, *J. Am. Chem. Soc.* **1998**, *120*, 3520.

8 K.R. Dress, L.J. Goossen, H. Liu, D.M. Jerina, K.B. Sharpless, *Tetrahedron Lett.* **1998**, *39*, 7669.

9 L.J. Goossen, H. Liu, K.R. Dress, K.B. Sharpless, *Angew. Chem. Int. Ed. Engl.* **1999**, *38*, 1080.

10 Selected examples: (a) K.C. Nicolaou, N.F. Jain, S. Natarajan, R. Hughes, M. E. Solomon, H. Li, J.M. Ramanjulu, M. Takayanagi, A.E. Koumbis, T. Bando, *Angew. Chem., Int. Ed.* **1998**, *37*, 2714; (b) K.C. Nicolaou, N. Takayanagi, N.F. Jain, S. Natarajan, A.E. Koumbis, T. Bando, J.M. Ramanjulu, *Angew. Chem., Int. Ed.* **1998**, *37*, 2717; (c) H. Sugiyama, T. Shiori, F. Yokokawa, *Tetrahedron Lett.* **2002**, *43*, 3489; (c) T.T. Upadhya, A. Sudalai, *Tetrahedron Asymmetry* **1997**, *8*, 3685.

11 For example: (a) S. Chandrasekhar, S. Mohapatra, *Tetrahedron Lett.* **1998**, *39*, 6415; (b) C.E. Masse, A.J. Morgan, J.S. Panek, *Org. Lett.* **2000**, *2*, 2571.

12 K.L. Reddy, K.B. Sharpless, *J. Am. Chem. Soc.* **1998**, *120*, 1207.

13 B. Tao, G. Schlingloff, K.B. Sharpless, *Tetrahedron Lett.* **1998**, *39*, 2507.

14 A.J. Morgan, J.S. Panek, *Org. Lett.* **1999**, *1*, 1949; (c) I.H. Kim, K.L. Kirk, *Tetrahedron Lett.* **2001**, *42*, 8401.

15 (a) G. Li, K.B. Sharpless, *Acta Chem. Scand.* **1996**, *50*, 649.

16 For later synthetic approaches, see: (a) M. Bruncko, G. Schlingloff, K.B. Sharpless, *Angew. Chem., Int. Ed. Engl.* **1997**, *36*, 1483; (b) see ref. 6; (c) C.E. Song, C.R. Oh, E.J. Roh, S.G. Lee, J.H. Choi, *Tetrahedron Asymmetry* **1999**, *10*, 671.

17 (a) L. Barboni, C. Lambertucci, G. Appendino, D.G. Van der Velde, R.H. Himes, E. Bombardelli, M. Wang, J.P. Snyder, *J. Med. Chem.* **2001**, *44*, 1576. (b) For a review: I. Ojima, S.N. Lin, T. Wang, *Curr. Med. Chem.* **1999**, *6*, 927.

18 S. Montiel-Smith, V. Cervantes-Mejía, J. Dubois, D. Guénard, F. Guéritte, J. Sandoval-Ramírez, *Eur. J. Org. Chem.* **2002**, 2260.

19 (a) D. Raatz, C. Innertsberger, O. Reiser, *Synlett* **1999**, 1907; (b) H.X. Zhang, P. Xia, W.S. Zhou, *Tetrahedron Asymmetry* **2000**, *11*, 3439; for AA on vinyl furans: (c) M.H. Haukaas, G.A. O'Doherty, *Org. Lett.* **2001**, *3*, 401; (d) P. Phukan, A. Sudalai, *Tetrahedron Asymmetry* **1998**, *9*, 1001; (e) M.L. Bushey, M.H. Haukaas, G.A. O'Doherty, *J. Org. Chem.* **1999**, *64*, 2984; for AA on vinyl indoles: (f) C.-G. Yang, J. Wang, X.-X. Tang, B. Jiang, *Tetrahedron Asymmetry* **2002**, *13*, 383.

20 (a) A.A. Thomas, K.B. Sharpless, *J. Org. Chem.* **1999**, *64*, 8379; (b) G. Cravotto, G.B. Giovenzana, R. Pagliarin, G. Palmisano, M. Sisti, *Tetrahedron Asymmetry* **1998**, *9*, 745.

21 (a) R. Angelaud, Y. Landais, K. Schenk, *Tetrahedron Lett.* **1997**, *38*, 1407; (b) R. Angelaud, O. Babot, T. Charvvat, Y. Landais, *J. Org. Chem.* **1999**, *64*, 9613.

22 H. Han, C.-W. Woo, K.D. Janda, *Chem. Eur. J.* **1999**, *5*, 1565.

23 N.S. Barta, D.R. Sidler, K.B. Somerville, S.A. Weissman, R.D. Larsen, P.J. Reider, *Org. Lett.* **2000**, *2*, 2821.

24 T.J. Donohoe, P.D. Johnson, M. Helliwell, M. Keenan, *Chem. Commun.* **2001**, 2078.

25 T.J. Donohoe, P.D. Johnson, A. Cowley, M. Keenan, *J. Am. Chem. Soc.* **2002**, *124*, 12934.327

26 A.E. Rubin, K.B. Sharpless, *Angew. Chem., Int. Ed. Engl.* **1997**, *36*, 2637.

27 W. Pringle, K.B. Sharpless, *Tetrahedron Lett.* **1999**, *40*, 5150.

28 V.V. Fokin, K.B. Sharpless, *Angew. Chem. Int. Ed.* **2001**, *40*, 3455.

29 (a) R. Criegee, *Liebigs Ann. Chem.* **1936**, *522*, 75; (b) R. Criegee, B. Marchand, H. Wannowius, *Liebigs Ann. Chem.* **1942**, *550*, 99.

30 H.C. Kolb, M.G. Finn, K.B. Sharpless, *Angew. Chem., Int. Ed.* **2001**, *40*, 2004.

31 A single example on the use of chloramine-M was included in the Baylis-Hilman adducts study; see ref. [27].

32 M.A. Andersson, R. Epple, V.V. Fokin, K.B. Sharpless, *Angew. Chem. Int. Ed.* **2002**, *41*, 490.

33 H. Han, J. Yoon, K.D. Janda, *J. Org. Chem.* **1998**, *63*, 2045.

34 S.-H. Lee, J. Yoon, S.-H. Chung, Y.-S. Lee, *Tetrahedron* **2001**, *57*, 2139.

35 (a) P. O'Brien, S.A. Osborne, D.D. Parker, *Tetrahedron Lett.* **1998**, *39*, 4099; (b) P. O'Brien, S.A. Osborne, D.D. Parker, *J. Chem. Soc., Perkin Trans I* **1998**, 2519.

36 Related diamine synthesis from racemic aziridinium compounds: (a) T.-H. Chuang, K.B. Sharpless, *Org. Lett.* **1999**, *1*, 1435; (b) T.-H. Chuang, K.B. Sharpless, *Org. Lett.* **2000**, *2*, 3555; (c) T.-H. Chuang, K.B. Sharpless, *Helv. Chim. Acta.* **2000**, *83*, 1734; (d) See also ref. 30.

37 For a general scheme on known olefin transformations, see Chapter 2.6.1.

38 (a) A.O. Chong, K. Oshima, K.B. Sharpless, *J. Am. Chem. Soc.* **1977**, *99*, 3420; (b) M.H. Schofield, T.P. Kee, J. T. Anhaus, R.R. Schrock, K.H. Johnson, W.M. Davis, *Inorg. Chem.* **1991**, *30*, 3595.

39 For stoichiometric metal-mediated, stepwise diamination of olefins, see: (a) V. Gómez Aranda, J. Barluenga, F. Aznar, *Synthesis* **1974**, 504; (b) J.-E. Bäckvall, *Tetrahedron Lett.* **1975**, 2225.

40 For catalytic achiral diamination of olefins, see: (a) J.U. Jeong, B. Tao, I. Sagasser, H. Henniges, K.B. Sharpless, *J. Am. Chem. Soc.* **1998**, *120*, 6844; (b) G. Li, H.-X. Wei, S.H. Kim, M. Carducci, *Angew. Chem., Int. Ed.* **2001**, *40*, 4277; (c) H.-X. Wei, S.H. Kim, G. Li, *J. Org. Chem.* **2002**, *67*, 4777.

41 K. Muñiz, *Eur. J. Org. Chem.* **2004**, 2243.

42 (a) K. Muñiz, M. Nieger, *Synlett* **2003**, 211; (b) K. Muñiz, A. Iesato, M. Nieger, *Chem. Eur. J.* **2003**, *9*, 5581.

43 D.V. Deubel, K. Muñiz, *Chem. Eur. J.* **2004**, *10*, 2475.

2.7
Epoxidations

2.7.1
Titanium-Catalyzed Epoxidation

Tsutomu Katsuki

2.7.1.1
Introduction

Titanium with an oxidation state of IV is stable, and various titanium(IV) complexes are readily available. Most of these are of low toxicity and show high catalytic performance for epoxidation. Accordingly, many titanium-mediated epoxidation reactions have been reported, and the reactions reported before 1997 have been summarized in the first edition of this book. Since then, several significant advancements have been made in this field, especially in heterogeneous epoxidation, and these are summarized in this chapter.

2.7.1.2
Epoxidation using Heterogeneous Catalysts

Titanium silicalite-1 (TS-1), which has an active $Ti(OSi{\equiv})_n$ site in a hydrophobic cavity, is one of the best heterogeneous catalysts for the epoxidation of alkenes with hydrogen peroxide [1], but it is less active toward epoxidation of bulky substrates such as branched and cyclic alkenes because of the diffusion restriction imposed by the medium-sized pore. In the mid-1990s, metal-containing mesoporous silicates, such as Ti-MCM-41 [2], Ti-beta [3], Ti-HMS [2b, 4], and amorphous titania-silica aerogel [5], showed high catalytic performance for selective oxidation of bulky substrates [6]. Since then, many studies on catalytic performance of various titanium-containing mesoporous materials have been implemented.

Despite their unique catalytic performance, mesoporous catalysts in general suffer a disadvantage, namely reduced hydrothermal and mechanical stability due to hydrophilicity caused by surface silanol groups.

Transition Metals for Organic Synthesis, Vol. 2, 2nd Edition.
Edited by M. Beller and C. Bolm

ISBN: 3-527-30613-7

To increase hydrophobicity, mesoporous mixed oxide (AM-Ti3) has been prepared by copolymerization of tetraethoxysilane and methyltriethoxysilane in the presence of titanium isopropoxide. AM-Ti3 shows high selectivity and high conversion in the epoxidation of a wide range of alkenes using *t*-butyl hydroperoxide (TBHP) (Scheme 1) [7]. The presence of a methylsilyl group makes the mesopore hydrophobic and improves its catalytic performance. It has also been reported that all-silica mesoporous MCM-41 with $Ti(OSiPh_3)_4$ grafted onto its internal surface [(Ph_3SiO)Ti-MCM-41] shows higher catalytic performance than (HO)Ti-MCM-41 [8]. Amorphous mesoporous titania-silica aerogel shows catalytic performance similar to that of AM-Ti3 [7]. It is noteworthy that the presence of an allylic hydroxy group enhances the epoxidation rate in homogeneous metal-mediated epoxidation, while it decreases the rate in heterogeneous epoxidation using a catalyst such as AM-Ti13 [7] and aerogel [6c]. The rate reduction in the heterogeneous reactions has been attributed to slower rates of diffusion of the more polar substrates in the catalyst pores. Despite this description, epoxidation of non-branched allylic alcohols using amorphous mesoporous titania-silica aerogel shows high selectivity as well as high chemical yield [5c]. It has also been reported that epoxidation of allylic and homoallylic alcohols is effected by using a modified amorphous mesoporous titania-silica as catalyst in ethyl acetate [9].

Ti-MCM-41 and Ti-MCM-48 have been modified by trimethylsilylation. The modified Ti-MCM-41 (sil) and Ti-MCM-48 (sil) show higher catalytic activity in the oxidation of alkenes using hydrogen peroxide than the parent Ti-MCM-41 and Ti-MCM-48, but epoxide selectivity is modest [10].

Hydrothermally stable SBA-12 is a highly ordered mesoporous silica possessing thick walls [11]. Ti-SBA-15 (sil) postsynthesized by the titanation of SBA-12 and subsequent trimethylsilylation has been reported to promote epoxidation of cyclohexene using TBHP with high epoxide selectivity (epoxide selectivity=97%, TON=843) [12].

Epoxidation of α,β-unsaturated ketones has also been studied. Epoxidation using a Ti-beta/aqueous H_2O_2 system in acetonitrile, a weak basic solvent, shows high epoxide selectivity, though conversion of the substrates are moderate (Scheme 2) [13]. On the other hand, epoxidation of cyclohexenone using an amorphous mesoporous titania-silica aerogel-TBHP system in toluene proceeds with moderate epoxide selectivity and modest substrate conversion [5c].

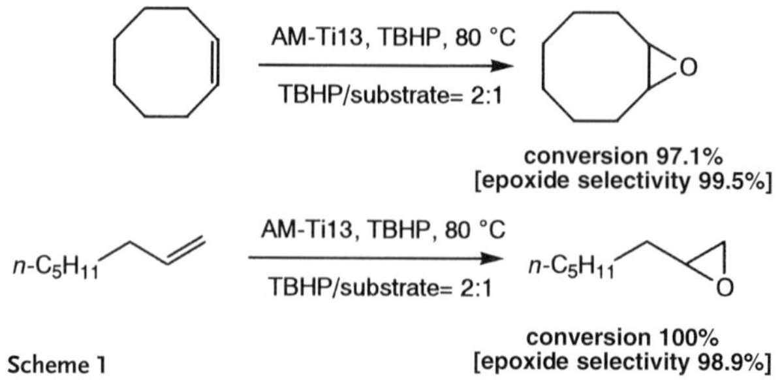

AM-Ti13, TBHP, 80 °C
TBHP/substrate= 2:1
O
conversion 97.1%
[epoxide selectivity 99.5%]

n-C_5H_{11}
AM-Ti13, TBHP, 80 °C
TBHP/substrate= 2:1
n-C_5H_{11}
O
conversion 100%
[epoxide selectivity 98.9%]

Scheme 1

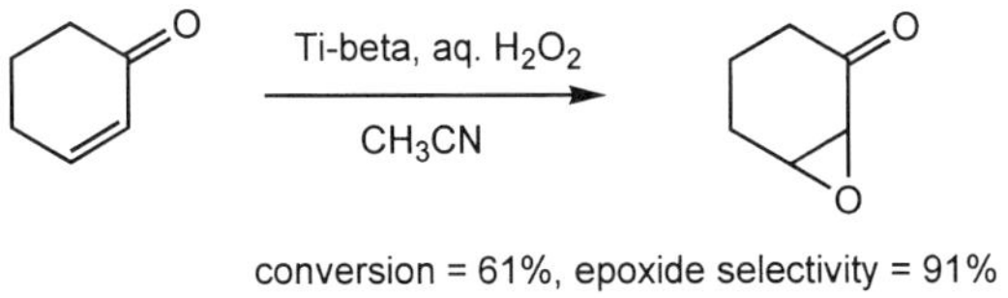

Ti-beta, aq. H_2O_2

CH_3CN

conversion = 61%, epoxide selectivity = 91%

Scheme 2

It has recently been reported that some heterogeneous catalysts other than titanium silicates are efficient for epoxidation using hydrogen peroxide as the oxidant.

Amphiphilic titanium-loaded zeolite (W/O-Ti-NaY) that is partly modified with hydrophobic alkylsilane shows unique catalytic properties for epoxidation [14]. W/O-Ti-NaY locates at the boundary between aqueous and organic phases because of its amphiphilic nature and catalyzes epoxidation with hydrogen peroxide smoothly without stirring. Although the substrates used are limited to terminal alkenes, epoxide selectivity is high (Scheme 3).

Titanium silsesquinoxanes [15, 16], which are soluble in organic solvents, are efficient catalysts for epoxidation using TBHP as the oxidant. Of these, titanium cyclopentylsilsesquinoxane (**1**) shows high catalytic activity together with high epoxide selectivity (Scheme 4) [16b].

It has been reported that titanium silsesquinoxane grafted onto three-dimensionally netted polysiloxane (**2**) shows catalytic activity similar to that of TS-1 [17]. Aqueous hydrogen peroxide can be used as the terminal oxidant for the epoxidation with **2**. It is noteworthy that bulky cyclic alkenes such as cyclooctene and cyclodecene can be selectively epoxidized using **2** as the catalyst (Scheme 5) [17].

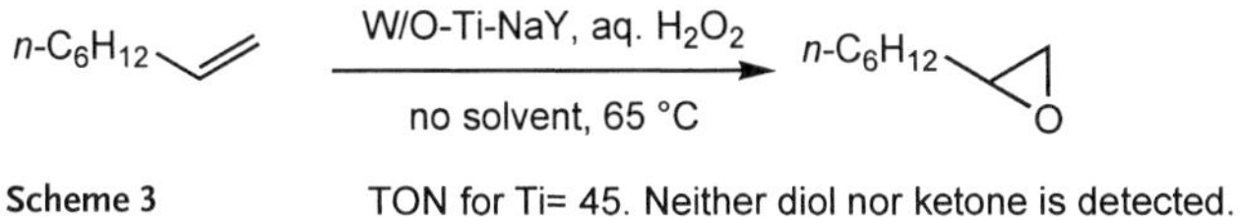

n-C_6H_{12} W/O-Ti-NaY, aq. H_2O_2 no solvent, 65 °C n-C_6H_{12}

Scheme 3 TON for Ti= 45. Neither diol nor ketone is detected.

1, TBHP

$CDCl_3$, 50 °C

conversion = 93%, epoxide selectivity = 98%

1: R= c-C_5H_9

titanium silsesquinoxane

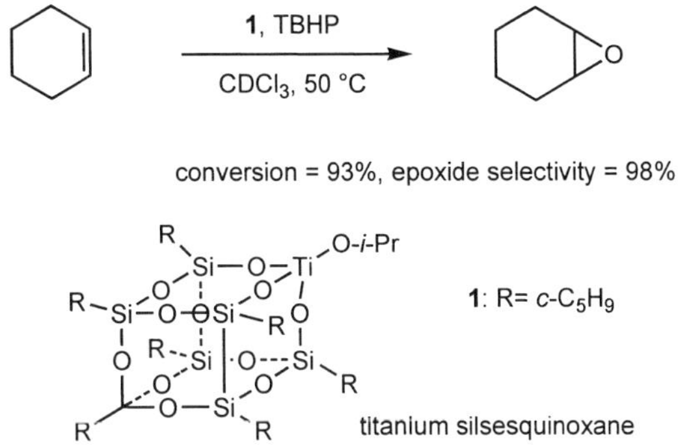

Scheme 4

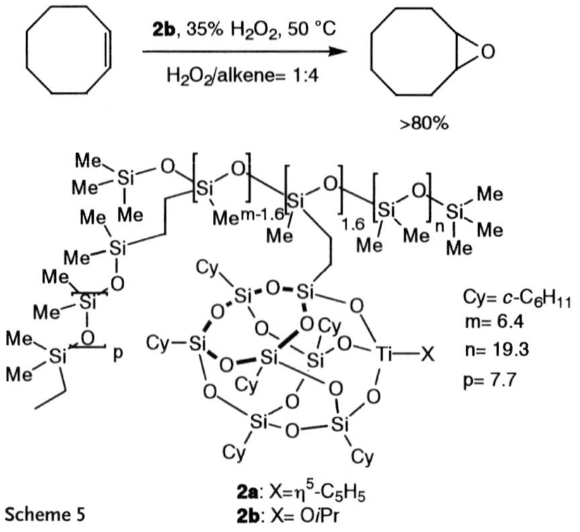

Scheme 5

Epoxidation with hydrogen peroxide in the presence of titanium-containing mesoporous silicates usually shows moderate epoxide selectivity [18].

2.7.1.3
Epoxidation using Homogeneous Catalyst

Ti(IV)-calix[4]arene complex **3** has been used as a catalyst for the epoxidation of allylic alcohols. As the exchange of chloro ligand with alcohol is slow, the catalytic activity of complex **3** itself is low. However, molecular sieves accelerate the ligand exchange, and the epoxidation using **3** as the catalyst in the presence of molecular sieves proceeds smoothly (Scheme 6) [19].

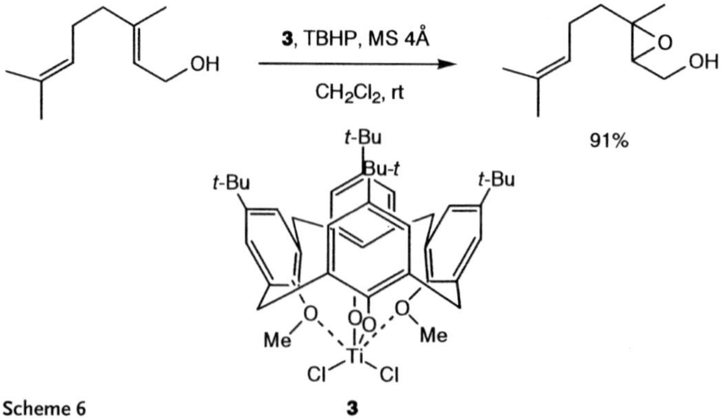

Scheme 6

2.7.1.4
Asymmetric Epoxidation

Since the discovery of asymmetric epoxidation of allylic alcohols using titanium tartrate catalyst [20], several efforts have been made to immobilize the catalyst. An early effort to develop a polymer-supported system met with a reduced enantioselectivity compared with that of the original homogeneous system [21]. Use of linear poly(tartrate ester) ligand **4** has recently been reported to give good chemical yield and to increase the enantioselectivity to 79% in the epoxidation of (*E*)-hex-2-en-1-ol, though the selectivity is still inferior to that of the original system (Scheme 7) [22]. It has also been reported that use of the gel-type crosslinked poly(tartrate ester) ligand **5** (a level of crosslinking ~ 3%, ligand : Ti = 2:1) further improves enantioselectivity up to 87% in the epoxidation of the same substrate (Scheme 7) [23]. However, enantioselectivity of the reaction is dependent on ligand : Ti ratio and the degree of crosslinking. Use of the ligand **5** with a degree of higher crosslinking reduces enantioselectivity. Under the optimized conditions, epoxidation of *E*-allylic alcohols proceeds with good to high enantioselectivity, but the epoxidation of geraniol shows moderate enantioselectivity.

High enantioselectivity and acceptable chemical yield have been quite recently achieved in the epoxidation of (*E*)-hex-2-en-1-ol by using soluble polymer-supported tartrate ester synthesized from tartaric acid and polyethylene glycol monomethyl ether as the ligand (Scheme 8) [24]. It is noteworthy that the polymer-supported tartrates **6a** and **6b** differ only in the length of the polymer units, but the sense of the asymmetric induction of each is opposite to that of the other. The origin of the reversal of asymmetric induction is unclear at present. Recovery of the polymer-supported tartrate is fairly simple, but only moderate enantioselectivity (**6b**: 49% *ee*) has been achieved by the epoxidation using the recovered catalyst.

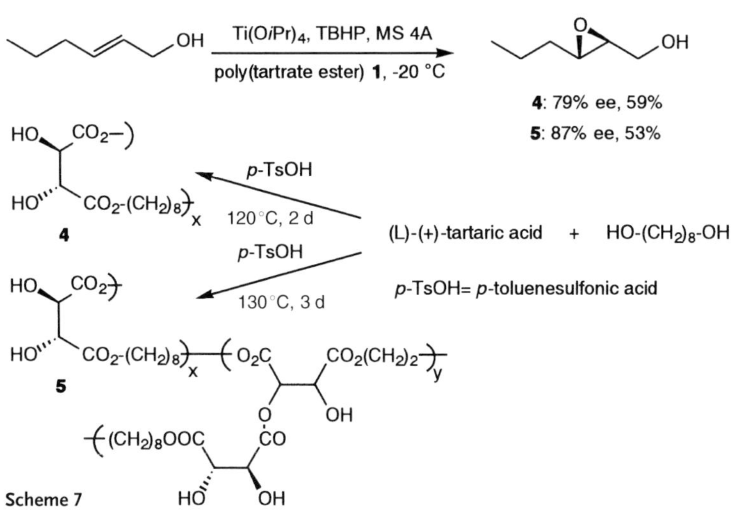

$Ti(OiPr)_4$, TBHP, MS 4A; poly(tartrate ester) **1**, -20 °C

4: 79% ee, 59%
5: 87% ee, 53%

(L)-(+)-tartaric acid + $HO\text{-}(CH_2)_8\text{-}OH$

p-TsOH, 120 °C, 2 d → **4**
p-TsOH, 130 °C, 3 d → **5**

p-TsOH= p-toluenesulfonic acid

Scheme 7

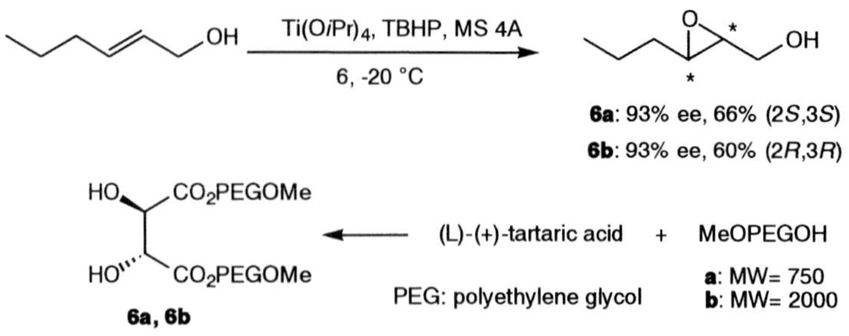

Scheme 8

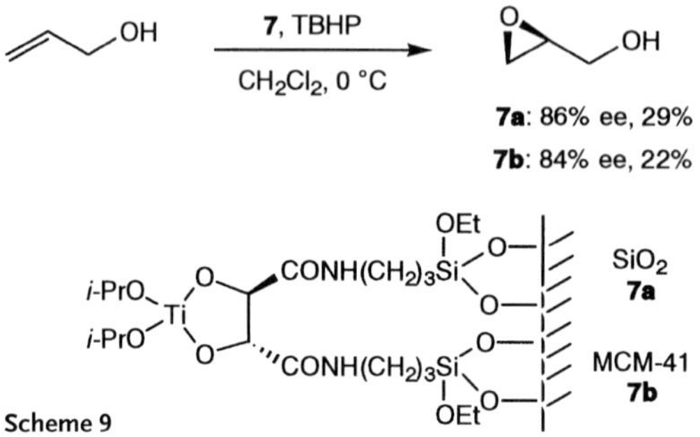

Scheme 9

Epoxidation of allylic alcohols other than (*E*)-hex-2-en-1-ol with these polymer-supported tartrates has not been reported.

Chiral tartramide derivatives 7 grafted onto inorganic supports SiO_2 and mesoporous MCM-41 have been successfully used as the chiral auxiliaries for the epoxidation of allyl alcohol (Scheme 9) [25]. Enantioselectivity of the epoxidation is al-

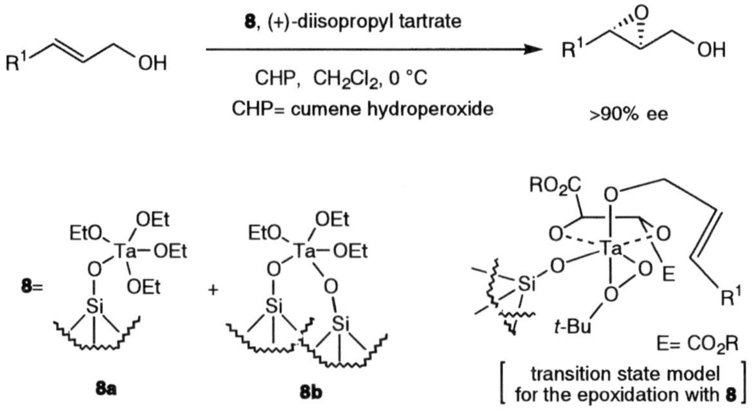

Scheme 10

most identical with that obtained by using homogeneous Ti/DET catalyst. An advantage of this system is that the catalyst is easily removed by simple filtration.

In connection with heterogeneous titanium tartrate catalyst, it is noteworthy that a combination of DAT and silica-supported tantalum alkoxides (**8a** and **8b**) prepared from $Ta(=CHCMe_3)(CH_2CMe_3)_3$ and $silica_{(500)}$ serves as an efficient catalyst for the epoxidation of *E*-allylic alcohols (Scheme 10), though homogeneous tantalum tartrate is a poor catalyst [26].

References

1 B. Notari, *Catal. Today* **1993**, *18*, 163–172.

2 (a) A. Corma, M. T. Navarro, J. Perez-Pariente, *J. Chem. Soc., Chem. Commun.* **1994**, 147–148. (b) P.T. Taneb, M. Chibwe, T. J. Pinnavaia, *Nature* **1994**, *368*, 321–323. (c) W. Zhang, M. Fröba, J. Wang, P.T. Tanev, J. Wong, T.J. Pinnavaia, *J. Am. Chem. Soc.* **1996**, *118*, 9164–9171. (d) S. Gontier, A. Tuel, *Appl. Catal. A: Gen.* **1996**, *143*, 125–135.

3 J.C. van der Waal, P. Lin, M.S. Rigutto, H. van Bekkum, *Stud. Surf. Sci. Catal.* **1997**, *105*, 1093–1100.

4 S. Gontiel, A. Tuel, *Zeolites* **1995**, *15*, 601–610.

5 (a) R. Hutter, T. Mallat, A. Baiker, *J. Catal.* **1995**, *153*, 177–189. (b) R. Hutter, T. Mallat, A. Baiker, *J. Catal.* **1997**, *157*, 665–675. (c) M. Dusi, T. Mallat, A. Baiker, *J. Mol. Catal. A: Chem.* **1999**, *138*, 15–23.

6 (a) R. Murugavel, H.W. Roesky, *Angew. Chem. Int. Ed. Engl.* **1997**, *36*, 477–479. (b) R.A. Sheldon, M. Wallau, I.W.C. A. Arends, U. Schuchardt, *Acc. Chem. Res.* **1998**, *31*, 485-493. (c) A.A. Sheldon, M.C.A. van Vliet, *Fine Chemicals through Heterogeneous Catalysis*, Wiley-VCH Verlag GmbH, Weinheim, **2001**, pp 473-490.

7 Y. Deng, W.F. Marier, *J. Catal.* **2001**, *199*, 115–122.

8 M.P. Attfield, G. Sankar, J.M. Thomas, *Catal. Lett.* **2000**, *70*, 155–158.

9 C. Berlini, G. Ferraris, M. Guidotti, G. Moretti, R. Psaro, N. Ravasio, *Microporous and Mesoporous Materials* **2001**, 595–602.

10 T. Tatsumi, K.A. Koyano, N. Igarashi, *Chem. Commun* **1998**, 325–326.

11 D. Zhao, Q. Huo, J. Feng, B.F. Chemelka, G.D. Stucky, *J. Am. Chem. Soc.* **1998**, *120*, 6024–6036.

12 P. Pu, T. Tatsumi, *Chem. Mater.* **2002**, *14*, 1657–1664

13 M. Sasidharan, P. Wu, T. Tatsumi, *J. Catal.* **2002**, *205*, 332–338.

14 (a) H. Nur, S. Ikeds, B. Ohtani, *Chem. Commun.* **2000**, 2325–2326. (b) H. Nur, S. Ikeds, B. Ohtani, *J. Catal.* **2001**, *204*, 402–408.

15 F.J. Feher, T.A. Budzichowski, *Polyhedron* **1995**, *14*, 3239–3253.

16 (a) T. Maschmeyer, M.C. Klunduk, C.M. Martin, D.S. Shephard, J.M. Thomas, B.F.G. Johnson, *Chem. Commun.* **1997**, 1847–1848. (b) P.P. Pescarmona, J.C. van der Waal, I.E. Maxwell, T. Maschmeyer, *Angew. Chem. Int. Ed.* **2001**, *40*, 740–743.

17 M.D. Skowronska-Ptasinska, M.L.W. Vortenbosch, A.A. van Santen, H.C. L. Abbenhuis, *Angew. Chem. Int. Ed.* **2002**, *41*, 637–639.

18 J.M. Fraile, J.I. Garcia, J.A. Mayoral, E. Vispe, *J. Catal.* **2001**, *204*, 145–156.

19 A. Massa, A. D'Ambrosi, A. Proto, A. Scettri, *Tetrahedron Lett.* **2001**, *42*, 1995–1998.

20 T. Katsuki, K.B. Sharpless, *J. Am. Chem. Soc.*, **1980**, *102*, 5974–5976.

21 M.J. Farrall, M. Alexis, M. Trecarten, *Nouv. J. Chim*,. **1983**, *7*, 449–451.

22 L. Canali, J.K. Karjalanien, D.C. Sherrington, O. Hormi, *Chem. Commun.* **1997**, 123–124.

23 J.K. Karjalanien, O.E.O. Hormi, D.C. Sherrington, *Tetrahedron Asymmetry*, **1998**, *9*, 1563–1569.

24 H. Guo, X. Shi, Z. Qiao, S. Hou, M. Wang, *Chem. Commun.* **2002**, 118–119.
25 S. Xiang, Y. Zhang, Q. Xin, C. Li, *Angew. Chem. Int. Ed.* **2002**, *41*, 821–824.
26 D. Meunier, A. Piechaczyk, A. Mallmann, *Angew. Chem. Int. Ed. Engl.* **1999**, *38*, 3540–3542.

2.7.2
Manganese-Catalyzed Epoxidations

Kilian Muñiz and Carsten Bolm

2.7.2.1
Introduction

The conversion of unfunctionalized olefins into epoxides has remained of general interest over many years. A wide range of transition metals are known to catalyze this transformation [1]. For asymmetric oxidations of such olefins [2], systems based on manganese have proved to be the most successful. Basically, three Mn-catalyzed epoxidation systems have been developed: salen-based complexes for enantioselective epoxidations with oxidants such as bleach, related manganese complexes for aerobic epoxidations, and triazacyclononane-based Mn complexes for epoxidations in the presence of hydrogen peroxide. In this chapter, we will discuss all three approaches, with special emphasis on the structural features which are required for efficient asymmetric epoxidations [3, 4].

2.7.2.2
Salen-based Manganese Epoxidation Complexes

After their studies on the use of salen chromium complexes [salen = *N,N*-ethylenebis(salicylidene aminato)] as catalysts for epoxidations of olefins [5], Kochi and coworkers searched for analogous metal complexes having higher catalytic reactivity. Thus, in 1986 they reported on cationic salen manganese(III) catalysts of type **1** and their capability to efficiently oxidize various types of olefins with iodosylbenzene as terminal oxidant [6]. Together with the X-ray crystal structure of a bis(pyridine) adduct of a salen manganese(III) complex, several important features of this new system, such as the stereospecificity of the epoxidation, the importance of axial donor ligands (D), the involvement of an intermediate oxomanganese(V) species (**2**), and a first discussion on possible radical intermediates were presented.

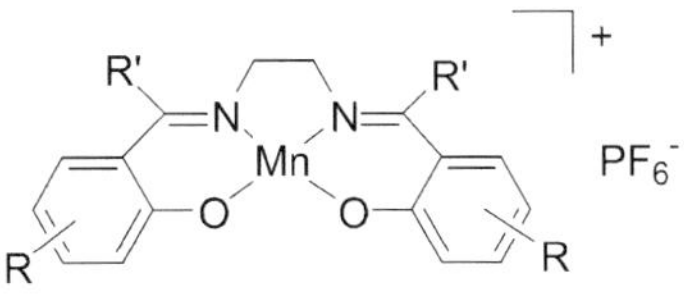

1 (R = H, MeO, Cl, NO_2, R' = H, Ph)

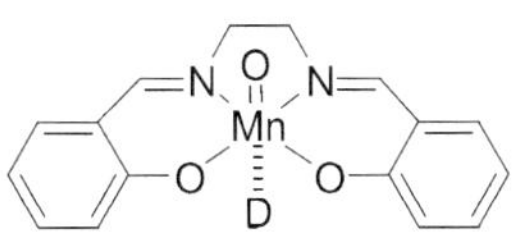

2

Only a few years after this key publication by Kochi, an asymmetric version of this olefin epoxidation was reported. In 1990, Jacobsen [7] and Katsuki [8] independently reported that the use of chiral salen manganese(III) catalysts resulted in the formation of optically active epoxides. The straightforward synthesis of the required enantiomerically pure salen ligands is conveniently carried out starting with optically active diamines and substituted salicyl aldehydes. In general, refluxing these starting materials in polar solvents, followed by reaction with $Mn(OAc)_2$, anion exchange with LiCl, and subsequent aerobic oxidation, leads to the readily available salen manganese(III) complexes [9]. The ease of this procedure allowed a rapid and extensive screening of various ligands with different electronic and steric patterns. The complexes themselves are usually obtained in form of dark, air-stable solids. Rational ligand optimization [10, 11] led to the development of two different salen manganese(III) type complexes **3** and **4** by Jacobsen and Katsuki, respectively.

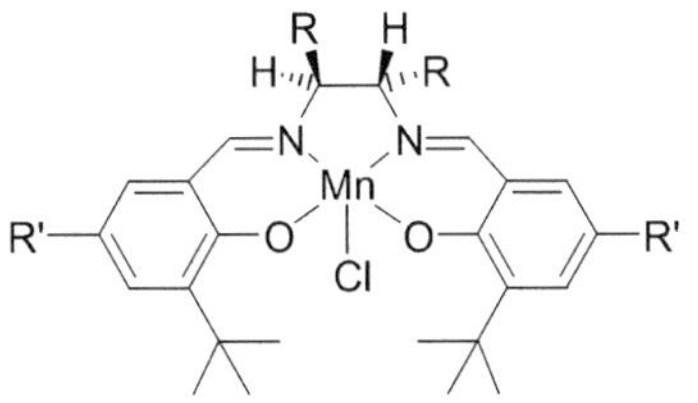

3a (R, R = -$(CH_2)_4$-, R' = *t*Bu)
3b (R, R = -$(CH_2)_4$-, R' = OMe)
3c (R, R = -$(CH_2)_4$-, R' = OSi*i*Pr_3)
3d (R = Ph, R' = Me)

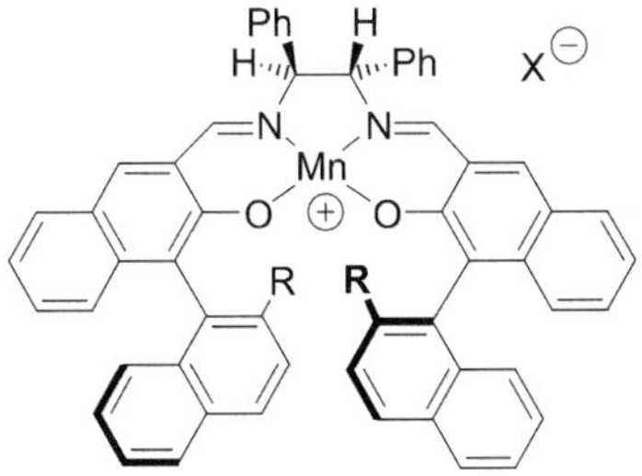

4a (R = Me, X = OAc)
4b (R = Ph, X = OAc)
4c (R = Ph, X = PF_6)

These chiral salen manganese(III) complexes are excellent catalysts in the enantioselective epoxidation of unfunctionalized olefins. Their use has already been intensively reviewed elsewhere [10, 12–17].

Epoxidations employing **3** or **4** are conveniently carried out in acetonitrile or dichloromethane with commercial bleach (NaOCl) or iodosylbenzene PhIO [18] as the oxygen source. Amine *N*-oxides such as *N*-methylmorpholine *N*-oxide (NMO), 4-phenylpyridine *N*-oxide or isoquinoline *N*-oxide serve as optional axial donor ligands and are proposed to stabilize the active oxo manganese species [19, 20]. In general, (*Z*)-disubstituted olefins give nearly perfect enantioselection [10], although in some cases attention has to be paid to electronic pattern and conjugated systems. For example, *α,β*-conjugated (*Z*)-olefins exhibit a remarkable isomerization during epoxidation, affording mixtures of *cis-* and *trans*-olefins [21, 22]. The analogous (*E*)-disubstituted olefins are less appropriate substrates. Special care has to be exercised for monosubstituted olefins. For example, high enantioselectivities in the epoxidation of styrene were only accomplished if the reaction was carried out at –78 °C in dichloromethane with *m*CPBA as oxidant in the presence of NMO, affording styrene oxide in 88% yield and 86% *ee* [23]. Tri- and tetrasubstituted olefins are generally good substrates as well [24, 25], although for the latter only chromene derivatives give enantioselectivities higher than 90% *ee*.

While the high efficiency of the Jacobsen-Katsuki epoxidation is widely acknowledged, the exact mechanism of the reaction remains a matter of debate [10, 16, 26–29]. Recently, significant work has been devoted to electron spray MS detection of the elusive key species, the oxomanganese(V) complex **2** [30]. Bolm, Bertagnolli and co-workers attempted to analyze this species by UV/Vis, Raman, XANES and EXAFS spectroscopy [31].

Various models (Scheme 1) have been proposed to explain the observed enantioselection. Concerning the path of the incoming olefin, reaction mechanisms involving a so-called "side-on approach" of the olefin parallel to the salen ligand were suggested (**A**) [10, 11, 16, 26]. The bulky substituents at the aryl groups, steric repulsions, and electronic interactions then control the approach of the olefin. The observed high enantiocontrol originates from a strict distinction between the two prochiral faces of the alkene by differentiation of the larger residue (R_L) from the smaller one (R_S). Although Jacobsen and Katsuki favor different approaches (**I** and **II**, respectively) both models are similar in general terms and provide a good

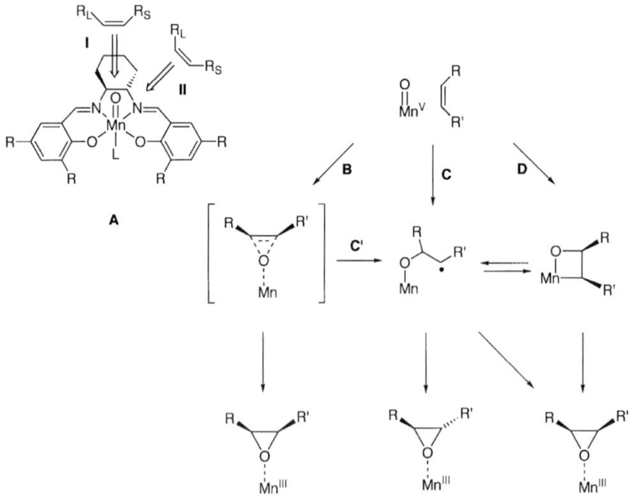

Scheme 1 Model for the "side-on approach" (**A**) and three distinct mechanisms for the oxygen transfer to olefins (**B**, **C**, **C′** and **D**).

explanation for the observed stereochemical outcome of the catalytic process. The subsequent step, the exact modality of the oxygen transfer, is still under debate. In principle, three different reaction sequences deserve attention: a concerted pathway (**B**), a route involving radical intermediates (**C**), and the formation of a metallaoxetane (**D**) [32].

Since alkyl-substituted (*Z*)-olefins yield the corresponding *cis*-oxiranes stereospecifically, the concerted pathway (**B**) is widely accepted for this type of substrate [10]. This is not the case for conjugated (*Z*)-olefins, which under standard reaction conditions are converted into a mixture of *cis*- and *trans*-epoxides. This result has been explained by assuming a reaction pathway (**C**) via a radical intermediate, which allows C-C bond rotation to give both *cis*- and *trans*-configurated products [20, 21, 33]. Contrary to this assumption, the epoxidation of various substituted vinylcyclopropanes revealed that, under the reaction conditions mentioned, neither epimerization nor cleavage of the cyclopropane occurred. Consequently, the reaction pathway involving radical intermediates (**C**) was rejected [28], and a different process via a metallaoxetane was postulated instead (**D**) [29]. In accord with a pathway of type **D** were results by Katsuki, who found a non-linear relationship between enantioselectivity and temperature, indicating the presence of a reversibly formed diastereomeric intermediate [34], which was suggested to be the manganaoxetane. However, Jacobsen did not find such a non-linear Eyring correlation for his salen manganese(III) system [20, 33], but both groups have pointed out the importance of entropic and enthalpic factors [33, 35]. As a consequence, Jacobsen has postulated a common early transition state via route **B** that either yields epoxides stereospecifically or leads to a radical intermediate (route **C′**). However, recent investigations point toward different reaction routes depending on the substrate class, and, in this context, Linde has presented a Hammet study that is inconsistent with **C′** [36]. Adam and Seebach have interpreted the degree of the final *cis*:*trans* ratio as a result of competitive concerted and radical-based mechanisms. These were correlated to an influence of both the terminal oxidant and the counterion in the salen Mn(III) complex [37, 38]. Moreover, the reaction course was analyzed by theoretical calculations [39–42], one of which indicated a relation between spin changes in the Mn oxo species and *cis*:*trans* ratios [39].

Since the oxo catalyst remains elusive, several discussions have been centered on the readily available Mn(III) salen catalyst precursors, and right from the beginning of asymmetric Mn salen epoxidation catalysis, several X-ray crystal structures became known. A series of them were compared in order to gain further insight into the relationship between the structures of the catalysts and their enantiofacial control [7, 43–45]. Very detailed structural elucidations stem from Katsuki and co-workers, who analyzed salen complexes (*R*,*S*)-**4c**$(OH_2)_2$ and (*R*,*S*)-**4c**(OH_2) (cyclopenteneoxide) and their respective diastereomeric (*R*,*R*)-counterparts with opposite absolute configuration in the chiral diamine backbone [16, 46–49]. The complexes with the apical cyclopentene oxide ligand formally represent the stereochemical scenery *after* occurrence of the oxo-transfer, and stereochemical conclusions must be drawn most carefully. Nevertheless, the stereochemical consequences are obvious: a conformation with the two phenyl substituents of the eth-

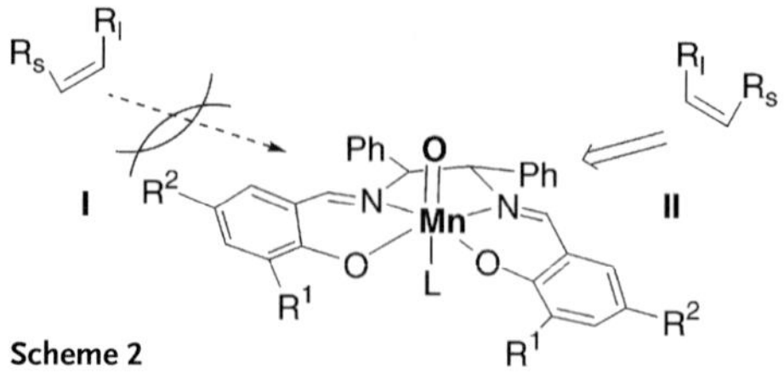

Scheme 2

ylene diamine moiety in equatorial position must be favored, thereby enforcing a non-planar structure of the aromatic biaryl groups. The resulting overall geometry for the preferential catalyst derived from (*R*,*S*)-**4c** is depicted in Scheme 2.

Obviously, a bulky aryl substitution pattern opens a single pathway for the approaching olefin (top right, identical to path II in Scheme 1), which has to be (*Z*)-configurated in order to minimize steric interactions with the ligand framework. As expected, the related complex with diastereomeric configuration displays a more pronounced folding, which retains an approximation of the incoming olefin. These X-ray structures were the first ones to unambiguously prove a non-pliable ligand conformation of hexa-coordinated salen manganese complexes [50].

As a further result of these structural insights, the oxidation of (*E*)-configurated olefins was re-examined [16]. Apparently, rigid and sterically crowded salen ligands such as the parent structure **4** suffer from unfavorable interactions between the incoming olefin and the ligand outer sphere (model **E**). In contrast, the catalyst precursor complex **5** (Scheme 3) leads to an oxo manganese(V) catalyst with a deeply folded structure and a significant decrease in steric hindrance (model **F**). Thus, oxidation of (*E*)-*β*-methyl styrene gives an epoxide with the high enantiomeric excess of 91% [49].

Owing to their non-pliable structure as well as to their conformational and configurational flexibility, oxomanganese(V) complexes with achiral salen ligands will exist as a racemic mixture of two enantiomers, thereby yielding racemic epoxides. However, significant enantiomeric excesses were obtained in the presence of non-racemic axial donor ligands such as sparteine (*ee* up to 73%) [51] and 2,2′-bipyri-

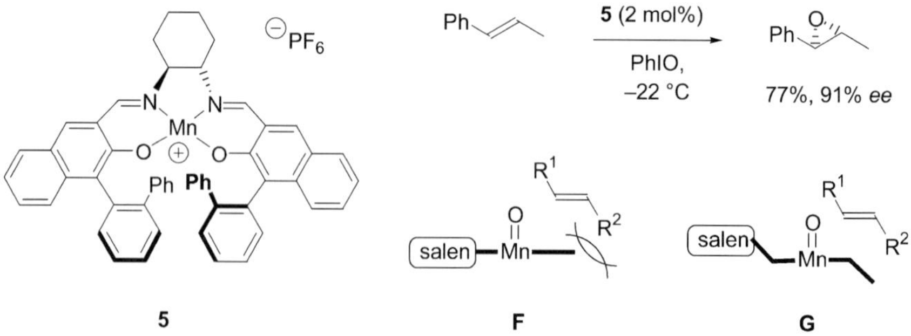

Scheme 3

dine *N,N'*-dioxide (*ee* up to 73%) [52]. Coordination of these ligands to manganese creates complexes of diastereomeric composition that enforce significant differences in equilibria, reaction rate, and enantioselectivities [53]. These examples were among the first in the area of asymmetric activation of configurationally flexible catalyst precursors [54, 55].

In addition to the catalysts mentioned above, various other salens or salen-like epoxidation systems have been reported, including a Katsuki-type salen ligand with intramolecular axial donor functionality [56], a C_1-symmetrical pentadentate salen-type ligand that employs hydrogen peroxide as terminal oxidant [57], and chiral binaphthyl Schiff bases [58], the latter system being particularly interesting for stereospecific epoxidation of (*Z*)-configurated olefins.

The applicability of the Jacobsen salen manganese catalysts for the synthesis of pharmaceutically important compounds, such as *Indinavir* [59], the TAXOL® side chain [60] or *BRL 55834* [61], has recently been demonstrated. Because of the impressive success of these asymmetric man-made catalysts, they have been compared positively with enzymes and catalytic antibodies [62].

A few attempts to prepare heterogeneous salen manganese catalysts [63] or membrane incorporation of Jacobsen-type catalysts [64] have also been reported, and alternative catalytic procedures developed so far include fluorinated chiral salen ligands for asymmetric epoxidation under biphasic conditions [65] as well as in ionic liquids [66].

2.7.2.3
Aerobic Epoxidation with Manganese Complexes

Mukaiyama reported the use of various manganese complexes in *aerobic* epoxidations [67, 68]. First, diastereoselective oxidations catalyzed by achiral (*β*-diketonato) manganese(II) complexes using cholesterol derivatives as test substrates were described [69]. Under 1 atm of molecular oxygen and in the presence of isobutyraldehyde, catalysis by bis(dipivaloylmethanato)manganese(II) [$Mn(dpm)_2$] afforded the corresponding *β*-epoxides with up to 82% *de*. This result was of particular interest because, unlike the case for epoxidations with *m*CPBA, this oxidation occurred preferentially from the more hindered *α*-face of the steroid, suggesting that the epoxidation with the manganese complex was not a process involving a simple carboxylic peracid generated from the aldehyde by autoxidation, but rather that an oxygenated metal complex was the reactive intermediate.

Soon after these studies, Mukaiyama reported *enantioselective* aerobic epoxidations [70]. Now salen manganese(III) complexes were used, and the combination of dioxygen and pivaldehyde gave the corresponding epoxides of several 1,2-dihydronaphthalenes in moderate to good yields. In order to achieve reasonably high enantioselectivities (up to 77% *ee*), the addition of *N*-methyl imidazole was essential. In its absence, chemical and optical yields were low, and the resulting epoxides had opposite absolute configuration. Additives of such a kind were suggested to act as additional ligands in the axial position to the metal center, and, by careful screening of various *N*-alkyl imidazoles, the enantiomeric excesses of the prod-

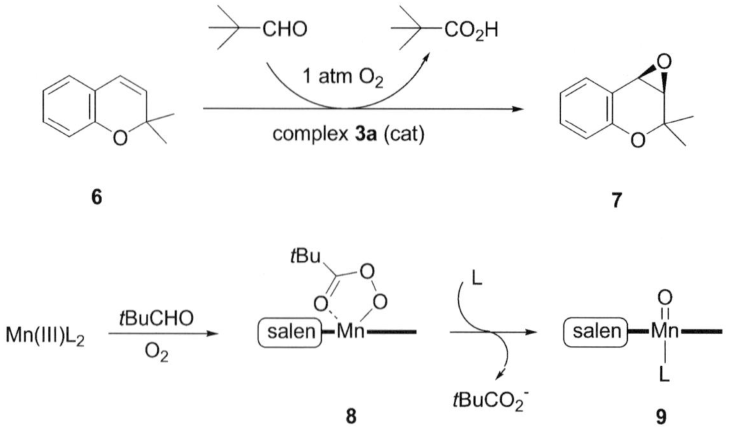

Scheme 4 Aerobic epoxidation under Mukaiyama conditions and postulated intermediates (L = *N*-octyl imidazole).

ucts were largely improved. For example, in the presence of *N*-octyl imidazole, 2,2-dimethyl-2*H*-chromene **6** was converted into the corresponding epoxide **7** in 37% chemical yield and with 92% *ee* [71, 72].

In order to rationalize these results, the following mechanism (Scheme 4) was proposed (with 1,2-dihydronaphthalene as the substrate). In a first step, an acylperoxomanganese complex (**8**) is formed from dioxygen, pivaldehyde, and the salen manganese(III) complex. In the absence of any axial ligand, **8** leads to the (1*R*, 2*S*)-epoxide of the olefin. However, in the presence of an *N*-alkyl imidazole (L), the acylperoxomanganese complex **8** is transformed into the oxomanganese complex **9**, an intermediate which is in accordance with the one proposed in the Jacobsen-Katsuki epoxidation. This oxomanganese complex selectively gives the (1*S*, 2*R*)-enantiomer of the epoxide, which, in turn, is identical to the one obtained under the usual Jacobsen-Katsuki oxidation conditions with iodosylbenzene or sodium hypochlorite as terminal oxidants.

Altering the ligand structure from salen derivatives to optically active ketoimine-type ligands gave the novel manganese catalysts **10**, which oxidize dihydronaphthalenes under aerobic conditions with moderate to good enantioselectivities [73–76].

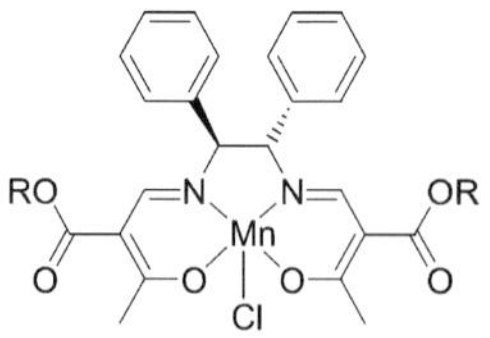

10 (with various R groups) 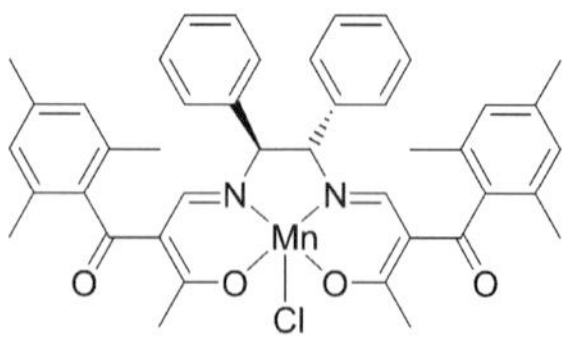 **11**

Further studies, including an X-ray crystal structure analysis [73], led to the rational design of the second-generation *β*-ketoiminato manganese(III) catalyst **11**, where the original *α*-ester groups of **10** were replaced by sterically more demanding mesitoyl moieties [74, 75]. Among other examples, oxidation of *cis*-*β*-methyl styrene yielded the optically active *cis*-epoxide with 80% *ee* in comparison to 67% *ee* obtained with the former ligand system. Most importantly, the enantiofacial selection in this aerobic epoxidation system again was the reverse of that obtained with a terminal oxidant such as sodium hypochlorite. It was deduced that the catalytically active species in this process must differ from the oxomanganese complex which is assumed for the epoxidation with terminal oxidants. Thus, the formation of an acylperoxomanganese complex like **8** from **11** was proposed.

2.7.2.4
Triazacyclononanes as Ligands for Manganese Epoxidation Catalysts

A different approach toward epoxidation of unfunctionalized olefins relates to biomimetic oxidations with manganese complexes [3]. Hage and co-workers from Unilever were interested in finding new catalysts for low-temperature bleaching [77], and in the course of these studies they also investigated the capability of manganese-containing systems with 1,4,7-triazacyclononanes (such as **12** or **13a**) as ligands to epoxidize styrenes. The activity of the resulting oxidation catalysts was remarkable. At an optimum pH of about 9.0 (buffered solution), a highly efficient epoxidation with hydrogen peroxide occurred, giving the corresponding epoxides with nearly quantitative conversion of the olefin.

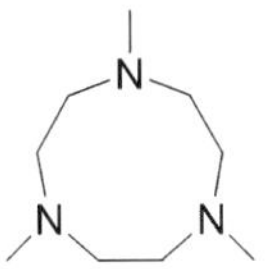

12 (tmtacn)

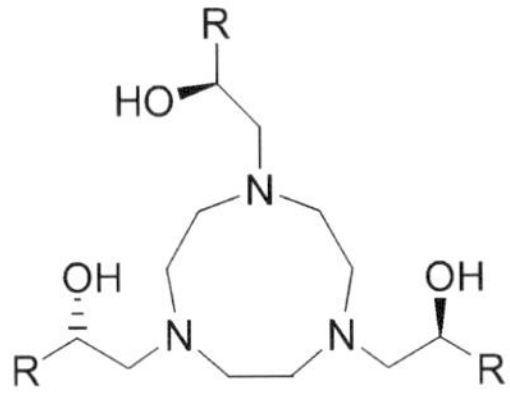

13a (R = H)
13b (R = Me)
13c (R = *i*Pr)

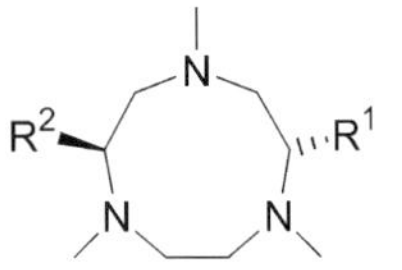

14a (R^1, R^2 = Me)
14b (R^1, R^2 = *i*Pr)
14c (R^1 = Me, R^2 = *i*Pr)

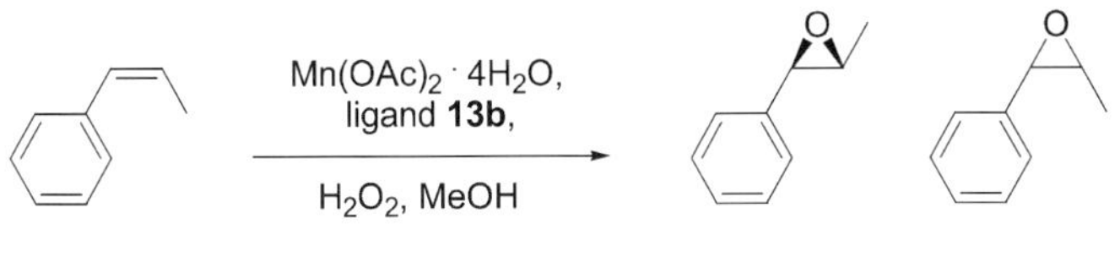

$Mn(OAc)_2 \cdot 4H_2O$, ligand **13b**, H_2O_2, MeOH

15 → *trans*-**16** + *cis*-**16**

Mn(OAc)$_3$, NaOAc,
NH_4PF_6, MeOH, H_2O

L_2Mn—O—MnL_2

17 **18**

Manganese complexes of 1,4,7-triazacyclononanes have been intensively studied by Peacock and Wieghardt [78]. Ligands with pendant arms bearing hydroxyl groups such as **13** are among the most interesting, because, upon complexation, the hydroxyls can either be deprotonated and coordinate to the central metal in an ionic manner or can remain protonated. However, the chemistry of the resulting complexes is highly complex and creates significant difficulties in the search for and design of defined triazacyclononane manganese complexes for epoxidation chemistry. However, these complexes are capable of overcoming the problem of hydrogen peroxide disproportionation, and after the original report by Hage [77] an optimization of the epoxidation protocol was described by various groups [79–82]. They found a dependence of the catalyst activity on temperature, solvent, and ligand structure.

It was only recently that applications of triazacyclononane manganese systems for enantioselective epoxidations were reported [83–89]. For example, Bolm described the use of an epoxidation catalyst formed *in situ* from manganese(II) acetate and enantiopure C_3-symmetric **13b** [83]. With hydrogen peroxide as the oxidant, epoxidation of (*Z*)-*β*-methylstyrene (**15**) yielded a 7:1 mixture of the corresponding isomeric epoxides *trans*-**16** with 55% *ee* and *cis*-**16** with 13% *ee*. This result favors the assumption that a stepwise radical mechanism is involved. For styrene oxide and 2,2-dimethyl-2*H*-chromene oxide (**7**), enantiomeric excesses of 43% and 38%, respectively, were observed. However, the enantiomeric excesses appeared to decrease upon longer reaction times, indicating that the catalytically active species presumably decomposes during the course of the catalysis [90]. Related chiral ligands with two stereogenic centers such as **14a–c** were described [86–89], but their manganese complexes led to inferior enantioselectivities (up to 23% *ee*). The interesting C_3-symmetric ligand **17** was synthesized by reduction of the corresponding L-proline cyclotripeptide [85]. Complexation of **17** to manganese afforded the bridged dimeric complex **18** that led to an insight into chiral enantiopure Mn TACN complexes. Among other results, cyclic voltammetry revealed a better stabilization for a higher oxidation state Mn in **18** than for the complex with the achiral ligand **12**. Consequently, asymmetric epoxidation with **18** enabled a high conversion (up to 88%), although the enantioselectivity did not exceed 26% *ee*. This result proves that further variations of the ligand system may be expected to have a significant effect on catalyst stability, activity, and enantioselectivity.

In addition, immobilizations of triazacyclononanes by either covalently attaching them to mesoporous siliceous support material (MCM-41) [91] or by incorporation into zeolites [92] have been described. The corresponding manganese complexes were then used as heterogeneous epoxidation catalysts [91–93]. Homogeneous polymer-bound systems (TACNs attached to ROMP-polymers) [94] and TACNs bearing fluoroponytail substituents [95] can also be used as ligands in manganese-catalyzed epoxidations.

2.7.2.5 Summary

In this review, several manganese-based epoxidation catalysts have been presented, with special focus on various ligand types and the resulting complexes. Extensive research has led to the discovery and development of a number of catalysts which now can be used for efficient olefin epoxidation. If it comes to stereochemical issues, however, significant problems remain to be solved. Thus, highly enantioselective transformations are still rare, and the discovery of appropriate catalysts in this field appears to be particularly difficult. Although several enantioselective catalysts are now known, most of these systems are either only applicable for a single specific class of olefins or do not satisfy the requirements in terms of extent of enantioselectivity and/or activity. Therefore, the ongoing search for new catalysts and the attempts at improving and further tailoring existing asymmetric catalytic systems are receiving close attention.

References

1 (a) K. A. Jørgensen in *Transition Metals for Organic Synthesis* (Eds. M. Beller, C. Bolm), Wiley-VCH, Weinheim, **1988**, *2*, p 157; (b) K. A. Jørgensen, *Chem. Rev.* **1989**, *89*, 431; (c) A. S. Rao in *Comprehensive Organic Synthesis* (Eds.: B. M. Trost, I. Fleming), Pergamon Press, Oxford **1991**, p. 357.

2 (a) V. Schurig, F. Betschinger, *Chem. Rev.* **1992**, *92*, 873; (b) S. Pedragosa-Moreau, A. Archelas, R. Furstoss, *Bull. Soc. Chim. Fr.* **1995**, *132*, 769; (c) P. Besse, H. Veschambre, *Tetrahedron* **1994**, *50*, 8885; (d) M. Bandini, P. G. Cozzi, A. Umani-Rochi, *Chem. Commun.* **2002**, 919; (e) C. Bonini, G. Righi, *Tetrahedron* **2002**, *58*, 4981.

3 For a review on biomimetic oxidations with manganese complexes: R. Hage, *Rec. Trav. Chim. Pays-Bas* **1996**, *115*, 385.

4 For reviews and leading references on oxidations with Mn porphyrins, see: (a) J. P. Collman, X. Zhang, V. J. Lee, E. S. Uffelman, J. I. Brauman, *Science* **1993**, *261*, 1404; (b) J. P. Collman, V. J. Lee, C. J. Kellen-Yuen, X. Zhang, J. A. Ibers, J. I. Brauman, *J. Am. Chem. Soc.* **1995**, *117*, 692; (c) R. L. Halterman, S. T. Jan, H. L. Nimmons, D. J. Standlee, M. A. Khan, *Tetrahedron* **1997**, *53*, 11257; (d) D. Dolphin, T. G. Traylor, L. Y. Xie, *Acc. Chem. Res.* **1997**, *30*, 251; (e) R. L. Halterman in *Transition Metals for Organic Synthesis* (Eds. M. Beller, C. Bolm), Wiley-VCH, Weinheim, **1988**, *2*, p. 300.

5 (a) E. G. Samsel, K. Srinivasan, J. K. Kochi, *J. Am. Chem. Soc.* **1985**, *107*, 7606; (b) K. Srinivasan, J. K. Kochi, *Inorg. Chem.* **1985**, *24*, 4671.

6 K. Srinivasan, P. Michaud, J. K. Kochi, *J. Am. Chem. Soc.* **1986**, *108*, 2309.

7 W. Zhang, J. L. Loebach, S. R. Wilson, E. N. Jacobsen, *J. Am. Chem. Soc.* **1990**, *112*, 2801.

8 R. Irie, K. Noda, Y. Ito, N. Matsumoto, T. Katsuki, *Tetrahedron Lett.* **1990**, *31*, 7345.

9 J. F. Larrow, E. N. Jacobsen, Y. Gao, Y. Hong, X. Nie, C. M. Zepp, *J. Org. Chem.* **1994**, *59*, 1939.

10 (a) E. N. Jacobsen in *Catalytic Asymmetric Synthesis* (Ed.: I. Ojima), VCH, New York, **1993**, p. 159; (b) T. Flessner, S. Doye, *J. Prakt. Chem.* **1999**, *341*, 436; (c) T. Kasuki in *Catalytic Asymmetric Synthesis*, 2nd edn (Ed.: I. Ojima), VCH, New York, **2000**, p. 287.

11 (a) N. Hosoya, A. Hatayama, R. Irie, H. Sasaki, T. Katsuki, *Tetrahedron* **1994**, *50*, 4311; (b) H. Sasaki, R. Irie, T. Hamada, K. Suzuki, T. Katsuki, *Tetrahedron* **1994**, *50*, 11827.

12 T. Katsuki, *Coord. Chem. Rev.* **1995**, *140*, 189.

13 T. Katsuki, *J. Mol. Catal.* **1996**, *113*, 87.

14 E. N. Jacobsen in *Stereoselective Reactions of Metal-Activated Molecules* (Eds.: H. Werner, J. Sundermeyer), Vieweg, Braunschweig Wiesbaden **1995**, p. 17.

15 Y. N. Ito, T. Katsuki, *Bull. Chem. Soc. Jpn.* **1999**, *72*, 603.

16 (a) T. Katsuki, *Adv. Synth. Catal.* **2002**, *344*, 131; (b) T. Katsuki, *Synlett* **2003**, 281.

17 T. Katsuki, *Curr. Org. Chem.* **2001**, *5*, 663.

18 A. Minatti, *Synlett* **2003**, 140.

19 D. L. Hughes, G. B. Smith, J. Liu, G. C. Dezeny, C. H. Senanayake, R. D. Larsen, T. R. Verhoeven, P. J. Reider, *J. Org. Chem.* **1997**, *62*, 2222.

20 N. S. Finney, P. J. Pospisil, S. Chang, M. Palucki, R. G. Konsler, K. B. Hansen, E. N. Jacobsen, *Angew. Chem. Int. Ed. Engl.* **1997**, *36*, 1720.

21 (a) N. H. Lee, E. N. Jacobsen, *Tetrahedron Lett.* **1991**, *32*, 6533; (b) E. N. Jacobsen, L. Deng, Y. Furukawa, L. E. Martínez, *Tetrahedron* **1994**, *50*, 4323; (c) S. Chang, N. H. Lee, E. N. Jacobsen, *J. Org. Chem.* **1993**, *58*, 6939; (d) H. Sasaki, R. Irie, T. Katsuki, *Synlett* **1994**, 356; (e) W. Zhang, N. H. Lee, E. N. Jacobsen, *J. Am. Chem. Soc.* **1994**, *116*, 425 and references cited therein.

22 K. G. Rasmussen, D. S. Thomsen, K. A. Jørgensen, *J. Chem. Soc., Perkin Trans. 1* **1995**, 2009.

23 (a) M. Palucki, P. J. Pospisil, W. Zhang, E. N. Jacobsen, *J. Am. Chem. Soc.* **1994**, *116*, 9333; (b) M. Palucki, G. J. McCormick, E. N. Jacobsen, *Tetrahedron Lett.* **1995**, *36*, 5457.

24 (a) B. D. Brandes, E. N. Jacobsen, *J. Org. Chem.* **1994**, *59*, 4378; (b) B. D. Brandes, E. N. Jacobsen, *Tetrahedron Lett.* **1995**, *36*, 5123.

25 T. Fukuda, R. Irie, T. Katsuki, *Synlett* **1995**, 197.

26 (a) T. Linker, *Angew. Chem. Int. Ed. Engl.* **1997**, *36*, 2060; (b) for a review on mechanisms in metal porphyrin oxidations see: D. Ostovic, T. C. Bruice, *Acc. Chem. Res.* **1992**, *25*, 314.

27 T. Hamada, T. Fukuda, H. Imanishi, T. Katsuki, *Tetrahedron* **1996**, *52*, 515.

28 C. Linde, M. Arnold, P.-O. Norrby, B. Åkermark, *Angew. Chem. Int. Ed. Engl.* **1997**, *36*, 1723.

29 P.-O. Norrby, C. Linde, B. Åkermark, *J. Am. Chem. Soc.* **1995**, *117*, 11035.

30 (a) D. Feichtinger, D. A. Plattner, *Angew. Chem. Int. Ed. Engl.* **1997**, *36*, 1718. (b) D. A. Plattner, D. Feichtinger, *Chem. Eur. J.* **2001**, *7*, 591. (c) D. Feichtinger, D. A. Plattner, *J. Chem. Soc., Perkin Trans. 2* **2000**, 1023.

31 M. P. Feth, C. Bolm, J. P. Hildebrand, M. Köhler, O. Beckmann, M. Bauer, R. Ramamonjisoa, H. Bertagnolli, *Chem. Eur. J.* **2003**, *9*, 1348.

32 (a) For a review on metallaoxetanes see: K. A. Jørgensen, B. Schiøtt, *Chem. Rev.* **1990**, *90*, 1483; for two recent important contributions on this topic, see: (b) K. P. Gable, E. C. Brown, *J. Am. Chem. Soc.* **2003**, *125*, 11018. (c) X. Chen, X. Zhang, P. Chen, *Angew. Chem. Int. Ed.* **2003**, *42*, 3798.

33 M. Palucki, N. S. Finney, P. J. Posipil, M. L. Güler, T. Ishida, E. N. Jacobsen, *J. Am. Chem. Soc.* **1998**, *120*, 948.

34 (a) H. Buschmann, H.-D. Scharf, N. Hoffmann, P. Esser, *Angew. Chem. Int. Ed. Engl.* **1991**, *30*, 477; (b) A. Gypser, P.-O. Norrby, *J. Chem. Soc., Perkin Trans. 2* **1997**, 939.

35 T. Nishida, A. Miyafuji, Y. N. Ito, T. Katsuki, *Tetrahedron Lett.* **2000**, *41*, 7053.

36 C. Linde, N. Koliaï, P.-O. Norrby, B. Åkermark, *Chem. Eur. J.* **2002**, *8*, 2568.

37 W. Adam, K. J. Roschmann, C. R. Saha-Möller, D. Seebach, *J. Am. Chem. Soc.* **2002**, *124*, 5068.

38 W. Adam, K. J. Roschmann, C. R. Saha-Möller, *Eur. J. Org. Chem.* **2000**, 3519.

39 C. Linde, B. Åkermark, P.-O. Norrby, M. Svensson, *J. Am. Chem. Soc.* **1999**, *121*, 5083.
40 T. Strassner, K.N. Houk, *Org. Lett.* **1999**, *1*, 419.
41 (a) H. Jacobsen, L. Cavallo, *Angew. Chem. Int. Ed.* **2000**, *39*, 589; (b) H. Jacobsen, L. Cavallo, *Chem. Eur. J.* **2001**, *7*, 800; (c) L. Cavallo, H. Jacobsen, *J. Org. Chem.* **2003**, *68*, 6202.
42 J. El-Bahraoui, O. Wiest, D. Feichtinger, D. A. Plattner, *Angew. Chem. Int. Ed.* **2001**, *40*, 2073.
43 M.T. Rispens, A. Meetsma, B.L. Feringa, *Rec. Trav. Chim. Pays-Bas* **1994**, *113*, 413.
44 P.J. Pospisil, D.H. Carsten, E.N. Jacobsen, *Chem. Eur. J.* **1996**, *2*, 974.
45 J.W. Yoon, T.-S. Yoon, S. W. Lee, W. Shin, *Acta Cryst.* **1999**, *C55*, 1766.
46 R. Irie, T. Hashihayata, T. Katsuki, M. Akita, Y. Moro-oka, *Chem. Lett.* **1998**, 1041.
47 T. Punniyamurthy, R. Irie, T. Katsuki, M. Akita, Y. Moro-oka, *Synlett* **1999**, 1049.
48 T. Hashihayata, T. Punniyamurthy, R. Irie, T. Katsuki, M. Akita, Y. Moro-oka, *Tetrahedron* **1999**, *55*, 14599.
49 H. Nishikori, C. Ohta, T. Katsuki, *Synlett* **2000**, 1557.
50 Such structural features have been confirmed by theoretical calculations, see ref. 39.
51 (a) T. Hashihayata, Y. Ito, T. Katsuki, *Synlett* **1996**, 1079; (b) T. Hashihayata, Y. Ito, T. Katsuki, *Tetrahedron* **1997**, *53*, 9541.
52 K. Miura, T. Katsuki, *Synlett* **1999**, 783.
53 K. Muiz, C. Bolm, *Chem. Eur. J.* **2000**, *6*, 2309.
54 K. Mikami, M. Terada, T. Korenaga, Y. Matsumoto, M. Ueki, R. Angelaud, *Angew. Chem. Int. Ed.* **2000**, *39*, 3532.
55 K. Mikami, K. Aikawa, Y. Yusa, J.J. Jodry, M. Yamanaka, *Synlett* **2002**, 1561.
56 Y.N. Ito, T. Katsuki, *Tetrahedron Lett.* **1998**, *39*, 4325.
57 (a) T. Schwenkreis, A. Berkessel, *Tetrahedron Lett.* **1993**, *34*, 4785; (b) A. Berkessel, M. Frauenkron, T. Schwenkreis, A. Steinmetz, *J. Mol. Cat.* **1997**, *117*, 339.
58 M.-C. Cheng, M.C.-W. Chan, S.-M. Peng, K.-K. Cheung, C.-M. Che, *J. Chem. Soc., Dalton Trans.* **1997**, 3479.
59 (a) C.H. Senanayake, G.B. Smith, K.M. Ryan, L.E. Fredenburgh, J. Liu, F.E. Roberts, D.L. Hughes, R.D. Larsen, T.R. Verhoeven, P.J. Reider, *Tetrahedron Lett.* **1996**, *37*, 3271; (b) D.L. Hughes, G.B. Smith, J. Liu, G.C. Dezeny, C.H. Senanayake, R.D. Larsen, T.R. Verhoeven, P.J. Reider, *J. Org. Chem.* **1997**, *62*, 2222; (c) P.J. Reider, *Chimia* **1997**, *51*, 306.
60 L. Deng, E.N. Jacobsen, *J. Org. Chem.* **1992**, *57*, 4320.
61 D. Bell, M.R. Davies, F.J.L. Finney, G.R. Geen, P.M. Kincey, I.S. Mann, *Tetrahedron Lett.* **1996**, *37*, 3895.
62 E.N. Jacobsen, N.S. Finney, *Chem. Biol.* **1994**, *1*, 85.
63 Review on homogeneous and supported chiral salen catalysts: L. Canali, D.C. Sherrington, *Chem. Soc. Rev.* **1999**, *28*, 85.
64 I.F.J. Vankelecom, D. Tas, R.F. Parton, V. Van der Vyver, P.A. Jacobs, *Angew. Chem. Int. Ed. Engl.* **1996**, *35*, 1346.
65 (a) M. Cavazzini, A. Manfredi, F. Montanari, S. Quici, G. Pozzi, *Chem. Commun.* **2000**, 2171. (b) M. Cavazzini, A. Manfredi, F. Montanari, S. Quici, G. Pozzi, *Eur. J. Org. Chem.* **2001**, 4639 and references cited therein.
66 (a) C.E. Song, E.J. Roth, *Chem. Commun.* **2000**, 837. (b) L. Gaillon, F. Bedioui, *Chem. Commun.* **2001**, 1458.
67 T. Mukaiyama, *Aldrichimica Acta* **1996**, *29*, 59.
68 T. Mukaiyama, T. Yamada, *Bull. Chem. Soc. Jpn.* **1995**, *68*, 17.
69 T. Yamada, K. Imagawa, T. Mukaiyama, *Chem. Lett.* **1992**, 2109.
70 T. Yamada, K. Imagawa, T. Nagata, T. Mukaiyama, *Chem. Lett.* **1992**, 2231.
71 K. Imagawa, T. Nagata, T. Yamada, T. Mukaiyama, *Chem. Lett.* **1994**, 527.
72 T. Yamada, K. Imagawa, T. Nagata, T. Mukaiyama, *Bull. Chem. Soc. Jpn.* **1994**, *67*, 2248.
73 T. Nagata, K. Imagawa, T. Yamada, T. Mukaiyama, *Inorg. Chim. Acta* **1994**, *220*, 283.

74 T. Mukaiyama, T. Yamada, T. Nagata, K. Imagawa, *Chem. Lett.* **1993**, 327.

75 T. Nagata, K. Imagawa, T. Yamada, T. Mukaiyama, *Chem. Lett.* **1994**, 1259.

76 T. Nagata, K. Imagawa, T. Yamada, T. Mukaiyama, *Bull. Chem. Soc. Jpn.* **1995**, *68*, 1455.

77 (a) R. Hage, J.E. Iburg, J. Kerschner, J.H. Koek, E.L.M. Lempers, R.J. Martens, U.S. Racherla, S.W. Russel, T. Swarthoff, M.R.P. van Vliet, J.B. Warnaar, L. van der Wolf, B. Krijnen, *Nature* **1994**, *369*, 637; (b) see also: J.H. Koek, E.W.M.J. Kohlen, S.W. Russell, L. van der Wolf, P.F. ter Steeg, J.C. Hellemons, *Inorg. Chim. Acta* **1999**, *295*, 189.

78 (a) A.A. Belal, P. Chaudhuri, I. Fallis, L.J. Farrugia, R. Hartung, N.M. Macdonald, B. Nuber, R.D. Peacock, J. Weiss, K. Wieghardt, *Inorg. Chem.* **1991**, *30*, 4397; (b) C. Stockheim, L. Hoster, T. Weyhermüller, K. Wieghardt, B. Nuber, *J. Chem. Soc., Dalton Trans.* **1996**, 4409; (c) K.P. Wainwright, *Coord. Chem. Rev.* **1997**, *166*, 35.

79 D.E. De Vos, T. Bein, *Chem. Commun.* **1996**, 917.

80 D.E. De Vos, T. Bein, *J. Organomet. Chem.* **1996**, *520*, 195.

81 A. Berkessel, C.A. Sklorz, *Tetrahedron Lett.* **1999**, *40*, 7965.

82 J. Brinksma, L. Schmieder, G. van Vliet, R. Boaron, R. Hage, D.E. de Vos, P.L. Alsters, B.L. Feringa, *Tetrahedron Lett.* **2002**, *43*, 2619.

83 (a) C. Bolm, D. Kadereit, M. Valacchi, *Synlett* **1997**, 687; (b) C. Bolm, D. Kadereit, M. Valacchi, DE 19720477.5, **1997**.

84 See also: M. Beller, A. Tafesh, W.R. Fischer, B. Scharbert (Hoechst AG) DE 19523891.5-44, **1995**.

85 (a) C. Bolm, N. Meyer, G. Raabe, T. Weyhermüller, T. Bothe, *Chem. Commun.* **2000**, 2435; (b) N. Meyer, dissertation at the RWTH Aachen, **2000**.

86 G. Argouarch, C.L. Gibson, G. Stones, D.C. Sherrington, *Tetrahedron Lett.* **2002**, *43*, 3795.

87 B.M. Kim, S.M. So, H.J. Choi, *Org. Lett.* **2002**, *4*, 949.

88 S.W. Golding, T.W. Hambley, G. Lawrence, S.M. Luther, M. Maeder, P. Turner, *J. Chem. Soc., Dalton Trans.* **1999**, 1975.

89 (a) J.E. W. Scheuermann, F. Ronketti, M. Motevalli, D.V. Griffiths, M. Watkinson, *New J. Chem.* **2002**, *26*, 1054; (b) J.E.W. Scheuermann, G. Ilashenko, D. V. Griffith, M. Watkinson, *Tetrahedron: Asymmetry* **2002**, *13*, 269.

90 C. Bolm, M. Valacchi, unpublished results.

91 (a) Y.V. Subba Rao, D.E. De Vos, T. Bein, P.A. Jacobs, *J. Chem. Soc. Chem. Commun.* **1997**, 355; (b) D.E. De Vos, S. de Wildeman, B.F. Sels, P.J. Grobet, P.A. Jacobs, *Angew. Chem. Int. Ed.* **1999**, *38*, 980.

92 D.E. De Vos, J.L. Meinershagen, T. Bein, *Angew. Chem.* **1996**, *108*, 2355; *Angew. Chem. Int. Ed.* **1996**, *35*, 2211.

93 Review on epoxidations with heterogeneous catalysts: D.E. De Vos, B.F. Sels, P.A. Jacobs, *Adv. Synth. Catal.* **2003**, *345*, 457.

94 (a) A. Grenz, S. Ceccarelli, C. Bolm, *Chem. Commun.* **2001**, 1726; (b) for oxidative CH-activations with this catalyst system, see: G.V. Nizova, C. Bolm, S. Ceccarelli, C. Pavan, G.B. Shul'pin, *Adv. Synth. Catal.* **2002**, *344*, 899.

95 (a) J.M. Vincent, A. Rabion, V.K. Yachandra, R.H. Fish, *Angew. Chem. Int. Ed.* **1997**, *36*, 2346; (b) review: R.H. Fish, *Chem. Eur. J.* **1999**, *5*, 1677.

2.7.3
Rhenium-Catalyzed Epoxidations

Fritz E. Kühn, Richard W. Fischer, and Wolfgang A. Herrmann

2.7.3.1
Introduction and Motivation

For a long time, significantly fewer efforts have been made to investigate and understand the chemistry of rhenium than have gone into exploring the chemistry of its neighboring elements, e.g., tungsten and osmium [1]. This situation, however, is changing, particularly with respect to high oxidation state organorhenium oxides, because of their outstanding catalytic activity in a surprisingly broad range of organic reactions [2]. The interest in organorhenium oxides was triggered by the discovery of the catalytic activity of methyltrioxorhenium(VII) in the late 1980s and early 1990s [2, 3]. Since then, the scope of application and the scientific interest in these complexes has dramatically widened [4].

2.7.3.2
Synthesis of the Catalyst Precursors

Methyltrioxorhenium(VII), nowadays usually abbreviated to MTO, was first synthesized in 1979 in a quite time-consuming (weeks) and small-scale (milligrams) synthesis [5a]. The breakthrough toward possible applications only came nearly 10 years later, when the first efficient synthetic route, starting from dirhenium heptoxide and tetramethyl tin, was reported [3a]. Several congeners of MTO were reported during the following years, most of these also prepared from dirhenium heptoxide and organo tin or organo zinc precursors [5b–5k]. The drawback of all these (otherwise excellent) approaches is the loss of half of the Re because of the formation of the low-reactivity trimethylstannyl perrhenate or zinc perrhenate, respectively. An improvement was made by using mixed esters of perrhenic and trifluoroacetic acid, avoiding the loss of rhenium [5l]. At the same time, the much

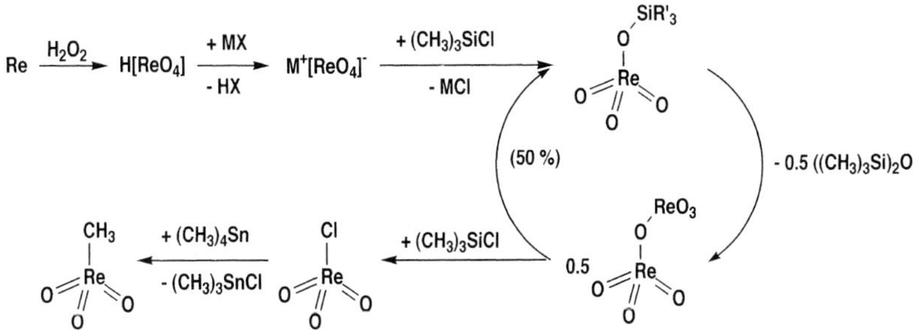

Scheme 1

less toxic tris(*n*-butyl)organyl tin was used for the selective organylation. For MTO, this route reached the laboratory pilot-plant stage in 1999 [5 m]. A further modification of the synthesis enables the use of the moisture-sensitive dirhenium heptoxide to be avoided as starting material and uses Re powder or perrhenates as the starting material [5 n]. This method is of particular interest since it allows the recyclization of catalyst decomposition products from reaction solutions. MTO is nowadays also commercially available from several producers [5 o]. Scheme 1 and Eq. (1) give an overview of the various routes to organorhenium(VII) oxides.

$$Re_2O_7 + RSnR'_3 \xrightarrow{THF} R{-}Re(=O)_3 + R'_3Sn{-}O{-}Re(=O)_3$$

$$2\,Re_2O_7 + ZnR_2 \xrightarrow{THF} 2\,R{-}Re(=O)_3 + Zn(THF)_2[ReO_4]_2$$

$$Cl{-}Re(=O)_3 + (nBu)_3Sn{-}R \xrightarrow{THF} R{-}Re(=O)_3 + (nBu)_3Sn{-}Cl \tag{1}$$

$$CF_3C(=O){-}O{-}Re(=O)_3 + (nBu)_3Sn{-}R \xrightarrow{CH_3CN} R{-}Re(=O)_3 + CF_3{-}C(=O){-}OSn(nBu)_3$$

2.7.3.3
Epoxidation of Olefins

Transition metal oxo complexes have already found applications as catalysts in industrial scale epoxidation reactions and other oxo transfer processes for several decades [6]. Especially molybdenum, titanium, and tungsten complexes have been under intense investigation, both to elucidate the catalytic mechanism and to broaden and optimize their field of application [7]. There is still a need for efficient, highly selective but broadly applicable and easily accessible catalysts activating cheap and safe oxidants, such as dilute hydrogen peroxide for olefin epoxidation. In addition, there is also a lack of epoxidation catalysts which are able to activate H_2O_2 without severely decomposing it. In the last decade, several new or improved epoxidation catalysts, based on the above-mentioned metals, emerged or were re-examined [8]. Additionally, other transition metal complexes also attracted considerable interest as oxidation catalysts, probably the most famous of them being the very versatile MTO and its derivatives. These complexes are highly efficient and selective epoxidation catalysts, activated by H_2O_2, as oxidant. Highly advantageously, rhenium systems show no H_2O_2 decomposition.

2.7.3.3.1 The Catalytically Active Species

The catalytic activity of MTO and some of its derivatives in the oxidation of olefins was noticed soon after these complexes were accessible in higher amounts [3]. However, the breakthrough in the understanding of the role of MTO in oxidation catalysis was the isolation and characterization of the reaction product of MTO with excess H_2O_2, i.e. a bisperoxo complex of stoichiometry $(CH_3)Re(O_2)_2O \cdot H_2O$ [9a]. This reaction takes place in any organic solvent or water (see Scheme 2).

In the solid state, it is isolated as a trigonal bipyramidal adduct with a donor ligand L ($L=H_2O$, $L=O=P[N(CH_3)_2]_3$) [9a, b], which is lost in the gas phase. The structures of $(CH_3)Re(O_2)_2O$ (electron diffraction), $(CH_3)Re(O_2)_2O \cdot H_2O$, and $(CH_3)Re(O_2)_2O \cdot (O=P[N(CH_3)_2]_3)$ (X-ray diffraction) were determined; the structure of the ligand-free complex $(CH_3)Re(O_2)_2O$ is known from the gas phase [9].

The adduct $(CH_3)Re(O_2)_2O \cdot H_2O$ melts at 56 °C and can be sublimed at room temperature in an oil pump vacuum. It reacts as a comparatively strong Brønsted acid in aqueous solution ($pK_s^{8\,°C}=6.1$, $pK_s^{20\,°C}=3.76$), whereas $(CH_3)Re(O_2)_2O \cdot (O=P[N(CH_3)_2]_3)$ melts at 65 °C and decomposes at ca. 75 °C. Both bis peroxo complex derivatives are explosive [9].

Experiments with the isolated bis(peroxo)complex $(CH_3)Re(O_2)_2O \cdot H_2O$ have shown beyond any reasonable doubt that it is an active species in olefin epoxidation catalysis and several other catalytic reactions [9a, 10]. *In situ* experiments show that the reaction of MTO with one equivalent of H_2O_2 leads to a monoperoxo complex of the likely composition $(CH_3)Re(O_2)O_2$ [10, 11]. $(CH_3)Re(O_2)O_2$ has never been isolated and exists only in equilibrium with MTO and $(CH_3)Re(O_2)_2O \cdot H_2O$. The monoperoxo complex is also catalytically active in oxidation processes. Kinetic experiments indicate that the rate constants for the transformation of most substrates into their oxidation products by catalysis with the mono and bisperoxo complex are of a comparable order of magnitude [11]. This result is supported by density functional calculations [12]. The transition states in the olefin epoxidation process starting from $(CH_3)Re(O_2)O_2$ and $(CH_3)Re(O_2)_2O \cdot H_2O$ are not different enough in energy to exclude one of these two catalysts totally from the catalytic process. The activation parameters for the coordination of H_2O_2 to MTO have also been determined. They indicate a mechanism involving nucleophilic attack. The protons lost in converting H_2O_2 to a coordinated O_2^{2-} ligand are transferred to one of the terminal oxygen atoms, which remains on the Re as the aqua ligand L. The rate of this reaction is

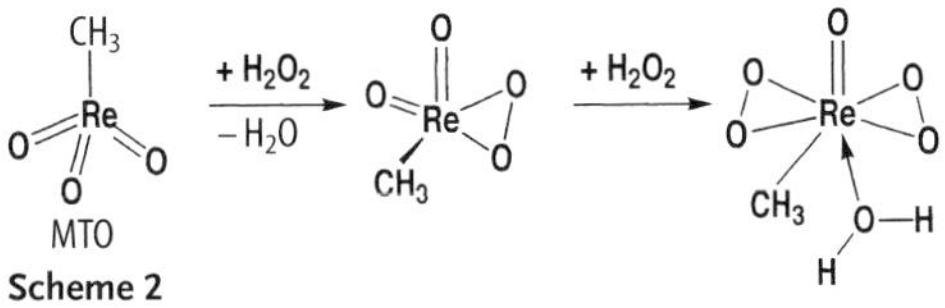

MTO $\xrightarrow[-H_2O]{+H_2O_2}$ $(CH_3)Re(O_2)O_2$ $\xrightarrow{+H_2O_2}$ $(CH_3)Re(O_2)_2O \cdot H_2O$

Scheme 2

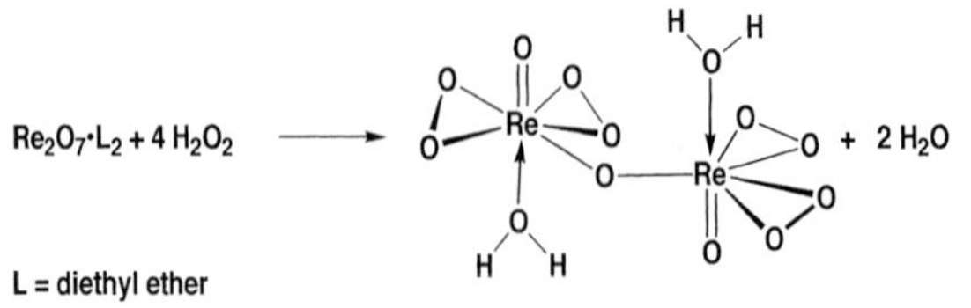

$Re_2O_7 \cdot L_2 + 4\ H_2O_2 \longrightarrow$ … $+\ 2\ H_2O$

L = diethyl ether

Scheme 3

not pH-dependent [11c]. More details about the reaction mechanism are discussed below.

As well as MTO and its derivatives, Re_2O_7 and ReO_3 form bisperoxo complexes when treated with excess H_2O_2 (Scheme 3). As in the case of MTO, a catalytically active species originating from the reaction of Re_2O_7 and four equivalents of H_2O_2 has been isolated and fully characterized, including X-ray crystallography of its diglyme adduct [13]. The red-orange, explosive compound of formula $H_4Re_2O_{13}$, containing two peroxo units per Re center, is the most oxygen-rich rhenium compound isolated to date. In contrast to $(CH_3)Re(O_2)_2O \cdot H_2O$, however, $O\{Re[O(O_2)_2]_2\} \cdot H_2O$ decomposes hydrolytically during the catalytic cycle and thus cannot compete in terms of catalytic activity in oxidation reactions involving H_2O_2. Anyway, it has also been demonstrated that with other oxidizing agents which do not produce H_2O as a by-product, such as bis(trimethylsilyl)peroxide (BTSP), Re_2O_7, ReO_3, and even $HReO_4$-derived catalysts act very efficiently [14]. The ease of their synthesis, however, is overshadowed by the price of the oxidizing agent. For special cases, these catalysts might nevertheless present interesting alternatives to established epoxidation systems. Perrhenic acid in combination with tertiary arsines is also reported to give versatile catalytic systems for epoxidation of alkenes with H_2O_2. The best results were obtained with dimethylarsine. A wide range of alkenes could be oxidized with aqueous H_2O_2 (60%) in 60–100% yields with substrate-to-catalyst ratios of up to 1000 [15].

2.7.3.3.2 **The Catalytic Cycles**

Two catalytic pathways for the olefin epoxidation may be described, corresponding to the concentration of the hydrogen peroxide used. If 85% hydrogen peroxide is used, only $(CH_3)Re(O_2)_2O \cdot H_2O$ appears to be responsible for the epoxidation activity (Scheme 4, cycle A). When a solution of 30 wt% or less H_2O_2 is used, the monoperoxo complex, $(CH_3)Re(O_2)O_2$, is also responsible for the epoxidation process, and a second catalytic cycle is involved as shown in Scheme 4, cycle B. For both cycles, a concerted mechanism is suggested in which the electron-rich double bond of the alkene attacks a peroxidic oxygen of $(CH_3)Re(O_2)_2O \cdot H_2O$. It has been inferred from experimental data that the system may involve a spiro arrangement [2, 4a, 12].

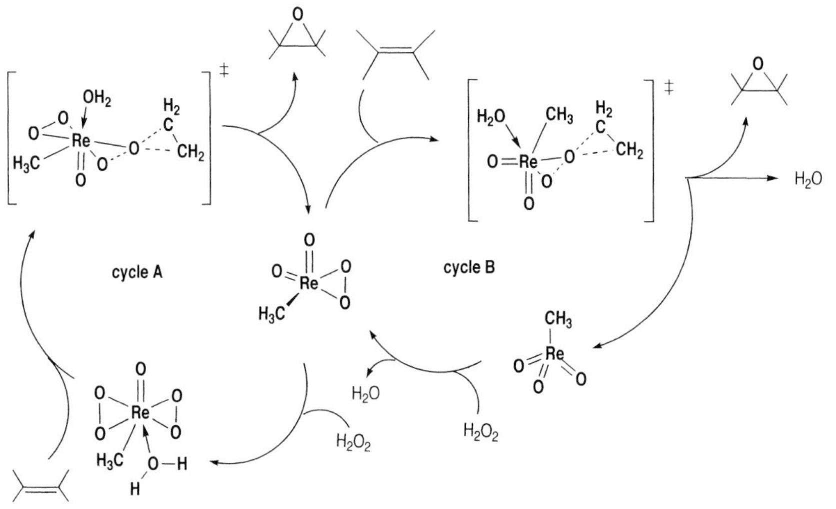

Scheme 4

2.7.3.3.3 **Catalyst Deactivation**

In spite of the extraordinarily strong Re-C bond [16], characteristic of MTO and its congeners, the cleavage of this bond plays a prominent role in the decomposition processes of these complexes [17]. Concerning MTO, the full kinetic pH profile for the base-promoted decomposition to CH_4 and ReO_4^- was examined. Spectroscopic and kinetic data give evidence for mono- and dihydroxo complexes of formulae $CH_3ReO_3(OH^-)$ and $CH_3ReO_3(OH^-)_2$ prior to and responsible for the decomposition process. In the presence of hydrogen peroxide, $(CH_3)Re(O_2)O_2$ and $(CH_3)Re(O_2)_2O \cdot H_2O$ decompose to methanol and perrhenate with a rate that is dependent on $[H_2O_2]$ and $[H_3O]^+$. The complex peroxide and pH dependencies are explained by two possible pathways: attack of either hydroxide on $(CH_3)Re(O_2)O_2$ or HO_2^- on MTO. The bisperoxo complex decomposes much more slowly to yield O_2 and MTO [17a]. Thus, critical concentrations of strong nucleophiles have to be avoided; a high excess of hydrogen peroxide stabilizes the catalyst. It turned out to be advantageous to keep the steady-state concentration of water during the oxidation reaction as low as possible to depress catalyst deactivation.

2.7.3.3.4 **The Role of Lewis Base Ligands**

The most important drawback of the MTO-catalyzed process is the concomitant formation of diols instead of the desired epoxides, especially in the case of more sensitive substrates [10]. It was quickly detected that the use of Lewis base adducts of MTO significantly decreases the formation of diols because of the reduced Lewis acidity of the catalyst system. However, while the selectivity increases, the conversion decreases [10, 18]. It was found that biphasic systems (water phase/organic phase) and the addition of a significant excess of pyridine as

Lewis base not only hamper the formation of diols but also increase the reaction velocity in comparison to MTO as catalyst precursor [19]. Additionally it was shown that 3-cyanopyridine and especially pyrazole as Lewis bases are even more effective and less problematic than pyridine itself, while pyridine *N*-oxides are less efficient [20]. From *in situ* measurements under one-phase conditions, it was concluded that both electronic and steric factors of the aromatic Lewis base involved play a prominent role during the formation of the catalytically active species. The Brønsted basicity of pyridines lowers the activity of hydronium ions, thus reducing the rate of opening of the epoxide ring [21].

MTO forms trigonal-bipyramidal adducts with pyridines and related Lewis bases (Formula I). Because of their obvious importance as catalyst precursors in olefin epoxidation, these complexes have been isolated and fully characterized [22a]. The complexes react with H_2O_2 to form mono- and bisperoxo complexes analogous to that of MTO, but coordinated by one Lewis base molecule instead of H_2O. From the Lewis-base-MTO complexes to the bisperoxo complexes a clear increase in electron deficiency at the Re center can be observed by spectroscopic methods. The activity of the bisperoxo complexes in olefin epoxidation depends on the Lewis bases, the redox stability of the ligands, and the excess of Lewis base used. Density functional calculations show that when the ligand is pyridine or pyrazole there are significantly stabilized intermediates and moderate energies of the transition states in olefin epoxidation. This ultimately causes an acceleration of the epoxidation reaction. Non-aromatic nitrogen bases as ligands were found to reduce the catalytic performance. The frontier orbital interaction between the olefin HOMO π(C-C) and orbitals with σ^*(O-O) character in the LUMO group of the Re-peroxo moiety controls the olefin epoxidation.

With bidentate Lewis bases, MTO forms octahedral adducts (Formula II), which also form very active and highly selective epoxidation catalysts. Peroxo complexes are generated, and one of the Re-N interactions is cleaved during this process. The peroxo complexes of the MTO Lewis bases are, in general, more sensitive to water than MTO itself [22b]. Furthermore, in the presence of olefins, which are not readily transformed to their epoxides, 2,2′-bipyridine can be oxidized to its *N*-oxide by the MTO/H_2O_2 system [23].

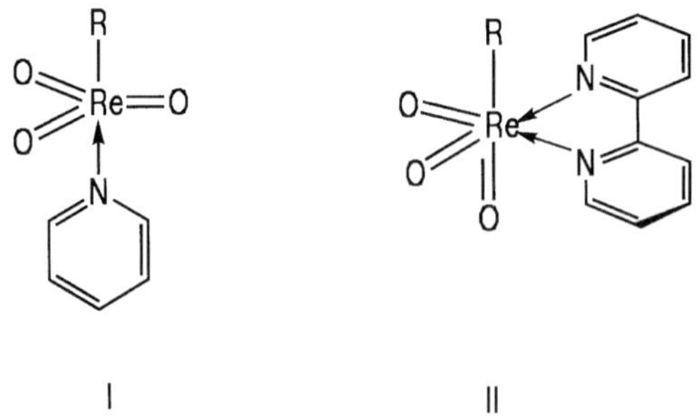

(I, II)

2.7.3.3.5 Heterogeneous Catalyst Systems

Alternative strategies to improve MTO-catalyzed oxidations have made use of host-guest inclusion chemistry [24]. It was found that a urea/hydrogen peroxide (UHP) complex is a very effective oxidant in heterogeneous olefin epoxidations and silane oxidations catalyzed by MTO [24 a, b, d]. Even stereoidal dienes have been successfully oxidized by the MTO/H_2O_2-urea system [24 g]. Using NaY zeolite as host for these reactions also resulted in high yields and excellent product selectivities [24 e]. MTO has also been supported on silica functionalized with polyether tethers [24 c]. In the absence of an organic solvent, this catalytic assembly catalyzed the epoxidation of alkenes with 30% H_2O_2 in high selectivity compared to the ring-opened products observed in homogeneous media. MTO has additionally been immobilized in the mesoporous silica MCM-41 functionalized with pendant bipyridyl groups of the type [4-($\equiv Si(CH_2)_4$)-4′-methyl-2,2′-bipyridine] [24 h]. Powder XRD and N_2 adsorption-desorption studies confirm that the regular hexagonal symmetry of the host is retained during the grafting reaction and that the channels remain accessible. The formation of a tethered Lewis base adduct of the type $CH_3ReO_3 \cdot (N\text{–}N)$ was confirmed. The XAFS results however indicated that not all the rhenium is present in this form, and this is consistent with elemental analysis which gave the Re:N ratio to be 1:1.1. It is likely that the excess rhenium is present as un-coordinated MTO molecules assembled in the MCM channels. Furthermore, novel heterogeneous derivatives of MTO were prepared with poly(4-vinylpyridine) and polystyrene as polymeric support [24 i]. In the case of poly(4-vinylpyridine)/MTO derivatives, a slightly distorted octahedral conformation of the metal's primary coordination sphere was observed. The Re-N bond was abnormally short in comparison to previously reported homogeneous MTO/pyridine complexes [22 a], showing a strong coordination of the MTO moiety to the surface. The reticulation grade of the polymer was a crucial factor for the morphology of the particle surface. The polymer-supported MTO proved to be an efficient and selective heterogeneous catalyst for the olefin epoxidation. The catalytic activity was reported to be maintained for at least five recycling experiments [24 i].

Rhenium oxides supported on zeolite Y (mixed silica-alumina and pure alumina) were prepared by impregnation of the supports with Re_2O_7 or NH_4ReO_4 [25]. These materials are also active catalysts in the epoxidation of cyclooctene and cyclohexene with anhydrous H_2O_2 in EtOAc. Catalyst stability with respect to metal leaching is closely correlated with the alumina content of the support, and almost no leaching was observed with ReO_4^- supported on pure alumina. Stable catalysts ReO_4-Al_2O_3 with ReO_4^- contents up to 12 wt% could be prepared. Higher contents result in extensive metal leaching and catalysis in the homogeneous phase. Selectivities for cyclooctene epoxide were ca. 96%; cyclohexanediol was obtained as the only product in cyclohexene epoxidation. Addition of pyridine in this latter case increased the epoxide amount from 0 to 67%. However, the conversion decreased significantly.

2.7.3.4
Summary: Scope of the Reaction

Epoxidations with the MTO/H_2O_2 catalytic system have received broad interest, both from industry and academics. MTO is easily available; active in low concentrations of both MTO (0.05 mol%) and H_2O_2 (<5 wt%), it works over a broad temperature range (–40 to +90 °C) and is stable in water under acidic conditions and in basic media in special cases. Furthermore, the MTO/H_2O_2 system works in a broad variety of solvents, ranging from highly polar solvents (e.g., nitromethane, water) to solvents with low polarity (e.g., toluene). However, the reactions between MTO/H_2O_2 and alkenes are approximately one order of magnitude faster in semi-

Tab. 1 Epoxidation of olefins, catalyzed by rhenium complexes. The data are taken from Refs. [10] (MTO/H_2O_2), [19] (MTO/H_2O_2/py) and (MTO/H_2O_2/pz), [14 a] (MTO/H_2O_2/cpy), [20 a] (Re_2O_7/BTSP), and [24 a] (MTO/UHP)

Catalyst/Oxidant	*Substrate*	*T (°C)*	*t (h)*	*Yield (%)*	*Selectivity (%)*
MTO/H_2O_2 [a)]	Cyclooctene	15	24	99	99
MTO/H_2O_2 [b)]	Cycloheptene	40	48	88	100
MTO/H_2O_2 [c)]	Cyclohexene	10	20	90	100
MTO/H_2O_2/py [d)]	Cyclooctene	25	2	99	>99
MTO/H_2O_2/py [d)]	Cycloheptene	25	3	99	>99
MTO/H_2O_2/py [d)]	Cyclohexene	25	6	96	>99
MTO/H_2O_2 [a)]	1-Decene	15	72	75	92
MTO/H_2O_2/py [d)]	1-Decene	25	48	82	>99
Re_2O_7/BTSP [e)]	1-Decene	25	14	94	
MTO/H_2O_2/cpy [f)]	1-Decene	25	17	94	>99
MTO/H_2O_2 [a)]	Styrene	25	3	60 (convers.)	0
MTO/UHP [g)]	Styrene	25	19	46	>95
MTO/H_2O_2/py [d)]	Styrene	25	5	70	>99
Re_2O_7/BTSP [e)]	Styrene	25	7	95	
MTO/H_2O_2/cpy [f)]	Styrene	25	5	85	>99
MTO/H_2O_2/pz [h)]	Styrene	25	5	>99	>99
MTO/H_2O_2 [i)]	*Cis*-1,4-dichloro-2-butene	25	48	73	96
MTO/H_2O_2 [j)]	4-Perfluoro-hexyl-1-butene	15	64	30	90

a) Solvent: *t*-BuOH, 7.68 mol olefin, 7.6 mmol MTO.
b) Solvent: *t*-BuOH, 0.17 mol olefin, 0.8 mmol MTO.
c) Solvent: 0.99 mol olefin, 1.6 mmol MTO.
d) Solvent: CH_2Cl_2, 2 mol/l olefin, 0.5 mol% MTO, 12 mol% pyridine, 1.5 equiv. 30% H_2O_2.
e) 10 mmol scale, 1.5 equiv. BTSP per double bond, 0.5 mol% Re_2O_7, solvent: CH_2Cl_2.
f) Equiv. olefin, 0.5 mol% MTO, 10 mol% 3-cyanopyridin, 30% H_2O_2, solvent: CH_2Cl_2.
g) MTO : UHP = 1 : 100 : 100, solvent: CH_2Cl_2.
h) MTO : H_2O_2:pyrazole = 0.5 : 200 : 12, solvent: CH_2Cl_2.
i) 0.14 mol olefin; 1.2 mmol MTO, solvent: *t*-BuOH.
j) 0.15 mol olefin, 0.8 mmol MTO, solvent: *t*-BuOH.

aqueous solvents (e.g., 85% H_2O_2) than in methanol. The rate constants for the reaction of MTO/H_2O_2 with aliphatic alkenes correlate closely with the number of alkyl groups on the alkene carbons. Theoretical calculations support these results [10a, b]. The reactions become significantly slower when electron-withdrawing groups such as -OH, -CO, -Cl, and -CN are present.

A major advantage of MTO and its derivatives is that hydrogen peroxide is not decomposed by the applied catalysts. This is in striking contrast to many other oxidation catalysts. Turnover numbers of up to 2500 (mol product per mol catalyst; reaction conditions: 0.1 mol% MTO, 5 mol% pyrazole, trifluoro ethanol as solvent [4d]) and turnover frequencies of up to 14000 (mol product per mol catalyst per hour; in fluorinated alcohols as solvent for cyclohexene at $<10\,^{\circ}C$ [4d]) have been reported, with typical MTO concentrations of 0.1–1.0 mol%. High selectivity (epoxide vs diol) can be adjusted by temperature control, trapping of water, or the use of certain additives such as aromatic Lewis-base ligands, which additionally accelerate the epoxidation reactions. Selectivities of >95% can be reached. Inorganic rhenium oxides, e.g., Re_2O_7 and ReO_3, in most cases display lower activity and selectivity. Table 1 gives a brief overview of the scope of olefins and both activity and selectivity of the catalytic systems used.

In comparison to the standard system for epoxidation, which uses *m*-chloroperoxybenzoic acid as oxidizing agent, the MTO/H_2O_2/aromatic Lewis base-system displays several advantages:

1. It is safer, but equal in price.
2. Because of the suppression of epoxide ring opening, it is much broader in scope.
3. Its selectivity is higher.
4. It is more reactive, requires less solvent, the product work-up is easier, and the only by-product formed is water.

References

1 (a) F. E. KÜHN, C. C. ROMÃO, W. A. HERRMANN in *Science of Synthesis: Houben-Weyl Methods of Molecular Transformations* (Eds.: T. IMAMOTO, D. BARBIER-BAUDRY), Vol. 2, Georg Thieme, Stuttgart **2002**; (b) C. C. ROMÃO in *Encyclopaedia of Inorganic Chemistry* (Ed.: R. B. KING), **1994**, 6, 3435, Wiley, Chichester; (c) K. A. JØRGENSEN, *Chem. Rev.*, **1989**, *89*, 447.

2 Recent reviews: (a) F. E. KÜHN, M. GROARKE in *Applied Homogeneous Catalysis with Organometallic Compounds*, 2nd edn (Eds.: B. CORNILS, W. A. HERRMANN), **2002**, *3*, 1304, Wiley-VCH, Weinheim; (b) F. E. KÜHN, W. A. HERRMANN, *Chemtracts-Organic Chemistry*, **2001**, *14*, 59; (c) F. E. KÜHN, W. A. HERRMANN in *Structure and Bonding* (Ed.: B. MEUNIER), **2000**, 97, 213, Springer, Heidelberg, Berlin; (d) W. ADAM, C. M. MITCHELL, C. R. SAHA-MÖLLER, O. WEICHOLD in Structure and Bonding (Ed.: B. MEUNIER), *97*, 237, Springer, Heidelberg, Berlin, **2000**; (e) G. S. OWENS, J. ARIAS, M. M. ABU-OMAR, *Catalysis Today* **2000**, *55*, 317; (f) F. E. KÜHN, R. W. FISCHER, W. A. HERRMANN, *Chem. Unserer Zeit* **1999**, *33*, 192; (g) J. H. ESPENSON, M. M. ABU-OMAR, *ACS Adv. Chem.* **1997**, *253*, 3507; (h) B. SCHMID, *J. Prakt. Chem.*, **1997**, *339*, 439; (i) C. C. ROMÃO, F. E. KÜHN, W. A. HERRMANN, *Chem. Rev.* **1997**, *97*, 3197; (j) S. N.

Brown, J.M. Mayer, *J. Am. Chem. Soc.* **1996**, *118*, 12119.

3 (a) W.A. Herrmann, J.G. Kuchler, J.K. Felixberger, E. Herdtweck, W. Wagner, *Angew. Chem. Int. Ed. Engl.*, **1988**, *27*, 394; (b) W.A. Herrmann, W. Wagner, U.N. Flessner, U. Volkhardt, H. Komber, *Angew. Chem. Int. Ed. Engl.* **1991**, *30*, 1636; (c) W. A. Herrmann, R.W. Fischer, D.W. Marz, *Angew. Chem. Int. Ed. Engl.* **1991**, *30*, 1638; (d) W.A. Herrmann, M. Wang, *Angew. Chem. Int. Ed. Engl.*, **1991**, *30*, 1641.

4 (a) W.A. Herrmann, F.E. Kühn, *Acc. Chem. Res.* **1997**, *30*, 169; (b) H. Rudler, J.R. Gregorio, B. Denise, J.M. Brégeault, A. Deloffre, *J. Mol. Catal. A. Chemical* **1998**, *133*, 255; (c) A.L.P. Villa D.E. Vos, C.C. de Montes, P.A. Jacobs, *Tetrahedron Lett.* **1998**, *39*, 8521; (d) M.C.A. van Vliet, I.W.C.E. Arends, R.A. Sheldon, *J. Chem. Soc., Chem. Commun.* **1999**, 821.

5 (a) J.R. Beattie, P.J. Jones, *Inorg. Chem.* **1979**, *18*, 2318; (b) W. A. Herrmann, M. Ladwig, P. Kiprof, J. Riede, *J. Organomet. Chem.* **1989**, *11*, C13; (c) W.A. Herrmann, C.C. Romão, R.W. Fischer, P. Kiprof, C. de Méric de Bellefon, *Angew. Chem. Int. Ed. Engl.* **1991**, *30*, 185; (d) W.A. Herrmann, M. Taillefer, C. de Méric de Bellefon, J. Behm, *Inorg. Chem.* **1991**, *30*, 3247; (e) C. de Méric de Bellefon, W.A. Herrmann, P. Kiprof, C.R. Whitaker, *Organometallics* **1992**, *11*, 1072; (f) W.A. Herrmann, F.E. Kühn, C.C. Romão, H. Tran-Huy, M. Wang, R.W. Fischer, W. Scherer, P. Kiprof, *Chem. Ber.* **1993**, *126*, 45; (g) W.A. Herrmann, F.E. Kühn, C. C. Romão, H. Tran Huy, *J. Organomet. Chem.* **1994**, *481*, 227; (h) F.E. Kühn, W.A. Herrmann, R. Hahn, M. Elison, J. Blümel, E. Herdtweck, *Organometallics* **1994**, *13*, 1601; (i) J. Sundermeyer, K. Weber, K. Peters, H.G. v. Schnering, *Organometallics* **1994**, *13*, 2560; (j) W.A. Herrmann, F.E. Kühn, C.C. Romão, *J. Organomet. Chem.* **1995**, *489*, C56; (k) W.A. Herrmann, F.E. Kühn, C.C. Romão, *J. Organomet. Chem.* **1995**, *495*, 209; (l) W.A. Herrmann, F.E. Kühn, R.W. Fischer, W.R. Thiel, C.C. Romão, *Inorg. Chem.* **1992**, *31*, 4431; (m) W.A. Herrmann in *Applied Homogeneous Catalysis with Organometallic Compounds*, 2nd edn (Eds.: B. Cornils, W.A. Herrmann), **2002**, *3*, 1319, Wiley-VCH, Weinheim; (n) W.A. Herrmann, R.M. Kratzer, R.W. Fischer, *Angew. Chem. Int. Ed. Engl.* **1997**, *36*, 2652; (o) Small amounts of MTO are commercially available from, e.g., Aldrich: 41,291-0 (100 mg, 500 mg); Fluka: 69489 (50 mg, 250 mg).

6 (a) R.A. Sheldon, in *Applied Homogeneous Catalysis with Organometallic Compounds*, (Eds.: B. Cornils, W.A. Herrmann), **2002**, *3*, 1304, Wiley-VCH, Weinheim; (b) H. Arzoumanian, *Coord. Chem. Rev.*, **1998**, *180*, 191; (c) R.H. Holm, *Chem. Rev.* **1987**, *87*, 1401; (d) Holm, R.H., *Coord. Chem. Rev.* **1990**, *100*, 183.

7 See also this book, Chapter 2.7.

8 See for example: (a) D.V. Deubel, J. Sundermeyer, G. Frenking, *J. Am. Chem. Soc.* **2000**, *122*, 10101; (b) G. Wahl, D. Kleinhenz, A. Schorm, J. Sundermeyer, R. Stowasser, C. Rummey, G. Bringmann, C. Fickert, W. Kiefer, *Chem. Eur. J.* **1999**, *5*, 3237; (c) F.E. Kühn, M. Groarke, É. Bencze, E. Herdtweck, A. Prazeres, A.M. Santos, M.J. Calhorda, C.C. Romão, I.S. Gonçalves, A.D. Lopes, M. Pillinger, *Chem. Eur. J.* **2002**, *8*, 2370; (d) F.E. Kühn, W.M. Xue, A. Al Ajlouni, A.M. Santos, S. Zhang, C.C. Romão, G. Eickerling, E. Herdtweck, *Inorg. Chem.* **2002**, in press; (e) D.E. de Voss, B.F. Sels, M. Reynaers, Y.V. Subba Rao, P.A. Jacobs, *Tetrahedron Lett.* **1998**, *39*, 3221; (f) A. Hroch, G. Gemmecker, W.R. Thiel, *Eur. J. Inorg. Chem.* **2000**, 1107.

9 (a) W.A. Herrmann, R.W. Fischer, W. Scherer, M.U. Rauch, *Angew. Chem. Int. Ed. Engl.* **1993**, *32*, 1157; (b) W.A. Herrmann, J.D.G. Correia, G.R.J. Artus, R.W. Fischer, C.C. Romão, *J. Organomet. Chem.* **1996**, *520*, 139; (c) H.S. Glenn, K.A. Lawler, R. Hoffmann, W. A. Herrmann, W. Scherer, R.W. Fischer, *J. Am. Chem. Soc.* **1995**, *117*, 3231.

10 (a) W.A. Herrmann, R.W. Fischer, M.U. Rauch, W. Scherer, *J. Mol. Catal.* **1994**, *86*, 243; (b) R.W. Fischer, Ph.D. thesis, Technische Universität München **1994**.

11 (a) A. Al-Ajlouni, H. Espenson, *J. Am. Chem. Soc.* **1991**, *117*, 9234; (b) S. Yamazaki, J.H. Espenson, P. Huston, *Inorg. Chem.* **1993**, *32*, 4683; (c) O. Pestovski, R. v. Eldik, P. Huston, J.H. Espenson, *J. Chem. Soc., Dalton Trans.* **1995**, 133; (d) J.H. Espenson, *J. Chem. Soc., Chem. Commun.* **1999**, 479; (e) W. Adam, C.R. Saha-Möller, O. Weichold, *J. Org. Chem.*, **2000**, *65*, 5001.

12 (a) P. Gisdakis, W. Antonczak, S. Köstlmeier, W.A. Herrmann, N. Rösch, *Angew. Chem. Int. Ed. Engl.* **1998**, *37*, 2211; (b) P. Gisdakis, N. Rösch, *Eur. J. Org. Chem.* **2001**, *4*, 719; (c) P. Gisdakis, I.V. Yudanov, N. Rösch, *Inorg. Chem.* **2001**, *40*, 3755; (d) C. di Valentin, R. Gandolfi, P. Gisdakis, N. Rösch, *J. Am. Chem. Soc.* **2001**, *123*, 2365.

13 W.A. Herrmann, J.D.G. Correia, F.E. Kühn, G.R.J. Artus, C.C. Romão, *Chem. Eur. J.* **1996**, *2*, 168.

14 (a) A.K. Yudin, K.B. Sharpless, *J. Am. Chem. Soc.* **1997**, *119*, 11536; (b) A.K. Yudin, J.P. Chiang, H. Adolfsson, C. Coperet, **2001**, *66*, 4713.

15 M.C.A. van Vliet, I.W.C.E. Arends, R.A. Sheldon, *J. Chem. Soc., Perkin Trans. 1* **2000**, 377.

16 (a) C. Mealli, J.A. Lopez, M.J. Calhorda, C.C. Romão, W.A. Herrmann, *Inorg. Chem.* **1994**, *33*, 1139; (b) A. Gobbi, G. Frenking, *J. Am. Chem. Soc.* **1994**, *116*, 9275.

17 (a) M.M. Abu-Omar, P.J. Hansen, J.H. Espenson, *J. Am. Chem. Soc.* **1996**, *118*, 4966; (b) G. Laurenczy, F. Lukács, R. Roulet, W.A. Herrmann, R.W. Fischer, *Organometallics* **1996**, *15*, 848; (c) J.H. Espenson, H. Tan, S. Mollah, R.S. Houk, M.D. Eager, *Inorg. Chem.* **1998**, *37*, 4621; (d) K.A. Brittingham, J.H. Espenson, *Inorg. Chem.* **1999**, *38*, 744.

18 W. Adam, C.M. Mitchell, C.R. Saha-Möller, *J. Org. Chem.* **1999**, *64*, 3699; (b) G.S. Owens, M. M. Abu-Omar, *J. Chem. Soc., Chem. Commun.* **2000**, 1165.

19 (a) J. Rudolph, K.L. Reddy, J.P. Chiang, K.B. Sharpless, *J. Am. Chem. Soc.* **1997**, *119*, 6189; (b) H. Adolfsson, A. Converso, K.B. Sharpless, *Tetrahedron Lett.* **1999**, *40*, 3991.

20 (a) C. Copéret, H. Adolfsson, K.B. Sharpless, *J. Chem. Soc., Chem. Commun.* **1997**, 1565; (b) W.A. Herrmann, R.M. Kratzer, H. Ding, H. Glas, W.R. Thiel, *J. Organomet. Chem.* **1998**, *555*, 293; (c) W.A. Herrmann, H. Ding, R.M. Kratzer, F.E. Kühn, J.J. Haider, R.W. Fischer, *J. Organomet. Chem.* **1997**, *549*, 319; (d) W.A. Herrmann, F.E. Kühn, M.R. Mattner, G.R.J. Artus, M. Geisberger, J.D.G. Correia, *J. Organomet. Chem.* **1997**, *538*, 203; (e) W.A. Herrmann, J.D.G. Correia, M.U. Rauch, G.R.J. Artus, F.E. Kühn, *J. Mol. Catal. A: Chemical* **1997**, *118*, 33.

21 W.D. Wang, J.H. Espenson, *J. Am. Chem. Soc.* **1998**, *120*, 11335.

22 (a) F.E. Kühn, A.M. Santos, P.W. Roesky, E. Herdtweck, W. Scherer, P. Gisdakis, I.V. Yudanov, C. Di Valentin, N. Rösch, *Chem. Eur. J.* **1999**, *5*, 3603; (b) P. Ferreira, W.M. Xue, É. Bencze, E. Herdtweck, F.E. Kühn, *Inorg. Chem.* **2001**, *40*, 5834.

23 M. Nakajima, Y. Sasaki, H. Iwamoto, S.I. Hashimoto, *Tetrahedron Lett.* **1998**, *39*, 87.

24 (a) W. Adam, C.M. Mitchell, *Angew. Chem. Int. Ed. Engl.* **1996**, *35*, 533; (b) T.R. Boehlow, C.D. Spilling, *Tetrahedron Lett.* **1996**, *37*, 2717; (c) R. Neumann, T.J. Wang, *J. Chem. Soc., Chem. Commun.* **1997**, 1915; (d) W. Adam, C.M. Mitchell, C.R. Saha-Möller, O. Weichold, *J. Am. Chem. Soc.* **1999**, *121*, 2097; (e) W. Adam, C.R. Saha-Möller, O. Weichold, *J. Org. Chem.* **2000**. *65*, 2897; (f) K. Dallmann, R. Buffon, *Catal· Commun.* **2000**, *1*, 9; (g) D. Sica, D. Musumeci, F. Zollo, S. de Marino, *Eur. J. Org. Chem.* **2001**, *19*, 3731; (h) C.D. Nunes, M. Pillinger, A.A. Valente, I.S. Gonçalves, J. Rocha, P. Ferreira, F.E. Kühn, *Eur. J. Inorg. Chem.* **2002**, 1100; (i) R. Saladino, V. Neri, A.R. Pelliccia, R. Caminiti, C. Sadun, *J. Org. Chem.* **2002**, *67*, 1423.

25 D. Mandelli, M.C.A. van Vliet, U. Arnold, R.A. Sheldon, U. Schuchardt, *J. Mol. Catal. A: Chemical* **2001**, *168*, 165.

2.7.4
Other Transition Metals in Olefin Epoxidation

W. R. Thiel

2.7.4.1
Introduction

During the last decade, continuing discussions on environmentally benign processes, ("green chemistry", etc.) resulted in intensified investigations on catalyzed reactions, since, by definition, catalyzed reactions generally show high atom efficiency. On the other hand, finding the right catalyst for a given reaction can be a long and stony way because of the special requirements of the substrate and the chemo-, regio- and stereoselectivities of the desired transformation.

However, olefin epoxidation is different. There are a few stoichiometric epoxidations using metal peroxo complexes, but also a multitude of transition metal compounds including almost every d-block element have found to be active in catalyzing this reaction. A few elements show pronounced effects, for example, titanium in the enantioselective epoxidation of allylic alcohols. While these systems are discussed separately in this book, the present chapter will focus on some aspects of other d- and f-block elements in olefin epoxidation. Probably the best way to organize these findings is by following the periodic table from group III to group XII (see below).

Before embarking on this, we first give a very brief overview of the general mechanistic aspects of oxidation reactions. For a long time, there has been discussion on the mechanisms of metal-catalyzed oxygen transfer to different substrates. By using hydroperoxides as the oxidizing agents, the reaction is allowed to proceed either via a peroxo or an oxo intermediate (Scheme 1).

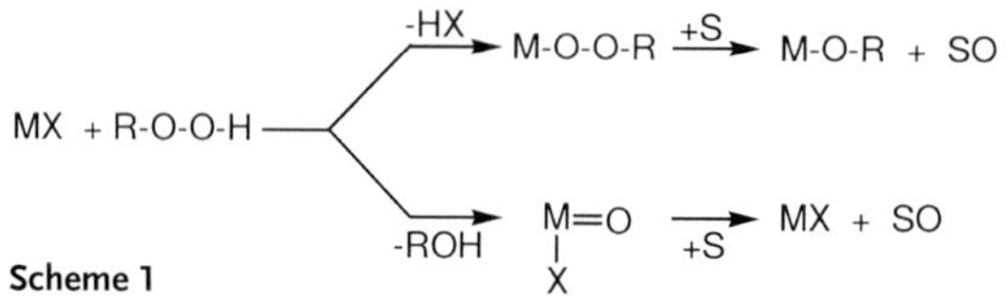

$$MX + R\text{-}O\text{-}O\text{-}H \xrightarrow{-HX} M\text{-}O\text{-}O\text{-}R \xrightarrow{+S} M\text{-}O\text{-}R + SO$$

$$MX + R\text{-}O\text{-}O\text{-}H \xrightarrow{-ROH} X\text{-}M{=}O \xrightarrow{+S} MX + SO$$

Scheme 1

Sheldon et al. have recently established an elegant mechanistic probe to distinguish between these two routes. The relative reactivities of *tert*-butyl hydroperoxide (TBHP) and pinane hydroperoxide (PHP) in metal-catalyzed oxidations were compared. When a rate-limiting oxygen transfer from a peroxometal species to the substrate is involved, huge differences in activity between TBHP and PHP were observed. In contrast, when the oxygen transfer from an oxometal species to the substrate is the rate-limiting step, little or no difference was found. Small but significant differences were observed when the reoxidation of the catalyst by the hydroperoxide to give the active oxometal species is the rate-limiting step. These findings can be explained by the different steric requirements of TBHP and PHP.

2.7.4.2
Group III Elements (Scandium, Yttrium, Lanthanum) and Lanthanoids

As well as some catalytic applications of heterogeneous systems and heteropolymetal acids containing Group III elements or members of the Lanthanoid family [2], a new and rapidly progressing field concerning the enantioselective epoxidation of conjugated, electron-deficient olefins has been opened up during recent years by the group of Shibasaki and others [3]. The catalyst (1–10 mol%) is generated *in situ* by mixing Group III or lanthanoid alkoxides and an enantiomerically pure BINOL ligand. An organic peroxide, ROOH, is used as the oxidizing agent. From the very beginning, high epoxide yields were reported. The enantiomeric excesses have been enhanced up to 99% by introducing sterically demanding oxidizing agents like cumene hydroperoxide instead of *tert*-butyl hydroperoxide, by some variations in the backbone of the BINOL ligand, by optimizing the central metal, and by the addition of triphenyl phosphine or arsine oxide. In an elegant spectroscopic study, Shibasaki et al. worked out the structure of the active catalyst and the epoxidation mechanism [4]. As depicted in Scheme 2 (epoxidation mechanism), the active species consists of a lanthanum BINOL unit coordinating the substrates and one additional donor molecule.

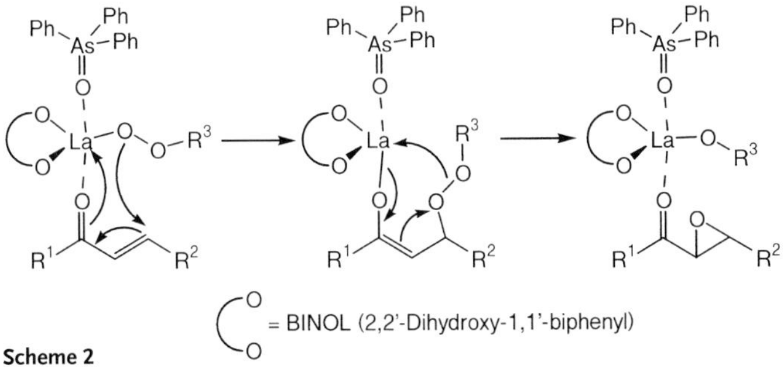

= BINOL (2,2'-Dihydroxy-1,1'-biphenyl)

Scheme 2

Up to now, the only critical requirement concerning the olefinic substrate is the conjugated enone moiety (O=C–HC=C), bearing no substituent at the α carbon atom. A whole variety of substituents at the carbonyl group and at the β carbon atom of the C=C double bond have been found to be uncritical for the performance of the reaction (Scheme 3) [4, 5].

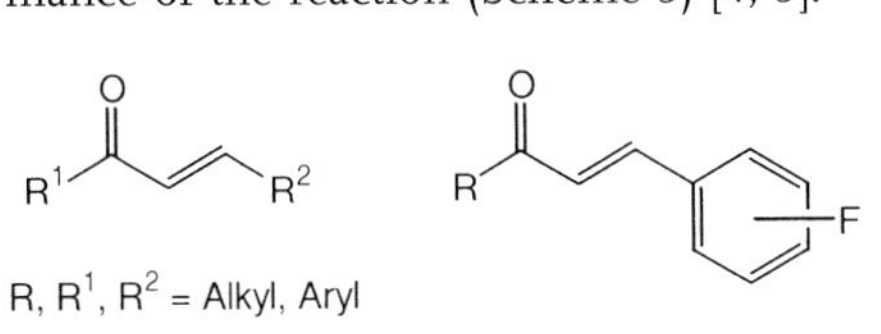

R, R^1, R^2 = Alkyl, Aryl

Scheme 3

A very interesting result is the asymmetric epoxidation of α,β-unsaturated carboxylic acid imidazolides (Scheme 4). This directly leads to the corresponding α,β-

epoxy peroxycarboxylic acid *tert*-butyl esters, which can be efficiently converted to chiral α,β-epoxy esters, amides, aldehydes, and γ,δ-epoxy β-keto esters [6].

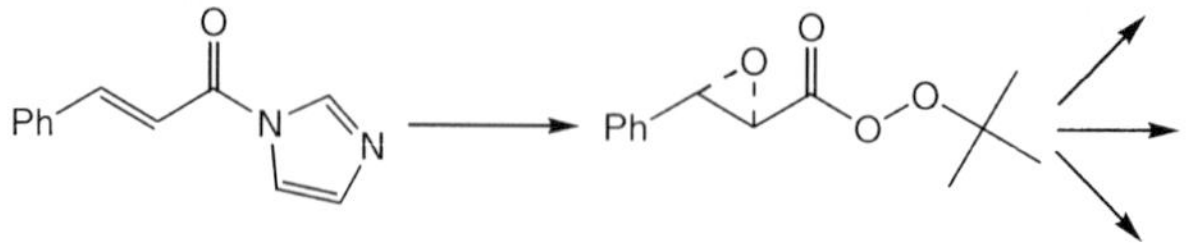

Scheme 4

This methodology has already reached application in drug synthesis, the novel PKC activator (+)-decursin and some derivatives having been obtained in high yields and enantiomeric excesses with $La(OiPr)_3$, BINOL and $O{=}AsPh_3$ (1:1:1) as the catalyst [7].

2.7.4.3
Group IV Elements (Zirconium, Hafnium)

Among group IV elements, titanium plays the dominant role as catalyst for olefin epoxidation. These systems are discussed elsewhere in this book. However, some new types of catalysts containing zirconium as the active metal have been developed during the last five years. They can be divided up into homogeneous and heterogeneous systems. For the latter, preparation of supported systems, characterization of the active species, and recovery of the catalytically active material have been the focus of attention [8].

Just a few, but nevertheless very interesting, reports on homogeneous applications of zirconium alcoholates as catalysts for olefin epoxidation have been published. Spivey et al. reported the synthesis of polyhydroxylated Celastraceae sesquiterpene cores using $Zr(OiPr)_4$ in a Sharpless type enantioselective epoxidation (>95% *ee*) of tertiary allylic alcohols, which are known to be notoriously poor substrates for the titanium reagent (Scheme 5) [9].

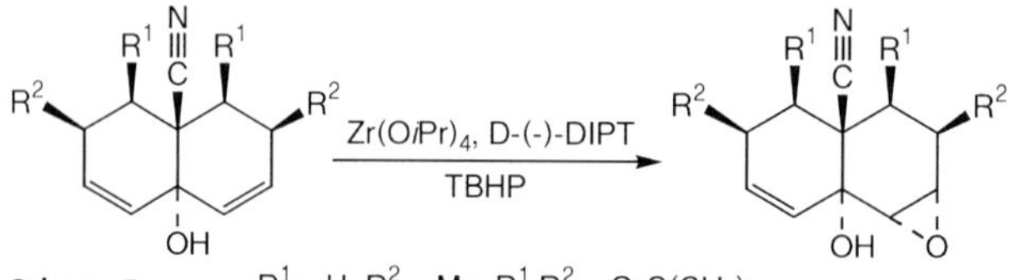

$Zr(OiPr)_4$, D-(-)-DIPT, TBHP

Scheme 5 $R^1 = H, R^2 = Me; R^1,R^2 = O_2C(CH_3)_2$

Shibasaki et al. worked out a protocol for the direct transformation of α-, internal, and cyclic olefins to the corresponding cyanhydrins, which must pass via the intermediate formation of an epoxide (for an example see Scheme 6) [10]. The reaction, which is tolerant of a whole variety of functional groups, requires $Zr(OtBu)_4$ (about 5–20 mol%) as catalyst, $(CH_3)_3Si\text{-}O\text{-}O\text{-}Si(CH_3)_3$ as the source of oxygen, $(CH_3)_3Si\text{-}CN$ as cyanide donor, some chelating 1,4-diols, and $Ph_3P{=}O$ as promoting ligand. By using a TADDOL derivative as chelator, moderate enantioselectivities of up to 62% *ee* have been observed.

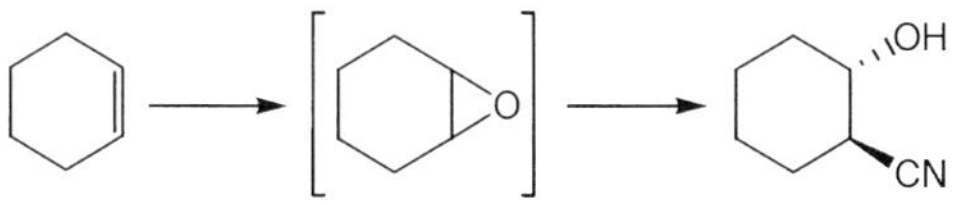

Scheme 6

2.7.4.4
Group V Elements (Vanadium, Niobium, Tantalum)

While heterogeneous vanadium oxidation catalysts have played an important role for a long time, almost nothing was known about applications of the heavier elements niobium and tantalum. One of the rare examples of high-performance enantioselective heterogeneous olefin epoxidation has been published by Basset et al. [11]. Supporting the carbene complex $Ta(=CHCMe_3)(CH_2CMe_3)_3$ on silica and reacting the product with ethanol gives a supported tetraethoxy tantalum(V) species which splits off two equivalents of ethanol when treated with 1.2–1.5 equiv. of diethyl or diisopropyl tartrate (Scheme 7). The resulting six-coordinated tantalum compound, obtained after the addition of TBHP and an allylic alcohol, is structurally closely related to the dimeric titanium species proposed to be the active component in the enantioselective epoxidation of allylic alcohols.

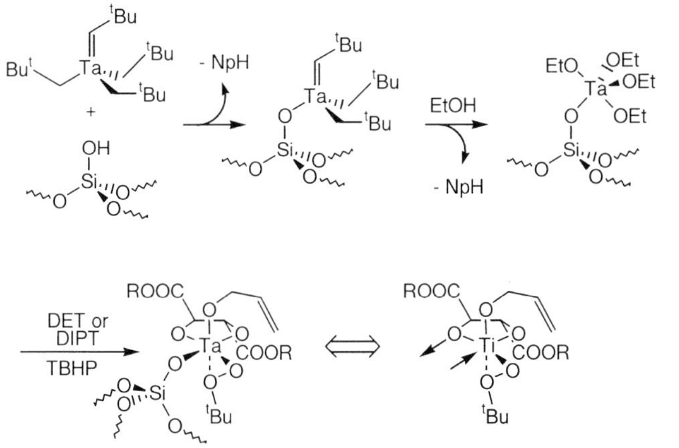

Scheme 7

For vanadium, known to catalyze the enantioselective epoxidation of allylic and homoallylic alcohols, recent investigations focussed on the fine tuning of chiral ligands for improved enantioselectivities and on the replacement of environmentally critical solvents. Olefin epoxidations were usually carried out in non-protic halogenated and/or aromatic solvents. These solvents, good for laboratory scale experiments, are not acceptable in industrial processes. Here, supercritical carbon dioxide can be the solvent of choice. Because of its low basicity it is an excellent solvent for epoxidations, which was demonstrated in a series of publications [12].

Ligand development mainly concentrates on systems bearing a hydroxamic acid group, a moiety which has given high stereochemical excesses in the past. Yamamoto et al. found an elegant and rapid access to this class of ligands starting from enantiomerically pure amino acids. A selection of systems with novel substitution patterns is presented in Scheme 8 [13].

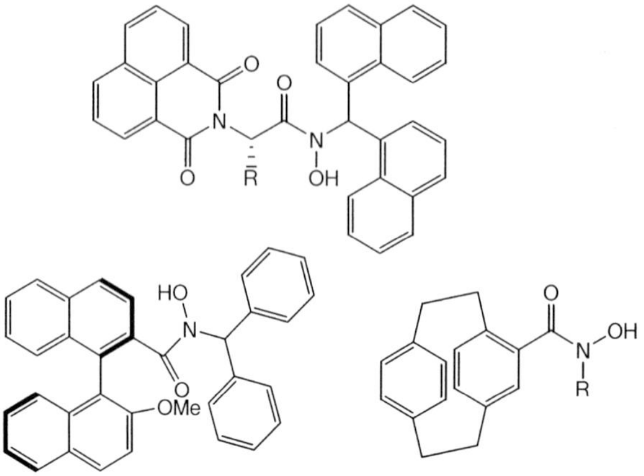

Scheme 8

Vanadium compounds are, as already mentioned, highly active catalysts for the enantioselective epoxidation of allylic and homoallylic alcohols. This has been impressively underlined in a series of syntheses of pharmaceutically active epoxies. Selected examples are given in Scheme 9 [14].

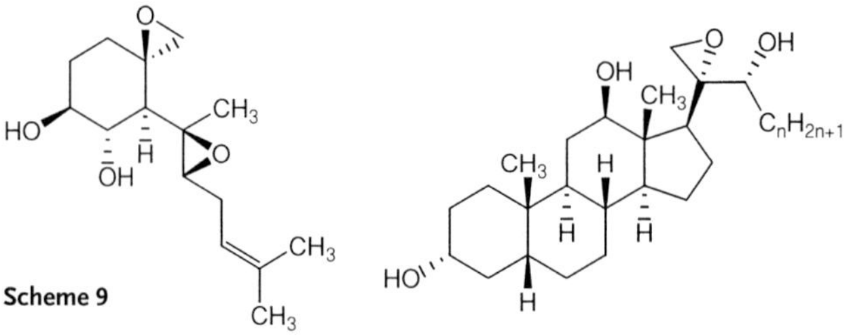

Scheme 9

2.7.4.5
Group VI Elements (Chromium, Molybdenum, Tungsten)

Olefin epoxidation with Group VI catalysts has been well established for a long time. The ARCO/HALCON process, one of the technical processes for the production of propylene oxide, runs with soluble molybdenum(VI) catalysts in combination with *tert*-butyl hydroperoxide or ethylbenzene hydroperoxide as the oxidizing agents. The side products, *tert*-butanol and 1-phenylethanol, are used for the production methyl of *tert*-butyl methyl ether and styrene. Therefore, investigations on

Group VI epoxidation catalysts have been focused on some special points of interest during recent years. One is the design of new ligand spheres, especially for the enantioselective epoxidation of unfunctionalized olefins. In contrast to the high reactivity of titanium and vanadium in the epoxidation of allylic and homoallylic olefins, molybdenum is known to show high activity for unfunctionalized substrates. Recent mechanistic studies proved that there is no direct interaction of the olefin with the metal center but only with the coordinated oxidizing agent, which prevents an efficient transfer of chirality. This was demonstrated again with two new types of chiral epoxidation catalysts (Scheme 10), which reached maximum enantiomeric excesses of about 40% [15].

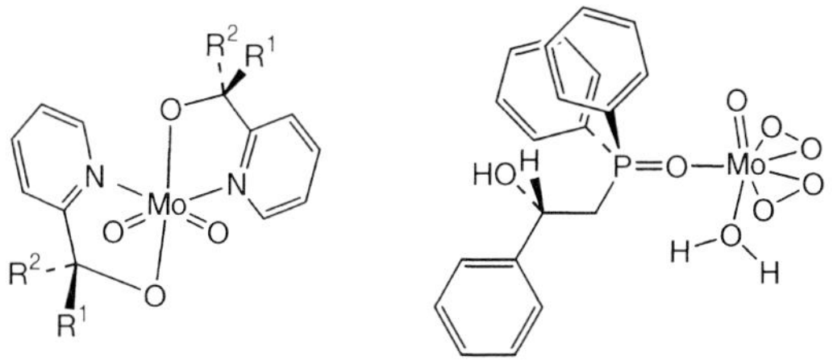

Scheme 10

Immobilization might be the right choice to overcome these problems. Che et al. showed that a chiral chromium(III) salen complex, supported on MCM-41, catalyzed the enantioselective epoxidation of styrenes with enantiomeric excesses of reaching >70% [16].

Additionally to these synthetic progresses, fundamental work aimed to obtain a better insight into mechanistic aspects of peroxide activation and oxygen transfer processes with Group VI elements was carried out. Limberg et al. and Ziegler et al. investigated the pathways of chromium-mediated oxidation reactions in a series of spectroscopic and theoretical studies [17]. Molybdenum- and tungsten-catalyzed olefin epoxidations were investigated mainly by theoretical methods [18].

2.7.4.6 Group VII Elements (Manganese, Technetium, Rhenium)

Manganese salen complexes and rhenium(VII) compounds of the type $RReO_3$ (R=alkyl, aryl) are widely used as catalysts for olefin epoxidation and are discussed elsewhere in this book.

2.7.4.7 Group VIII Elements (Iron, Ruthenium, Osmium)

Quantitative recovery of noble metal catalysts can be performed either by supporting these systems on ceramic or polymeric materials or by dissolving them in fluorocarbons. The latter procedure allows a simple phase separation but requires ligands equipped with fluorocarbon side chains. Fluorocarbon solvents are advantageous for oxidation reactions because of their inertness against oxidative degradation and because of the high solubility of dioxygen in these phases. This was proved by performing olefin epoxidations in triphasic or biphasic systems using a ruthenium catalyst [19].

Ruthenium under aerobic conditions, however, applied in common organic solvents, has some applications in the epoxidation of natural products. A novel ruthenium(II) bisoxazoline complex shows high activity and selectivity in steroid epoxidation (Scheme 11) [20].

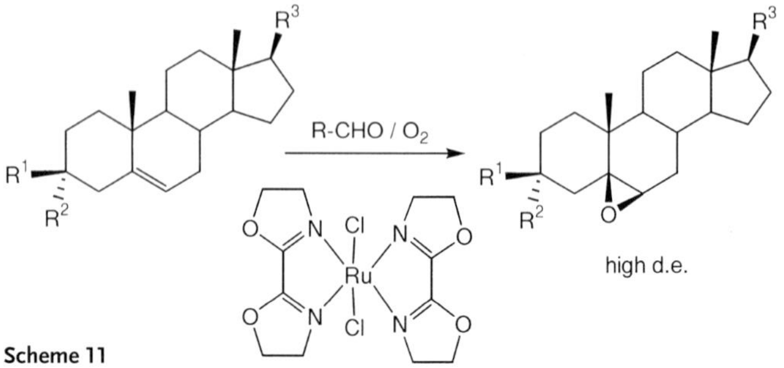

R-CHO / O_2

high d.e.

Scheme 11

Pfaltz et al. published a series of novel ruthenium complexes bearing chiral bis(dihydrooxazolylphenyl)oxalamide ligands, which catalyze the epoxidation of (*E*)-stilbene and (*E*)-1-phenylpropene with moderate enantioselectivities (up to 70%) using $NaIO_4$ as the oxidant [21]. An even higher chiral induction of up to 94% *ee* for the epoxidation of *trans*-*β*-methyl styrene was observed with the enantiomerically pure *Δ*-isomer of $[(bipy)_2RuCl(R\text{-}CH_3S(O)(p\text{-}C_6H_4CH_3)]^+$ and $PhI(OAc)_2$ as the oxidant [22].

Iron is the central metal of a series of proteins catalyzing epoxidations in nature. In synthetic processes it has not found much application because of the fa-

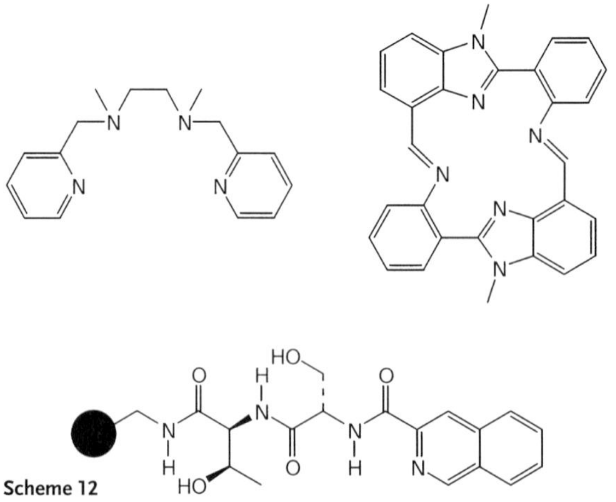

Scheme 12

cile generation of radical species in the presence of active oxygen compounds, thus leading to unselective oxidation reactions. However, ligand design can help to overcome these problems. Scheme 12 shows three ligands which are responsible for a dramatic enhancement of epoxidation selectivity and activity when combined with iron [23]. The polymer-bound substituted peptide (bottom) system was developed by methods of combinatorial chemistry.

2.7.4.8
Late Transition Metals

Late transition metals have shown activity as catalysts in the so-called Mukaiyama epoxidation. Here, an aldehyde is treated with dioxygen to form a peracid. Especially cobalt- and nickel-based systems have been used in this process. Ligand development and process improvement including the implementation of supported systems for such reactions is still ongoing [24].

Adding to these well-known procedures, zinc has found a new application in olefin epoxidation. Pu et al. have synthesized BINOL polymers which are able to activate zinc alkyl peroxo species (generated from either TBHP or Et_2Zn+O_2) for the asymmetric epoxidation of α,β-unsaturated ketones. Up to 81% *ee* has been achieved. A very interesting positive cooperative effect of the catalytic sites in the polymer chain is observed, which leads to greatly increased enantioselectivity compared with that achieved with the corresponding monomeric ligands.

References

1 H. E. B. Lempers, A. Ripollès i Garcia, R. A. Sheldon, *J. Org. Chem.* **1998**, *63*, 1408–1413.

2 (a) S. C. Grice, W. R. Flavell, A. G. Thomas, S. Warren, P. G. D. Marr, D. E. Jewitt, N. Khan, P. M. Dunwoody, S. A. Jones, *Int. J. Mol. Sci.* **2001**, *2*, 197–210. (b) W. P. Griffith, N. Morley-Smith, H. I. S. Nogueira, A. G. F. Shoair, M. Suriaatmaja, A. J. P. White, D. J. Williams, *J. Organometal. Chem.* **2000**, *607*, 146–155. (c) Y. Kera, Y. Mochizuki, S. Yamaguchi, J. Ichihara, H. Kominami, *Kidorui* **1998**, *32*, 308–309. (d) R. Shiozaki, A. Inagaki, A. Ozaki, H. Kominami, S. Yamaguchi, J. Ichihara, Y. Kera, *J. Alloys Comp.* **1997**, *261*, 132–139. (e) A. Inagaki, K. Satoh, H. Kominami, Y. Kera, S. Yamaguchi, J. Ichihara, *Kidorui* **1997**, *30*, 288–289. (f) A. Inagaki, K. Satoh, H. Kominami, Y. Kera, S. Yamaguchi, J. Ichihara, *Kidorui* **1997**, *30*, 286–287.

3 (a) M. Bougauchi, S. Watanabe, T. Arai, H. Sasai, M. Shibasaki, *J. Am. Chem. Soc.* **1997**, *119*, 2329–2330. (b) K. Daikai, M. Kamura, T. Hanamoto, I. Junji, *Kidorui* **1998**, *32*, 298–299. (c) K. Daikai, M. Kamura, I. Junji, *Tetrahedron Lett.* **1998**, *39*, 7321–7322. (d) S. Watanabe, Y. Kobayashi, T. Arai, H. Sasai, M. Bougauchi, M. Shibasaki, *Tetrahedron Lett.* **1998**, *39*, 7353–7356.

4 T. Nemoto, T. Ohshima, K. Yamaguchi, M. Shibasaki, *J. Am. Chem. Soc.* **2001**, *123*, 2725–2732.

5 (a) R. Chen, C. Qian, J. G. de Vries, *Tetrahedron* **2001**, *57*, 9837–9842. (b) T. Kagawa, A. Tanaka (Tosoh Corp., Japan), JP 2001233869.

6 T. Nemoto, T. Ohshima, M. Shibasaki, *J. Am. Chem. Soc.* **2001**, *123*, 9474–9475.

7 T. Nemoto, T. Ohshima, M. Shibasaki, *Tetrahedron Lett.* **2000**, *41*, 9569–9574.

8 (a) A. Choplin, B. Coutant, C. Dubuisson, P. Leyrit, C. McGill, F. Quignard, R. Teissier, *Stud. Surf. Sci. Catal.* **1997**, *108*, 353–360. (b) S. Gontier, A. Tuel, *Stud. Surf. Sci. Catal.* **1997**, *105B*, 1085–1092. (c) F. Quignard, A. Choplin, R. Teissier, *J. Mol. Catal. A: Chem.* **1997**, *120*, L27–L31. (d) S. Imamura, T. Yamashita, K. Utani, H. Kanai, K. Hamada, *React. Kinet. Catal. Lett.* **2001**, *72*, 11–20. (e) H. Kanai, Y. Okumura, K. Utani, K. Hamada, S. Imamura, *Catal. Lett.* **2001**, *76*, 207–211. (f) M.S. Wong, H.C. Huang, J.Y. Ying, *Chem. Mater.* **2002**, *14*, 1961–1973.

9 A.C. Spivey, S.J. Woodhead, M. Weston, B.I. Andrews, *Angew. Chem. Int. Ed.* **2001**, *40*, 769–771.

10 S. Yamasaki, M. Kanai, M. Shibasaki, *J. Am. Chem. Soc.* **2001**, *123*, 1256–1257.

11 D. Meunier, A. Piechaczyk, A. de Mallmann, J.-M. Basset, *Angew. Chem. Int. Ed.* **1999**, *38*, 3540–3542.

12 (a) G.R. Haas, J.W. Kolis, *Tetrahedron Lett.* **1998**, *39*, 5923–5926. (b) B.A. Stradi, J.P. Kohn, M.A. Stadtherr, J.F. Brennecke, *J. Supercrit. Fluids* **1998**, *12*, 109–122. (c) D.R. Pesiri, D.K. Morita, W. Tumas, W. Glaze, *Chem. Commun.* **1998**, 1015–1016. (d) D.R. Pesiri, D.K. Morita, T. Walker, W. Tumas, *Organometal.* **1999**, *18*, 4916–4924. (e) B.A. Stradi, M.A. Stadtherr, J.F. Brennecke, *J. Supercrit. Fluids* **2001**, *20*, 1–13.

13 (a) Y. Hoshino, H. Yamamoto, *J. Am. Chem. Soc.* **2000**, *122*, 10452–10453. (b) Y. Hoshino, N. Murase, M. Oishi, H. Yamamoto, *Bull. Chem. Soc. Jpn.* **2000**, *73*, 1653–1658. (c) C. Bolm, T. Kuhn, *Synlett* **2000**, 899–901. (d) N. Murase, Y. Hoshino, M. Oishi, H. Yamamoto, *J. Org. Chem.* **1999**, *64*, 338–339. (e) B. Traber, Y.-G. Jung, T.K. Park, J.-I. Hong, *Bull. Kor. Chem. Soc.* **2001**, *22*, 547–548.

14 (a) S. Amano, N. Ogawa, M. Ohtsuka, N. Chida, *Tetrahedron* **1999**, *55*, 2205–2224. (b) H. Asanuma, H. Wada, Y. Yamada (Taisho Pharmaceutical Co. Ltd, Japan), JP 10251294 (**1998**).

15 (a) W.A. Herrmann, J.J. Haider, J. Fridgen, G.M. Lobmaier, M. Spiegler, *J. Organometal. Chem.* **2000**, *603*, 69–79. (b) A.A. Valente, I.S. Goncalves, A.D. Lopes, J.E. Rodriguez-Borges, M. Pillinger, C.C. Romão, J. Rocha, X. Garcia-Mera, *New J. Chem.* **2001**, *25*, 959–963. (c) F.E. Kuhn, A.M. Santos, A.D. Lopes, I.S. Goncalves, J.E. Rodriguez-Borges, M. Pillinger, C.C. Romão, *J. Organometal. Chem.* **2001**, *621*, 207–217. (d) R.J. Cross, P.D. Newman, R.D. Peacock, D. Stirling, *J. Mol. Catal. A: Chem.* **1999**, *144*, 273–284.

16 X.-G. Zhou, X.-Q. Yu, J.-S. Huang, S.-G. Li, L.-S. Li, C.-M. Che, *Chem. Commun.* **1999**, 1789–1790.

17 (a) C. Limberg, S. Cunskis, A. Frick, *Chem. Commun.* **1998**, 225–226; (b) T. Wistuba, C. Limberg, P. Kircher, *Angew. Chem. Int. Ed.* **1999**, *38*, 3037–3039. (c) C. Limberg, R. Koeppe, *Inorg. Chem.* **1999**, *38*, 2106–2116. (d) M. Torrent, L. Deng, M. Duran, M. Sola, T. Ziegler, *Can. J. Chem.* **1999**, *77*, 1476–1491. (e) M. Torrent, L. Deng, T. Ziegler, *Inorg. Chem.* **1998**, *37*, 1307–1314.

18 (a) D.V. Deubel, G. Frenking, J. Sundermeyer, H.M. Senn, *Chem. Commun.* **2000**, 2469–2470. (b) D.V. Deubel, J. Sundermeyer, G. Frenking, *J. Am. Chem. Soc.* **2000**, *122*, 10101–10108. (c) I. V. Yudanov, C. Di Valentin, P. Gisdakis, N. Rösch, *J. Mol. Catal. A: Chem.* **2000**, *158*, 189–197. (d) A. Hroch, G. Gemmecker, W.R. Thiel, *Eur. J. Inorg. Chem.* **2000**, 1107–1114. (e) D.V. Deubel, J. Sundermeyer, G. Frenking, *Inorg. Chem.* **2000**, *39*, 2314–2320. (f) C. Di Valentin, P. Gisdakis, I.V. Yudanov, N. Rösch, *J. Org. Chem.* **2000**, *65*, 2996–3004. (g) G. Wahl, D. Kleinhenz, A. Schorm, J. Sundermeyer, R. Stowasser, C. Rummey, G. Bringmann, C. Fickert, W. Kiefer, *Chem. Eur. J.* **1999**, *5*, 3237–3251. (h) P. Macchi, A.J. Schultz, F.K. Larsen, B.B. Iversen, *J. Phys. Chem. A* **2001**, *105*, 9231–9242. (i) D.V. Deubel, J. Sundermeyer, G. Frenking, *Eur. J. Inorg. Chem.* **2001**, 1819–1827. (j) P. Gisdakis, I.V. Yudanov, N. Rösch, *Inorg. Chem.* **2001**, *40*, 3755–3765. (k) D.V. Deubel, *J. Phys. Chem. A* **2001**, *105*, 4765–4772.

19 (a) I. Klement, H. Lutjens, P. Knochel, *Angew. Chem. Int. Ed. Engl.* **1997**, *36*, 1454–1456. (b) S. Quici, M. Cavazzini, S. Ceragioli, F. Montanari, G. Pozzi,

GIANLUCA, *Tetrahedron Lett.* **1999**, *40*, 3647–3650.

20 (a) V. KESAVAN, S. CHANDRASEKARAN, *J. Chem. Soc. Perkin Trans. 1* **1997**, 3115–3116. (b) V. KESAVAN, S. CHANDRASEKARAN, *J. Org. Chem.* **1998**, *63*, 6999–7001.

21 (a) N. END, A. PFALTZ, *Chem. Commun.* **1998**, 589–590. (b) N. END, L. MACKO, M. ZEHNDER, A. PFALTZ, *Chem. Eur. J.* **1998**, *4*, 818–824.

22 F. PEZET, H. AIT-HADDOU, J.-C. DARAN, I. SASAKI, G.G.A. BALAVOINE, *Chem. Commun.* **2002**, 510–511.

23 (a) M.C. WHITE, A.G. DOYLE, E.N. JACOBSEN, *J. Am. Chem. Soc.* **2001**, *123*, 7194–7195. (b) P. PAYRA, S.-C. HUNG, W.H. KWOK, D. JOHNSTON, J. GALLUCCI, M.K. CHAN, *Inorg. Chem.* **2001**, *40*, 4036–4039. (c) M.B. FRANCIS, E.N. JACOBSEN, *Angew. Chem. Int. Ed.* **1999**, *38*, 937–941.

24 (a) J. ESTRADA, I. FERNANDEZ, J.R. PETRO, X. OTTENWAELDER, R. RAFAEL, Y. JOURNAUX, *Tetrahedron Lett.* **1997**, *38*, 2377–2380. (b) R.I. KURESHY, N.H. KHAN, S.H.R. ABDI, P. IYER, A.K. BHATT, *J. Mol. Catal. A: Chem.* **1998**, *130*, 41–50. (c) B.B. WENTZEL, S.-M. LEINONEN, S. THOMSON, D.C. SHERRINGTON, M.C. FEITERS, R.J.M. NOLTE, *Perkin 1* **2000**, 3428–3431. (d) N. KOMIYA, T. NAOTA, Y. ODA, S.-I. MURAHASHI, *J. Mol. Catal. A: Chem.* **1997**, *117*, 21–37.

25 H.-B. YU, X.-F. ZHENG, Z.-M. LIN, Q.-S. HU, W.-S. HUANG, L. PU, *J. Org. Chem.* **1999**, *64*, 8149–8155.

2.8 Wacker-Type Oxidations

Lukas Hintermann

2.8.1 Introduction

The oxidative functionalization of alkenes by means of palladium-catalyzed reactions is important for both large-scale industrial processes and research-scale synthetic organic chemistry [1–3]. Many of these reactions proceed according to the following general principle:

- An oxygen-nucleophile (ROH) attacks an olefin coordinated to palladium(II) (*Oxypalladation*)
- The intermediary alkyl-palladium species undergoes β-H-elimination, releasing the oxygenated organic product and H–Pd–X
- The loss of HX from H–Pd–X results in Pd(0), which must be re-oxidized to palladium(II) in order to start a new catalytic cycle.

The first and foremost of these reactions is the Wacker-Hoechst acetaldehyde process, where ethylene and oxygen react to acetaldehyde by means of a catalyst composed of $PdCl_2$ and $CuCl_2$ in aqueous HCl [4, 5]. Pd(II) is the actual catalytic reagent for the oxygenation of the olefin (Scheme 1a), whereas Cu(II) is a co-catalytic reoxidant for Pd(0) (b), and elemental oxygen is the terminal oxidant (c) in the overall process.

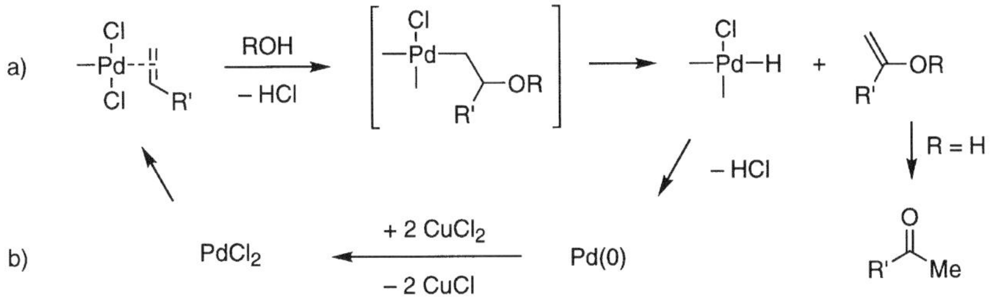

a) ROH, – HCl; – HCl; R = H

b) $PdCl_2 \xleftarrow[-2\,CuCl]{+2\,CuCl_2} Pd(0)$

c) $2\,CuCl + 0.5\,O_2 + 2\,HCl = 2\,CuCl_2 + H_2O$

Scheme 1

Transition Metals for Organic Synthesis, Vol. 2, 2nd Edition.
Edited by M. Beller and C. Bolm

ISBN: 3-527-30613-7

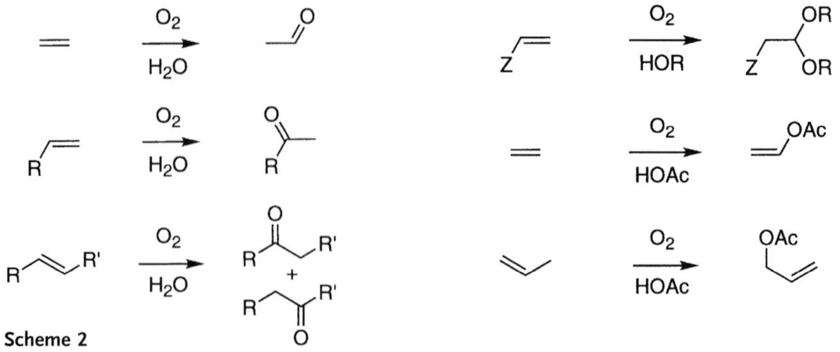

Scheme 2

Related processes of industrial use include the acetoxylation of ethane to vinyl acetate and of propene to allyl acetate, and the diacetoxylation of butadiene to 1,4-diacetoxy-2-butene [6]. The application of "Wacker"-Chemistry to higher olefins has been most successful for the conversion of terminal alkenes to methyl ketones, whereas the oxidation of internal olefins is sometimes complicated by olefin isomerization and lack of regioselectivity. Other reactions with broad synthetic scope include the synthesis of oxa-heterocycles by cyclization reactions of alkenols, the allylic acetoxylation of olefins, and the 1,4-diacetoxylation of dienes.

The oxypalladation of an olefin, because it yields an alkyl-palladium species, can also serve as entry point into a variety of C–C coupling reactions, including olefin-insertion or carbonylation.

Scheme 2 gives an overview of the types of reactions treated in this chapter.

2.8.2
The Wacker-Hoechst Acetaldehyde Synthesis

The original process was communicated from the research organization of Wacker-Chemie in 1959 [7–9]. According to the patent application, a gaseous mixture of olefin, oxygen, and water (or hydrogen) was passed over a heterogeneous catalyst, which consisted of platinum-metal compounds (preferably of Pd) and certain metal salts (e.g., of Cu, Fe, Mn, or V) precipitated on activated carbon. The industrial Wacker-Hoechst process for the oxidation of ethylene to acetaldehyde by oxygen is, however, a homogeneous catalysis in the aqueous phase, based on a soluble $PdCl_2/CuCl_2$ catalyst [4, 5]. In a variant, called the "two-stage" process, the oxidation of ethylene is performed by catalytic Pd and stoichiometric Cu(II), while the reoxidation of Cu(I) to Cu(II) is carried out in a separate reactor using air. The catalyst solution is cycled between the two reactors. Similarly, propene is oxidized to acetone in a two-stage process.

2.8.3
The Wacker-Tsuji Reaction

Terminal alkenes are fairly reliably oxidized to methyl ketones under Wacker conditions in the presence of a wide range of functionality, including internal alkenes, aldehydes, carboxylic acids, esters, alcohols, MOM-ethers, acetals, bromides, amines, etc. [1, 10, 11]. In terms of synthesis planning, a terminal alkene can thus be regarded as a masked methyl ketone.

2.8.3.1
Reaction Conditions

Many synthetic applications rely on the reagent combination of $PdCl_2/CuCl/O_2$ in DMF containing some water. For typical experimental procedures, see [11, 12]. Numerous variations of these conditions have been reported [2]. Concerning solvents, alcohols tend to give faster reactions, but also tend to speed up olefin isomerization. With cyclodextrins as phase-transfer catalysts, reactions have also been performed in water [13]. In terms of oxidants/reoxidants, $CuCl_2$ can be used instead of $CuCl/O_2$, but chlorinated side products are sometimes formed. Thus, halide-free reaction systems may be desirable, and some of the most successful include Pd salts in DMSO solvent with O_2 as the direct oxygen source [14] or $Pd(OAc)_2$ with benzoquinone (BQ) as the stoichiometric oxidant [15]. BQ has been applied co-catalytically with electrochemical or Co(salen)/O_2-mediated and other reoxidation processes [2, 16]. Further oxidants for Wacker reactions include H_2O_2 in *t*-BuOH or HOAc [17], alkyl nitrites, polyoxo-heterometallates/O_2 [18], or Pd complexes of redox-active polymers [19].

2.8.3.2
Synthetic Applications

The terminal alkene → methyl ketone conversion has been used in several synthetic sequences toward functionalized carbonyl compounds. A general access to 1,4-diketones consists in the allylation of ketone enolates, followed by Wacker oxidation. Base-induced cyclization of 1,4-diketones affords cyclopentenones, and this overall synthetic sequence has found use for the annulation of 5-rings (Scheme 3a) [11].

Likewise, 1,5-diketones are obtained from the Wacker oxidation of 5-alkenyl-carbonyl compounds. These emerge, among others, from the conjugate allylation of alkenones, the butenylation of ketone enolates, or from 3,3-sigmatropic rearrangements (oxy-Cope). Tsuji has highlighted the use of 6-ring annulation sequences for steroid synthesis, based on the Wacker oxidation of α-3-butenyl-cycloalkanones [11, 20] (Scheme 3b).

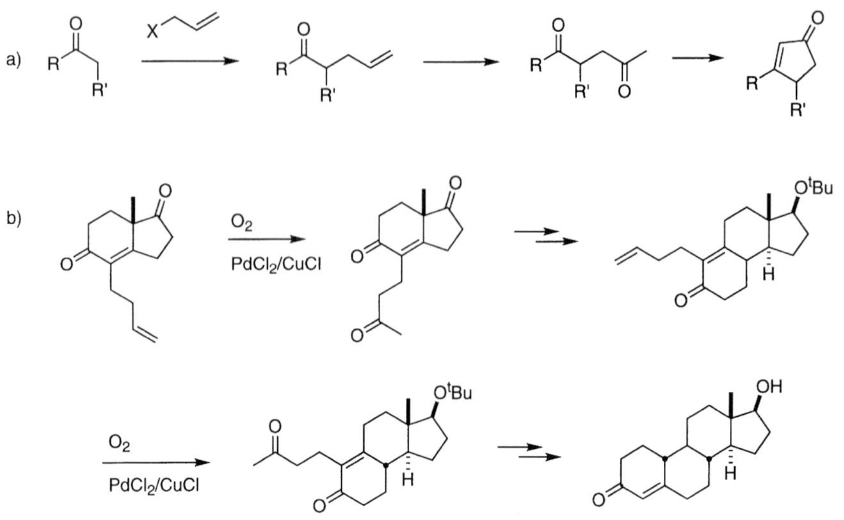

Scheme 3

2.8.3.2.1 Inversion of Regioselectivity: Oxidation of Terminal Olefins to Aldehydes and Lactones

Considerable aldehyde formation sometimes occurs under standard Wacker conditions, and this is usually connected to the presence of directing functional groups coordinating to Pd(II) in a substrate [21, 22]. A modified catalytic system based on $[PdCl(NO_2)(MeCN)_2]/CuCl_2$ in *t*-BuOH has been reported to generally yield aldehydes as major products, independently of the presence of directing groups [23]. Anti-Markovnikov selectivity is also observed in an oxidative cyclization of silylated homo-propargylic alcohols to butyrolactones under modified Wacker conditions [24] (Scheme 4). Butyrolactones are also obtained via homoallyl alcohol cyclization to hemiacetals and subsequent oxidation [25, 26].

2.8.3.2.2 Oxidation of Internal Alkenes

The Wacker oxidation of simple internal olefins is slow and not generally regioselective. However, substrates with certain functionalization patterns undergo regioselective oxidation because of a combination of electronic influences and coordination to Pd(II). Allyl ethers, either as open-chain substrates [1, 27] or as α-vinyl-tetrahydropyranes [28], yield β-alkoxy-ketones in fairly high selectivity (Scheme 5a).

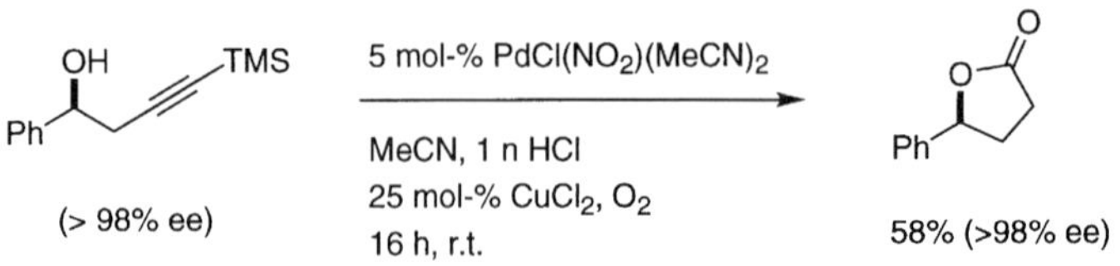

Scheme 4

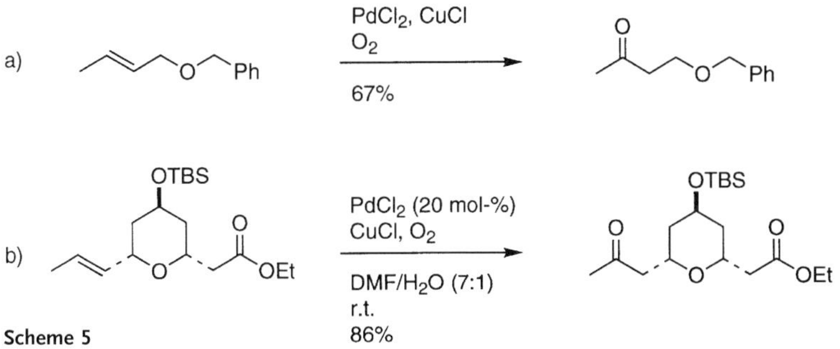

Scheme 5

This selective conversion was put to good use in the synthesis of a building block for a complex natural product (Leucascandrolide A [29]) (Scheme 5b).

Further examples of regioselective Wacker oxidation of internal olefins include the conversion of esters of 3-alken-1-ols to 4-acyloxy-ketones [27], of 3-butenoic derivatives to 1,4-dicarbonyl compounds [30], and of 4-alken-1-ones or 4-alkenoic derivatives to 1,5-dicarbonyl compounds [1]. The electronic influence in α,β-unsaturated ketones directs the nucleophilic attack to the β-position. A catalytic system based on Na_2PdCl_4/*t*-BuOOH converts those substrates to 1,3-dicarbonyl compounds [31].

2.8.4 Addition of ROH with β-H-Elimination to Vinyl or Allyl Compounds

In Wacker-type reactions with oxygen nucleophiles other than water (carboxylic acids, alcohols), initial alkoxy or acyloxy palladation is followed by β-H-elimination to either vinyl or allyl compounds, with liberation of $[PdH(X)(L)_n]$.

2.8.4.1 Synthesis of Vinyl Ethers and Acetals

The reaction of acetic acid with ethylene and oxygen to yield vinyl acetate (see Scheme 2), introduced by Moiseev [15], is performed industrially. It requires the presence of some NaOAc. Catalysis by Pd clusters, as an alternative to Pd(II)-salts, was proposed to proceed with altered reaction characteristics [32].

Olefins bearing electron-withdrawing substituents, such as α,β-unsaturated carbonyl compounds, are regioselectively converted to acetals in alcoholic solvent [33]. Acrylates are converted to 3,3-dialkoxy-propionates by a $Pd(OAc)_2$-NPMoV/C (molybdo-vanadophosphate on carbon) catalyst in acidic medium [34], or by a traditional Wacker catalyst ($PdCl_2/CuCl/O_2$) in supercritical CO_2 [35]. Further examples include the preparation of 3,3-dimethoxypropionitrile from acrylonitrile ($PdCl_2$/MeONO/MeOH) [36] (Scheme 6a) or a stereoselective acetal synthesis with an Evans-type chiral auxiliary (Scheme 6b) [37]. Intramolecular acetal formation

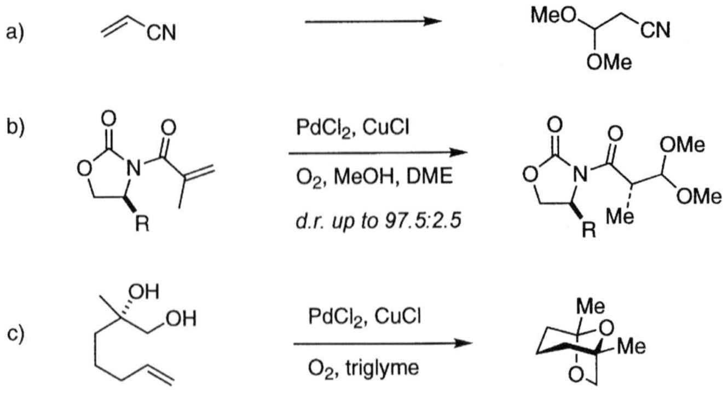

Scheme 6

occurs on oxidation of certain alkene diols, and this has been applied to pheromone synthesis (Scheme 6c) [38].

2.8.4.2
Allyl Ethers by Cyclization of Alkenols

The oxidative cyclization of alkenols is a mild method for the synthesis of oxacycles [39, 40]. Tetrahydrofurans are obtained from the 5-exo-trig cyclization of 4-alkenols, and tetrahydropyrans from the exo-cyclization of 5-alkenols (Scheme 7a) [14].

Ortho-allylphenols cyclize to benzo[*b*]furans and benzo[*b*]pyrans, depending on the reaction conditions and choice of Pd-source [41]. Under classical Wacker conditions, the cyclization to benzofurans is favored [42]. In an asymmetric version of this reaction, a benzo-dihydrofuran with an exocyclic double bond is formed in high *ee* (Scheme 7b) [43].

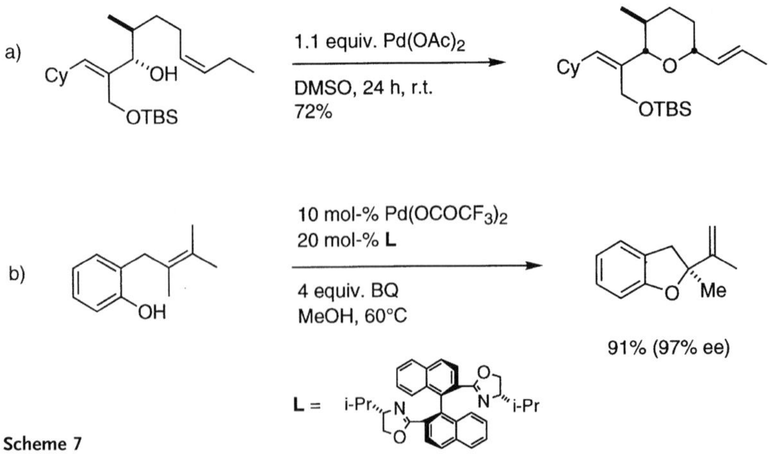

Scheme 7

2.8.4.3
Synthesis of Allyl Esters from Olefins

The palladium-catalyzed oxidative acetoxylation of olefins is particularly suitable for the synthesis of cyclic allylic acetates (Scheme 8a), whereas open-chain alkenes usually give regioisomeric mixtures [32, 44–46]. Depending on the conditions, this reaction may or may not proceed according to an allylic substitution mechanism rather than an acyloxy-palladation/β-elimination sequence [47]. Intramolecular versions of this reaction are known and applied to the synthesis of 5- and 6-membered lactones mediated by Li_2PdCl_4 and $[PdCl_2(MeCN)_2]$, and for the preparation of *iso*-coumarins and phthalides from allylbenzoic acids using catalytic amounts of $Pd(OAc)_2$ in DMSO with O_2 [3, 48, 49].

The related di-acetoxylation of dienes certainly involves an intermediary π-allyl complex. In the case of cyclic substrates such as 1,3-cyclohexadiene, this intermediate allyl complex reacts to either *cis*- or *trans*-1,4-diacetoxylated product, depending on whether halide ions are present as co-ligands (Scheme 8b) [50]. The acetoxylation reactions are usually performed using $Pd(OAc)_2$ as a catalyst in the presence of co-catalytic benzoquinone (BQ) with MnO_2 as terminal oxidant, or with a Co(salen) complex for BQ regeneration and O_2 as terminal oxidant [16].

2.8.5
Further Reactions Initiated by Hydroxy-Palladation

The lifetime of the β-oxy-palladium species, which is formed from the attack of ROH on an olefin, is usually limited by fast β-H-elimination. However, under suitable conditions, further palladium-organic chemistry may take place, notably in-

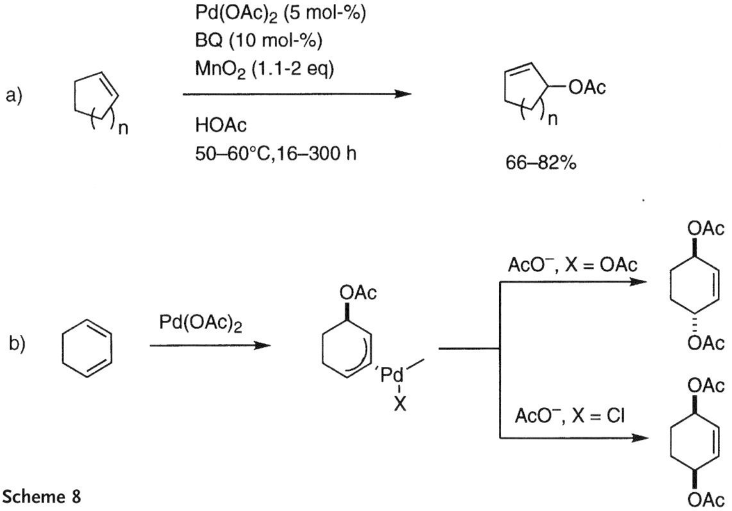

Scheme 8

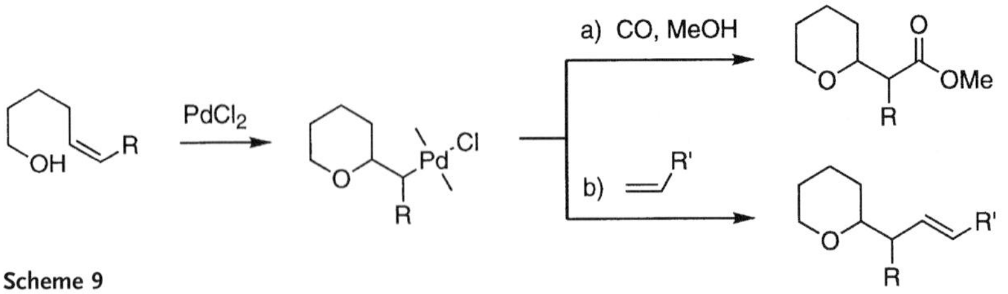

Scheme 9

sertion reactions of carbon monoxide. If the cyclization reactions of alkenols are performed in the presence of CO and excess $CuCl_2$ in alcoholic solvents, alkoxycarbonylation takes place and tetrahydrofuranyl or pyranyl esters are obtained (Scheme 9a) [3, 40]. The insertion of olefins is also possible (Scheme 9b) [51].

At high chloride concentrations, $CuCl_2$ cleaves the carbon-palladium bond, introducing chlorine with inversion of the configuration at carbon [52]. This process leads to undesired side-products (2-chloroethanol) in the industrial Wacker oxidation [53], but it has also been worked out to a useful catalytic and asymmetric synthesis of chlorohydrins [54, 55].

2.8.6
Palladium-Catalyzed Addition Reactions of Oxygen Nucleophiles

In Wacker-type reactions, the addition of an oxygen nucleophile to an olefin is coupled to a redox reaction of the reaction intermediate with Pd(II). The simple, redox-neutral addition of ROH to olefins is only realized with α,β-unsaturated substrates [56, 57] as far as palladium catalysis is concerned (Scheme 10a).

Olefinic substrates bearing a leaving group X in an allylic position may undergo alkoxy palladation followed by fast -X-elimination [58]. This mechanistic pathway

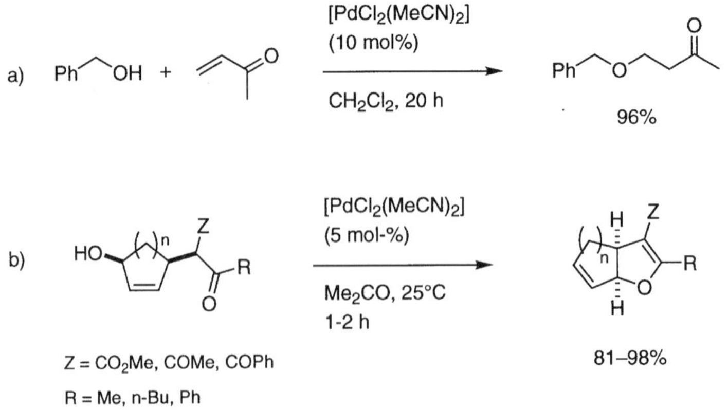

Scheme 10

might occur in the dehydrative cyclization of some cyclo-alkenols [59] and alk-3-en-1,2-diols (Scheme 10b) [60].

2.8.7 Conclusion

The Wacker oxidation represents a catalytic, atom-economic conversion of olefins with oxygen (air) to carbonyl compounds, and has a proven potential for large-scale applications. Several variants of the Wacker reaction for the selective oxidative functionalization of alkenes are regularly applied in synthesis, while others may need some improvement in terms of selectivity and catalyst turnover before they will find broader use. It is desirable to improve the performance of several of the re-oxidation systems and to generally switch to O_2 as the terminal oxidant. By meeting these goals, Wacker-type oxidations will continue to be textbook examples of clean oxidation processes by transition metal catalysis.

References

1 J. Tsuji in *Comprehensive Organic Synthesis* (Eds.: B. M. Trost, I. Fleming), Pergamon Press, Oxford, **1991**, *7*, 449.
2 J. Tsuji, *Palladium Reagents and Catalysts*, Wiley, New York, **1995**.
3 *Handbook of Organopalladium Chemistry for Organic Synthesis* (Ed.: E.-I. Negishi), John Wiley & Sons, New York, **2002**, *2*, 2119–2192.
4 R. Jira in *Applied Homogeneous Catalysis with Organometallic Compounds*, 2nd edn (Eds.: B. Cornils, W. A. Herrmann), Wiley-VCH, Weinheim, **2002**, *1*, 386–405.
5 *"Acetaldehyde"* in *Ullmann's Encyclopedia of Industrial Chemistry*, 6th edn., **2001**.
6 K. Weissermel, H.-J. Arpe, *Industrial Organic Chemistry*, 3rd edn., VCH, Weinheim, **1997**.
7 J. Smidt, W. Hafner, R. Jira, R. Rüttinger, DP 1049845, **1959**.
8 J. Smidt, W. Hafner, R. Jira, J. Sedlmeier, R. Sieber, R. Rüttinger, H. Kojer, *Angew. Chem.* **1959**, *71*, 176.
9 J. Smidt, W. Hafner, R. Jira, R. Sieber, J. Sedlmeier, A. Sabel, *Angew. Chem.* **1962**, *74*, 93.
10 T. Takahashi, K. Kasuga, J. Tsuji, *Tetrahedron Lett.* **1978**, 4917.
11 J. Tsuji, *Synthesis* **1984**, 369.
12 J. Tsuji, H. Nagashima, H. Nemoto, *Org. Synth.* **1984**, *62*, 9.
13 E. Monflier, E. Blouet, Y. Barbaux, A. Montreux, *Angew. Chem. Int. Ed. Engl.* **1994**, *33*, 2100.
14 M. F. Semmelhack, C. R. Kim, W. Dobler, M. Meier, *Tetrahedron Lett.* **1989**, *30*, 4925.
15 I. I. Moiseev, M. N. Vargaftik, Y. K. Syrkin, *Dokl. Akad. Nauk SSSR* **1960**, *133*, 377; *Chem. Abstr.* **1960**, *54*, 127953.
16 J.-E. Bäckvall, R. B. Hopkins, H. Grennberg, M. M. Mader, A. K. Awasthi, *J. Am. Chem. Soc.* **1990**, *112*, 5160.
17 M. Roussel, H. Mimoun, *J. Org. Chem.* **1980**, *45*, 5387.
18 S. F. Davison, B. E. Mann, P. M. Maitlis, *J. Chem. Soc., Dalton Trans.* **1984**, 1223.
19 M. Higuchi, S. Yamaguchi, T. Hirao, *Synlett* **1996**, 1213.
20 I. Shimizu, Y. Naito, J. Tsuji, *Tetrahedron Lett.* **1980**, *21*, 487.
21 H. Pellissier, P.-Y. Michellys, M. Santelli, *Tetrahedron* **1997**, *53*, 10733.
22 T. Hosokawa, S. Aoki, M. Takano, T. Nakahira, Y. Yoshida, S. Murahashi, *J. Chem. Soc., Chem. Commun.* **1991**, 1559.

23 B. L. Feringa, *J. Chem. Soc., Chem. Commun.* **1986**, 909.

24 P. Compain, J.-M. Vatèle, J. Goré, *Synlett* **1994**, 943.

25 J. Nokami, H. Ogawa, S. Miyamoto, T. Mandai, S. Wakabayashi, J. Tsuji, *Tetrahedron Lett.* **1988**, *29*, 5181.

26 T. M. Meulemans, N. H. Kiers, B. L. Feringa, P. W. N. M. van Leeuwen, *Tetrahedron Lett.* **1994**, *35*, 455.

27 J. Tsuji, H. Nagashima, K. Hori, *Tetrahedron Lett.* **1982**, *23*, 2679.

28 E. Keinan, K. K. Seth, R. Lamed, *J. Am. Chem. Soc.* **1986**, *108*, 3474.

29 A. Fettes, E. M. Carreira, *Angew. Chem. Int. Ed. Engl.* **2002**, *41*, 4098.

30 H. Nagashima, K. Sakai, J. Tsuji, *Chem. Lett.* **1982**, 859.

31 J. Tsuji, N. Nagashima, K. Hori, *Chem. Lett.* **1980**, 257.

32 I. I. Moiseev, M. N. Vargaftik in *Applied Homogeneous Catalysis with Organometallic Compounds*, 2nd edn (Eds.: B. Cornils, W. A. Herrmann), Wiley-VCH, Weinheim, **2002**, *1*, 406–412.

33 T. Hosokawa, S.-I. Murahashi, *Acc. Chem. Res.* **1990**, *23*, 49.

34 A. Kishi, S. Sakaguchi, Y. Ishii, *Org. Lett.* **2000**, *4*, 523.

35 L. Jia, H. Jiang, J. Li, *Chem. Commun.* **1999**, 985.

36 A. Iwayama, K. Matsui, S. Uchimumi, T. Umeza, Ube-Industries, EP 55108, **1982**.

37 T. Hosokawa, T. Yamanaka, M. Itotani, S.-I. Murahashi, *J. Org. Chem.* **1995**, *60*, 6159.

38 T. Hosokawa, Y. Makabe, T. Shinohara, S.-I. Murahashi, *Chem. Lett.* **1985**, 1529.

39 T. Hosokawa, S.-I. Murahashi in *Handbook of Organopalladium Chemistry for Organic Synthesis* (Ed.: E.-I. Negishi), John Wiley & Sons, New York, **2002**, *2*, 2169–2192.

40 T. Hosokawa, S.-I. Murahashi, *Heterocycles* **1992**, *33*, 1079.

41 R. C. Larock, L. Wei, T. R. Hightower, *Synlett* **1998**, 522.

42 A. I. Roshchin, S. M. Kel'chevski, N. A. Bumagin, *J. Organomet. Chem.* **1998**, *560*, 163.

43 Y. Uozumi, K. Kato, T. Hayashi, *J. Am. Chem. Soc.* **1997**, *119*, 5063.

44 A. Heumann, B. Åkermark, *Angew. Chem. Int. Ed. Engl.* **1984**, *23*, 453.

45 S. E. Byström, E. M. Larsson, B. kermark, *J. Org. Chem.* **1990**, *55*, 5674.

46 S. Hansson, A. Heumann, T. Rein, B. Åkermark, *J. Org. Chem.* **1990**, *55*, 975.

47 H. Grennberg, J.-E. Bäckvall, *Chem. Eur. J.* **1998**, *4*, 1083.

48 D. E. Korte, L. S. Hegedus, R. K. Wirth, *J. Org. Chem.* **1977**, *42*, 1329.

49 R. C. Larock, T. R. Hightower, *J. Org. Chem.* **1993**, *58*, 5298.

50 J. E. Bäckvall, S. E. Byström, R. E. Nordberg, *J. Org. Chem.* **1984**, *49*, 4619.

51 M. F. Semmelhack, W. R. Epa, *Tetrahedron Lett.* **1993**, *34*, 7205.

52 J.-E. Bäckvall, *Tetrahedron Lett.* **1977**, 467.

53 H. Stangl, R. Jira, *Tetrahedron Lett.* **1970**, 3589.

54 J.-Y. Lai, F.-S. Wang, G.-Z. Guo, L.-X. Dai, *J. Org. Chem.* **1993**, *58*, 6944.

55 O. Hamed, P. M. Henry, *Organometallics* **1998**, *17*, 5184.

56 T. Hosokawa, T. Shinohara, Y. Ooka, S.-I. Murahashi, *Chem. Lett.* **1989**, 2001.

57 K. J. Miller, T. T. Kitagawa, M. M. Abu-Omar, *Organometallics* **2001**, *20*, 4403.

58 J. W. Francis, P. M. Henry, *Organometallics* **1992**, *11*, 2832.

59 A. Tenaglia, F. Kammerer, *Synlett* **1996**, 576.

60 S. Saito, T. Hara, N. Takahashi, M. Hirai, T. Moriwake, *Synlett* **1992**, 237.

2.9
Catalyzed Asymmetric Aziridinations

Christian Mößner and Carsten Bolm

2.9.1
Introduction

Enantiopure aziridines have attracted considerable interest because of their potential use as intermediates for the synthesis of complex molecules [1] on the one hand and the interesting biological activities of aziridine-containing alkylating agents or natural products [2] on the other. For their synthesis, various strategies have been developed, which either make use of compounds of the chiral pool such as amino acids [3] or, alternatively, involve stereoselective transformations of simple substrates such as olefins or imines [4]. This account focuses on catalytic asymmetric methods for the preparation of aziridines with well-defined stereochemistry.

2.9.2
Olefins as Starting Materials

2.9.2.1
Use of Chiral Copper Complexes

2.9.2.1.1 Nitrene Transfer with Copper Catalysts bearing Bis(Oxazoline) Ligands

In 1991, Evans reported on the copper-catalyzed aziridination of olefins using (*N*-(*p*-toluenesulfonyl)imino)phenyliodane (**1**) as the nitrene source (Eq. 1) [5]. Efficient systems involved 5–10 mol% of a soluble copper salt such as copper triflate or copper perchlorate and a polar aprotic solvent such as acetonitrile. Solvents with a higher polarity led to both increased reaction rate and enhanced efficiency. In contrast to copper-catalyzed cyclopropanations, copper(II) salts were also suitable catalyst precursors. Mostly, the nitrene source was used as the limiting reagent with a 3- to 5-fold excess of the olefin. Both electron-rich and electron-poor olefins gave good to high yields in the range 23–95%.

Transition Metals for Organic Synthesis, Vol. 2, 2nd Edition.
Edited by M. Beller and C. Bolm

ISBN: 3-527-30613-7

$$\underset{\text{3-5 eq.}}{R^1R^2C{=}CHR^3} \xrightarrow[\text{CH}_3\text{CN, rt}]{\text{1 eq. of PhINTs (1), Cu(I) or Cu(II) salt (5-10 mol\%)}} \underset{\text{23-95\% yield}}{\text{N-Ts aziridine}} \qquad R^1, R^2, R^3 = \text{aryl, alkyl} \tag{1}$$

In most cases, aziridination reactions of this type proceed stereospecifically. In transformations of conjugated *cis*-olefins such as *cis*-stilbene or *cis*-methylstyrene, however, partial isomerization and formation of the *trans*-aziridines has been observed. The absence of ring-opened products in experiments with vinylcyclopropane as a hypersensitive radical trap indicates a concerted mechanism [5 b]. From results of UV measurements, Evans concluded that the active copper catalyst had the oxidation state two [5 b, 6].

Subsequently, Evans investigated asymmetric aziridinations with copper complexes bearing 4,4′-disubstituted bis(oxazolines) **2** and **3**.

2: R = *t*-Bu
3: R = Ph

4

5

6: X = NHPh
7: X = OH

The catalysis was highly dependent on experimental details, and for each substrate the reaction conditions had to be optimized. Generally, the system involved 5 mol% of the metal source (preferentially CuOTf), 6 mol% of the ligand, and the olefin as the limiting reagent [6]. With cinnamate esters, excellent enantioselectivities (up to 97% *ee*) could be obtained. Simple olefins such as styrene and *trans*-methylstyrene gave products with significantly lower *ee* values (63 and 70%, respectively, with **2** as ligand).

Andersson showed that the nitrene donor had an influence on the catalyst performance and that substituents at the aromatic ring of the sulfonyl moiety affected yield and enantioselectivity [7]. Predictions, however, remained difficult, and consistent rules could not be deduced. Since some nitrene precursors were not isolable or difficult to prepare, Dodd reported on an *in situ* formation starting from iodosylbenzene and free sulfonamides in the presence of molecular sieves [8].

Other oxazoline derivatives have also been applied as ligands. For example, the aziridination of styrene with camphor-derived BOX ligand **4** gave very high enantioselectivities, as reported by Masamune [9]. Stremp [10] and Andersson [11] introduced tartrate-based, anionic *N,N*- and *N,O*-bidentate oxazoline ligands **5**, **6** and **7**, respectively. With none of them, however, was the enantioselectivity improved.

Hutchings performed a detailed study on the heterogeneous aziridination of styrene with copper-exchanged zeolite in the presence of chiral BOX ligands such as

2 and **3** [12]. Under optimized conditions using a slight excess of PhINNs as nitrene precursor the best yield was 88% and the enantioselectivity reached 95% *ee*.

2.9.2.1.2 Nitrene Transfer with Copper Catalysts bearing Schiff Base Ligands

Besides bis(oxazolines), Schiff bases have successfully been applied as ligands in copper-catalyzed asymmetric aziridinations. Jacobsen introduced chiral copper-salen complexes. Tetradentate ligands, which were suitable for manganese-catalyzed epoxidations and copper-catalyzed cyclopropanations, proved ineffective in aziridination reactions [13]. Changing to neutral bidentate ligand **8**, however, led to success.

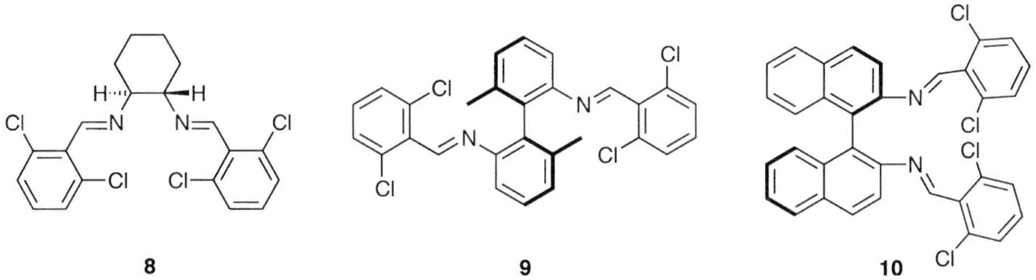

8 **9** **10**

Generally, the reactions were carried out in the presence of 10 mol% of catalyst in dichloromethane at –78 °C. As in asymmetric epoxidations and cyclopropanations, *cis*-olefins were more suitable substrates than *trans*-olefins. The highest enantioselectivity (> 98% *ee*) was obtained with 6-cyano-2,2-dimethylchromene (**11**) as substrate (Eq. 2). Use of simple olefins such as indene and 1,2-dihydronaphthalene afforded products with lower *ee* (58% and 87%, respectively).

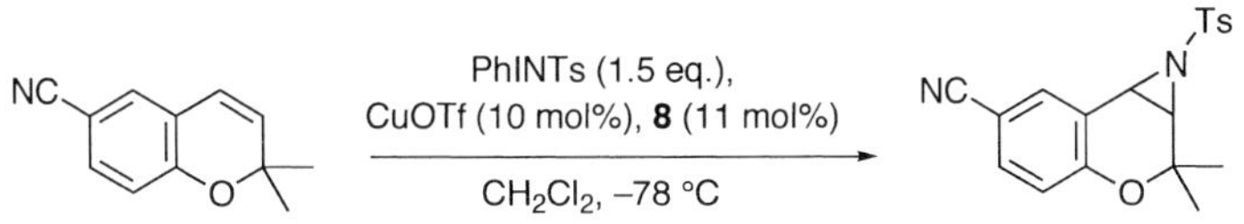

PhINTs (1.5 eq.), CuOTf (10 mol%), **8** (11 mol%); CH_2Cl_2, –78 °C (2)

11 **12** (>98% ee)

In analogy to epoxidations, aziridination reactions with metal-salen complexes are non-stereospecific. Thus, reaction of *cis*-methylstyrene affords a 3 : 1 mixture of the *cis/trans*-isomers, indicating a non-concerted mechanism for the nitrene addition to the olefin.

Scott developed chiral Schiff base ligands derived from biphenyldiamines. The substitution pattern was found to be crucial for the catalyst performance. Ligands lacking 2,6-substituents led to the formation of bimetallic $Cu_2L_2^{2+}$ complexes, which showed no catalytic activity in aziridination reactions. *Ortho*-disubstitution was required to get active monometallic CuL^+ catalysts. As in the Jacobsen system, compound **9**, derived from 2,6-dichlorobenzaldehyde, led to the most effective catalyst [14].

Most interestingly, use of **9** gave excellent results in aziridinations of both *cis*- and *trans*-olefins. For example, starting from cinnamate ester **13**, aziridine **14** with 98% *ee* was obtained in 59% yield (in CH_2Cl_2 at –40 °C; Eq. 3). Again, conversions of chromene derivatives such as **11** gave enantioselectivities near to perfection (up to 99% *ee*) [14–16].

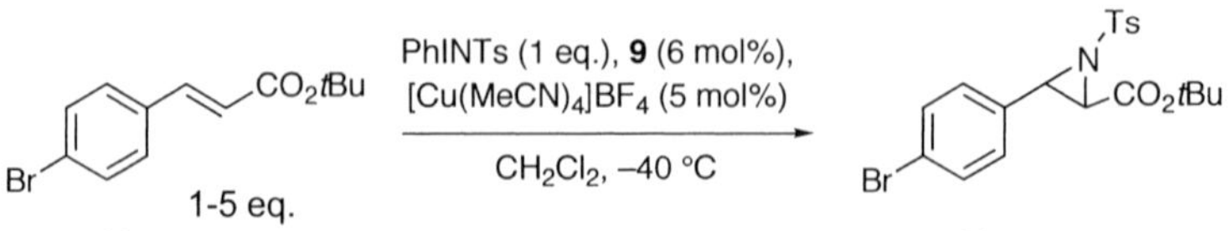

(3)

Chan introduced bis(naphthyldiimine)-derived ligand **10**, which also allowed aziridinations of cinnamate esters to be performed with very high enantioselectivities (up to 97% *ee*) [17].

Two mechanisms have been proposed for copper-catalyzed aziridinations with ArINTs (Scheme 1). The first (pathway **A**) involves a classical copper-nitrene complex and a complete release of ArI before the selectivity-determining step. On pathway **B**, the metal catalyst only serves as a Lewis acid for the activation of ArINTs.

Mechanistic investigations of aziridinations with Jacobsen's diimine-based catalysts provided evidence for a Cu(I)/Cu(III) catalytic cycle. The reaction of photochemically generated tosylnitrene in the presence of the copper complex with **8** as ligand led to the same results as the reaction with PhINTs. Moreover, use of sterically bulky nitrene source **15** had almost no influence on the catalyst performance (Scheme 2). Both facts indicate that a discrete copper-nitrene species is involved and that a Lewis acid mechanism is unlikely [18].

Detailed DFT studies from Norrby and Andersson on the Jacobsen system supplied further evidence for a Cu(I)/Cu(III) catalytic cycle. The calculations revealed that for simple systems both singlet and triplet metallonitrene have quite similar

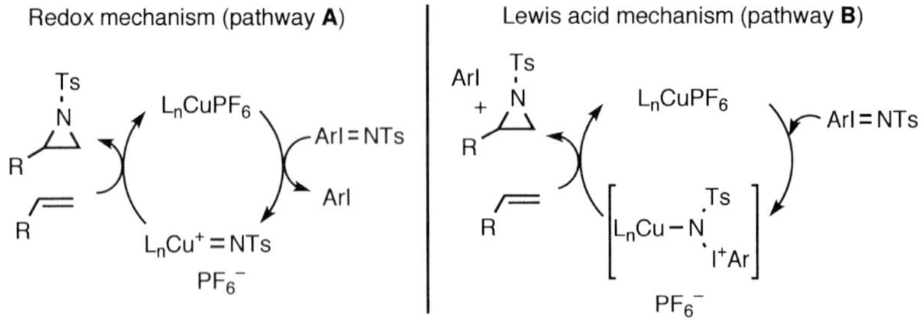

Scheme 1 Proposed mechanisms of copper-catalyzed aziridinations.

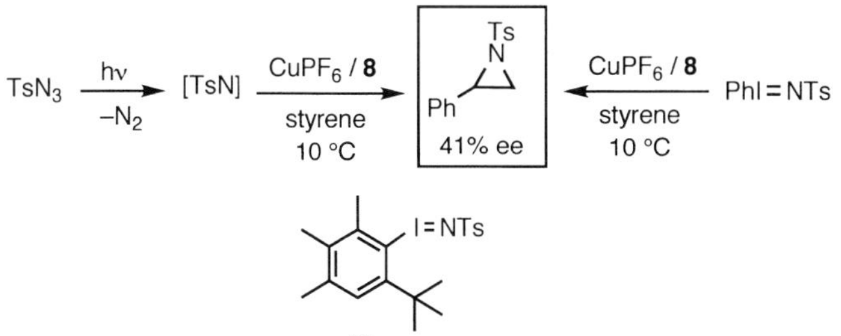

Scheme 2 Copper-catalyzed aziridinations with various nitrene precursors.

energies. For more complex systems, the triplet state is energetically favored, thus explaining the observed *cis/trans* isomerization with some substrates [19].

2.9.2.1.3 Miscellaneous Ligands

Kim investigated the use of bis(ferrocenyldiamine) **16** in combination with a copper(II) salt as a catalyst for aziridinations. The reactions were performed at room temperature with 10 mol% of catalyst, and they showed respectable yields and enantioselectivities (74% and 70% *ee* for styrene and 1-hexene, respectively) [20].

Tanner and Andersson applied C_2-symmetric bis(aziridines) such as **17** as ligands in various metal-catalyzed asymmetric transformations. In enantioselective aziridinations, low enantioselectivities were observed [21].

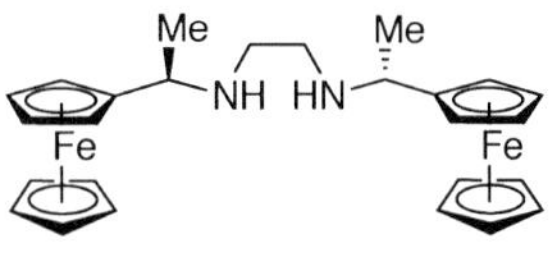

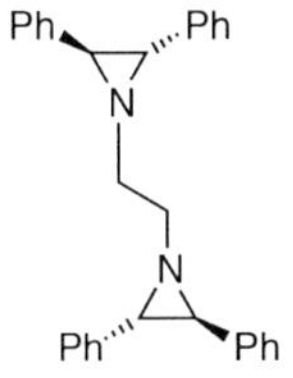

Arndtsen investigated the influence of a chiral anion on the enantioselectivity in copper-catalyzed aziridinations [22]. However, with Binol-derived borate **18** as counterion, only low enantioselectivities (up to 7% *ee* in benzene) were achieved.

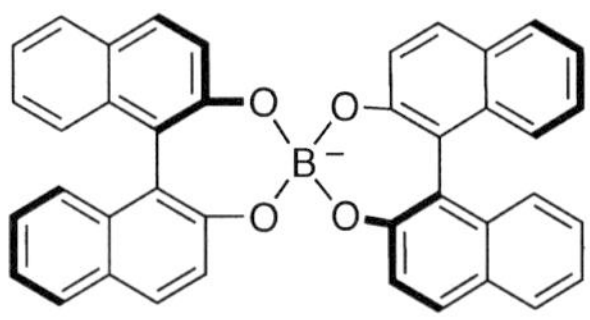

2.9.2.2 Rh-Catalyzed Aziridinations

Rhodium-catalyzed aziridinations have been studied by Müller in great detail [4]. They were found to be highly dependent on the nitrene donor [23, 24]. Whereas the use of PhINTs gave only modest yields (up to 59% with styrene as substrate), optimized conditions with PhINNs led to up to 85% yield. In terms of the substrate scope, copper-based systems appear to be superior to the ones based on rhodium. For example, in contrast to copper catalyses, electron-rich olefins such as 4-

methoxystyrene give no aziridines under rhodium catalysis, and only rearranged pyrrolidines arising from cycloaddition reactions between ring-opened aziridines and remaining olefins are obtained. Furthermore, copper catalysts convert *trans*-methylstyrene stereospecifically to *trans*-aziridines, whereas rhodium-based systems often lead to allylic sulfonamides stemming from allylic C-H insertion [24].

Rhodium complexes bearing various carboxamidate ligands, which proved effective as cyclopropanation catalysts, have also been tested in asymmetric aziridinations (Eq. 4). Good results could be obtained with $[Rh_2\{(R)\text{-}(-)\text{-bnp}\}_4]$ (**19**), which led to the corresponding products of styrene and *cis*-methylstyrene with high yields and enantioselectivities of 55 and 73% *ee*, respectively [24].

PhINNs; $[Rh_2\{(R)\text{-bnp}\}_4]$

R = H: 74% yield; 55% ee
R = Me: 80% yield; 73% ee

19 [(*R*)-bnp]

(4)

Attempts to trap the photochemically generated free nitrene from NsN_3 in the presence of Rh-complex **19** remained unsuccessful. Only marginal yields and enantioselectivities were obtained [24].

2.9.2.3 Other Metals in Aziridinations

Manganese- and ruthenium-based catalysts have also been used in aziridination reactions. Stoichiometric approaches with isolated nitridomanganese complexes have been reported by a number of groups [25]. Most of them utilized salen- or porphyrin-type ligands. In 1984, Mansuy described the first catalytic aziridination of alkenes with iron- and manganese-porphyrin complexes, which were found to catalyze the nitrogen transfer from PhINTs to alkenes with reasonable yields [26].

2.9.2.3.1 Nitrene Transfer with Salen Complexes

Burrows tested various metal-salen complexes (**20**) (with M = Mn, Co, Fe, Rh, etc.) in catalytic epoxidation and aziridination reactions [27]. Only Mn(salen)Cl catalyzed the reaction to give the desired aziridines, while all other metal complexes hydrolyzed PhINTs to give phenyl iodide and tosyl sulfonamide. Whereas an enantiodifferentiation was observed in epoxidation reactions with complexes **21**, aziridination reactions failed to give any asymmetric induction.

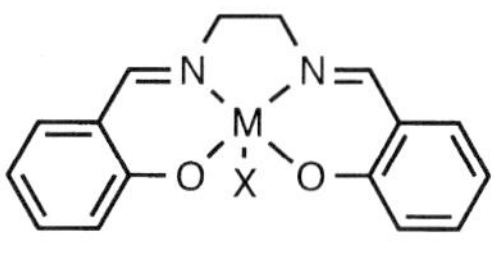

20

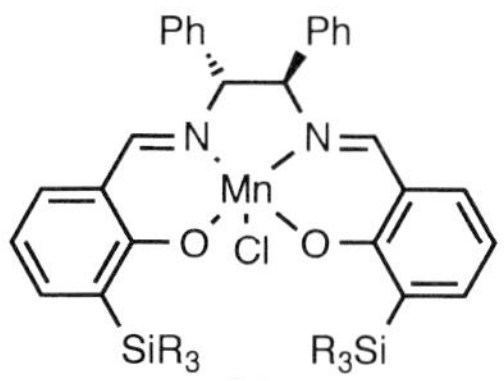

21

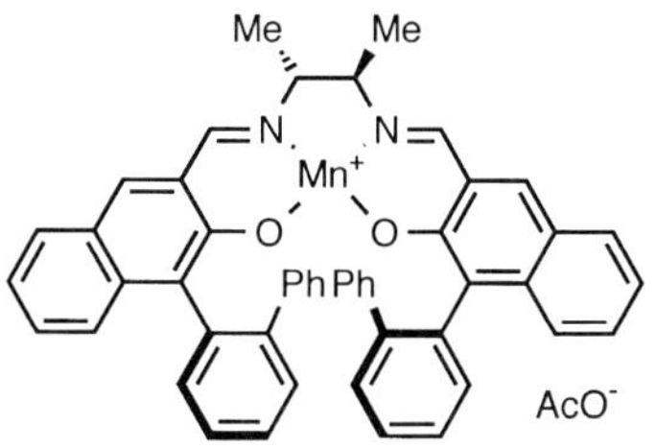

22

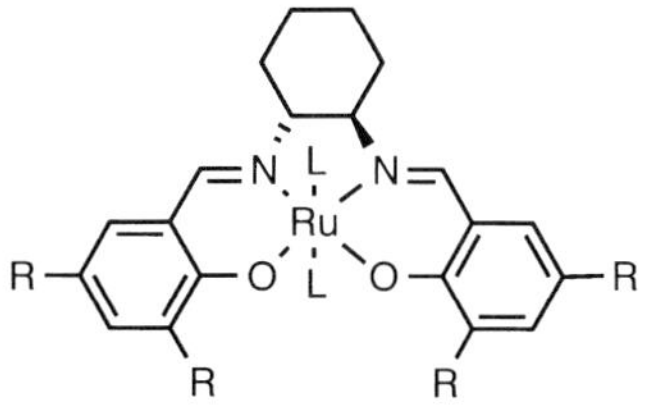

L = PPh_3, R = NO_2, I, Br

23

Katsuki reported similar results [28]. Modifications of the Mn-salen system led to complex **22**, which gave high yields and enantioselectivities in the aziridination of styrene in the presence of 4-phenyl-pyridine *N*-oxide (Eq. 5; Ar = Ph: 76% yield, 94% *ee*). Other styrenic olefins such as indene showed significantly lower enantioselectivities (50% *ee* and 10% yield) [29].

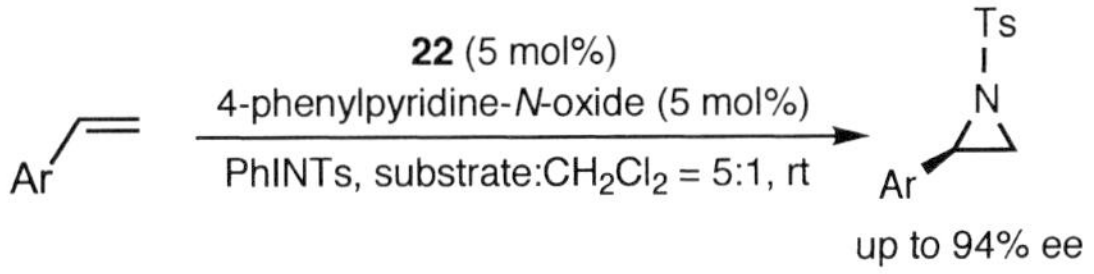

22 (5 mol%)
4-phenylpyridine-*N*-oxide (5 mol%)
PhINTs, substrate:CH_2Cl_2 = 5:1, rt

up to 94% ee

(5)

Catalyses with 12.5 mol% of Ru-salen complexes **23** led to low conversions, but in some cases the enantioselectivities were high. Thus, up to 83% *ee* was achieved in the aziridination of olefines and up to 97% *ee* in the nitrene transfer on silyl enol ethers to afford α-amino ketones [30].

2.9.2.3.2 Nitrene Transfer with Porphyrin Complexes

After the early work by Groves and Mansuy, who had first shown a nitrogen transfer from nitrido manganese(V) porphyrin complexes to olefins [25 a, b, 26], Che investigated the use of D_4-symmetric complexes **24** and **25** in catalytic asymmetric aziridinations and amidations of olefins [31, 32].

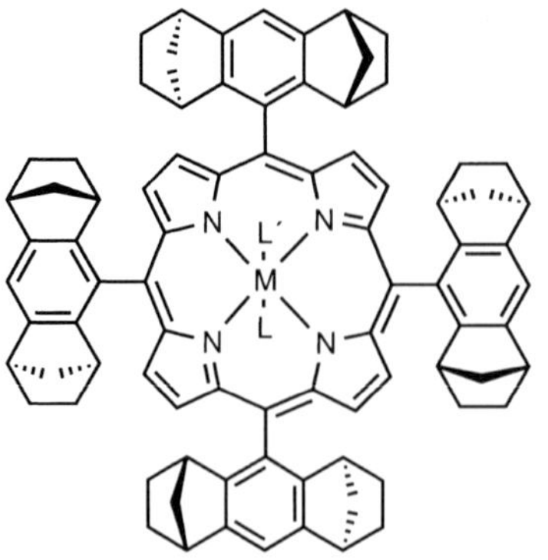

24: M = Mn, L/L′ = OH/MeOH
25: M = Ru, L/L′ = CO/EtOH

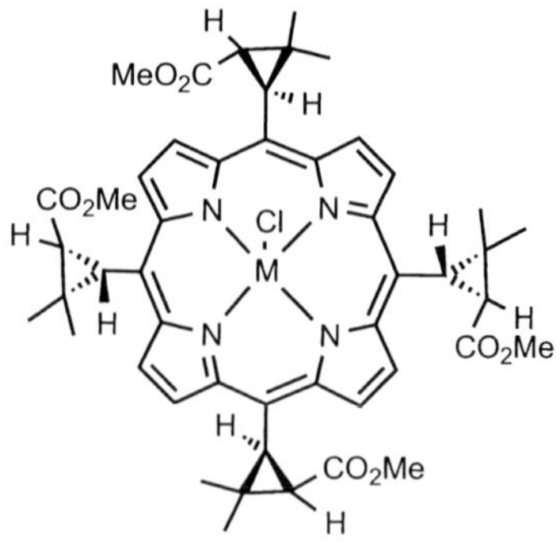

26: M = Mn
27: M = Fe

With only 0.5–1.3 mol% of complex **24** in dichloromethane at 40 °C, aziridines were obtained in high yields (43–94%). However, the enantioselectivities remained rather moderate (up to 68% *ee*). Decreasing the catalyst loading to 0.05 mol% afforded products (from styrene) with up to 42% *ee* in 58% yield, which corresponds to a TON of 1160. For the nitrogen atom transfer to olefins this is the highest TON achieved so far. Under the same conditions, asymmetric amidations of saturated benzylic C-H bonds could be performed (ee_{max}=56%) [32]. Ruthenium complex **25** proved less efficient (up to 29% and 47% *ee* in aziridinations and amidations, respectively).

Marchon reported the use of tetramethylchiroporphyrins **26** and **27** with Mn(III) or Fe(III) as central metal, respectively. In the aziridination of styrene, enantiomeric excesses of 57% (for **26**) and 28% (for **27**) were achieved. Interestingly, both complexes led to products with opposite absolute configuration. Since the structures of the two complexes were very similar, the stereochemical reversal was attributed to electronic effects leading to different reaction pathways [33].

A PEG-supported achiral Ru-porphyrin catalyst (with loadings up to 0.143 mmol/g) led to aziridines from olefins with up to 88% yield [34].

2.9.3
Imines as Starting Materials

In analogy to their well-established transfer to olefins, affording cyclopropanes, carbenes can be added to imines giving aziridines. With diazo compounds as starting materials [1], two reaction pathways can be distinguished. The first (**A**) involves an initial decomposition of the diazo compound to give a metal-carbene complex, which transfers its carbon fragment to the imine in a more or less concerted [2+1]-cycloaddition. On the second pathway (**B**) the imine is activated by a Lewis acid and is subsequently attacked by the diazo compound with loss of dinitrogen (Scheme 3).

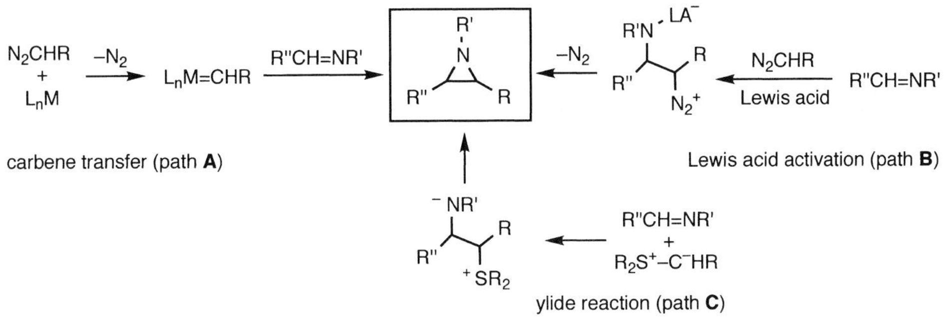

Scheme 3 Aziridination pathways starting from imines.

A third reaction pathway (**C**) utilizes sulfonium ylides. These can also add to imines, and intramolecular ring closure with loss of sulfide then affords aziridines.

2.9.3.1
Use of Metal Complexes

Aziridinations of imines and iminoesters with ethyl diazoacetate (EDA, **29**) and diiodomethane were first described by Baret [35, 36]. Jørgensen then expanded the scope of this reaction, utilizing catalytic amounts of $Cu(OTf)_2$ and various *N*-protected imines. The corresponding aziridines were formed in good yields with diethylmaleate and diethylfumarate as by-products [37]. Both yield and diastereoselectivity were highly dependent on the nitrogen substituent. By employing chiral bisoxazolines as ligands, the yields decreased and aziridines with low enantiomeric excesses were obtained. Jacobsen investigated the use of BOX ligands in asymmetric aziridinations of *N*-aryl-aldimines with EDA (Eq. 6) [38]. With imine **28** as substrate and a copper(I) complex bearing **2** as ligand, a diastereoselectivity in the formation of **30** of 60% *de* (*cis*:*trans*=4:1) was observed. The enantioselectivities were 44% and 35% *ee* for *cis*- and *trans*-**30**, respectively, and both isomers were obtained in a combined yield of 37%. As a minor product, racemic pyrrolidine **31** was formed. In reactions of α-iminoesters such as **33** with trimethylsilyldiazomethane (**34**) (Eq. 7) catalyzed by a copper(I) complex bearing $(Tol)_2$P-Binap (**32**) as ligand, Jørgensen achieved enantioselectivities in the formation of *cis*-**35** of up to 72% *ee* (*cis*:*trans* ratio of 19:1) [39].

28 + N_2CHCO_2Et (**29**) → [**2** (15 mol%), $[Cu(CH_3CN)_4PF_6]$ (10 mol%), CH_2Cl_2, rt] → *cis*-**30** + *trans*-**30**

rac-**31** **32** (6)

33 + $N_2CHSiMe_3$ (**34**) → [**32** (10 mol%), $CuClO_4$ (10 mol%), THF, –78 °C] → **35** (7)

Once more, rhodium-based systems were less suitable than copper-based ones, and both yield and enantioselectivity were low in the reaction of **28** with methyl diazoacetate (MDA). Müller attributed this result to a lower ylide-coordinating capability of the Rh(II)-catalysts [40].

2.9.3.2
Use of Lewis Acids

A wide range of transition metal and main group Lewis acids (such as ZnI_2, BF_3, $AlCl_3$, $TiCl_4$, $SnCl_4$, $Zn(OTf)_2$, $Ln(OTf)_3$, and MTO) have been employed in the reaction of imines with diazo compounds [41–48]. Usually, *cis*-aziridines are the main products, and the *cis*:*trans* ratios depend on the catalyst, the substrate, and the solvent. In no case have products from carbene-coupling reactions been observed. Jørgensen studied asymmetric aziridinations using various C_2-symmetric ligands in the presence of $Zn(OTf)_2$ and $Yb(OTf)_3$. Only marginal enantioselectivities (5–15% *ee*) were achieved [44]. Kobayashi reported on $Yb(OTf)_3$-catalyzed aziridine formations by three-component couplings, in which imines were first formed *in situ* and then reacted with EDA [45].

In 1999, Wulff introduced highly enantioselective aziridinations applying boron-based Lewis acids prepared from axial-chiral VAPOL (**38**) and VANOL (**39**) (Eq. 8). With only 2.5–10 mol% of the catalyst, aromatic or aliphatic benzhydryl imines **36** in combination with EDA (**29**) gave *cis*-aziridines **37** in high yields and with excellent stereoselectivities (*cis/trans* ratios of up to >50:1 and *ee* values of up to 99%) [49].

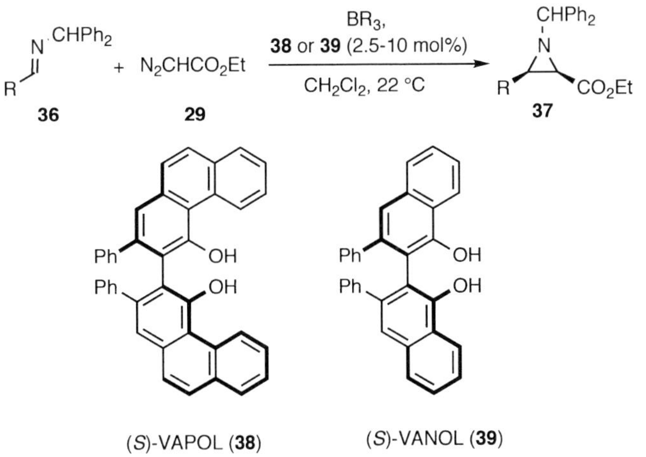

(8)

The use of trialkyl- or triarylborates instead of borane as the boron source resulted in analogous or even higher diastereo- and enantioselectivities [50].

2.9.3.3
Ylide Reactions

In contrast to ylide-mediated asymmetric epoxidations and cyclopropanations [51], the corresponding aziridinations have attracted much less attention. On the basis of his established epoxidation method, Aggarwal generated sulfonium benzylides *in situ* from sulfides and diazo compounds under rhodium catalysis, and, upon their reaction with imines, aziridines were formed (Scheme 4) [52]. Copper(II) salts could also be used, but generally rhodium catalysts gave better yields and – in asymmetric versions of this process using a chiral sulfide – slightly higher enantioselectivities.

In order to avoid the potentially hazardous handling of diazo compounds Aggarwal introduced an extension of his protocol which involves readily available tosyl hydrazones as diazo precursors. For example, the reaction of trimethylsilylethylsulfonyl-protected aldimine **40** with deprotonated tosyl hydrazone **41** in the presence of 20 mol% of chiral sulfide **43**, an ammonium salt as phase transfer catalyst (10 mol%), and 1 mol% of $Cu(acac)_2$ or $[Rh_2(OAc)_4]$, afforded aziridine **42** in 75% yield (Eq. 9). As is most common for these reactions, the diastereomer ratio was

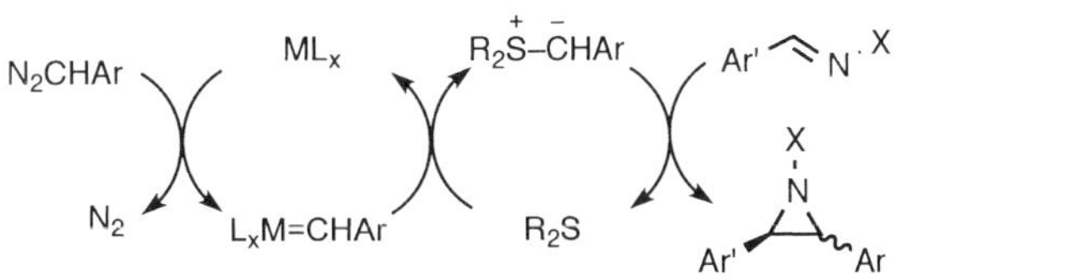

Scheme 4 Aziridination formation starting from imines via sulfur ylides.

1:2.5 in favor of the *trans* isomer, and, in this particular case, *trans*-**42** had an *ee* of 94% [53].

(9)

In addition to the SES group, other electron-withdrawing *N*-substituents such as tosyl-, Boc-, and DPP (diphenylphosphino) were found to be suitable for this reaction.

2.9.4 Conclusion

For a long time, catalytic approaches toward aziridines appeared to be less developed than analogous reactions leading to other three-membered (hetero)cycles such as epoxides or cyclopropanes. Recently, however, great advances have been made, which now allow us to prepare aziridines in an efficient manner, affording products in excellent yields. Major progress has also been achieved in catalyzed asymmetric aziridinations. Various synthetic strategies have been investigated, and several substrates can now be converted to products with high diastereoselectivities and outstanding enantioselectivities. In terms of substrate scope and catalyst activity, however, more general protocols are still desirable. Their future development is close, and, once they are found, the importance of catalytic asymmetric aziridinations as a key transformation of organic synthesis will be further highlighted.

References

1 (a) D. Tanner, *Angew. Chem.* **1994**, *106*, 625; *Angew. Chem. Int. Ed. Engl.* **1994**, *33*, 599. (b) W. McCoull, F.A. Davies, *Synthesis* **2000**, 1347.

2 (a) M. Kasai, M. Kono, *Synlett* **1992**, 778. b) J. Sweeney, *Chem. Soc. Rev.* **2002**, *31*, 247.

3 H.M. Osborn, J. Sweeney, *Tetrahedron: Asymmetry* **1997**, *8*, 1693.

4 P. Müller, C. Fruit, *Chem. Rev.* **2003**, *103*, 2905.

5 (a) D.A. Evans, M.M. F.aul, M.T. Bilodeau, *J. Org. Chem.* **1991**, *56*, 6744. (b) D.A. Evans, M.M. Faul, M.T. Bilodeau, *J. Am. Chem. Soc.* **1994**, *116*, 2742.

6 (a) D.A. Evans, K.A. Woerpel, M.M. Hinman, M.M. Faul, *J. Am. Chem. Soc.* **1991**, *113*, 726. (b) D.A. Evans, M.M. Faul, M.T. Bilodeau, B.A. Anderson,

D. M. Barnes, *J. Am. Chem. Soc.* **1993**, *115*, 5328.

7 M. J. Södergren, D. A. Alonso, P. G. Andersson, *Tetrahedron: Asymmetry* **1997**, *8*, 3563.

8 P. Dauban, L. Sanière, A. Tarrade, R. H. Dodd, *J. Am. Chem. Soc.* **2001**, *123*, 7707.

9 R. E. Lowenthal, S. Masamune, *Tetrahedron Lett.* **1991**, *32*, 7373.

10 A. M. Harm, J. G. Knight, G. Stemp, *Synlett* **1996**, 677.

11 S. K. Bertilsson, L. Tedenborg, D. A. Alonso, P. G. Andersson, *Organometallics* **1999**, *18*, 1281.

12 (a) C. Langham, P. Piaggio, D. Bethell, D. F. Lee, P. McMorn, P. C. Bulman-Page, D. J. Willock, C. Sly, F. E. Hancock, F. King, G. J. Hutchings, *Chem. Commun.* **1998**, 1601. (b) S. Taylor, J. Gullick, P. McMorn, D. Bethell, D. F. Lee, P. C. Bulman-Page, F. E. Hancock, F. King, G. J. Hutchings, *J. Chem. Soc., Perkin Trans. 2* **2001**, 1714. (c) S. Taylor, J. Gullick, P. McMorn, D. Bethell, D. F. Lee, P. C. Bulman-Page, F. E. Hancock, F. King, G. J. Hutchings, *J. Chem. Soc., Perkin Trans. 2* **2001**, 1724.

13 (a) Z. Li, K. R. Conser, E. N. Jacobsen, *J. Am. Chem. Soc.* **1993**, *115*, 5326. (b) W. Zhang, N. H. Lee, E. N. Jacobsen, *J. Am. Chem. Soc.* **1994**, *116*, 425.

14 C. J. Sanders, K. M. Gillespie, D. Bell, P. Scott, *J. Am. Chem. Soc.* **2000**, *122*, 7132.

15 K. M. Gillespie, C. J. Sanders, P. O'Shaughnessy, I. Westmoreland, C. P. Thickitt, P. Scott, *J. Org. Chem.* **2002**, *67*, 3450.

16 For X-ray structure determinations and DFT calculations, see: K. M. Gillespie, E. J. Crust, R. J. Deeth, P. Scott, *Chem. Commun.* **2001**, 785.

17 M. Shi, C.-J. Wang, A. S. C. Chan, *Tetrahedron: Asymmetry* **2001**, *12*, 3105.

18 Z. Li, R. W. Quan, E. N. Jacobsen, *J. Am. Chem. Soc.* **1995**, *117*, 5889.

19 P. Brandt, M. J. Södergren, P. G. Andersson, P.-O. Norrby, *J. Am. Chem. Soc.* **2000**, *122*, 8013.

20 D.-J. Cho, S.-J. Jeon, H.-S. Kim, C.-S. Cho, S.-C. Shim, T.-J. Kim, *Tetrahedron: Asymmetry* **1999**, *10*, 3833.

21 D. Tanner, P. G. Andersson, A. Harden, P. Somfai, *Tetrahedron Lett.* **1994**, *35*, 4631.

22 D. B. Llewllyn, D. Adamson, B. A. Arndtsen, *Org. Lett.* **2000**, *2*, 4165.

23 P. Müller, C. Baud, Y. Jacquier, *Tetrahedron* **1996**, *52*, 1543.

24 P. Müller, C. Baud, Y. Jacquier, *Can. J. Chem.* **1998**, *76*, 738.

25 (a) J. T. Groves, T. Takahashi, *J. Am. Chem. Soc.* **1983**, *105*, 2073. (b) J. T. Groves, T. Takahashi, W. M. Butler, *Inorg. Chem.* **1983**, *22*, 884. (c) S. Minakata, T. Ando, M. Nishimura, I. Ryu, M. Komatsu, *Angew. Chem.* **1998**, *110*, 3596; *Angew. Chem. Int. Ed. Engl.* **1998**, *37*, 3392. (d) M. Nishimura, S. Minakata, T. Takahashi, Y. Oderaotoshi, M. Komatsu, *J. Org. Chem.* **2002**, *67*, 2101. (e) M. Nishimura, S. Minakata, S. Thongchant, I. Ryu, M. Komatsu, *Tetrahedron Lett.* **2000**, *41*, 7089. (f) J. Du Bois, J. Hong, E. Carreira, M. W. Day, *J. Am. Chem. Soc.* **1996**, *118*, 915. (g) J. Du Bois, C. S. Tomooka, J. Hong, E. Carreira, M. W. Day, *J. Am. Chem. Soc.* **1997**, *119*, 3179. (h) J. Du Bois, C. S. Tomooka, J. Hong, E. Carreira, M. W. Day, *Acc. Chem. Res.* **1997**, *30*, 364. (i) J. Du Bois, C. S. Tomooka, J. Hong, E. Carreira, M. W. Day, *Angew. Chem.*, **1997**, *109*, 1722; *Angew. Chem. Int. Ed.* **1997**, *36*, 1645.

26 (a) D. Mansuy, J.-P. Mahy, A. Dueault, G. Bedi, P. Battioni, *J. Chem. Soc., Chem. Commun.* **1984**, 1161. (b) J.-P. Mahy, G. Bedi, P. Battioni, D. Mansuy, *J. Chem. Soc., Perkin Trans. 2* **1988**, 1517.

27 K. J. O'Connor, S.-J. Wey, C. J. Burrows, *Tetrahedron Lett.* **1992**, *33*, 1001.

28 K. Noda, N. Hosoya, R. Irie, Y. Ito, T. Katsuki, *Synlett* **1993**, 469.

29 H. Nishikori, T. Katsuki, *Tetrahedron Lett.* **1996**, *37*, 9245.

30 J.-L. Liang, X.-Q. Yu, C.-M. Che, *Chem. Commun.* **2002**, 124.

31 T.-S. Lai, H.-L. Kwong, C.-M. Che, S.-M. Peng, *Chem. Commun.* **1997**, 2373.

32 J.-L. Liang, J.-S. Huang, X.-Q. Yu, N. Zhu, C.-M. Che, *Chem. Eur. J.* **2002**, *8*, 1563.

33 J.-P. Simonato, J. Pécaut, W. R. Scheidt, J.-C. Marchon, *Chem. Commun.* **1999**, 989.

34 J.-L. Zhang, C.-M. Che, *Org. Lett.* **2002**, *4*, 1911.

35 P. Baret, H. Buffet, J.-L. Pierre, *Bull. Chem. Soc. Fr.* **1972**, 825.

36 P. Baret, H. Buffet, J.-L. Pierre, *Bull. Chem. Soc. Fr.* **1972**, 2493.

37 K.G. Rasmussen, K.A. Jørgensen, *J. Chem. Soc., Chem. Commun.* **1995**, 1401.

38 K.B. Hansen, N.S. Finney, E.N. Jacobsen, *Angew. Chem.* **1995**, *107*, 750; *Angew. Chem. Int. Ed. Engl.* **1995**, *35*, 1720.

39 K. Juhl, R.G. Hazell, K.A. Jørgensen, *J. Chem. Soc., Perkin Trans. 1* **1999**, 2293.

40 M. Moran, G. Bernardinelli, P. Müller, *Helv. Chim. Acta* **1996**, *78*, 2048.

41 R. Bartnik, G. Mloston, *Synthesis* **1983**, 924.

42 L. Casarrubios, J.A. Pérez, M. Brookhart, J.L. Templeton, *J. Org. Chem.* **1996**, *61*, 8358.

43 H.-J. Ha, K.-H. Kang, J.-M. Suh, Y.-G. Ahn, *Tetrahedron Lett.* **1996**, *37*, 7069.

44 K.G. Rasmussen, K.A. Jørgensen, *J. Chem. Soc., Perkin Trans. 1* **1997**, 1287.

45 S. Nagayama, S. Kobayashi, *Chem. Lett.* **1998**, 685.

46 W. Xie, J. Fang, J. Li, P.G. Wang, *Tetrahedron* **1999**, *55*, 12929.

47 Z. Zhu, J.H. Espenson, *J. Am. Chem. Soc.* **1996**, *118*, 9901.

48 J.C. Antilla, W.D. Wulff, *J. Am. Chem. Soc.* **1999**, *121*, 5099.

49 J.C. Antilla, W.D. Wulff, *Angew. Chem.* **2000**, *112*, 4692; *Angew. Chem. Int. Ed.* **2000**, *39*, 4518.

50 (a) A.-H. Li, L.-X. Dai, V.K. Aggarwal, *Chem. Rev.* **1997**, *97*, 2341. (b) V.K. Aggarwal, *Synlett* **1998**, 329.

51 V.K. Aggarwal, A. Thompson, R.V.H. Jones, M.C.H. Standen, *J. Org. Chem.* **1996**, *61*, 8368.

52 V.K. Aggarwal, E. Alonso, G. Fang, M. Ferrara, G. Hynd, M. Porcelloni, *Angew. Chem.* **2001**, *113*, 1482; *Angew. Chem. Int. Ed.* **2001**, *40*, 1433.

2.10
Catalytic Amination Reactions of Olefins and Alkynes

Matthias Beller, Annegret Tillack, and Jaysree Seayad

2.10.1
Introduction

Industrially important catalytic reactions are often refinement reactions of olefins. Here, the catalytic formation of carbon-carbon or carbon-hydrogen bonds is particularly important in hydrogenations, telomerisations, hydroformylations, hydrocyanations, etc. On the other hand, the atom-efficient formation of carbon-heteroatom bonds from olefins, e.g., carbon-nitrogen bonds, is comparatively rare in natural product synthesis and fine or bulk chemical production. This methodological gap is somewhat surprising if one considers the importance of amines and their derivatives in organic chemistry. For instance, most amines, enamines, and imines are useful as pharmaceutically and biologically active substances, dyes, and fine chemicals [1]. Typical methods for the synthesis of amines include alkylation, nitration of aromatics followed by reduction, reductive amination of carbonyl compounds, hydrocyanation of alkenes, etc. [2]. Among these, aside from the reductive amination of carbonyl compounds, atom-efficient synthetic routes to amines are rare. Thus, there is considerable interest in the development of new and improved synthetic protocols for the construction of carbon-nitrogen bonds. Here, the catalytic hydroamination of olefins and alkynes appears to be a particularly "green" method (Scheme 1) [3]. The procedure is in principle environmentally friendly, i.e. each atom from the starting material is present in the product and

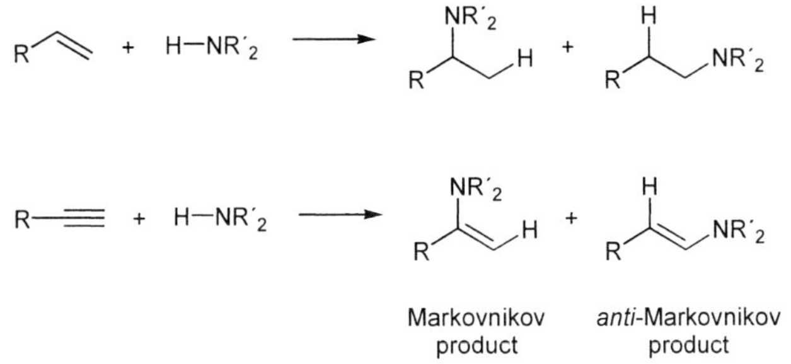

Markovnikov product   *anti*-Markovnikov product

Scheme 1 Hydroamination of olefins or alkynes.

Transition Metals for Organic Synthesis, Vol. 2, 2nd Edition.
Edited by M. Beller and C. Bolm

ISBN: 3-527-30613-7

no by-products can be formed. Furthermore, olefins, selected alkynes, and amines are both inexpensive and readily available feedstocks.

Even though considerable progress in catalytic amination reactions has been made (see below), an efficient general catalytic process for the intermolecular amination of non-activated or neutral unsaturated systems still remains a challenge. In particular, the efficient hydroamination of aliphatic alkenes is not yet possible and remains an important goal for future catalysis research.

2.10.2
The Fundamental Chemistry

Amination of unsaturated systems can take place either as hydroamination, which constitutes the formal addition of an N–H bond across the C–C multiple bond (Scheme 1) or an oxidative amination reaction whereby an imine or enamine is formed (Scheme 2).

In theory, terminal olefins provide two regioisomeric amines, the Markovnikov and the anti-Markovnikov product [4]. In acid-catalyzed reactions the Markovnikov regioisomer is usually favored because of the higher stability of the intermediate carbocation.

Though the direct nucleophilic addition of amines across multiple bonds seems to be simple, the negative entropy balance of the hydroamination reaction does not permit the use of high reaction temperatures. Thus, a catalyst is required for successful transformations. On the other hand, it is clear that the direct nucleophilic addition of amines proceeds easily to electron-deficient (activated) π-systems containing neighboring functional groups such as keto, ester, nitrile, sulfoxide, or nitro, usually leading to the anti-Markovnikov products (Michael addition) [5].

2.10.3
Catalysts

In the presence of Brønsted or Lewis acid catalysts (e.g., zeolites), aliphatic as well as most aromatic olefins react with amines, usually forming the Markovnikov product, because of the higher stability of the intermediate carbocation. This type of reaction has mainly been studied in industry. For example, propene [6] and isobutene [7] are reported to react with ammonia over zeolites forming *iso*-propyl-

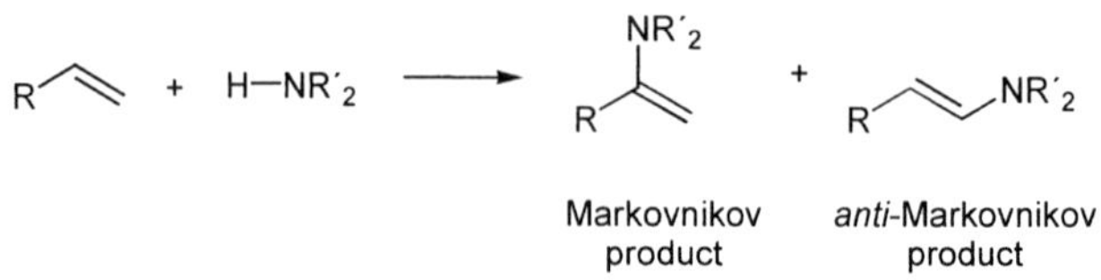

R + H–NR′₂ → (R)C(NR′₂)=CH₂ + R–CH=CH–NR′₂

Markovnikov product   *anti*-Markovnikov product

Scheme 2 Oxidative amination of olefins.

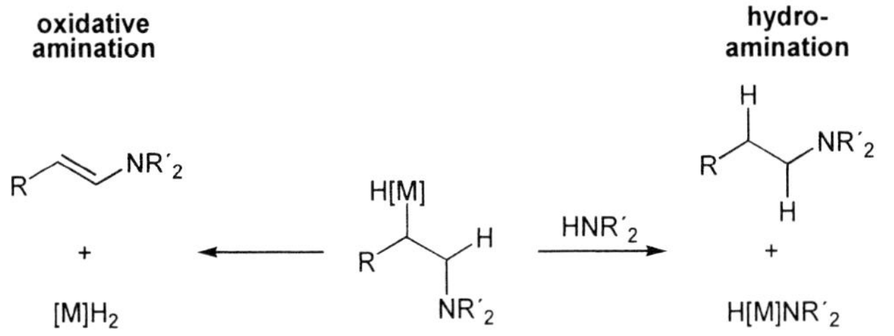

Scheme 3 Formation of hydroamination or oxidative amination products.

amine (90–100% selectivity) and *tert*-butylamine (up to 99% selectivity) respectively. BASF has commercialized a *tert*-butylamine process based on this reaction [8].

Despite the fact that reacting two nucleophilic compounds with each other can be a problem, the direct nucleophilic addition of amines to inert, non-activated unsaturated systems is known. This reaction is promoted by alkali metals [9], early [10] or late [11] transition metals, lanthanides, and actinide complexes [12], which activate either the amine or the olefin (alkyne or allene) for the coupling process. Alkenes or alkynes are susceptible to nucleophilic attack by amines by coordination to an electrophilic transition metal center [13]. β-Hydride elimination from the resulting 2-aminoalkylmetal complex leads to the oxidative amination product, and protonolysis leads to the hydroamination product (Scheme 3).

The amine can be activated by oxidative addition to a transition metal [14], which allows insertion of the alkene into the M–N or M–H bond. In addition, deprotonation of the amine in the presence of a strong base forms a more nucleophilic amide, which is able to react with certain olefins (e.g. ethylene, styrene, or 1,3-butadiene) at higher temperatures. Early transition metal complexes can also activate the amine by converting it into the coordinated imide M=NR and can enable the reaction of C–C multiple bonds with the M–N bond. The successful intra- and intermolecular amination of alkynes and alkenes using bases, early transition metals, and f-block element catalysts demonstrates the feasibility of these different approaches. However, the main disadvantage of these catalysts is that they are highly air and moisture sensitive. Thus, the development of less sensitive late transition metal catalysts for the amination of olefins is of interest. Unfortunately, the catalytic activation of π-systems using late transition metals is rather difficult to achieve in the presence of amines because of the strong coordination of the amines to electrophilic metal centers, which will replace rather than attack the π-coordinated compound. Despite these problems, a few late transition metal-catalyzed oxidative amination processes have been realized (see below) by careful variation of catalysts, substrates, and reaction conditions.

2.10.4
Oxidative Aminations

Pioneering work on catalytic oxidative aminations has been done by Hegedus and co-workers [15]. They synthesized indole and substituted indoles by Pd(II)-catalyzed regiospecific cyclization of *o*-vinyl or *o*-allyl aniline in the presence of benzoquinone and lithium chloride. The catalytic cyclization proceeds well with a number of allyl as well as styryl systems under mild and neutral conditions. Similarly, Pd(II)-catalyzed intramolecular oxidative aminations of aromatic as well as aliphatic amino olefins to give *N*-vinylic [16] and *N*-allylic heterocycles [17] were realized by converting the amino group to the corresponding *p*-toluenesulfonamide. Here, the reaction proceeds catalytically, mainly because of the decreased basicity of aromatic and tosylated amines and the stability of the cyclized products. In another example, the catalytic intermolecular oxidative amination of acrylates (activated olefins) with lactams was reported to proceed smoothly under oxygen in the presence of $PdCl_2(MeCN)_2$ and CuCl in DME at 60 °C [18]. Brunet and co-workers reported the intermolecular oxidative amination of styrene and hexene with aniline catalyzed by a rhodium complex $[Rh(PEt_3)_2Cl]_2$ in the presence of lithium anilide, forming the corresponding imines. However, only low turnover numbers (TON = 21) and turnover frequencies (TOF = < 0.07 h^{-1} at 70 °C) were achieved [19].

Later, we discovered that cationic rhodium complexes of the type $[RhL_4]X$ (L = olefin or phosphine, X = BF_4^-) catalyze the selective formation of anti-Markovnikov enamines [20]. For instance, $[Rh(cod)_2]BF_4$ in the presence of triphenylphosphine as ligand catalyzes the reaction between styrene and piperidine (Scheme 4), forming *N*-styrylpiperidine, without yielding even traces of the Markovnikov product. The formation of ethylbenzene, the hydrogenation product of styrene, is observed as another major product.

In general, the more nucleophilic cyclic and acyclic aliphatic monofunctional amines are more reactive than the less nucleophilic aromatic monofunctional amines.

2.10.5
Transition Metal-Catalyzed Hydroaminations

Similarly to oxidative aminations, intramolecular hydroaminations of alkenes and alkynes are thermodynamically more favored and more easily performed than the corresponding intermolecular reactions. Hence, the catalytic cyclization of amino-

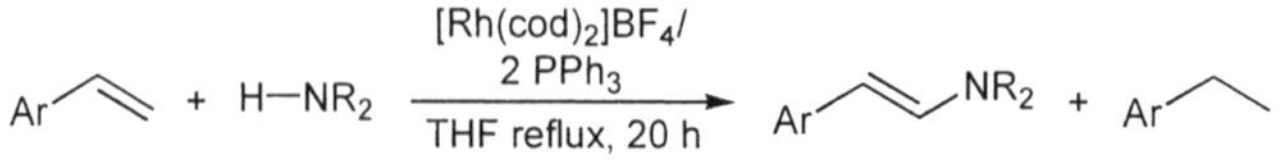

Ar–CH=CH₂ + H–NR₂ —[Rh(cod)₂]BF₄/2 PPh₃, THF reflux, 20 h→ Ar–CH=CH–NR₂ + Ar–CH₂CH₃

$$\text{ArCH=CH}_2 + \text{H–NR}_2 \xrightarrow[\text{THF reflux, 20 h}]{[Rh(cod)_2]BF_4/\ 2\ PPh_3} \text{ArCH=CH–NR}_2 + \text{ArCH}_2\text{CH}_3$$

Scheme 4 Oxidative amination of styrenes.

alkenes, aminoalkynes, and aminoallenes to give nitrogen-containing heterocycles was studied preferentially in the past using a variety of catalysts. In addition to the early transition metal complexes and organolanthanides, Pd, Ni, Au, and Ag compounds were used for this conversion. For example, $Ni(CO)_2(PPh_3)_2$ [21] and simple $PdCl_2$ were found to catalyze the cyclization of 1-amino-3-alkynes and 1-amino-4-alkynes to give 1-pyrrolines in 40–67% yield [22]. Similarly, $NaAuCl_4 \cdot 2H_2O$ and $PdCl_2(MeCN)_2$ catalyze the cyclization of 1-amino-5-alkynes, giving almost quantitative yields of corresponding tetrahydropyridines [23]. Recent studies by Müller et al. [24] have demonstrated the efficient regioselective intramolecular hydroamination of aminoalkynes of the type $RC \equiv C(CH_2)_nNH_2$ (n=3; R=H, Ph; n=4, R=H) and 2-(phenylethynyl)aniline to pyrrolidines, piperidines bearing an alkylidine functionality, and 2-phenylindole, respectively, using transition metal complexes of groups 7 to 12 ($TOF \geq 1600\ h^{-1}$). A number of metal complexes with d^8 and d^{10} configurations were found to be suitable for this reaction. Among these, the most active complexes were $[Cu(CH_3CN)_4]PF_6$, $Zn(CF_3SO_3)_2$, and $[Pd(tripos)](CF_3SO_3)_2$. Mechanistic investigations into these catalytic systems suggested that the reaction is more likely to proceed by the activation of the alkyne rather than the oxidative addition of the N–H bond. A cationic Rh complex of the type $[Rh((mim)_2CH_2)(CO)_2]BPh_4$ (mim = *N*-methylimidazol-2-yl) was also identified to act as an efficient catalyst for the intramolecular hydroamination of both terminal and non-terminal aminoalkynes ($TOF > 220\ h^{-1}$) [25]. More recently, low-valent Ru complexes with π-acidic ligands, such as $Ru(\eta^6\text{-cot})(dmfm)_2$ (cot = 1,3,5-cyclooctatriene, dmfm = dimethyl fumarate) and $Ru_3(CO)_{12}$ were reported to be active for the intramolecular hydroamination of aminoalkynes [26]. In this reaction, which is highly regioselective, the nitrogen atom is selectively attached to the internal carbon of alkynes to form several five-, six- and seven-membered heterocycles as well as indoles in good to high yields.

The intramolecular hydroamination of aminoalkenes is more difficult than the corresponding reaction of aminoalkynes. A special case of this reaction was reported by Westcott and co-workers, who demonstrated that Pd and Pt compounds catalyze the intramolecular hydroamination of aminopropyl vinylether to tetrahydro-2-methyl-1,3-oxazine [27]. An example describing late transition metal-catalyzed intermolecular hydroaminations of alkynes includes the use of $Pd(PPh_3)_4$ and benzoic acid as a catalyst system for the intermolecular hydroamination of certain aromatic acetylenes with secondary amines [28]. For example, 1-phenyl-1-propyne reacts with dibenzylamine in dioxane at 100 °C to give a 98% yield of the corresponding allylamine. Wakatsuki et al. introduced an $Ru_3(CO)_{12}$/additive (NH_4PF_6 or HBF_4/OEt_2) catalyst system permitting the conversion of terminal phenyl acetylenes with anilines to the corresponding Markovnikov imines in high yields (88–95%) [29]. $Ru_3(CO)_{12}$ was also found to be active, in the absence of any additives, for the hydroamination of alkynes with *N*-methylaniline, forming *N*-methyl-*N*-(α-styrylamines) in high yields (76–88%) [30]. Non-activated aliphatic alkynes react smoothly with anilines even at room temperature in the presence of a cationic rhodium catalyst $[Rh(cod)_2]BF_4/2\ PCy_3$. Here, the regioselective formation of branched imines was observed (Scheme 5) [31].

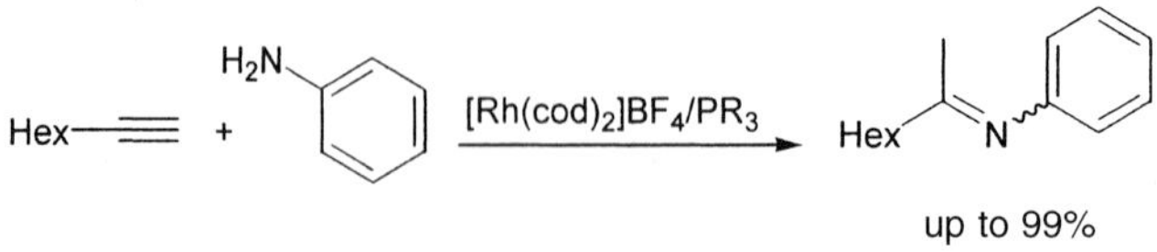

Hex—≡ + H_2N–Ph —[Rh(cod)$_2$]BF$_4$/PR$_3$→ Hex–C(=N–Ph)–CH$_3$

up to 99%

Scheme 5 Amination of 1-octyne in the presence of cationic rhodium complexes.

In the last 5 years the intermolecular hydroamination of alkynes using early transition metal catalyst systems has been developed into a convenient tool for the synthesis of imines. Notable advances have been published by the groups of Bergmann, Doye, Odom, and others [32]. Nowadays, a variety of different titanium complexes can be used as catalysts for this type of reaction, an interesting example being the first anti-Markovnikov hydroamination of terminal alkynes, which was achieved in the presence of $Cp_2Ti(\eta^2\text{-}Me_3SiC{\equiv}CSiMe_3)$ (Rosenthal's catalyst) (Scheme 6) [33].

Related to the hydroamination of alkynes is the amination of allenyl compounds to allylic amines, which is achieved using a palladium-based catalyst system consisting of $Pd_2(dba)_3 \cdot CHCl_3$-dppf-acetic acid [34].

The first transition metal-catalyzed intermolecular hydroamination of olefins was introduced by DuPont for the reaction of ethylene with secondary amines using Rh and Ir salts [35]. For example, piperidine was converted to *N*-ethylpiperidine in 70% yield using $RhCl_3 \cdot 3H_2O$ as the catalyst. However, this process was limited to ethylene and highly basic amines. We later showed that hydroamination of styrenes and vinylpyridines with aliphatic amines and anilines is possible in the presence of cationic rhodium catalysts [36]. Even activated aliphatic alkenes, e.g., norbornadiene, can be hydroaminated with cationic rhodium complexes (Scheme 7) [37].

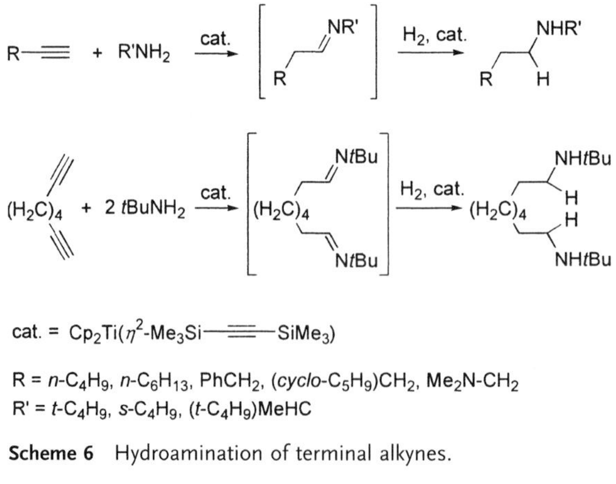

R—≡ + R'NH$_2$ —cat.→ [R–CH$_2$–CH=NR'] —H$_2$, cat.→ R–CH$_2$–CH$_2$–NHR'

(H$_2$C)$_4$(C≡CH)$_2$ + 2 *t*BuNH$_2$ —cat.→ [(H$_2$C)$_4$(CH$_2$CH=N*t*Bu)$_2$] —H$_2$, cat.→ (H$_2$C)$_4$(CH$_2$CH$_2$NH*t*Bu)$_2$

cat. = $Cp_2Ti(\eta^2\text{-}Me_3Si{\equiv}SiMe_3)$

R = *n*-C_4H_9, *n*-C_6H_{13}, $PhCH_2$, (*cyclo*-C_5H_9)CH_2, Me_2N-CH_2
R' = *t*-C_4H_9, *s*-C_4H_9, (*t*-C_4H_9)MeHC

Scheme 6 Hydroamination of terminal alkynes.

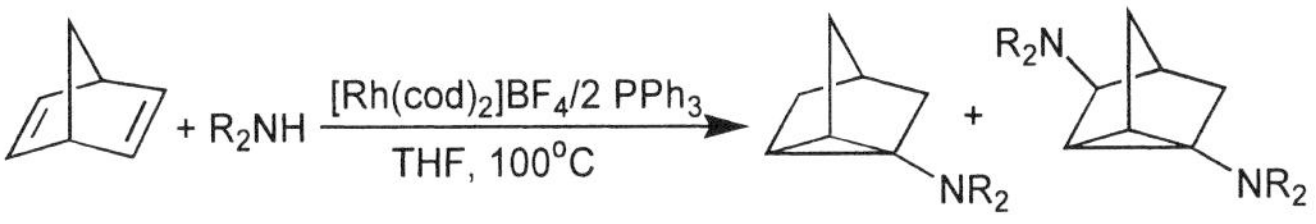

+ R_2NH $\xrightarrow[\text{THF, 100°C}]{[Rh(cod)_2]BF_4/2\ PPh_3}$ (NR_2) + (R_2N, NR_2)

Scheme 7 Amination of norbornadiene.

Milstein and co-workers [38] have established the feasibility of intermolecular hydroamination of norbornene with aniline using an electron-rich iridium complex catalyst $[Ir(PEt_3)_2(C_2H_4)_2Cl]$ in combination with $ZnCl_2$ as the co-catalyst to form exo-(2-phenylamino)norbornane as the product. This was the first successful demonstration of hydroamination of an olefin by the transition metal-catalyzed N–H activation mechanism. Stable *cis*-anilidohydride iridium complexes, resulting from the oxidative addition of aniline to iridium complexes of the type $Ir(PMe_3)_3(C_8H_{14})Cl$, $[Ir(PEt_3)_2Cl)]$, and $[Ir(PMe_3)_4PF_6]$, were synthesized and characterized. The actual catalytic species in this case is a $14e^-$ species $[Ir(PEt_3)_2Cl)]$ formed by the liberation of ethylene ligands from the precatalyst complex. Based on Milstein's work, Togni and co-workers elegantly demonstrated the possibility of catalytic asymmetric hydroaminations [39]. In a study on the intermolecular hydroamination of norbornene with aniline using Ir complexes containing chiral ligands [(R)-(S)-Josiphos, BINAP, Biphemp], they have shown that high yields (81%) and enantioselectivities (up to 95% *ee*) of exo-(2-phenylamino)norbornane (R) can be achieved using "naked" fluoride ions as co-catalyst (TON up to 80, TOF up to 3.4 h^{-1}). However, the precise role of the fluoride ions is not yet known.

Recently, a highly enantioselective palladium-catalyzed hydroamination of vinylarenes with anilines was reported by Hartwig and co-workers [40]. They found that aniline and styrene react in the presence of a catalyst system consisting of $Pd(PPh_3)_4$ or $Pd(OC(O)CF_3)_2$/dppf and triflic acid to form the Markovnikov addition product in high yields (>99%). Here, the major role was played by the acid co-catalyst, which presumably oxidizes the Pd(0) species to an active Pd(II) species. When chiral phosphine ligands were used, non-racemic amine products were obtained with good *ees*. For instance, the reaction of aniline with trifluoromethylstyrene and vinylnaphthalene catalyzed by $[\{(R)\text{-BINAP}\}Pd(OSO_2CF_3)_2]$ at 25 °C yielded the addition products in quantitative yields and 81% and 64% enantioselectivities respectively. From the same group, the enantioselective amination of 1,3-dienes is also reported using a $[Pd(\pi\text{-allyl})Cl]_2$ complex, along with optically active phosphines, which provides high yields (up to 94%) and *ees* (up to 95%) for a variety of aryl amines [41]. Also, the feasibility of transition metal-catalyzed amination of acrylic acid derivatives using a high-throughput colorimetric assay was demonstrated by Hartwig and co-workers [42].

2.10.6
Base-Catalyzed Hydroaminations

An important advantage of base-catalyzed hydroaminations [43] over the above-mentioned transition metal-based reactions is the lower price of alkali metal catalyts. Generally alkyl lithium reagents, lithium and sodium amides, NaH, and KO*t*Bu are used as catalysts. However, base-catalyzed hydroaminations often proceed (although they have been less studied) in the presence of simple alkali metals. The feasibility of base-catalyzed hydroaminations on an industrial scale is demonstrated by the Takasago process for (–)-menthol. (In 1996 more than 2000 tons of (–)-menthol and other terpenes were produced.) The key intermediates of the process, *N,N*-diethylgeranylamine and *N,N*-diethylnerylamine, are synthesized in high yields from myrcene or isoprene, respectively, by treatment with diethylamine and a catalytic quantity (1 mol%) of lithium diethylamide (Scheme 8) [44].

The amination of isoprene to *N,N*-diethylnerylamine (telomerization) using *n*-BuLi or PhLi catalysts is also an important step in the synthesis of other industrially important acyclic monoterpenes such as linalool, hydroxylinalool, and citronellol (Scheme 9) [45].

In addition to these reactions, the base-catalyzed hydroamination of simple styrene derivatives and 1,3-butadiene with primary and secondary amines is easily done. Especially β-arylethylamines and amphetamines are accessible by the base-catalyzed anti-Markovnikov hydroamination of the corresponding styrene derivatives [46].

In addition to the simple hydroamination reaction, KO*t*Bu is also useful for the domino hydroamination-aryne cyclization reaction of 2-halostyrenes with anilines, forming the corresponding indoles (Scheme 10) [47].

In fact, the cyclization of 2-chlorostyrene with aniline in the presence of three equivalents of KO*t*Bu in toluene at 135 °C provided *N*-phenyl-2,3-dihydroindole in 53% yield.

Base-catalyzed hydroaminations of alkynes have also been realized. For example, Knochel and co-workers reported the reaction of phenylacetylene with diphe-

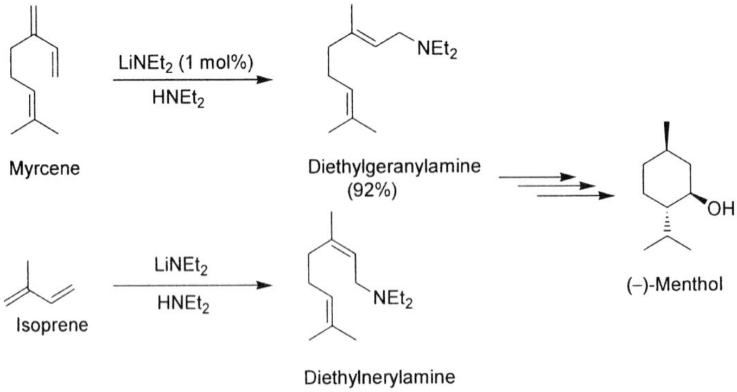

Scheme 8 Takasago (–)-menthol process.

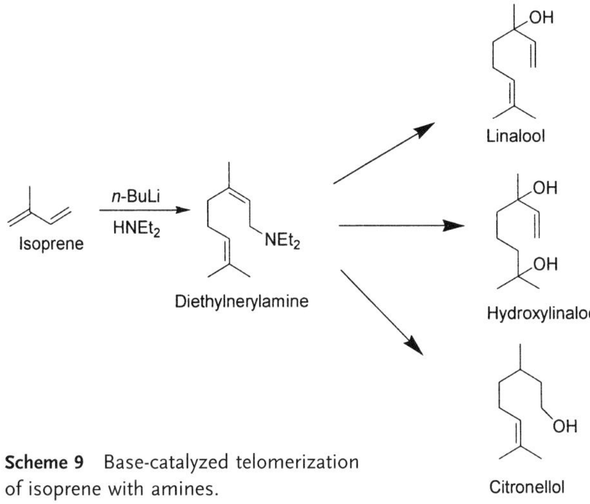

Scheme 9 Base-catalyzed telomerization of isoprene with amines.

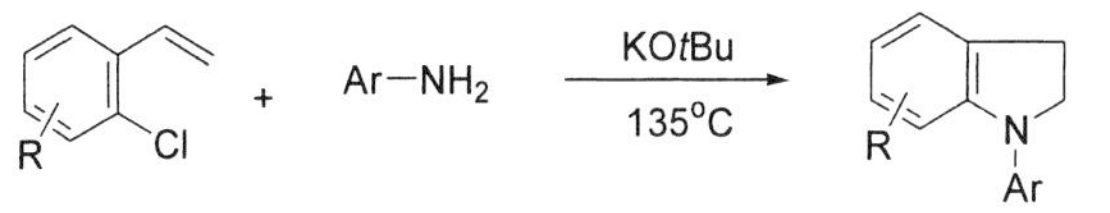

Scheme 10 Domino hydroamination-aryne cyclization reaction.

nylaniline and *N*-methylaniline in the presence of catalytic amounts of CsOH in NMP at 90–120 °C, leading to corresponding enamines in 82% and 46% yield, respectively [48]. Under similar conditions, pyrrole, imidazole, indole, and benzimidazole add to phenylacetylene giving 65–83% yield. The same group reported an elegant base-catalyzed intramolecular hydroamination of 2-(2-alkynyl)anilines to form substituted indoles in 61–90% yield (Scheme 11) [49]. The cyclization reaction is fast even at room temperature and tolerates several functional groups such as hydroxy, acetal, amino, nitro, and alkyne, enabling a variety of polyfunctional

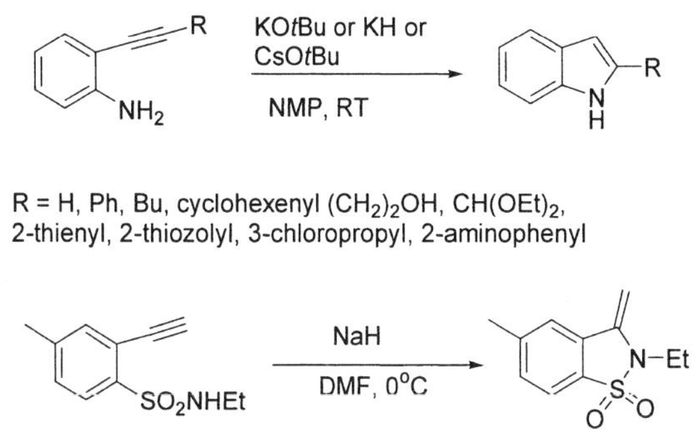

Scheme 11 Base-catalyzed intramolecular hydroamination-substituted alkynes.

indoles to be prepared. This cyclization reaction was also extended to various heterocyclic amines such as aminopyridines.

2.10.7
Conclusions

Catalytic hydroamination of olefins and alkynes offers a simple and atom-efficient access to a variety of amines and their derivatives. Apart from strongly activated olefins, 1,3-dienes and styrene derivatives can be used in a more general way. Nevertheless, a number of problems in this area await solution. Clearly, a general catalytic method for the hydroamination of simple aliphatic olefins is an important goal, and a general procedure for catalytic asymmetric aminations of olefins would be of high value for fine chemical synthesis. Here, new transition metal-catalyzed reactions are especially likely to open up new possibilities.

References

1 (a) For general references see March, J. *Advanced Organic Chemistry*, 4th edn., Wiley: New York, **1992**; p. 768 and references therein. (b) Collman, J. P., Trost, B. M., Veroeven, T. R. in *Comprehensive Organometallic Chemistry*, Wilkinson, G., Stone, F. G. A., Eds., Pergamon Press: Oxford, **1982**, 8, 892 and references therein. (c) Gibson, M. S. in *The Chemistry of Amino Group*, Patai, S., Ed., Interscience, New York, **1968**; p. 61.

2 (a) Hartwig, J. F. *Synlett* **1997**, 329. (b) Roundhill, D. M. *Chem. Rev.* **1992**, *92*, 1.

3 For leading reviews of hydroamination, see: (a) Müller, T. E., Beller, M. *Chem. Rev.* **1998**, *98*, 675. (b) Haak, E., Doye, S. *Chem. Unserer Zeit* **1999**, *33*, 296. (c) Brunet, J. J., Neibecker, D., Niedercorn, F. *J. Mol. Catal.* **1989**, *49*, 235.

4 (a) Markovnikov, V. V., *Ann. Chem. Pharm.* **1870**, *153*, 228. (b) Markovnikov, V. V., *C. R. Acad. Sci.* **1875**, *85*, 668.

5 (a) Bozel, J. J., Hegedus, L. S. *J. Org. Chem.* **1981**, *46*, 2561. (b) Suminov, S. I., Kost, A. N. *Russ. Chem. Rev.* **1969**, *38*, 884. (c) Larock, R. C., Leong, W. W. *Compr. Org. Synth.* **1991**, *4*, 269. (d) Jung, M. E. *Compr. Org. Synth.* **1991**, *4*, 1.

6 (a) Deeba, M., Ford, M. E., Johnson, T. A. *J. Chem. Soc. Chem. Commun.* **1987**, 562. (b) Deeba, M., Ford, M. E. *J. Org. Chem.* **1988**, *53*, 4594.

7 (a) Tabata, M., Mizuno, N., Iwamoto, M. *Chem. Lett.* **1991**, 1027. (b) Mizuno, N., Tabata, M., Uematsu, T., Iwamoto, M. *J. Catal.* **1994**, *146*, 249.

8 Chauvel, A., Delmon, B., Hölderich, W. F. *Appl. Catal. A: Gen.* **1994**, *115*, 173.

9 (a) Howk, B. W., Little, E. L., Scott, S. L., Whitman, G. M. *J. Am. Chem. Soc.* **1954**, *76*, 1899. (b) Wollensak, J., Closson, R. D. *Org. Synth.* **1963**, *43*, 45. (c) Pez, G. P., Galle, J. E. *Pure. Appl. Chem.* **1985**, *57*, 1917. (d) Steinborn, D., Thies, B., Wagner, I., Taube, R. *Z. Chem.* **1989**, *29*, 333. (e) Beller, M., Breindl, C. *Tetrahedron* **1998**, *54*, 6359. (f) Hartung, C. G., Breindl, C., Tillack, A., Beller, M. *Tetrahedron* **2000**, *56*, 5175.

10 (a) McGrane, P. L., Livinghouse, T. *J. Am. Chem. Soc.* **1993**, *115*, 11485. (b) McGrane, P. L., Livinghouse, T. *J. Org. Chem.* **1992**, *57*, 1323. (c) McGrane, P. L., Jensen, M., Livinghouse, T. *J. Am. Chem. Soc.* **1992**, *114*, 5459. (d) Walsh, P. J., Baranger, A. M., Bergman, R. G. *J.

Am. Chem. Soc. **1992**, *114*, 1708. (e) Barranger, A.M., Walsh, P.J., Bergman, R.G. *J. Am. Chem. Soc.* **1993**, *115*, 2753. (f) Haak, E., Bytschkov, I., Doye, S. *Angew. Chem.* **1999**, *111*, 3584; *Angew. Chem. Int. Ed.* **1999**, *38*, 3389. (g) Haak, E., Siebeneicher, H., Doye, S. *Org. Lett.* **2000**, *2*, 1935.

11 (a) Brunet, J.J. *Gazz. Chim. Ital.* **1997**, *127*, 111. (b) Beller, M., Eichberger, M., Trauthwein, H. In *Catalysis of Organic Reactions*, Herkes, F.E., Ed., Marcel Dekker Inc.: New York, **1998**, p. 319. (c) Schaffrath, H., Keim, W. *J. Mol. Catal. A: Chem.* **2001**, *168*, 9.

12 (a) Hong, S., Marks, T.J. *J. Am. Chem. Soc.* **2002**, *124*, 7886. (b) Kim, Y.K., Livinghouse, T. *Angew. Chem. Int. Ed.* **2002**, *41*, 3645; *Angew. Chem.* **2002**, *114*, 3797. (c) Li, Y., Marks, T.J. *J. Am. Chem. Soc.* **1998**, *120*, 1757. (d) Eisen, M.S., Straub, T., Haskel, A. *J. Alloys Compd.* **1998**, *271–273*, 116. (e) Haskel, A., Straub, T., Eisen, M.S. *Organometallics* **1996**, *15*, 3773. (f) Li, Y., Marks, T.J. *J. Am. Chem. Soc.* **1996**, *118*, 9295. (g) Li, Y., Marks, T.J. *J. Am. Chem. Soc.* **1996**, *118*, 707. (h) Li, Y., Marks, T.J. *Organometallics* **1996**, *15*, 3770. (i) Giardell, M.A., Conticello, V.P., Brard, L., Gagné, M.R., Marks, T.J. *J. Am. Chem. Soc.* **1994**, *116*, 10241. (j) Giardell, M.A., Conticello, V.P., Brard, L., Sabat, M., Rheingold, A.L., Stern, C.L., Marks T.J. *J. Am. Chem. Soc.* **1994**, *116*, 10212. (k) Li, Y., Fu, P.F., Marks, T.J. *Organometallics* **1994**, *13*, 439. (l) Gagné, M.R., Stern, C.L., Marks, T.J. *J. Am. Chem. Soc.* **1992**, *114*, 275. (m) Gagné, M.R., Stern, C.L., Gagné, M.R., Nolan, S.P., Marks, T.J. *Organometallics* **1990**, *9*, 1716. (n) Gagné, M.R., Marks, T.J. *J. Am. Chem. Soc.* **1989**, *111*, 4108.

13 (a) Hegedus, L.S., Åkermark, B., Zetterberg, K., Olsson, L.F. *J. Am. Chem. Soc.* **1984**, *106*, 7122. (b) Hegedus, L.S. *Angew. Chem.* **1988**, *100*, 1147; *Angew. Chem. Int. Ed. Engl.* **1988**, *27*, 1113. (c) Eisenstein, O., Hoffmann, R. *J. Am. Chem. Soc.* **1981**, *103*, 4308. (d) Selington, A.L., Trogler, W.C. *Organometallics*, **1993**, *12*, 744.

14 (a) Selington, A.L., Cowan, R.L., Trogler, W.C. *Inorg. Chem.* **1991**, *30*, 3371. (b) Driver, M.S., Hartwig, J.F. *J. Am. Chem. Soc.* **1996**, *118*, 4206. (c) Fryzuk, M.D., Montgomery, C.D. *Coord. Chem. Rev.* **1989**, *95*, 1. (d) Casalnuovo, A.L., Calabrese, J.C., Milstein, D. *Inorg. Chem.* **1987**, *26*, 971. (e) Schulz, M., Milstein, D. *J. Chem. Soc. Chem. Commun.* **1993**, 318.

15 (a) Hegedus, L.S., Allen, G.F., Waterman, E.L. *J. Am. Chem. Soc.* **1976**, *98*, 2674. (b) Hegedus, L.S., Allen, G.F., Bozell, J.J., Waterman, E.L. *J. Am. Chem. Soc.* **1978**, *100*, 5800.

16 Hegedus, L.S., McKearin, J.M. *J. Am. Chem. Soc.* **1982**, *104*, 2444.

17 Larock, R.C., Hightower, T.R., Hasvold, L.A., Peterson, K.P. *J. Org. Chem.* **1996**, *61*, 3584.

18 Hosokawa, T., Takano, M., Kuroki, Y., Murahashi, S. *Tetrahedron Lett.* **1992**, *33*, 6643.

19 Brunet, J.J., Neibecker, D., Philippot, K. *Tetrahedron Lett.* **1994**, *34*, 3877.

20 (a) Beller, M., Eichberger, M., Trauthwein, H. *Angew. Chem.* **1997**, *109*, 2306; *Angew. Chem. Int. Ed. Engl.* **1997**, *36*, 222. (b) Beller, M., Trauthwein, H., Eichberger, M., Breindl, C., Müller, T.E., Zapf, A. *J. Organomet. Chem.* **1998**, *566*, 277.

21 Campi, E.M., Jackson, W.R. *J. Organomet. Chem.* **1996**, *523*, 205.

22 Utimoto, K. *Pure Appl. Chem.* **1983**, *55*, 1845.

23 Fukuda, Y., Utimoto, K., Nozaki, H. *Heterocycles* **1987**, *25*, 297.

24 (a) Müller, T.E., Pleier, A.-K. *J. Chem. Soc. Dalton. Trans.* **1999**, 583. (b) Müller, T.E., Grosche, M., Herdtweck, E., Pleier, A.-K., Walter, E., Yan, Y.K. *Organometallics*, **2000**, *19*, 170.

25 Burling, S., Field, L.D., Messerle, B.A. *Organometallics* **2000**, *19*, 87.

26 Kondo, T., Okada, T., Suzuki, T., Mitsudo, T. *J. Organomet. Chem.* **2001**, *622*, 149.

27 Vogels, C.M., Hayes, P.G., Shaver, M., Westcott, S.A. *Chem. Commun.* **2000**, 51.

28 Kadota, I., Shibuya, A., Lutete, L.M., Yamamoto, Y. *J. Org. Chem.* **1999**, *64*, 4570.

29 Tokunaga, M., Eckert, M., Wakatsuki, Y. *Angew. Chem.* **1999**, *111*, 3417; *Angew. Chem. Int. Ed.* **1999**, *38*, 3222.

30 Uchimaru, Y. *Chem. Commun.* **1999**, 1133.

31 Hartung, C.G., Tillack, A., Trauthwein, H., Beller, M. *J. Org. Chem.* **2001**, *66*, 6339.

32 Recent examples: (a) Johnson, J.S., Bergman, R.G. *J. Am. Chem. Soc.* **2001**, *123*, 2923. (b) Pohlki, F., Doye, S. *Angew. Chem.* **2001**, *113*, 2361; *Angew. Chem. Int. Ed.* **2001**, *40*, 2305. (c) Bytschkov, I., Doye, S. *Eur. J. Org. Chem.* **2003**, 935. (d) Pohlki, F., Doye, S. *Chem. Soc. Rev.* **2003**, *32*, 104. (e) Shi, Y., Ciszewski, J.T., Odom, A.L. *Organometallics* **2001**, *20*, 3967. (f) Cao, C., Ciszewski, J.T., Odom, A.L. *Organometallics* **2001**, *20*, 5011. (g) Shi, Y., Hall, C., Ciszewski J.T., Cao, C., Odom, A.L. *Chem. Commun.* **2003**, 586. (h) Ong, T.-G; Yap, G.P.A., Richeson, D.S. *Organometallics* **2002**, *21*, 2839.

33 Tillack, A., Garcia Castro, I., Hartung, C.G., Beller, M. *Angew. Chem. Int. Ed.* **2002**, *41*, 2541; *Angew. Chem.* **2002**, *114*, 2646.

34 Al-Masum, M., Meguro, M., Yamamoto, Y. *Tetrahedron Lett.* **1997**, *38*, 6071.

35 Coulson, D.R. *Tetrahedron Lett.* **1971**, *12*, 429.

36 (a) Beller, M., Trauthwein, H., Eichberger, M., Breindl, C., Herwig, J., Müller, T.E., Thiel, O.R. *Chem. Eur.J.* **1999**, *5*, 1306. (b) Tillack, A., Trauthwein, H., Hartung, C.G., Eichberger, M., Pitter, S., Jansen, A., Beller, M. *Monatsh. Chem.*, **2000**, *131*, 1327. (c) Beller, M., Trauthwein, H., Eichberger, M., Breindl, C., Müller, T.E. *Eur.J. Inorg. Chem.* **1999**, 1121. (d) Beller, M., Thiel.O.R., Trauthwein, H., Hartung, C.G. *Chem. Eur. J.* **2000**, *6*, 2513.

37 Trauthwein, H., Tillack, A., Beller, M. *Chem. Commun.* **1999**, 2029.

38 Casalnuovo, A.L., Calabrese, J.C., Milstein, D. *J. Am. Chem. Soc.* **1988**, *110*, 6738.

39 Dorta, R., Egli, P., Zürcher F., Togni A. *J. Am. Chem. Soc.* **1997**, *119*, 10857.

40 Kawatsura, M., Hartwig, J.F. *J. Am. Chem. Soc.* **2000**, *122*, 9546.

41 Löber, O., Kawatsura, M., Hartwig, J.F. *J. Am. Chem. Soc.* **2001**, *123*, 4366.

42 Kawatsura, M., Hartwig, J.F. *Organometallics* **2001**, *20*, 1960.

43 Seayad, J., Tillack, A., Hartung, C.G.,M. Beller, *Adv. Synth. Catal.* **2002**, *344*, 795.

44 (a) Akutagawa, S. in *Chirality in Industry* (Eds.: Collins, A.N., Sheldrake, G.N., Crosby, J.), John Wiley and Sons, England, **1995**, pp. 313. (b) Akutagawa, S., Tani, K. in *Catalytic Asymmetric Synthesis* (Ed.: Ojima, I.), VCH, Weinheim, **1993**, pp. 43. (c) Inoue, S.I., Takaya, H., Tani, K., Otsuka, S., Sato, T., Noyori, R. *J. Am. Chem. Soc.* **1990**, *112*, 4897.

45 (a) Takabe, K., Katagiri, T., Tanaka, J. *Tetrahedron Lett.* **1975**, 3005. (b) Tani, K., Yamagata, T., Otsuke, S., Akutagawa, S., Komubayashi, H., Taketomi, T., Takaya, H., Aiyashita, A., Noyori, R. *Chem. Commun.* **1982**, 600.

46 Beller, M., Breindl, C. *Chemosphere* **2001**, *43*, 21.

47 Beller, M., Breindl, C., Riermeier, T.H., Eichberger, M., Trauthwein, H. *Angew. Chem.* **1998**, *110*, 3571; *Angew. Chem. Int. Ed.* **1998**, *37*, 3389.

48 Tzalis, D., Koradin, C., Knochel, P. *Tetrahedron Lett.* **1999**, *40*, 6193.

49 Rodriguez, A., Koradin, C., Dohle, W., Knochel, P. *Angew. Chem. Int. Ed.* **2000**, *39*, 2488; *Angew. Chem.* **2000**, *112*, 2607.

2.11
Polyoxometalates as Catalysts for Oxidation with Hydrogen Peroxide and Molecular Oxygen

Ronny Neumann

2.11.1
Introduction

The requirement for sustainable chemical processes combining environmentally acceptable or "green" syntheses under economically viable conditions is a key area of activity in present-day research in organic synthesis. This need has led to great emphasis on research into the use of ecologically friendly oxidants such as hydrogen peroxide [1] and molecular oxygen [2] in place of classic stoichiometric or super-stoichiometric oxidants. Linked with this general research direction is the desire to develop practical synthetic methods that can be carried out in non-noxious solvents, preferably water, or without solvent. In order to make the use of hydrogen peroxide or molecular oxygen a viable option for fine chemical synthesis, the development of practical catalysts is necessary. The basic catalyst requirements are that (a) the catalyst should be able to activate the oxidants selectively, (b) the catalyst should be stable to strongly oxidizing conditions, (c) catalyst recycle should be a simple and quanti-

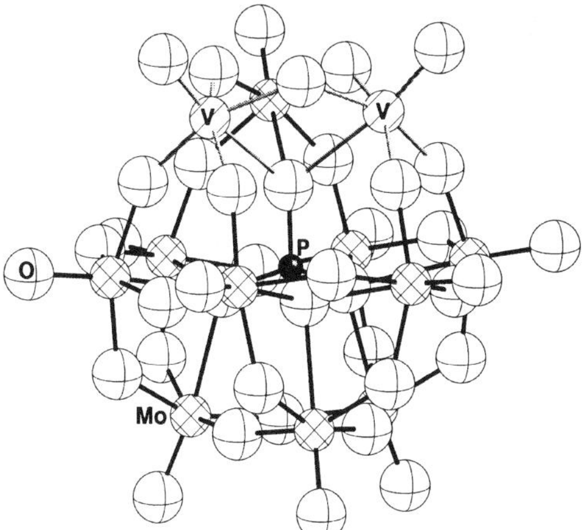

Fig. 1 Polyoxometalates with the Keggin structure, e.g., $[XM_{10}V_2O_{40}]^{(3+x)-}$.

Transition Metals for Organic Synthesis, Vol. 2, 2nd Edition.
Edited by M. Beller and C. Bolm

ISBN: 3-527-30613-7

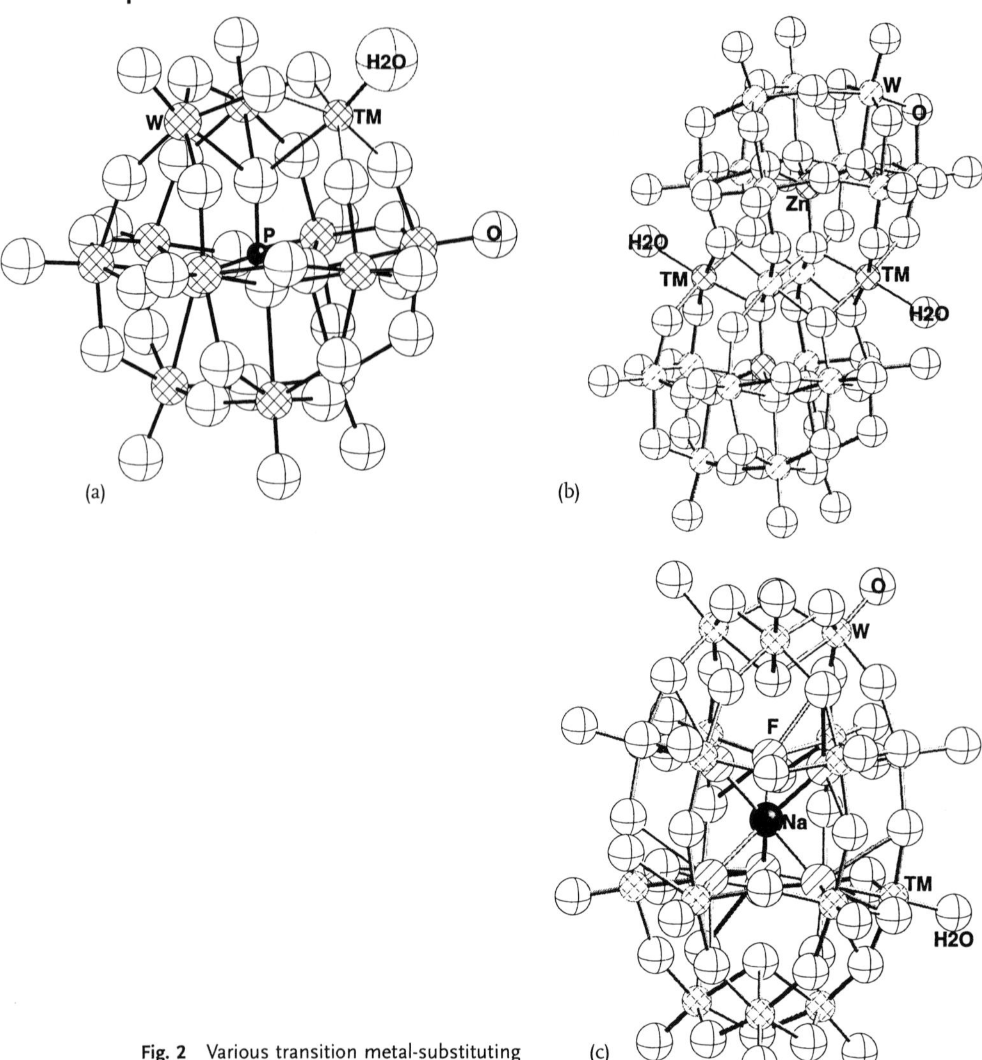

Fig. 2 Various transition metal-substituting polyoxometalates.

tative procedure, (d) in the case of hydrogen peroxide there should be minimal non-productive decomposition to water and oxygen, and (e) for dioxygen, methods must be found to prevent non-selective catalytic autoxidation.

A basic premise behind the use of polyoxometalates in homogeneous oxidation chemistry is the fact that polyoxometalates are oxidatively stable. This, *a priori*, leads to the conclusion that for practical purposes polyoxometalates would have distinct advantages over widely investigated organometallic compounds that are vulnerable to decomposition due to oxidation of the ligand bound to the metal center. In general, polyoxometalates, also called heteropolyanions, can be described by the general formula $[X_xM_mO_y]^{q-}$ ($x \leq m$), where X is defined as the heteroatom and M are the addenda atoms. Polyoxometalates with the Keggin structure, $[XM_{12\text{-}x}M'_xO_{40}]^{(3+x)-}$, especially where (X=P, M=Mo, M′=V, and x=0, 1 or 2) (Fig. 1) represent a major subclass of polyoxometalates and are often used for catalysis.

An important family of polyoxometalate derivatives comprises those compounds in which a transition metal, typically TM=Co, Mn, Fe, Cu, Ru, etc., substitutes an M=O moiety at the polyoxometalate surface. In such compounds the transition metal is pentacoordinated by the "parent" polyoxometalate, with a sixth (labile) ligand, L, usually water. This lability allows the interaction of the transition metal atom with a reaction substrate and/or oxidant, leading to reaction at the transition metal center; the rest of the polyoxometalate acts as an inorganic ligand. Many structural variants of such transition metal-substituting polyoxometalates are known, for example, (a) the transition metal-substituting "Keggin" type compounds, $[XTM(L)M_{12}O_{39}]^{q-}$ (X=P, Si, M=Mo, W, Fig. 2a), (b) the so-called "sandwich" type polyoxometalates, $\{[(WZnTM_2(H_2O)_2][(ZnW_9O_{34})_2]\}^{q-}$ (Fig. 2b), having a ring of transition metals between two truncated Keggin "inorganic ligands", and (c) the polyfluorooxometalates (Fig. 2c), of a quasi Wells-Dawson structure. Especially these latter two compounds classes often (i.e. normally) have superior catalytic activity and stability, as will be shown below.

In the Sections below, the use of polyoxometalates as catalysts in liquid phase synthetic oxidative applications using the environmentally and economically favored hydrogen peroxide and dioxygen will be surveyed and discussed. Readers interested in the use of more "exotic" oxidants, e.g., iodosobenzene [3], nitrous oxide [4], ozone [5], sulfoxides [6], gas phase applications, acid-catalyzed reactions, more catalytically oriented research, and other related subjects are encouraged to go to some of the published comprehensive reviews elsewhere [7].

2.11.2
Oxidation with Hydrogen Peroxide

The fact that polyoxometalates are also a subclass of oxotungstates or oxomolybdates with high-valent d^0 tungsten or molybdenum atoms make them excellent candidates for the heterolytic activation of hydrogen peroxide through formation of inorganic peroxo or hydroperoxo intermediates. A little more than fifteen years ago, Ishii and his co-workers were the first to describe a procedure where the hexadecylpyridinium quaternary ammonium salt of Keggin compound $[PM_{12}O_{40}]^{3-}$ (M=Mo or W) was used to catalyze the oxidation of numerous types of organic substrates using aqueous 30–35% hydrogen peroxide as oxidant. Transformations were typical of reactions of hydrogen peroxide in the presence of tungsten-based catalysts and included epoxidation of allylic alcohols [8] and alkenes [9] with yields generally above 90% using an approximately 50% excess of H_2O_2. Under more acidic conditions and at higher temperatures there is hydrolysis of the epoxide and formation of vicinal diols followed by oxidation to keto-alcohols or α,β-diketones [10], or carbon-carbon bond cleavage to yield carboxylic acids and ketones. The phosphotungstate polyoxometalate was also effective for oxidation of secondary alcohols to ketones, while primary alcohols were not reactive, allowing for the high-yield regioselective oxidation of non-vicinal diols to the corresponding keto-alcohols; α,ω-diols did, however, react to give lactones (e.g., γ-butyrolactone from 1,4-butanediol) with high yields [11]. Additional research showed that alkynes [12], amines [13], and sulfides [14] could be oxidized efficiently to ketones, *N*-oxides, and sulfoxides and sulfones, respectively. Various quinones were also synthesized from active arene precursors [15].

For researchers unfamiliar with the field of polyoxometalate-catalyzed oxidation, it is very important to point out that there was originally much disagreement about the identity of the true catalyst in the reactions using aqueous hydrogen peroxide as oxidant and Keggin type compounds as catalysts. It was first Brégeault and his co-workers [16] and at about the same time also the groups of Griffith [17] and Hill [18] who suggested and convincingly proved that the heteropolyanion in the Ishii system for alkene epoxidation was only a precursor of the true catalyst, $\{PO_4[MO(O_2)_2]\}^{3-}$ (M = W, Mo), the so-called Venturello compound independently synthesized and used in similar catalytic oxidation reactions at about the same time [19]. The lack of solvolytic stability was attributed to decomposition of the Keggin compound by *aqueous* hydrogen peroxide. These results were nicely supported both by isolation of the compound and solution spectroscopic studies. It has become clear that of all the possible "real" catalysts in the Keggin-plus-H_2O_2 system the Venturello complex is the most active. This is not to say, however, that other active intermediate peroxo species, some catalytically active, are not also present. The true identity of the active species in each case is probably a function of a combination of factors including the oxidizability of the substrate, the solvent, the temperature used, and the rate of decomposition of the Keggin heteropolyanion under reaction conditions.

Following the research on the so-called Venturello-Ishii catalytic systems, polyoxometalates that were solvolytically stable to aqueous hydrogen peroxide were sought and investigated. It was observed that, in general, larger polyoxometalalates, specifically polyoxotungstates of various "sandwich" type structures, were solvolytically stable toward hydrogen peroxide. These "sandwich" type structures generally have low-valent transition metals substituting into the polyoxometalate structure (see for example Fig. 2b). Notably, the substituting transition metal often catalyzes fast decomposition of hydrogen peroxide, leading to low reaction yields and non-selective reactions of little synthetic value. However, there is now a considerable body of research into several types of transition metal-substituted polyoxometalates that are synthetically useful. Hill and co-workers have reported on a number of iron-containing polyoxometalates that have shown good activity for alkene oxidation with only moderate non-productive decomposition of hydrogen peroxide [20]. Mizuno and co-workers have also reported the use of metal-substituted Keggin compounds, although catalyst stability was not definitely determined in every case [21]. We have also observed that transition metal-substituted polyfluorooxometalates, especially the nickel-substituted compound (Fig. 2c), were also very active and stable oxidation catalysts for epoxidation with H_2O_2 [22]. We have found that the $\{[(WZnTM_2(H_2O)_2][(ZnW_9O_{34})_2]\}^{q-}$ polyoxometalates were far more catalytically active. Originally we observed that, among this class of compounds, the manganese and analogous rhodium derivatives were uniquely active when reactions were carried out in biphasic systems, preferably 1,2-dichloroethane-water [23]. Significantly, at low temperatures, highly selective epoxidation could be carried out even on cyclohexene, which is normally highly susceptible to allylic oxidation. Non-productive decomposition of hydrogen peroxide at low temperatures was minimal but increased with temperature. The rhodium compound was preferable in terms of H_2O_2 dismutation, but of course is more expensive. In a further kinetic and mechanistic study, it was shown that the catalyst was stable under turnover conditions and tens of thousands of turnovers could be attained with little H_2O_2 decomposition [24].

After the initial discovery of the $\{[(WZnMn(II)_2(H_2O)_2][(ZnW_9O_{34})_2]\}^{12-}$ polyoxometalate as a catalyst for hydrogen peroxide activation, the synthetic utility of the reaction was studied for a variety of substrates [25]. Allylic primary alcohols were oxidized selectively to the corresponding epoxides in high yields and >90% selectivity. Allylic secondary alcohols were oxidized to a mixture of -unsaturated ketones (major product) and epoxides. Secondary alcohols were oxidized to ketones and sulfides to a mixture of sulfoxides and sulfones. The reactivity of simple alkenes is inordinately affected by the steric bulk of the substrate. Despite the tendency toward higher reactivity upon substitution at the double bond, which increases its nucleophilicity (e.g., 2,3-dimethyl-2-butene was more reactive than 2-methyl-2-heptene), substrates such as 1-methylcyclohexene were less reactive than cyclohexene. This led, for example, to unusual reaction selectivity in limonene epoxidation, where both epoxides were formed in equal amounts, in contrast to the usual situation where epoxidation at the *endo* double bond is highly preferred. In these catalytic systems, high turnover conditions can be easily achieved, but sometimes, for less reactive substrates such as terminal alkenes, yields are low. This can remedied by continuous or semi-continuous addition of hydrogen peroxide and removal of spent aqueous phases. Another problem in these systems is the use of organic solvents, which reduces the environmental attractiveness of the use of hydrogen peroxide. As one way of compensating for this problem, we have shown that functionalized silica catalytic assemblies containing a polyoxometalate attached or adsorbed onto a silica surface can be prepared. The catalytic activity is essentially the same as that in the traditional bi-phasic liquid-liquid reaction medium, but now an organic solvent is not required and the solid catalyst particles were easily recoverable [26].

Recently, Adam and Neumann et al. have begun to reinvestigate the use of "sandwich" type polyoxometalates. Thus $\{[(WZnTM_2(H_2O)_2][(ZnW_9O_{34})_2]\}^{q-}$ compounds were active catalysts for the epoxidation of allylic alcohols [27]. The identity of the transition metal did not affect the reactivity, chemoselectivity, or stereoselectivity of the allylic alcohol epoxidation by hydrogen peroxide. These selectivity features support a conclusion that a tungsten peroxo complex rather than a high-valent transition-metal-oxo species operates as the key intermediate in the "sandwich" type POMs-catalyzed epoxidations. The marked enhancement of reactivity and selectivity of allylic alcohols versus simple alkenes was explained by a template formation in which the allylic alcohol is coordinated through metal-alcoholate bonding and the hydrogen-peroxide oxygen source is activated in the form of a peroxo tungsten complex. 1,3-Allylic strain expresses a high preference for the formation of the *threo* epoxy alcohol, whereas 2-allylic strain expresses a preference for the *erythro* diastereomer. In contrast to acyclic allylic alcohol, the $\{[(WZnTM_2(H_2O)_2][(ZnW_9O_{34})_2]\}^{q-}$-catalyzed oxidation of the cyclic allylic alcohols by H_2O_2 gives significant amounts of enone.

2.11.3
Oxidation with Molecular Oxygen

Commonly, molecular oxygen tends to react in the liquid phase via autoxidation pathways. One way to utilize this type of reactivity is to oxidize a hydrocarbon in the presence of a reducing agent. In the most synthetically interesting case, a polyoxometalate may initiate a radical chain reaction between oxygen and an aldehyde. The initial product of this reaction is an acylperoxo radical or an acylhydroperoxide (peracid). These active intermediate species may then be used for the epoxidation of alkenes, the oxidation of alkanes to ketones and alcohols, and the Baeyer-Villiger oxidation of ketones to esters. This has been demonstrated using both vanadium ($H_5PV_2Mo_{10}O_{40}$) and cobalt ($Co(II)PW_{11}O_{39}^{5-}$) containing Keggin type polyoxometalates as catalysts with *iso*-butyraldehyde as the preferred peracid precursor [28]. Significant yields at very high selectivities were obtained in most examples. The catalytic effect is probably mostly in the peracid generation step, but catalysis of the substrate oxygenation cannot be ruled out.

It is of course also possible to use transition metal-substituted polyoxometalates in a more straightforward manner as autooxidation catalysts. In this way, the trisubstituted Keggin compound, $M_3(H_2O)_3PW_9O_{37}^{6-}$ [M = Fe(III) and Cr(III)] and $Fe_2M(H_2O)_3PW_9O_{37}^{7-}$ [M = Ni(II), Co(II), Mn(II) and Zn(II)] were used in the autooxidation of alkanes such as propane and isobutane to acetone and *tert*-butyl alcohol [29]. Later $Fe_2Ni(OAc)_3PW_9O_{37}^{10-}$ was prepared and used to oxidize alkanes such as adamantane, cyclohexane, ethylbenzene and *n*-decane, where the reaction products (alcohol and ketone) and regioselectivities were typical for metal-catalyzed autooxidations [30]. An interesting recent application of such an autooxidation is the oxidation of 3,5-di-*tert*-catechol by iron- and/or vanadium-substituted polyoxometalates [31]. In this reaction there is a very high turnover number, >100 000. In this case the polyoxometalates are excellent mimics of catechol dioxygenase.

The activation of substrates, both organic and inorganic, by polyoxometalates, in a redox type interaction involving electron transfer followed by re-oxidation of the reduced polyoxometalate with molecular oxygen is the oldest and possibly the most developed of all the applications of polyoxometalates in homogeneous oxidation chemistry.

$$\text{Substrate} + \text{POM}_{ox} \longrightarrow \text{Product} + \text{POM}_{red}$$

$$\text{POM}_{red} + O_2 \longrightarrow \text{POM}_{ox}$$

The most commonly used catalysts for this reaction are the phosphovanandomolybdates, $PV_xMo_{12-x}O_{40}^{(3+x)-}$, especially but not exclusively when x=2. In fact, $H_5PV_2Mo_{10}O_{40}$ was first described as a co-catalyst in the Wacker reaction [Pd(0)=substrate], a reaction which best epitomizes this type of mechanism, as a substitute for the chloride-intensive $CuCl_2$ system, which is both corrosive and forms chlorinated side-products [32]. In the 1990s the oxidative hydration of ethylene to acetaldehyde was significantly improved by Grate and co-workers at Catalytica [33]. Another inorganic application was the aerobic oxidation of gaseous HBr, which was utilized for the *in situ* selective bromination of phenol to 4-bromophenol [34]. Another early interest in the catalytic chemistry of $H_{3+x}PV_xMo_{12-x}O_{40}$ was its use in the oxidation of sulfur-containing compounds in the purification of industrial waste and natural gas, including the oxidation of H_2S to elemental sulfur, sulfur dioxide to sulfur trioxide (sulfuric acid), mercaptans to disulfides, and sulfides to sulfoxides and sulfones [35]. Hill and his group have continued the investigation of the oxidation chemistry of sulfur compounds [36]. Investigation of the use of $PV_2Mo_{10}O_{40}^{5-}$ for the oxidation of hydrocarbon substrates led to the finding that cyclic dienes could be oxidatively dehydrogenated to the corresponding aromatic derivatives [37]. Later, this polyoxometalate compound was used in other oxydehydrogenation reactions such as the selective oxydehydrogenation of alcohol compounds to aldehydes with no over-oxidation to the carboxylic acids [38]. Significantly, autooxidation of the aldehyde to the carboxylic acid was strongly inhibited; in fact, at the catalyst concentrations used, 1 mol% $PV_2Mo_{10}O_{40}^{5-}$ can be considered an autooxidation inhibitor. An important observation in these systems was that active carbon as a support was unique in its function. Ishii and his group later repeated many of these oxydehydrogenation reactions using a similarly supported $PV_6Mo_6O_{40}^{9-}$ on carbon. The scope of the reactions was extended to include oxidative dehydrogenations of allylic alcohols to allylic aldehydes [39]. A subsequent study led to the supposition that quinones, possibly formed on the active carbon surface, might play a role as an intermediate oxidant [40]. Thus, a catalytic cycle may be postulated, whereby a surface quinone oxidizes the alcohol to the aldehyde and is reduced to the hydroquinone, which is reoxidized in the presence of the catalyst and molecular oxygen. Similarly to alcohol dehydrogenation to aldehydes, amines may be dehydrogenated to intermediate and unstable imines. In the presence of water, aldehyde is formed, and this may then immediately further react with the initial amine to yield a Schiff base. Since the Schiff base is formed under equilibrium conditions, aldehydes are eventually the sole products. In the judicious absence of water, the intermediate imine was dehydrogenated to the corresponding nitrile.

Another reaction of practical interest studied by several groups including our own is the oxidation of phenols to quinones. For example, the oxidation of 2,5,6-trimethylphenol in acetic acid [41] gave 2,5,6-trimethylbenzoquinone as the main product along with a small amount of coupled biphenol as by-product. Addition of water lowered reaction selectivity, and more biphenol was formed. Reactions in alcohol on the other hand gave the monomeric benzoquinone as sole product [42]. However, oxidation of 2,6-substituted phenols in alcohol solvents yielded only oxidative dimerization of the activated phenols to the corresponding diphenoquinones as sole products. Unfortunately, under these mild conditions, the less reactive phenol did not react. An interesting extension of this work is the oxidation of 2-methyl-1-naphthol to 2-methyl-1,4-naphthaquinone (Vitamin K_3, menadione) in fairly high selectivities, ~83% at atmospheric O_2 [43]. This work could lead to a new environmentally favorable process to replace the stoichiometric CrO_3 oxidation of 2-methylnaphthalene used today.

Another interesting set of reactions is described by Brégeault and co-workers. Here, $H_5PV_2Mo_{10}O_{40}$ was used in combination with dioxygen to oxidatively cleave vicinal diols [44] and ketones [45]. Only vanadium-containing heteropoly compounds appear to be active, and the acidic site seems to be a prerequisite for the catalytic reaction. For example, 1-phenyl-2-propanone can be cleaved to benzaldehyde (benzoic acid) and acetic acid, ostensibly through the *α,β*-diketone intermediate, 1-phenyl-1,2-propane dione. Similarly, cycloalkanones can be cleaved to keto-acids and di-acids. In general, the conversions and selectivities are very high. It would be interesting to carry out the oxidative cleavage of diols, also under non-acidic conditions, as a possible pathway to the formation of a chiral pool from natural sources. Iodomolybdates have been found to show some activity in these reactions [46].

Using *α*-terpinene as a model substrate, extensive mechanistic research utilizing kinetic and spectroscopic tools was carried out to decipher $PV_2Mo_{10}O_{40}^{5-}$ polyoxometalate-catalyzed oxydehydrogenations [47]. Dehydrogenation was explained by a series of fast electron and proton transfers. Interestingly, there were clear indications that the re-oxidation of the reduced polyoxometalate with molecular oxygen proceeded via an inner sphere mechanism, presumably via formation of a *μ*-peroxo intermediate. Subsequent research has given conflicting and inconclusive evidence that the re-oxidation might occur via an outer sphere mechanism [48]. An additional, effective, and general method for the aerobic selective oxidation of alcohols to aldehydes or ketones is by the use of nitroxide radicals and $PV_2Mo_{10}O_{40}^{5-}$ as cocatalysts. Typically, quantitative yields were obtained for aliphatic, allylic, and benzylic alcohols [49]. Based mostly on kinetic evidence and some spectroscopic support, a reaction scheme was formulated as follows. The polyoxometalate oxidizes the nitroxyl radical to the nitrosium cation. The latter oxidizes the alcohol to the ketone/aldehyde and is reduced to the hydroxylamine, which is then reoxidized by $PV_2Mo_{10}O_{40}^{5-}$.

Another important example of the use of polyoxometalates in a two-step redox type mechanism is the technology proposed by Hill and Weinstock for the delignification of wood pulp [50]. In the first step, lignin is oxidized preferentially compared to cellulose and the polyoxometalate is reduced. The now solubilized lignin component is separated from the whitened pulp and mineralized with oxygen to CO_2 and H_2O. During the mineralization process, the polyoxometalate is re-oxidized and can be used for an additional process cycle. A closer examination of all the reactions presented above reveals that in all the examples given the oxidation reaction proceeds by transfer of electrons (and protons) without oxygenation or oxygen transfer from the catalyst or molecular oxygen to the organic substrate. A more general question therefore arose – can there also be oxygen transfer reactions in reactions catalyzed by $PV_2Mo_{10}O_{40}^{5-}$ or other polyoxometalates? This subject is relevant to an important area of classical heterogeneous reactions in which, through catalysis by a metal oxide compound at high temperature, oxygen is transferred from the lattice of the oxide to a hydrocarbon substrate hydrocarbon. This type of mechanism was originally proposed by Mars and van Krevelen and is important in several industrial applications such as the oxidation of propene to acrolein and butane to maleic anhydride. It was shown that, in the case of the $PV_2Mo_{10}O_{40}^{5-}$ catalyst, oxygenation was possible via an initial activation of a hydrocarbon by electron transfer even at temperatures of 25–60 °C [51]. Substrates oxygenated in this manner included polycyclic aromatic compounds such as anthracene and alkyl aromatic compounds with activated benzylic positions such as xanthene. The use of $^{18}O_2$ and isotopically labeled polyoxometalates as well as carrying out stoichiometric reactions under anaerobic conditions provided strong evidence for a homogeneous Mars-van Krevelen type mechanism and clearly provided evidence against autooxidation and oxidative nucleophilic substitution as alternative possibilities. Evidence of the activation of the hydrocarbon by electron transfer was provided by the excellent correlation of the reaction rate with the oxidation potential of the substrate. For anthracene the intermediate cation radical was observed by ESR spectroscopy, whereas for xanthene the cation radical quickly underwent additional electron and proton transfer, yielding a benzylic cation species observed by ^1H NMR.

An additional mode of oxygen activation is via a "dioxygenase type" mechanism. Such an activation of molecular oxygen is possible by use of a ruthenium-substituted polyoxometalate with a "sandwich" structure [52]. Evidence for such a mechanism for hydroxylation of adamantane and alkene epoxidation was obtained by showing that there are no autooxidation reactions and that the reaction stoichiometery was substrate/O_2=1. In addition, a ruthenium-oxo intermediate was isolated and shown to viably transfer oxygen in a quantitative and stereoselective manner. The catalytic cycle was also supported by kinetic data.

2.11.4 Conclusion

In a short period of only about fifteen years, the synthetic applications of polyoxometalates as oxidation catalysts have shown that these compounds have considerable potential. Additional synthetic procedures are just around the corner, as a very wide variety of polyoxometalates can be prepared. Although the catalysts are of high molecular weight, efficient methods of catalyst recycle such as nanofiltration and the use of supports (heterogeneous catalysts) are already available, making these compounds an attractive solution for the replacement of environmentally damaging stoichiometric oxidants.

References

1 G. Strukul, *Catalytic Oxidations with Hydrogen Peroxide as Oxidant*, Kluwer Academic, The Netherlands, **1992**.

2 L.I. Simandi, *Catalytic Activation of Dioxygen by Metal Complexes*, Kluwer Academic, The Netherlands, **1992**.

3 C.L. Hill, R.B. Brown, *J. Am. Chem. Soc.* **1986**, *108*, 536. D. Mansuy, J.F. Bartoli, P. Battioni, D.K. Lyon, R.G. Finke, *J. Am. Chem. Soc.* **1991**, *113*, 7222. H. Weiner, Y. Hayashi, R.G. Finke, *Inorg. Chem.* **1999**, *38*, 2579.

4 R. Ben-Daniel, L. Weiner, R. Neumann, *J. Am. Chem. Soc.* **2002**, *124*, 8788. R. Ben-Daniel, R. Neumann, *Angew. Chem. Int. Ed.* **2003**, *42*, 92.

5 R. Neumann, A.M. Khenkin, *Chem. Commun.* **1998**, 1967.

6 A.M. Khenkin, R. Neumann, *J. Am. Chem. Soc.* **2002**, *124*, 4198. A.M. Khenkin, R. Neumann, *J. Org. Chem.* **2002**, *67*, 7075.

7 M.T. Pope, *Isopoly and Heteropoly Anions*, Springer, Berlin, Germany, **1983**. A. Müller, *Polyoxometalate Chemistry*, Kluwer Academic, Dordrecht, The Netherlands, **2001**. I.V. Kozhevnikov, *Catalysis by Polyoxometalates*, Wiley, Chichester, England, **2002**. C.L. Hill, C.M. Prosser-McCartha, *Coord. Chem. Rev.* **1995**, *143*, 407. N. Mizuno, M. Misono, *Chem. Rev.* **1998**, *98*, 171. R. Neumann, *Prog. Inorg. Chem.* **1998**, *47*, 317.

8 Y. Matoba, Y. Ishii, M. Ogawa, *Synth. Commun.* **1984**, *14*, 865.

9 Y. Ishii, K. Yamawaki, T. Ura, H. Yamada, T. Yoshida, M. Ogawa, *J. Org. Chem.* **1988**, *53*, 3587. T. Oguchi, Y. Sakata, N. Takeuchi, K. Kaneda, Y. Ishii, M. Ogawa, *Chem. Lett.* **1989**, 2053. M. Schwegler, M. Floor, H. van Bekkum, *Tetrahedron Lett.* **1988**, *29*, 823.

10 Y. Sakata, Y. Katayama, Y. Ishii, *Chem. Lett.* **1992**, 671. Y. Sakata, Y. Ishii, *J. Org. Chem.* **1991**, *56*, 6233; T. Iwahama, S. Sakaguchi, Y. Nishiyama, Y. Ishii, *Tetrahedron Lett.* **1995**, *36*, 1523.

11 Y. Ishii, K. Yamawaki, T. Yoshida, M. Ogawa, *J. Org. Chem.* **1988**, *53*, 5549.

12 F.P. Ballistreri, S. Failla, E. Spina, G.A. Tamaselli, *J. Org. Chem.* **1989**, *54*, 947.

13 S. Sakaue, Y. Sakata, Y. Nishiyama, Y. Ishii, *Chem. Lett.* **1992**, 289.

14 Y. Ishii, H. Tanaka, Y. Nishiyama, *Chem. Lett.* **1994**, 1.

15 H. Orita, M. Shimizu, T. Haykawa, K. Takehira, *React. Kinet. Catal. Lett.* **1991**, *44*, 209. L.A. Petrov, N.P. Lobanova, V.L. Volkov, G.S. Zakharova, I.P. Kolenko, L.Yu. Buldakova, *Izv. Akad. Nauk SSSR, Ser. Khim.* **1989**, 1967. M. Shimizu, H. Orita, T. Hayakawa, K. Takehira, *Tetrahedron Lett.* **1989**, *30*, 471.

16 L. Salles, C. Aubry, F. Robert, G. Chottard, R. Thouvenot, H. Ledon, J.-M. Brégault, *New J. Chem.* **1993**, *17*, 367. C. Aubry, G. Chottard, N. Platzer, J.-M. Brégault, R. Thouvenot, F. Chauveau, C. Huet, H. Ledon, *Inorg. Chem.* **1991**, *30*, 4409. L. Salles, C. Aubry, R. Thouvenot, F. Robert, C. Dorémieux-Morin, G. Chottard, H. Ledon, Y. Jeannin, J.-M. Brégault, *Inorg. Chem.* **1994**, *33*, 871.

17 A.C. Dengel, W.P. Griffith, B.C. Parkin, *J. Chem. Soc., Dalton Trans.* **1993**, 2683. A.J. Bailey, W.P. Griffith, B.C. Parkin, *J. Chem. Soc., Dalton Trans.* **1995**, 1833.

18 D.C. Duncan, R.C. Chambers, E. Hecht, C.L. Hill, *J. Am. Chem. Soc.* **1995**, *117*, 681.

19 C. Venturello, R. D'Aloiso, J.C. Bart, M. Ricci, *J. Mol. Catal.* **1985**, *32*, 107.

20 A.M. Khenkin, C.L. Hill, *Mendeleev Commun.* **1993**, 140. X. Zhang, Q. Chen, D.C. Duncan, R.J. Lachicotte, C.L. Hill, *Inorg. Chem.* **1997**, *36*, 4381. X. Zhang, Q. Chen, D.C. Duncan, C.F. Campana, C.L. Hill, *Inorg. Chem.* **1997**, *36*, 4208. X. Zhang, T.M. Anderson, Q. Chen, C.L. Hill, *Inorg. Chem.* **2001**, *40*, 418.

21 Y. Seki, J.S. Min, M. Misono, N. Mizuno, *J. Phys. Chem. B* **2000**, *104*, 5940. N. Mizuno, Y. Seki, Y. Nishiyama, I. Kiyoto, M. Misono, *J. Catal.* **1999**, *184*, 550.

22 R. Ben-Daniel, A.M. Khenkin, R. Neumann, *Chem. Eur. J.* **2000**, *6*, 3722.

23 R. Neumann, M. Gara, *J. Am. Chem. Soc.* **1994**, *116*, 5509. R. Neumann, A.M. Khenkin, *J. Mol. Catal.* **1996**, *114*, 169.
24 R. Neumann, M. Gara, *J. Am. Chem. Soc.* **1995**, *117*, 5066.
25 R. Neumann, D. Juwiler, *Tetrahedron* **1996**, *47*, 8781. R. Neumann, A.M. Khenkin, D. Juwiler, H. Miller, M. Gara, *J. Mol. Catal.* **1997**, *117*, 169.
26 R. Neumann, H. Miller, *J. Chem. Soc., Chem. Commun.* **1995**, 2277. R. Neumann, M. Cohen, *Angew. Chem.* **1997**, *109*, 1810.
27 W. Adam, P.L. Alsters, R. Neumann, C.R. Saha-Möller, D. Sloboda-Rozner, R. Zhang, *Synlett* **2002**, 2011. W. Adam, P.L. Alsters, R. Neumann, C.R. Saha-Möller, D. Sloboda-Rozner, R. Zhang, *J. Org. Chem.* **2003**, *68*, 1721.
28 M. Hamamoto, K. Nakayama, Y. Nishiyama, Y. Ishii, *J. Org. Chem.* **1993**, *58*, 6421. N. Mizuno, T. Hirose, M. Tateishi, M. Iwamoto, *Chem. Lett.* **1993**, 1839. N. Mizuno, M. Tateishi, T. Hirose, M. Iwamoto, *Chem. Lett.* **1993**, 1985. N. Mizuno, T. Hirose, M. Tateishi, M. Iwamoto, *Stud. Surf. Sci. Catal.* **1994**, *82*, 593. A.M. Khenkin, A. Rosenberger, R. Neumann, *J. Catal.* **1999**, *182*, 82.
29 J.E. Lyons, P.E. Ellis, V.A. Durante, *Stud. Surf. Sci. Catal.* **1991**, *67*, 99.
30 N. Mizuno, T. Hirose, M. Tateishi, M. Iwamoto, *J. Mol. Catal.* **1994**, *88*, L125. N. Mizuno, M. Tateishi, T. Hirose, M. Iwamoto, *Chem. Lett.* **1993**, 2137.
31 H. Weiner, R.G. Finke, *J. Am. Chem. Soc.* **1999**, *121*, 9831.
32 K.I. Matveev, *Kinet. Catal.* **1977**, *18*, 716. K.I. Matveev, I.V. Kozhevnikov, *Kinet. Catal.* **1980**, *21*, 855.
33 J.R. Grate, D.R. Mamm, S. Mohajan, *Mol. Eng.* **1993**, *3*, 205. J.R. Grate, D.R. Mamm, S. Mohajan in *Polyoxometalates: From Platonic Solids to Anti-Retroviral Activity*, M.T. Pope, A. Müller (eds.), Kluwer, Dordrecht, The Netherlands, **1993**, 27.
34 R. Neumann, I. Assael, *J. Chem. Soc., Chem. Commun.* **1998**, 1285.
35 I.V. Kozhevnikov, V.I. Simagina, G.V. Varnakova, K.I. Matveev, *Kinet. Catal.* **1979**, *20*, 506. B.S. Dziiumakaeva, V.A. Golodov, *J. Mol. Catal.* **1986**, *35*, 303.
36 M.K. Harrup, C.L. Hill, *Inorg. Chem.* **1994**, *33*, 5448. M.K. Harrup, C.L. Hill, *J. Mol. Catal.* **1996**, *106*, 57.
37 R. Neumann, M. Lissel, *J. Org. Chem.* **1989**, *54*, 4607.
38 R. Neumann, M. Levin, *J. Org. Chem.* **1991**, *56*, 5707.
39 K. Nakayama, M. Hamamoto, Y. Nishiyama, Y. Ishii, *Chem. Lett.* **1993**, 1699.
40 A.M. Khenkin, I. Vigdergauz, R. Neumann, *Chem. Eur. J.* **2000**, *6*, 875.
41 O.A. Kholdeeva, A.V. Golovin, I.V. Kozhevnikov, *React. Kinet. Catal. Lett.* **1992**, *46*, 107. O.A. Kholdeeva, A.V. Golovin, R.A. Maksimovskaya, I.V. Kozhevnikov, *J. Mol. Catal.* **1992**, *75*, 235.
42 M. Lissel, H. Jansen van de Wal, R. Neumann, *Tetrahedron Lett.* **1992**, *33*, 1795.
43 K.I. Matveev, E.G. Zhizhina, V.F. Odyakov, *React. Kinet. Catal. Lett.* **1995**, *55*, 47.
44 J.-M. Brégeault, B. El Ali, J. Mercier, J. Martin, C. Martin, *C.R. Acad. Sci. II* **1989**, *309*, 459.
45 B. El Ali, J.-M. Brégeault, J. Martin, C. Martin, *New J. Chem.* **1989**, *13*, 173. B. El Ali, J.-M. Brégeault, J. Mercier, J. Martin, C. Martin, O. Convert, *J. Chem. Soc., Chem. Commun.* **1989**, 825. A. Atlamsani, M. Ziyad, J.-M. Brégeault, *J. Chim. Phys., Phys.-Chim. Biol.* **1995**, *92*, 1344.
46 A.M. Khenkin, R. Neumann, *Adv. Syn. Catal.* **2002**, *344*, 1017.
47 R. Neumann, M. Levin, *J. Am. Chem. Soc.* **1992**, *114*, 7278.
48 D.C. Duncan, C.L. Hill, *J. Am. Chem. Soc.* **1997**, *119*, 243.
49 R. Ben-Daniel, P.L. Alsters, R. Neumann, *J. Org. Chem.* **2001**, *66*, 8650.
50 I.A. Weinstock, R.H. Atalla, R.S. Reiner, M.A. Moen, K.E. Hammel, C.J. Houtman, C.L. Hill, *New J. Chem.* **1996**, *20*, 269. I.A. Weinstock, R.H. Atalla, R.S. Reiner, M.A. Moen, K.E. Hammel, C.J. Houtman, C.L. Hill, M. K. Harrup, *J. Mol. Catal. A-Chem.* **1997**, *116*, 59. I.A. Weinstock, R.H. Atalla, R.S. Reiner, C.J. Houtman, C.L. Hill, *Holzforschung* **1998**, *52*, 304.

51 A. M. Khenkin, R. Neumann, *Angew. Chem. Int. Ed.* **2000**, *39*, 4088. A. M. Khenkin, L. Weiner, Y. Wang, R. Neumann, *J. Am. Chem. Soc.* **2001**, *123*, 8531.

52 R. Neumann, M. Dahan, *Nature* **1997**, *388*, 353. R. Neumann, M. Dahan, *J. Am. Chem. Soc.* **1998**, *120*, 11969.

2.12 Oxidative Cleavage of Olefins

Fritz E. Kühn, Richard W. Fischer, Wolfgang A. Herrmann, and Thomas Weskamp

2.12.1 Introduction and Motivation

Oxidative cleavage of olefins is one of the paramount reactions developed in organic chemistry. The plethora of oxidative pathways discussed in the literature can be broken down into two main methodologies:

1. Transformation of olefins into 1,2-diols followed by oxidative cleavage [1a].
2. Direct cleavage into a variety of functionalized products dependent on the condition applied [1b].

The oxidation of olefinic double bonds using transition metals as catalysts is often limited to epoxidations and the consecutive hydrolysis of the primarily formed oxiranes to the corresponding vicinal diols. From an economic point of view, the oxidative transformation of long-chain olefins like waxes or fatty acid derivatives performed with the aid of transition metal catalysts has a high technical potential in the emerging field of natural resources.

The standard method for the direct oxidative cleavage of olefins is ozonolysis. This reaction has been well developed and yields aldehydes or carboxylic acids upon reductive or oxidative workup, respectively. As important as ozonolysis has proved to be in synthetic chemistry, there are relatively few alternative reactions that duplicate this transformation, i.e., the direct cleavage of olefins without the intermediacy of 1,2-diols [1c,d]. The drawback, however, is the application of the stoichiometric amounts of the expensive oxidant ozone and the need for a consecutive oxidative treatment of the ozonides, formed as intermediates, to yield the desired carboxylic acids. Also, a major issue with ozonolysis is safety [1e,f]. Thus, such oxidative conversions using ozone as the reactant will be reserved for the pharmaceutical and high-price specialty chemicals industry. It should be noted, however, that an organometallic ozonolysis, applying an osmium tetroxide-promoted catalytic oxidative cleavage of olefins with acid yields ranging from 80 to 97%, has been reported recently [1g].

With respect to the afore-mentioned general drawbacks of the ozonolysis reactions, we will focus here on a selection of catalytic systems for the oxidative cleavage of C=C double bonds.

Transition Metals for Organic Synthesis, Vol. 2, 2nd Edition.
Edited by M. Beller and C. Bolm

ISBN: 3-527-30613-7

- Two-step syntheses of carboxylic acids via carbonyl compounds or diols, respectively.
- One-step oxidations applying ruthenium catalysts using peroxo compounds as oxidants.
- One step oxidations of C=C double bonds to aldehydes or carboxylic acids by organo rhenium(VII) catalyst systems.

2.12.2
Two-Step Synthesis of Carboxylic Acids from Olefins

2.12.2.1
Formation of Keto-Compounds from Olefinic Precursors – Wacker-Type Oxidations

The well-known Wacker Hoechst process [2a, b], still on stream for the synthesis of acetaldehyde from ethylene using a bimetallic palladium dichloride/copper dichloride catalyst system, can also be efficiently applied for the conversion (69–100%) of higher terminal olefins or their derivatives into keto acids [2c] with high selectivities (90–100%) and in fair yields (41–73%) on a preparative scale, all dependent on the substrate. For example, 1-octene can be converted into 2-octanone, and 9-decenoic acid into 9-oxo-decenoic acid. Warwel reports a small change to the Wacker system: the copper(II) species is formed here from CuCl during reaction to serve as *in situ*-generated re-oxidant for the Pd(I)/Pd(II) cycle [2a–c]. As well as this, $RhCl_3/FeCl_3$ [2d] proved to be a more selective but less active catalyst. The reaction is highly dependent on the solvent and also restricted to a temperature range of 25–60 °C. Highly polar solvents are advantageous. The best results were obtained in DMF and tetramethyl urea [2e]. Since catalytic oxidation of higher, functional, and cyclic olefinic compounds with $PdCl_2$ in the presence of cupric chloride often results in high amounts of chlorinated by-products, chlorine-free oxidants can be used in order to avoid such chlorinating reactions [2a].

To avoid the use of corrosive additives, such as large amounts of copper salts, chlorides and acid to maintain the catalytic cycle, co-catalysts such as the heteropolyacid $H_3PMo_6V_6O_{40}$ [2e] or a combination of benzoquinone with either iron(II) phthalocyanine [2f] or heteropolyacids [2g] have been developed. Water-soluble palladium(II) bathophenanthroline is another novel, stable, and recyclable catalyst for the selective aerobic oxidation of terminal olefins to the corresponding 2-alkanones in a biphasic liquid-liquid system, the active catalyst being a homogeneous mononuclear species according to kinetic measurements [2h, i]. Additional experiments with chloride-free reactants have been reported on the oxidation of C_2H_4 and C_3H_6 using an [alkene/Pd black (anode)/H_3PO_4/graphite (cathode)/NO + O_2] gas cell at 353 K. The co-feed of NO with O_2 at the cathode dramatically enhanced the Wacker-type oxidations of the alkenes. The enhancement is ascribed to the acceleration in the rate of electrochemical oxidation of Pd(0) to Pd(II) due to the formation of NO_2 at the cathode [2j].

2.12.2.2
Cleavage of Keto-Compounds and vic-Diols into Carboxylic Acids

The oxidative transformation of aliphatic aldehydes or ketones into their corresponding carboxylic acids is standard in industrial chemistry [3]. Especially, short-chain aldehydes and ketones like acetaldehyde or methyl-ethyl ketone are used as efficient co-oxidants because of their high capacity to form peroxy acid radicals in the presence of a transition metal catalyst and oxygen or air as the oxidizing agent. Thus, it is state of the art to oxidize C_2–C_5 aldehydes and ketones applying manganese salts (i.e., acetate, stearate, acac) as catalysts in a concentration of about 1–2 mol% at atmospheric pressure or up to 30 bar. In the case of longer-chain substrates, as in fatty acid chemistry (C_9–C_{15}), or the oxidation of keto derivatives of waxes (C_{30+}), selectivity drops and chain degradation occurs. Methylketo fatty esters are cleaved mainly at the carbonyl group to yield dicarboxylic acids that are one or two C-atom units shorter than the starting material. For example 9-oxo-decanoic acid will be converted to cork (C_8) and azelaic acid (C_9), 10-oxo-tetradecanoic acid to azelaic and sebacic acid (C_{10}), respectively. 13-Oxo-tetradecenoic acid methylester oxidized with oxygen (from air), mediated by simple, easy-to-recycle manganese catalyst, is cleaved into dicarboxylic monomethyl esters with corresponding chain lengths of C_{13} (60%), C_{12} (29%), C_{11} (5%), C_{10} (2%), C_9, C_8 (1%) and C_7, C_4 (traces). C_8–C_{13} and higher mono- and dicarboxylic acids find application in the production of special plasticizers and ester-based lubricants.

Besides the oxidative cleavage of keto-compounds, aliphatic vic-diols derived from rhenium [4a–e] and manganese [4f], molybdenum [5], tungsten [6], or osmium [7]-catalyzed epoxidations or hydroxylations can be cleaved by Co(III)-catalyzed aerobic oxidations or by application of W(VI), Mo(VI) or Os(VIII) catalyst systems [4, 7, 8].

2.12.3
One-Step Oxidative Cleavage Applying Ruthenium Catalysts and Percarboxylic Acids as Oxidants

2.12.3.1
General Aspects

In spite of the broad industrial application of Wacker-type oxidations of terminal olefins to aldehydes and manganese-catalyzed formations of carboxylic acids from carbonyl group-containing compounds, the above-mentioned two-step synthesis sequence suffers from some disadvantages: low overall selectivity and yield in the case of long-chain substrates, application of two different catalyst systems, and the need to isolate the intermediate products before further transformation. To overcome these drawbacks, catalysts containing only one metal species have to be applied. One of the few transition metals suitable as a catalyst for a single-step cleavage of olefinic double bond systems is ruthenium. Like osmium tetraoxide, ruthe-

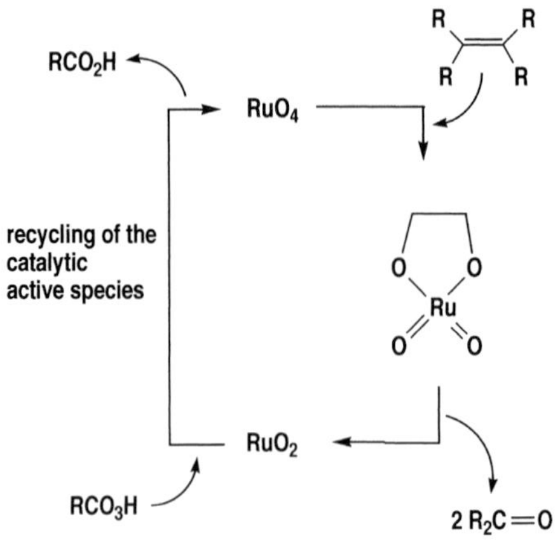

Scheme 1

nium tetraoxide reacts with C=C double bonds to give cyclic ruthenate(VI) esters. Oxidative work-up, likewise with peracids, yields bond cleavage to give mainly keto compounds (Scheme 1) [9].

To substitute the highly reactive and aggressive RuO_4, easy-to-handle catalyst precursors like $RuCl_3$, Ru(acac), or RuO_2 are commonly used. Besides these simple ruthenium compounds, a wide variety of different ruthenium complexes like dimeric $(Cp^*RuCl_2)_2$, $CpRu(PPh_3)_2Cl$, $Ru_3(CO)_{12}$, $Ru(CO)_2Cl(OEt_2)$, or $Ru(CF_3SO_3)$, as well as systems such as $RuCl_3$/Oxone/$NaHCO_3$ can serve as efficient catalyst precursors showing similar activity to that observed for the above-mentioned systems or for RuO_4 itself, suggesting that under catalytic conditions (presence of peroxyacetic acid) the precursors are transformed into the same or closely related catalytic active species. Thus, with oxidants like sodium periodate, bleach (NaOCl), cerium(VI) salts, or organic peracids (mostly peracetic acid, other peroxy acids being less suitable because of increased epoxide formation), the low-valent ruthenium compounds are transformed into the active species "RuO_4" [10].

2.12.3.2
Optimized Catalyst Systems and Reaction Conditions

The system "RuO_4"/peracetic acid (on-site-formed peroxyacetic acid from H_2O_2, acetic acid, and H_2SO_4 as catalyst) reacts with olefins like 1-octene to give two different primary products, i.e., 1-heptanal and formic aldehyde, mediated by the ruthenium catalyst, and 1-octenoxide, formed by direct epoxidation of the olefin by peracetic acid. The aldehydes formed are easily transformed by the oxidant into the corresponding acids. The octene oxide is solvolyzed into the glycol (further oxidized to the corresponding acids by the peracid), the vicinal hydroxyacetate, and the diacetate, respectively. Finally these last two products can also be cleaved oxidatively into their corresponding acids. Because of some degradation, small amounts of hexanoic acid are formed, the total by-products amounting to approxi-

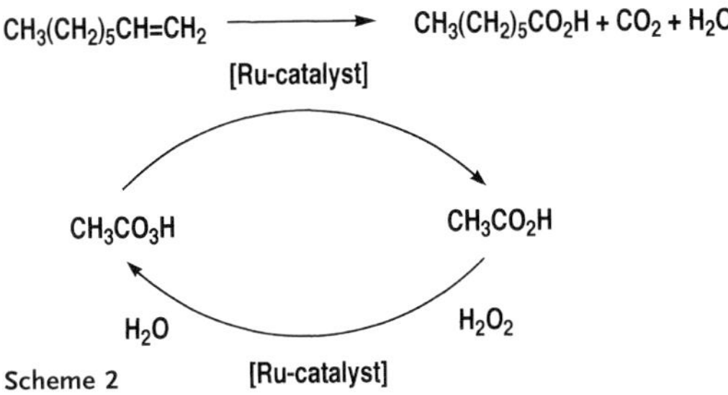

Scheme 2

mately 10%. The molar ratio of olefin to peroxyacetic acid needed is generally greater than the theoretical value of 1/4, i.e. 1/5 to 1/6. An excess of peroxyacetic acid will increase the yield of the desired carboxylic acid at an optimized pH of ca. 2 (Na_2CO_3/H_2SO_4 buffer) [2]. In general, the normally used ratio of catalyst to substrate amounts to 1/1000, yielding the best selectivities. However, the remarkable activity of ruthenium-containing catalyst can be followed up to an $Ru(acac)_3$/olefin ratio as low as 1/20000. Under such conditions, heptanoic acid is obtained from 1-octene in 62% yield [2]. Even at Ru/olefin ratios of 1/60000, a 15% yield of heptanoic acid has been reported. Thus, the enormous catalytic activity of ruthenium catalysts confers the outstanding advantage of this catalyst metal for the oxidative cleavage of C=C double bonds. In practice, however, for reasons of selectivity as well as space-time yield, the applied concentrations are somewhat higher. The best solvents are water/*n*-hexane mixtures (two-phase system). Strong coordinating or even complexing solvents like DMF, acetonitrile, or THF are less suitable. Ruthenium-based systems suitable for oxidative cleavage of olefinic double bonds, also in the case of long chain waxes, are quite well optimized [9, 10]. However, they often suffer from complex reaction conditions like the application of various solvent systems, auxiliary reactants, or expensive oxidants like organic peroxy acids. The major drawback of ruthenium catalysts is their high activity in decomposing hydrogen peroxide. Simple ruthenium compounds like RuO_2 or $RuCl_3$ show rapid H_2O_2 decomposition, three orders of magnitude faster than that with MnO_2 [12a]. Donor ligand-substituted ruthenium compounds like $RuCl_3(PPh)_3$ or $RuCl_3(dmp)_2$ show significantly lower decomposition rates for H_2O_2 than RuO_4 or $RuCl_3$. Thus, by applying these complexes in acetic acid as solvent it is possible to activate hydrogen peroxide without rapid decomposition and with an acid formation selectivity (1-octene to heptanoic acid) of 46% at 100% conversion [12a]. It is reasonable to assume that *in situ*-formed peracetic acid is acting as the primary oxidant during the catalytic cycle (Scheme 2).

2.12.4
Selective Cleavage of Olefins Catalyzed by Alkylrhenium Compounds

2.12.4.1
Rhenium-Catalyzed Formation of Aldehydes from Olefins

Alkylrhenium oxides are known as highly efficient and selective oxidation catalysts, especially in the field of epoxidation reactions [4]. The advantage of the rhenium catalysts is their ability to activate hydrogen peroxide as a cheap and environmentally friendly oxidant without any H_2O_2 decomposition, independently of the concentration of the hydrogen peroxide used (5–85 wt%). Compared to the $RuCl_3$ mentioned above, in the presence of methyltrioxorhenium (CH_3ReO_3, MTO) the half-life of H_2O_2 is 20000 times higher, analogously compared to MnO_2 it is higher by a factor of 50, Na_2WO_4 by a factor of 20, and even Re_2O_7 by a factor of 2 [12a]. In this light, MTO appears as a first class catalyst for the efficient activation of hydrogen peroxide. Dependent on the reaction conditions, alkylrhenium oxides can be turned into epoxidation catalysts (low temperature, presence of co-ligands, correct stoichiometry of oxidant and olefin), dihydroxylation catalysts (ambient temperature and higher, presence of water), or catalysts for the cleavage of C=C double bonds. The latter can be achieved under nearly water-free conditions, the right choice of solvent, and a defined excess of hydrogen peroxide (olefin/oxidant/catalyst=1/>4/0.01). To turn alkylrhenium oxides like MTO into a C=C double-bond-cleaving catalyst it is necessary to trap the formed water with $MgSO_4$, ortho-esters, or by azeotropic distillation during the course of the oxidation reaction to increase activity and to avoid catalyst hydrolysis (see Tab. 1). This increases the catalyst lifetime at the required higher reaction temperature of 60 °C. Under these conditions, olefins are converted into aldehydes in fair to high yields. Special aprotic solvents like *t*-butyl methyl ether (mtbe) allow higher water concentrations without it being essential to trap H_2O_2 from the reaction mixture.

Tab. 1 Oxidation of *n*-octene-1 with MTO/H_2O_2 [a] [12b]

Drying agent	*Solvent*	*Aldehyde yield (%)* [b]	*Diol yield (%)* [b]
$MgSO_4$	MTBE	65	35
$MgSO_4$	*t*-BuOH	48	52
$MgSO_4$	CH_3CN	32	68
$MgSO_4$	di-*n*-butyl ether	10	90
Na_2SO_4	*t*-BuOH	27	73
$HC(OEt)_3$	*t*-BuOH	23	77
–	*t*-BuOH	0	70 [c]
–	CH_3CN	0	41 [c]

a) Reaction conditions: 10.0 mmol olefin, 0.1 mol catalyst, 60 mmol H_2O_2 (30% in organic solvent), T=60 °C, t=7 h.
b) All given yields are GC yields.
c) Conversion of olefin.

The addition of non-coordinating Brønsted acids like HBF_4 or $HClO_4$ as co-catalysts increases the yield of aldehyde from olefin oxidation from 68 to 85% under two-phase conditions, e.g., with chloroform as the organic phase [12b].

2.12.4.2
Acid Formation from Olefins with Rhenium/Co-Catalyst Systems

A strong solvent effect was discovered by the application of the oxidation system mtbe/H_2O_2/HBF_4/MTO, which oxidizes the primarily formed aldehydes further to their corresponding carboxylic acids in 60% selectivity at complete conversion. In contrast, in mtbe as solvent without addition of HBF_4, only aldehyde formation is observed.

The right combination of aprotic solvent (mtbe) and co-catalyst (HBF_4) leads to a one-pot transformation of olefinic double bonds to aliphatic as well as aromatic carboxylic acids. Thus, according to the reaction conditions used, MTO and the homologous alkylrhenium catalysts can be freely tuned, depending on the desired reaction pathway (Scheme 3). Besides simple olefins, long-chain olefins, waxes, and fatty acid derivatives can also be cleaved in the aforesaid manner to the aldehydes and carboxylic acids, applying co-catalyst systems. A C_{30} wax fraction (MW 564 g/mol; chain lengths between C_{26} and C_{54}; Chevron) is cleaved by MTO (0.5 mol%)/H_2O_2 to 57% aldehydes and 43% *vic*-diols at full conversion of the substrate mixture. 2-Alkyl-1-alkene compounds are oxidized to the ketones. Under HBF_4 conditions, further oxidation to the carboxylic acid is observed. At reaction temperature, the wax is completely soluble in mtbe, and after cooling to ambient temperature the catalyst solution (mtbe/MTO/H_2O_2) is easily separated from the solid reaction products by filtration [12b].

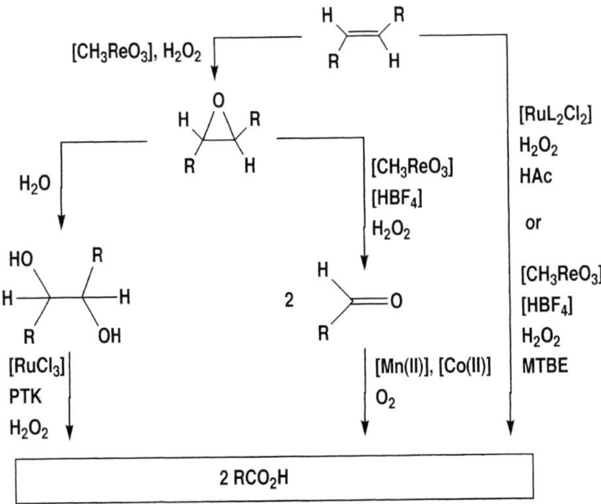

Scheme 3

Furthermore, the MTO/H_2O_2 system catalyzes the oxidation of cyclic *β*diketones to carboxylic acids [12c]. Conversions are usually above 85%, and the product selectivity is almost quantitative. The reaction is performed in a 1:1 water-acetonitrile solution at room temperature. It has been assumed that enolic forms which exist in solution are initially epoxidized. After a rearrangement step, the C-C bond is cleaved and an oxygen atom is inserted. Then, an *α*diketone intermediate forms, and this is finally oxidized to the carboxylic acid [12c].

In summary, the rhenium catalyzed olefin cleavage has several advantages: the use of hydrogen peroxide as oxidant, broad applicability at various reaction conditions, and a multi-purpose catalyst system tuned solely by oxidant concentration, solvent system, and reaction temperature. Scheme 3 gives an overview of some of the most efficient transition metal catalysts for the cleavage of olefins to aldehydes and carboxylic acids.

2.12.5 Other Systems

During recent years, several other efficient systems for C=C bond cleavage have been found and described [13]. Among them is Re_2O_7 in 70% *t*-butyl hydroperoxide, which acts as a comparatively mild and efficient catalyst for the carbon-carbon bond cleavage of ketones to the corresponding carboxylic acids [13a].

The use of tungstic acid or tris(cetylpyridinium) 12-tungstophosphate under homogeneous conditions (*t*-BuOH as the solvent) for the production of carboxylic acids from alkenes has limited practical value, as it requires long reaction times (24 h at 80 °C) and affords moderate to low yields of acids with *α*-olefins [13b–e]. The oxidative cleavage of alkenes to carboxylic acids with 40% w/v hydrogen peroxide catalyzed by methyltrioctylammonium tetrakis(oxodiperoxotungsto)phosphate(3-), however, is reported to occur in high yields and selectivities under two-phase conditions in the absence of organic solvents [13d]. Two main reaction pathways leading to acids have been recognized, one involving the perhydrolysis, the other the hydrolysis of the epoxide initially formed. The perhydrolytic reaction pathway appears to play a primary role in the oxidation of medium- and long-chain alkenes to acids, while it intervenes to a rather limited extent in the oxidation of arylalkenes and C_5–C_7 cycloalkenes. Hydrogen peroxide concentration appears to exert a remarkable influence on medium acidity and thereby affects the reaction efficiency [13d].

Reaction of alkenes with aqueous hydrogen peroxide and catalytic quantities of heteropolyacids of Mo and W, both in free form and adsorbed onto magnesium, aluminum or zinc oxide leads in some cases to complete, rapid cleavage of the alkene to give carbonyl compounds [13f–i].

References

1 (a) K.T.M. Shing in *Comprehensive Organic Synthesis* (Eds.: B.M. Trost, I. Flemming), **1991**, *7*, 703–716, Pergamon Press, Oxford, **1991**; (b) C.R. Larock in *Comprehensive Organic Transformations*, 2nd edn, Wiley-VCH, **1999**, pp 1213–1215, Wiley-VCH, New York; (c) L. Albarella, F. Giordano, M. Lasalvia, V. Picialli, D. Sica, *Tetrahedron Lett.* **1995**, *36*, 5267; (d) D. Yang, C. Zhang, *J. Org. Chem.* **2001**, 66, 4814; (e) K. Koike, G. Inoue, T. Fukuda, *J. Chem. Eng. Jpn.* **1999**, *32*, 295; (f) R.A. Ogle, J.L. Schumacher, *Process Saf. Prog.* **1998**, *17*, 127; (g) B.R. Travis, R.S. Narayan, B. Borhan, *J. Am. Chem. Soc.* **2002**, *124*, 3824.

2 (a) R. Jira in *Applied Homogeneous Catalysis with Organometallic Compounds*, 2nd edn (Eds.: B. Cornils, W.A. Herrmann), **2002**, *1*, 386-405, Wiley-VCH, Weinheim, and references cited therein; (b) C. Elschenbroich, A. Salzer, *Organometallics – A Concise Introduction*, 2nd edn, 425–427, VCH, Weinheim, **1992**; (c) S. Warwel, M. Sojka, M. Rüsch, *Top. Curr. Chem.*, **1993**, *164*, 83; (d) F.J. McQuillin, D.G. Parker, *J. Chem. Soc., Perkin Trans. 1*, **1975**, 2092; (e) J.H. Grate, D.R. Hamm, S. Mahajan, *Mol. Eng.*, **1993**, *3*, 205; (f) S. Srinivasan, W.T. Ford, *J. Mol. Catal.* **1991**, *64*, 291; (g) Y. Kim, H. Kim, J. Lee, K. Sim, Y. Han, H. Paik, *Applied Catalysis A: General* **1997**, *155*, 15; (h) G.J. ten Brink, I.W. C.E. Arends, G. Papadogianakis, R.A. Sheldon, *Applied Catalysis A: General* **2000**, *194/195*, 435; (i) G.J. ten Brink, I. W.C.E. Arends, R.A. Sheldon, *Science*, **2000**, *287*, 1636; (j) K. Otsuka, I. Yamanaka, A. Nishi, *J. Electrochem. Soc.* **2001**, *148*, D4.

3 K. Weissermel, H.J. Arpe, *Industrielle Organische Chemie*, 5th edn, Wiley-VCH, Weinheim, **1998**.

4 Recent reviews: (a) F.E. Kühn, M. Groarke in *Applied Homogeneous Catalysis with Organometallic Compounds*, 2nd edn (Eds.: B. Cornils, W.A. Herrmann), **2002**, *3*, 1304-1318, Wiley-VCH, Weinheim; (b) F.E. Kühn, W.A. Herrmann, *Chemtracts-Organic Chemistry*, **2001**, *14*, 59; (c) F.E. Kühn, W.A. Herrmann in *Structure and Bonding* (Ed.: B. Meunier), **2000**, *97*, 213, Springer, Heidelberg Berlin; (d) W. Adam, C.M. Mitchell, C.R. Saha-Möller, O. Weichold in *Structure and Bonding* (Ed.: B. Meunier), **2000**, *97*, 237, Springer, Heidelberg, Berlin; (e) G.S. Owens, J. Arias, M.M. Abu-Omar, *Catalysis Today*, **2000**, *55*, 317; (f) J. Brinksma, L. Schmieder, G. van Vliet, R. Boaron, R. Hage, D.E. de Vos, P.L. Alsters, B.L. Feringa, *Tetrahedron Lett.* **2002**, *43*, 2619.

5 Recent reviews: (a) R.A. Sheldon in *Applied Homogeneous Catalysis with Organometallic Compounds*, 2nd edn (Eds: B. Cornils, W.A. Herrmann), **2002**, 412-426, VCH, Weinheim; (b) H. Arzoumanian, *Coord. Chem. Rev.*, **1998**, *180*, 191; (c) R.H. Holm, *Chem. Rev.* **1987**, *87*, 1401; (c) R.H. Holm, *Coord. Chem. Rev.* **1990**, *100*, 183.

6 (a) D.E. de Vos, B.F. Sels, P.A. Jacobs, *Adv. Catal.* **2002**, *46*, 1; (b) I.A. Weinstock, E.M.G. Barbuzzi, M.W. Wemple, J.J. Cowan, R.S. Reiner, D.M. Sonnen, R.A. Heintz, J.S. Bond, C.L. Hill, *Nature* **2001**, *414*, 191; (c) J. Ichihara, *Tetrahedron Lett.* **2001**, *42*, 695; (d) D.V. Deubel, *J. Phys. Chem. A* **2001**, *105*, 4765; (e) D. Hoegaerts, B.F. Sels, D.E. de Vos, F. Verpoort, P.A. Jacobs, *Catal. Today* **2000**, *60*, 209; (f) K. Vassilev, R. Stamenova, C. Tsvetanov, *React. Funct. Polym.* **2000**, *46*, 165.

7 (a) B.M. Choudary, N.S. Chowdari, K. Jyothi, M.L. Kantam, *J. Am. Chem. Soc.* **2002**, *124*, 5341; (b) J. Muldoon, S.N. Brown, *Org. Lett.* **2002**, *4*, 1043; (c) C. Dobler, G.M. Mehltretter, U. Sundermeier, M. Eckert, H.C. Militzer, M. Beller, *Tetrahedron Lett.* **2001**, *42*, 8447; (d) C. Dobler, G.M. Mehltretter, U. Sundermeier, M. Beller, *J. Organomet. Chem.* **2001**, *621*, 70; (e) T. Sammakia, T.B. Hurley, D.M. Sammond, R.S. Smith, S.B. Sobolov, T.R. Öschger, *Tetrahedron Lett.* **1996**, *37*, 4427.

8 (a) Co(III)-oxidation: R.W. Fischer, F. Röhrscheid in *Applied Homogeneous Catalysis with Organometallic Compounds*,

2nd edn (Eds.: B. Cornils, W. A. Herrmann), **2002**, *3*, 448–449, Wiley-VCH, Weinheim; W(VI): C. Venturello, M. Ricci, *J. Org. Chem.*, **1986**, *54*, 1599; (c) F. di Furia in *Dioxygen Activation and Homogeneous Catalytic Oxidation*, Elsevier, Amsterdam, **1991**, p. 375.

9 (a) C. Derjassi, R. Engel, *J. Am. Chem. Soc.* **1953**, *75*, 3838; (b) L. M. Berkowitz, P. N. Rylander, *J. Am. Chem. Soc.* **1958**, *80*, 6682.

10 (a) R. A. Sheldon, J. K. Kochi, *Metal Catalyzed Oxidations of Organic Compounds*, Academic Press, New York **1981**, pp 162, 297; (b) E. A. Seddon, K. R. Seddon in *The Chemistry of Ruthenium* (Ed.: R. J. H. Clark), Elsevier, Amsterdam, p. 52; (c) M. Hudlicky, *Oxidations in Organic Chemistry*, Washington, **1990**, p 82; (d) T. Mitsudo, T. Kondo, *Synlett* **2001**, 309; (e) T. Kondo, *J. Synth. Org. Chem. Jpn.* **2001**, *59*, 170; (f) D. Yang, C. Zhang, *J. Org. Chem.* **2001**, *66*, 4814; (g) R. H. Jih, K. Y. King, *Curr. Sci.* **2001**, *81*, 1043.

11 The catalytic system ruthenium compound/peracid as an oxidation system for bond cleavage has been known for more than 30 years: (a) P. H. Washecheck (Continental Oil Co.) Ger. Offen. 2046034, **1991**; (b) P. N. Sheng (Atlantic Richfield Co.), US 3839375, **1974**; (c) S. To, K. Aihara, M. Matsumoto, *Tetrahedron Lett.* **1983**, *24*, 5249; (d) R. Neumann, C. Abu-Gnim, *J. Chem. Soc., Chem. Commun.* **1989**, 1324; (e) K. A. Keblys, M. Dubeck (Ethyl Corporation), US Pat. 3409649, **1968**; (f) MacLean, A. Fiske, Ger. Offen. 1568346, **1970**; (g) T. A. Foglia, P. A. Barr, A. J. Malloy, *J. Am. Oil Chem. Soc.* **1977**, *54*, 858A; (h) A. Fiske, A. L. Stautzenberger, Ger. Offen. 1568363, **1970**; (i) S. Wolfe, S. K. Hasan, J. R. Campbell, *J. Chem. Soc., Chem. Commun.* **1970**, 1420; (j) K. Kaneda, S. Haruna, T. Imanaka, K. Kawamoto, *J. Chem. Soc., Chem. Commun.* **1990**, 1467.

12 (a) T. Weskamp, Diploma Thesis, Technische Universität München, **1996**, p 15; (b) W. A. Herrmann, T. Weskamp, J. P. Zoller, R. W. Fischer, *J. Mol. Catal.* **2000**, *153*, 49; (c) M. M. Abu-Omar, J. H. Espenson, *Organometallics* **1996**, *15*, 3543.

13 (a) S. Gurunath, A. Sudalai, *Synlett* **1999**, 559; (b) Y. Ishii, K. Yamawaki, T. Ura, H. Yamada, T. Yoshida, M. Ogawa, *J. Org. Chem.* **1988**, *53*, 3587; (c) T. Oguchi, T. Ura, Y. Ishii, M. Ogawa, *Chem. Lett.* **1989**, 857; (d) E. Antonelli, R. D.'Aloisio, M. Gambaro, T. Fioriani, C. Venturello, *J. Org. Chem.* **1998**, *63*, 719; (e) K. Sato, M. Aoki, J. Tagaki, K. Zimmermann, R. Noyori, *Bull. Chem. Soc. Jpn.* **1999**, *72*, 2287; (f) C. D. Brooks, L. C. Huang, M. McCarron, R. A. W. Johnstone, *J. Chem. Soc., Chem. Commun.* **1999**, 37; (g) Y. M. A. Yamada, M. Ichinohe, H. Takahashi, S. Ikegami, *Org. Lett.* **2001**, *3*, 1837; (h) M. Hashimoto, K. Itoh, K. Y. Lee, M. Misono, *Top. Catal.* **2001**, *15*, 265; (i) J. M. Brégeault, F. Launay, A. Atlamsani, *C. R. Acad. Sci. Ser. II Fasc. Chim.* **2001**, *4*, 11; (j) K. Sato, M. Aoki, R. Noyori, *Science* **1998**, *281*, 1646; (k) U. Schuchardt, D. Cardoso, R. Sercheli, R. Perreira, R. S. de Cruz, M. C. Guerreiro, D. Mandelli, E. V. Spinace, E. L. Fires, *Appl. Catal. A-General* **2001**, *211*, 1.

2.13 Aerobic, Metal-Catalyzed Oxidation of Alcohols

István. E. Markó, Paul R. Giles, Masao Tsukazaki, Arnaud Gautier, Raphaël Dumeunier, Kanae Doda, Freddi Philippart, Isabelle Chellé-Regnault, Jean-Luc Mutonkole, Stephen M. Brown, and Christopher J. Urch

2.13.1 Introduction

The oxidation of alcohols (**1**) into aldehydes and ketones (**2**) is a ubiquitous transformation in Organic Chemistry (Fig. 1). The plethora of reagents available to accomplish this key reaction is a testimony to its importance, both in large-scale processes and in the manufacture of fine chemicals [1]. Unfortunately, most of these oxidants are required at least in stoichiometric quantities and are either toxic or hazardous or both. Moreover, the purification of the reaction products is often demanding and laborious. To circumvent these problems, a number of catalytic oxidation processes based upon the combination of a salt of a metal, e.g., V, Mo, W, Ru, and Co and stoichiometric oxidants such as NMO, *t*BuOOH, PhIO, NaOCl and H_2O_2 have been devised and are now routinely used [2].

From an economical and environmental viewpoint, catalytic oxidation processes are thus extremely valuable. Among these procedures, catalytic systems employing molecular oxygen or air are particularly attractive. Indeed, they employ the cheapest and most readily available stoichiometric oxidant (air or O_2) and are ecologically friendly since they only release H_2O as the by-product. However, while the petrochemical-based industry already takes advantage of aerobic oxidations for the preparation of epoxides, diols, ketones, and acids at the ton scale, few efficient, catalytic aerobic processes are known that are amenable to the preparation of fine chemicals [3].

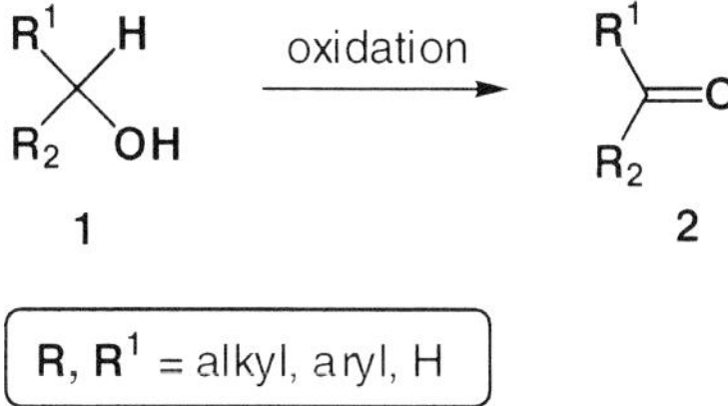

Fig. 1

Transition Metals for Organic Synthesis, Vol. 2, 2nd Edition.
Edited by M. Beller and C. Bolm

ISBN: 3-527-30613-7

In this Chapter, we shall briefly review the most pertinent aerobic oxidation systems described so far in the literature and discuss in greater detail our own contribution to this area.

2.13.2 General Survey

Probably the oldest catalytic aerobic oxidation of alcohols is the aqueous platinum-based process [4]. This system has been continuously refined over the past century, and problems pertaining to catalyst deactivation and to the use of water-insoluble substrates have been partially resolved. The oxidations are usually performed under mild conditions (20–90 °C, 1 atm O_2) with a catalyst-to-substrate ratio in the range 0.2–0.005 wt%. Unfortunately, the yields of carbonyl compounds depend strongly upon the pH of the solution, and the optimum pH has to be determined for every reaction [5]. Moreover, whilst oxidation of benzylic and allylic alcohols affords the corresponding aldehydes in good yields, aliphatic primary alcohols are rapidly oxidized to the acids, especially under basic conditions. The corresponding symmetric esters are also produced in significant quantities. Furthermore, catalyst deactivation is frequently encountered, necessitating a high catalyst/substrate loading. Finally, the explosion risk in the case of readily dehydrogenating substrates must be particularly stressed. In this context, a recent contribution by Baiker demonstrated that aerobic oxidation of alcohols using platinum-based catalysts could be efficiently accomplished in supercritical CO_2, thus overcoming this stringent limitation [6].

$PdCl_2$ in combination with NaOAc has been reported to oxidize alcohols to carbonyl compounds under 1 atm of O_2. However, the reactions are particularly slow, proceed at best with moderate yields, and can only be performed in a limited number of solvents (e.g., sulfolane or ethylene carbonate) [7]. Moreover, while the procedure is efficient for the transformation of secondary aliphatic alcohols into ketones, it is not compatible with the presence of olefinic linkages or unhindered amines (Fig. 2).

Nonetheless, these initial results have triggered, over the past few years, a resurgence of interest in palladium-catalyzed aerobic oxidations. Eschavarren reported that $Pd(PPh_3)_4$, in the presence of NH_4PF_6, was a competent catalyst for the selective oxidation of allylic alcohols into enals and enones [8]. The conditions are, however, rather harsh, requiring prolonged reflux at 110 °C under an oxygen at-

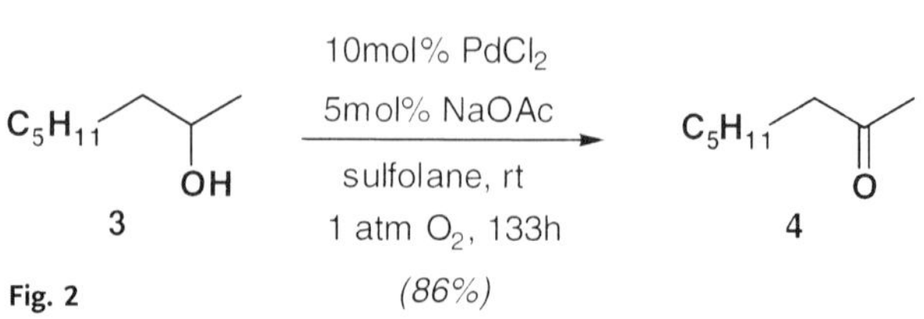

C_5H_{11} (OH) 3 — 10mol% $PdCl_2$, 5mol% NaOAc, sulfolane, rt, 1 atm O_2, 133h (86%) → C_5H_{11} (O) 4

Fig. 2

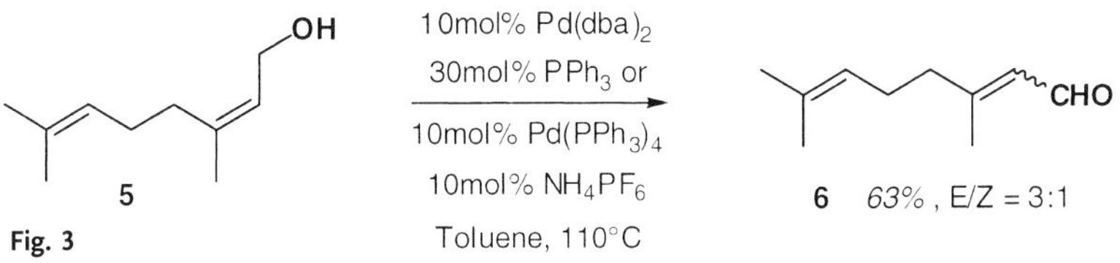

Fig. 3

mosphere, and typically resulting in the production of a mixture of (*E*)- and (*Z*)-enals (Fig. 3).

In 1998, Uemura [9] and Larock [10] published simultaneously the use of $Pd(OAc)_2$ for the aerobic oxidation of various classes of alcohols to form the corresponding carbonyl derivatives. While Uemura employs pyridine, in toluene at 80 °C and in the presence of 3 Å MS, Larock recommends the use of $NaHCO_3$ (or no base at all) in DMSO at 80 °C. Under Larock's conditions, primary and secondary benzylic alcohols are transformed in good yields into aromatic aldehydes and ketones, but allylic substrates usually give modest yields. Uemura's system appears broader ranging. Indeed, not only primary and secondary benzylic alcohols are smoothly oxidized into the corresponding carbonyl derivatives, but primary and secondary aliphatic alcohols also afford the desired products in excellent yields (Tab. 1).

Again, allylic alcohols are poor substrates, and the catalyst does not tolerate the presence of strongly coordinating functions. A few examples of resilient substrates are depicted in Fig. 4.

The proposed mechanism of this aerobic oxidation is depicted in Fig. 5. The reaction begins with a ligand exchange between $Pd(OAc)_2Py_2$ and the alcohol **7**, generating intermediate **14**, which undergoes a β-hydride elimination, affording the carbonyl derivative **8** and the hydrido complex **15**. Reaction of **15** with molecular oxygen leads to the peroxide **16**, which, after addition of alcohol **7** and release of hydrogen peroxide, regenerates the loaded complex **14**. A new catalytic cycle then ensues. The liberated H_2O_2 is then decomposed by the 3 Å MS.

Subsequently to Uemura's work, Sheldon reported the use of water-soluble palladium(II) complexes for the aerobic oxidation of alcohols [11]. Modified, water-soluble phenanthroline ligands were appended onto $Pd(OAc)_2$, and, after adjusting the pH to 11.5, the oxidation was carried out at 100 °C and an oxygen pres-

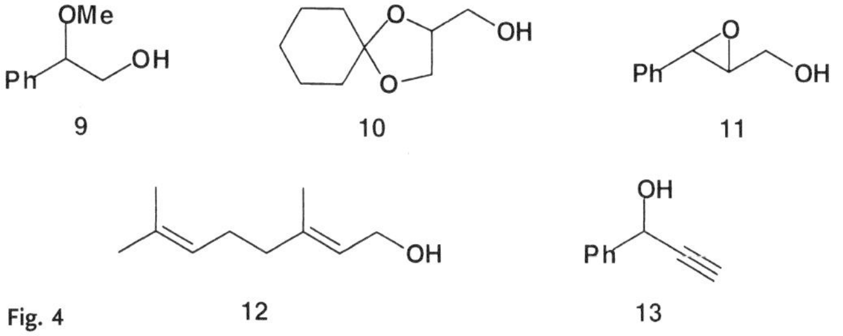

Fig. 4

Tab. 1 Palladium-catalyzed aerobic oxidation (Uemura/Larock)

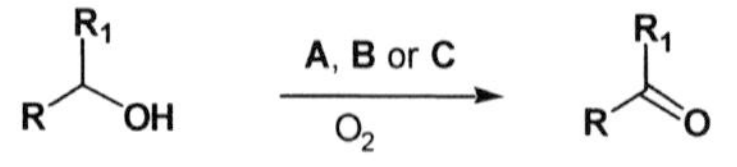

A = 5 mol% $Pd(OAc)_2$, 20 mol% Py, 3 Å MS, 80 °C, Toluene
B = 5 mol% $Pd(OAc)_2$, 2 eq $NaHCO_3$, DMSO, 80 °C
C = 5 mol% $Pd(OAc)_2$, DMSO, 80 °C

Entry	Condition	Substrate	Product	Yield
1	A			95%
2	C			92%
3	A	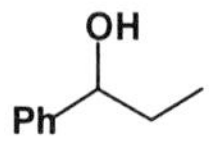	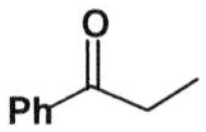	94%
4	B	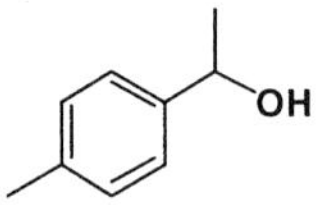	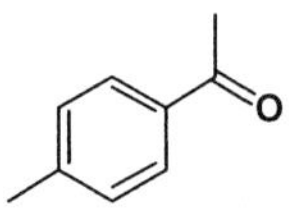	81%
5	A			87%
6	B			67%
7	A			86%
8	C			69%
9	A	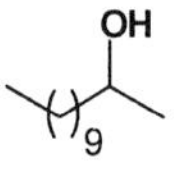	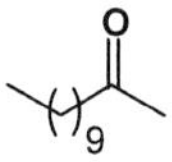	97%
10	A	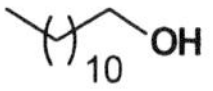	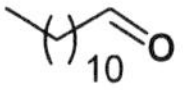	93%

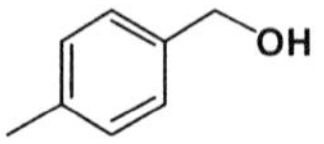

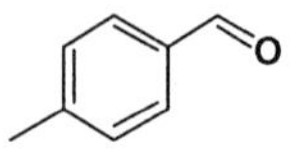

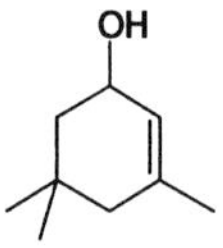

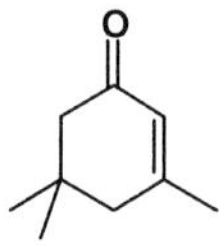

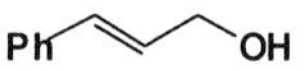

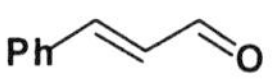

sure of 30 bar. Under these conditions, secondary aliphatic alcohols provide the corresponding ketones but primary aliphatic substrates are directly oxidized to carboxylic acids (Tab. 2).

Whilst the substrate/catalyst ratio can be as low as 200:1 to 400:1 and recycling is possible, this system does not tolerate S, N, and coordinating functions. Furthermore, the strongly basic conditions preclude the use of base-sensitive alcohols or carbonyls. The proposed mechanism differs from the one previously postulated by Uemura (Fig. 6).

In the continuation of his studies, Sheldon found that the addition of TEMPO (4 equiv. per palladium) led to the selective formation of aldehydes (Tab. 2, En-

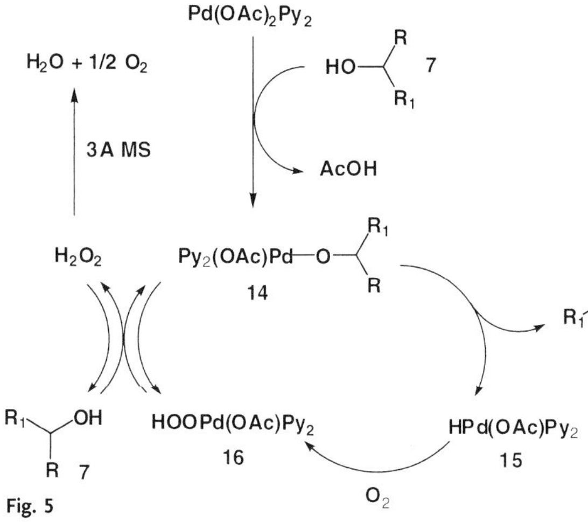

Fig. 5

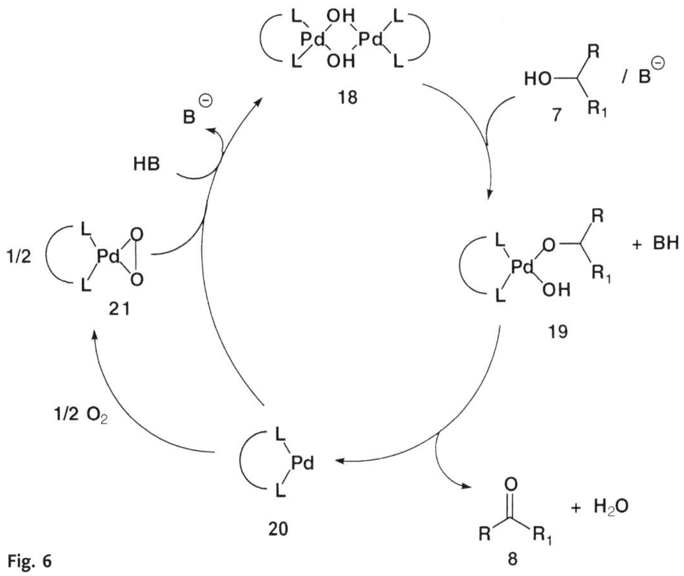

Fig. 6

try 5). For lipophilic substrates, the use of a co-solvent or other additives such as alkanesulfonates becomes mandatory. In some cases, Wacker oxidation of the double bond can be a important side reaction. Finally, Sigman reported an interesting modification of the Uemura/Larock protocol [12]. By switching from pyridine to Et_3N, he found that the aerobic oxidation of a variety of alcohols could be efficiently performed at room temperature instead of 80–100 °C. Unfortunately, the limitations pertaining to the previous palladium-based procedures still apply with this system.

Tab. 2 Palladium-catalyzed aerobic oxidation (Sheldon)

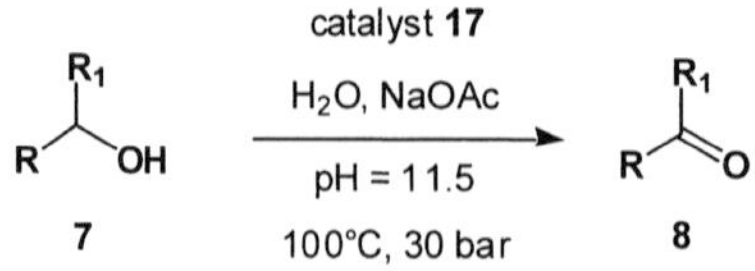

R(R$_1$)CH–OH **7** → catalyst **17**, H_2O, NaOAc, pH = 11.5, 100°C, 30 bar → R(R$_1$)C=O **8**

Entry	*Substrate*	*Product*	*Yield*
1	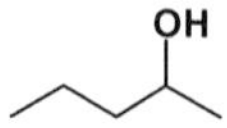	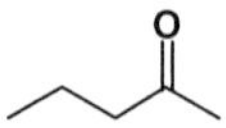	90%
2	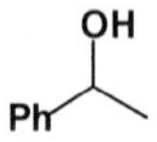 Ph	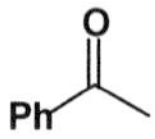 Ph	85%
3	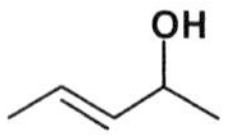	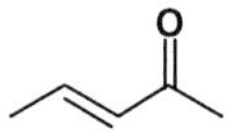	79%
4	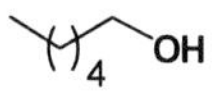 ()$_4$ OH	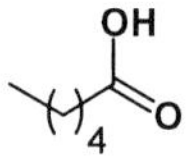 ()$_4$ (OH)C=O	80%
5	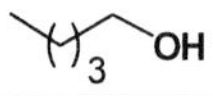 ()$_3$ OH	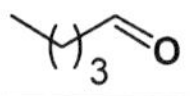 ()$_3$ CHO	90% [a)]

a) 4 equivalents of TEMPO were added

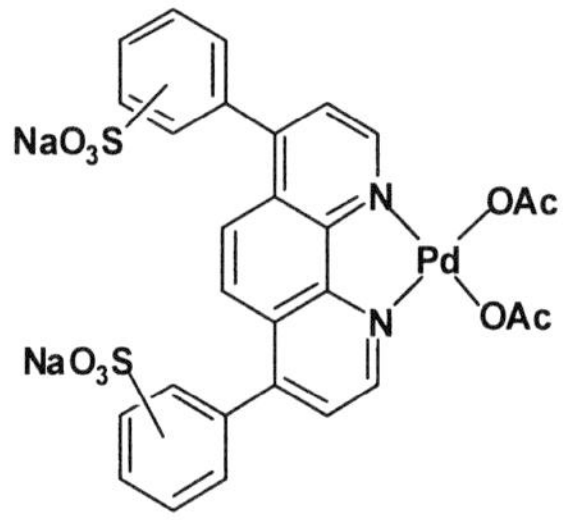

17

An important breakthrough in the Pd-catalyzed aerobic oxidation of alcohols was disclosed by Stoltz [13] and Sigman [14] simultaneously. Both research groups found that, in the presence of (–)-sparteine (4 equiv. per Pd), kinetic resolution of a range of benzylic (and one allylic) alcohols took place, affording the unconsumed starting material **22** in high enantiomeric purity (Tab. 3).

The X-ray structure analysis of the Pd-(–)-sparteine complex was obtained, but this organometallic reagent proved to be inert under the reaction conditions. The addition of excess (–)-sparteine was required for catalysis to occur. Desymmetrization of meso diols also proceeded with good *ee*'s (Fig. 7).

Tab. 3 Palladium-catalyzed aerobic kinetic resolution (Stoltz/Sigman)

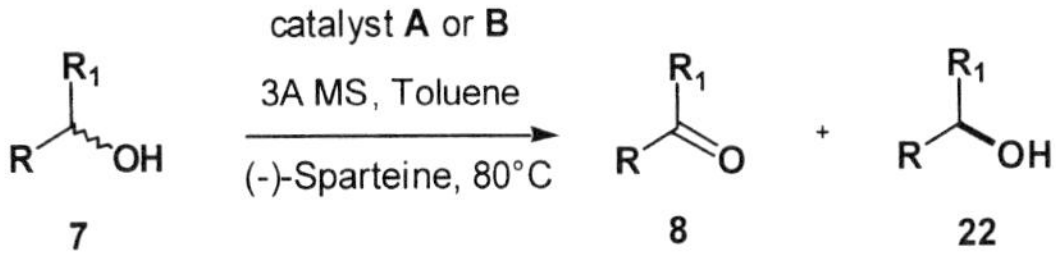

catalyst **A** or **B**, 3A MS, Toluene, (-)-Sparteine, 80°C: **7** → **8** + **22**

A = $PdCl_2$(nbd); **B** = $Pd(OAc)_2$ or $PdCl_2(MeCN)_2$

Entry	Condition	Product	Conversion	ee
1	A	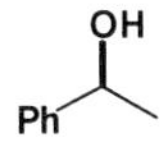 1-phenylethanol	60%	98.7%
2	B		66%	98.2%
3	A	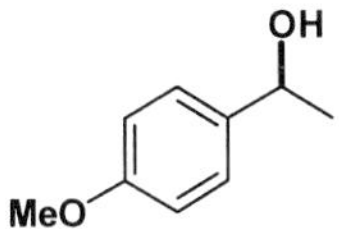 1-(4-methoxyphenyl)ethanol	66.6%	98.1%
4	B		67.2%	99%
5	A	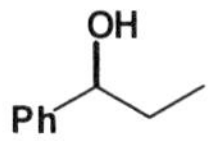 1-phenylpropanol	59.3%	98%
6	B		57.2%	88.5%
7	A	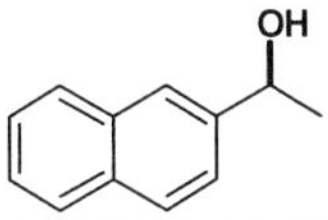 1-(2-naphthyl)ethanol	55.2%	99%
8	B		65.7%	96%

In order to overcome some of the stringent limitations pertaining to the use of palladium salts in the aerobic oxidation of alcohols, Kaneda investigated the utilization of heterogeneous Pd catalysts. He discovered that the cluster $Pd_4Phen_2(CO)(OAc)_4$, in the presence of small amounts of acetic acid, smoothly effected the transformation of a number of primary allylic alcohols into the corresponding enals in good yield (Tab. 4) [15].

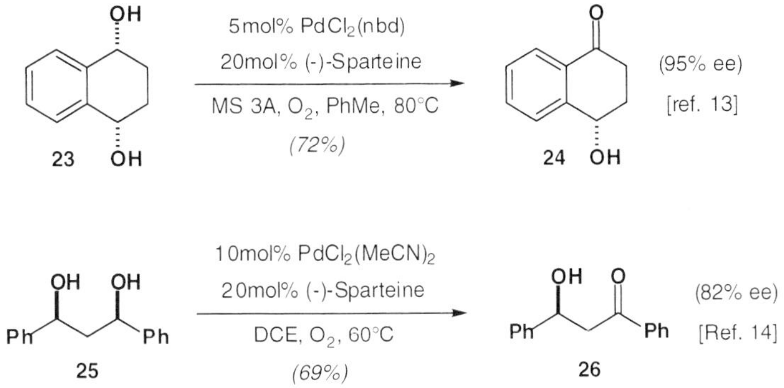

23 → **24**: 5mol% $PdCl_2$(nbd), 20mol% (-)-Sparteine, MS 3A, O_2, PhMe, 80°C (72%); (95% ee) [ref. 13]

25 → **26**: 10mol% $PdCl_2(MeCN)_2$, 20mol% (-)-Sparteine, DCE, O_2, 60°C (69%); (82% ee) [Ref. 14]

Fig. 7

Tab. 4 Palladium-cluster-catalyzed aerobic oxidation (Kaneda)

R, R_1, OH **7** → 4 mol% $Pd_4Phen_2(CO)(OAc)_4$ / C_6H_6, AcOH, O_2, 50°C → R, R_1, O **8**

Entry	*Substrate*	*Product*	*Yield*
1	Ph, OH	Ph, O	93%
2	Ph, OH	Ph, O	89%
3	Ph, Ph, OH	Ph, Ph, O	98% [a)]
4	OH	O	91%
5	OH	O	79% [a)]
6	OH	O	92%

a) Only 83% conversion in this case.

No reaction was observed with aliphatic alcohols, and weak activity was noticed for secondary allylic and primary and secondary benzylic substrates. The giant Pd cluster $Pd_{561}Phen_{60}(OAc)_{180}$ displayed similar activity [16]. An advantage of these heterogeneous systems is their ease of recycling, though activity was gradually lost over time.

In a similar manner, Uemura reported that hydrotalcite, $Mg_6Al_2(OH)_{16}CO_3 \cdot 4H_2O$, was a good support for Pd(II) salts [17]. In the presence of 5 mol% of this heterogeneous catalyst and variable amounts of pyridine, the aerobic oxidation of a variety of alcohols occurred smoothly, affording the corresponding carbonyl derivatives in good yields (Tab. 5).

In the case of geraniol and nerol, up to 5 equivalents of pyridine are required. The catalyst can be recycled, but the activity declines sharply after the second run (run 1: 98%, run 2: 93%, run 3: 77%). It is interesting to note that some diols can be mono-oxidized with good selectivity using this protocol. Unfortunately, the limitations of this heterogeneous system are similar to those observed in the case of $Pd(OAc)_2$. Finally, Uozumi recently reported the use of amphiphilic resin dispersion of Pd nanoparticles in the aerobic oxidation of alcohols in water [18]. Apart from primary aliphatic alcohols, which are directly converted to carboxylic

Tab. 5 Pd(II)-hydrotalcite-catalyzed aerobic oxidation (Uemura)

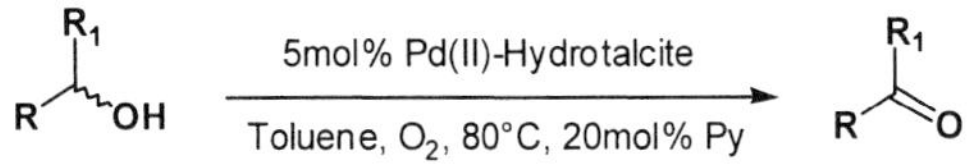

7 → 8 (5mol% Pd(II)-Hydrotalcite; Toluene, O_2, 80°C, 20mol% Py)

Entry	Substrate	Product	Yield
1	 Ph–OH	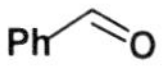 Ph–O	>99%
2	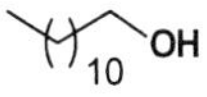 OH (10)	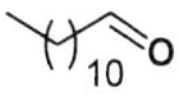 O (10)	86%
3	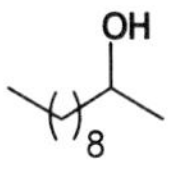 OH (8)	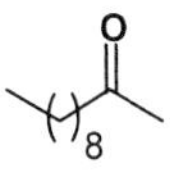 O (8)	93% [a)]
4	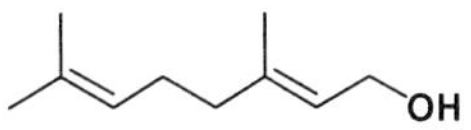 OH	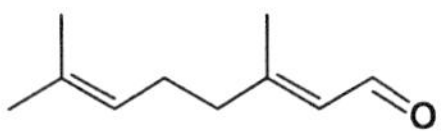 O	91% [b)]

a) 1 equivalent of pyridine employed.
b) 5 Equivalents of pyridine employed.

acids, benzylic and secondary aliphatic substrates give the desired carbonyl derivatives in good yields. In this case, recycling of the catalyst occurs without loss of activity.

Cobalt-based catalysts have also enjoyed wide popularity. In 1981, Tovrog, Diamond, and Mares [19] reported the oxidation of benzylic and secondary alcohols to the corresponding aldehydes and ketones using catalytic pyCo(saloph)NO_2 or pyCo(TPP)NO_2 in the presence of $BF_3 \cdot Et_2O$ or $LiPF_6$. The Lewis acid is crucial. No reaction is observed in its absence, and H-bonding solvents are required for catalytic activity (no reaction in benzene). Later, Iqbal [20] showed that the Co-Schiff base complex **27** oxidized a range of alcohols to the corresponding carbonyl derivatives in the presence of 2 equiv. of 2-oxocyclopentanecarboxylate **28**. The yields are usually moderate but the oxidation could be highly chemoselective (Fig. 8).

Ishii and co-workers found that the combination *N*-hydroxy-phthalimide/$Co(acac)_3/O_2$ was an efficient system for the production of ketones from secondary alcohols and acids from primary hydroxyl compounds [21]. Conversions are usually good, but the catalyst does not tolerate many functional groups (e.g., double bonds are cleaved). During subsequent studies on this system, Ishii discovered that the addition of small quantities of organic acids led to a significant improvement in the yield and rate of oxidation (Tab. 6) [22].

Benzylic and secondary aliphatic alcohols are good substrates, but primary aliphatic alcohols are directly oxidized to the corresponding carboxylic acids. Moreover, in the case of some allylic derivatives, moderate yields are obtained because

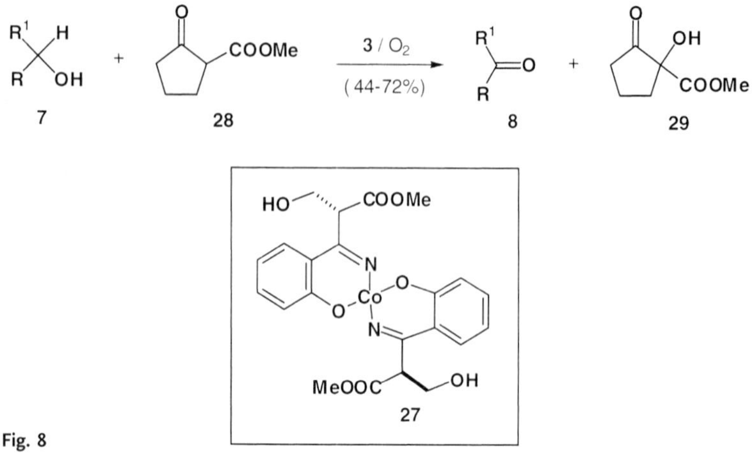

Fig. 8

Tab. 6 Cobalt-catalyzed aerobic oxidation (Ishii)

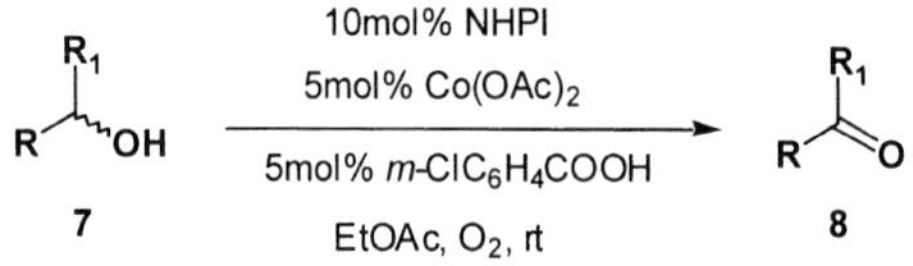

Entry	Substrate	Product	Conversion	Yield
1	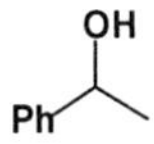	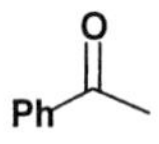	100%	98%
2	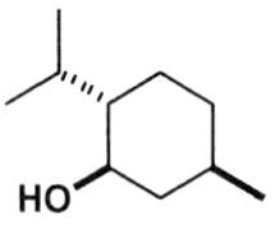	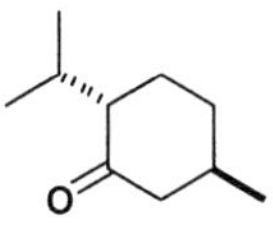	59%	47%
3	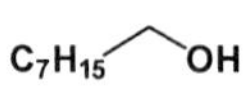 C_7H_{15} OH	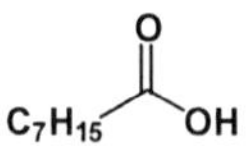 C_7H_{15} COOH	79%	78%
4	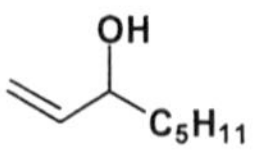	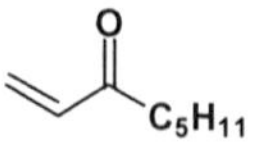	79%	67%
5	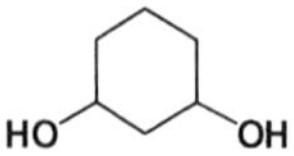	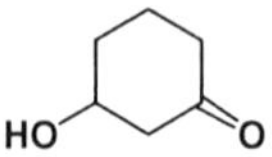	80%	80%
6	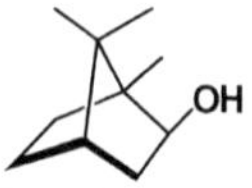	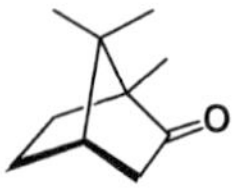	91%	91%

of the competing addition of some radical intermediate onto the C-C double bond. The proposed mechanism is illustrated in Fig. 9.

The role of the added organic acid is not clearly understood, but it appears to involve its coordination to the cobalt catalyst, generating a complex, which rapidly decomposes peroxide **33**.

An interesting bimetallic oxidant based upon Os/Cr was shown by Sharpley to oxidize alcohols in the presence of oxygen. Unfortunately, the conversions are very poor [23]. In this context, it is worth mentioning the report of Neumann and Levin, who employed a supported Mo/V heteropolyanion salt to oxidize alcohols and amines to aldehydes and imines, respectively [24]. The process is, however, severely limited to benzylic substrates. More recently, the elegant work of Bäckvall, who uses a combination of Co and Ru catalysts for the oxidation of some allylic and benzylic alcohols, is a notable contribution to this area of research [25]. In the context of bimetallic catalysis, Osborn reported that the combination OsO_4/CuCl generated a species capable of selectively oxidizing benzylic alcohols and some allylic ones. However, the yields are rather modest, and the catalyst appears to be particularly sensitive to steric hindrance [26]. Aliphatic substrates barely react under these conditions. Subsequently, Brown described a modification of the Osborn protocol in which the complex $OsO_4 \cdot$ quinuclidine, in conjunction with $Cu(II)_2$-ethylhexanoate and ethyl allylether, was employed to catalyze the aerobic oxidation of a variety of benzylic and allylic alcohols (Tab. 7) [27].

Unfortunately, aliphatic substrates are essentially inert under these conditions. Quinuclidine is an important component, and its absence leads to a 10-fold de-

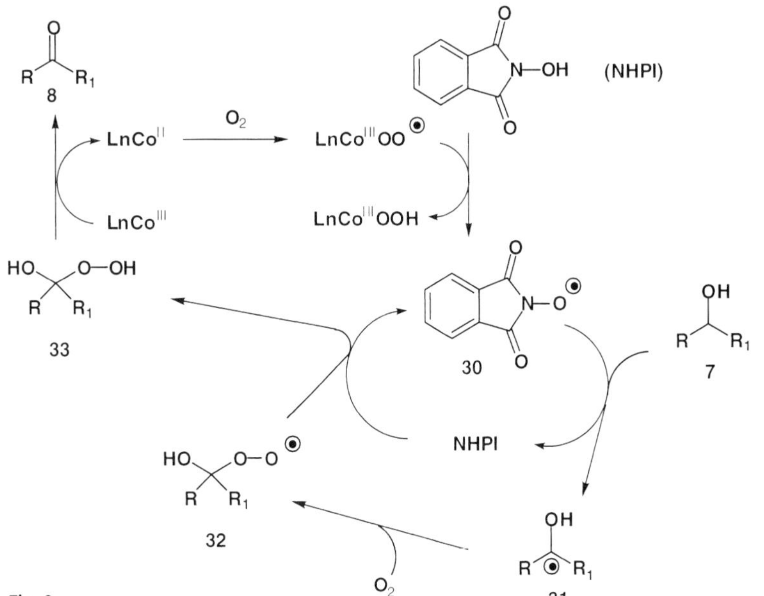

Fig. 9

Tab. 7 Osmium-catalyzed aerobic oxidation (Brown)

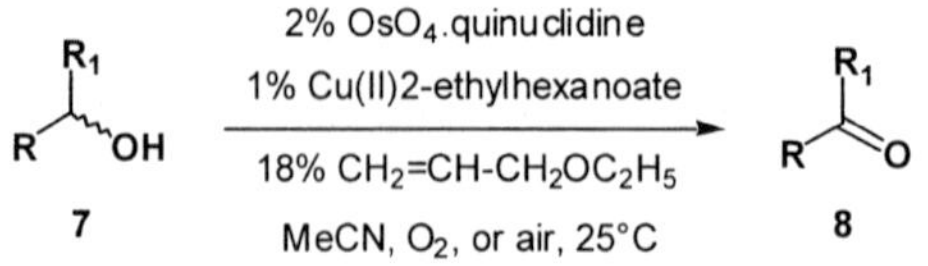

Entry	Substrate	Product	Yield
1	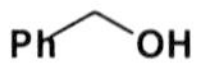	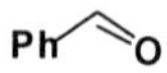	98%
2	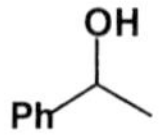	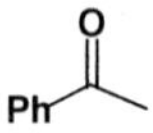	97%
3	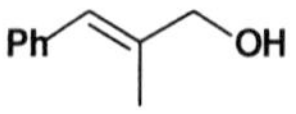	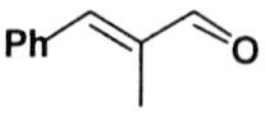	97%
4	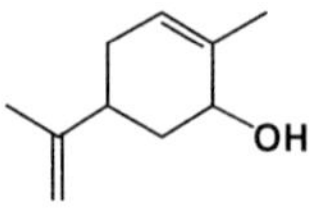	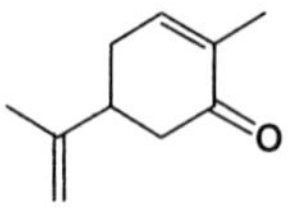	98%
5	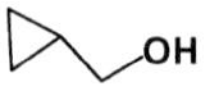	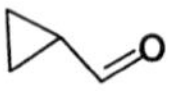	15%

crease in reaction rate. Remarkably, no dihydroxylation of C-C double bonds is observed using this catalytic system. However, it is important to note that 1,2- and 1,3-diols strongly inhibit the reaction. The proposed mechanism is described in Fig. 10.

The oxidation of aldehydes to acids using Ni catalysts was reported by Mukaiyama [28]. The reaction presumably proceeds via the Ni-peracyl derivative. Such a combination of O_2/aldehyde/metal catalyst was subsequently employed by this author and many others to effect asymmetric epoxidation and Baeyer-Villiger reactions. Finally, the use of ruthenium catalysts has also been investigated in some depth. Tang has reported that $RuCl_3$ catalyzed the aerobic oxidation of secondary

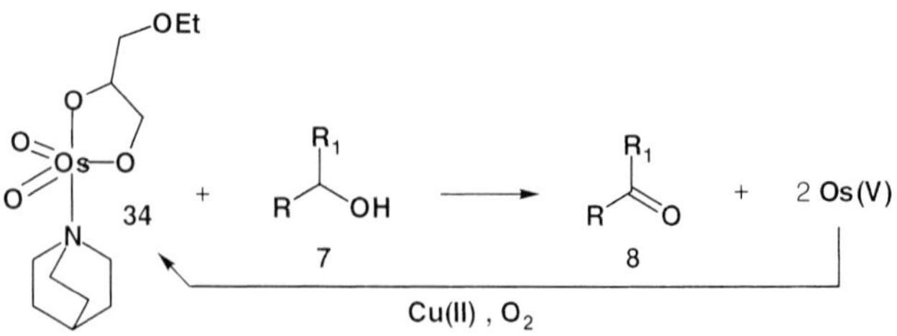

Fig. 10

alcohols to ketones, although in modest yield [29], and Matsumoto has shown that RuO_2 hydrate conveniently transforms allylic alcohols into enals and enones and thus could serve as a useful replacement for MnO_2 [30].

Perhaps the most successful ruthenium-based systems described so far are the trinuclear complexes reported by Drago. These catalysts oxidize a variety of alcohols into aldehydes and ketones under 40 psi pressure of O_2 [31]. More recently, Chang described the aerobic oxidation of a variety of alcohols in the presence of 3 mol% of $[RuCl_2(p\text{-cymene})]_2$ at 100 °C in toluene [32]. Excellent yields are obtained in the transformation of benzylic and allylic substrates, and good conversions are realized in the case of secondary aliphatic alcohols. However, in this last case, up to 13 mol% of catalyst are required, in addition to 1 equiv. of Cs_2CO_3, to reach a good yield of ketone. No example of the oxidation of primary aliphatic alcohol has been reported.

Katsuki, employing the chiral ruthenium-salen complex **35**, has shown that the kinetic resolution of secondary allylic, benzylic, and propargylic alcohols could be efficiently carried out in the presence of air and light (Fig. 11) [33].

After 60–65% conversion, the recovered starting material displayed remarkably high levels of enantioselectivity. Irradiation by fluorescent light is a prerequisite to activate catalyst **35**, but the oxidation does not appear to involve a Ru=O species since no competing epoxidation is observed.

During the course of his work on the aerobic oxidation of alcohols catalyzed by ruthenium complexes, Sheldon observed that the addition of TEMPO remarkably altered not only the rate of these transformations but also the scope of the oxida-

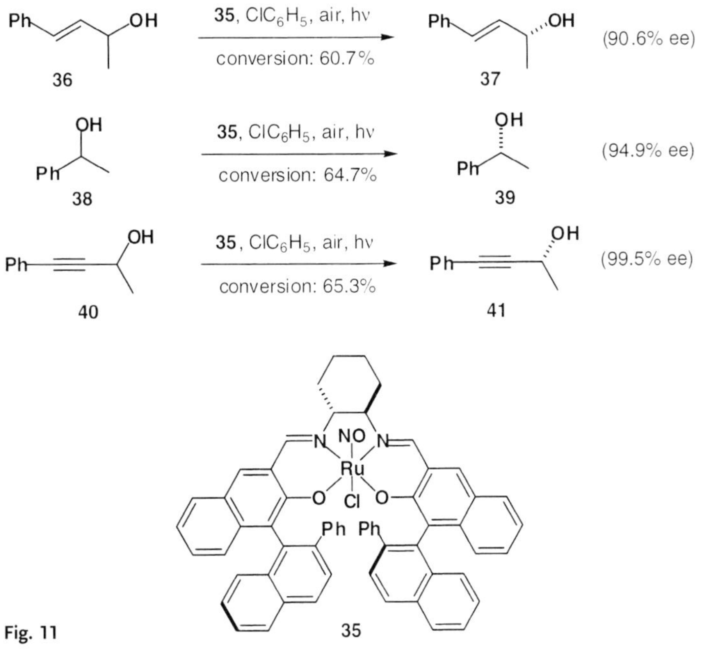

Fig. 11

Tab. 8 Ruthenium-catalyzed aerobic oxidation (Sheldon)

R(R$_1$)CH–OH **7** → (1 mol% $RuCl_2(PPh_3)_3$, 3 mol% TEMPO, ClC_6H_5, 100 °C, O_2, 10 bar) → R(R$_1$)C=O **8**

Entry	*Substrate*	*Product*	*Yield*
1	OH, ( )$_4$	O, ( )$_4$	98%
2	( )$_6$ OH	( )$_6$ O	97%
3	Ph OH	Ph O	97%
4	OH	O	98%

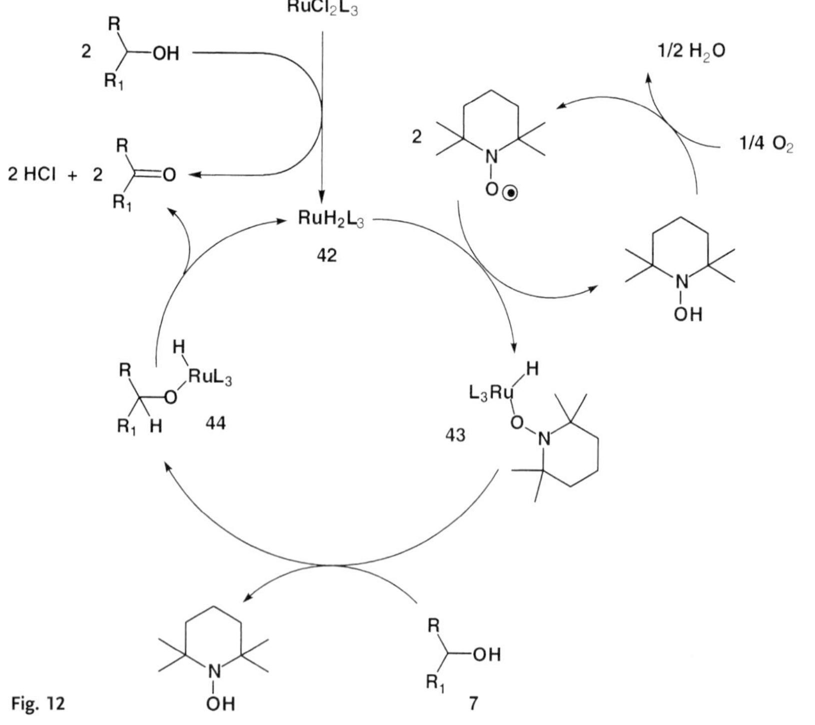

Fig. 12

Tab. 9 Ruthenium-alumina catalyzed aerobic oxidation (Mizuno)

$R(R_1)CH$–OH (**7**) $\xrightarrow[\text{PhCF}_3,\ 356°\text{K},\ 1\ \text{atm}\ O_2]{2.5-5\text{mol\%}\ \text{Ru/Al}_2\text{O}_3}$ $R(R_1)C$=O (**8**)

Entry	*Substrate*	*Product*	*Conversion*
1	Ph, OH	Ph, O	>99%
2	OH, Ph	O, Ph	>99%
3	OH	O	84%
4	6, OH	6, O	87%[a)]
5	OH	O	90%
6	S, OH	S, O	>99%
7	N, OH	N, O	93%[b)]

a) 5 mol% Ru/Al_2O_3 + 5 mol% hydroquinone.
b) 5 mol% Ru/Al_2O_3.

tions [34]. In the presence of 1 mol% $RuCl_2(PPh_3)_3$ and 3 mol% TEMPO in chlorobenzene at 100 °C and 10 bar pressure, a variety of alcohols could be efficiently converted into the corresponding carbonyl derivatives in excellent yields (Tab. 8).

The Ru-TEMPO catalyst displays some preference for primary alcohols over secondary ones, and selective oxidations are sometimes possible. Unfortunately, this system is inhibited by the presence of coordinating functions such as sulfides, amines, ethers, and acids in the substrate. The mechanism has been thoroughly studied and has revealed the key role of ruthenium as the oxidant. TEMPO acts as a hydrogen acceptor, which is continuously recycled by oxygen (Fig. 12).

A similar system employing copper instead of ruthenium has also been described, but the scope and limitations are similar [35]. Finally, immobilized TEMPO can be employed in these oxidations, and recycling of the catalyst is possible [36].

The use of heterogeneous ruthenium-based catalysts for the aerobic oxidation of alcohols has been studied for a number of years. Kaneda recently reported that hydrotalcites, containing ruthenium incorporated in their cationic brucite layer, oxidize allylic and benzylic alcohols in the presence of oxygen [37]. The yields are generally good, and the catalytic system can be recycled several times. Mizuno described the use of Ru/Al_2O_3 for the same reaction [38]. In the presence of 2.5 mol% of the catalyst, a variety of alcohols could be smoothly and efficiently converted into the corresponding carbonyl derivatives (Tab. 9).

In contrast to the previously mentioned system, this heterogeneous oxidant is equally competent for the transformation of aliphatic, allylic, and benzylic substrates. Primary alcohols are oxidized faster than secondary ones and no radical intermediates are involved in this reaction.

2.13.3 Copper-Based Aerobic Oxidations

The reaction of oxygen with Cu(I) and Cu(II) complexes has been thoroughly investigated, especially with regard to the understanding of the biological mode of action of hemocyanins, a widespread class of oxygen-carrying enzymes present in molluscs and arthropods [39]. The kinetics of oxygen binding to dinuclear copper complexes and the mechanism of subsequent reactions of the initially generated peroxy dicopper species have been studied in depth.

Several copper-oxygen complexes have been isolated, and X-ray diffraction analyses have revealed that dioxygen binds to the dinuclear copper system either in an h2 fashion [40] or as a μ-peroxide [41]. However, the ability of copper complexes to oxidize alcohols to carbonyl compounds has not received the same attention. Rivière and Jallabert [42] were probably the first to report that a CuCl · amine (Phen or bipy) complex, in the presence of excess base (K_2CO_3) in benzene under reflux and under a stream of O_2, was able to convert benzyl alcohol into benzaldehyde (Tab. 10). Unfortunately, two equivalents of CuCl · Phen were required to obtain a good yield of the aldehyde. The reaction was also strongly limited to benzylic alcohols. Indeed, aliphatic and allylic alcohols gave poor yields of aldehydes or ketones and β-phenethylalcohol only afforded benzaldehyde, resulting from C-C bond cleavage.

Subsequently, Semmelback reported that catalytic amounts of CuCl (10 mol%) in conjunction with TEMPO (10 mol%) and molecular oxygen efficiently oxidized a variety of primary alcohols to aldehydes [43]. A base is necessary to remove the HCl formed, and CaH_2 was typically used. Optimized conditions employ 20 mol% TEMPO, 22 mol% CuCl and 300 mol% CaH_2 (Tab. 11).

Since this reaction proceeds very poorly with secondary alcohols, the chemoselective discrimination between a primary and a secondary alcohol can be efficiently realized. Thus, diol **45** affords a 19:1 mixture of aldehyde **46** and ketone **47** (Fig. 13).

Subsequent attempts to improve the synthetic utility of the Cu/O_2 system came from Maumy and Capdevielle [44]. They investigated the influence of ligand, solvent, catalyst, temperature, and substrate on the rate of oxidation. In the oxidation

Tab. 10 Riviere and Jallabert type aerobic oxidations

$$\underset{\mathbf{7}}{R\text{–}CH(R_1)\text{–}OH} \xrightarrow[\text{2 eqs } K_2CO_3,\ O_2,\ 80\,^\circ C]{\text{2 eqs CuCl.Phen, } C_6H_6} \underset{\mathbf{8}}{R\text{–}C(R_1)\text{=}O}$$

Entry	*Substrate*	*Product*	*Yield*
1	$PhCH_2OH$	$PhCHO$	86%
3	$PhCH{=}CHCH_2OH$	$PhCH{=}CHCHO$	83%
4	$C_6H_{13}CH(Et)OH$	$C_6H_{13}C(Et){=}O$	18%
5	$C_7H_{15}CH_2OH$	$C_7H_{15}CHO$	22%

Tab. 11 Aerobic oxidations using TEMPO and CuCl

$$\underset{\mathbf{7}}{R\text{–}CH(R_1)\text{–}OH} \xrightarrow[\text{20 mol\% TEMPO, 3 eqs } CaH_2]{\text{22 mol\% CuCl}} \underset{\mathbf{8}}{R\text{–}C(R_1)\text{=}O}$$

Entry	*Substrate*	*Product*	*Yield*
1	pMeOPhCH_2OH	pMeOPhCHO	96%
3	$PhCH_2OH$	$PhCHO$	94%
4	$PhCH{=}CHCH_2OH$	$PhCH{=}CHCHO$	93%

of benzhydrol, they found that the best ligand was 2,2′-bipyridyl (100 mol%) in conjunction with CuCl (10 mol%) in MeCN at 60 °C for 24 h. Complete conversion was observed, and benzophenone was isolated in up to 99% yield. Again, the experimental procedure is limited to activated benzylic alcohols, 1-hexanol giving only 15% conversion after 7 h and hexane-2-ol 10% conversion after the same

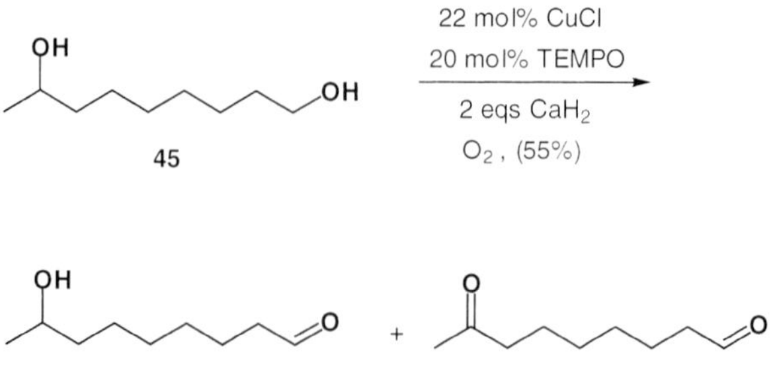

Fig. 13

time. The mechanism of the reaction was also investigated in some detail, and it was shown that copper alkoxides were intermediates (Fig. 14).

Reaction of *in situ*-generated copper alkoxides **48** with molecular oxygen affords the μ-peroxodicopper derivative **49**, which undergoes fragmentation to the Cu(III)-oxo species **50**. β-Hydrogen elimination generates the carbonyl compounds and CuOH. Independently prepared copper alkoxide **48** followed the same dehydrogenation route. It is noteworthy that β-scission is a significant side reaction when stabilized radicals could be generated, as in the case of β-phenethylalcohol and α-ketols, which led solely to carboxylic acids.

In a recent study, Sawyer demonstrated that a combination of bis(dipyridyl)copper(II) salts and 2 equiv. of base in acetonitrile (1 atm of O_2) dehydrogenated benzylic and allylic alcohols to the corresponding carbonyl compounds [45]. Water, generated during the reaction, deactivates the catalyst by reducing Cu(II) to the Cu(I) state. Again, aliphatic primary and secondary alcohols are poor substrates for this interesting oxidation system.

Chaudhury reported the use of the copper complex **54** for the oxidation of benzyl alcohol and ethanol under aerobic conditions (Fig. 15) [46].

No reaction is observed when methanol or isopropanol are employed as substrate. The postulated mechanism appears to be similar to that of galactose oxidase and proceeds by single-electron transfer.

More recently, Knochel employed the fluorinated bipyridyl ligand **55** and $CuBr \cdot Me_2S$ in the presence of TEMPO to perform the aerobic oxidation of a vari-

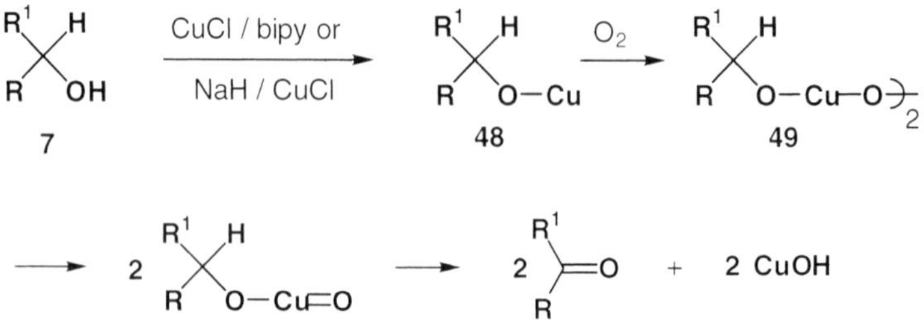

Fig. 14

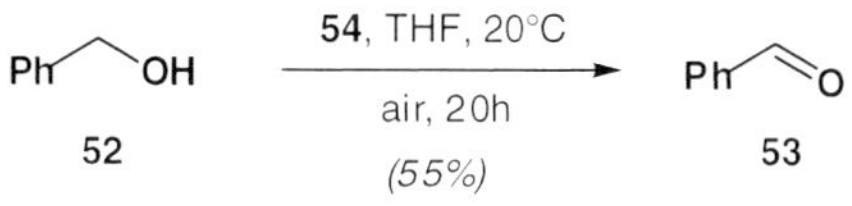

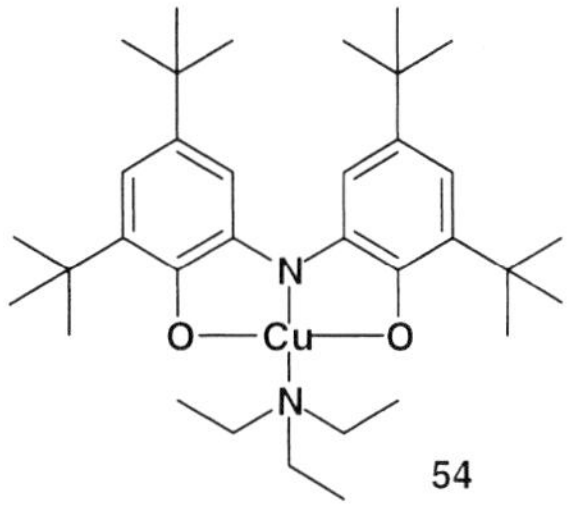

Fig. 15

ety of alcohols using the fluorophase principle [47]. By simple separation of the two liquid layers, the catalyst can easily be recovered and recycled. After 8 runs, the yield of *p*-nitro benzaldehyde averages 86% (Tab. 12).

It is noteworthy that this catalyst appears to be particularly sensitive to steric hindrance, as shown by the poor yields obtained in Entry 6.

Our fascination for the Rivière and Jallabert procedure prompted us to reinvestigate this system and to modify various parameters in the hope of achieving good catalyst turnover and establishing a useful and efficient aerobic protocol for the oxidation of all classes of alcohols into carbonyl derivatives.

Our initial experiments were performed on *p*-chlorobenzyl alcohol and employed two equivalents of CuClr · Phen. It was rather disappointing to find that, aside from NaOAc, all the other bases tested were far less efficient than K_2CO_3 [48]. However, during the course of these optimization studies, a dramatic influence of the solvent on the reaction rate was uncovered. For example, a 3- to 4-fold acceleration was obtained when toluene was substituted for benzene. In contrast, replacing benzene with *m*- or *p*-xylene resulted in a decrease in the rate of the reaction. Although it is difficult to offer a rational explanation for the profound effect displayed by minute changes in the structure of the solvent, it is quite reasonable to assume that the coordinating properties of these aromatic solvents may significantly alter the stability and reactivity of the copper complexes [49]. Finally, it was also discovered that molecular oxygen could be replaced by air, a more readily available and inexpensive stoichiometric oxidant [50].

But the real breakthrough was achieved when it was decided to lower the amount of the catalyst (Fig. 16). Under the original Rivière and Jallabert conditions (2 equiv. CuClr · Phen, benzene), any attempt at decreasing the concentration of the catalyst resulted in a disastrous curtailment in the reaction conversion. However, in toluene, reducing the quantity of the CuClr · Phen complex did not impair the oxidation of the benzylic alcohol. Although the reaction took longer to reach completion, quantitative formation of *p*-chlorobenzaldehyde could be accomplished using as little as 0.05 equiv. of the catalyst.

Tab. 12 Copper-catalyzed aerobic oxidation (Knochel)

R_1, R, OH — **7** → (2mol% **55**, 2mol% $CuBr.Me_2S$; 3.5mol% TEMPO, O_2, 90°C; C_8F_{18}, C_6H_5Cl) → R_1, R, O — **8**

Entry	*Substrate*	*Product*	*Yield*
1	O_2N, OH	O_2N, O	93%
2	Ph, OH	Ph, O	79%
3	OH	O	76%
4	8, OH	8, O	73%
5	OH, 8	O, 8	71%
6	OH, 8	O, 8	31%

C_8F_{17} ( )4 — 4( ) C_8F_{17}, N, N — **55**

Conversion vs Amount of CuCl.phen

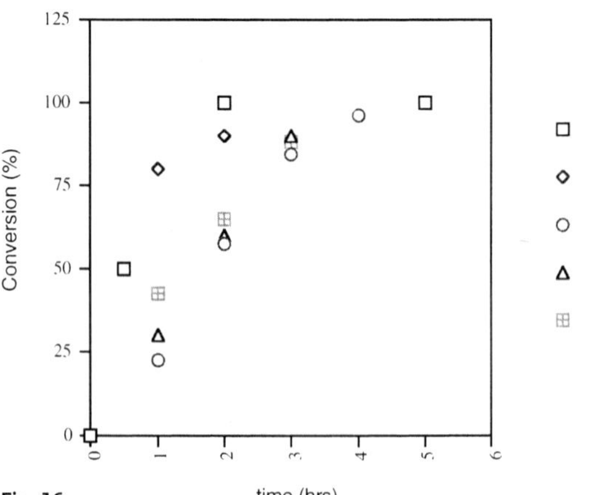

Fig. 16

Unfortunately, this initial catalytic system proved to be (among other things) severely restricted to benzylic alcohols. Based upon previous work in the biochemistry of hemocyanins and tyrosinases [39], a reasonable mechanism for this aerobic oxidation could be envisioned, in which the μ_2-peroxide **59** occupies a cardinal position (Fig. 29). This intermediate **59** can be formed by two different pathways, either (1) by the displacement of the chloride ion in complex **56** by the alcohol nucleophile [51] followed by dimerization in the presence of O_2, or (2) by the initial formation of a chloro bis-copper peroxide **57** followed by the exchange of the chloride substituent for the alcohol ligand. The loaded μ_2-peroxide **59** can then undergo homolytic cleavage of the labile O-O bond and generate the reactive species **60**. Intramolecular hydrogen abstraction leads to the copper-bound carbonyl derivative **61** with concomitant reduction of Cu(II) [or Cu(III)] to Cu(I). Finally, ligand exchange with the starting alcohol and release of H_2O completes the catalytic cycle (Fig. 17).

Such a simple mechanistic proposal accommodated the observation that highly activated, benzylic alcohols were good substrates, because of the enhanced lability of their α-hydrogen atoms. In contrast, aliphatic alcohols are far less reactive toward H-radical abstraction and, accordingly, poor conversions should ensue. How-

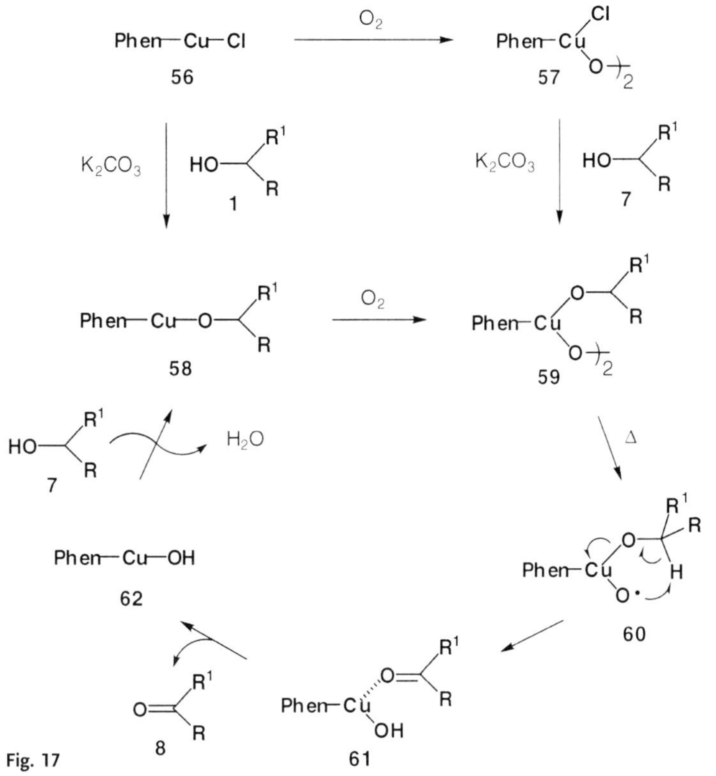

Fig. 17

ever, it was rather disturbing to note that allylic alcohols, such as geraniol and nerol, displayed poor reactivity in this system.

Furthermore, it was observed that the aerobic oxidation of aliphatic alcohols invariably resulted in the rapid formation of a green copper(II) salt, with concomitant deactivation of the catalyst. This observation strongly suggested that the regeneration of the active copper(I) species was an urgent requirement in the oxidation of aliphatic alcohols. It was therefore decided to test the effect of various reductants in this aerobic oxidation reaction. Naturally, we turned to the hydrazine family of reducing agents (Tab. 13) [52].

Remarkably, addition of hydrazine or *N,N*-dimethylhydrazine (20 mol%) to the reaction mixture resulted in a significant enhancement in the rate of the oxidation reaction. The presence of electron-withdrawing groups on the hydrazine led to an even more dramatic improvement in both yield and reaction rate, the oxidation of **63** being virtually complete within 15 min using DEAD-H_2 (Tab. 13, Entry 3). Although the efficiency of the hydrazine additive depended to a small extent on steric hindrance, it was largely affected by electronic factors. For example, whereas a small methyl ester substituent proved less efficient than the bulkier ethyl group, a more sterically demanding isopropyl ester only slightly reduced the rate of the reaction, complete conversion being observed in 30 min (Tab. 13, Entries 2, 3 and 4). More importantly, if the ester substituent was replaced by an acyl function, such as acetyl or benzoyl, virtually no oxidation took place, regardless of the s-*cis* or s-*trans* conformation of the acyl group (Tab. 13, Entries 5–7). Having found that optimum conversions could be achieved using as little as 25 mol% of DEAD-H_2, we then applied these conditions to the oxidation of a range of representative alcohols. Some pertinent results are collected in Tab. 14.

Tab. 13 Effect of the hydrazine additives

63 → 64 (5 mol% CuCl.Phen, toluene; H_2O; 90°C, 200 mol% K_2CO_3, 25 mol% **additive**)

Entry	*Additive*	*Conversion (%)* [a] 15 min	*Conversion (%)* [a] 30 min
1	Me_2NNH_2	10	39
2	$(MeO_2CNH\text{-})_2$	31	56
3	**$(EtO_2CNH\text{-})_2$ (DEAD-H_2)**	**98**	**>99**
4	$(^iPrO_2CNH\text{-})_2$ (DIAD-H_2)	70	99
5	$(MeCONH\text{-})_2$	5	5
6	$(PhCONH\text{-})_2$	<1	<1
7	phthalhydrazide	<1	<1

a) The conversions were determined by ^1H NMR spectroscopy and/or capillary GC analysis.

Tab. 14 Copper-catalyzed aerobic oxidation of alcohols using DEAD-H_2

R^1RCH–OH **7** → (5 mol% CuCl.Phen; toluene; 90 °C / 25 mol% DEAD-H_2; 2 eqs K_2CO_3; O_2) → R^1RC=O **8**

Entry	*Substrate*	*Product*	*Conversion* [a]
1	2-thienylmethanol (S, OH)	thiophene-2-carbaldehyde (S, CHO)	100%
2	Ph–CH(Me)–OH	Ph–C(Me)=O	100%
3	Ph–CH(CF_3)–OH	Ph–C(CF_3)=O	100%
4	Ph–CH=CH–CH_2OH	Ph–CH=CH–CHO	100%
5	geraniol	geranial (CHO)	80% [b]
6	Ph–CO–CH(OH)–Ph	Ph–CO–CO–Ph	100%
7	^{t}Bu–cyclohexyl–OH	^{t}Bu–cyclohexanone (=O)	40% [c]
8	C_9H_{19}–CH(OH)–CH_3	C_9H_{19}–CO–CH_3	44% [c]

a) The conversions were determined by ^1H NMR spectroscopy and/or capillary GC analysis.
b) Neral was not detected in this reaction.
c) 20 mol% CuCl · Phen was employed in this reaction.

As can be seen from Tab. 14, both benzylic and allylic alcohols underwent smooth and quantitative transformation into the corresponding aldehyde or ketone within 1–4 h. It is noteworthy that the catalyst tolerates sulfur heterocycles. The stereochemical integrity of the C-C double bond of the starting allylic alcohols is also retained in the final products, with geraniol giving solely geranial (Tab. 14, Entry 5). Remarkably, trifluoromethyl alcohols and α-ketols are excellent substrates, affording the corresponding trifluoromethyl ketone and α-diketone, respectively, in high yield (Tab. 14, Entries 3 and 6). Interestingly, the corresponding azo dicarboxylate (DEAD) could be replaced by the hydrazide derivative (DEAD-H_2) with equal efficiency [53]. Unfortunately, both primary and secondary aliphatic alcohols proved to be poor substrates, and only modest conversions could be achieved under these conditions, even when a larger amount of the CuCl · Phen

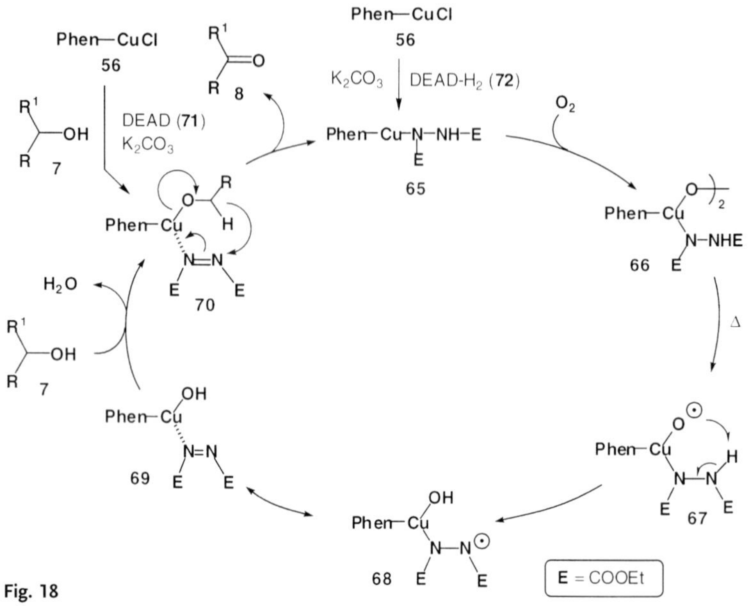

Fig. 18

catalyst was employed (Tab. 14, Entries 7 and 8). A plausible mechanism, involving both the azo and hydrazide derivatives, can be formulated as shown in Fig. 18.

In the presence of DEAD-H_2 (**72**) and base (K_2CO_3), displacement of the chloride ligand by the hydrazide nucleophile takes place, affording the hydrazino copper(I) complex **65** [54]. A rapid reaction with oxygen then ensues, leading to the –$_2$-peroxo-bis-copper(II) derivative (**66**). It is believed that, upon heating, this complex undergoes homolytic cleavage of the labile O-O peroxidic bond, resulting in the generation of the copper-alkoxy radical **67**. Intramolecular hydrogen atom abstraction ensues, affording the capto-datively stabilized nitrogen-centered radical **68** [55], which is none other than the azo-substituted copper(I) hydroxyl species **69**. This particular sequence is thus responsible for the reduction of the copper(II) species to the catalytically active copper(I) complex. Ligand exchange with the alcohol and concomitant release of H_2O then results in the formation of the ternary loaded catalyst **70**. In this complex, both the alcohol and azo-substituents are held together in close proximity by coordination to the copper center. An intramolecular hydrogen shift, akin to the Meerwein-Pondorff-Verley-Oppenauer reaction [56], then takes place, affording transiently a carbonyl-bound hydrazido-copper derivative [57].

The aldehyde or ketone can now desorb, leading to the initial copper(I) hydrazide complex **65**, which re-enters the catalytic cycle. The replacement of DEAD-H_2 (**72**) by DEAD (**71**) can be easily understood when considering this catalytic cycle. Indeed, several entries to the main catalytic cycle are possible, either via the hydrazino copper species **65** or via the direct formation of the ternary loaded complex **70** from the azo-derivative **71**, Phen · CuCl (**56**) and the alcohol **7**. The key role played by the hydrazine or azo compounds can also be readily appreciated when considering the proposed mechanistic rationale. The hydrazide not only helps in reducing the copper(II) salt to the copper(I) state but, by virtue of its easy passage into the azo derivative, it also acts as a hydrogen acceptor, allowing the efficient oxidation of the alcohol into the carbonyl compound.

Moreover, we believe that the azo form helps in stabilizing several of the reactive copper complexes involved in this catalytic cycle, such as the hydroxy copper complex (**69**). Thus, we surmise that this novel catalytic, aerobic procedure for oxidizing alcohols into carbonyl derivatives proceeds via a dehydrogenation mechanism and relies on the effective role of hydrazine or azo compounds as hydrogen shuttles and stabilizing ligands for the various copper complexes [58].

Further evidence for the occurrence of this dehydrogenation mechanism can be gathered from the following experiments. The hydrazido-copper complex **65** can be prepared independently, by reacting Phen · CuCl (**56**) with the sodium salt of DEAD-H_2. Addition of an alcohol in the absence of O_2 results in no oxidation to the corresponding carbonyl compound. However, when oxygen is admitted into the reaction medium, rapid and quantitative conversion into the desired product is achieved. Moreover, combining an alcohol with Phen · CuCl and DEAD under anaerobic condition led to the rapid oxidation of the starting material and the simultaneous generation of equimolar amounts of DEAD-H_2. The proportion of aldehyde/ketone and hydrazine formed is equivalent to the quantity of starting azo derivative. Independent reaction of the alcohol with DEAD and K_2CO_3 in the absence of air and copper salts results in the quantitative recovery of the starting alcohol [59]. Although we have not yet been able to obtain direct evidence for some of the intermediates postulated in this catalytic scheme, we believe that the abovementioned experiments lend credence to the involvement of complex **65**, **66**, and **70**, and strongly support our proposed mechanism.

However, two major observations still need to be accounted for: the lack of reactivity of aliphatic substrates and the need for a 5-fold excess of DEAD or DEAD-H_2 over CuCl · Phen to achieve quantitative oxidations of benzylic and allylic alcohols. An interesting clue to these questions was provided when monitoring the fate of the reagents and products involved in the aerobic catalytic oxidation of undecanol to undecanal (Fig. 19) [60].

Whereas the decay of the alcohol follows the expected kinetic course, the formation of the aldehyde shows an abnormal behavior. In the early part of the reaction, the aldehyde formation matches almost perfectly the disappearance of the alcohol. However, after ca. 50% conversion, it reaches a maximum and then slowly begins to decrease. Clearly, a side reaction is consuming the alcohol substrate as soon as the aldehyde concentration attains a critical value. The fate of the DEAD-H_2 additive is even more interesting. In stark contrast to our expectation, the concentration of DEAD-H_2 does not remain constant throughout the course of the reaction but gradually decreases over time. The disappearance of DEAD-H_2 corresponds exactly to the point of maximum aldehyde formation. Thus, the destiny of the aldehyde and the DEAD-H_2 additive are intimately linked, and, as the hydra-

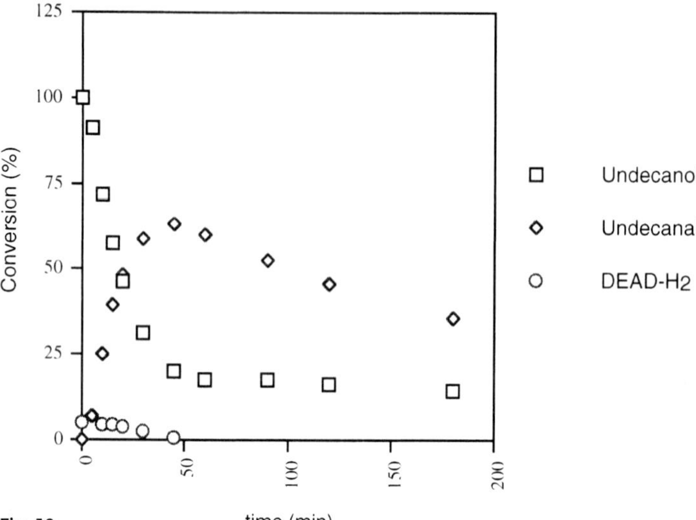

Fig. 19

zide is removed from the reaction mixture, the formation of the aldehyde simultaneously stops and its concentration decreases after the hydrazide has been totally consumed. A similar pattern is observed for the related DEAD compound, though in this case an initial and extremely rapid transformation of DEAD into DEAD-H_2 takes place. Only the hydrazide can be observed later in the reaction medium. Closer examination of the reaction by-products using stoichiometric DEAD under anaerobic conditions led to the isolation of an unexpectedly large quantity of the mixed carbonate **72** (Fig. 20).

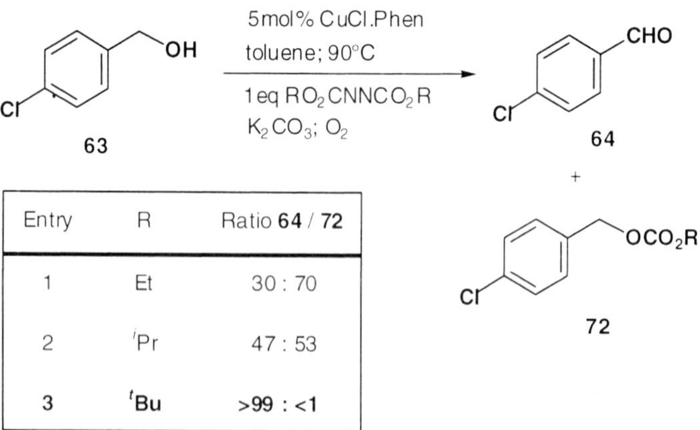

Entry	R	Ratio **64** / **72**
1	Et	30 : 70
2	^{i}Pr	47 : 53
3	**^{t}Bu**	**>99 : <1**

Fig. 20

Thus, the hydrazide or its azo analog not only plays a key role in the catalytic cycle as a hydrogen acceptor and a reductant for the copper catalyst, but it also acts as an acyl transfer reagent, generating competitively the undesired mixed carbonate **72**. This by-product presumably originates from the inter- or intra-molecular nucleophilic attack of the alcohol on either the copper hydrazide or azo complexes **65** or **70**, respectively, resulting ultimately in the deactivation of the catalyst. To minimize this undesired *trans*-acylation reaction, sterically demanding azo derivatives were tested (Fig. 20). While di-isopropyl azodicarboxylate (DIAD) produced a more favorable aldehyde/carbonate ratio (Fig. 20, Entry 2), we were gratified to find that the corresponding di-*tert*-butyl azo-dicarboxylate (DBAD) led solely to the formation of the desired oxidation product, with no trace of the mixed carbonate contaminant (Fig. 20, Entry 3).

Under these optimized conditions, the aerobic oxidation of alcohols can be efficiently achieved using as little as 5 mol% of the DBAD or DBAD-H_2 additives (Tab. 15).

Using this improved protocol, a variety of allylic, benzylic, and secondary alcohols are now smoothly oxidized to the corresponding carbonyl derivatives in high yield. Unfortunately, primary aliphatic alcohols still appear to be poor substrates, and a conversion of only 65% can be achieved before catalyst deactivation (Tab. 15, Entry 9).

Careful analysis of the reaction products revealed the absence of a mixed carbonate akin to **72**, even though gradual decomposition of DBAD was again observed [61]. This observation suggested that the rather basic reaction conditions might be responsible for the degradation of the azo derivative and that a decrease in the amount of K_2CO_3 is mandatory if better conversions and longer catalyst lifetimes are to be attained. The stringent requirement for 2 equivalents of K_2CO_3 in toluene was puzzling, and we initiated some studies in order to understand the role(s) of this heterogeneous base. In particular, we wondered if suitable reaction conditions might be found in which smaller quantities of base could be employed in order to transform our original system into a more ecologically friendly protocol.

Therefore, a variety of other bases (Na_2CO_3, Li_2CO_3, Na_2HPO_4, NaH_2PO_4, Al_2O_3, NaOAc, KOAc, KOH, and $CuCO_3$) were tested in this aerobic oxidation system. Surprisingly, none proved to be as efficient as K_2CO_3 [62]. Examination of the postulated mechanism of this transformation (Fig. 18) suggested a number of possible roles for K_2CO_3 [63]. Firstly, K_2CO_3 should act as a base and react with the HCl formed during the initial replacement of the chloride ligand of **56** by the alcohol (**7**) or the hydrazine (**72**). However, if this was the sole purpose of K_2CO_3, then only 5 mol% should actually be necessary in the reaction to fulfil the requirement for catalyst formation.

Secondly, examination of the oxidation in toluene revealed its heterogeneous nature. Filtration of the dark-brown suspension gave a filtrate devoid of oxidizing activity and a solid material which, once re-suspended in toluene, smoothly oxidized alcohols to the corresponding ketones and aldehydes. It thus appears that K_2CO_3 may also serve as a solid support on which the copper catalyst can be adsorbed.

Tab. 15 Copper-catalyzed aerobic oxidation of alcohols using DBAD-H_2

R^1RCH–OH **7** → (5 mol% CuCl.Phen; toluene; 90 °C / 5 mol% DBAD-H_2; 2 eqs K_2CO_3; O_2) → R^1RC=O **8**

Entry	*Substrate*	*Product*	*Yield* [a]
1	2-thienyl–CH_2OH	2-thienyl–CHO	85%
2	3-pyridyl–CH_2OH	3-pyridyl–CHO	81%
3	4-MeS–C_6H_4–CH_2OH	4-MeS–C_6H_4–CHO	92%
4	Ph–CH=CH–CH_2OH	Ph–CH=CH–CHO	89%
5	geraniol	geranial (CHO)	71% [b]
6	nerol	neral (CHO)	73% [c]
7	${}^tBu$-cyclohexanol (OH)	${}^tBu$-cyclohexanone (=O)	84% [d]
8	C_9H_{19}CH(OH)CH_3	C_9H_{19}C(=O)CH_3	88% [e]
9	C_9H_{19}–CH_2OH		65%

a) All yields refer to pure, isolated compounds.
b) Neral was not detected in this reaction.
c) Geranial was not detected in this experiment.
d) 10 mol% CuCl · Phen and 10 mol% DBAD were used in this reaction.
e) 5 mol% DBAD was employed instead of DBAD-H_2.

Finally, since water is released during the oxidation process, K_2CO_3 might also be acting as a water scavenger.

The importance of the dehydrating properties of K_2CO_3 was clearly revealed by performing the oxidation reaction using only 10 mol% K_2CO_3 in the presence of an excess of 4 Å MS. Although 4 Å MS proved to be less efficient than K_2CO_3 in trapping the released water (larger loading and longer reaction time are required), the oxidation went smoothly to completion.

A major breakthrough was accomplished while studying the influence of the solvent on the amount of K_2CO_3 required for the aerobic oxidation of 2-undecanol into 2-undecanone (Tab. 16) [64].

Whereas, in toluene, 2 equivalents of K_2CO_3 are necessary to achieve complete conversion of alcohol **73** into ketone **74** (Tab. 16, Entries 1–3), we were quite surprised to find that only a mediocre yield of the desired ketone was obtained in fluorobenzene (Tab. 16, Entry 4) under comparable conditions [65]. Unexpectedly, lowering the amount of K_2CO_3 dramatically increased the conversion of **73** into **74**, reaching a 100% conversion even when only 25 mol% of the base was employed (Tab. 16, Entry 7). Under these conditions, 2-undecanone (**74**) could be isolated in up to 99% yield. These optimized conditions were then applied to the aerobic oxidation of a variety of structurally representative alcohols. The results are summarized in Tab. 17.

It is quite remarkable that essentially every type of alcohol is smoothly oxidized into the corresponding carbonyl derivative in high yield and with good to complete conversion. Under these conditions, primary aliphatic, allylic, and benzylic alcohols afford the expected aldehydes, and secondary alcohols are smoothly transformed into ketones. Geraniol produces geranial and nerol produces neral, with no detectable loss of the geometric integrity of the C-C double bond (Tab. 17, Entries 5 and 6). The catalyst tolerates both sulfur and nitrogen substituents (Tab. 17, Entries 2, 4 and 8). Protected β-amino-alcohols are smoothly converted into the corresponding aldehydes without detectable racemization (Tab. 17, Entries 4 and 8) [66]. It is also noteworthy that the reaction conditions are sufficiently mild to be compatible with the Boc-protecting group (Tab. 17, Entry 8).

Tab. 16 Solvent and base effects in the aerobic oxidation of 2-undecanol

$C_9H_{19}CH(OH)CH_3$ (**73**) → [5mol% CuCl.Phen; 5mol% DBAD; O_2; K_2CO_3; solvent; 80°C] → $C_9H_{19}C(O)CH_3$ (**74**)

Entry	*Solvent*	*K_2CO_3*	*Conversion* [a)]	*Yields* [b]
1	$CH_3C_6H_5$	2 eqs	90%	88%
2	$CH_3C_6H_5$	1 eq	70%	61% [c)]
3	$CH_3C_6H_5$	0.5 eq	<20%	11%
4	FC_6H_5	2 eqs	50%	44% [d)]
5	FC_6H_5	1 eq	100%	94%
6	FC_6H_5	0.5 eq	100%	>99%
7	*FC_6H_5*	*0.25 eq*	*100%*	>99%
8	FC_6H_5	0.1 eq	60%	58% [e)]

a) The % conversions were measured by ^1H NMR spectroscopy and by capillary GC analysis.
b) All yields are for pure, isolated compounds.
c) The reaction was stopped after 5 h.
d) The reaction was stopped after 3 h.
e) The reaction was stopped after 8 h.

Tab. 17 Aerobic, catalytic oxidation of alcohols in fluorobenzene

Entry	*Substrate*	*Product*	*Yield*[a]
1	$C_9H_{19}CH_2OH$	$C_9H_{19}CHO$	65%[b]
2	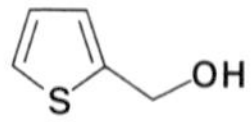 (2-thienyl)methanol (S, OH)	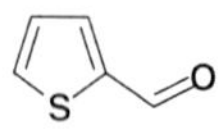 thiophene-2-carbaldehyde (S, O)	85%
3	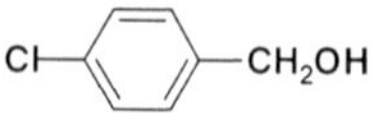 Cl–C₆H₄–CH_2OH	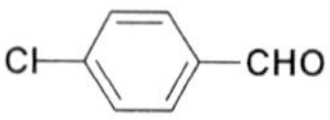 Cl–C₆H₄–CHO	87%
4	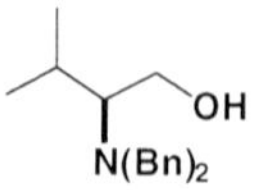 N(Bn)$_2$, OH	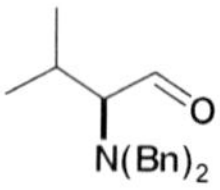 N(Bn)$_2$, O	85%[c]
5	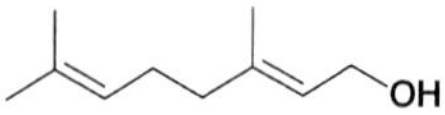 OH	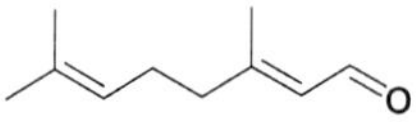 O	86%[d]
6	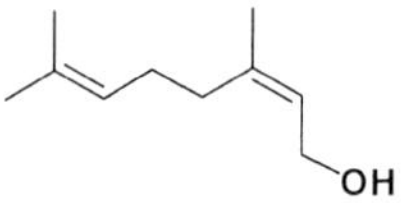 OH	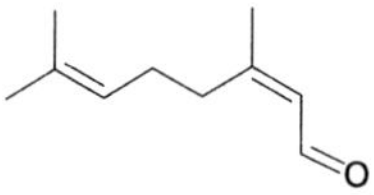 O	80%[d]
7	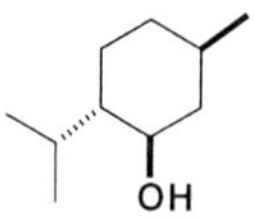 OH	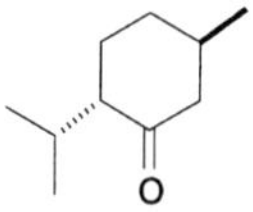 O	75%
8	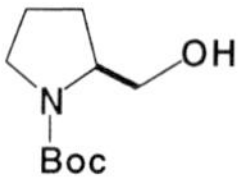 N-Boc, OH	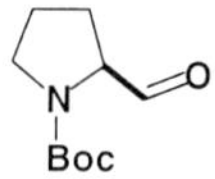 N-Boc, O	80%[e]
9	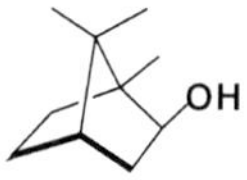 OH	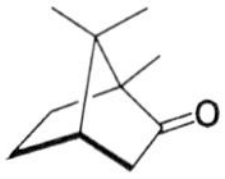 O	93%
10	 OH	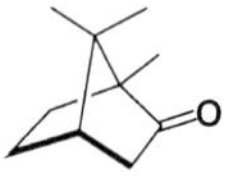 O	90%

a) All yields are for pure, isolated compounds. Complete conversions were achieved unless specified.

b) In this case, the reaction was stopped after 70% conversion. The recovered alcohol amounted to ~ 30%.

c) No racemization was observed in this oxidation reaction. The aldehyde was reduced to the alcohol ($LiAlH_4$ in THF) and the *ee* of the resulting amino alcohol was measured by HPLC analysis: Daicel chiralpak column; 2% ^{i}PrOH in hexane; 1 mL/min; $T = 20\,°C$; $\lambda = 254$ nm; (*R*)-Bn_2Valinol: 11.16 min; (*S*)-Bn_2Valinol: 12.70 min; *ee* > 99%.

d) No double bond isomerization took place under these conditions.

e) No racemization was observed in this oxidation reaction. The *ee* was measured by chiral GC (CP-Chiral-Dex CB, 25 m; ∅ = 0.25 mm, DF = 0.25 μ, 130 °C for 12 min then 1 °C per min) of the derived *bis*-Boc-prolinol obtained by $LiAlH_4$ reduction of Boc-prolinal followed by derivatization with Boc_2O (R_t(*R*)-enantiomer: 43.1 min, R_t(*S*)-enantiomer: 43.6 min).

Interestingly, both *endo-* and *exo*-borneol are oxidized to camphor at the same rate, despite the enormous difference in the steric environment of these two alcohols (Tab. 17, Entries 9 and 10).

Using our new protocol, only 25 mol% of K_2CO_3 is required for optimum activity. This unexpected breakthrough thus provides us with a novel system, which is completely catalytic in all its ingredients. Such a low loading of the heterogeneous base appears to be highly specific to fluorobenzene as the solvent (compare Tab. 16, Entries 1–3 with Entries 4–8). The property of fluorobenzene which is responsible for its unequalled behavior is not yet known [67, 68], although we believe that it is a combination of factors, rather than a single one, that gives fluorobenzene its uniqueness.

When less than 25 mol% of K_2CO_3 was employed in this protocol (Tab. 16, Entry 8), the reaction became rather sluggish and proved to be difficult to transpose to other alcohols. The search for an alternative to K_2CO_3 then became one of our prime objectives.

After unsuccessfully screening a number of different additives, we were gratified to find that *t*BuOK uniquely satisfied our requirements. Interestingly, we also noticed that the mode of addition of the various reaction partners played a crucial role in the success of this new procedure (Fig. 21).

Thus, it appeared that addition of the base to the pre-formed CuCl · Phen/DBAD complex resulted in rapid deactivation of the system, as demonstrated by the poor conversion of 2-undecanol (**73**) into the corresponding ketone **74** (Fig. 21, Entry 1) [69]. On the other hand, adding *t*BuOK to CuCl · Phen in the presence of 2-undecanol, followed by the addition of DBAD and heating under a gentle stream of oxygen, led to complete conversion of **73** into **74** (Fig. 21, Entry 2). This efficient, catalytic procedure was then applied to a range of representative alcohols. Some selected examples are shown in Tab. 18.

As can be seen from Tab. 18, secondary aliphatic, allylic, and benzylic alcohols are all quantitatively converted into the corresponding carbonyl derivatives. It is interesting to note that no epimerization of menthone takes place under these conditions (Entry 3). Furthermore, fairly hindered decalin derivatives (Entry 5) are also smoothly oxidized.

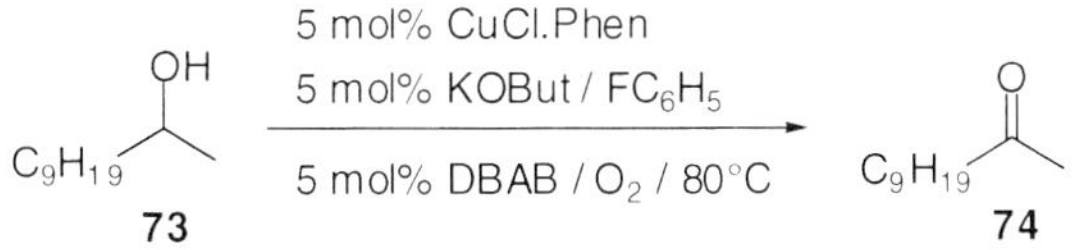

$C_9H_{19}CH(OH)CH_3$ **73** → $C_9H_{19}COCH_3$ **74**

5 mol% CuCl.Phen
5 mol% KOBut / FC_6H_5
5 mol% DBAB / O_2 / 80°C

Order of addition	Conversion
CuCl.Phen / DBAD / tBuOK / **73** / Δ	40%
CuCl.Phen / **73** / tBuOK / DBAD / Δ	100%

Fig. 21

Tab. 18 Aerobic oxidation of alcohols usin $^{t}BuOK$ [a)]

Entry	*Substrate*	*Product*	*Yield* [b, c)]
1	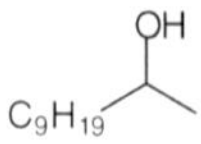 C_9H_{19}, OH	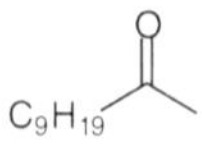 C_9H_{19}, O	90%
2	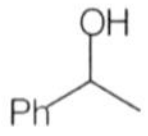 Ph, OH	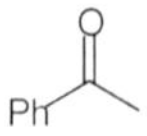 Ph, O	93%
3	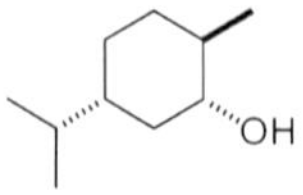 OH	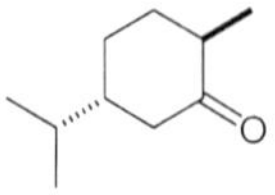 O	92%
4	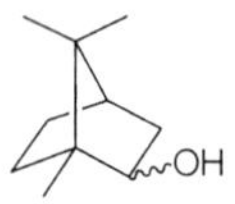 OH	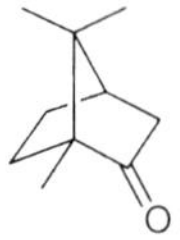 O	93% [d)]
5	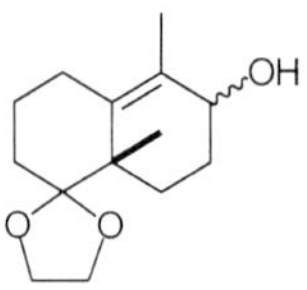 OH, O, O	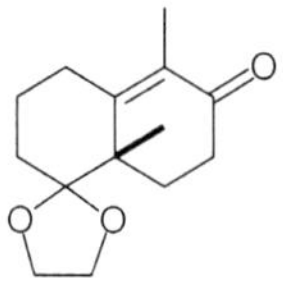 O, O, O	84% [e, f)]
6	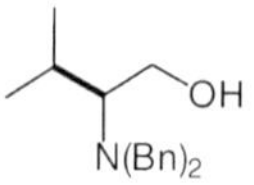 OH, N(Bn)2	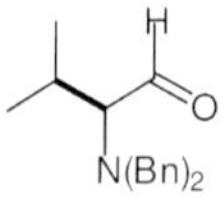 H, O, N(Bn)2	84% [g, h)]
7	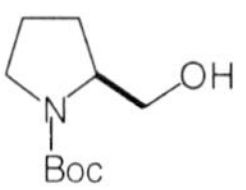 OH, N, Boc	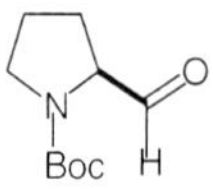 O, N, Boc, H	97% [h)]

a) The reaction conditions are described in [32].
b) All yields refer to pure, isolated products.
c) Unless otherwise stated, all the conversions are quantitative.
d) The oxidation was performed on an 80/20 mixture of borneol and *iso*-borneol.
e) The oxidation was effected on a 30/70 mixture of axial and equatorial isomers.
f) The conversion amounted to 95% in this case.
g) After silica gel column chromatography.
h) No racemization was detected.

These observations imply that the Cu oxidant is little sensitive to the steric surroundings of the hydroxyl function. The scope of the reaction can be further extended to protected primary β-amino alcohols with equal efficiency. The oxidation of dibenzyl valinol (Entry 6), which contains a tertiary nitrogen atom, proceeds in excellent yield. Moreover, the involvement of a neutral medium is ideally demonstrated by the lack of racemization of both dibenzyl valinal and Boc-prolinal (Entries 6 and 7). Purification of this latter product, which was prepared on a gram scale, necessitated only a simple filtration [70].

It is important to note that this new protocol operates under completely neutral conditions. Indeed, addition of *t*BuOK to the copper chloride · Phen/alcohol mixture generates the corresponding copper alkoxide. From that point onward, the oxidation proceeds under neutral conditions, since all the base has been consumed. It is noteworthy that sensitive substrates do not undergo epimerization or racemization.

Unfortunately, even using this optimized procedure, we were not able to improve the conversion of primary alcohols into the corresponding aldehydes. However, close examination of the oxidation behavior of several primary aliphatic alcohols revealed intriguing features (Tab. 19). While poor conversion of 1-decanol to decanal was achieved (Tab. 19, Entry 1), dibenzyl leucinol and Boc-prolinol were quantitatively transformed into the corresponding aldehydes (Tab. 19, Entries 2 and 3). The enhanced reactivity of these two primary alcohols could be due to either an increased steric effect at the α-carbon center or an electronic influence of the α-nitrogen substituent or a combination of both. To test the importance of steric hindrance, the aerobic oxidation of cyclohexane methanol and adamantane methanol was carried out. Much to our surprise, oxidation of cyclohexane methanol afforded the corresponding aldehyde in 70% conversion (Tab. 19, Entry 4), and transformation of adamantane methanol proceeded with 80% conversion

Tab. 19 Copper-catalyzed aerobic oxidation of selected primary alcohols

Entry	*Substrate*	*Product*	*Yield*[a, b]
1	 C_9H_{19} OH	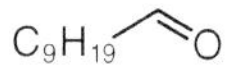 C_9H_{19} O	(60%) 51%
2	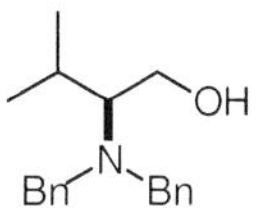 OH N Bn Bn	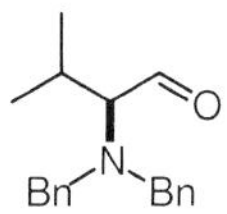 O N Bn Bn	(100%) 84%
3	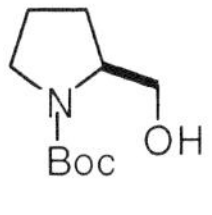 N Boc OH	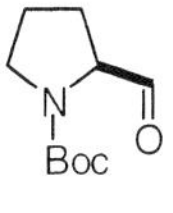 N Boc O	(100%) 97%
4	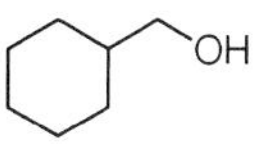 OH	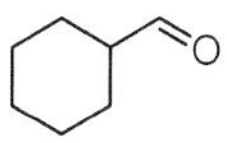 O	(70%) 64%
5	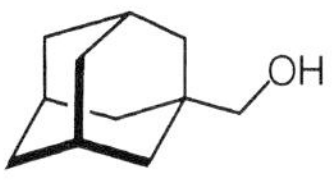 OH	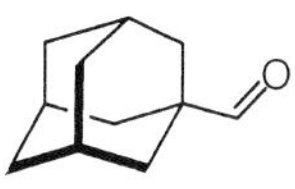 O	(80%) 77%

a) Values in parentheses refer to the percentage conversion of the starting material.
b) Yields of isolated, pure product.

(Tab. 19, Entry 5). Clearly, increased substitution at the α-position favors the oxidation of primary aliphatic alcohols, though the conversions are still not optimum.

In order to improve this transformation, a variety of selected additives were tested in the aerobic oxidation of 1-decanol (**75**). The high affinity of heterocyclic amines for copper salts, coupled with their ubiquitous presence as ligands in biologically active copper-containing proteins [39], prompted us to investigate them initially. Some selected results are collected in Tab. 20.

As can be seen from Tab. 20, the conversion of 1-decanol (**75**) to the desired aldehyde **76** proceeded poorly in the absence of additive (Tab. 20, Entry 1). In the presence of 5 mol% of 4-DMAP (4-dimethylaminopyridine), a significant increase in the transformation of **75** to **76** was observed (Tab. 20, Entry 2), and complete conversion was eventually reached using 10 mol% of 4-DMAP (Tab. 20, Entry 3). Interestingly, only 7 mol% of NMI (*N*-methyl imidazole) was required to transform **75** completely into **76** (Tab. 20, Entry 4).

These conditions were next applied to the aerobic oxidation of a variety of primary alcohols. A selection of pertinent examples is displayed in Tab. 21.

As can be seen from Tab. 21, all the primary alcohols employed were quantitatively converted into the corresponding aldehydes with 100% selectivity. It is noteworthy that no trace of carboxylic acid was observed under these aerobic conditions. The reaction tolerates both simple aliphatic primary alcohols (Tab. 21, Entry 1) and more hindered derivatives (Tab. 21, Entries 2 and 3) as well as various protecting groups (Tab. 21, Entries 4 and 8). Simple alkenes are unaffected (Tab. 21, Entry 5), and base-sensitive substrates are smoothly oxidized (Tab. 21, Entry 6). It is interesting to note that, under these neutral conditions, highly acid-sensitive substrates are also quantitatively converted into the corresponding aldehydes (Tab. 21, Entry 7). Finally, a sig-

Tab. 20 Influence of additives on the aerobic oxidation of 1-decanol

$C_9H_{19}CH_2OH$ (**75**) → [5 mol% Phen.CuCl, 5 mol% DBAD, FC_6H_5; 5 mol% *t*BuOK, O_2, **additive**, 70-80°C] → $C_9H_{19}CHO$ (**76**)

Entry	*Additive*	*Amount*	*Conversion*
1	none	none	60%
2	4-DMAP (structure)	5 mol%	80%
3	4-DMAP (structure)	10 mol%	100%
4	NMI (structure)	7 mol%	100%

a) The conversions were measured by capillary gas chromatography using the internal standard method.

Tab. 21 Efficient, aerobic, catalytic oxidation of primary alcohols

Entry	*Substrate*	*Product*	*Yield* [a]
1	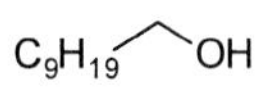 C_9H_{19} OH	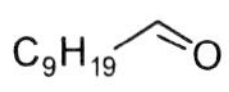 C_9H_{19} O	95%
2	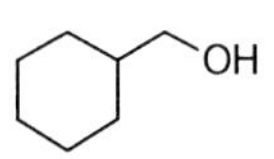 OH	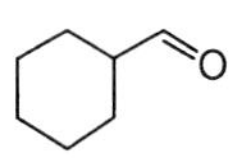 O	93%
3	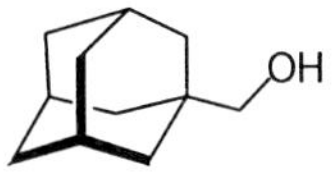 OH	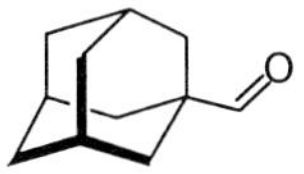 O	95%
4	 TBSO OH	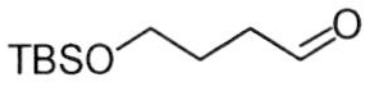 TBSO O	94%
5	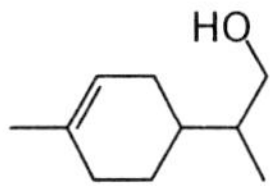 HO	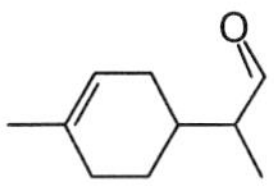 O	94%
6	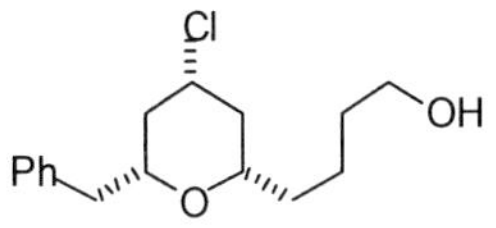 Cl Ph O OH	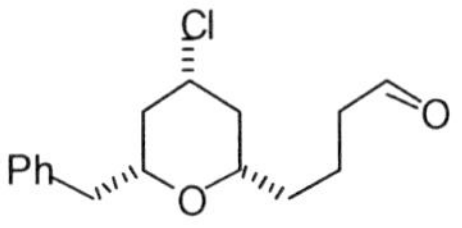 Cl Ph O O	83%
7	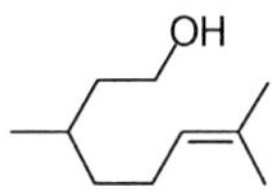 OH	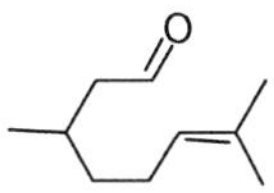 O	82%
8	 BnO OH	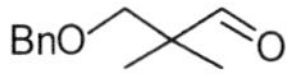 BnO O	97%
9	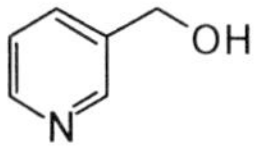 N OH	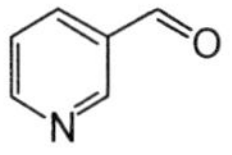 N O	93%
10	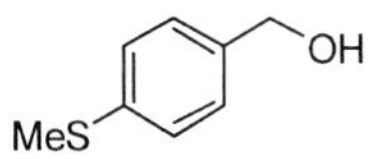 MeS OH	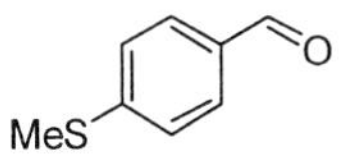 MeS O	95%

a) All yields are for pure, isolated products.

nificant impediment pertaining to all the other reported aerobic oxidation protocols is their inability to oxidize alcohols possessing a chelating function, a nitrogen atom, or a sulfur substituent. Such is not the case for the copper catalyst, which transforms strongly coordinating substrates quantitatively into the aldehyde (Tab. 21, Entry 8) and tolerates both heteroatoms (Tab. 21, Entries 9 and 10).

The remarkable effect of 4-DMAP and NMI on the ability of the copper catalyst to oxidize efficiently a wide range of primary alcohols is surprising, and the origin of this effect was investigated, initially using the mechanistically simpler anaerobic system. In the absence of oxygen and NMI, 1-decanol was smoothly and quantitatively oxidized to decanal. Addition of 7 mol% NMI did not improve either the conversion or the rate of the reaction; rather, NMI had a slightly retarding effect [71].

In order to reconcile these observations with the previously established catalytic cycle for the aerobic oxidation of alcohols using the CuCl · Phen/DBAD system, a new catalytic manifold has to be operative in the presence of NMI (Fig. 22).

The productive catalytic cycle begins with the ternary loaded complex **75**. Intramolecular hydrogen transfer from the alkoxy substituent to the azo ligand generates copper(I) hydrazide (**76**). Subsequent release of the aldehyde produces complex **77**, which is rapidly captured by oxygen, affording Cu(II) hydrazide derivative **78**. Reorganization of **78** under the thermal conditions of the reaction leads to the hydroxy copper(I) species **79**. Finally, ligand exchange and elimination of water regenerates the active, loaded complex **75**, and a new catalytic cycle ensues. Among the various active species involved in this system, complex **77**, bearing an empty coordination site, appears to be the most likely candidate to suffer a competitive deactivation by the primary alcohols [72].

Indeed, while **77** usually reacts rapidly with oxygen, it can occasionally undergo competitive coordination to an alcohol, producing the copper derivative **80**, which might undergo hydrogen transfer and loss of the hydrazine substituent, resulting in the inactive complex **81** [73, 74].

In the case of secondary alcohols, competitive coordination of the OH function and oxygen to **77** largely favors the latter, and the bis-copper peroxide **78** is formed. However, when primary aliphatic alcohols are employed, coordination of

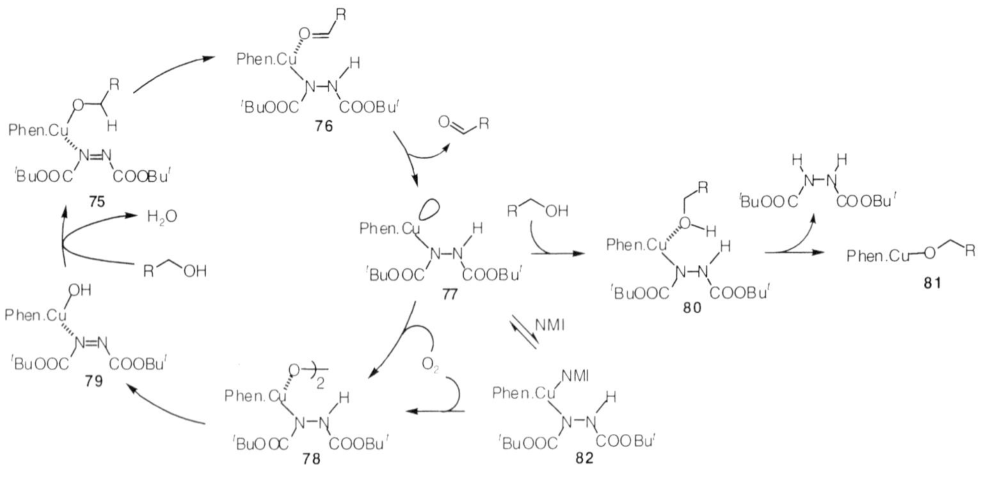

Fig. 22

the less hindered OH group now becomes competitive. The formation of inactive complex **77** gradually depletes the catalytic cycle in the active oxidizing species, and the reaction grinds to a halt. This mechanistic proposal also explains the observed increased conversions when employing more hindered aliphatic primary alcohols.

The role of NMI and 4-DMAP would thus be to bind rapidly to copper complex **77**, generating intermediate **82**, which is probably in equilibrium with **77**. Such coordination would preclude the competitive addition of the alcohol and suppress the undesired formation of the inert derivative **81** [75].

In summary, we have established a simple and environmentally friendly, catalytic aerobic protocol for the efficient oxidation of a wide variety of alcohols into aldehydes and ketones. This novel catalytic system uses oxygen or air as the stoichiometric oxidant and releases water as the sole by-product. We have also shown that the use of the simple and inexpensive additive NMI strongly modified the course of the copper-catalyzed aerobic oxidation of primary aliphatic alcohols. Under these novel conditions, a wide range of primary substrates could be transformed efficiently into the corresponding aldehydes with no trace of over-oxidized carboxylic acids being detected. Moreover, the neutral conditions employed are compatible with base- and acid-sensitive substrates. Furthermore, these results have shed some light on an unsuspected decomposition pathway, the inhibition of which held the key to a highly successful aerobic oxidation procedure for primary alcohols.

Although much still remains to be done, we believe that, through the combined research effort of several groups throughout the world, a genuine leap has been realized in the establishment of mild, functionally tolerant, and ecologically benign catalytic systems for the oxidation of alcohols into carbonyl derivatives.

Acknowledgements

Financial support was provided by Zeneca Limited through the Zeneca Strategic Research Fund. IEM is grateful to Zeneca for his appointment to the Zeneca Fellowship (1994–1997) and the 2003 Astra-Zeneca European Lectureship.

References

1 For general reviews on oxidation reactions, see: (a) Larock, R. C. in *Comprehensive Organic Transformations*; VCH Publishers Inc.: New York, **1989**, 604. (b) Procter, G. in *Comprehensive Organic Synthesis*, Ley, S. V. (Ed.), Pergamon: Oxford, **1991**, *7*, 305. (c) Ley, S. V., Madin, A. in *Comprehensive Organic Synthesis*, vol. 7; Trost, B. M., Fleming, I. (Eds.), Pergamon: Oxford, **1991**, 251. (d) Lee, T. V. in *Comprehensive Organic, Synthesis*, Trost, B. M., Fleming, I. (Eds.), Pergamon: Oxford, **1991**, *7*, 291. (e) Trahanovsky, W. S. in *Oxidation in Organic Chemistry*, Blomquist, A. T., Wasserman, H. (Eds.); Part A–D, Acad. Press. (f) Noyori, R., Hashiguchi, S. *Acc. Chem. Res.* **1997**, *30*, 97 and references cited therein.

2 (a) Sheldon, R.A., Kochi, J. K. In *Metal-Catalyzed Oxidations of Organic Compounds*; Academic Press, New York, **1981**. (b) Ley, S.V., Norman, J., Griffith, W. P., Marsden, S.P. *Synthesis* **1994**, 639. (c) Murahashi, S.-I., Naota, T., Oda, Y., Hirai, N. *Synlett* **1995**, 733. (d) Krohn, K., Vinke, I., Adam, H. *J. Org. Chem.* **1996**, *61*, 1467. (e) Strukul, G. in *Catalytic Oxidations with Hydrogen Peroxide as Oxidant*, Kluwer Academic Publishers, London, **1992**. (f) Sato, K., Takagi, J., Aoki, M., Noyori, R. *Tetrahedron Lett.* **1998**, *39*, 7549. (g) Sato, K., Aoki, M., Noyori, R. *Science* **1998**, *281*, 1646. (h) Berkessel, A., Sklorz, C.A. *Tetrahedron Lett.* **1999**, *40*, 7965.

3 (a) Sheldon, R.A. in *Dioxygen Activation and Homogeneous Catalytic Oxidation*; Simandi, L.L. (Ed.), Elsevier: Amsterdam, **1991**, p. 573. (b) James, B.R. in *Dioxygen Activation and Homogeneous Catalytic Oxidation*; Simandi, L.L. (Ed.), Elsevier, Amsterdam, **1991**, p. 195. (c) Sheldon, R.A., Arends, I.W.C.E., Dijksman, A. *Catal. Today* **2000**, 57, 157. (d) Matsumoto, M., Ito, S. *J. Chem. Soc., Chem. Commun.* **1981**, 907. (e) Hinzen, B., Lenz, R., Ley, S.V. *Synthesis* **1998**, 977. (k) Bleloch, A., Johnson, B.F.G., Ley, S.V., Price, A.J., Shephard, D.S., Thomas, A.W. *Chem. Commun.* **1999**, 1907. (l) Hallman, K., Moberg, C. *Adv. Synth. Catal.* **2001**, *343*, 260.

4 Jia, C.-G., Jing, F.-Y., Hu, W.-D., Huang, M.-Y., Jiang, Y.-Y. *J. Mol. Catal.* **1994**, *91*, 139.

5 Mallat, T., Baiker, A. *Catal. Today* **1994**, *19*, 247.

6 Jenzer, G., Sueur, D., Mallat, T., Baiker, A. *Chem. Comun.* **2000**, 2247.

7 Blackburn, T.F., Schwartz, J. *J. Chem. Soc., Chem. Commun.* **1977**, 157.

8 Gomez-Bengoa, E., Noheda, P., Echavarren, A.M. *Tetrahedron Lett.* **1994**, *35*, 7097.

9 Peterson, K.P., Larock, R. C. *J. Org. Chem.* **1998**, *63*, 3185.

10 (a) Nishimura, T., Onoue, T., Ohe, K., Uemura, S. *Tetrahedron Lett.* **1998**, *39*, 6011. (b) Nishimura, T., Onoue, T., Ohe, K., Uemura, S. *J. Org. Chem.* **1999**, *64*, 6750. (c) Kakiuchi, N., Maeda, Y., Nishimura, T., Uemura, S. *J. Org. Chem.* **2001**, *66*, 6620.

11 ten Brink, G.-J., Arends, I.W.C.E., Sheldon, R.A. *Science* **2000**, *287*, 1636.

12 Schultz, M.J., Park, C.C., Sigman, M. S. *Chem. Commun.* **2002**, 3034.

13 Ferreira, E.M., Stoltz, B.M. *J. Am. Chem. Soc.* **2001**, *123*, 7725.

14 Jensen, D.R., Pugsley, J.S., Sigman, M.S. *J. Am. Chem. Soc.* **2001**, *123*, 7475.

15 Kaneda, K., Fujii, M., Morioka, K. *J. Org. Chem.* **1996**, *61*, 4502.

16 Kaneda, K., Fujie, Y., Ebitani, K. *Tetrahedron Lett.* **1997**, *38*, 9023.

17 (a) Nishimura, T., Kakiuchi, N., Inoue, M., Uemura, S. *Chem. Commun.* **2000**, 1245. (b) Kakiuchi, N., Maeda, Y., Nishimura, T., Uemura, S. *J. Org. Chem.* **2001**, *66*, 6620.

18 Uozumi, Y., Nakao, R. *Angew. Chem. Int. Ed.* **2003**, *42*, 194.

19 Tovrog, B.S., Diamond, S.E., Mares, F., Szalkiewicz, A. *J. Am. Chem. Soc.* **1981**, *103*, 3522.

20 (a) Punniyamurthy, T., Iqbal, J. *Tetrahedron Lett.*, **1994**, *35*, 4007. (b) Mandal, A.K., Iqbal, J. *Tetrahedron* **1997**, *53*, 7641.

21 (a) Iwahama, T., Sakaguchi, S., Nishiyama, Y., Ishii, Y. *Tetrahedron Lett.* **1995**, *36*, 6923. (b) Iwahama, T., Sukaguchi, S., Nishiyama, Y. Ishii, Y. *Tetrahedron Lett.* **1998**, 36, 6923.

22 Iwahama, T., Yoshino, Y., Keitoku, T., Sakaguchi, S., Ishii, Y., *J. Org. Chem.* **2000**, *65*, 6502.

23 Zhang, N., Mann, C.M., Shapley, P.A. *J. Am. Chem. Soc.* **1988**, *110*, 6591.

24 Neumann, R., Levin, M. *J. Org. Chem.* **1991**, *56*, 5707.

25 (a) Bäckvall, J.-E., Chowdhury, R.L., Karlsson, U. *J. Chem. Soc., Chem. Commun.* **1991**, 473. (b) Wang, G.-Z., Andreasson, U., Bäckvall, J.E. *J. Chem. Soc., Chem. Commun.* **1994**, 1037.

26 Coleman, K.S., Coppe, M., Thomas, C., Osborn, J.A. *Tetrahedron Lett.* **1999**, *40*, 3723.

27 Muldoon, J., Brown, S.N. *Org. Lett.* **2002**, *4*, 1043.

28 Yamada, T., Rhode, O., Takai, T., Mukaiyama, T. *Chem. Lett.* **1991**, 5.

29 Tang, R., Diamond, S. E., Neary, N., Mares, F. J. Chem. Soc., *Chem. Commun.* **1978**, 562.
30 Matsumoto, M., Watanabe, N. *J. Org. Chem.* **1984**, *49*, 3436.
31 Bilgrien, C., Davis, S., Drago, R. S. *J. Am. Chem. Soc.*, **1987**, *109*, 3786.
32 Lee, M., Chang, S. *Tetrahedron Lett.* **2000**, *41*, 7507.
33 Masutani, K., Uchida, T., Irie, R., Katsuki, T. *Tetrahedron Lett.* **2000**, *41*, 5119.
34 (a) Dijksman, A., Arends, I. W. C. E., Sheldon, R. A. *Chem. Commun.* **1999**, 1591. (b) Dijksman, A., Marino-González, A., i Payeras, A. M., Arends, I. W. C. E., Sheldon, R. A. *J. Am. Chem. Soc.* **2001**, *123*, 6826.
35 Sheldon, R. A., Arends, I. W. C. E., ten Brink, G.-J., Dijksman, A. *Acc. Chem. Res.* **2002**, *35*, 774.
36 Dijksman, A., Arends, I. W. C. E., Sheldon, R. A. *Synlett* **2001**, 102.
37 Kaneda, K., Yamashita, T., Matsushita, T., Ebitani, K. *J. Org. Chem.* **1998**, *63*, 1750.
38 Yamaguchi, K., Mizuno, N. *Angew. Chem. Int. Ed* **2002**, *41*, 4538.
39 For excellent reviews on the formation, isolation and reactions of dinuclear copper(II) peroxides, see: (a) Karlin, K. D., Gultneh, Y. *Prog. Inorg. Chem.* **1987**, *35*, 219–327. (b) Zuberbühler, A. D. in *Copper Coordination Chemistry: Biochemical and Inorganic Perspectives* (Eds.: Karlin, K. D., Zubieta, J.) Adenine, Guilderland, New York, **1983**. (c) Sakharov, A. M., Skibida, I. P., *Kinet. Catal.* **1988**, *29*, 96-102. (d) Fox, S., Nanthakumar, A., Wikstrom, M., Karlin, K. D.,Blackburn, N. J. *Kinet. Catal.* **1996**, *118*, 24–34. (e) Solomon, E. I., Sundaram, U. M., Machonkin, T. E. *Chem. Rev.* **1996**, *96*, 2563–2605.
40 Tyleklar, Z., Jacobson, R. R., Wei, N., Murthy, N. N., Zubieta, J. ; Karlin, K. D. *J. Am. Chem. Soc.* **1993**, *115*, 2677–2689.
41 Kitajima, N., Fujisawa, K., Fujimoto, C., Moro-oka, Y., Hashimoto, S., Kitagawa, T., Toriumi, K., Tatsumi, K., Nakamura, A. *J. Am. Chem. Soc.* **1992**, *114*, 1277–1291.
42 (a) Jallabert, C., Rivière, H. *Tetrahedron Lett.*, **1977**, 1215. (b) Jallabert, C., Lapinte, C., Rivière, H. *J. Mol. Catal*, **1980**, *7*, 127. (c) Jallabert, C., Rivière, H. *Tetrahedron* **1980**, *36*, 1191. (d) Jallabert, C., Lapinte, C., Rivière, H. *J. Mol. Catal.* **1982**, *14*, 75. For other pertinent studies on aerobic oxidation of alcohols using copper complexes, see for example: (a) Munakata, M., Nishibayashi, S., Sakamoto, H. *J. Chem. Soc., Chem. Commun.* **1980**, 219. (b) Bhaduri, S., Sapre, N. Y. *J. Chem. Soc., Dalton Trans.* **1981**, 2585.
43 Semmelhack, M. F., Schmid, C. R., Cortes, D. A., Chon, C. S. *J. Am. Chem. Soc.* **1984**, *106*, 3374.
44 (a) Capdevielle, P., Sparfel, D., Baranne-Lafont, J., Cuong, N. K., Maumy, M. *J. Chem. Research* (S) **1993**, 10 and references cited therein. (b) Capdevielle, P., Audebert, P., Maumy, M. *Tetrahedron Lett.* **1984**, *25*, 4397.
45 Jiu, X., Qiu, A., Sawyer, D. T. *J. Am. Chem. Soc.*, **1993**, *115*, 3239.
46 Chaudhury, P., Hess, M., Weyhermüller, T., Wieghardt, K. *Angew. Chem. Int. E.* **1999**, *38*, 1095.
47 Betzemeier, B., Cavazzini, M., Quici, S., Knochel, P. *Tetrahedron Lett.* **2000**, *41*, 4343. For a related Pd-catalyzed oxidation, see: Nishimura, T., Maeda, Y., Kakiuchi, N., Uemura, S. *J. Chem. Soc., Perkin Trans. 1* **2000**, 4301.
48 Other bases tested include e.g., Na_2CO_3, Li_2CO_3, Na_2HPO_4, NaH_2PO_4, Al_2O_3, NaOAc, KOAc, KOH and $CuCO_3$. Only KOBut appears to act as an efficient base in the catalytic oxidation process.
49 Solomon, R. G., Kochi, J. K. *J. Am. Chem. Soc.* **1973**, *95*, 3300.
50 The use of air instead of oxygen results in a slower reaction rate. The oxidation can be increased by passing the air through a porous glass frit which creates microbubbles. Under these conditions, the speed of the catalytic oxidation of alcohols using air matches the one employing oxygen.
51 The preparation of Copper(I) alkoxides and their reactivity towards O_2 has been reported in the literature. See for example: Capdevielle, P., Audebert, P., Mau-

MY, M. *Tetrahedron Lett.* **1984**, *25*, 4397-4400.

52 Stoichiometric amounts of substituted azo compounds have been used to oxidize magnesium alkoxides to the corresponding carbonyl compounds: NARASAKA, K., MORIKAWA, A., SAIGO, K., MUKAIYAMA, T. *Bull. Chem. Soc. Jpn.* **1977**, *50*, 2773. The decomposition mechanism of hydrazines in the presence of copper complexes has been reported: (a) ERLENMEYER, H., FLIERL, C., SIGEL, H. *J. Am. Chem. Soc.* **1969**, *91*, 1065. (b) ZHONG, Y., LIM, P. K. *J. Am. Chem. Soc.* **1989**, *111*, 8398.

53 (a) MARKÓ, I. E., GILES, P. R., TSUKAZAKI, M., BROWN, S. M., URCH C. J. *Science* **1996**, *274*, 2044. (b) MARKÓ, I. E., GILES, P. R., TSUKAZAKI, M., CHELLÉ-REGNAUT, I., URCH C. J., BROWN, S. M. *J. Am. Chem. Soc.* **1997**, *119*, 12661. (c) MARKÓ, I. E., TSUKAZAKI, M., GILES, P. R., BROWN, S. M., URCH C. J. *Angew. Chem. Int. Ed., Engl.* **1997**, *36*, 2208. (d) MARKÓ, I. E., GILES, P. R., TSUKAZAKI, M., BROWN, S. M., URCH, C. J. in *Transition Metals for Organic Synthesis*, BELLER, M., BOLM, C. (Eds.) **1998**, *2*, Chapter 2.12, 350. (e) MARKÓ, I. E., GAUTIER, A., CHELLÉ-REGNAUT, I., GILES, P. R., TSUKAZAKI, M., URCH, C. J., BROWN, S. M. *J. Org. Chem.* **1998**, 63, 7576. (f) MARKÓ, I. E., GAUTIER, A., MUTONKOLE, J.-L., DUMEUNIER, R., ATES, A., URCH, C. J., BROWN, S. M. *J. Organomet. Chem.* **2001**, *624*, 344. For an independent report of the aerobic TPAP-catalyzed oxidation of alcohols, see: LENZ, R., LEY, S. V. *J. Chem. Soc., Perkin I* **1997**, 3291.

54 The intermediacy of complex **65** in the aerobic oxidations was supported by the following observations: (1) independently generated complex **65** (CuCl · Phen/DBADH$_2$/NaH) proved to be unreactive under anaerobic conditions; (2) passing O_2 through the reaction mixture containing **65** and alcohol **63** restored the catalytic activity and good yields of aldehyde **64** were again obtained.

55 (a) SUSTMANN, R., MÜLLER, W., MIGNANI, S; MERÉNYI, R., JANOUSEK, Z., VIEHE, H. G., *New J. Chem.*, **1989**, *13*, 557. (b) DE BOECK, B., JANOUSEK, Z., VIEHE, H. G. *Tetrahedron* **1995**, *51*, 13239–13246.

56 For general reviews on Oppenauer-type oxidations, see: (a) DE GRAAUW, C. F., PETERS, J. A., VANDEKKUM, H., HUSKENS, J. *Synthesis* **1994**, 1007–1017. (b) DJERASSI, C. *Org. React. (N.Y.)* **1951**, 6, 207–212. (c) KROHN, K., KNAUER, B., KUPKE, J., SEEBACH, D., BECK, A. K., HAYAKAWA, M. *Synthesis* **1996**, 1341–1344.

57 The use of stoichiometric amounts of dipiperidinyl azodicarboxamide to oxidize magnesium alkoxides to the corresponding carbonyl compounds has been described: NARASAKA, K., MORIKAWA, A., SAIGO, K., MUKAIYAMA, T. *Bull. Chem. Soc. Jpn.* **1977**, *50*, 2773. No reaction is observed under our catalytic anaerobic conditions if DBAD is replaced by the azodicarboxamide derivative.

58 Another argument against the oxo-transfer mechanism in our catalytic, aerobic oxidation protocol is the lack of formation of sulfoxides from sulfides, *N*-oxides from amines, and phosphine oxides from phosphines. Alkenes also proved to be inert toward oxidation; no epoxide formation could be detected under our reaction conditions.

59 The oxidation of alcohols using azodicarboxylates has been previously reported (YONEDA, F., SUZUKI, K., NITTA, Y. *J. Org. Chem.* **1967**, *32*, 727–729.). Control experiments were therefore performed to establish the need for copper salts in our oxidation procedure. Thus, under our reaction conditions, no aldehyde or ketone could be detected in the absence of the CuCl · Phen catalyst, even if phenanthroline was added as an activating base. Moreover, certain reactive alcohols were oxidized partially by CuCl · Phen in the absence of the azo-derivative **71**, though only in moderate yields. These control experiments thus clearly establish the key role of the copper ion in these oxidations.

60 The oxidation reactions were monitored by GC (Permabond SE-52-DF-0.25; 25 m×0.25 mm ID), using tetradecane as the internal standard.

61 The decomposition appears to result from the activation of the azo derivative by the copper complex, in conjunction with the deprotonation of the *tert*-butyl substituent by the base, resulting in the loss of CO_2 and isobutene.

62 One rare exception appears to be KOBut. For example, the aerobic oxidation of 2-undecanol (5 mol% CuCl · Phen, 5 mol% KOBut, toluene, 80–90 °C) afforded 2-undecanone in almost quantitative yields. However, this system appears, so far, to be limited to secondary alcohol oxidations.

63 For a discussion of the possible mechanism of this reaction, see: MARKÓ, I. E., TSUKAZAKI, M., GILES, P. R., BROWN, S. M., URCH, C. J. *Angew. Chem. Int. Ed. Engl.* **1997**, *36*, 2208.

64 It is interesting to note that other solvents gave repeatedly poorer conversions (benzene, xylenes) or destroyed the catalyst activity (CH_2Cl_2, $CHCl_3$, $ClCH_2CH_2Cl$, DMF, and MeCN).

65 It is interesting to note that fluorobenzene was also used successfully by Mukaiyama and co-workers as a solvent in their Mn(salen)-catalyzed epoxidation of alkenes using the O_2/aldehyde protocol: YAMADA, T., IMAGAWA, K., NAGATA, T., MUKAIYAMA, *T. Chem. Lett.* **1992**, *11*, 2231.

66 A small amount of racemization was observed during the oxidation of Boc-prolinol.

67 Fluorobenzene possesses some remarkable properties. For example, the solubility of O_2 in FC_6H_5 is greater than that for other alkylbenzene or monohalobenzene derivatives. The relative solubility of O_2 in toluene is 8.77 as compared to 15.08 for FC_6H_5 (NAUMENKO, N. V., MUKHIN, N. N., ALESKOVIKII, V. B., *Zh. Prikl. Khim. (Leningrad)* **1969**, *42*, 2522). Furthermore, fluorobenzene possesses unusual solvent property parameters and is more polar than toluene.

Parameters[a]	***Toluene***	***Fluorobenzene***
Gutmann donor number	0.1	3.00
Dipole moment	1.0	4.90
Dielectric constant	2.38	5.42
ET(30)	33.9	37
Solvatochromic p*	0.54	0.62

a) These data were measured at Zeneca Ltd.

Like most aromatic solvents, fluorobenzene is highly flammable (Fp=–12 °C). It is irritant to the skin und can cause serious damage to the eyes. It is only weakly toxic by inhalation (rat: LC_{50}=27 mg/L) and even less by ingestion (rat: LC_{50}=4000 mg/L). On large-scale experiments, it can be easily recycled by drying and distillation.

68 It is possible that the greater polarity of fluorobenzene, which can lead to a higher concentration of soluble base, might be responsible in part for the improved yields and rate of reaction observed in this medium. Moreover, the amount of oxygen dissolved in boiling fluorobenzene might be greater than in toluene, leading to a more efficient reoxidation of the active copper species. In this regard, it is noteworthy that finely divided oxygen or air bubbles (obtained by passing the gas through a glass frit) result in enhanced reaction rate.

69 The deactivation of the catalyst could arise from base-catalyzed decomposition of copper-coordinated DBAD by *t*BuOK in the absence of added alcohol.

70 Aerobic oxidation of Boc-prolinol. 1,10-Phenanthroline (45 mg, 0.25 mmol, 5 mol%) was added to 45 mL of dry FC_6H_5, and this was followed by solid CuCl (25 mg, 0.25 mmol, 5 mol%). After stirring for 5 min at room temperature, L-Boc-prolinol (1.0 g, 4.97 mmol) was added followed by solid KOBu (28 mg, 0.25 mmol, 5 mol%). The resulting yellowish solution was stirred at room temperature for 10 min before DBAD (57.5 mg, 0.25 mmol, 5 mol%) was added. The reaction mixture was refluxed

under a gentle stream of O_2 for 4.5 h. After cooling to 20 °C, celigel (1 g of 80/20 w/w mixture of celite and silica gel) was added, and stirring was continued for 2 min. Filtration, washing off of the solid residue with 100 mL ether, and evaporation of the solvents *in vacuo* afforded pure L-Boc-prolinal as a colorless oil (960 mg, 97%). ^{1}H NMR ($CDCl_3$, 200 MHz): d=9.55 (brs, 1H, rotamer 1), 9.45 (brd, J=3 Hz, 1H, rotamer 2), 4.3 (m, 1H, rotamer 1), 4.0 (m, 1H, rotamer 2), 3.6–3.3 (m, 2H), 2.2–1.8 (m, 4H), 1.45 (brs, 9H, rotamer 1), 1.40 (brs, 9H, rotamer 2). ^{13}C NMR ($CDCl_3$, 75 MHz): d=199.5, 199.3, 79.5, 64.4, 46.2, 28.1, 27.6, 24.4, 23.8. The *ee* was measured by chiral GC (CP-Chiral-Dex CB, 25 m; F=0.25 mm, 130 °C for 12 min then 1 °C per min) of the derived bis-Boc-prolinol obtained by $LiAlH_4$ reduction of Boc-prolinal followed by derivatization with Boc2O (tR (R)-enantiomer, 43.1 min; tR (S)-enantiomer, 43.6 min).

71 Whereas quantitative conversion of **75** into **76** occurred, under anaerobic conditions, in the absence and presence of 7 mol% of NMI, the oxidation of **75** proceeded more slowly in the presence of this additive. The coordination of NMI to copper results in a slower exchange with the excess DBAD and hence in a longer reaction time.

72 Studies performed on the anaerobic version of this catalytic system revealed that aliphatic primary alcohols were oxidized with the same efficiency as that of all the other classes of alcohols, thus ruling out complexes **75**, **76**, and **79** as the culprit for the decomposition pathway. While we could not experimentally eliminate complex **78**, coordination of an alcohol to **78** should involve the participation of a pentacoordinated copper species. Although these are not uncommon, their formation requires a higher activation energy than the coordination to **77**.

73 This hydrogen transfer is essentially an intramolecular acid-base reaction. The hydrogen of the coordinated alcohol function is acidified by coordination to the copper center, while the hydrazine ligand possesses basic properties. The elimination of the hydrazine substituent is irreversible under these neutral conditions. Indeed, in the absence of excess base, $DBADH_2$ is unable to displace the alkoxide ligand from the copper complex **81**.

74 We have previously demonstrated that **81** was not a competent catalyst in the aerobic oxidation protocol when R=alkyl.

75 Markó, I. E., Tsukazaki, M., Giles, P. R., Brown, S. M., Urch, C. J. *Angew. Chem. Int. Ed. Engl.* **1997**, *36*, 2208.

2.14
Catalytic Asymmetric Sulfide Oxidations

H. B. Kagan and T. O. Luukas

2.14.1
Introduction

Sulfoxides have an asymmetric center at the sulfur atom, and chiral sulfoxides have generated a lot of interest as auxiliaries in asymmetric synthesis [1–4]. The preparations of enantiopure sulfoxides are variously based on resolution, transformation of a chiral sulfinate into a sulfoxide, and asymmetric oxidation of a sulfide. The Andersen method was for a long time (and still is) the most practical way to prepare chiral sulfoxides. It was mainly dedicated to the compounds Ar-S(O)-R [5], but a variation involving sugar sulfinates has recently been used to produce various types of alkyl sulfoxides [4, 6]. Asymmetric oxidation of sulfides R-S-R′ is a very general approach to chiral sulfoxides, since wide variations in the nature of the R and R′ groups are available. However, for a long time this route gave quite small *ee*s (<10%), the oxidant being a chiral peracid. Only in the last 15 years have significant results (*ee* >80%) been obtained by using *stoichiometric chiral reagents,* namely oxaziridines [7], hydroperoxides in combination with some chiral titanium complexes [8, 9], or oxidants in presence of BSA [10, 11]. New developments are presently being introduced by the use of *asymmetric catalysts* of sulfoxidation (see reviews in [12–14]). It is the purpose of the present chapter to summarize the main achievements in this area by focusing on *asymmetric organometallic catalysis* (enzymatic processes are excluded).

2.14.2
Sulfoxidation Catalyzed by Chiral Titanium Complexes

2.14.2.1
Diethyl Tartrate as Ligand

Oxidation of sulfides by *t*-butyl hydroperoxide (TBHP) in the presence of stoichiometric amounts of some chiral complexes has been shown to lead to the formation of aryl methyl sulfoxides with a quite good *ee* (up to 90%) by a suitable modification of the Sharpless reagent using chiral tartrate (DET) as the ligand [8, 9]. The combi-

Transition Metals for Organic Synthesis, Vol. 2, 2nd Edition.
Edited by M. Beller and C. Bolm

ISBN: 3-527-30613-7

nations $Ti(OiPr)_4/(R,R)$-DET/H_2O=1:2:1 and $Ti(OiPr)_4/(R,R)$-DET=1:4 have been respectively used by the author in Orsay and by another research group in Padua [8, 9]. The main results are detailed in a review article [12]. A significant improvement in *ees* was afforded by the replacement of TBHP by cumene hydroperoxide. This can be used as a preparative method to obtain highly enantioenriched sulfoxides [15]. The structures of the Padua reagent and the Orsay reagent were not established, although in the latter case the molecular weight in solution was indicative of a dimeric structure [8]. Some comparisons have been made between the reactions of the two systems [16]. Aggregation has been confirmed by a strong negative nonlinear effect [17]. NMR studies of the two systems have been carried out, and these show a strong similarity [18] (see below).

Very careful control of the experimental conditions in the preparation of the chiral titanium complex allowed us to optimize the sulfoxidation by CHP, and some results are indicated in Fig. 1. Especially impressive is the high enantiomeric excess, reaching values of 99% in several cases [19a, b].

The decrease in the amount of the combination $Ti(OiPr)_4/(R,R)$-DET/H_2O=1:2:1 (Orsay reagent) drops the enantioselectivity once there is less than 50 mol% of the titanium complex. However it was found that methyl *p*-tolyl sulfoxide (85% *ee*) could be produced with 20 mol% of the titanium complex (instead of 96% *ee*) in the oxidation by CHP. This moderate but significant catalytic sulfoxidation has been achieved in the presence of molecular sieves [20].

It was discovered in the author's laboratory that the combination $Ti(OiPr)_4/(R,R)$-DET/*i*PrOH=1:4:4 may be used in acceptable catalytic conditions (10 mol%) in presence of some 4 Å molecular sieves [21a, b]. Enantiomeris excesses of up to 95% were observed for methyl *p*-tolyl sulfoxide or various aryl methyl sulfoxides.

Von Unge et al. prepared a highly potent gastric acid secretion inhibitor – esomeprazole – on a multi-kilogram scale by using 4 mol% of modified Orsay re-

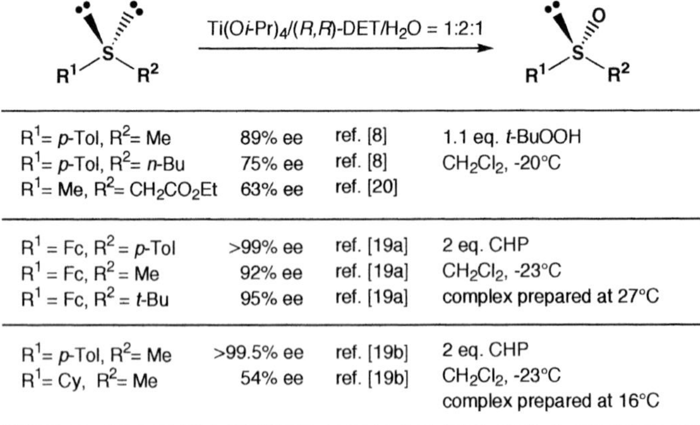

R^1= *p*-Tol, R^2= Me	89% ee	ref. [8]	1.1 eq. *t*-BuOOH
R^1= *p*-Tol, R^2= *n*-Bu	75% ee	ref. [8]	CH_2Cl_2, -20°C
R^1= Me, R^2= CH_2CO_2Et	63% ee	ref. [20]	
R^1 = Fc, R^2 = *p*-Tol	>99% ee	ref. [19a]	2 eq. CHP
R^1 = Fc, R^2 = Me	92% ee	ref. [19a]	CH_2Cl_2, -23°C
R^1 = Fc, R^2 = *t*-Bu	95% ee	ref. [19a]	complex prepared at 27°C
R^1= *p*-Tol, R^2= Me	>99.5% ee	ref. [19b]	2 eq. CHP
R^1= Cy, R^2= Me	54% ee	ref. [19b]	CH_2Cl_2, -23°C
			complex prepared at 16°C

Fig. 1 Some examples of sulfide oxidation by *tert*-butyl hydroperoxide and by cumyl hydroperoxide in the presence of a water-modified chiral titanium complex ("Orsay" reagent, see text). See also ref. 73.

agent $Ti(O\textit{i}Pr)_4$/(*S,S*)-DET/H_2O = 1:2:0.3 giving an *ee* of 91% [22]. They modified the original procedure, preparing the reagent in the presence of the sulfide. The reagent was equilibrated at an elevated temperature, and the oxidation was performed in the presence of an amine, preferably *N,N*-diisopropylethylamine. However, the authors observed a decrease in reproducibility when less than 30 mol% of reagent was used.

The mechanisms of the asymmetric sulfoxidations involving the above various combinations of $Ti(O\textit{i}Pr)_4$/(*R,R*)-DET and some additives are not yet well understood. It is necessary to take into account the diversity of the titanium complexes which are produced by tartrates and which may interconvert in solution [23]. Recently, Potvin and Fieldhouse studied titanium-tartrate mixtures by NMR spectroscopy [18]. The authors stabilized titanium-tartrate complexes by disulfonamides, and with the aid of the ^{1}H- and ^{13}C-NMR spectra they proposed a structure (**A**) for the titanium complex (Fig. 2). However, the interpretation of spectra for the titanium-tartrate complex prepared from $Ti(O\textit{i}Pr)_4$/DIPT 1:2 was more complicated, and the authors proposed that the active systems are more likely to be mixtures of **A**- and **B**-like complexes. In all the asymmetric sulfoxidations promoted or catalyzed by various chiral titanium complexes, it is very reasonable to assume that the hydroperoxide reacts to give a peroxotitanium species (**1**) (Fig. 3). This is well supported by the recent X-ray crystal structure of **3** produced from the reaction of (diethylamino)titanatrane (**2**) with TBHP [24]. It is interesting to note that peroxo complex **3** cleanly oxidizes benzyl methyl sulfide into benzyl methyl sulfoxide at 0 °C in dichloromethane.

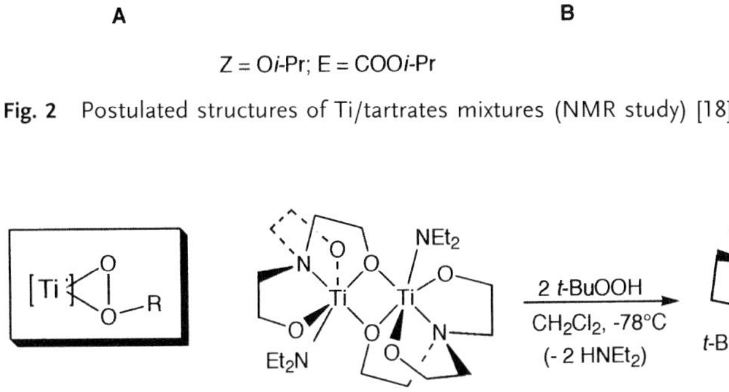

A **B**

Z = O*i*-Pr; E = COO*i*-Pr

Fig. 2 Postulated structures of Ti/tartrates mixtures (NMR study) [18].

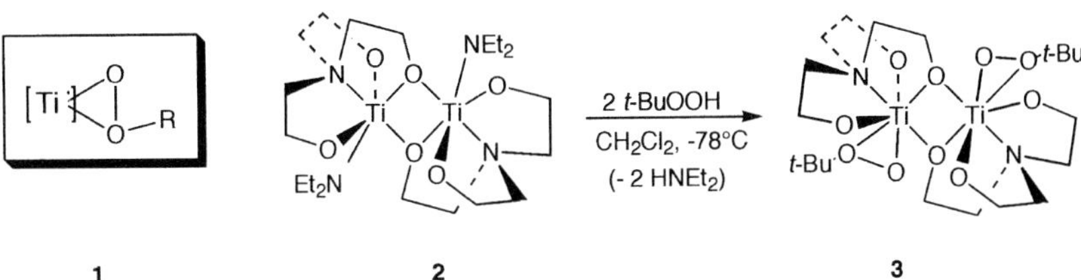

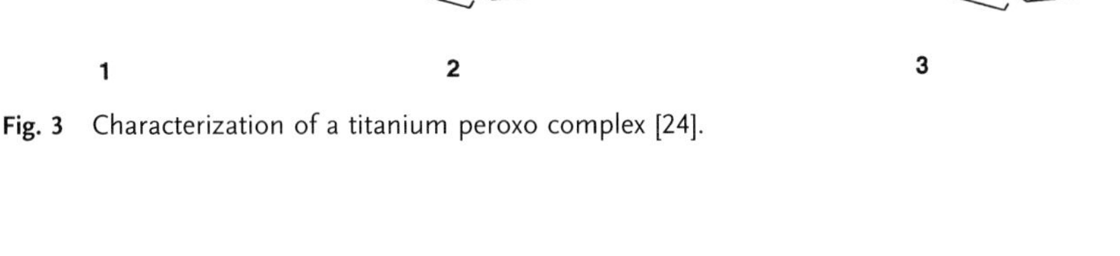

Fig. 3 Characterization of a titanium peroxo complex [24].

2.14.2.2
1,2-Diarylethane 1,2-Diols as Ligands

The replacement of diethyl tartrate by some chiral 1,2-diols in the Orsay water-modified reagent has been studied [25]. Methyl *p*-tolyl sulfoxide could be formed with an *ee* of up to 84% (Fig. 4) using diol **4** as ligand. Interestingly, there is a reversal of absolute configuration for the sulfoxide with the *para*-substituted ligand **5**. A similar inversion of configuration occurred in the oxidation of methyl benzyl sulfide: **4** and **5** gave respectively 6% *ee* (*S*)- and 43% *ee* (*R*)-methyl benzyl sulfoxide.

Catalytic conditions have been developed using a water-modified titanium complex having 1,2-diphenylethane 1,2-diol as ligand [26]. The authors found experimental conditions which avoid the overoxidation to sulfones and decomposition of **6** into various products. The reaction of aryl methyl sulfides was performed at 0 °C with 2 equiv. of TBHP in presence of 5 mol% of the combination Ti(O*i*Pr)$_4$/**6**/H_2O=1:2:20. This catalytic method allowed to reach an *ee* of 99% for benzyl phenyl sulfoxide (Fig. 4).

2.14.2.3
Binol as Ligand

Uemura et al. investigated the replacement of diethyl tartrate by 2,2′-dihydroxy-1,1′-binaphthyl (binol) in the water-modified Sharpless reagent (see Section 2.14.2.1) [27 a]. They developed a titanium catalyst (10 mol%) which had the composition Ti(O*i*Pr)$_4$/(*R*)-binol/H_2O=1:2:20. The reaction was performed in CCl_4 with TBHP (70% in water) at 20 °C. In these conditions, methyl (*R*)-*p*-tolyl sulfoxide (53% *ee*) was produced in 80% yield [27 a, b]. For a useful application see ref. [27 c]. In the initial report, higher *ee* has been noticed in slightly different experimental conditions (see Section 2.14.9.2) [27 a]. In the absence of water, enantioselectivity was very low. The authors assumed the formation of a mononuclear titanium complex with two binaphthyl ligands, in which water affects the structure and rate of formation of this complex. A nonlinear effect was also indicative of complexes with several chiral ligands.

Reetz et al. prepared (*R*)-octahydrobinol (**7**) and its dinitro derivative **8** (Fig. 5) [28]. This last compound was an excellent titanium catalyst when used in the right conditions [27 a]. Methyl *p*-tolyl sulfoxide was obtained with 86% *ee* (kinetic resolution may occur) and with (*S*)-configuration, which is the opposite of the one given by (*R*)-binol.

Recently, Bolm and Dabard reported a catalytic oxidation of sulfides with a novel type of steroid- derived binol analog **9** prepared from equilenine [29]. When the authors used this diol under the conditions described by Uemura, they found an improvement in both applicability and catalytic efficiency. In the presence of 10 mol% of titanium catalyst prepared from Ti(O*i*Pr)$_4$, water, and diol **9**, oxidation of phenyl methyl sulfide by TBHP was performed in THF with high enantioselectivity, giving up to 92% *ee* in 76% chemical yield. It is interesting to note that the oxidation in THF has a much higher enantioselectivity (92% *ee*) than the reaction carried out in DCM (49% *ee*).

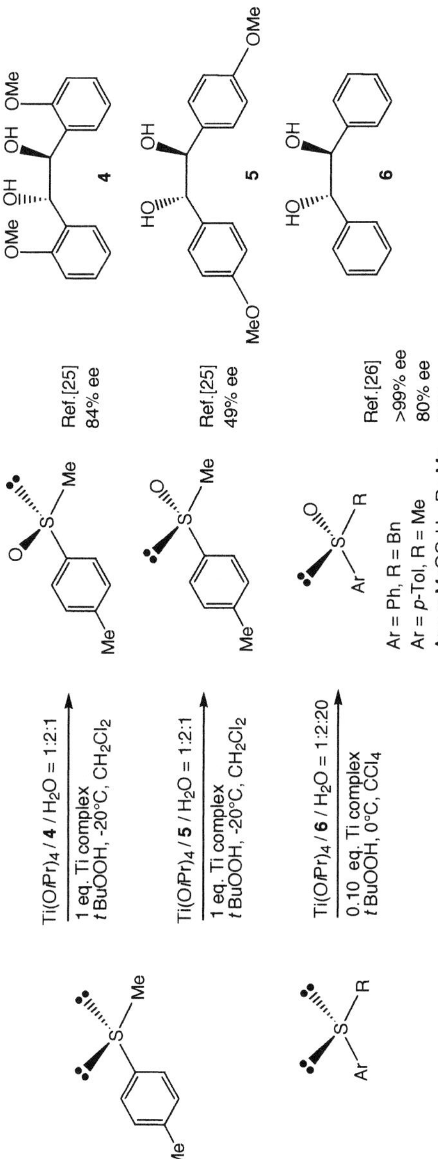

Fig. 4 Water-modified titanium complexes prepared from chiral 1,2-diols.

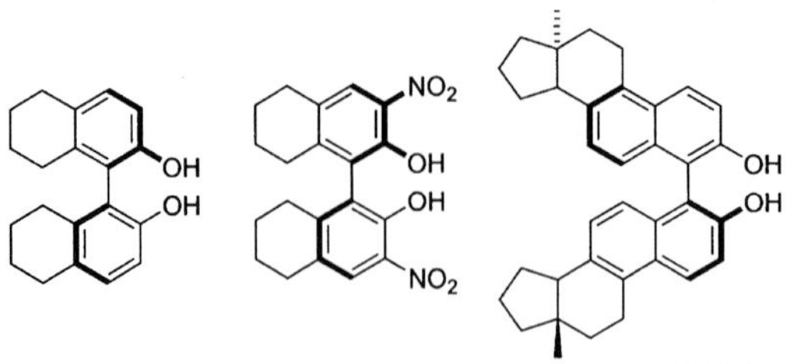

(*R*)-**7** (*R*)-**8** (*S,S*)-**9**

Fig. 5 Titanium/binol derivatives for titanium complexes.

2.14.2.4
Trialkanolamines as Ligands

The reaction between $Ti(OiPr)_4$ and trialkanolamines **10** (Fig. 6) has been studied [30]. Tetradentate titanium complex **11** was characterized by ^1H NMR in $CDCl_3$. Further addition of *t*-BuOOH afforded the peroxo complex **12a** and **11a** (equilibrium constant=3.5 at 22 °C in $CDCl_3$). Catalytic reactions were performed in 1,2-dichloroethane using CHP. A preliminary screening showed that sulfide gave a mixture of sulfoxide and sulfone, the best *ee*s being given by ligand **10b**. 10 mol% of catalyst was routinely used with 0.5 mol/ equiv. of CHP with respect to sulfide. In these conditions the overall chemical yields were excellent (sulfoxide + sulfone). Methyl *p*-tolyl sulfide gave a mixture of (*S*)-sulfoxide (45% *ee*) and sulfone (62:38), while benzyl phenyl sulfide provided (*S*)-sulfoxide (84% *ee*) and sulfone (77:23). Sulfone is produced at the very beginning of the reaction. Some kinetic resolution working in the same direction as asymmetric sulfoxidation has been demonstrated.

2.14.2.5
Chiral Schiff Bases as Ligands

Chiral Schiff bases are easily prepared from chiral amines and may give rise to a wide diversity of structures. In 1986 Pasini et al. prepared oxotitanium complexes **13** (Fig. 7), which are highly active for the oxidation of methyl phenyl sulfide with

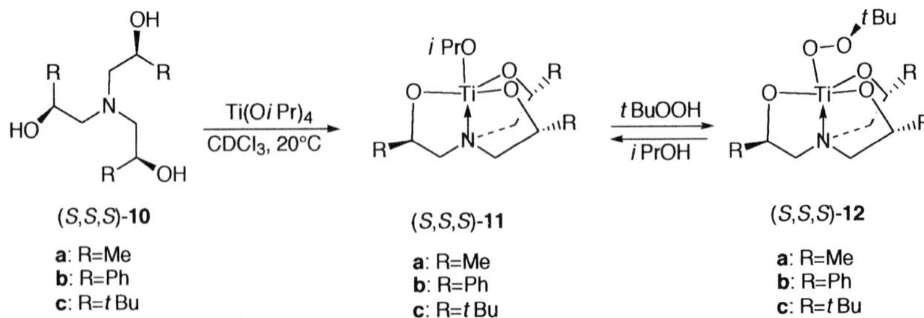

Fig. 6 Chiral titanium complexes prepared from aminotriols [30].

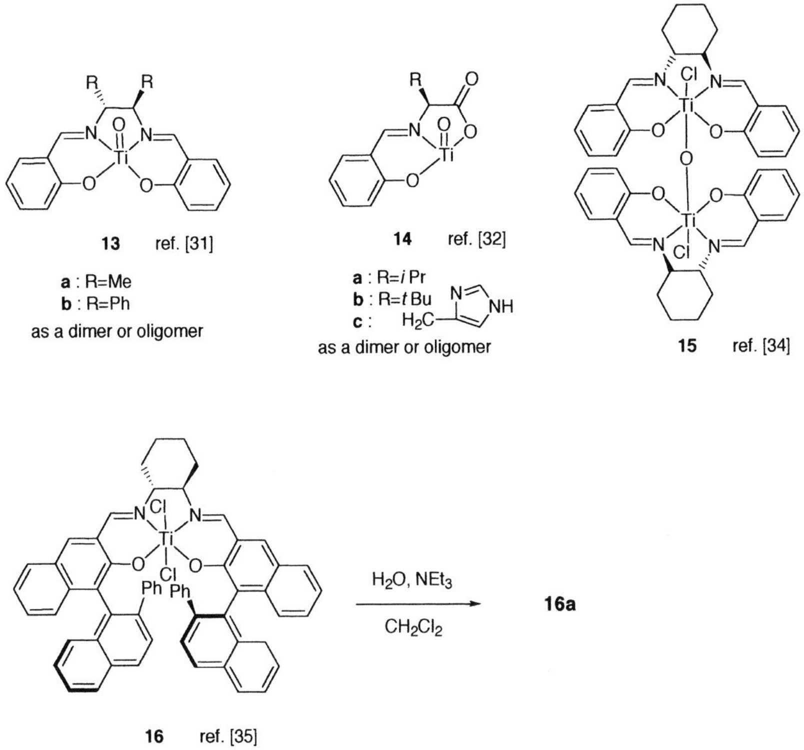

13 ref. [31]

a : R=Me
b : R=Ph
as a dimer or oligomer

14 ref. [32]

a : R=*i* Pr
b : R=*t* Bu
c : H_2C–
as a dimer or oligomer

15 ref. [34]

H_2O, NEt_3 / CH_2Cl_2 → 16a

16 ref. [35]

Fig. 7 Some chiral salen titanium complexes for sulfoxidation.

35% H_2O_2 in methanol or dichloromethane (0.1 mol% catalyst) [31]. However, the enantioselectivity is not higher than 20% and some sulfone is also produced. The authors favored a mechanism with the precoordination of sulfide on titanium followed by the external attack of hydrogen peroxide.

The titanium complex **14** was prepared by Colonna et al. in 1987 from Schiff bases of *α*-aminoacids [32]. They can be used as catalysts (10 mol%) in the oxidation of methyl *p*-tolyl sulfide and various sulfides with *t*-BuOOH. Reactions were performed at room temperature in benzene but gave sulfoxides in *ee*s lower than 25%.

The bis-salen titanium complex **15** has been obtained by serendipity [33]. The chiral salen (salen*) was mixed with $TiCl_4$ in pyridine. Instead of the expected complex (salen* $TiCl_2$), **15** was isolated, whose structure was established by X-ray crystallography [34]. Moisture in pyridine presumably hydrolyzed one Ti-Cl bond and gave rise to the oxo bridge between two titanium atoms. The isolated complex is catalytically active (4 mol%) for the asymmetric sulfoxidation by trityl hydroperoxide in methanol at 0 °C. (*R*)-methyl phenyl sulfoxide was formed in good yield and moderate *ee* (53%). Other peroxides (TBHP or CHP) gave inferior enantioselectivities. The authors assumed that complex **15** is modified in solution in order to generate the catalytically active species.

Recently Saito and Katsuki prepared aryl alkyl sulfoxides in excellent enantiomeric purities by using 2 mol% Ti(salen)-catalyst **16a** (prepared by controlled hydrolysis of **16**) and UHP (urea hydrogen peroxide) as oxidant [35a, b]. They recovered phenyl methyl sulfoxide in 98% *ee* with 78% yield. The reaction could be extended to various sulfides. Interestingly, the authors observed a positive nonlinear effect. As the reactions were carried out in methanol at 0 °C, the authors believed that the homomeric species (*R,R*)- and (*S,S*)- di–oxo complexes were well solubilized and the racemic di-μ-oxo complex was less solubilized thus forming a non-active reservoir.

2.14.3
Sulfoxidation Catalyzed by Chiral Salen Vanadium Complexes

Fujita et al. prepared the salen oxovanadium (IV) complexes **17a**–**17c** (Fig. 8) and used them as catalysts (10 mol%) in asymmetric sulfoxidation [34]. Reactions were performed at room temperature in dichloromethane with CHP. Chemical yields in aryl methyl sulfoxides are excellent but enantioselectivities are lower than 40% *ee* (methyl phenyl sulfoxide).

Bolm and Bienewald greatly improved the catalytic sulfoxidation catalyzed by asymmetric vanadium complexes [36]. They prepared an *in situ* catalyst (1 mol%) by the 1:1 combination of $VO(acac)_2$ and Schiff base **18** (Fig. 8). The oxidations were performed in CH_2Cl_2 by aqueous H_2O_2 (30%, 1.1 equiv. added slowly at room temperature). These conditions minimized the sulfone formation. The best enantioselectivities are shown in Fig. 6; 85% *ee* has been reached in the monooxidation of a dithioacetal. A screening of the structural features of salen **18** established that a sterically demanding group *ortho* to the phenolic hydroxyl enhances the enantioselectivity. A *para*-nitro substituent was also generally beneficial. ^{51}V NMR spectroscopic investigations showed that several species are formed in the conditions of the reaction. Ligands **18** seem well devised for asymmetric sulfoxidation, while the related complexes **17** are quite inefficient [36]. Recently, the asymmetric oxidation of the di-*t*-butyl disulfide to form the corresponding *t*-butyl *t*-butanethiosulfinate was very successful (91% *ee*) [37]. The reaction has been scaled up (1 mole scale) with excellent results using 0.25 mol% of the catalyst [38]. The thiosulfinate is a good precursor of *t*-butanesulfinyl compounds by nucleophilic substitution with full inversion of stereochemistry. Ellman et al. checked Schiff bases prepared from various β-aminoalcohols, the best ligand, **18a**, being derived from *t*-leucinol [38]. Some progress has been made toward elucidating the mechanism of the reaction, which is in competition with a non-selective oxidation route [39]. Many analogs of the Schiff base ligands **18** have been prepared by Berkessel et al. [40]. Introduction of an additional chiral fragment led to match-mismatch effects. The best combination was the compound **18** with X=CH_3 and R=2-*exo*-(*S*)-bornyl, giving 78% *ee* in the oxidation of *o*-bromothioanisole.

Skarzewski et al. screened several Schiff bases deriving from (*S*)-valinol in the oxidation of thioanisol and acyclic disulfides [41]. The enantioselectivity was within the range of those reached by Bolm et al. A bis-sulfoxide of 95% *ee* and 60% *de* has been obtained in 41% yield with ligand **18e**. The high *ee* is the result of the known amplification arising from the two identical asymmetric reactions on a substrate with two prochiral centers [42–45]. Ligand **18e** was also efficient (with the Bolm protocol) for the preparation of a sulfoxide (70% *ee*) from the corresponding γ,δ-unsaturated sulfide [46].

Katsuki et al. tried to improve Bolm's procedure with new Schiff base tridentate ligands [47]. The best ligand was **19**, which gave 87% *ee* in methyl phenyl sulfoxide (for 1 mol% catalyst).

Because of the high activity of the chiral vanadium catalysts and the quite good *ee*s obtained, calculations by a density functional method have been carried out [48]. Various hydroperoxo and peroxo vanadium complexes have been explored as well as the possible transition states in the disulfide oxidation.

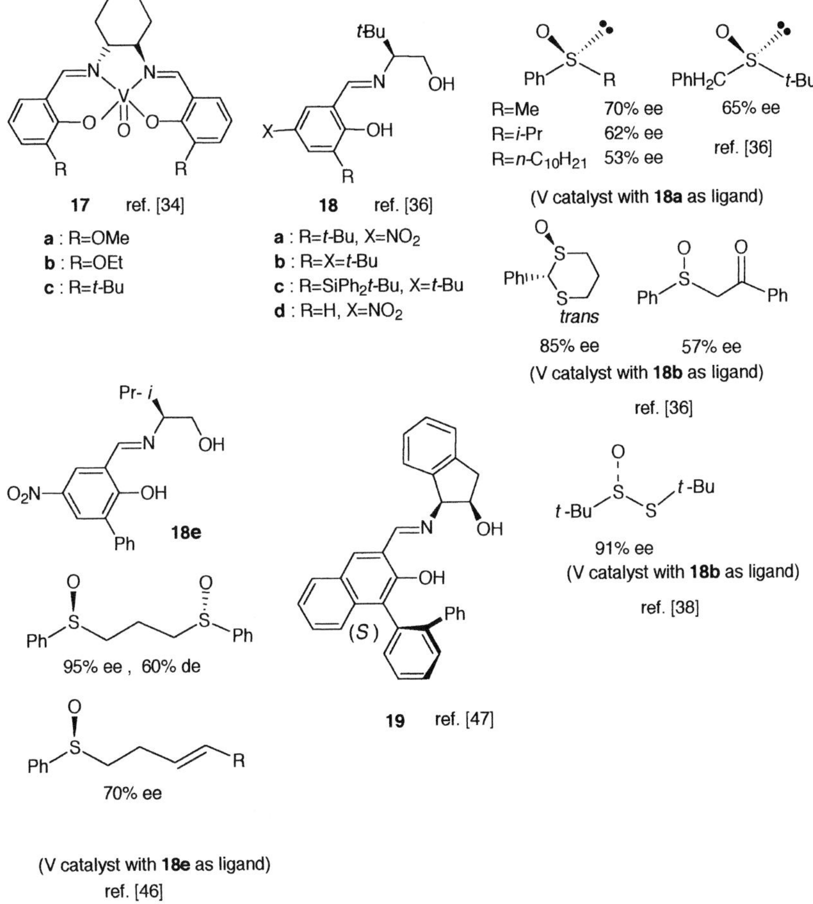

Fig. 8 Chiral vanadium complexes for sulfoxidation.

2.14.4
Sulfoxidation Catalyzed by Chiral Salen Manganese(III) Complexes

Chiral (salen)Mn(III) complexes are excellent catalysts for asymmetric epoxidation of isolated double bonds. Jacobsen et al. found that complex **20** (Fig. 9) catalyzes (2–3 mol%) the asymmetric oxidation of aryl alkyl sulfides with unbuffered 30% hydrogen peroxide in acetonitrile. The maximum enantioselectivity was 68% *ee* (for methyl *o*-bromophenyl sulfoxide) [49]. Katsuki et al. used salen manganese complexes **20** or **21** as catalysts (9 mol%) for sulfoxidation by iodosylbenzene [50a, b]. The reactions were performed at –20 °C in acetonitrile and gave up to 90% *ee* (methyl *o*-nitrophenyl sulfoxide) with catalyst **21 b**. In these conditions the formation of sulfones is almost suppressed. A comparison of the efficiency and enantioselectivity of catalysts **20** and **21** (1 mol%) has been realized by oxidation of methyl phenyl sulfide by 2 mol equiv. of PhIO in acetonitrile. The chemical yields of the methyl phenyl sulfoxide are similar, but there were strong differences in *ee* for **21 a, b**, **22 a, b**, (3% *ee*, 29% *ee*, 20% *ee* and 62% *ee* respectively). Modified Mn(salen) complexes with an additional source of chirality (binaphthyl fragments) have been investigated. Complex **23** was the most efficient, allowing the formation of various alkylsulfoxides (*ee*s around 90%) [51 a]. Unfortunately, io-

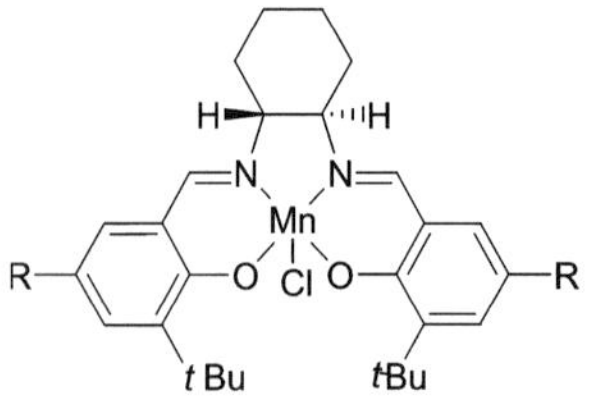

20 ref. [49]
R=OMe

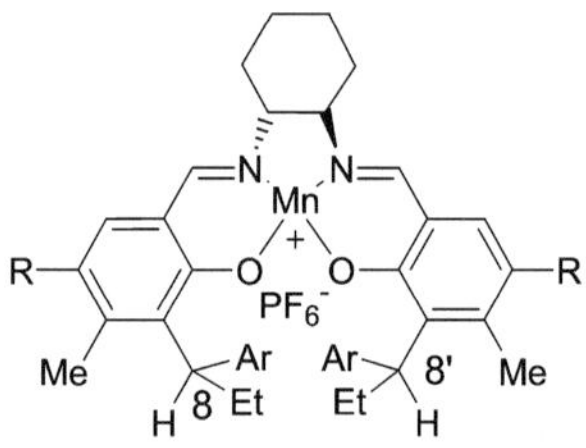

21 ref. [50]
a : R=H; 8R,8'R
b : R=OMe; 8S,8'S
Ar=4-*t*-butylphenyl

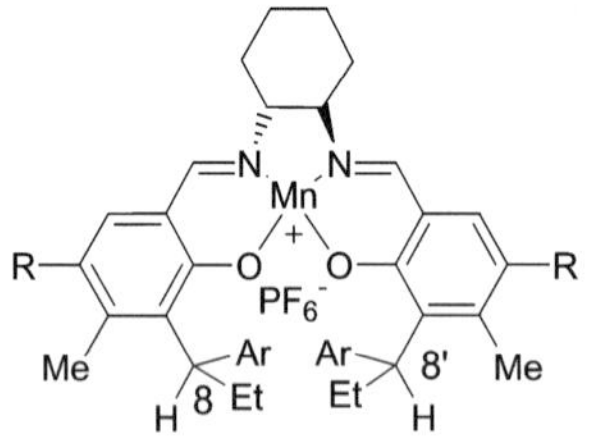

22 ref. [50]
a : R=H; 8R,8'R
b : R=OMe; 8S,8'S
Ar=4-*t*-butylphenyl

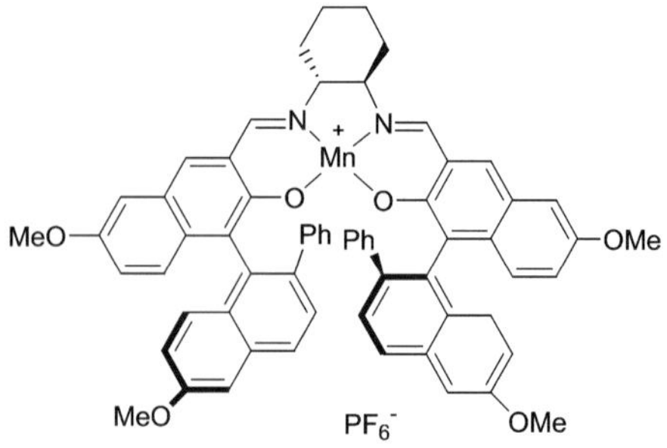

23

Fig. 9 Some chiral salen manganese complexes for sulfoxidation.

dosylbenzene has to be used as terminal oxidant. It is interesting to point out that an achiral salen ligand in combination with (–)-sparteine as an axial coligand enabled thioanisole to be oxidized with 25% *ee* [51 b].

2.14.5
Sulfoxidation Catalyzed by Chiral β-Oxo Aldiminatomanganese(III) Complexes

Mukaiyama et al. developed a new family of manganese catalysts (**24**) for the asymmetric epoxidation of isolated double bonds by the combination RCHO/molecular oxygen. They applied this oxidant system to asymmetric sulfoxidation [52, 53]. The β-oxo aldiminato Mn(III) complex **24a** catalyzes the oxidation of methyl *o*-bromophenyl sulfoxide in toluene at room temperature. The *ee* of methyl *o*-bromophenyl sulfoxide was dependent on the nature of the aldehyde: *t*-BuCHO (52% *ee*), *i*-PrCHO (46% *ee*), *n*-PrCHO (42% *ee*). This has been taken by authors as evidence that oxidation goes through an acylperoxomanganese complex (**25**). Pivalaldehyde has been selected for asymmetric sulfoxidation (some results are listed in Fig. 10) using **24b** as the catalyst (which gave higher *ee*s than **24a**). The chemical yields are satisfactory (60–90%); only in a few cases has sulfone been detected.

2.14.6
Sulfoxidation Catalyzed by Iron or Manganese Porphyrins

In 1990, Groves and Viski prepared binaphthyl iron(III)-tetraphenyl porphyrin [54]. This compound is an active catalyst (0.1 mol%) in the asymmetric oxidation of sulfides with iodosylbenzene. Enantioselectivities up to 48% *ee* (methyl *p*-tolyl sulfoxide) were achieved. The active species is presumably an oxoironporphyrin intermediate.

Naruta et al. simultaneously described asymmetric sulfoxidations catalyzed by the chiral "twin coronet" iron porphyrin [55]. Quite high catalytic activity could be observed in CH_2Cl_2 using iodosylbenzene as oxidant and 1-methylimidazole (which acts as an axial ligand of iron). The reaction was performed at –15 °C in the following

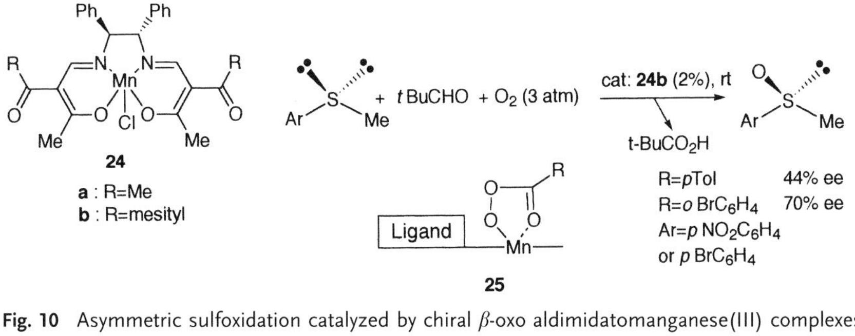

Fig. 10 Asymmetric sulfoxidation catalyzed by chiral β-oxo aldimidatomanganese(III) complexes [52, 53].

conditions: Ar-S-Me/PhIO/porphyrin/1-methylimidazole= 2:1:0.002:0.02. The *ees* were 46% (Ar=Ph), 54% (Ar=*p*-Tol), and 73% (Ar=C_6F_5) with turnover numbers (based on the amount of isolated sulfoxides) of 139, 144, and 55 respectively. In the absence of 1-methylimidazole, phenyl methyl sulfoxide is formed with only 31% *ee*. The authors proposed a mechanism for explaining asymmetric induction [56]. It is based on the steric approach control of the sulfide to the oxo iron center in the molecular cavity.

The two previous examples deal with C_2-symmetric iron-porphyrins. Haltermann et al. catalyzed the oxidation of some sulfides by a D_4-symmetric manganese-tetraphenylporphyrin complex [57]. The reaction was performed at 20 °C in the stoichiometry sulfide/PhIO/porphyrin=2:1:0.005. Methyl phenyl sulfoxide and methyl *o*-bromophenyl sulfoxide were obtained with 55% *ee* and 68% *ee* respectively.

2.14.7
Sulfoxidation Catalyzed by Iron Non-Porphyrinic Complexes

Fontecave et al. prepared the binuclear iron(III) complex **26** (Fig. 11), and found that it catalyzed the oxidation of aryl methyl sulfides by hydrogen peroxide (*ees* of up to 40%) [58]. It was established that the active species is the peroxo adduct of the complex. Recently the authors compared the properties of **26** with an analogous mononuclear iron(III) complex which was less enantioselective [59]. Bolm and Legros developed a new catalyst system based on the combination of [Fe(acac)$_3$] and ligands **18** [60]. The authors used 30% aqueous hydrogen peroxide as the oxidant and 2 mol% catalyst. The yields are usually around 40%, with *ees* of up to 90% (oxidation of methyl *p*-nitrophenyl sulfide). The most promising ligand is **18** (X=R=I).

2.14.8
Sulfoxidation Catalyzed by Chiral Ruthenium or Tungsten Complexes

A new approach to catalytic sulfoxidation has been proposed by Fontecave et al. [61]. It is based on the use of "chiral-at-metal" octahedral Ru(III), bearing only achiral ligands. An enantioselectivity of 18% (oxidation of *p*-bromophenyl methyl sulfide by hydrogen peroxide) was obtained.

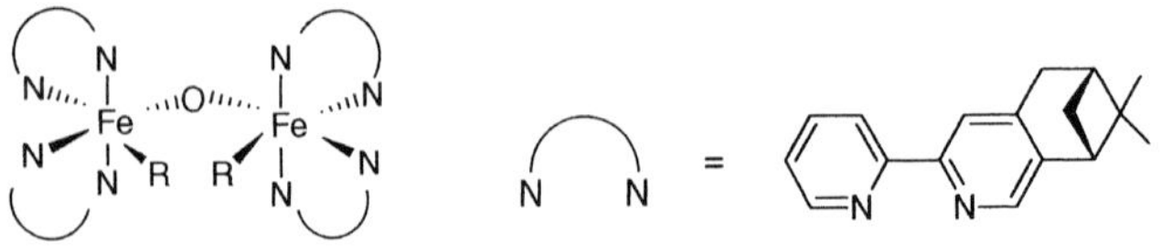

R = OMe

26

Fig. 11 A chiral iron catalyst [58].

A heterogeneous catalytic system (WO_3-L*-30% aq. H_2O_2-THF-0 °C or 25 °C) was recently reported by Sudalai and Thakur [62]. The chiral ligand L* is easily available since it is a cinchona alkaloid such as (–)-quinine or the alkaloid derivatives which are used in the Sharpless asymmetric dihydroxylation of alkenes. For example, benzyl phenyl sulfide has been transformed at 25 °C into the corresponding (*R*)-sulfoxide (53% *ee*) in the presence of 5 mol% of WO_3 and 10 mol% of $(DHQD)_2$-PYR. An interesting application is the asymmetric synthesis of (*R*)-Lanoprazole (84% yield, 88% *ee*), an anti-ulcer drug. The oxidation was performed at 0 °C on the corresponding sulfide, with $(DHQD)_2$-PYR as the chiral auxiliary.

2.14.9 Kinetic Resolution

Two kinds of processes may occur by asymmetric oxidation at sulfur: firstly the kinetic resolution of a *racemic sulfide,* giving a mixture of enantioenriched sulfide and sulfoxide, and secondly the kinetic resolution of a *racemic sulfoxide* with formation of a mixture of enantioenriched sulfoxide and sulfone. This reaction has incidentally been observed in asymmetric sulfoxidation, amplifying the *ee* of the sulfoxide initially obtained.

2.14.9.1 Kinetic Resolution of a Racemic Sulfide

There are a few reports of kinetic resolution using chiral titanium reagents. The Orsay reagent $Ti(OiPr)_4/(R,R)$-DET/H_2O, in combination with TBHP, has been used to resolve racemic sulfides **27** [63, 64], **28** [65], and **29** [66] (Fig. 12) with stereoselectivity factors $s = k_R/k_S$ of 12, 4.5, and 7.0 respectively.

2.14.9.2 Kinetic Resolution of a Racemic Sulfoxide

Uemura et al. developed a $Ti(OiPr)_4/(R)$-binol/H_2O catalyst (see Section 2.14.2.3) which gave kinetic resolution of racemic sulfoxides [27 a, 27 b]. The process amplifies the *ee*s initially obtained in asymmetric sulfoxidation. For example, it was estimated that the asymmetric oxidation generates methyl *p*-tolyl sulfoxide with 53% *ee*, but this sulfoxide may be obtained in 44% yield with 96% *ee* by oxidation of methyl *p*-tolyl sulfide. There is an enhancement of *ee* by overoxidation to sulfone (with a faster oxidation of the minor sulfoxide). This is a general phenomenon often observed in asymmetric syntheses occurring by group selection at a prochiral center. It is detailed in Fig. 12 (**30** → **31** + **32**). Uemura et al. also established that selectivity factor $s = k_R/k_S$ is around 2.2 for the kinetic resolution of methyl *p*-tolyl sulfoxide, with some asymmetric amplification when the binol is not enantiomerically pure [27 b].

Imamoto et al. catalytically oxidized methyl *p*-tolyl sulfide into sulfoxide using 2,2,5,5-tetramethyl-3,4-hexanediol as a ligand. The initial oxidation gave 40% *ee*

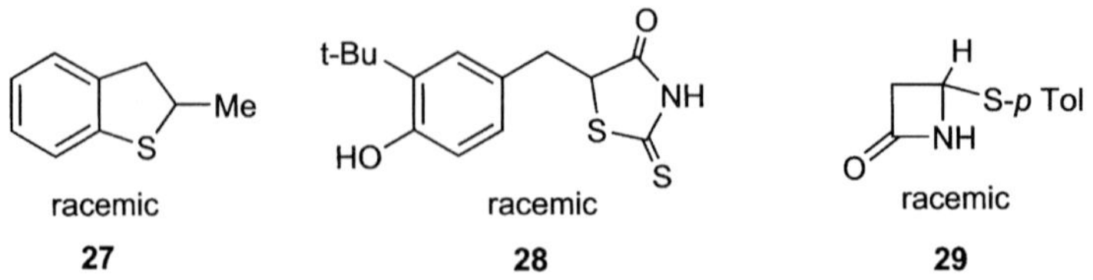

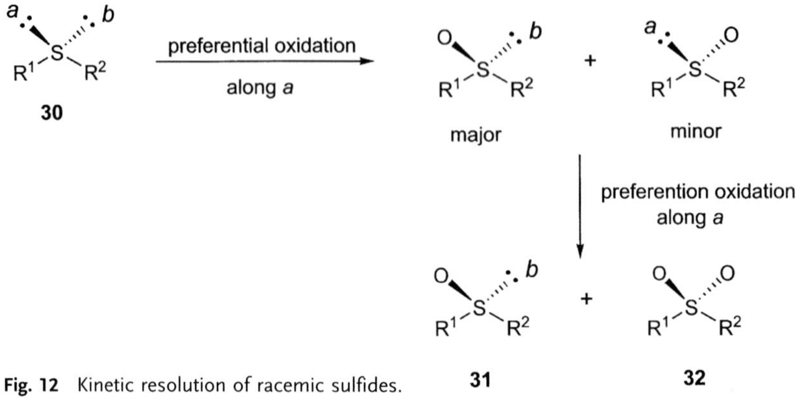

racemic
27

racemic
28

racemic
29

preferential oxidation along *a*

major

minor

preferention oxidation along *a*

31

32

Fig. 12 Kinetic resolution of racemic sulfides.

(20% yield), but, because of the kinetic resolution at the end of the reaction, sulfoxide was obtained in 42% yield with 95% *ee* ($s=3.0$) [67].

The Orsay reagent $Ti(O\textit{i}Pr)_4/(R,R)$-DET/$H_2O$ with DET of various *ee*'s catalyzes the kinetic resolution of racemic methyl *p*-tolyl sulfoxide during its oxidation into sulfone ($s=2.2$) [74]. Also, the chiral salen manganese(III) catalyst **24b** gave a poor kinetic resolution of methyl phenyl sulfoxide, and the *β*-oxo aldiminato manganese(III) complex **18** catalyzed oxidation of methyl *p*-nitrophenyl sulfoxide ($s=2.0$) into sulfone [55, 56].

Kinetic resolution of racemic sulfoxides Ar-S(O)-Me by CHP and the Padova reagent $Ti(O\textit{i}Pr)_4/(R,R)$-DET$=1:4$ at $-23\,^{\circ}C$ in CH_2Cl_2 gave significant results. Thus, at 65% conversion, sulfoxides (*R*)-ArS(O)Me with Ar$=$*p*-Tol, Ph, *p*-ClC_6H_4 were isolated with 83% *ee*, 87% *ee*, and 94% *ee* respectively [68]. Some reduced kinetic resolution occurs if the titanium complex is used in catalytic amount (20%), as established for R$=$*p*-ClC_6H_4 at room temperature (1 equiv. Ti: 64% *ee*, 0.2 equiv. Ti: 41% *ee*).

It was shown that the heterogeneous catalytic system [WO_3-cinchona alkaloids]-30% aq. H_2O_2-THF give some kinetic resolution of racemic sulfoxides [62].

2.14.9.3 Kinetic Resolution of Racemic Hydroperoxides during Asymmetric Sulfoxidation

Scretti et al. discovered that racemic furyl hydroperoxides such as **33** may be used instead of CHP for the asymmetric oxidation of methyl *p*-tolyl sulfide in the presence of 1 equiv. of the Padova reagent $Ti(O\textit{i}Pr)_4/(R,R)$-DIPT$=1:4$ [68, 69]. The de-

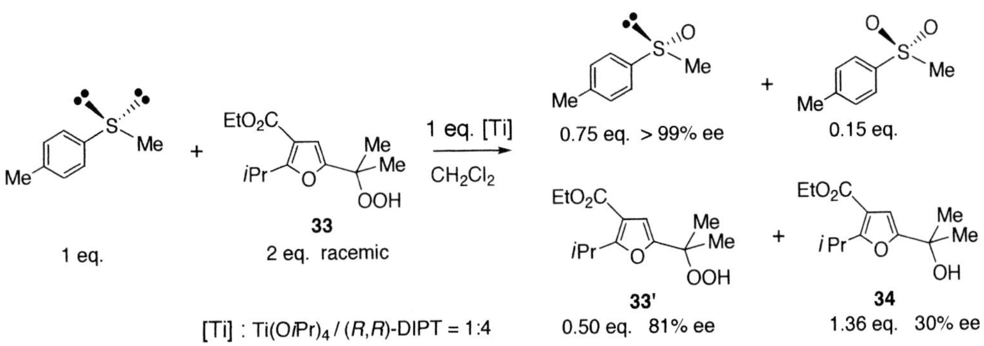

Fig. 13 Kinetic resolution of racemic hydroperoxides.

tails of one experiment are given in Fig. 13. The furyl alcohol **34** (30% *ee*) derives from hydroperoxide **33** which has reacted in the sulfoxidation. From these data one can estimate that the kinetic resolution of **33** occurred with $s = 2.0$.

2.14.10 Conclusion

Asymmetric sulfoxidation mediated or catalyzed by chiral organometallic species may give very high enantioselectivities, although mainly related to structures such as Ar-S(O)-Me. This approach has been used on a multikilogram scale in industry [22, 70–72]. There is growing interest in the area of catalytic sulfoxidation, and respectable enantioselectivities have been achieved. However, chiral catalysts combining both high catalytic activity and high enantioselectivity have yet to be found. The problem of avoiding overoxidation to sulfone has been solved in many cases. Kinetic resolution of racemic sulfides or sulfoxides by asymmetric oxidation has so far met with only moderate success ($s < 12$), and further work to improve this situation is needed.

References

1 G. Solladié, *Synthesis* **1981**, 185–196.
2 G.H. Posner in *The Chemistry of Sulfones and Sulfoxides* (Eds.: S. Patai, Z. Rappoport, C.J.M. Sterling), J. Wiley and Sons, Chichester, UK, **1988**, Chapter 16.
3 M.C. Carreno, *Chem. Rev.* **1995**, *95*, 1717–1760.
4 I. Fernández, N. Khiar, *Chem. Rev.* **2003**, *103*, 3651–3705.
5 K.K. Andersen in *The Chemistry of Sulfones and Sulfoxides* (Eds.: S. Patai, Z. Rappoport, C.J.M. Sterling), J. Wiley and Sons, Chichester, UK, **1988**, Chapter 3.
6 I. Fernandez, N. Khiar, J.M. Lhera, F. Alcudia, *J. Org. Chem.* **1992**, *57*, 6789–6796.

7 F. A. Davis, J. P. McCauley Jr., M. E. Harakal, *J. Org. Chem.* **1984**, *49*, 1465–1467.

8 P. Pitchen, M. Deshmukh, E. Dunach, H. B. Kagan, *J. Am. Chem. Soc.* **1984**, *106*, 8188–8193.

9 F. Furia, G. Modena, R. Seraglia, *Synthesis* **1984**, 325–326.

10 T. Sugimoto, T. Kokubo, J. Miyazaki, S. Tanimoto, M. Okano, *J. Chem. Soc. Chem. Commun.* **1989**, 1052–1053.

11 S. Colonna, S. Banfi, M. Sommaruga, *J. Org. Chem.* **1985**, *50*, 769–771.

12 H. B. Kagan in *Catalytic Asymmetric Synthesis* (Ed.: I. Ojima), J. Wiley and Sons, NY, **2000**, Chapter 6C, 325–354.

13 C. Bolm, K. Muniz, J. P. Hildebrand, *Comprehensive Asymmetric Catalysis* (Eds.: E. N. Jacobsen, A. Pfaltz, H. Yamamoto), Springer, Berlin Heidelberg New York, **1999**, 697–710.

14 K. Katsuki, *Adv. Synth. Catal.* **2002**, *344*, 131–147.

15 S. Zhao, O. Samuel, H. B. Kagan, *Org. Synth.* **1989**, *68*, 49–56.

16 V. Conte, F. Di Furia, G. Licini, G. Modena, G. Sbampato in *Dioxygen Activation and Homogeneous Catalytic Oxidation* (Ed.: L. J. Simandi), Elsevier Science Publishers, Amsterdam, **1991**, 385–394.

17 C. Puchot, O. Samuel, E. Dunach, S. Zhao, C. Agami, H. B. Kagan, *J. Am. Chem. Soc.* **1986**, *108*, 2353–2357.

18 P. G. Poitvin, B. G. Fieldhouse, *Tetrahedron: Asymmetry* **1999**, *10*, 1661–1672.

19 (a) P. Diter, O. Samuel, S. Taudien, H. B. Kagan, *Tetrahedron: Asymmetry* **1994**, *5*, 549–552. (b) J. M. Brunel, P. Diter, M. Deutsch, H. B. Kagan, *J. Org. Chem.* **1995**, *60*, 8086–8088.

20 S. Zhao, O. Samuel, H. B. Kagan, *Tetrahedron* **1987**, *43*, 5135–5144.

21 (a) J. M. Brunel, H. B. Kagan, *Synlett* **1996**, 404–406. (b) J. M. Brunel, H. B. Kagan, *Bull. Soc. Chim. Fr.* **1996**, *133*, 1109–1115.

22 H. Cotton, T. Elebring, M. Larsson, L. Li, H. Sörensen, S. von Unge, *Tetrahedron: Asymmetry* **2000**, *11*, 3819–3825.

23 D. J. Berrisford, K. B. Sharpless, C. Bolm, *Angew. Chem. Int. Ed. Engl.* **1995**, *34*, 1059–1070.

24 G. Boche, K. Möbus, K. Harms, M. Marsch, *J. Am. Chem. Soc.* **1996**, *118*, 2770–2771.

25 K. Yamamoto, H. Ands, T. Shuetaka, H. Chikamatsu, *J. Chem. Soc. Chem. Commun.* **1989**, 754–755.

26 S. Superchi, C. Rosini, *Tetrahedron: Asymmetry* **1997**, *8*, 349–352.

27 (a) N. Komatsu, M. Hashizume, T. Sugita, S. Uemura, *J. Org. Chem.* **1993**, *58*, 4529–4533. (b) N. Komatsu, M. Hashizume, T. Sugita, S. Uemura, *J. Org. Chem.* **1993**, *58*, 7624–7626. (c) M. M. Capozzi, C. Cardellicchio, G. Fracchiolla, F. Naso, P. Tortorella, *J. Am. Chem. Soc.* **1999**, *121*, 4708–4709.

28 M. T. Reetz, C. Merk, G. Naberfeld, J. Rudolph, N. Griebenow, R. Goddard, *Tetrahedron Lett.* **1997**, *38*, 5273–5276.

29 C. Bolm, O. A. G. Dabard, *Synlett* **1999**, *3*, 360–362.

30 F. Di Furia, G. Licini, G. Modena, R. Motterle, W. A. Nugent, *J. Org. Chem.* **1996**, *61*, 5175–5177.

31 A. Colombo, G. Marturano, A. Pasini, *Gazz. Chim. Ital.* **1986**, *116*, 35–40.

32 S. Colonna, A. Manfredi, M. Spadoni, L. Casella, M. Gulloti, *J. Chem. Soc. Perkin Trans I* **1987**, 71–73.

33 K. Nakajima, C. Sasaki, M. Kojima, T. Aoyama, S. Ohba, Y. Saito, J. Fujita, *Chem. Lett.* **1987**, 2189–2192.

34 K. Nakajima, M. Kojima, J. Fujita, *Chem. Lett.* **1986**, 1483–1486.

35 (a) B. Saito, T. Katsuki, *Tetrahedron Lett.* **2001**, *42*, 3874–3876. (b) B. Saito, T. Katsuki, *Tetrahedron Lett.* **2001**, *42*, 8333–8336.

36 C. Bolm, F. Bienewald, *Angew. Chem. Int. Ed. Engl.* **1995**, *34*, 2640–2642.

37 G. Liu, D. A. Cogan, J. Ellman, *J. Am. Chem. Soc.* **1997**, *119*, 9913–9914.

38 D. A. Cogan, G. Liu, K. Kim, B. J. Backes, J. A. Ellman, *J. Am. Chem. Soc.* **1998**, *120*, 8011–8019.

39 S. A. Blum, R. G. Bergman, J. A. Ellman, *J. Org. Chem. Soc.* **2003**, *68*, 150–155.

40 A. H. Vetter, A. Berkessel, *Tetrahedron Lett.* **1998**, *39*, 1741–1744.

41 J. Skarzewski, E. Ostrycharz, R. Siedlecka, *Tetrahedron: Asymmetry* **1999**, *10*, 3457–3461.

42 T. R. Hoye, J. C. Suhadonik, *J. Am. Chem. Soc.* **1985**, *107*, 5312–5313.

43 K. Soai, H. Hori, M. Kawahara, *J. Chem. Soc. Chem. Commun.* **1994**, 106.

44 V. Rautenstrauch, *Bull. Soc. Chim. Fr.* **1994**, *131*, 515–524.

45 S. El Baba, K. Sartor, J. C. Poulin, H. B. Kagan, *Bull. Soc. Chim. Fr.* **1994**, *131*, 525–533.

46 J. Skarzewski, E. Wojaczynska, I. Turowska-Tyrk, *Tetrahedron: Asymmetry* **2002**, *13*, 369–375.

47 C. Ohta, H. Shimieu, A. Kondo, T. Katsuki, *Synlett* **2002**, 161–163.

48 B. Balcells, F. Maseras, A. Lledo, *J. Org. Chem.* **2002**, *67*, 161–163.

49 M. Palucki, P. Hanson, E. N. Jacobsen, *Tetrahedron Lett.* **1992**, *33*, 7111–7114.

50 (a) K. Noda, N. Hosoya, K. Yanai, R. Irie, T. Katsuki, *Tetrahedron Lett.* **1994**, *35*, 1887–1890. (b) K. Noda, N. Hosoya, R. Irie, Y. Yamashita, T. Katsuki, *Tetrahedron* **1994**, *50*, 9609–9618.

51 (a) C. Kokubo, T. Katsuki, *Tetrahedron* **1996**, *52*, 13895–13900. (b) T. Hashihayata, Y. Ito, T. Katsuki, *Tetrahedron Lett.* **1997**, *38*, 9541–9544.

52 K. Imagawa, T. Nagata, T. Yamada, T. Mukaiyama, *Chem. Lett.* **1995**, 335–336.

53 T. Nagata, K. Imagawa, T. Yamada, T. Mukaiyama, *Bull. Chem. Soc. Jpn.* **1995**, *68*, 3241–3246.

54 J. T. Groves, P. Viski, *J. Org. Chem.* **1990**, *55*, 3628–3634.

55 Y. Naruta, F. Tani, K. Maruyama, *J. Chem. Soc. Chem. Commun.* **1990**, 1378–1380.

56 Y. Naruta, F. Tani, K. Maruyama, *Tetrahedron: Asymmetry* **1991**, *2*, 533–542.

57 R. L. Haltermann, S. T. Jan, H. L. Nimmens, *Synlett* **1991**, 791–792.

58 C. Duboc-Toia, S. Ménage, C. Lambeaux, M. Fontecave, *Tetrahedron Lett.* **1997**, *38*, 3727–3730.

59 Y. Mekmouche, H. Hummel, R. N. Y. Ho, L. Que Jr., V. Schünemann, F. Thomas, A. X. Trautwein, C. Lebrun, K. Gorgy, J.-C. Leprêtre, M.-N. Collomb, A. Deronzier, M. Fontecave, S. Ménage, *Chem. Eur. J.* **2002**, *8*, 1195–1204.

60 J. Legros, C. Bolm, *Angew. Chem. Int. Ed.* **2003**, *42*, 5487–5489.

61 M. Chavarot, S. Ménage, O. Hamelin, F. Chanay, J. Pecaut, M. Fontecave, *Inorg. Chem.* **2003**, *42*, 4810–4816.

62 V. V. Thakur, A. Sudalai, *Tetrahedron: Asymmetry* **2003**, *14*, 407–410.

63 T. Takata, W. Ando, *Tetrahedron Lett.* **1986**, *27*, 1591–1594.

64 W. Ando, L. Huang, *Tetrahedron Lett.* **1986**, *27*, 3391–3394.

65 M. I. Phillips, D. M. Berry, J. A. Panetta, *J. Org. Chem.* **1992**, *57*, 4047–4049.

66 C. Nemecek, H. B. Kagan, *Pol. J. Chem.* **1994**, *68*, 2467–2475.

67 Y. Yamanoi, T. Imamoto, *J. Org. Chem.* **1997**, *62*, 8560–8564.

68 A. Scretti, F. Bonadies, A. Lattanzi, A. Senatore, A. Soriente, *Tetrahedron: Asymmetry* **1996**, *7*, 657–658.

69 A. Scretti, F. Bonadies, A. Lattanzi, *Tetrahedron: Asymmetry* **1996**, *7*, 629–632.

70 P. Pitchen, C. J. France, I. M. McFarlane, C. G. Newton, D. M. Thompson, *Tetrahedron Lett.* **1994**, *35*, 485–488.

71 P. J. Hogan, P. A. Hopes, W. O. Moss, G. E. Robinson, I. Patel, *Org. Proc. Res. Dev.* **2002**, 6, 225–229.

72 Sun Pharmaceutical Industries Limited (India), International patent WO 03/089408.

73 M. M. Capozzi, C. Cardellicchio, F. Naso, P. Tortorella, *J. Org. Chem.* **2000**, *65*, 2843–2846.

74 T. O. Luukas, C. Girard, D. Denwick, H. B. Kagan, *J. Am. Chem. Soc.* **1999**, *121*, 9299–9306.

2.15
Amine Oxidation

Shun-Ichi Murahashi and Yasushi Imada

2.15.1
Introduction

Amines can be oxidized readily; however, selective oxidations are generally very difficult. To accomplish selective oxidation of amines, generation of single oxidizing species is essential. Metabolism of amines is controlled selectively by various enzymes such as amine oxidase, flavoenzyme, and cytochrome P-450. The functions of these enzymes are classified by oxidase and oxygenase, i.e. dehydrogenation and the oxygen atom transfer reactions, respectively. The transition metal-catalyzed reactions of amines with various oxidizing reagents may correspond to these reactions. In this chapter, these two types of catalytic oxidation reactions of amines will be described.

The oxidation of amines with stoichiometric amounts of metal salts has been reviewed recently [1, 2].

2.15.2
Low-Valent Transition Metals for Catalytic Dehydrogenative Oxidation of Amines

Activation of amines with low-valent transition metal catalysts gives two types of key intermediates. The reaction of an amine which has an N–H bond gives an imine metal complex (**1**) [3–5], while that of an amine without an N–H bond gives an iminium ion complex (**2**) (Scheme 1) [6]. Using these intermediates, various catalytic transformations of amines can be explored.

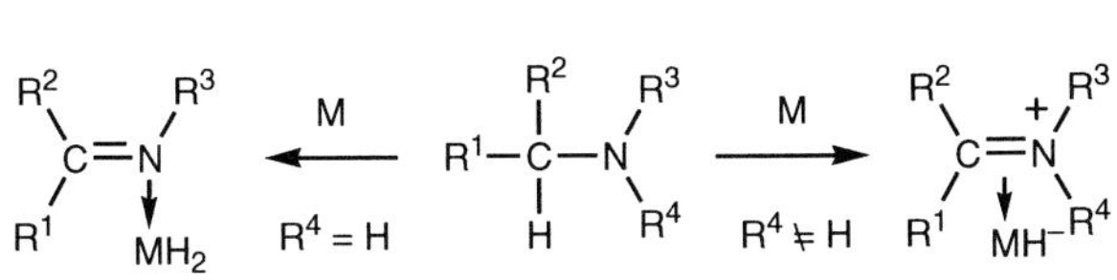

Scheme 1. The key intermediate of catalytic dehydrogenative oxidation of amines.

Transition Metals for Organic Synthesis, Vol. 2, 2nd Edition.
Edited by M. Beller and C. Bolm

ISBN: 3-527-30613-7

2.15.2.1
Oxidation of Primary and Secondary Amines

The study of the generation of iminium ions by activating amines with transition metal catalysts led to the discovery of catalytic transalkylations of amines [3–5]. Pd black is an excellent catalyst, although other heterogeneous and homogeneous transition metal catalysts can be used similarly. Variation of the exchange reaction can open up convenient processes for the synthesis of tertiary amines, diamines, polyamines (Eq. 1), and heterocyclic amines.

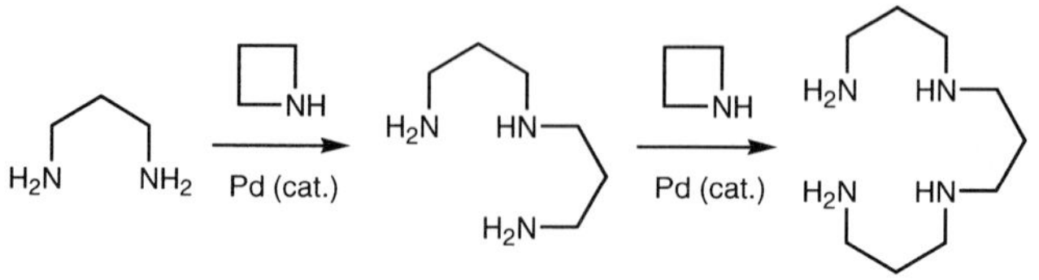

(1)

The key intermediate for the reaction is an imine metal hydride complex (**1**), which is derived from oxidative addition of a low-valent metal to the N–H bond and subsequent β-metal hydride elimination. Nucleophilic addition of a second molecule of amine to **1** gives **3**, and intramolecular reductive cleavage of **3** with the metal hydride provides amines with exchanged substituents (Scheme 2).

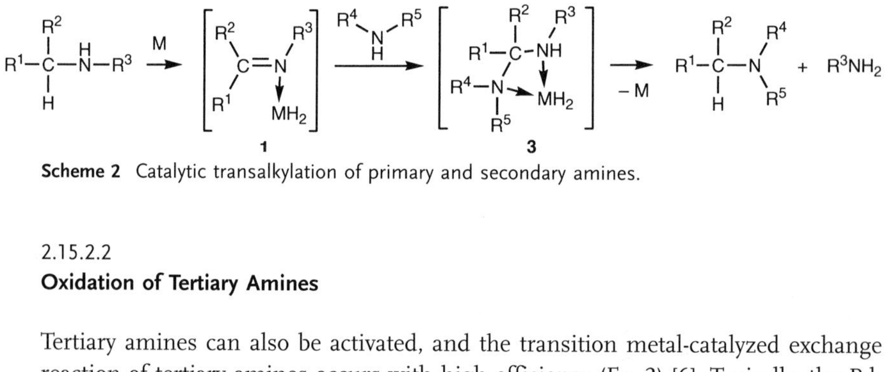

Scheme 2 Catalytic transalkylation of primary and secondary amines.

2.15.2.2
Oxidation of Tertiary Amines

Tertiary amines can also be activated, and the transition metal-catalyzed exchange reaction of tertiary amines occurs with high efficiency (Eq. 2) [6]. Typically, the Pd-catalyzed reaction of dibutylhexylamine gave a mixture of tributylamine (26%), dibutylhexylamine (37%), butyldihexylamine (24%), and trihexylamine (3%); the alkyl groups are distributed statistically in these tertiary amines. This process may provide a convenient method for the synthesis of unsymmetrical tertiary amines.

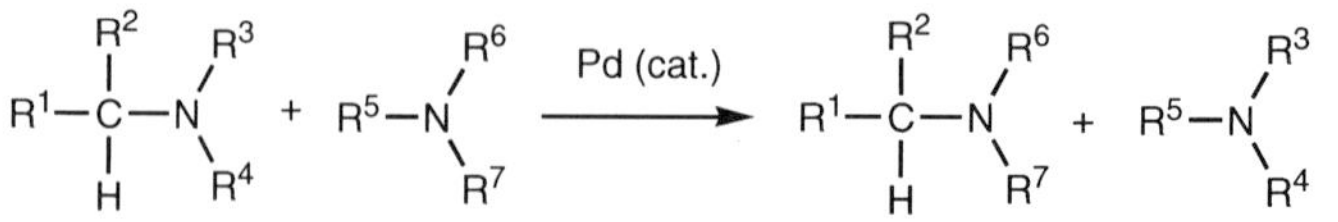

$$R^1R^2CH{-}NR^3R^4 + R^5{-}NR^6R^7 \xrightarrow{\text{Pd (cat.)}} R^1R^2CH{-}NR^6R^7 + R^5{-}NR^3R^4 \quad (2)$$

This reaction can be rationalized by assuming a mechanism which involves iminium ion–palladium complex **2**. The transition metal coordinates to nitrogen and

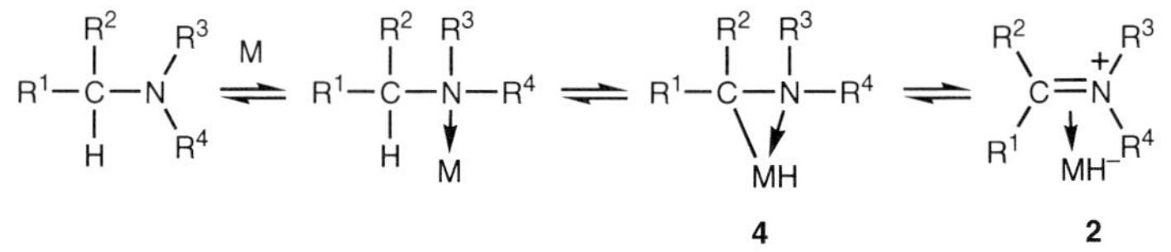

Scheme 3 Activation of tertiary amines by metal catalyst.

inserts into the adjacent C–H bond to give **4**, which is in equilibrium with a key intermediate, the iminium ion complex **2** (Scheme 3) [6]. The nucleophilic attack of a second molecule of tertiary amine on the extremely electrophilic **2** and subsequent reductive cleavage gives products. This is the pioneering work of heteroatom-induced α-C–H bond activation with metals or metal complexes.

The Rh-catalyzed asymmetric isomerization of allylamines (**5**) to enamines (**6**) (Eq. 3), which is one of the key steps of the industrial synthesis of menthol, is initiated by C–H activation to form the iminium–RhH^- π-complex similar to **2** [7].

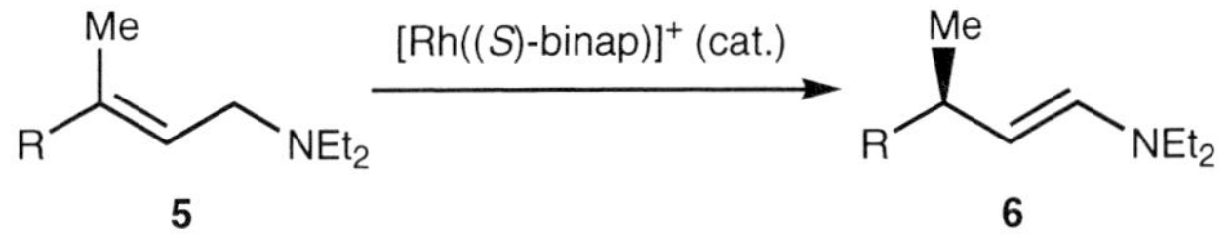

(3)

The iminium ion metal complex **2** can be trapped with an external nucleophile. Thus, palladium-catalyzed hydrolysis of tertiary amines can be performed upon treatment with Pd black in the presence of water (Eq. 4) [8]. The reaction proceeds via nucleophilic attack of water on **2** followed by cleavage.

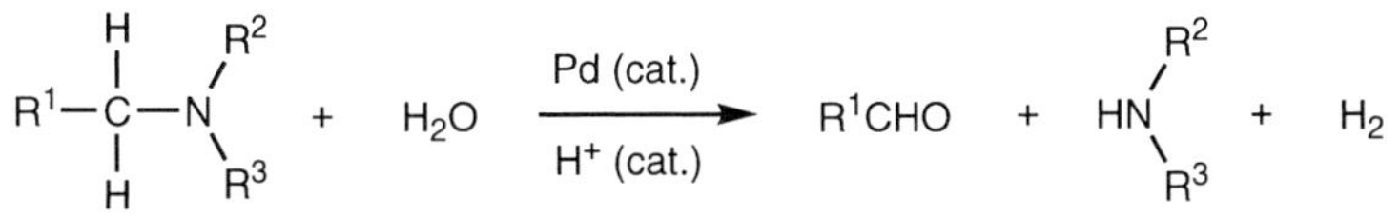

$$R^1CH_2NR^2R^3 + H_2O \xrightarrow[H^+\ (cat.)]{Pd\ (cat.)} R^1CHO + HNR^2R^3 + H_2 \qquad (4)$$

Similar catalytic reactions proceed in the presence of the homogeneous cluster catalysts such as $Rh_6(CO)_{16}$, $Ru_3(CO)_{12}$, and $Os_3(CO)_{12}$ [9–11]. (η^1-Ylide)palladium complexes [12] and unusual amino–carbene cluster complexes [13] have been isolated as key intermediates in these reactions.

2.15.3
Metal Hydroperoxy and Peroxy Species for Catalytic Oxygenation of Amines

Reaction of early transition metals such as V, Mo, W, and Ti with H_2O_2 gives metal hydroperoxy or peroxy species, of which peroxygens have an electrophilic nature with respect to H_2O_2. Oxygen transfer from these species to a nitrogen atom takes place readily to perform oxygenation of amines.

2.15.3.1
Oxygenation of Secondary Amines

Hydroperoxytungstate, which is generated by the reaction of tungstate with H_2O_2, is an excellent reagent for the oxygenation of secondary amines. Direct oxidative transformation of secondary amines to nitrones was discovered in 1984. Thus, the treatment of secondary amines with 30% H_2O_2 in the presence of Na_2WO_4 catalyst in either MeOH or water gives the corresponding nitrones (**7**) in good yields (Eq. 5) [14–16]. This single-step synthesis of nitrones from secondary amines is extremely useful, since the products are highly valuable synthetic intermediates and spin-trapping reagents. The oxidation of secondary amines with hydroperoxytungstate gives hydroxylamines, which undergo further oxidation followed by dehydration to give nitrones (**7**). An important variation of this reaction is the decarboxylative oxidation of *N*-alkyl-*α*-amino acids to give the nitrones regioselectively (Eq. 6) [17, 18].

$$R^1-CH(R^2)-NH-R^3 \xrightarrow[Na_2WO_4\ (cat.)]{H_2O_2} R^1R^2C{=}N^+(R^3)-O^- \quad (\mathbf{7}) \tag{5}$$

$$R^1-CH(CO_2H)-NH-R^2 \xrightarrow[Na_2WO_4\ (cat.)]{H_2O_2} R^1(H)C{=}N^+(R^2)-O^- \tag{6}$$

SeO_2-catalyzed oxidation of secondary amines in acetone is also convenient for the synthesis of water-soluble nitrones [19]. Since the above two methods were discovered, similar transformations of secondary amines to nitrones have been reported using peroxotungstophosphate (PCWP) [20, 21], titanium silicate (TS-1) [22], and $MeRe(O)_3$ catalysts [23–25]. Urea–H_2O_2 complex (UHP) is also used as an alternative oxidant in the presence of Na_2WO_4, Na_2MoO_4, and SeO_2 catalysts [26]. Peroxo species, such as $L_nW(O_2)$ [21], $Ti(\mu\text{-}O_2)$ [22], and $MeRe(O)(O_2)_2$ [23–25] are proposed as active oxidants.

An assembled catalyst of phosphotungstate and non-crosslinked amphiphilic polymer is a highly active immobilized catalyst for organic solvent-free oxidations with H_2O_2 [27]. Alkyl hydroperoxides can be used for the oxidation under anhydrous conditions in the presence of trialkanolamine-bound Ti complex catalyst [28].

The tungstate-catalyzed reaction can be used for the oxidation of various substrates. The oxidation of tetrahydroquinolines provides a convenient method for the synthesis of hydroxamic acids (**8**) (Eq. 7) [29]. The reaction is rationalized by assuming the formation of nitrones and subsequent addition of H_2O_2 to give 2-hydroperoxy-*N*-hydroxylamines, which undergo dehydration to give **8**.

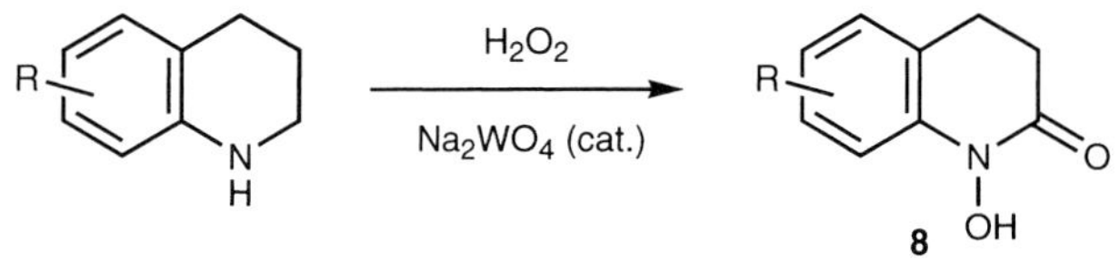

$$\text{R-tetrahydroquinoline} \xrightarrow[\text{Na}_2\text{WO}_4\text{ (cat.)}]{\text{H}_2\text{O}_2} \text{N-hydroxy-R-3,4-dihydroquinolin-2-one } (\mathbf{8}) \tag{7}$$

2.15.3.2
Oxygenation of Primary Amines

Oxygen transfer from metal peroxides to primary amines results in a wide variety of oxidized products, depending on the oxidant and reaction conditions employed. Scheme 4 outlines the oxygenation of primary amines, which gives nitro compounds by way of hydroxylamines and nitroso compounds. Further, nitroso compounds are rather reactive intermediates, which undergo condensation with amines (**9**) or hydroxylamines (**10**) to give azo or azoxy compounds, and nitrosoalkanes having an α-hydrogen are readily rearranged to the oximes.

Primary amines having no α-hydrogen, such as anilines and *tert*-butylamine, are oxidized to the nitro compounds upon treatment with *t*-BuOOH in the presence of a catalytic amount of Mo and V complexes [30] and chromium silicate (CrS-2) [31] at elevated temperature. The oxidation of anilines with *t*-BuOOH in the presence of Ti complex gives azoxybenzenes [32].

Nitroso compounds are synthetically useful reagents; however, selective oxidation of primary amines to nitroso compounds is difficult because of overoxidation and formation of coupling products. Selective, catalytic oxidations of anilines with H_2O_2 to nitrosobenzenes can be performed in the presence of (dipic)Mo(O)(O_2) (hmpa) (dipic = pyridine-2,6-dicarboxylato) [33] and Mo(O)$(O_2)_2(H_2O)$(hmpa) [34] (Eq. 8).

$$\text{ArNH}_2 \xrightarrow[\text{(dipic)Mo(O)(O}_2\text{)(tmpa) (cat.) or Mo(O)(O}_2\text{)(H}_2\text{O)(hmpa) (cat.)}]{\text{H}_2\text{O}_2} \text{ArNO} \tag{8}$$

The oxidation of anilines catalyzed by peroxotungstophosphate (PCWP) can give some different oxidized products, depending on the reaction conditions employed. Thus, PCWP-catalyzed oxidation of anilines gives nitrosobenzenes selectively

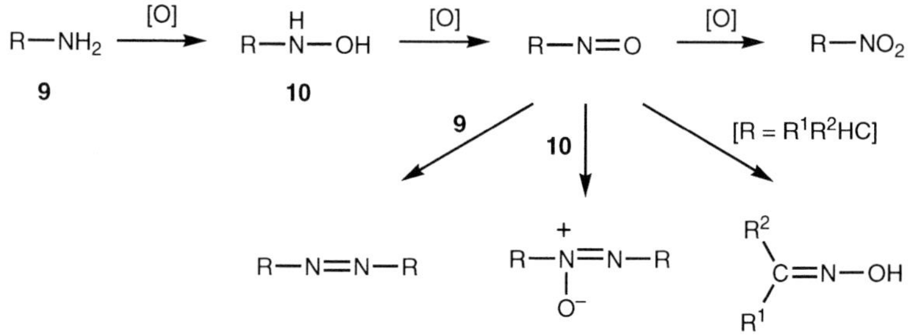

$$\underset{\mathbf{9}}{\text{R—NH}_2} \xrightarrow{[\text{O}]} \underset{\mathbf{10}}{\text{R—N(H)—OH}} \xrightarrow{[\text{O}]} \text{R—N=O} \xrightarrow{[\text{O}]} \text{R—NO}_2$$

$$\text{R—N=O} \xrightarrow{\mathbf{9}} \text{R—N=N—R}; \quad \text{R—N=O} \xrightarrow{\mathbf{10}} \text{R—N}^+(\text{O}^-)\text{=N—R}; \quad \text{R—N=O} \xrightarrow{[\text{R} = \text{R}^1\text{R}^2\text{HC}]} \text{R}^1\text{R}^2\text{C=N—OH}$$

Scheme 4 Metal-catalyzed oxygenation of primary amines.

upon treatment with 35% H_2O_2 at room temperature, while similar oxidations at high temperature and those with diluted H_2O_2 afford nitrobenzenes and azoxybenzenes, respectively [35]. The $MeRe(O)_3$-catalyzed H_2O_2 oxidations of primary amines which have no α-hydrogen also afford nitroso [36] or nitro [37] compounds, depending on the reaction conditions employed.

Nitrosoalkanes possessing α-hydrogens undergo prototopic rearrangement to give oximes. Typically, cyclohexanone oxime, which is an intermediate for nylon-6, can be obtained by the oxidation of cyclohexylamine with H_2O_2 in the presence of Na_2MoO_4, Na_2WO_4 [38], PCWP [20], $Mo(O)(O_2)(H_2O)$(hmpa) [39], $MeRe(O)_3$ [25], and amphiphilic-polymer-bound phosphotungstate [27] catalysts.

2.15.3.3
Oxygenation of Tertiary Amines

The oxidation of tertiary amines is simple in comparison to those of secondary and primary amines. *N*-Oxides are the only products derived from the oxygen transfer from metal peroxides to a nitrogen atom. Thus, tertiary amines are readily oxidized to the corresponding amine *N*-oxides with catalytic systems such as Mo, V, or Ti/ROOH [40, 41] and Na_2WO_4 [42], $MeRe(O)_3$ [36, 43], Mn–porphyrin [44], TS-1 [45], or tungstate-exchanged Mg–Al-layered double hydroxide/H_2O_2 [46] (Eq. 9). Molecular oxygen can be used as an alternative oxidant in the presence of Ru catalyst [47].

$$R^1-N(R^2)(R^3) \xrightarrow[\text{M (cat.)}]{\text{ROOH or } H_2O_2} R^1-\overset{+}{N}(R^2)(R^3)-O^- \qquad (9)$$

2.15.4
Metal Oxo Species for Catalytic Oxygenation of Amines

Since oxidative *N*-dealkylation of tertiary amines mediated by oxoiron species (Fe=O) is an important cytochrome P-450-specific reaction, model reactions for *N*-demethylation of tertiary methylamines using Fe porphyrins have been reported [48–51]. The reaction may involve the iminium ion intermediates, which are derived by transfer of an electron from nitrogen to oxoiron species followed by transfer of hydrogen. Generation of metal oxo species by the reaction of transition metals with monooxygen donors will provide a new type of oxygenation of amines.

2.15.4.1
Oxygenation of Tertiary Amines

Ruthenium(II) complex-catalyzed oxidation of tertiary amines with *t*-BuOOH gives the α-(*tert*-butyldioxy)alkylamines **11** with high efficiency (Eq. 10) [52]. Benzylic and allylic positions and carbon–carbon double bonds tolerate the oxidation.

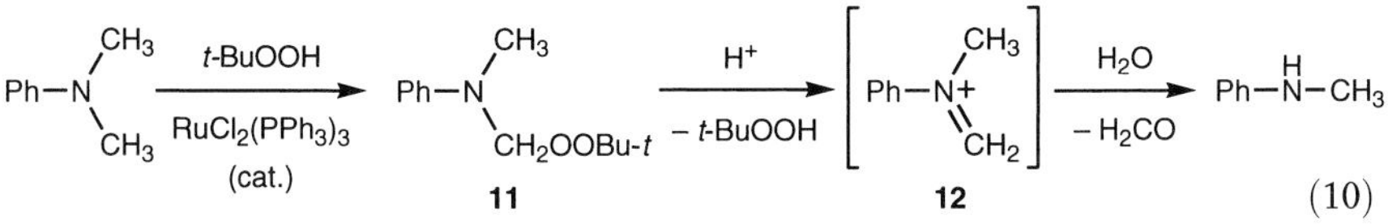

$$\text{Ph-N(CH}_3)_2 \xrightarrow[\text{RuCl}_2(\text{PPh}_3)_3 \text{ (cat.)}]{t\text{-BuOOH}} \underset{\mathbf{11}}{\text{Ph-N(CH}_3)\text{CH}_2\text{OOBu-}t} \xrightarrow[-\,t\text{-BuOOH}]{\text{H}^+} \underset{\mathbf{12}}{\left[\text{Ph-N}^+(\text{CH}_3)\text{=CH}_2\right]} \xrightarrow[-\,\text{H}_2\text{CO}]{\text{H}_2\text{O}} \text{Ph-NH-CH}_3 \quad (10)$$

Selective *N*-demethylation of tertiary methylamines is performed by this Ru-catalyzed oxidation and subsequent hydrolysis of **11** with an aqueous HCl solution [52]. This is the first synthetically practical method for the *N*-demethylation of tertiary methylamines. The reaction involves protonation, removal of *t*-BuOOH, and hydrolysis of iminium ion intermediate **12**. Generation of the iminium ion **12** also provides novel methods for the construction of piperidine structures via an olefin–iminium ion cyclization reaction [52].

The oxidation reaction can be rationalized by assuming the cytochrome P-450-type mechanism (Scheme 5). Ru^{IV}=O complex is generated by the reaction of Ru^{II} complex with *t*-BuOOH. Tertiary amines react with Ru^{IV}=O species by electron transfer followed by proton transfer to give iminium ion complex **13**. Nucleophilic attack of *t*-BuOOH on **13** gives **14**, water, and Ru^{II} species to complete the catalytic cycle [52].

Oxoruthenium species can be generated with other monooxygen donors, and iminium ion complex **13** thus obtained can be trapped with other nucleophiles. The Ru-catalyzed oxidation of tertiary methylamines with H_2O_2 in MeOH gives α-methoxymethylamines (**15**) with high efficiency (Eq. 11) [53]. The reaction also provides an efficient method for selective *N*-demethylation of tertiary methylamines and construction of quinoline skeletons from tertiary methylamines. The Ru-catalyzed oxidation with H_2O_2 in the presence of NaCN gives α-cyanomethylamines (**16**), which are readily hydrolyzed giving α-amino acids (Eq. 12) [54].

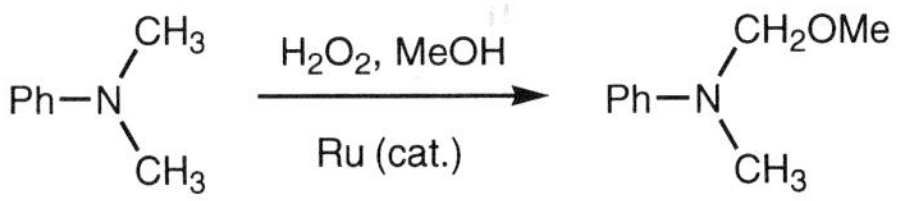

$$\text{Ph-N(CH}_3)_2 \xrightarrow[\text{Ru (cat.)}]{\text{H}_2\text{O}_2,\ \text{MeOH}} \underset{\mathbf{15}}{\text{Ph-N(CH}_3)\text{CH}_2\text{OMe}} \quad (11)$$

$$\text{R}^1\text{R}^2\text{CH-NR}^3\text{R}^4 \xrightarrow{\text{Ru}^{IV}\text{=O}} \underset{\mathbf{13}}{\left[\text{R}^1\text{R}^2\text{C=N}^+\text{R}^3\text{R}^4 \quad \text{Ru}^{II}(\text{OH})\right]} \xrightarrow[-\,\text{Ru}^{II},\ -\,\text{H}_2\text{O}]{t\text{-BuOOH}} \underset{\mathbf{14}}{\text{R}^1\text{R}^2\text{C}(t\text{-BuOO})\text{-NR}^3\text{R}^4}$$

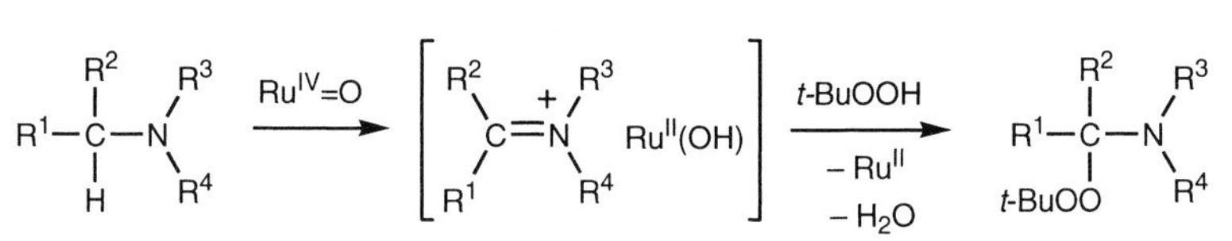

Scheme 5 Ru-catalyzed oxidation of teriary amines with *t*-BuOOH.

$$\text{Ph-N(CH}_3)_2 \xrightarrow[\text{Ru (cat.)}]{H_2O_2,\ \text{NaCN}} \underset{\mathbf{16}}{\text{Ph-N(CH}_3)\text{CH}_2\text{CN}} \xrightarrow{H_3O^+} \text{Ph-N(CH}_3)\text{CH}_2\text{CO}_2\text{H} \quad (12)$$

Aerobic oxidation of *N,N*-disubstitutred anilines in the presence of Fe(salen) [55] or $CoCl_2$ [56] proceeds to give *N*-substituted anilines along with *N*-formyl derivatives. Catalytic *α*-cyanation of tertiary arylamines has been reported to proceed using the $FeCl_3/O_2$–PhCOCN [57] or $RuCl_3/O_2$–NaCN system [58] to give **16**.

2.15.4.2
Oxygenation of Secondary and Primary Amines

Treatment of secondary amines with *t*-BuOOH in the presence of $RuCl_2(PPh_3)_3$ catalyst at room temperature gives the corresponding imines **17** in high yields (Eq. 13) [59]. The reaction proceeds via iminium ion complex **15** (R^4=H), which undergoes decomposition to give imines. This is the first catalytic oxidative transformation of secondary amines to imines.

$$R^1\text{-C(R}^2\text{)(H)-NH-}R^3 \xrightarrow[\text{RuCl}_2(\text{PPh}_3)_3\ (\text{cat.})]{t\text{-BuOOH}} \underset{\mathbf{17}}{R^1R^2\text{C=N-}R^3} \quad (13)$$

Secondary amines can be transformed into either imines or nitrones by changing the active oxidizing species. Thus, the $RuCl_2(PPh_3)_3$-catalyzed oxidation of secondary amines with *t*-BuOOH gives imines **18** via oxometal species (M=O) (Eq. 14) [59], while the Na_2WO_4-catalyzed oxidation with H_2O_2 gives nitrones **19** via hydroperoxymetal species (MOOH) (Eq. 15) [15].

$$\text{tetrahydroisoquinoline (BnO, MeO; NH)} \xrightarrow[\text{RuCl}_2(\text{PPh}_3)_3\ (\text{cat.})\ [\text{Ru=O}]]{t\text{-BuOOH}} \mathbf{18}\ (\text{BnO, MeO; C=N}) \quad (14)$$

$$\text{tetrahydroisoquinoline (BnO, MeO; NH)} \xrightarrow[\text{Na}_2\text{WO}_4\ (\text{cat.})\ [\text{WOOH}]]{H_2O_2} \mathbf{19}\ (\text{BnO, MeO; CH=N}^+\text{-O}^-) \quad (15)$$

After the catalytic oxidation of secondary amines to imines was demonstrated [58], similar transformations were reported recently by using catalytic systems such as

$RuCl_2(PPh_3)_3$/PhIO [60, 61], Co(salen)/O_2 [62], Co(salen)/*t*-BuOOH [63], Mo–V heteropolyoxometalate/O_2 [64, 65], $NiSO_4/K_2S_2O_8$ [66], Pr_4NRuO_4/*N*-methylmorpholine *N*-oxide [67], and hydroxyapatite-bound Ru/O_2 [68].

Oxidation of primary amines having an α-CH_2 group gives the corresponding nitriles using catalytic systems of $K_2RuO_4/K_2S_2O_8$ [69], $RuCl_3/O_2$ [70], and hydroxyapatite-bound Ru/O_2 [68]. Primary amines can be converted to nitriles in the presence of *trans*-$[Ru^{VI}(tmp)(O)_2]$ (tmp = tetramesitylporphyrin) under air (Eq. 16) [71].

$$PhCH_2NH_2 \xrightarrow[\textit{trans}\text{-}[Ru^{VI}(tmp)(O)_2]\ (cat.)]{O_2} PhCN \qquad (16)$$

2.15.5 Conclusion

Catalytic oxidative transformation of secondary amines either to nitrones (**7**) or to imines (**17**), both of which react readily with various nucleophiles affording α-substituted hydroxylamines and amines in a diastereo- or an enantioselective manner, is extremely useful. Oxidative transformations of tertiary *N*-methylarylamines to α-oxygenated amines (**14**), which generate highly reactive iminium ions upon treatment with acid and react readily with various nucleophiles, is also important.

References

1 *Organic Syntheses by Oxidation with Metal Compounds* (Eds.: W. J. Mijs, C. R. H. I. de Jonge), Plenum Press, New York, **1986.**
2 T. L. Gilchrist in *Comprehensive Organic Synthesis* (Eds.: B. M. Trost, I. Fleming), Vol. 7, Pergamon Press, London, **1991**, 735–756.
3 S.-I. Murahashi, N. Yoshimura, T. Tsumiyama, T. Kojima, *J. Am. Chem. Soc.* **1983**, *105*, 5002–5011.
4 N. Yoshimura, I. Moritani, T. Shimamura, S.-I. Murahashi, *J. Am. Chem. Soc.* **1973**, *95*, 3038–3039.
5 N. Yoshimura, I. Moritani, T. Shimamura, S.-I. Murahashi, *J. Chem. Soc., Chem. Commun.* **1973**, 307–308.
6 S.-I. Murahashi, T. Hirano, T. Yano, *J. Am. Chem. Soc.* **1978**, *100*, 348–350.
7 S. Inoue, H. Takaya, K. Tani, S. Otsuka, T. Sato, R. Noyori, *J. Am. Chem. Soc.* **1990**, *112*, 4897–4905.
8 S.-I. Murahashi, T. Watanabe, *J. Am. Chem. Soc.* **1979**, *101*, 7429–7430.
9 R. M. Laine, D. W. Thomas, L. W. Cary, S. E. Buttrill, *J. Am. Chem. Soc.* **1978**, *100*, 6527–6528.
10 Y. Shvo, R. M. Laine, *J. Chem. Soc., Chem. Commun.* **1980**, 753–754.
11 R. B. Wilson Jr., R. M. Laine, *J. Am. Chem. Soc.* **1985**, *107*, 361–369.
12 R. McCrindle, G. Ferguson, G. J. Arsenault, A. J. McAlees, *J. Chem. Soc., Chem. Commun.* **1983**, 571–572.
13 R. D. Adams, H.-S. Kim, S. Wang, *J. Am. Chem. Soc.* **1985**, *107*, 6107–6108.
14 H. Mitsui, S. Zenki, T. Shiota, S.-I. Murahashi, *J. Chem. Soc., Chem. Commun.* **1984**, 874–875.

15 S.-I. Murahashi, H. Mitsui, T. Shiota, T. Tsuda, S. Watanabe, *J. Org. Chem.* **1990**, *55*, 1736–1744.
16 S.-I. Murahashi, T. Shiota, Y. Imada, *Org. Synth.* **1991**, *70*, 265–271.
17 S.-I. Murahashi, Y. Imada, H. Ohtake, *J. Org. Chem.* **1994**, *59*, 6170–6172.
18 H. Ohtake, Y. Imada, S.-I. Murahashi, *Bull. Chem. Soc. Jpn.* **1999**, *72*, 2737–2754.
19 S.-I. Murahashi, T. Shiota, *Tetrahedron Lett.* **1987**, *28*, 2383–2386.
20 S. Sakaue, Y. Sakata, Y. Nishiyama, Y. Ishii, *Chem. Lett.* **1992**, 289–292.
21 F.P. Ballistreri, U. Chiacchio, A. Rescifina, G.A. Tomaselli, R.M. Toscano, *Tetrahedron* **1992**, *48*, 8677–8684.
22 R. Joseph, A. Sudalai, T. Ravindranathan, *Synlett* **1995**, 1177–1178.
23 A. Goti, L. Nanneli, *Tetrahedron Lett.* **1996**, *37*, 6025–6028.
24 R.W. Murray, K. Iyanar, *J. Org. Chem.* **1996**, *61*, 8099–8102.
25 S. Yamazaki, *Bull. Chem. Soc. Jpn.* **1997**, *70*, 877–883.
26 E. Marcantoni, M. Petrini, O. Polimanti, *Tetrahedron Lett.* **1995**, *36*, 3561–3562.
27 Y.M.A. Yamada, H. Tabata, H. Takahashi, S. Ikegami, *Synlett* **2002**, 2031–2034.
28 M. Forcato, W.A. Nugent, G. Licini, *Tetrahedron Lett.* **2003**, *44*, 49–52.
29 S.-I. Murahashi, T. Oda, T. Sugahara, Y. Masui, *J. Org. Chem.* **1990**, *55*, 1744–1749.
30 G.R. Howe, R.R. Hiatt, *J. Org. Chem.* **1970**, *35*, 4007–4012.
31 B. Jayachandran, M. Sasidharan, A. Sudalai, T. Ravindranathan, *J. Chem. Soc., Chem. Commun.* **1995**, 1523–1524.
32 K. Kosswig, *Justus Liebigs Ann. Chem.* **1971**, *749*, 206–208.
33 E.R. Møller, K.A. Jørgensen, *J. Am. Chem. Soc.* **1993**, *115*, 11814–11822.
34 S. Tollari, M. Cuscela, F. Porta, *J. Chem. Soc., Chem. Commun.* **1993**, 1510–1511.
35 S. Sakaue, T. Tsubakino, Y. Nishiyama, Y. Ishii, *J. Org. Chem.* **1993**, *58*, 3633–3638.
36 Z. Zhu, J.H. Espenson, *J. Org. Chem.* **1995**, *60*, 1326–1332.
37 R.W. Murray, K. Iyanar, J. Chen, J.T. Wearing, *Tetrahedron Lett.* **1996**, *37*, 805–808.
38 K. Kahr, *Angew. Chem.* **1960**, *72*, 135–137.
39 S. Tollari, F. Porta, *J. Mol. Catal.* **1993**, *84*, L137–L140.
40 L. Kuhnen, *Chem. Ber.* **1966**, *99*, 3384–3386.
41 M.N. Sheng, J.G. Zajacek, *J. Org. Chem.* **1968**, *33*, 588–590.
42 P. Burckard, J.P. Fleury, F. Weiss, *Bull. Soc. Chim. Fr.* **1965**, 2730–2733.
43 C. Copéret, H. Adolfsson, T.-A.V. Khuong, A.K. Yudin, K.B. Sharpless, *J. Org. Chem.* **1998**, *63*, 1740–1741.
44 A. Thellend, P. Battioni, W. Sanderson, D. Mansuy, *Synthesis* **1997**, 1387–1388.
45 M.R. Prasad, G. Kamalakar, G. Madhavi, S.J. Kulkarni, K.V. Raghavan, *Chem. Commun.* **2000**, 1577–1578.
46 B.M. Choudary, B. Bharathi, C.V. Reddy, M.L. Kantam, K.V. Raghavan, *Chem. Commun.* **2001**, 1736–1737.
47 S.L. Jain, B. Sain, *Chem. Commun.* **2002**, 1040–1041.
48 P. Shannon, T.C. Bruice, *J. Am. Chem. Soc.* **1981**, *103*, 4580–4582.
49 N. Miyata, H. Kiuchi, M. Hirobe, *Chem. Pharm. Bull.* **1981**, *29*, 1489–1492.
50 J.R. Lindsay-Smith, D.N. Mortimer, *J. Chem. Soc., Perkin Trans. 2* **1986**, 1743–1749.
51 K. Fujimori, S. Fujiwara, T. Takata, S. Oae, *Tetrahedron Lett.* **1986**, *27*, 581–584.
52 S.-I. Murahashi, T. Naota, K. Yonemura, *J. Am. Chem. Soc.* **1988**, *110*, 8256–8258.
53 S.-I. Murahashi, T. Naota, N. Miyaguchi, T. Nakato, *Tetrahedron Lett.* **1992**, *33*, 6991–6994.
54 S.-I. Murahashi, N. Komiya, JP 11255729, **1999** [*Chem. Abstr.* **1999**, *131*, 214088].
55 S. Murata, M. Miura, M. Nomura, *J. Org. Chem.* **1989**, *54*, 4700–4702.
56 S. Murata, A. Tamatani, K. Suzuki, M. Miura, M. Nomura, *Chem. Lett.* **1990**, 757–760.
57 S. Murata, K. Teramoto, M. Miura, M. Nomura, *Bull. Chem. Soc. Jpn.* **1993**, *66*, 1297–1298.

58 S.-I. Murahashi, N. Komiya, H. Terai, T. Nakae, *J. Am. Soc.* **2004**, *125*, 15312–15313.

59 S.-I. Murahashi, T. Naota, H. Taki, *J. Chem. Soc., Chem. Commun.* **1985**, 613–614.

60 P. Müller, D.M. Gilabert, *Tetrahedron* **1988**, *44*, 7171–7175.

61 F. Porta, C. Crotti, S. Cenini, G. Palmisano, *J. Mol. Catal.* **1989**, *50*, 333–341.

62 A. Nishinaga, S. Yamazaki, T. Matsuura, *Tetrahedron Lett.* **1988**, *29*, 4115–4118.

63 K. Maruyama, T. Kusukawa, Y. Higuchi, A. Nishinaga, *Chem. Lett.* **1991**, 1093–1096.

64 R. Newmann, M. Levin, *J. Org. Chem.* **1991**, *56*, 5707–5712.

65 K. Nakayama, M. Hamamoto, Y. Nishiyama, Y. Ishii, *Chem. Lett.* **1993**, 1699–1702.

66 S. Yamazaki, *Chem. Lett.* **1992**, 823–826.

67 A. Goti, M. Romani, *Tetrahedron Lett.* **1994**, *35*, 6567–6570.

68 K. Mori, K. Yamaguchi, T. Mizugaki, K. Ebitani, K. Kaneda, *Chem. Commun.* **2001**, 461–462.

69 M. Schröder, W.P. Griffith, *J. Chem. Soc., Chem. Commun.* **1979**, 58–59.

70 R. Tang, S.E. Diamond, N. Neary, F. Mares, *J. Chem. Soc., Chem. Commun.* **1978**, 562.

71 A.J. Bailey, B.R. James, *Chem. Commun.* **1996**, 2343–2344.

3
Special Topics

Transition Metals for Organic Synthesis, Vol. 2, 2nd Edition.
Edited by M. Beller and C. Bolm

ISBN: 3-527-30613-7

3.1 Two-Phase Catalysis

D. Sinou

3.1.1 Introduction

Homogeneous organometallic catalysts have many advantages over their heterogeneous counterparts. Generally, higher activities and selectivities can be achieved under the reaction conditions. However, one of the major problems in homogeneous catalysis, and particularly for industrial applications, is the separation of the products from the catalyst, the latter generally being a costly and toxic transition metal. A possible solution to this problem is the use of a liquid-liquid two-phase system. Aqueous-organic systems have been successfully applied, and this is because of the easy and quantitative recovery of the catalyst in active form by simple phase separation and also the environmental benefits of the use of water. The use of such a system could also give selectivities different than those generally found in an organic medium. Although this methodology has been extensively studied since its discovery in 1975 [1, 2], other systems based on, e.g., perfluorohydrocarbons or ionic liquids have been proposed as the non-aqueous phase. Some reviews have appeared in the literature on the applications of water-soluble phosphines in catalysis [3–9], and this article covers developments since 1990 on aqueous-organic two-phase catalysis and other two-phase systems, with emphasis on the actual developments in the field of applications in organic synthesis. Since we define a two-phase system as a system with two liquid phases, reactions performed in water only have not been considered, although in many cases the substrates themself are not soluble in water and form a different phase. The use of the two-phase systems perfluorocarbon-organic solvent and ionic liquid-organic solvent will not be discussed, since Chapters 3.2 and 3.4 are devoted to these two subjects.

Transition Metals for Organic Synthesis, Vol. 2, 2nd Edition.
Edited by M. Beller and C. Bolm

ISBN: 3-527-30613-7

3.1.2
Catalysis in an Aqueous-Organic Two-Phase System

3.1.2.1
Hydrogenation of Unsaturated Substrates

The hydrogenation of unsaturated compounds has been predominantly catalyzed by rhodium and ruthenium complexes associated with various water-soluble ligands such as tppms, tppts, or amphos (Scheme 1). The ruthenium and rhodium complexes are active in the hydrogenation of alkenes [10], cycloalkenes [10], and unsaturated carbonyl compounds [11–13].

One of the most valuable applications is the selective reduction of α,β-unsaturated aldehydes to unsaturated alcohols or saturated aldehydes, depending on the nature of the metal used. For example, 3-methyl-2-buten-1-al or prenal was selectively reduced to prenol with selectivity up to 97% using the catalyst $RuCl_3$/tppts prepared *in situ* in a biphasic medium, water-toluene at 35 °C and 20 bar hydrogen [11, 12], while the use of Rh(I)/tppts as the catalyst at 80 °C and 20 bar hydrogen cleanly gave the saturated aldehyde with selectivity higher than 90% (Eq. 1). Such a selectivity was also observed for other unsaturated aldehydes such as (*E*)-cinnamaldehyde, 2-butenal, and 3,7-dimethyl-2,6-octadien-1-al or citral. In all cases, catalyst recycling was possible without loss of activity and selectivity.

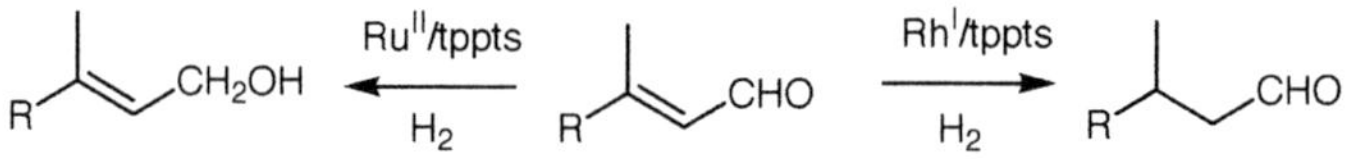

$$\mathrm{R(CH_3)C{=}CH{-}CH_2OH} \xleftarrow[\mathrm{H_2}]{\mathrm{Ru^{II}/tppts}} \mathrm{R(CH_3)C{=}CH{-}CHO} \xrightarrow[\mathrm{H_2}]{\mathrm{Rh^{I}/tppts}} \mathrm{R(CH_3)CH{-}CH_2{-}CHO} \quad (1)$$

$Ph_nP(C_6H_4SO_3Na)_{3-n}$: n = 0 **tppts**; n = 2 **tppms**

$Ph_2P\text{-}(CH_2)_2\text{-}NMe_3^+I^-$ **amphos**

PTA

norbos-Na Ar = C_6H_4-*p*-SO_3Na

bisbis-Na

binas-Na

Ar = C_6H_4-*m*-SO_3Na; n = 0, 1

Xantphos$_{DS}$

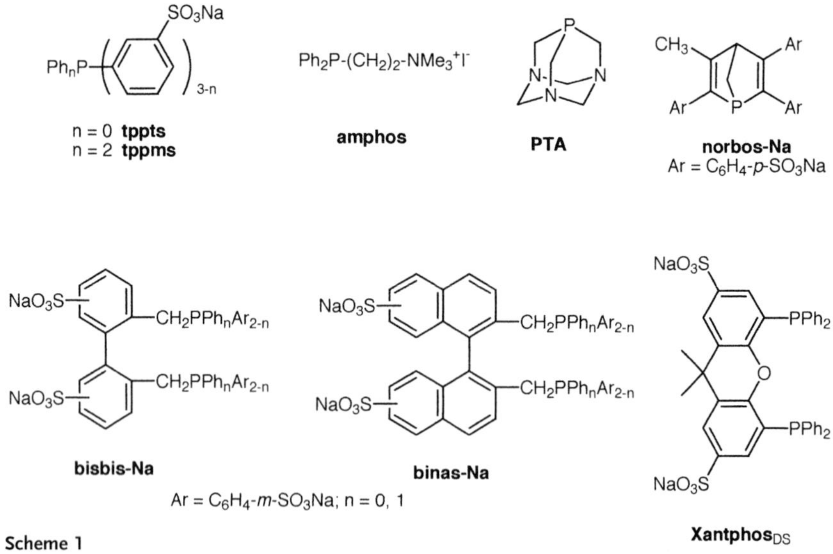

Scheme 1

Dimethyl itaconate was also reduced in a continuous flow reactor using $[Rh(COD)Cl]_2$ + tppts as the catalyst [14].

The same behavior was observed using $RuCl_2(PTA)_4$ or $RhCl(PTA)_3$ as the catalyst; the first of these systems catalyzed the reduction of unsaturated aldehydes to unsaturated alcohols under 28 bar of hydrogen at 80 °C, while the second was very active in the hydrogenation to saturated aldehyde [15].

Series of carboxylated phosphines of the type $Ph_2P\text{-}(CH_2)_n\text{-}CO_2Na$ or $Ph_2P\text{-}C_6H_4\text{-}CO_2Na$ [16], and sulfonated phosphines $NaO_3S(C_6H_4)CH_2C(CH_2PPh_2)_3$ [17], have been used as the ligand of rhodium and ruthenium complexes in the reduction of various alkenes. Hydrogenation of unsaturated bonds in polybutadiene, styrene-butadiene, and nitrile-butadiene polymer emulsions was catalyzed by the complex $[RhCl(Ph_2P\text{-}(CH_2)_n\text{-}CO_2Na)_2]_2$ in an aqueous-toluene biphasic medium, the results being similar to those obtained using $RhCl(tppms)_3$ as the catalyst.

The water-soluble Cp*-ruthenium complexes $[Cp^*Ru(CO)Cl(tppts)]CF_3SO_3$ (**1**) [18] and $[Cp^*Ru(\eta^4\text{-}MeCH{=}CHCH{=}CH\text{–}CO_2H)]CF_3SO_3$ (**2**) [19] are very effective catalysts for the selective hydrogenation of sorbic acid to *trans*-hex-3-enoic acid and *cis*-hex-3-enoic acid, respectively, and of sorbic alcohol to *cis*-hex-3-en-1-ol in various two-phase systems (Eq. 2).

$$\text{CH}_3\text{CH=CHCH=CHCO}_2\text{H} \xrightarrow[\text{cat.}]{H_2} \text{CH}_3\text{CH}_2\text{CH=CHCH}_2\text{CO}_2\text{H} \tag{2}$$

cat. **1** H_2O/*n*-heptane 85% *(E)*-isomer
cat. **2** ethyleneglycol/MTBE 96% *(Z)*-isomer

The regioselectivity in the reduction of di-unsaturated acids using the Wilkinson catalyst was reversed going from the benzene solution to the benzene/H_2O biphasic system (Eq. 3) [20].

$$CH_2{=}CH(CH_2)_3CH{=}CH\text{-}CH_2CO_2H \xrightarrow[H_2]{RhCl(PPh_3)_3} C_5H_{11}CH{=}CHCH_2CO_2H + CH_2{=}CH(CH_2)_6CO_2H \tag{3}$$

	$C_5H_{11}CH{=}CHCH_2CO_2H$	$CH_2{=}CH(CH_2)_6CO_2H$
in benzene	18%	70%
in benzene/H_2O	72%	15%

The water-soluble ruthenium-benzene complexes $[Ru(\eta^6\text{-}C_6H_6)(CH_3CN)]BF_4$ [21] and $Ru_2Cl_4(\eta^6\text{-}C_6H_6)_2$ [22] hydrogenated alkenes and α,β-unsaturated carbonyl compounds, and benzene, respectively.

Hydrogenation could also be performed in a biphasic aqueous-organic medium with formate as the hydrogen source. For example, $RuCl_3$/tppts [12] or $RuCl_2(PTA)_4$ [15] reduced the unsaturated aldehydes to unsaturated alcohols, whereas $[Rh(PTAH)(PTA)_2Cl]Cl$ afforded under the same conditions the saturated aldehyde with high selectivity [23].

The use of chiral water-soluble phosphines (Scheme 2) allowed the enantioselective reduction of some prochiral compounds in an aqueous-organic system [24]. Asymmetric hydrogenation of some α-amino acid precursors (Eq. 4) occurred using chiral sulfonated phosphines derived from Chiraphos (**3**), BDPP (**4**), and Cyclobutanediop (**5**) [25, 26], and also ligands **7–9**, whose water solubility is due to the quaternization of the nitrogen atom [27–29]. However, the effect of water on the enantioselectivity varies widely according to the system used. Generally, rhodium complexes of Chiraphos derivatives **3** and **7** retained their high enantioselectivities, up to 96% *ee*, when complexes of BDPP (**4**) and (**8**), or Cyclobutanediop (**5**) or Diop (**9**) gave lower enantioselectivities (up to 71% *ee* for BDPP (**4**) and (**8**), and 34% *ee* for Cyclobutanediop (**5**) or Diop (**9**). Hydrogenation of these α-amino acid precursors using a rhodium complex associated with tetrasulfonated Binap (**6**) gave 70% enantioselectivity, while the ruthenium complex gave *ee* up to 88% [30, 31].

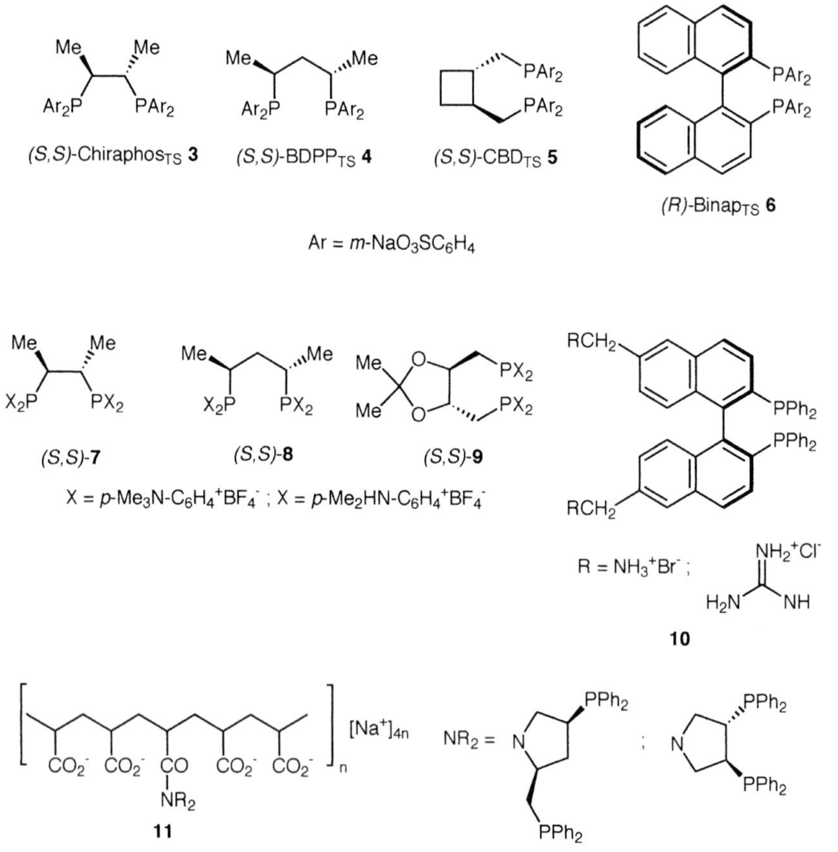

(*S,S*)-Chiraphos$_{TS}$ **3** (*S,S*)-BDPP$_{TS}$ **4** (*S,S*)-CBD$_{TS}$ **5** (*R*)-Binap$_{TS}$ **6**

Ar = *m*-$NaO_3SC_6H_4$

(*S,S*)-**7** (*S,S*)-**8** (*S,S*)-**9**

X = *p*-Me_3N-$C_6H_4{}^+BF_4{}^-$; X = *p*-Me_2HN-$C_6H_4{}^+BF_4{}^-$

R = $NH_3{}^+Br^-$; H_2N–C(=$NH_2{}^+Cl^-$)–NH

10

$[Na^+]_{4n}$ NR_2 =

11

Scheme 2

$$\text{R}^1\text{CH=C(CO}_2\text{R}^2\text{)NHCOR}^3 \xrightarrow[\text{H}_2\text{O/Organic solvent}]{[\text{Rh}] \text{ or } [\text{Ru}]/\mathbf{3\text{-}11},\ \text{H}_2} \text{R}^1\text{CH}_2\text{CH(CO}_2\text{R}^2\text{)NHCOR}^3 \quad (4)$$

Bisphosphinites derived from α,α- or β,β-trehalose are also very effective ligands in the rhodium-catalyzed reduction of methyl α-acetamidocinnamate, enantioselectivities of up to 98% *ee* being reached in the mixture of solvents $H_2O/CH_3OH/$ AcOEt [32, 33].

The reduction was also extended to dehydropeptides using $[Rh(COD)Cl]_2$ associated with ligands (**4**) and **5** [34]. In this case, the diastereomeric excess depends strongly on the absolute configuration of the substrate; for example, with the tetrasulfonated BDPP **4**, a *de* value of 72% was obtained in the reduction of Ac-Δ-Phe-(*R*)-Ala-OCH_3, while a *de* value of only 10% was obtained for Ac–Phe-(*S*)-Ala-OCH_3.

α,β-Unsaturated acids and esters, such as itaconic acid and its dimethyl ester, have also been reduced using rhodium or ruthenium complexes associated with chiral water-soluble ligands. An interesting application is the reduction of 2-(6′-methoxy-2′-naphthyl)-acrylic acid to give Naproxen with *ee* values of 81% and 77% using Binap (**6**) or PEG-bound Binap as the water-soluble ligand, respectively (Eq. 5) [35, 36].

$$\text{MeO-C}_{10}\text{H}_6\text{-C(=CH}_2\text{)CO}_2\text{H} \xrightarrow{\text{H}_2/[\text{Ru}]} \text{MeO-C}_{10}\text{H}_6\text{-CH(CH}_2\text{)CO}_2\text{H} \quad (5)$$

ligand **6**	81% ee
SAPC/**6**	70% ee
PEG-bound Binap	77% ee

The hydrogenation of various β-keto esters proceeded also in a two-phase system in the presence of ruthenium complexes associated with the ligands **10**; the β-hydroxy esters were obtained with enantioselectivities of up to 94% *ee*, the catalyst being reused with no loss of enantioselectivity [37, 38].

A drastic influence of the degree of sulfonation of chiral BDPP on the enantioselectivity in the reduction of prochiral imines was observed [39–41]. Hydrogenation of various imines in a two-phase system AcOEt/H_2O with the catalyst obtained by mixing $[Rh(COD)Cl]_2$ and the monosulfonated BDPP yielded the corresponding amines with *ee* up to 94%, while the reduction using the di- or the trisulfonated BDPP as the ligand proceeded with quite low enantioselectivity (Eq. 6).

$$\mathrm{C_6H_5C(=NBn)CH_3} \xrightarrow[\text{AcOEt/H}_2\text{O/70 atm}]{\text{[Rh]/ligand/H}_2} \mathrm{C_6H_5CH(NHBn)CH_3} \tag{6}$$

$BDPP_{MS}$ 94% ee
$BDPP_{DS}$ 2% ee
$BDPP_{TS}$ 63% ee

The binding of the chiral ligands PPM and pyrphos to a water-soluble polymer such as polyacrylic acid gave macroligands **11** (Scheme 2) [42–44], which were used in the reduction of *α*-acetamidocinnamic acid; enantioselectivities up to 83% *ee* were obtained using EtOAc/H_2O (1/1) as the solvents.

The mechanistic role of water in the hydrogenation reaction was investigated by the groups of Sinou [45, 46] and Joo [47]. Rhodium-catalyzed hydrogenation (or deuteration) of methyl *α*-acetamidocinnamate in AcOEt/D_2O (or H_2O) indicated a 75% regiospecific incorporation of deuterium (or hydrogen) at the *α*-position to the acetamido and the ester groups.

3.1.2.2 Hydroformylation

After the discovery of tppts at Rhône-Poulenc Ind. in Lyon in 1975, its use in hydroformylation as an industrial process was developed by RuhrChemie AG [9, 48, 49]. Concerning the hydroformylation of propene (Eq. 7), the capacity of the plants is actually 300 000 ta^{-1} of *n*-butyraldehyde (<4% *iso*-butyraldehyde) under typical conditions. Because of the success of this process, there was a need to develop new water-soluble ligands with better efficiency. Examples of such novel ligands are bisbis-Na, norbos-Na, and binas-Na (Scheme 1) [50, 51]. In the biphasic hydroformylation of propene, the sulfonated ligands bisbis-Na, norbos-Na, and bisnas-Na showed very high activities and productivies at low phosphine/rhodium ratios compared to tppts (relative activities: tppts/bisbis-Na/norbos-Na/binas-Na= 1/5.6/7.4/11.1). Furthermore, ligands bisbis-Na and bisnas-Na gave *n/iso* ratios of 97/3 and 98/2, respectively, in the resulting butyraldehyde. The concept of a large "natural" bite angle in chelating diphosphines has been extended to two-phase alkene hydroformylation by the use of the water-soluble diphosphine $Xantphos_{DS}$; the rhodium catalyst system obtained from $Rh(CO)_2(acac)$ and this ligand is very selective for the formation of linear aldehydes [52].

$$\mathrm{CH_2{=}CHCH_3} + \mathrm{CO} + \mathrm{H_2} \xrightarrow{\text{[Rh]/ligand}} \underset{\textit{linear}}{\mathrm{CH_3CH_2CH_2CHO}} + \underset{\textit{iso}}{\mathrm{(CH_3)_2CHCHO}} \tag{7}$$

Binuclear rhodium complexes $[Rh(\mu\text{-}SR)(CO)(tppts)]_2$ showed the usual activity in the hydroformylation of hex-1-ene, but an unusual *n/iso* ratio of 95/5 [53, 54]. It is to be noted that in this case the hydroformylation can be carried out without hydrogen, since water can function as the hydrogen source via the water gas shift reaction. More recently, a rhodium complex associated with a water-soluble dendrimer exhibited high catalytic activity and high selectivity (*iso*-aldehyde for the styrene derivatives, and *n*-aldehyde for the aliphatic alkenes) [56].

In the hydroformylation of acrylate esters in a two-phase system in the presence of water-soluble rhodium complexes of tppts as the catalyst, an increase in the reaction rate relative to that observed in a homogeneous system by a factor of 2–14 was observed [57, 58].

Hydroformylation of various alkenes (1-hexene, 2-pentene, etc.) using cobalt catalysts associated with tppts afforded the corresponding oxo products in good yields, an *n/iso* ratio of up to 70/30 being obtained [59].

Because of mass transfer limitations, low catalytic activity was obtained in the hydroformylation of higher olefins in a two-phase system using the rhodium/tppts catalyst. One way to circumvent this problem in the case of water-insoluble substrates and to improve reaction rate is to use rhodium catalysts modified with $PPh_2C_6H_4CO_2Na$, tppms, or tppts, and to add transfer agents or surfactants such as $PhCH_2N^+$*n*-Bu_3Cl^- or $C_{12}H_{25}N^+Me_3Br^-$ [60, 61]. Under these conditions, dodec-1-ene and hexadec-1-ene were converted to *n*-aldehydes with high conversion and *n/iso* selectivities up to 22. An alternative method is to use surface-active phosphines [62–69]; for example, oct-1-ene hydroformylation occurred more efficiently at lower ligand/rhodium ratios than with tppts in a two-phase system, and with a better selectivity (*n/iso*=8–9.5 compared to 3.6 for tppts).

Polyether-substituted triphenylphosphines demonstrate an inverse temperature-dependent solubility in water and have been used as thermoregulated phase transfer ligands; the rhodium complexes of these ligands are very active hydroformylation catalysts for extremely water-immiscible alkenes in an aqueous/organic two-phase system [70–74].

Another approach used the notion of promoter ligand for the hydroformylation of such olefins [75]. A rate enhancement by a factor of 10–50 was observed in the hydroformylation of oct-1-ene using the catalyst $[Rh(COD)Cl]_2$/tppts in a two-phase system when PPh_3 was added in the organic phase. The rate of hydroformylation of the water-soluble allyl alcohol was increased by a factor of 5 using $[Rh(COD)Cl]_2/PPh_3$ as the catalyst and by adding tppts as a promoter.

Very low enantioselectivities were obtained in the hydroformylation of styrene and analogs in a two-phase system, and this still remains an unresolved problem [76, 77].

3.1.2.3
Alkylation and Coupling Reaction

Among the organometallic catalysts used for the alkylation and coupling reaction, palladium has a predominant role. Palladium catalysts are effectively used in a large number of useful transformations in organic chemistry. During the last decade, the excellent compatibility of palladium catalysts containing water-soluble phosphines has considerably increased their potential in organic synthesis.

Casalnuovo and Calabrese were the first to investigate the cross-coupling reaction of various aryl and vinyl iodides and bromides with terminal alkynes (the so-called Sonogashira reaction), and phenylboronic acids and esters (the so-called Suzuki-Stille coupling) in an aqueous medium (water/acetonitrile) in the presence of $Pd(tppms)_3$ as the catalyst [78]. The reaction occurred with quite good yields and tolerated a broad range of functional groups including those present in unprotected nucleosides and amino acids. Vinylation or arylation of activated alkenes (Heck reaction) occurred also in an aqueous or a biphasic medium using the same catalyst. Even arylation of ethylene occurred in quite good yield using $PdCl_2(tppms)_2$ as the catalyst, leading to functionalized styrene derivatives [79].

The catalyst obtained *in situ* from $Pd(OAc)_2$ and tppts has been used with success in aqueous media in many coupling reactions under very mild conditions: Sonogashira coupling (Eq. 8), Suzuki and Stille coupling (Eq. 9), inter- and intramolecular Heck reaction (Eq. 10), cycloisomerization of enynes (Eq. 11), and Tsuji-Trost reaction (Eq. 12), high chemical yields being generally obtained. A review of this subject has appeared recently [80]. However, most of these reactions were performed in a monophasic system H_2O/CH_3CN, although the recycling of the catalyst was possible by extraction of the organic product using an organic solvent.

OHC–C₆H₄–I + H–C≡C–SiMe₃ → [Pd(OAc)₂/tppts, Et₃N; MeCN/H₂O/25 °C; 80%] → OHC–C₆H₄–C≡C–SiMe₃ (8)

Bu–CH=CH–B(OH)₂ + 3-iodocyclohex-2-enone → [Pd(OAc)₂/tppts, *i*-Pr₂NH; MeCN/H₂O/25 °C; 95%] → 3-(hex-1-enyl)cyclohex-2-enone (Bu) (9)

H₂N–C₆H₄–I + CH₂=CH–CO₂Et → [Pd(OAc)₂/tppts, Et₃N; MeCN/H₂O/37 °C; 98%] → H₂N–C₆H₄–CH=CH–CO₂Et (10)

enyne ether (Ph) → [PdCl₂/tppts/80 °C; dioxane/H₂O; 85%] → tetrahydrofuran product (H, OH, Ph) (11)

$$\text{R}\!-\!\text{CH}\!=\!\text{CH}\!-\!\text{CH}_2\text{OCOR}^1 \xrightarrow[\text{Pd(OAc)}_2\text{/tppts/50 °C, RCN/H}_2\text{O}]{\text{NuH}} \text{R}\!-\!\text{CH}\!=\!\text{CH}\!-\!\text{CH}_2\text{Nu} \qquad (12)$$

$R^1 = CH_3, OCH_3$; NuH = $CH_2(CO_2Me)_2$, $CH_2(COMe)_2$, $CH_2(NO_2)CO_2Et$, HNR_2;
RCN: CH_3CN, C_3H_7CN, C_6H_5CN

The first palladium-catalyzed reaction performed in a true two-phase system was the allylic substitution [81–83]. Carbon nucleophiles as well as heteronucleophiles were used in this reaction, giving the products of alkylation in yields of up to 95%. It should be noted that the observed regio- and stereoselectivities in this reaction were analogous to the selectivities found in a usual organic medium; however the use of primary amines led very cleanly to the formation of the secondary amine by monoalkylation, in contrast to the mono- and dialkylation products observed in an organic medium. Recycling of the catalyst without formation of metallic palladium was possible.

However, although the mixture H_2O/CH_3CN is a homogeneous phase, performing the palladium-catalyzed reaction in this medium can drastically change the selectivity of the reaction. For example, the intramolecular Heck-type reaction, carried out in water/acetonitrile in the presence of $Pd(OAc)_2$/tppts, afforded the *endo* cyclized product, instead of the usual *exo* product formed in an organic medium [84].

Another example is the allylation of uracils and 2-thiouracils [85, 86]. The palladium-catalyzed allylation of uracils and 2-thiouracils in an organic medium led to a complex mixture resulting from allylation at N-1, N-2 and sulfur. When the reaction was performed with H_2O/CH_3CN as the solvent and $Pd(OAc)_2$/tppts as the catalyst, regioselective allylation occurred at nitrogen for uracils and at sulfur for 2-thiouracils (Eq. 13). It was shown that reaction in this medium was kinetically controlled.

(13)

A very interesting application of this allylic substitution using $Pd(OAc)_2$/tppts as the catalyst is the removal of allyl protecting groups under very mild conditions using the monophasic system CH_3CN/H_2O or a biphasic system [80]. For example, the use of the biphasic system butyronitrile/H_2O allowed an easy separation of the deprotected substrate and the recycling of the catalyst (Eq. 14); chemoselective removal (e.g., allyl versus dimethylallyl) was also observed.

$Pd(OAc)_2$/tppts
25 °C/30 min
n-PrCN/H_2O, $HNEt_2$
100%
10 recyclings
OH
(14)

Polar hydrophilic phosphines containing mono- and disaccharide moieties associated with $Pd(OAc)_2$ exhibited superior catalytic performance compared to the tppts ligand in Suzuki and Heck reactions in a two-phase system [87, 88]. The complex obtained from $[Pd(\eta^3\text{-}C_3H_5)Cl]_2$ and an amphiphilic ligand derived from D-glucosamine was found to be an efficient catalyst for asymmetric allylic substitution of 1,3-diphenyl-2-propenyl acetate in a toluene/H_2O mixture, enantioselectivity of up to 80% *ee* being obtained, and recycling of the catalyst being possible [89].

3.1.2.4
Other Reactions

Hydrocarboxylation of alkenes to carboxylic acids in the presence of carbon monoxide and water appears as an attractive process for the synthesis of carboxylic acids. This reaction was performed in a two-phase system using the water-soluble palladium complex of tppts in association with a Brønsted acid as promoter [90, 91], and was recently extended to the asymmetric hydrocarboxylation of vinyl arenes, enantioselectivities of up to 43% being obtained [92].

Substituted phenyl acetic acids were also obtained in quite good yields by palladium-catalyzed carboxylation of benzylic alcohols [93] or benzyl chlorides [94].

The telomerization of butadiene has been intensively investigated with compounds containing active hydrogen such as alcohols, amines, phenols, acids, etc. in two-phase systems. Addition of ethyl acetoacetate or other active methylene compounds to asymmetrical dienes such as myrcene occurred in a 1,4-fashion with regioselectivity up to 99% at high conversion in the presence of Rh(I)/tppts as the catalyst [95–97]; actually this is an industrial route to vitamin E. The telomerization of butadiene with ammonia afforded the corresponding primary octadienylamines with selectivities of up to 88% in the presence of $Pd(OAc)_2$/tppts, provided that a two-phase toluene-water medium was used [98].

3.1.3
Other Methodologies

3.1.3.1
Supported Aqueous Phase Catalyst

In 1989, Davis's group introduced the concept of Supported Phase Catalysts (or SAPC) [99]. In such a system, the water-soluble organometallic catalyst is dissolved in a film of water which is supported on a high surface-area hydrophilic solid. Hydroformylation of various alkenes such as hex-1-ene, oct-1-ene, dec-1-ene, or oleic alcohol occurred using $RhH(CO)(tppts)_3$ immobilized on silica [100–102]. Quantitative conversions were obtained, the *n/iso* ratio depending on the water content of the catalyst; no leaching of rhodium was observed. Recently, the supported catalyst $RhH(CO)_2(Xantphos_{DS})$ with a large P-M-P bite angle allowed the hydroformylation of higher olefins with a very high regioselectivity (*n/iso*=40:1), the catalyst being reused in numerous consecutive catalytic cycles [103]. The catalytic rate of hydroformylation of acrylate esters was also greatly improved using this methodology [57, 104].

The SAP methodology was also extended to the alkylation of allylic carbonates with various nucleophiles (ethyl acetoacetate, dimethyl malonate, morpholine, phenol, 2-mercaptopyridine) using $Pd(OAc)_2$/tppts supported on silica [105–107]; this catalyst is drastically more active using C_6H_5CN rather than CH_3CN as the organic solvent. Higher activities were obtained using polysaccharides such as chitosan or cellulose as the support instead of silica [108, 109].

The enantioselective hydrogenation of 2-(6′methoxy-2′-naphthyl)-acrylic acid with a ruthenium catalyst associated with sulfonated Binap **6** and supported on silica gave Naproxen with *ee* values of up to 77%, using ethyl acetate containing 3% water as the solvent [110]. This enantioselectivity increased to 96% by using ethyleneglycol instead of water and a cyclohexane-chloroform mixture as the solvent [111].

3.1.3.2
Inverse Phase Catalysis

Recently, Okano et al. introduced the notion of “counter phase catalysis” [112], where the water-soluble catalyst transferred the hydrophobic organic reactants into the water phase and catalyzed their transformations into the aqueous phase. Most of the reactions studied were the catalytic transformation of organic halides using $PdCl_2(tppms)_2$ as the catalyst, and they exhibited a high efficiency and selectivity compared to the reaction performed in the presence of $PdCl_2(PPh_3)_2$ as the catalyst. For example, carbonylation of benzylic halides or aryl halides (Eq. 15), as well as cyanation of aryl iodides, occurred very efficiently in the presence of $PdCl_2(tppms)_2$ [113–115].

$$C_6H_5\text{-}I + CO + NaOH \xrightarrow[H_2O/\text{heptane}]{PdCl_2L_2} C_6H_5\text{-}CO_2H \qquad (15)$$

L = tppms 88%
L = P(*o*-tolyl)$_3$ 5%

Another approach, reported by Harada [116], used cyclodextrins as the counter phase catalyst. This approach was used for the transformation of long-chain alkenes; the cyclodextrins transfer the alkenes, whose solubilities are limited in the aqueous phase, across the phase interface. For example, oxidation of higher olefins (C_8–C_{16}) to ketones occurred in the presence of oxygen in a two-phase system with high yields (>90%) using $PdSO_4/H_9PV_6Mo_6O_{40}/CuSO_4$ in the presence of per(2,6-di-*O*-methyl)-β-cyclodextrin [117–120].

This concept was extended to hydroformylation of various water-insoluble terminal olefins, giving the corresponding aldehydes with high yields and selectivities [121, 122], hydrocarboxylation of higher α-olefins [123], hydrogenation of aldehydes [124, 125], as well as cleavage of allylic protecting groups [126].

A quite new approach concerns the concept of covalently connecting a catalytically active transition metal center to a cyclodextrin [127].

3.1.4 Conclusion

A survey of the literature of the past few years concerning the use of two-phase catalysis in organic synthesis shows that this methodology has been extended to various organic transformations including hydrogenation, hydroformylation, and carbon-carbon bond formation, and also to various metal complexes including rhodium, ruthenium, and palladium. Although the first goal was the easy separation of the catalyst from the products of the reaction for its eventual recycling, the use of water as a co-solvent in a two-phase system or even in a homogeneous system could also exhibit a quite different selectivity than that observed in an organic phase. However, some problems remain to be solved, for example, routes to more efficient and easily accessible achiral and chiral water-soluble ligands have to be found, and efficient asymmetric hydroformylation or hydroxycarbonylation in a two-phase system is still a problem. Some results with immobilized catalysts, particularly on natural supports, are also very promising.

References

1 Rhône-Poulenc Recherche (E. Kuntz), FR 2314910, **1975**.
2 F. Joo, M.T. Beck, *React. Kin. Catal. Lett.* **1975**, *2*, 357–363.
3 F. Joo, Z. Tòth, *J. Mol. Catal.* **1980**, *8*, 369–383.
4 E.G. Kuntz, *Chemtech* **1987**, *17*, 570–575.
5 D. Sinou, *Bull. Soc. Chim. Fr.* **1987**, 480–486.
6 T.G. Southern, *Polyhedron* **1989**, *8*, 407–413.
7 M. Barton, J.D. Atwood, *J. Coord. Chem.* **1991**, *24*, 43–67.
8 P. Kalck, F. Monteil, *Adv. Organomet. Chem.* **1992**, *34*, 219–284.
9 W.A. Herrmann, C.W. Kohlpaintner, *Angew. Chem., Int. Ed. Engl.* **1993**, *32*, 1524–1544.
10 A. Andrillo, A. Bolivar, F.A. Lopez, D.E. Paez, *Inorg. Chim. Acta* **1995**, *238*, 187–192.
11 J.M. Grosselin, C. Mercier, *J. Mol. Catal.* **1988**, *63*, L25–27.
12 J.M. Grosselin, C. Mercier, G. Allmang, F. Grass, *Organometallics* **1991**, *10*, 2126–2133.
13 M. Hernandez, P. Kalck, *J. Mol. Catal. A Chem.* **1997**, *116*, 131–146.

14 C. de Bellefon, N. Tanchoux, S. Caravieilhes, D. Schweich, *Catal. Today* **1999**, *48*, 211–219.

15 D.J. Darensbourg, F. Joo, M. Kannisto, A. Katho, J.H. Reibenspies, *Organometallics* **1992**, *11*, 1990–1993.

16 D.C. Mudalige, G.L. Rempel, *J. Mol. Catal. A: Chem.* **1997**, *116*, 309–316.

17 I. Rojas, F.L. Linares, N. Valencia, C. Bianchini, *J. Mol. Catal. A: Chem.* **1999**, *144*, 1–6.

18 B. Driessen-Hölscher, J. Heinen, *J. Organomet. Chem.* **1998**, *570*, 141–146.

19 S. Steines, U. Englert, B. Driessen-Hölscher, *Chem. Commun.* **2000**, 217–218.

20 T. Okano, M. Kaji, S. Isotani, J. Kiji, *Tetrahedron Lett.* **1992**, *33*, 5547–5550.

21 W.-C. Chan, C.-P. Lau, L. Cheng, Y.-S. Leung, *J. Organomet. Chem.* **1994**, *464*, 103–106.

22 E.G. Fidalgo, L. Plasseraud, G. Süss-Fink, *J. Mol. Catal. A: Chem.* **1998**, *132*, 5–12.

23 D.J. Darensbourg, N.W. Stafford, F. Joo, J.H. Reibenspies, *J. Organomet. Chem.* **1995**, *488*, 99–108.

24 D. Sinou, *Adv. Synth. Catal.* **2002**, *344*, 221–237.

25 F. Alario, Y. Amrani, Y. Coleuille, T. P. Dang, J. Jenck, D. Morel, D. Sinou, *J. Chem. Soc., Chem. Commun.* **1986**, 202–203.

26 Y. Amrani, L. Lecomte, D. Sinou, J. Bakos, I. Toth, B. Heil, *Organometallics* **1989**, *8*, 542–547.

27 I. Toth, B.E. Hanson, *Tetrahedron: Asymmetry* **1990**, *1*, 895–912.

28 I. Toth, B.E. Hanson, *Tetrahedron: Asymmetry* **1990**, *1*, 913–930.

29 I. Toth, B.E. Hanson, M.E. Davis, *Catal. Lett.* **1990**, *5*, 183–188.

30 K.T. Wan, M.E. Davis, *J. Chem. Soc., Chem. Commun.* **1993**, 1262–1264.

31 K.T. Wan, M.E. Davis, *Tetrahedron: Asymmetry* **1993**, *4*, 2461–2468.

32 K. Yonehara, T. Hashizume, K. Mori, K. Ohe, S. Uemura, *J. Org. Chem.* **1999**, *64*, 5593–5598.

33 S. Shin, T.V. RajanBabu, *Org. Lett.* **1999**, *1*, 1229–1232.

34 M. Laghmari, D. Sinou, A. Masdeu, C. Claver, *J. Organomet. Chem.* **1992**, *438*, 213–216.

35 K.-T. Wan, M.E. Davies, *J. Catal.* **1994**, *148*, 1–8.

36 Q.-H. Fan, G.-J. Deng, X.-M. Chen, W.-C. Xie, D.-Z. Jiang, D.-S. Liu, A.S.C. Chan, *J. Mol. Catal. A: Chem.* **2000**, *159*, 37–43.

37 T. Lamouille, C. Saluzzo, R. ter Halle, F. Le Guyader, M. Lemaire, *Tetrahedron Lett.* **2001**, *42*, 663–664.

38 P. Guerreiro, V. Ratovelomanana-Vidal, J.-P. Genêt, P. Dellis, *Tetrahedron Lett.* **2001**, *42*, 3423–3426.

39 J. Bakos, A. Orosz, B. Heil, M. Laghmari, P. Lhoste, D. Sinou, *J. Chem. Soc., Chem. Commun.* **1991**, 1684–1685.

40 C. Lensink, J.G. de Vries, *Tetrahedron: Asymmetry* **1992**, *3*, 235–238.

41 C. Lensink, E. Rijnberg, J.G. de Vries, *J. Mol. Catal. A: Chem.* **1997**, *116*, 199–207.

42 T. Malström, C. Andersson, *Chem. Commun.* **1996**, 1135–1136.

43 T. Malström, C. Andersson, *J. Mol. Catal. A: Chem.* **1999**, *139*, 259–270.

44 T. Malström, C. Andersson, *J. Mol. Catal. A: Chem.* **2000**, *157*, 79–82.

45 M. Laghmari, D. Sinou, *J. Mol. Catal.* **1991**, *66*, L15–28.

46 J. Bakos, R. Karaivanov, M. Laghmari, D. Sinou, *Organometallics* **1994**, *13*, 2951–2956.

47 F. Joo, E. Papp, A. Katho, *Top. Catal.* **1998**, *5*, 113–124.

48 B. Cornils, E. Wiebus, *Chemtech* **1995**, 33–38.

49 B. Cornils, E.G. Kuntz, *J. Organomet. Chem.* **1995**, *502*, 177–186.

50 W.A. Herrmann, C.W. Kohlpaintner, H. Bahrman, W. Konkol, *J. Mol. Catal.* **1992**, *73*, 191–199.

51 W.A. Herrmann, C.W. Kohlpaintner, R.B. Manetsberger, H. Bahrman, H. Kottmann, *J. Mol. Catal.* **1995**, *97*, 65–72.

52 M.S. Goedheijt, P.C.J. Kamer, P.W.N. M. van Leeuwen, *J. Mol. Catal A: Chem.* **1998**, *134*, 243–249.

53 P. Escaffre, A. Thorez, P. Kalck, *J. Chem. Soc., Chem. Commun.* **1987**, 6146–6147.

54 P. Escaffre, A. Thorez, P. Kalck, *New J. Chem.* **1987**, *11*, 6601–6604.

55 P. Kalck, P. Escaffre, F. Serein-Spirau, A. Thorez, *New J. Chem.* **1988**, *12*, 687–690.

56 A. Gong, Q. Fan, Y. Chen, H. Liu, C. Chen, F. Xi, *J. Mol. Catal A: Chem.* **2000**, *159*, 225–232.

57 G. Fremy, E. Monflier, Y. Castanet, J. F. Carpentier, A. Mortreux, *Angew. Chem., Int. Ed. Engl.* **1995**, *34*, 1474–1476.

58 G. Fremy, E. Monflier, J. F. Carpentier, Y. Castanet, A. Mortreux, *J. Mol. Catal. A: Chem.* **1998**, *129*, 35–40.

59 M. Beller, J. G. E. Krauter, *J. Mol. Catal. A: Chem.* **1999**, *143*, 31–39.

60 M. J. H. Russell, *Platinum Met. Rev.* **1988**, *32*, 179–186.

61 H. Chen, Y. Li, J. Cehn, P. Cheng, X. Li, *Catal. Today* **2002**, *74*, 131–135.

62 T. Bartik, B. Bartik, B. E. Hanson, *J. Mol. Catal.* **1994**, *88*, 43–56.

63 H. Ding, B. E. Hanson, T. Bartik, B. Bartik, *Organometallics* **1994**, *13*, 3761–3763.

64 T. Bartik, H. Ding, B. Bartik, B. E. Hanson, *J. Mol. Catal.* **1995**, *98*, 117–122.

65 H. Ding, B. E. Hanson, *J. Chem. Soc., Chem. Commun.* **1994**, 2747–2748.

66 H. Ding, J. Kang, B. E. Hanson, C. W. Kohlpaintner, *J. Mol. Catal. A: Chem.* **1997**, *124*, 21–28.

67 B. E. Hanson, H. Ding, C. W. Kohlpaintner, *Catal. Today* **1998**, *42*, 421–429.

68 E. A. Karakhanov, Y. S. Kardasheva, E. A. Runova, V. A. Semernina, *J. Mol. Catal. A: Chem.* **1999**, *142*, 339–347.

69 S. Bischoff, M. Kant, *Catal. Today* **2001**, *66*, 183–189.

70 Z. Jin, X. Zheng, B. Fell, *J. Mol. Catal. A: Chem.* **1997**, *116*, 55–58.

71 X. Zheng, J. Jiang, X. Liu, Z. Jin, *Catal. Today* **1998**, *44*, 175–182.

72 R. Chen, X. Liu, Z. Jin, *J. Organometal. Chem.* **1998**, *571*, 201–204.

73 J. Jiang, Y. Wang, C. Liu, F. Han, Z. Jin, *J. Mol. Catal. A: Chem.* **1999**, *147*, 131–136.

74 R. Chen, J. Jiang, Y. Wang, Z. Jin, *J. Mol. Catal. A: Chem.* **1999**, *149*, 113–117.

75 R. V. Chaudhari, B. M. Bhanage, R. M. Desphande, H. Delmas, *Nature* **1995**, *373*, 501–503.

76 R. W. Eckl, T. Priermeier, W. A. Herrmann, *J. Organometal. Chem.* **1997**, *532*, 243–249.

77 M. D. Miquel-Serrano, A. M. Masdeu-Bultò, C. Claver, D. Sinou, *J. Mol. Catal. A: Chem.* **1999**, *143*, 49–55.

78 A. L. Casalnuovo, J. C. Calabrese, *J. Am. Chem. Soc.* **1990**, *112*, 4324–4330.

79 J. Kiji, T. Okano, T. Hasegawa, *J. Mol. Catal.* **1995**, *97*, 73–77.

80 J. P. Gent, M. Savignac, *J. Organometal. Chem.* **1999**, *576*, 305–317.

81 M. Safi, D. Sinou, *Tetrahedron Lett.* **1991**, *32*, 2025–2028.

82 E. Blart, J. P. Genêt, M. Safi, M. Savignac, D. Sinou, *Tetrahedron* **1994**, *50*, 505–514.

83 S. Sigismondi, D. Sinou, *J. Mol. Catal. A: Chem.* **1997**, *116*, 289–296.

84 S. Lemaire-Audoire, M. Savignac, C. Dupuis, J. P. Gent, *Tetrahedron Lett.* **1996**, *37*, 2003–2006.

85 S. Sigismondi, D. Sinou, M. Perez, M. Moreno-Maas, R. Pleixats, M. Villarroya, *Tetrahedron Lett.* **1994**, *35*, 7085–7088.

86 C. Goux, S. Sigismondi, D. Sinou, M. Perez, M. Moreno-Maas, R. Pleixats, M. Villarroya, *Tetrahedron* **1996**, *52*, 9521–9534.

87 M. Beller, J. G. E. Krauter, A. Zapf, *Angew. Chem., Int. Ed. Engl.* **1997**, *36*, 772–774.

88 M. Beller, J. G. E. Krauter, A. Zapf, S. Bogdanovic, *Catal. Today* **1999**, *48*, 279–290.

89 T. Hashizume, K. Yonehara, K. Ohe, S. Uemura, *J. Org. Chem.* **2000**, *65*, 5197–5201.

90 S. Tilloy, E. Monflier, F. Bertoux, Y. Castanet, A. Mortreux, *New J. Chem.* **1997**, *21*, 529–531.

91 F. Bertoux, S. Tilloy, E. Monflier, Y. Castanet, A. Mortreux, *J. Mol. Catal. A: Chem.* **1999**, *138*, 53–57.

92 M. D. Miquel-Serrano, A. Aghmiz, M. Diéguez, A. M. Masdeu-Bultò, C. Claver, D. Sinou, *Tetrahedron: Asymmetry* **1999**, *10*, 4463–4467.

93 G. Verspui, G. Papadogianakis, R.A. Sheldon, *Catal. Today* **1998**, *42*, 449–458.

94 C.W. Kohlpaintner, M. Beller, *J. Mol. Catal. A: Chem.* **1997**, *116*, 259–267.

95 G. Mignani, D. Morel, Y. Colleuille, *Tetrahedron Lett.* **1985**, *26*, 6337–6340.

96 C. Mercier, G. Mignani, M. Aufrand, G. Allmang, *Tetrahedron Lett.* **1991**, *32*, 1433–1436.

97 C. Mercier, P. Chabardes, *Pure Appl. Chem.* **1994**, *66*, 1509–1518.

98 T. Prinz, W. Keim, B. Driessen-Hölscher, *Angew. Chem., Int. Ed. Engl.* **1996**, *35*, 1708–1710.

99 J.P. Arhancet, M.E. Davis, J.S. Merola, B.E. Hanson, *Nature* **1989**, *339*, 454–455.

100 J.P. Arhancet, M.E. Davis, B.E. Hanson, *Catal. Lett.* **1991**, *11*, 129–136.

101 J.P. Arhancet, M.E. Davis, J.S. Merola, B.E. Hanson, *J. Catal.* **1990**, *121*, 327–339.

102 J.P. Arhancet, M.E. Davis, B.E. Hanson, *J. Catal.* **1991**, *129*, 100–105.

103 A.J. Sandee, V.F. Slagt, J.N.H. Reek, P.C.J. Kamer, P.W.N.M. van Leeuwen, *Chem. Commun.* **1999**, 1633–1634.

104 G. Fremy, E. Monflier, J.-F. Carpentier, Y. Castanet, A. Mortreux, *J. Catal.* **1996**, *162*, 339–348.

105 S. Dos Santos, Y. Tong, F. Quignard, A. Choplin, D. Sinou, J.P. Dutasta, *Organometallics* **1998**, *17*, 78–89.

106 A. Choplin, S. Dos Santos, F. Quignard, S. Sigismondi, D. Sinou, *Catal. Today* **1998**, *42*, 471–478.

107 S. Dos Santos, F. Quignard, D. Sinou, A. Choplin, *Top. Catal.* **2000**, *13*, 311–318.

108 F. Quignard, A. Choplin, A. Domard, *Langmuir* **2000**, *16*, 9106–9108.

109 F. Quignard, A. Choplin, *Chem. Commun.* **2001**, 21–22.

110 K.T. Wan, M.E. Davies, *Nature* **1994**, *370*, 449–450.

111 K.T. Wan, M.E. Davies, *J. Catal.* **1995**, *152*, 25–30.

112 T. Okano, Y. Moriyama, H. Konishi, J. Kiji, *Chem. Lett.* **1986**, 1463–1466.

113 J. Kiji, T. Okano, W. Nishiumi, H. Konishi, *Chem. Lett.* **1988**, 957–960.

114 T. Okano, I. Uchida, T. Nakagaki, H. Konishi, J. Kiji, *J. Mol. Catal.* **1989**, *54*, 65–71.

115 T. Okano, T. Hayashi, J. Kiji, *Bull. Chem. Soc. Jpn* **1994**, *67*, 2339–2341.

116 A. Harada, *Syn. Org. Chem. Jpn* **1990**, *48*, 517–521.

117 E. Monflier, E. Blouet, Y. Barbaux, A. Mortreux, *Angew. Chem., Int. Ed. Engl.* **1994**, *33*, 2100–2102.

118 E. Monflier, S. Tilloy, G. Fremy, Y. Barbaux, A. Mortreux, *Tetrahedron Lett.* **1995**, *36*, 387–388.

119 E. Monflier, S. Tilloy, E. Blouet, Y. Barbaux, A. Mortreux, *J. Mol. Catal.* **1996**, *109*, 27–35.

120 E. Karakhanov, A. Maximov, A. Kirillov, *J. Mol. Catal. A: Chem.* **2000**, *157*, 25–30.

121 E. Monflier, G. Fremy, Y. Castanet, A. Mortreux, *Angew. Chem., Int. Ed. Engl.* **1995**, *34*, 2269–2271.

122 E. Monflier, S. Tilloy, G. Fremy, Y. Castanet, A. Mortreux, *Tetrahedron Lett.* **1995**, *36*, 9481–9484.

123 E. Monflier, S. Tilloy, F. Bertoux, Y. Castanet, A. Mortreux, *New J. Chem.* **1997**, *21*, 857–859.

124 E. Monflier, S. Tilloy, Y. Castanet, A. Mortreux, *Tetrahedron Lett.* **1998**, *39*, 2959–2960.

125 S. Tilloy, H. Bricout, E. Monflier, *Green Chem.* **2002**, *4*, 188–193.

126 R. Widehem, T. Lacroix, H. Thibaut, H. Bricoud, E. Monflier, *Synlett* **2000**, 722–724.

127 M. *T.* Reetz, *Catal. Today* **1998**, *42*, 399–411.

3.2 Transition Metal-Based Fluorous Catalysts

Rosenildo Corrêa da Costa and J. A. Gladysz

3.2.1 Brief Introduction to Fluorous Catalysis

Fluorous catalysis is a recently developed technique for catalyst/product separation and recycling that exploits the temperature-dependent miscibility of organic and fluorous solvents [1–3]. Since the seminal 1994 publication of Horváth and Rábai [1], this protocol has seen steadily increasing application. The original formulation, shown in Scheme 1, involves an organic and a fluorous solvent such as a perfluoroalkane. Catalysts are derivatized with "pony tails" or $(CH_2)_m(CF_2)_{n-1}CF_3$ segments (abbreviated $(CH_2)_mR_{fn}$) that provide high fluorous-solvent affinities. Reactions can be conducted under homogeneous conditions at the one-phase, high-temperature limit. Products normally have much greater affinities for the non-fluorous solvent, and are easily separated at the two-liquid phase (low-temperature limit).

As detailed elsewhere, several modified procedures have been developed [3]. For example, $CF_3C_6H_5$ [(trifluoromethyl)toluene] effectively solubilizes both fluorous and non-fluorous solutes, which can subsequently be separated by extraction or chromatography [4]. Also, a number of fluorous compounds exhibit highly tem-

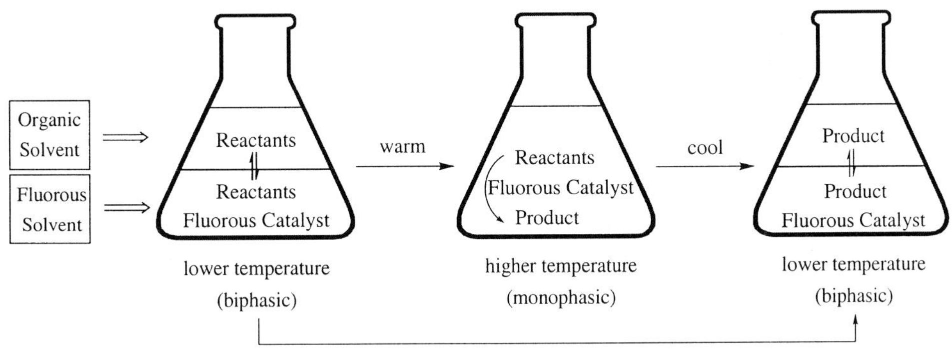

Scheme 1 Most common protocol for fluorous biphase catalysis and catalyst recycling.

Transition Metals for Organic Synthesis, Vol. 2, 2nd Edition.
Edited by M. Beller and C. Bolm

ISBN: 3-527-30613-7

perature-dependent solubilities in organic solvents [5]. Such "thermomorphic" behavior allows homogeneous reaction conditions at higher temperatures, with catalyst recovery via liquid/solid-phase separation at lower temperatures. Other physical characteristics of fluorous solvents and compounds have been outlined in reviews [2, 3], and the electronic properties of "pony tails" have been studied as a function of structure [6].

This brief entry-level review covers *molecular* transition metal-based fluorous catalysts that have been applied to organic synthesis, irrespective of the solvent system. All catalysts to date constitute modified versions of non-fluorous catalysts, which are extensively treated in the other chapters of this compendium. Thus, the first step or "getting started" involves the synthesis of a fluorous version of an established ligand. There are extensive series of papers dealing only with ligand synthesis and some papers dealing only with catalyst synthesis. However, for this review only papers that describe catalyst applications are cited.

In the other chapters in this compendium, catalysts are evaluated on the basis of product yields, selectivities, and turnover numbers. With fluorous catalysts, there is the additional criterion of recoverability. While it is important to minimize catalyst leaching and/or decomposition, these themes will not be critically treated in this review. Enough partition coefficients have now been measured to allow researchers to "dial in" the desired degree of fluorophilicity into a target molecule. However, a number of subtle points or traps deserve emphasis, including the facts that (1) the fluorous properties of the catalyst rest state are the most critical for recovery, and (2) yield data do not constitute good measures of recyclability [7]. For example, some fluorous palladacycles described below serve as nothing more than steady-state sources of catalytically active palladium nanoparticles [8]. After several cycles, the palladacycle catalyst precursors completely decompose and activity ceases.

3.2.2 Alkene Hydroformylation

Since alkene hydroformylation was the first application investigated by Horváth and Rábai [1], it is treated first. The combination of a rhodium(I) source – typically $Rh(acac)(CO)_2$ – and fluorous phosphorus donor ligands gives highly active catalysts for the hydroformylation of terminal alkenes [9]. As shown in Scheme 2, typical ligands include trialkylphosphine (**1**), triarylphosphine (**2**), and triarylphosphite (**3**). The recyclability of the first system is especially high. The normal/branched aldehyde ratios differ somewhat from those for analogous non-fluorous catalysts, but there are no dramatic surprises. Fluorous polymers that are soluble in fluorous solvents and contain pendant $-PPh_2$ groups have also been employed [9d].

3.2.3
Alkene Hydrogenation

As in the previous section, all examples to date feature rhodium(I) catalyst precursors [10]. The $ClRhL_3$ species in Scheme 3, **4–7**, were isolated before use [10a, d]. Others are conveniently generated *in situ* [10b]. Complexes **6** and **7** feature the silicon-containing triarylphosphines $P(4\text{-}C_6H_4SiMe_2CH_2CH_2R_{fn})_3$. Cationic rhodium catalyst precursors of the types **8–10**, which have chelating fluorous silicon-containing diphosphines, are also highly effective [10f, g]. Note that the silicon substituents

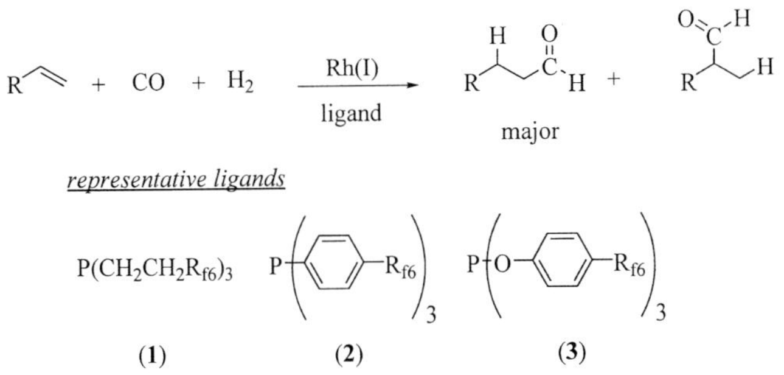

Scheme 2 Fluorous catalysts for alkene hydroformylation.

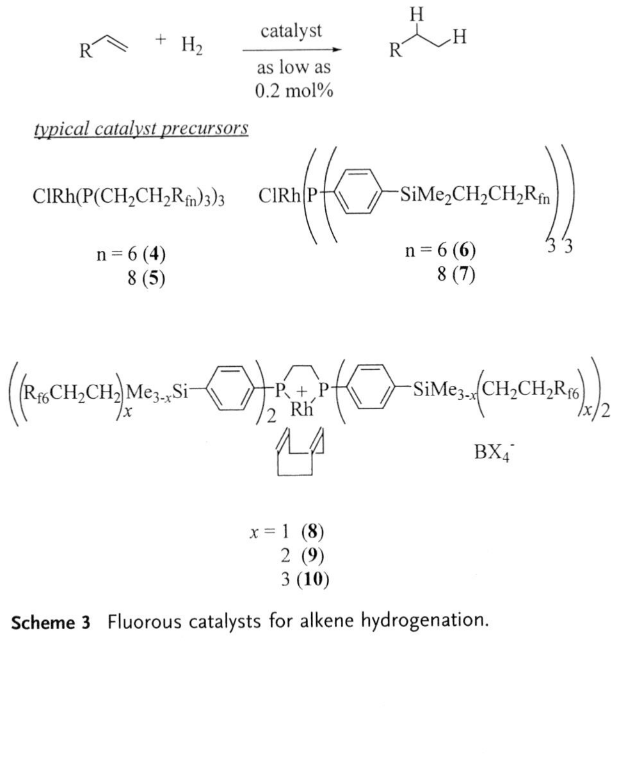

Scheme 3 Fluorous catalysts for alkene hydrogenation.

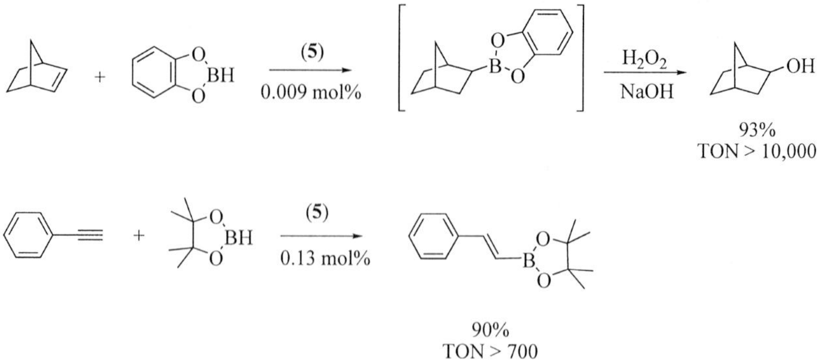

Scheme 4 Fluorous catalysts for the hydroboration of alkenes and alkynes.

can be used to append as many as three pony tails per aromatic ring. This provides a high degree of fluorophilicity. As would be expected, **8–10** also catalyze the hydrogenation of alkynes. With all catalysts, some C=C isomerization can be observed.

3.2.4
Alkene/Alkyne Hydroboration and Alkene/Ketone Hydrosilylation

The fluorous rhodium complexes **4** and **5** are also effective catalyst precursors for additions of boron-hydrogen bonds to alkenes and alkynes, and representative examples are given in Scheme 4 [11]. The catalysts are easily separated before the customary oxidation step, which would lead to decomposition. Complexes **6** and **11** similarly catalyze additions of silicon-hydrogen bonds to alkenes, as illustrated in Scheme 5 [12]. Related reactions have been conducted with **6** in an ionic liquid phase that was rendered fluorophilic by means of a fluorous anion [12b]. Complex **5** efficiently catalyzes the hydrosilylation of ketones [13]. As exemplified in Scheme 5, α,β-unsaturated ketones predominantly undergo 1,4-addition.

3.2.5
Reactions of Diazo Compounds

A number of dirhodium complexes of the formula $Rh_2(O_2CR)_4$, where R is a fluorous alkyl or aryl moiety, have been prepared [14, 15]. A partial list is shown in Scheme 6. These are active catalysts for the cyclopropanation of alkenes by diazo compounds as well as for intramolecular and intermolecular carbon-hydrogen bond insertion reactions.

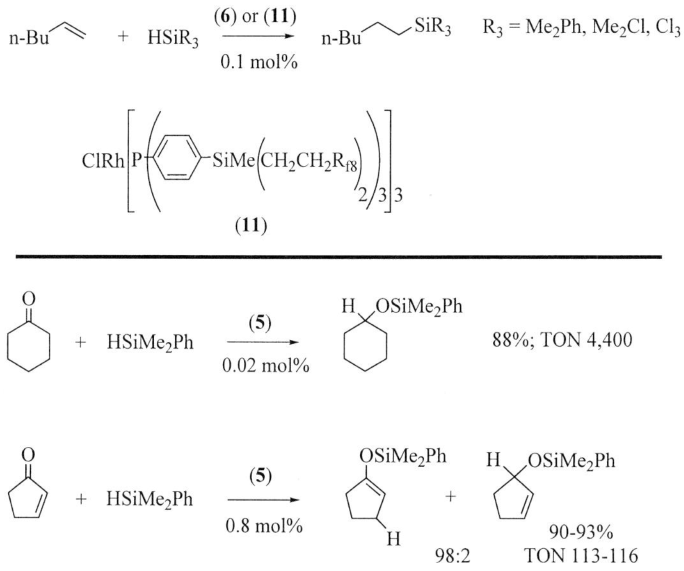

n-Bu–CH=CH$_2$ + HSiR$_3$ —(6) or (11), 0.1 mol%→ n-Bu–CH$_2$CH$_2$–SiR$_3$ R$_3$ = Me$_2$Ph, Me$_2$Cl, Cl$_3$

ClRh[P(C$_6$H$_4$-SiMe(CH$_2$CH$_2$R$_{f8}$)$_2$)$_3$]$_3$

(11)

cyclohexanone + HSiMe$_2$Ph —(5), 0.02 mol%→ H, OSiMe$_2$Ph (cyclohexyl)   88%; TON 4,400

cyclopentenone + HSiMe$_2$Ph —(5), 0.8 mol%→ OSiMe$_2$Ph, H (silyl enol ether) + H, OSiMe$_2$Ph (cyclopentenyl)   98:2   90-93% TON 113-116

Scheme 5 Fluorous catalysts for the hydrosilylation of alkenes and ketones.

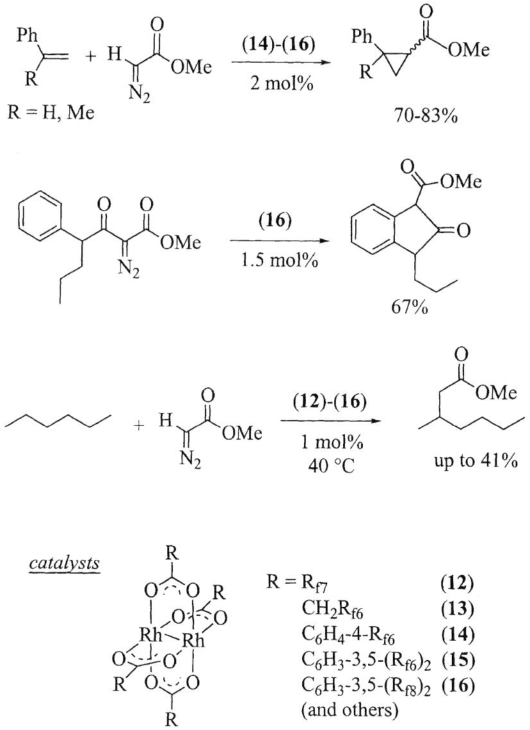

Ph(R)C=CH$_2$ + H–C(=N$_2$)–C(=O)OMe —(14)-(16), 2 mol%→ Ph, R cyclopropane–C(=O)OMe   70-83%

R = H, Me

PhCH(propyl)–C(=O)–C(=N$_2$)–C(=O)OMe —(16), 1.5 mol%→ indanone, C(=O)OMe, =O   67%

hexane + H–C(=N$_2$)–C(=O)OMe —(12)-(16), 1 mol%, 40 °C→ CH$_2$C(=O)OMe   up to 41%

catalysts

Rh$_2$(O$_2$CR)$_4$

R = R$_{f7}$ **(12)**
CH$_2$R$_{f6}$ **(13)**
C$_6$H$_4$-4-R$_{f6}$ **(14)**
C$_6$H$_3$-3,5-(R$_{f6}$)$_2$ **(15)**
C$_6$H$_3$-3,5-(R$_{f8}$)$_2$ **(16)**
(and others)

Scheme 6 Reactions of diazo compounds.

3.2.6
Palladium-Catalyzed Carbon-Carbon Bond-Forming Reactions of Aryl Halides

Fluorous catalyst precursors have been employed in the classical palladium-catalyzed reactions of aryl halides, such as the Heck [8, 16], Suzuki [8, 17, 18], Stille [19], Sonogashira [18, 20], and organozinc [21] cross-coupling reactions. As shown in Scheme 7, palladium complexes of fluorous phosphines, and N-donor and S-donor palladacycles (**18**, **19**), have been utilized for Heck reactions. There is compelling evidence that **18** and **19** do not give molecular catalysts, but rather palladium nanoparticles with high affinities for *non*-fluorous phases [8]. However, enantioselective Heck reactions have been effected with the fluorous BINAP ligand **20** (Scheme 7) [16b]. The *ee* values are as high as 93%, and often exceed those obtained with BINAP. These exceptional enantioselectivities indicate a molecular catalyst.

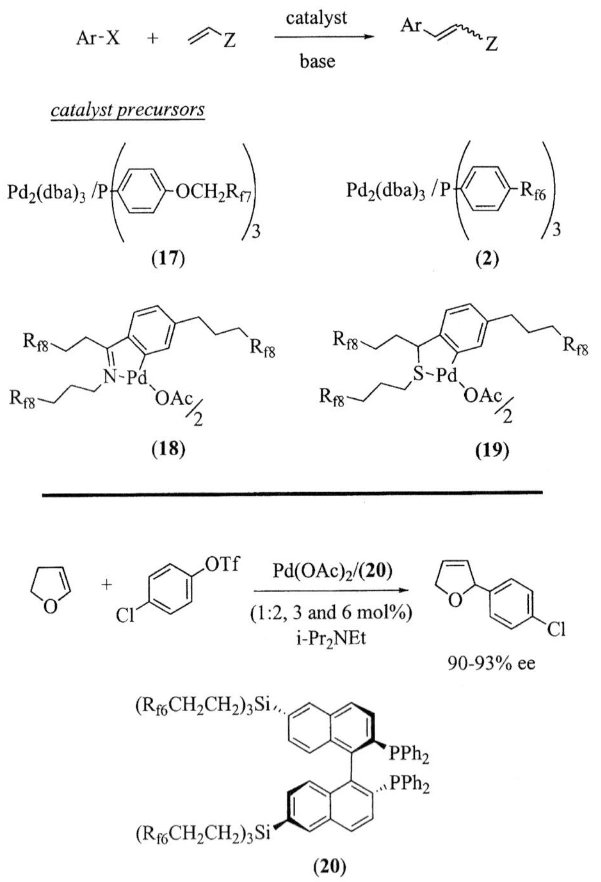

Ar-X + Z → (catalyst, base) Ar–Z

catalyst precursors

$Pd_2(dba)_3$ /P(C_6H_4-OCH_2R_{f7})$_3$ **(17)**

$Pd_2(dba)_3$ /P(C_6H_4-R_{f6})$_3$ **(2)**

(18) (R_{f8}, N-Pd, OAc, ½)

(19) (R_{f8}, S-Pd, OAc, ½)

(dihydrofuran) + Cl-C_6H_4-OTf → ($Pd(OAc)_2$/(**20**), (1:2, 3 and 6 mol%), i-Pr_2NEt) product, 90-93% ee

(20) ($(R_{f6}CH_2CH_2)_3Si$, PPh_2, PPh_2, $(R_{f6}CH_2CH_2)_3Si$)

Scheme 7 Fluorous catalysts for the Heck reaction.

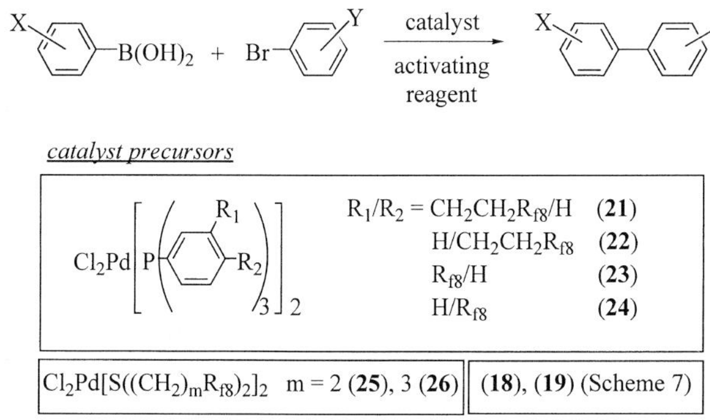

$$X\text{-}C_6H_4\text{-}B(OH)_2 + Br\text{-}C_6H_4\text{-}Y \xrightarrow[\text{activating reagent}]{\text{catalyst}} X\text{-}C_6H_4\text{-}C_6H_4\text{-}Y$$

catalyst precursors

$Cl_2Pd[P(C_6H_3R_1R_2)_3]_2$

R_1/R_2	
$CH_2CH_2R_{f8}/H$	(**21**)
$H/CH_2CH_2R_{f8}$	(**22**)
R_{f8}/H	(**23**)
H/R_{f8}	(**24**)

$Cl_2Pd[S((CH_2)_mR_{f8})_2]_2$ m = 2 (**25**), 3 (**26**) | (**18**), (**19**) (Scheme 7)

Scheme 8 Fluorous catalysts for the Suzuki reaction.

Fluorous catalyst precursors for the Suzuki reaction are summarized in Scheme 8. The complexes **25** and **26**, which feature fluorous thioethers, also appear to give palladium nanoparticle catalysts [17b]. The catalysts **21–24** (Scheme 8), and closely related species, have also been used for Stille and Sonogashira reactions (not depicted). A recent report describes the adsorption of **21**, **22**, and analogous $-OCH_2R_{f7}$-substituted species onto fluorous silica gel [18]. The resulting free-flowing powders are effective catalyst precursors for Suzuki and Sonogashira reactions, which are conducted at 80–100 °C in the customary organic solvents. They are recovered by decantation at 0 °C with <2% palladium loss and reused.

3.2.7 Other Palladium-Catalyzed Carbon-Carbon Bond-Forming Reactions

Fluorous catalyst precursors have also been employed in palladium-catalyzed alkylations of allylic acetates [22]. In the most recent efforts, chiral ligands have been utilized, and these are summarized in Scheme 9 [22b–d]. While good enantioselectivities can be achieved with the binaphthyl (**27**) and bisoxazolines (**30–31**), they do not exceed those of non-fluorous homologs. Also, the closely related binaphthyls **28** and **29** give much lower *ee* values. Thus, enantioselectivity can be a sensitive function of structure. Finally, the cyclodimerization of conjugated enynes to vinyl arenes (not depicted) is efficiently catalyzed by $Pd_2(dba)_3$ and fluorous triaryl phosphines [23].

3.2.8 Zinc-Catalyzed Additions of Dialkylzinc Compounds to Aldehydes

A variety of chiral amino alcohols and amino thiols react with dialkylzinc compounds to give chelate complexes that are highly enantioselective catalysts for additions of dialkylzinc compounds to aldehydes. As illustrated in Scheme 10, sev-

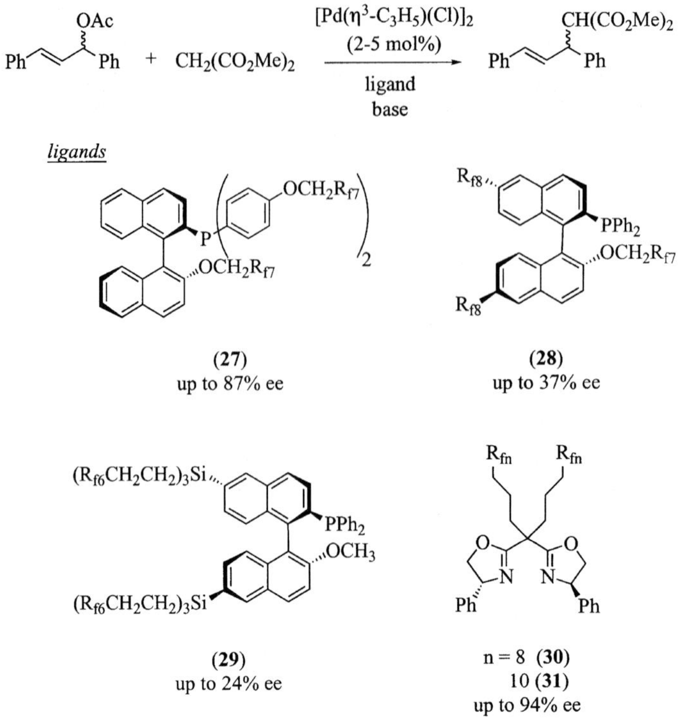

OAc
Ph
Ph
+ CH2(CO2Me)2
[Pd(η3-C3H5)(Cl)]2
(2-5 mol%)
ligand
base
CH(CO2Me)2
ligands
OCH2Rf7
P
OCH2Rf7
2
Rf8
PPh2
OCH2Rf7
(27)
up to 87% ee
(28)
up to 37% ee
(Rf6CH2CH2)3Si
PPh2
OCH3
Rfn
O
N
(29)
up to 24% ee
n = 8 (30)
10 (31)
up to 94% ee

Scheme 9 Fluorous catalysts for the enantioselective alkylation of allylic acetates.

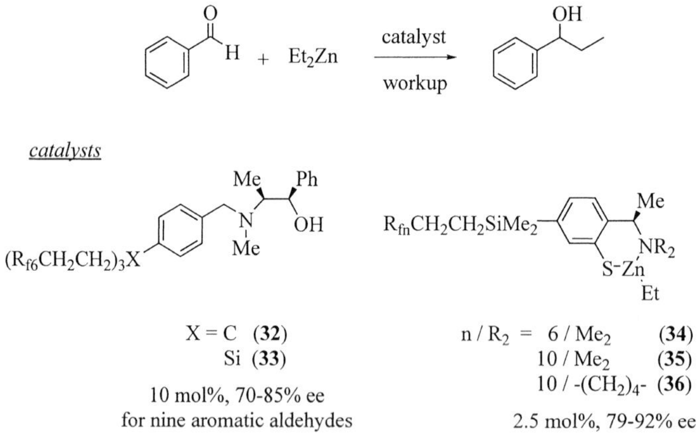

O
H
+ Et2Zn
catalyst
workup
OH
catalysts
Me
Ph
N
OH
Me
(Rf6CH2CH2)3X
RfnCH2CH2SiMe2
NR2
S-Zn
Et
X = C (32)
Si (33)
10 mol%, 70-85% ee
for nine aromatic aldehydes
n / R2 = 6 / Me2 (34)
10 / Me2 (35)
10 / -(CH2)4- (36)
2.5 mol%, 79-92% ee

Scheme 10 Zinc-catalyzed enantioselective addition of dialkylzinc compounds to aldehydes.

eral types of fluorous ligands have been applied in these reactions [24]. Among the amino alcohols, **32** and **33** are the most effective. The former features quaternary carbon to which three pony tails are attached. In the case of amino thiols, the zinc thiolates **34–36** were first isolated and characterized [24a]. These also gave very high *ee* values, exceeding those of non-fluorous analogs.

3.2.9
Titanium-Catalyzed Additions of Carbon Nucleophiles to Aldehydes

A variety of chiral binaphthols have been shown to react with $Ti(O\text{-}i\text{-}Pr)_4$ to give chelate complexes that are highly enantioselective catalysts for additions of dialkylzinc compounds to aldehydes. Such reactions are best conducted with some $Ti(O\text{-}i\text{-}Pr)_4$ as co-catalyst. As illustrated in Scheme 11, several types of fluorous binaphthols have been so applied [16c, 25]. The performance characteristics of **37–38** are particularly noteworthy. The ligand **39** has furthermore been used for the analogous addition of triethyl aluminum to aromatic aldehydes (63–38% *ee*) [25a]. Titanium adducts of other chiral fluorous binaphthols have recently been shown to catalyze the allylation of aldehydes by allyl tri(n-butyl)tin (88–51% *ee*) [25c]. Platinum complexes of fluorons phosphines are also competent catalyst precursors and have been applied in parallel synthesis [26].

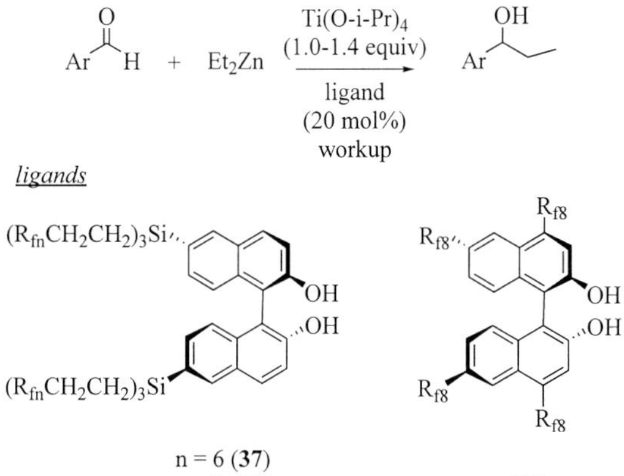

Ar(C=O)H + Et_2Zn → [$Ti(O\text{-}i\text{-}Pr)_4$ (1.0-1.4 equiv); ligand (20 mol%); workup] → ArCH(OH)Et

ligands

$(R_{fn}CH_2CH_2)_3Si$ ... OH, OH ... $(R_{fn}CH_2CH_2)_3Si$

n = 6 (**37**)
8 (**38**)

78-91% ee
for nine aryl groups

R_{f8}, R_{f8} ... OH, OH ... R_{f8}, R_{f8}

(**39**)

54% ee, Ar =
C_6H_5, p-ClC_6H_4

similar additions of $H_2C{=}CHCH_2Sn(n\text{-}Bu)_3$
using other fluorous binaphthols and 20 mol% $Ti(O\text{-}i\text{-}Pr)_4$
(51-88% ee for eight aryl groups)

Scheme 11 Titanium-mediated addition of carbon nucleophiles to aldehydes.

3.2.10 Oxidations

A variety of transition metal-based fluorous oxidizing systems have been reported [27–34]. However, many fluorous oxidizing systems that do not contain a metal have also been developed. Given the widely publicized applicability of some fluorous media as blood substitutes, it would be natural to surmise that they might hold promise for oxidations. However, on molarity concentration scales, the solubility of oxygen and other gaseous molecules in fluorous solvents is only 2–3 times greater than that in typical organic solvents (the increase versus water is much greater because of the strong hydrogen bond network) [2g, 35]. Accordingly, no special rate accelerations have been found.

3.2.10.1 Alkene Epoxidation [27, 28]

Three achiral catalyst systems are shown in Scheme 12. The first consists of the fluorous cobalt porphyrin **40**, oxygen and isobutyraldehyde [27a], the second of $RuCl_3$, a fluorous bipyridine (**41–43**), and $NaIO_4$ [27c], and the third of the fluorous ruthenium acetylacetonate complex **44**, oxygen, and isobutyraldehyde [28]. Not surprisingly, recent emphasis has focused on enantioselective epoxidation. The first-generation chiral fluorous salen(manganese) systems gave *ee* values that were usually much lower than non-fluorous analogs [27b,d]. However, the second-generation complexes **45–46** yielded markedly improved results, as summarized in Scheme 12 [27e,f].

3.2.10.2 Other Oxidations of Alkenes and Alkanes [29–31]

The fluorous cyclic polyamines **47** and **48**, shown in Scheme 13, are components of catalyst systems for the oxidation of cyclohexene [29, 30]. Both *tert*-butyl hydroperoxide and molecular oxygen are required. The former is needed for initiation (hydrogen atom abstraction), and the latter for free radical chain propagation. Cyclohexane and cyclooctane can similarly be oxidized, although not as efficiently. The fluorous palladium(II) acetylacetonate complex **49** (Scheme 13) catalyzes a number of preparatively useful Wacker oxidations of alkenes, using *tert*-butyl hydroperoxide [31].

3.2.10.3 Oxidations of Other Functional Groups [28, 32–34]

Two very effective fluorous catalysts for the aerobic oxidation of alcohols are shown in Scheme 14 [32, 33]. One uses a palladium complex of the pyridine **50**, and the other a copper adduct of the bipyridine **51**. High yields have been obtained for a broad spectrum of alcohols. The nickel analog of palladium complex

1. achiral catalysts

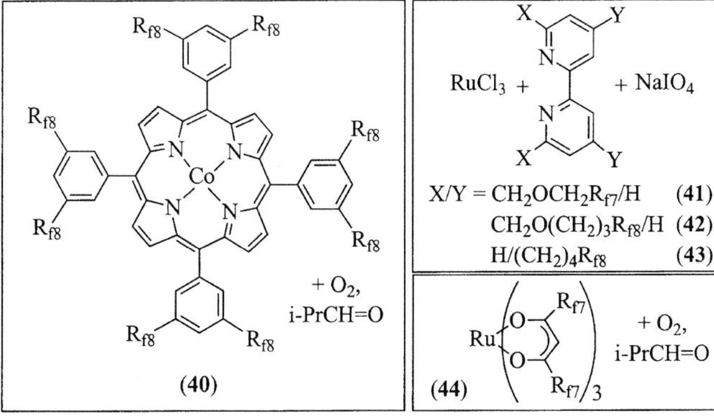

2. chiral catalysts

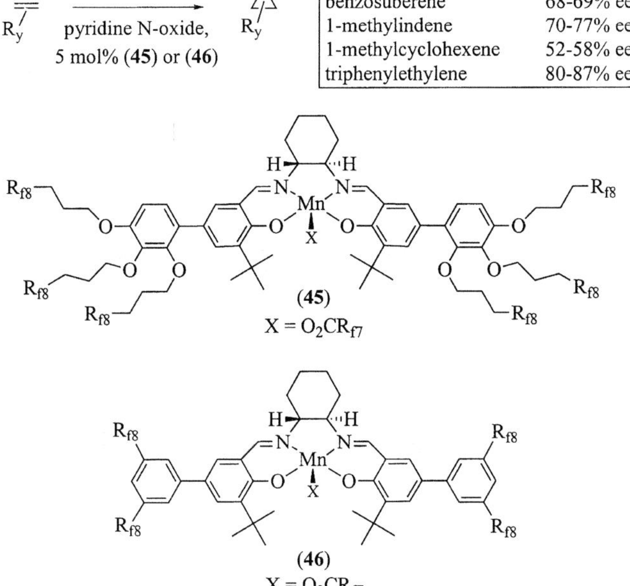

Scheme 12 Fluorous catalysts for the epoxidation of alkenes.

49 (Scheme 13), **52**, catalyzes the reaction of aldehydes and oxygen to give carboxylic acids [28]. As illustrated in Scheme 14, this same system also catalyzes the aerobic oxidation of sulfides to sulfoxide and/or sulfones. Stoichiometric quantities of isobutyraldehyde are required, the amount of which determines the degree of oxidation. The cobalt porphyrin **40** and fluorous cobalt phthalocyanine **53** effect similar oxidations of sulfides [34].

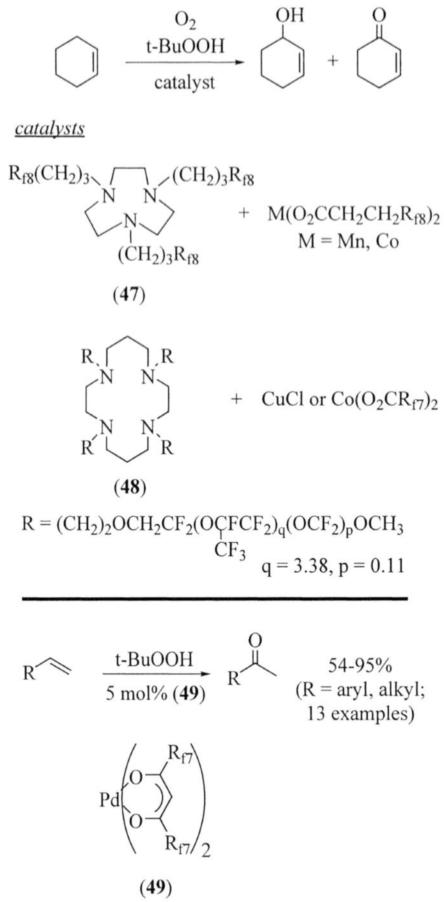

O_2
t-BuOOH
catalyst
OH
O
+

catalysts

$R_{f8}(CH_2)_3$ $(CH_2)_3R_{f8}$
N N
N
$(CH_2)_3R_{f8}$
+ $M(O_2CCH_2CH_2R_{f8})_2$
M = Mn, Co

(**47**)

R R
N N
N N
R R
+ CuCl or $Co(O_2CR_{f7})_2$

(**48**)

$R = (CH_2)_2OCH_2CF_2(OCFCF_2)_q(OCF_2)_pOCH_3$
CF_3
q = 3.38, p = 0.11

R
t-BuOOH
5 mol% (**49**)
O
R
54-95%
(R = aryl, alkyl;
13 examples)

R_{f7}
O
Pd
O
R_{f7} 2

(**49**)

Scheme 13 Other oxidations of alkenes.

3.2.11
Other Metal-Catalyzed Reactions

As illustrated in Scheme 15, complexes formed between $[Ir(COD)(Cl)]_2$ and the chiral fluorous diimine and diamine ligands **54–59** are catalysts for enantioselective hydrogen transfer reductions of ketones, using isopropanol as the hydrogen donor [36]. Also, chiral fluorous salen(cobalt) complexes are able to catalyze the kinetic resolution of racemic terminal epoxides via hydrolytic ring opening [37]. Compound **60** (Scheme 15) was the most effective of several tested. As shown in Scheme 16, the fluorous nickel pincer complexes **61–62** catalyze the addition of CCl_4 to methyl methacrylate (Kharasch reaction) [38]. Copper(I) complexes of the acyclic fluorous polyamines **63–64** catalyze a related intramolecular reaction [39], as well or living radical polymerizations of acrylates [40]. Finally, a perfluorinated polyether has been derivatized with an acetylacetonate moiety, a nickel complex of which catalyzes the oligomerization of ethylene [41].

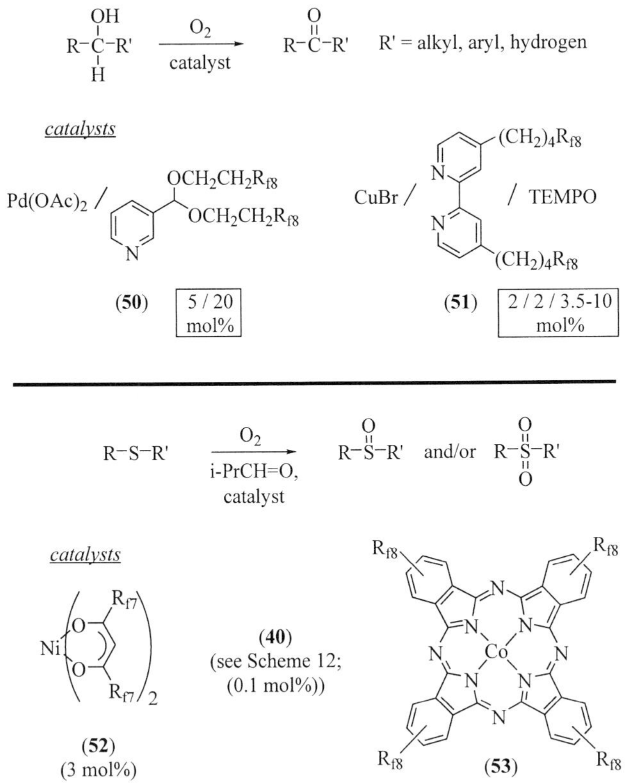

$$R-CH(OH)-R' \xrightarrow[\text{catalyst}]{O_2} R-C(=O)-R' \quad R' = \text{alkyl, aryl, hydrogen}$$

catalysts

$Pd(OAc)_2$ / 3-pyridyl-$CH(OCH_2CH_2R_{f8})_2$ (50) 5 / 20 mol%

CuBr / 4,4'-bis($(CH_2)_4R_{f8}$)-2,2'-bipyridine / TEMPO (51) 2 / 2 / 3.5-10 mol%

$$R-S-R' \xrightarrow[\text{i-PrCH=O, catalyst}]{O_2} R-S(=O)-R' \text{ and/or } R-S(=O)_2-R'$$

catalysts

Ni(R_{f7}COCHCOR_{f7})$_2$ (52) (3 mol%)

(40) (see Scheme 12; (0.1 mol%))

Co phthalocyanine with four R_{f8} (53)

Scheme 14 Aerobic oxidations of alcohols and sulfides.

3.2.12 Related Methods

Several topics formally outside the scope of this article or compendium merit brief mention. First, unsurprisingly, fluorous catalysis has been extended to lanthanide metal-containing species [42]. Second, fluorous compounds generally have excellent solubilities in supercritical CO_2, and there is an extensive parallel (sometimes overlapping) catalysis literature involving this medium. Finally, it is also possible to render catalytically active palladium nanoparticles fluorophilic and recoverable by the protocol in Scheme 1 [43].

3.2.13 Summary and Outlook

In looking back over the preceding chemistry, it is evident that fluorous analogs of many transition metal-based catalysts are now available. In most respects, these seem to function just like their non-fluorous counterparts, with the major advan-

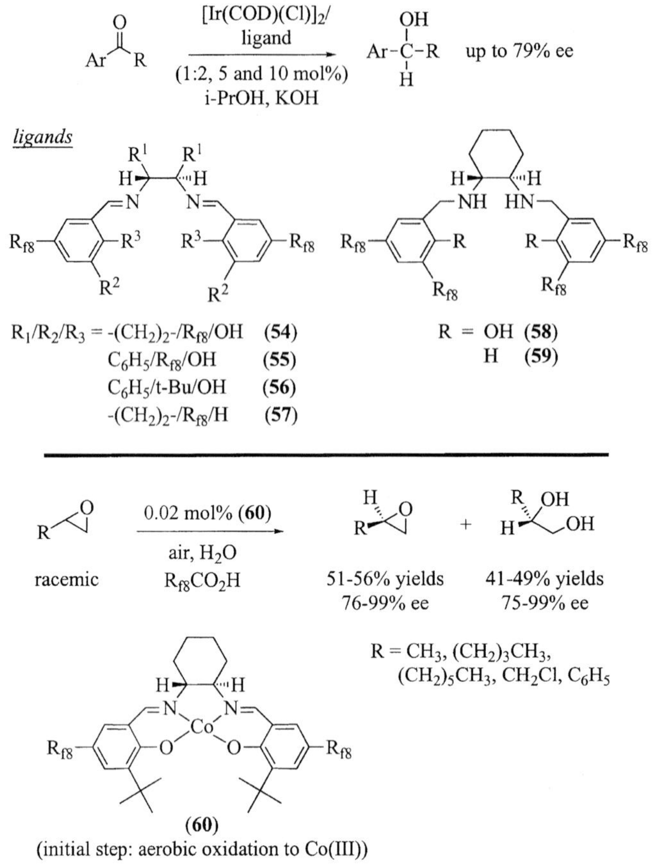

$[Ir(COD)(Cl)]_2$/ ligand (1:2, 5 and 10 mol%) i-PrOH, KOH

Ar–C(OH)(H)–R up to 79% ee

ligands

$R_1/R_2/R_3$ = $-(CH_2)_2-/R_{f8}/OH$ (**54**)
$C_6H_5/R_{f8}/OH$ (**55**)
C_6H_5/t-Bu/OH (**56**)
$-(CH_2)_2-/R_{f8}/H$ (**57**)

R = OH (**58**)
H (**59**)

racemic

0.02 mol% (**60**)
air, H_2O
$R_{f8}CO_2H$

51-56% yields
76-99% ee

41-49% yields
75-99% ee

R = CH_3, $(CH_2)_3CH_3$, $(CH_2)_5CH_3$, CH_2Cl, C_6H_5

(**60**)
(initial step: aerobic oxidation to Co(III))

Scheme 15 Other reactions of chiral metal-containing fluorous catalysts.

tage of recoverability. Although increasing numbers of metal-mediated carbon-fluorine bond activation processes are being discovered [44], no side reactions involving the pony tails of fluorous catalysts have ever been observed.

Although fluorous catalysts can give somewhat different chemo-, regio- and stereoselectivities than those of their non-fluorous counterparts, no dramatic reversals have been reported to date. The enantioselectivities of first-generation chiral fluorous catalysts are often lower. However, it should be kept in mind that comparisons are usually being made to non-fluorous catalysts that have been highly optimized through years of work. Furthermore, pony tails can exert substantial electronic effects [6], with the potential to influence selectivities.

A decade ago, fluorous catalysts or reagents of any description were unknown. In the upcoming decade, many types of extensions can be anticipated. One will certainly be the development of additional metal-based catalysts. However, fluorous versions of metal-containing reagents such as zirconium hydrides or osmium

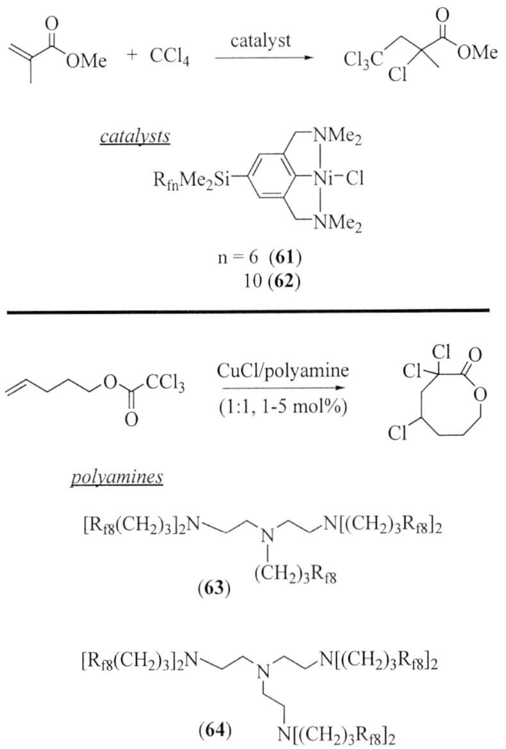

Scheme 16 Fluorous catalysts for atom transfer radical additions.

oxidants can also be expected. Another active front will be the optimization of all of these reactions from the standpoint of catalyst/reagent recoverability. Efforts will also be vigorously directed at modified procedures that minimize or eliminate the need for fluorous solvents [5].

Acknowledgement

The authors thank the Bundesministerium für Bildung und Forschung (BMBF) for support.

References

1 I.T. Horváth, J. Rábai, *Science* **1994**, *266*, 72.

2 (a) I.T. Horváth, *Acc. Chem. Res.* **1998**, *31*, 641. (b) E. de Wolf, G. van Koten, B.-J. Deelman, *Chem. Soc. Rev.* **1999**, *28*, 37. (c) M. Cavazzini, F. Montanari, G. Pozzi, S. Quici, *J. Fluorine Chem.* **1999**, *94*, 183. (d) E.G. Hope, A.M. Stuart, *J. Fluorine Chem.* **1999**, *100*, 75. (e) J. Yoshida, K. Itami, *Chem. Rev.* **2002**, *102*, 3693. (f) A.P. Dobbs, M.R. Kimberly, *J. Fluorine Chem.* **2002**, *118*, 3. (g) Survey of

practical considerations and underlying physical principles: L. P. Barthel-Rosa, J. A. Gladysz, *Coord. Chem. Rev.* **1999**, *190–192*, 587.

3 Handbook of Fluorons Chemistry, J. A. Gladysz, D. P. Curran, I. T. Horváth, Eds.; Wiley/VCH, Weinheim, **2004**.

4 J. J. Maul, P. J. Ostrowski, G. A. Ublacker, B. Linclaw, D. P. Curran, in *Topics in Current Chemistry* ("Modern Solvents in Oragnic Synthesis") Vol. 206 (P. Knuchel, Ed.). Springer, Berlin, **1999**, p. 80.

5 M. Wende, J. A. Gladysz, *J. Am. Chem. Soc.* **2003**, *125*, 5861.

6 H. Jiao, S. Le Stang, T. Soós, R. Meier, P. Rademacher, K. Kowski, L. Jafarpour, J.-B. Hamard, S. P. Nolan, J. A. Gladysz, *J. Am. Chem. Soc.* **2002**, *124*, 1516.

7 (a) J. A. Gladysz, *Pure Appl. Chem.* **2001**, *73*, 1319. (b) J. A. Gladysz, *Chem. Rev.* **2002**, *102*, 3214.

8 (a) C. Rocaboy, J. A. Gladysz, *Org. Lett.* **2002**, *4*, 1993. (b) C. Rocaboy, J. A. Gladysz, *New J. Chem.* **2003**, *27*, 39.

9 (a) I. T. Horváth, G. Kiss, R. A. Cook, J. E. Bond, P. A. Stevens, J. Rábai, E. J. Mozeleski, *J. Am. Chem. Soc.* **1998**, *120*, 3133. (b) D. F. Foster, D. Gudmunsen, D. J. Adams, A. M. Stuart, E. G. Hope, D. J. Cole-Hamilton, G. P. Schwarz, P. Pogorzelec, *Tetrahedron* **2002**, *58*, 3901, and earlier work cited therein. (c) T. Mathivet, E. Monflier, Y. Castanet, A. Mortreux, J.-L. Couturier, *Tetrahedron* **2002**, *58*, 3877. (d) W. Chen, L. Xu, Y. Hu, A. M. Banet Osuna, J. Xiao, *Tetrahedron* **2002**, *58*, 3889.

10 (a) D. Rutherford, J. J. J. Juliette, C. Rocaboy, I. T. Horváth, J. A. Gladysz, *Catal. Today* **1998**, *42*, 381. (b) E. G. Hope, R. D. W. Kemmitt, D. R. Paige, A. M. Stuart, *J. Fluorine Chem.* **1999**, *99*, 197. (c) D. E. Bergbreiter, J. G. Franchina, B. L. Case, *Org. Lett.* **2000**, *2*, 393. (d) B. Richter, A. L. Spek, G. van Koten, B.-J. Deelman, *J. Am. Chem. Soc.* **2000**, *122*, 3945. (e) T. Soós, B. L. Bennett, D. Rutherford, L. P. Barthel-Rosa, J. A. Gladysz, *Organometallics* **2001**, *20*, 3079. (f) E. de Wolf, A. L. Spek, B. W. M. Kuipers, A. P. Philipse, J. D. Meeldijk, P. H. H. Bomans, P. M. Frederik, B.-J. Deelman, G. van Koten, *Tetrahedron* **2002**, *58*, 3911. (g) J. van den Broeke, E. de Wolf, B.-J. Deelman, G. van Kuten, *Adv. Synth. Cat.* **2003**, *345*, 625. (h) D. Sinou, D. Maillard, A. Aghmiz, A. M. Masden i-Bultó, *Adv. Synth. Cat.* **2003**, *345*, 603.

11 (a) J. J. J. Juliette, I. T. Horváth, J. A. Gladysz, *Angew. Chem., Int. Ed. Engl.* **1997**, *36*, 1610; *Angew. Chem.* **1997**, *109*, 1682. (b) J. J. J. Juliette, D. Rutherford, I. T. Horváth, J. A. Gladysz, *J. Am. Chem. Soc.* **1999**, *121*, 2696.

12 (a) E. de Wolf, E. A. Speets, B.-J. Deelman, G. van Koten, *Organometallics* **2001**, *20*, 3686. (b) J. van den Broeke, F. Winter, B.-J. Deelman, G. van Koten, *Org. Lett.* **2002**, *4*, 3851.

13 L. V. Dinh, J. A. Gladysz, *Tetrahedron Lett.* **1999**, *40*, 8995.

14 A. Endres, G. Maas, *Tetrahedron* **2002**, *58*, 3999, and references therein.

15 A. Endres, G. Maas, *J. Organomet. Chem.* **2002**, *643/644*, 174.

16 (a) J. Moineau, G. Pozzi, S. Quici, D. Sinou, *Tetrahedron Lett.* **1999**, *40*, 7683. (b) Y. Nakamura, S. Takeuchi, S. Zhang, K. Okumura, Y. Ohgo, *Tetrahedron Lett.* **2002**, *43*, 3053. (c) Y. Nakamura, S. Takeuchi, Y. Ohgo, *J. Fluorine Chem.* **2003**, *120*, 121.

17 (a) S. Schneider, W. Bannwarth, *Helv. Chim. Acta* **2001**, *84*, 735. (b) C. Rocaboy, J. A. Gladysz, *Tetrahedron* **2002**, *58*, 4007.

18 C. C. Tzschucke, C. Markert, H. Glatz, W. Bannwarth, *Angew. Chem., Int. Ed.* **2002**, *41*, 4500; *Angew. Chem.* **2002**, *114*, 4678.

19 S. Schneider, W. Bannwarth, *Angew. Chem., Int. Ed.* **2000**, *39*, 4142; *Angew. Chem.* **2000**, *112*, 4293.

20 C. Markert, W. Bannwarth, *Helv. Chim. Acta* **2002**, *85*, 1877.

21 B. Betzemeier, P. Knochel, *Angew. Chem., Int. Ed. Engl.* **1997**, *36*, 2623; *Angew. Chem.* **1997**, *109*, 2736.

22 (a) R. Kling, D. Sinou, G. Pozzi, A. Choplin, F. Quignard, S. Busch, S. Kainz, D. Koch, W. Leitner, *Tetrahedron Lett.* **1998**, *39*, 9439. (b) M. Cavazzini, G. Pozzi, S. Quici, D. Maillard, D. Sinou, *Chem. Commun.* **2001**, 1220. (c) D. Mail-

LARD, J. BAYARDON, J. D. KURICHIPARAMBIL, C. NGUEFACK-FOURNIER, D. SINOU, *Tetrahedron: Asymmetry* **2002**, *13*, 1449. (d) J. BAYARDON, D. SINOU, *Tetrahedron Lett.* **2003**, *44*, 1449.

23 S. SAITO, Y. CHOUNAN, T. NOGAMI, O. OHMORI, Y. YAMAMOTO, *Chem. Lett.* **2001**, 444.

24 (a) H. KLEIJN, E. RIJNBERG, J. T. B. H. JASTRZEBSKI, G. VAN KOTEN, *Org. Lett.* **1999**, *1*, 853. (b) Y. NAKAMURA, S. TAKEUCHI, K. OKUMURA, Y. OHGO, *Tetrahedron* **2001**, *57*, 5565.

25 (a) Y. TIAN, Q. C. YANG, T. C. W. MAK, K. S. CHAN, *Tetrahedron* **2002**, *58*, 3951, and earlier work cited therein. (b) Y. NAKAMURA, S. TAKEUCHI, K. OKUMURA, Y. OHGO, D. P. CURRAN, *Tetrahedron* **2002**, *58*, 3963, and earlier work cited therein. (c) Y.-Y. YIN, G. ZHAO, Z.-S. QIAN, W.-X. YIN, *J. Fluorine Chem.* **2003**, *120*, 117.

26 Q. ZHANG, Z. LUO, D. P. CURRAN, *J. Org. Chem.* **2000**, *65*, 8866.

27 (a) G. POZZI, F. MONTANARI, S. QUICI, *Chem. Commun.* **1997**, 69. (b) G. POZZI, F. CINATO, F. MONTANARI, S. QUICI, *Chem. Commun.* **1998**, 877. (c) S. QUICI, M. CAVAZZINI, S. CERAGIOLI, F. MONTANARI, G. POZZI, *Tetrahedron Lett.* **1999**, *40*, 3647. (d) G. POZZI, M. CAVAZZINI, F. CINATO, F. MONTANARI, S. QUICI, *Eur. J. Org. Chem.* **1999**, 1947. (e) M. CAVAZZINI, A. MANFREDI, F. MONTANARI, S. QUICI, G. POZZI, *Chem. Commun.* **2000**, 2171. (f) M. CAVAZZINI, A. MANFREDI, F. MONTANARI, S. QUICI, G. POZZI, *Eur. J. Org. Chem.* **2001**, 4639.

28 I. KLEMENT, H. LÜTJENS, P. KNOCHEL, *Angew. Chem., Int. Ed. Engl.* **1997**, *36*, 1454; *Angew. Chem.* **1997**, *109*, 1605.

29 (a) J.-M. VINCENT, A. RABION, V. K. YACHANDRA, R. H. FISH, *Angew. Chem., Int. Ed. Engl.* **1997**, *36*, 2346; *Angew. Chem.* **1997**, *109*, 2438. (b) J.-M. VINCENT, A. RABION, V. K. YACHANDRA, R. H. FISH, *Can. J. Chem.* **2001**, *79*, 888.

30 G. POZZI, M. CAVAZZINI, S. QUICI, S. FONTANA, *Tetrahedron Lett.* **1997**, *38*, 7605.

31 B. BETZEMEIER, F. LHERMITTE, P. KNOCHEL, *Tetrahedron Lett.* **1998**, *39*, 6667.

32 T. NISHIMURA, Y. MAEDA, N. KAKIUCHI, S. UEMURA, *J. Chem. Soc., Perkin Trans. 1* **2000**, 4301.

33 (a) B. BETZEMEIER, M. CAVAZZINI, S. QUICI, P. KNOCHEL, *Tetrahedron Lett.* **2000**, *41*, 4343. (b) G. RAGAGNIN, B. BETZEMEIER, S. QUICI, P. KNOCHEL, *Tetrahedron* **2002**, *58*, 3985.

34 S. COLONNA, N. GAGGERO, F. MONTANARI, G. POZZI, S. QUICI, *Eur. J. Org. Chem.* **2001**, 181.

35 M.-A. GUILLEVIC, C. ROCABOY, A. M. ARIF, I. T. HORVÁTH, J. A. GLADYSZ, *Organometallics* **1998**, *17*, 707.

36 D. MAILLARD, G. POZZI, S. QUICI, D. SINOU, *Tetrahedron* **2002**, *58*, 3971.

37 (a) M. CAVAZZINI, S. QUICI, G. POZZI, *Tetrahedron* **2002**, *58*, 3943. (b) I. SHEPPERSON, M. CAVAZZINI, G. POZZI, S. QUICI, *J. Fluor. Chem.* **2004**, *125*, 175.

38 H. KLEIJN, J. T. B. H. JASTRZEBSKI, R. A. GOSSAGE, H. KOOIJMAN, A. L. SPEK, G. VAN KOTEN, *Tetrahedron* **1998**, *54*, 1145.

39 F. DE CAMPO, D. LASTÉCOUÈRES, J.-M. VINCENT, J.-B. VERLHAC, *J. Org. Chem.* **1999**, *64*, 4969.

40 D. M. HADDLETON, S. G. JACKSON, S. A. F. BON, *J. Am. Chem. Soc.* **2000**, *122*, 1542.

41 W. KEIM, M. VOGT, P. WASSERSCHEID, B. DRIESSEN-HÖLSCHER, *J. Mol. Cat. A* **1999**, *139*, 171.

42 (a) A. G. M. BARRETT, N. BOULOC, D. C. BRADDOCK, D. CATTERICK, D. CHADWICK, A. J. P. WHITE, D. J. WILLIAMS, *Tetrahedron* **2002**, *58*, 3835. (b) K. MIKAMI, Y. MIKAMI, H. MATSUZAWA, Y. MATSUMOTO, J. NISHIKIDO, F. YAMAMOTO, H. NAKAJIMA, *Tetrahedron* **2002**, *58*, 4015.

43 (a) M. MORENO-MAÑAS, R. PLEIXATS, S. VILLARROYA, *Organometallics* **2001**, *20*, 4524. (b) V. CHECHIK, R. M. CROOKS, *J. Am. Chem. Soc.* **2000**, *122*, 1243.

44 T. G. RICHMOND, *Angew. Chem., Int. Ed.* **2000**, *39*, 3241; *Angew. Chem.* **2000**, *112*, 3378, and references cited therein.

3.3
Organic Synthesis with Transition Metal Complexes using Compressed Carbon Dioxide as Reaction Medium

Giancarlo Franciò and Walter Leitner

3.3.1
Carbon Dioxide as Reaction Medium for Transition Metal Catalysis

Transition metal-catalyzed syntheses are generally conducted as solution phase processes requiring a suitable solvent system to ensure intimate contact of the reagents and catalysts. The molecular structure of the solvent can play a crucial role for the stabilization of reactive intermediates and has often a decisive influence on the rate and selectivity of a particular reaction. Interactions with catalytically active species through temporary binding at open coordination sites are in many cases inferred as key steps in catalytic cycles. In addition to molecular interactions, physico-chemical properties affecting aspects such as solubility of gases and mass transfer can have pronounced effects on the performance of catalysts. Environmental and safety issues resulting from toxicity, flammability, or ecological burden associated with certain solvents need to be considered also during the planning of a synthetic procedure. Isolation of the pure solvent-free product(s) and recycling of reagents or catalysts can be a major difficulty in solution phase synthesis even on a laboratory scale. These considerations become even more pressing for the commercial production of fine chemicals and biologically active compounds. Therefore, the solvent is a strategic parameter for the planning of a synthesis already in the initial stage.

In recent years, several new solvent concepts have received systematic attention, expanding the traditional scope of reaction media available for chemical synthesis. Compressed carbon dioxide, in either its liquid ($liqCO_2$) or supercritical ($scCO_2$) state, is of particular interest as a "green" solvent for catalysis. The supercritical state is reached when CO_2 is heated and compressed beyond its critical temperature ($T_c = 31.0\,^\circ C$) and pressure ($p_c = 73.7$ bar). It is a particularly attractive option offering a whole range of unique properties that can lead to improved synthetic performance, or may be exploited in advanced separation schemes (Tab. 1) [1]. The properties to be considered are often different from classical solvent parameters such as polarity, E_T-values, coordination strength, etc., that form the basis of the synthetic chemist's experience and intuition with organic solvents [2]. Although some of the properties summarized in Table 1 are associated strictly speaking with the supercritical state only, most of them already become signifi-

Transition Metals for Organic Synthesis, Vol. 2, 2nd Edition.
Edited by M. Beller and C. Bolm

ISBN: 3-527-30613-7

Tab. 1 Potential reaction benefits and corresponding physico-chemical properties for compressed CO_2 as reaction medium in transition metal-catalyzed organic synthesis

Potential benefit	***Physico-chemical property***
Higher rates	Miscibility with gases, rapid mass transfer
Different selectivities	Weak coordination, pressure tuning
Fewer reaction steps	*In situ* protection of amines
Additional safety	No toxicity, inertness, good heat transfer
Enhanced separation	Tunable solvent properties, multiphase systems
Continuous flow operation	Multiphase systems, mass transfer

cant in the so-called "near critical region", i.e. at temperatures and pressures in the vicinity of the critical point. At room temperature, CO_2 is liquefied at pressures above ca. 55 bar, and most of the effects will be largely retained under these conditions also.

The high miscibility of $scCO_2$ with reaction gases at liquid-like solvatation can be advantageous for many important transformations such as hydrogenation, oxidation, or carbonylation reactions. The balance between reactivity and inertness of the CO_2 molecule provides another significant difference from classical solvent systems if exploited in the proper way. The possibility to modify solvent properties widely through comparatively small variations in system pressure provides an additional parameter for controlling rate and selectivity in certain cases. In more general terms, this pressure tuning also enables the solubility in $scCO_2$ to be controlled, and novel separation schemes can be envisaged on the basis of this phenomenon. The combination with CO_2-insoluble materials opens up interesting strategies for multiphase catalysis, which are facilitated by the gas-like mass transfer properties of the supercritical state. These themes have emerged individually or in combination for a large number of transition metal-catalyzed reactions in $scCO_2$ reaction media in the last ten years [3, 4]. In the present chapter, we discuss selected recent examples arranged according to the main reaction types, thus providing an overview of synthetic methodologies that have been successfully implemented in this medium.

3.3.2
Reaction Types and Catalytic Systems for Organic Synthesis with Transition Metal Complexes in Compressed Carbon Dioxide

3.3.2.1
Hydrogenation and Related Reactions

The pioneering work on hydrogenation of CO_2 itself to yield formic acid or derivatives such as formamides under supercritical conditions demonstrated that the high solubility of hydrogen in the $scCO_2$ phase can be exploited to achieve very

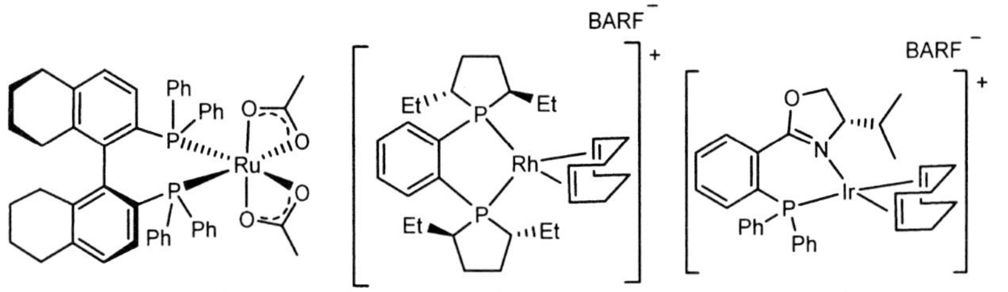

1 **2** **3**

BARF = tetrakis[3,5-bis(trifluoromethyl)phenyl]borate

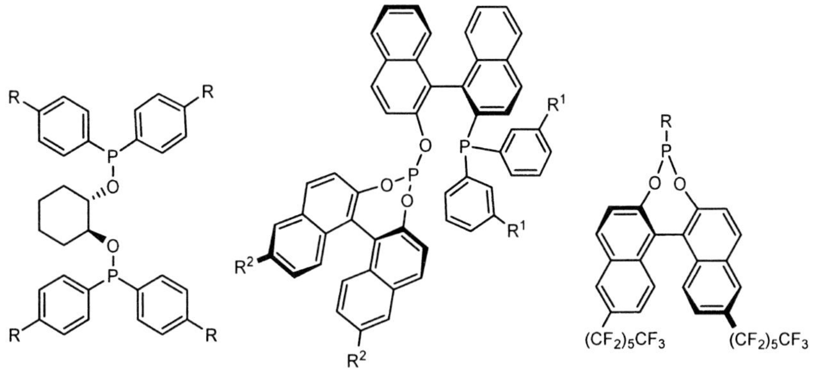

4, R = $(CH_2)_2(CF_2)_5CF_3$

5a, $R^1 = (CH_2)_2(CF_2)_5CF_3$, R^2 = H
5b, R^1 = H, $R^2 = (CH_2)_3(CF_2)_5CF_3$

6a, R = NMe_2
6b, R = O–C_6H_4–$(CF_2)_5CF_3$

Chart 1 Selected ligands and catalysts for homogeneously catalyzed asymmetric hydrogenation in $scCO_2$. For leading references see **1**: [6], **2**: [7], **3**: [8], **4**: [9], **5a**: [10], **5b**: [11], and **6a, b**: [12].

high turnover frequencies with soluble metal catalysts [5]. This also spurred interest in asymmetric hydrogenation in this medium, and a selection of sufficiently "CO_2-philic" chiral ligands and catalysts is summarized in Chart 1. Whereas the ruthenium catalyst **1** still required the presence of a co-solvent for optimum performance [6], the cationic catalysts **2** [7] and **3** [8] could be used in pure CO_2 when combined with the highly fluorinated BARF anion. These conditions led to a significantly higher *ee* when the rhodium catalyst **2** was used for the hydrogenation of dehydroamino acids such as **7** (Eq. 1) and to a 20-fold reduction in the reaction time required for 100% conversion of imine **9** with iridium complex **3** (Eq. 2).

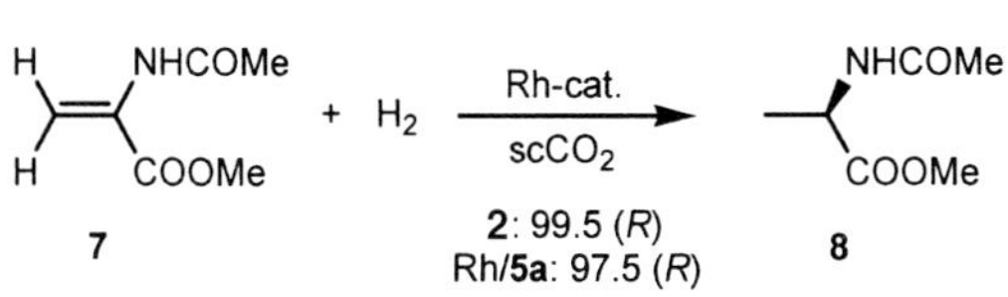

(1)

$$9 + H_2 \xrightarrow[\text{scCO}_2]{\text{Ir-cat.}} 10 \qquad \textbf{3}: 81.0\ (R) \tag{2}$$

Introduction of perfluoroalkyl groups into the ligand framework has proved to be an efficient methodology to render organometallic catalysts highly CO_2-philic [13]. However, the nature and position of the substituent can have pronounced effects on the performance of the catalyst. The ligands **4** [9] and **5a** [10] bear a $-(CH_2)_2-$ spacer through which the perfluorinated alkyl chain is attached at the PPh_2 moiety of the parent ligand. Both ligands showed enantioselectivities that were fully compatible with those of their unsubstituted congeners. In fact, the BINAPHOS derivative **5a** was found to be an excellent ligand for hydrogenation in $scCO_2$ even though it had been originally designed for hydroformylation (see below). In contrast, the substitution pattern of ligands **6a,b** [12] did not lead to catalytic systems that could be operated in $scCO_2$ with similar efficiency to that in conventional solvents.

Fluorinated phosphines have been investigated also for Ru-catalyzed chemoselective hydrogenation of α,β-unsaturated aldehydes in $scCO_2$ [14]. Perfluoroalkyl-substituted ligands were employed successfully in transformations related to hydrogenation such as the hydrosilylation [15] and hydroboration [16] of olefins. A remarkable CO_2 effect was observed for the hydroboration of **11** with a catalyst formed *in situ* from **13** and ligand **14**, leading to greatly enhanced chemo- and regioselectivities toward the desired branched boronic ester **15** as compared to conventional liquid reaction media (Eq. 3).

$$\textbf{11} + \textbf{12} \xrightarrow[\text{scCO}_2\text{, 40 °C, 193 bar, 5 h}]{[(\text{hfacac})\text{Rh}(\text{coe})_2]\ (\textbf{13}\text{, 2 mol\%}),\ Cy_2P(CH_2)_2(CF_2)_5CF_3\ (\textbf{14}\text{, 4 mol\%})} \textbf{15}\text{, quant.} \tag{3}$$

Catalysts bearing a large number of perfluoroalkyl group [17] or extended fluorous polymers (cf. Chart 2) [18] reach a size where they can be efficiently retained by membrane filtration. Hydrogenation of butane as a model reaction has been carried out successfully in a continuous membrane reactor using a highly fluorous derivative of Wilkinson's catalyst [17, 19]. Hydrogenation catalysts can also be immobilized for use with $scCO_2$ by covalent linking to solid supports, as demonstrated, for example, in CO_2 hydrogenation [20, 21]. Although this technique has not yet been applied to asymmetric hydrogenation, recent very positive results with a resin-bound version of BINAPHOS in hydroformylation are encouraging (cf. Section 3.3.2.3) [22].

A conceptually different way of catalyst immobilization is the use of liquid phases that are not soluble in $scCO_2$ to dissolve and retain the catalyst. Water is only sparingly soluble in $scCO_2$ and can be applied as a catalyst compartment with typical sulfonated water-soluble phosphine ligands [23]. Specific "CO_2-philic" surfactants have been used to generate emulsion and micro-emulsion type reaction mixtures for enhanced mass transfer in these systems [24]. High-molecular-weight polyethyleneglycol (PEG) is yet another catalyst immobilization phase for organometallic hydrogenation catalysts [25]. Although this catalyst phase would be a solid at the reaction temperature, it melts in the presence of $scCO_2$ as the compressed gas dissolves in the polymer. The intriguing feature of this system is that no catalyst modification is necessary with triphenylphosphine-based catalysts, suggesting that many chiral catalysts might also operate under the same conditions.

Ionic liquids (ILs) form biphasic mixtures with $scCO_2$, exhibiting a unique phase behavior that can be exploited for catalyst immobilization [26]. The hydrogenation of supercritical CO_2 to formamides was used to investigate the possibility of controlling consecutive reaction pathways in such biphasic mixtures [27]. Asymmetric hydrogenation with ruthenium-BINAP catalysts similar to **1** in ILs enabled the chiral catalyst to be recycled after extraction of the product with $scCO_2$ [28]. The ionic nature of the catalyst phase seems to make this approach particularly suited for cationic catalysts. Indeed, the biphasic system was shown to have clear advantages over either of the two solvent systems separately for the hydrogenation of **9** with catalysts of type **3** [29].

3.3.2.2
Hydroformylation and Carbonylation Reactions

As in the case of hydrogenation, the design of homogeneous and multiphase systems has been a major research focus for rhodium-catalyzed hydroformylation in $scCO_2$ (Eq. 4, Chart 2). Asymmetric hydroformylation of vinyl arenes and acrylate esters to give the corresponding chiral branched aldehydes of type **18** using BINAPHOS-type ligands **5a,b** [10, 11] and **27** [22] allows for an interesting comparison. The introduction of the CO_2-philic side groups in **5a-b** led to homogeneous catalyst systems that gave comparable or even higher overall stereoselectivities than those associated with the parent ligand. Separation of products and catalyst was demonstrated with **5a** in a batch-wise procedure relying on the tunable solvent properties of the supercritical medium (catalysis and extraction using supercritical solutions, CESS). Conversion and regioselectivity remained stable, but the enantioselectivity of the catalytic system decreased after several cycles [10b]. The resin-bound solid catalyst **27** gave initially slightly lower enantioselectivities than its homogeneous counterpart. It could be implemented, however, in a very efficient continuous-flow reactor system allowing for consecutive highly selective transformations of a variety of substrates with the same catalyst loading [22]. This example nicely highlights the interplay between molecular design and reaction engineering, which can lead to novel generic methodologies for transition metal catalysis.

$$\underset{\mathbf{16}}{\text{R-CH=CH}_2} + CO + H_2 \xrightarrow[\text{scCO}_2]{\text{Rh/Lig}} \underset{\mathbf{17}}{\text{R-CH}_2\text{CH}_2\text{CHO}} + \underset{\mathbf{18}}{\text{R-CH(CHO)CH}_3} \tag{4}$$

The excellent mass transfer properties of $scCO_2$ also lead to better results than those obtained in conventional liquid solvents when other solid phase-bound hydroformylation catalysts are used [38, 39]. On the basis of the ionic liquid/supercritical biphasic system, a particularly effective continuous flow system has been developed for hydroformylation of long-chain olefins [37]. On the molecular level, the choice of an appropriate counterion (imidazolium vs Na^+) in ligand **25** was again a crucial design factor for the successful implementation of the engineering concept in this case.

Highly polar substrates and/or products cannot be processed under the single-phase or multiphase conditions so far discussed because of their low solubility in $scCO_2$. An "inverted" biphasic system $H_2O/scCO_2$ has been designed, where a rhodium catalyst based on **19a** is contained in the $scCO_2$ phase and the substrate is in the aqueous phase [40]. A norbornene carboxylic acid derivative was hydro-

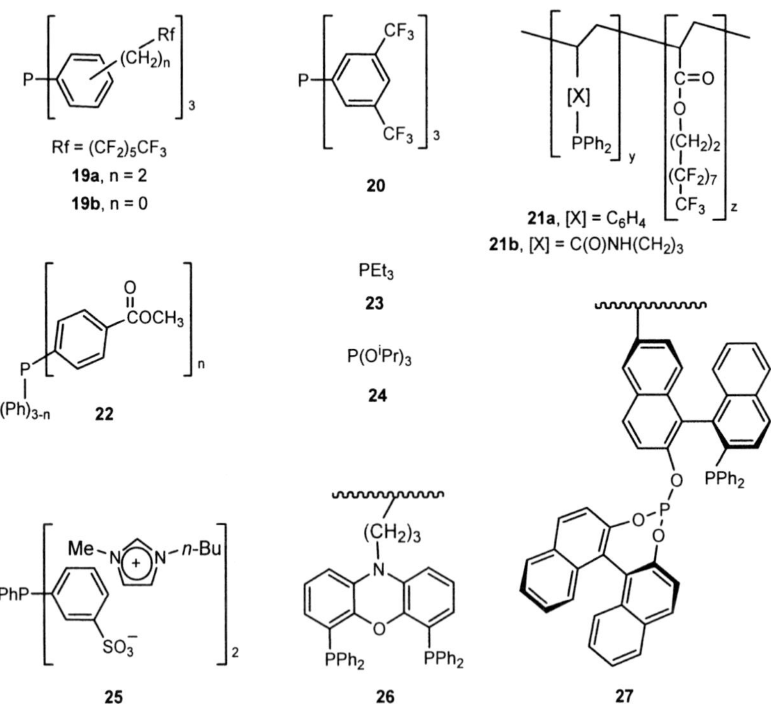

Chart 2 Selected ligands for rhodium-catalyzed hydroformylation in $scCO_2$ under single-phase (**19–24**) or multiphase (**25–27**) conditions. For leading references see **19a**: [13, 30], **19b**: [31], **20**: [32], **21a**: [33], **21b**: [34], **22**: [35], **23**, **24**: [36], **25**: [37], **26**: [38], and **27**: [22].

formylated efficiently as a test substrate, and catalyst recycling was possible at sufficiently high catalyst loadings.

The hydroformylation/hydrogenation sequence shown in Eq. (5) leads to the cyclic amide **32** as the major product in conventional solvents such as dioxane. In sharp contrast, the saturated heterocycles **30** and **31** are formed preferentially in $scCO_2$ with high selectivity [41]. High-pressure NMR spectroscopy revealed that this results from *in situ* protection of the amine functionality in **28** by reversible reaction with the solvent to give the corresponding carbamic acid. High-pressure multinuclear NMR was also used to probe the active intermediates during rhodium-catalyzed hydroformylation, with **5a** [10b] and $P(p\text{-}CF_3C_6H_4)_3$ [42] as ligands. As in hydrogenation, the key intermediates of the catalytic cycle in $scCO_2$ were found to be largely identical to those observed in conventional media.

[(hfacac)Rh(cod)] (**29**, 1 mol%)
19a (3 mol%)
$scCO_2$ (0.78 g mL^{-1})
80 °C, 44 h, conv. 77%

28 + CO/H_2 (40 bar) → **30**, 76%; **31**, 13%; **32**, 7%; **33**, 4%

(5)

The high miscibility of $scCO_2$ with carbon monoxide has also stimulated research into other carbonylation reactions including methanol [43], aryl halides [44], olefins [45], and even alkanes [46] as substrates. The catalytic Pauson-Khand reaction occurs very efficiently in $scCO_2$ (Eq. 6), yielding substituted cyclopentenones such as **36** in excellent yields [47]. Interestingly, cobalt carbonyl-catalyzed hydroformylation using **35** as catalyst precursor was the first homogeneous organometallic catalytic reaction studied in $scCO_2$ [48].

34 + CO (30 bar) → **36**, 82%
$Co_2(CO)_8$ (**35**, 2.5 mol%)
$scCO_2$, 90 °C, 200 bar, 24 h

(6)

3.3.2.3
C-C Bond Formation Reactions

Palladium-catalyzed C-C coupling reactions can be carried out in $scCO_2$ with good yields and reaction rates under single-phase conditions. Several standard test reactions of the various coupling types have been examined, aiming at the development of "CO_2-philic" catalyst systems [49–54]. Enhanced product selectivities or alternative reaction pathways were explicitly addressed in several cases [55–57]. In terms of synthetic application, the coupling of solid-phase-bound substrates is particularly noteworthy as exemplified by the Suzuki reaction of **37** with boronic esters **38a,b** (Eq. 7) [58]. The concept of combining solid-phase synthesis with supercritical reaction media seems highly attractive, as swelling properties and mass transfer between the polymer and the continuous phase will be significantly different from those observed in the case of conventional solvent systems.

37 + **38a**, R = H; **38b**, R = CH_3 → [$[Pd(OAc)_2]$ (**39**, 5 mol%), $P(t\text{-}Bu)_3$ (**40**, 10 mol%), $N(i\text{-}Pr)_2Et$, $scCO_2$, 80 °C, 207 bar, 16 h] → **41a**, 67%; **41b**, 70% (7)

The nickel-catalyzed co-dimerization of olefins (hydrovinylation, Eq. 8) has been studied under single-phase conditions [59] and in an IL/$scCO_2$ biphasic system [60]. The nature of the anion used to activate the nickel precursor **43** by chloride abstraction was found to be of decisive influence on activity and selectivity in both cases. Very high reaction rates were achieved under homogeneous conditions, but rapid deactivation was observed during batch-wise recycling. Under biphasic continuous flow conditions, the highly sensitive active species is embedded in the IL and a constant supply of substrate and removal of product can be assured with $scCO_2$ as the mobile phase. Thus, deactivation pathways can be successfully reduced, leading to continuous production of **44** with good conversion and an *ee* of over 56%.

42 + ethylene (10 bar) → [**43**, IL/liqCO_2, 0 °C, 80 bar, 60 h continuous flow, conv. 65-78%, *ee* 56-66% (*S*)] → **44** (8)

Olefin metathesis has been developed into one of the most versatile catalytic processes for C-C bond formation in recent years, and $scCO_2$ is an attractive alternative solvent, especially for ring-closing processes where inter- and intramolecular reaction pathways are competing [61]. In addition, $scCO_2$ can be used as a protective medium in olefin metathesis, as shown in Eq. (9). The secondary amine functionality of substrate **45** leads to irreversible deactivation of catalyst **46** in conventional solvents such as methylene chloride. In $scCO_2$, the NH group is "masked" in the form of the carbamic acid, and cyclization occurs smoothly. Under ambient conditions, the carbamic acid spontaneously releases CO_2, and the desired amine **47** is isolated after depressurization of the reactor without the need for protecting group manipulation.

O N H **45** → [Cl, Cl, Ru, PCy_3, PCy_3, Ph, Ph] (**46**, 1 mol%), 40 °C, 72 h, CH_2Cl_2: 0%, $scCO_2$: 74 % (*Z*:*E* = 1 : 2.4) → O O NH **47** (9)

Cyclotrimerizations of acetylenes [62] and isocyanates [63] were studied in $scCO_2$ using $[cpCo(CO)_2]$ as the catalyst precursor. Although the organometallic complex was found to be highly soluble in $scCO_2$, lower yields than those obtained under neat conditions were observed. This was explained by a reduced rate of reaction caused by the dilution effect. In contrast, no reduction of turnover frequencies has been observed in the palladium-catalyzed dimerization of methylacrylate under biphasic IL/$scCO_2$ conditions, even though investigation of the substrate and product distribution between the two phases indicated that there is a lower substrate-to-palladium ratio in the IL phase [64].

Lanthanide complexes are versatile Lewis acid catalysts for highly selective C-C bond-forming reactions. $ScCO_2$ provides an interesting reaction medium for such processes because it has a very low Lewis basicity, and solvent cage effects can be dramatically different from organic solvents typically used for this type of process. A recent study reported on the diastereoselective Diels-Alder reaction using various rare-earth catalysts in $scCO_2$ (Eq. 10) [65]. Although phase behavior and solubility of catalysts such as **50** were not addressed explicitly, high yields and selectivities of **51** were observed in several cases. Earlier studies of related Diels-Alder reactions had revealed that the catalyst performance increased with increasing length of the perfluoroalkyl chains in triflate type anions of scandium catalysts [66, 67].

48 + 49 → 51 (10)

Yb(OTf)$_3$ (**50**, 10 mol%)
scCO$_2$, 40 °C, 100 bar, 0.5 h
yield: 91%
endo/exo : 76/24
de (*endo*) : 50%

The Mannich reaction shown in Eq. (11) occurred smoothly in $scCO_2$ for a remarkably broad range of substrates using lanthanide complexes as catalysts [68]. In this case, PEG of low molecular weight (PEG-400) was found to be a necessary additive, and the formation of emulsion type reaction mixtures was confirmed to be crucial for optimum performance. In line with the assumption that PEG acts as a surfactant under these conditions, high-molecular-weight PEG was not effective, although it can form biphasic mixtures with hydrogenation catalysts [25].

52 + **53** (1.2 eq) → **54** (11)

50 (5 mol%)
scCO$_2$,150 bar, 50 °C, 3 h
without additives: 10 %
with PEG 400: 72 %

3.3.2.4
Oxidation Reactions

Carbon dioxide in its liquid or supercritical state is a very attractive reaction medium for oxidation reactions, combining good solubility for most oxidants including molecular oxygen with the additional safety of a totally inert environment. In addition, side products from oxidation of the solvent are *a priori* avoided. Despite theses obvious advantages offered by compressed CO_2, the number of synthetically useful oxidation reactions in this medium still remains very low. This largely reflects a lack of practical catalytic systems that would be compatible with the solvent properties rather than shortcomings of the methodology [69]. In an attempt to obtain a compromise between the catalyst solubilities in organic solvents and the mass transfer and safety features of CO_2, the concept of so-called "CO_2-expanded solvents" has been developed and successfully applied to reactions based on porphyrin and salen-type catalysts such as **56** (Eq. 12) [70].

$$\mathbf{55} + O_2 \ (1\ \text{bar}) \xrightarrow[\substack{CO_2/CH_3CN\ (V/V_0 = 2) \\ 60\ °C,\ 80\ \text{bar} \\ \text{TOF = ca. } 30\ h^{-1}}]{\mathbf{56}} \mathbf{57},\ 80\% + \mathbf{58},\ 20\% \tag{12}$$

Another difficulty in aerobic oxidation reactions in $scCO_2$ can arise from background reactions not involving transition metal oxo mechanisms. Radical processes are usually very rapid in CO_2 media, and the presence of stainless steel from the reactor walls can initiate such processes quite efficiently [71]. Using O_2 in the presence of aldehydes as the terminal oxidant, this has been exploited to develop synthetically potentially useful methods for the epoxidation of olefins [72] and the Bayer-Villiger oxidation of ketones such as **59 a–c** (Eq. 13) [73].

$$\mathbf{59} + O_2\ (20\ \text{bar}) \xrightarrow[\substack{\text{stainless steel} \\ \text{liq}CO_2,\ \text{rt},\ 18\ h}]{PhCHO \rightarrow PhCO_2H} \mathbf{60} \tag{13}$$

59a, R = Ph, n = 1 → **60a**, 82%
59b, R = H, n = 2 → **60b**, 70%
59c, R = H,n = 3 → **60c**, 75%

Palladium-catalyzed oxidations of the Wacker type have been investigated both in $scCO_2$ [74, 75] and in IL/$scCO_2$ mixtures [76]. The latter conditions seem to be most promising, as the catalyst systems can be expected to be insoluble in compressed CO_2. In accord with this consideration, selectivity and catalyst stability for the oxidation of 1-hexene was highest under biphasic conditions as compared to the use of individual solvents alone or to neat conditions [76].

Truly homogeneously metal-catalyzed epoxidation reactions using hydroperoxides as oxidants have been demonstrated in $scCO_2$ with molybdenum catalysts under various conditions [77–79]. The epoxidation of allylic and homoallylic alcohols **61 a, b** with *tert*-butylhydroperoxide (**62**) in compressed CO_2 was performed using the soluble vanadium-based catalyst **63** (Eq. 14) and in asymmetric fashion under Sharpless conditions [80]. High stereoselectivity was obtained in the latter case, opening up a large area of potential synthetic applications.

$$\mathbf{61} + t\text{-BuOOH}\ (\mathbf{62},\ 2.4\ \text{eq}) \xrightarrow[\text{liq}CO_2,\ 25\ °C,\ 103\ \text{bar},\ 24\ h]{VO(O\textit{i}\text{-}Pr)_3\ (\mathbf{63},\ 3.5\ \text{mol\%})} \mathbf{64},\ 98\ \% \tag{14}$$

3.3.3
Conclusion and Outlook

The examples discussed in this chapter substantiate the principle of the compatibility of compressed CO_2 as solvent with transition metal-catalyzed reactions. Various reaction types can be performed effectively in CO_2, and improved catalyst performance is observed in certain cases compared to the use of conventional organic solvents. However, it is also clear that there are still a vast number of catalytic transformations that have yet to be explored in CO_2 or under multiphase conditions based on this medium. The examples covered in the present chapter reflect only a small fraction of the synthetic methodologies summarized throughout the other sections of this book. As a growing number of examples become available, some general patterns emerge and help to identify reactions that are likely to benefit from the unique properties of carbon dioxide media. The effects summarized in Table 1 and discussed on the basis of the present examples are only a first and crude compilation of such criteria, and exciting new possibilities are yet to be added to this provisional list.

References

1 P. G. Jessop, W. Leitner (Eds.) *Chemical Synthesis Using Supercritical Fluids*, Wiley-VCH, Weinheim, **1999**.

2 C. Reichardt, *Solvent Effects in Organic Chemistry*, 3rd ednWiley-VCH, Weinheim, **2002**.

3 For early reviews see (a) P. G. Jessop, T. Ikariya, R. Noyori, *Science* **1995**, *269*, 1065; (b) E. Dinjus, R. Fornika, M. Scholz in *Chemistry under Extreme or Non-Classical Conditions* (Eds.: R. van Eldik, C. D. Hubbard), Wiley, New York, **1996**, pp. 219; (c) P. G. Jessop, T. Ikariya, R. Noyori, *Chem. Rev.* **1999**, *99*, 475; (d) W. Leitner, *Top. Curr. Chem.* **1999**, *206*, 107.

4 For recent reviews see (a) R. S. Oakes, A. A. Clifford, C. M. Rayner, *J. Chem. Soc. Perkin Trans. 1* **2001**, 917; (b) W. Leitner, *Acc. Chem. Res.* **2002**, *35*, 746; (c) D. J. Cole-Hamilton, *Science* **2003**, *299*, 1702.

5 P. G. Jessop, Y. Hsiao, T. Ikariya, R. Noyori, *J. Am. Chem. Soc.* **1994**, *116*, 8851.

6 J. Xiao, S. C. A. Nefkens, P. G. Jessop, T. Ikariya, R. Noyori, *Tetrahedron Lett.* **1996**, *37*, 2813.

7 M. J. Burk, S. Feng, M. F. Gross, W. Tumas, *J. Am. Chem. Soc.* **1995**, *117*, 8277.

8 S. Kainz, A. Brinkmann, W. Leitner, A. Pfaltz, *J. Am. Chem. Soc.* **1999**, *121*, 6421.

9 S. Lange, A. Brinkmann, P. Trautner, K. Woelk, J. Bargon, W. Leitner, *Chirality* **2000**, *12*, 450.

10 (a) G. Franciò, W. Leitner, *Chem. Commun.* **1999**, 1663; (b) G. Franciò, K. Wittmann, W. Leitner, *J. Organomet. Chem.* **2001**, *621*, 130.

11 D. Bonafoux, Z. H. Hua, B. H. Wang, I. Ojima, *J. Fluor. Chem.* **2001**, *112*, 101.

12 D. J. Adams, W. P. Chen, E. G. Hope, S. Lange, A. M. Stuart, A. West, J. L. Xiao, *Green Chem.* **2003**, *5*, 118.

13 S. Kainz, D. Koch, W. Baumann, W. Leitner, *Angew. Chem. Int. Ed.* **1997**, *36*, 1628.

14 F. Zhao, Y. Ikushima, M. Chatterjee, O. Sato, M. Arai, *J. Supercrit. Fluids* **2003**, *27*, 65.

15 L.N. He, J.C. Choi, T. Sakakura, *Tetrahedron Lett.* **2001**, *42*, 2169.

16 C.A.G. Carter, R.T. Baker, S.P. Nolan, W. Tumas, *Chem. Commun.* **2000**, 347.

17 L.J.P. van den Broeke, E.L.V. Goetheer, A.W. Verkerk, E. de Wolf, B.-J. Deelman, G. van Koten, J.T.F. Keurentjes, *Angew. Chem. Int. Ed.* **2001**, *40*, 4473.

18 I. Kani, M.A. Omary, M.A. Rawashdeh-Omary, Z.K. Lopez-Castillo, R. Flores, A. Akgerman, J.P. Fackler, *Tetrahedron* **2002**, *58*, 3923.

19 E.L.V. Goetheer, A.W. Verkerk, L.J.P. van den Broeke, E. de Wolf, B.-J. Deelman, G. van Koten, J.T.F. Keurentjes, *J. Catal.* **2003**, *219*, 126.

20 L. Schmid, O. Krocher, R.A. Koppel, A. Baiker, *Micropor. Mesopor. Mater.* **2000**, *35/36*, 181.

21 Y. Kayaki, Y. Shimokawatoko, T. Ikariya, *Adv. Synth. Catal.* **2003**, *345*, 175.

22 F. Shibahara, K. Nozaki, T. Hiyama, *J. Am. Chem. Soc.* **2003**, *125*, 8555.

23 B.M. Bhanage, M. Shirai, M. Arai, Y. Ikushima, *Chem. Commun.* **1999**, 1277.

24 G.B. Jacobson, C.T. Lee, K.P. Johnston, W. Tumas, *J. Am. Chem. Soc.* **1999**, *121*, 11902.

25 D.J. Heldebrant, P.G. Jessop, *J. Am. Chem. Soc.* **2003**, *125*, 5600.

26 L.A. Blanchard, D. Hâncu, E.J. Beckman, J.F. Brennecke, *Nature* **1999**, *399*, 28.

27 F.C. Liu, M.B. Abrams, R.T. Baker, W. Tumas, *Chem. Commun.* **2001**, 433.

28 R.A. Brown, P. Pollet, E. McKoon, C. A. Eckert, C.L. Liotta, P.G. Jessop, *J. Am. Chem. Soc.* **2001**, *123*, 1254.

29 M. Solinas, P. Wasserscheid, W. Leitner, A. Pfaltz, *Chem. Ing. Tech.* **2003**, *75*, 1153.

30 D. Koch, W. Leitner, *J. Am. Chem. Soc.* **1998**, *120*, 13398.

31 A.M.B. Osuna, W.P. Chen, E.G. Hope, R.D.W. Kemmitt, D.R. Paige, A.M. Stuart, J.L. Xiao, L.J. Xu, *J. Chem. Soc. Dalton Trans.* **2000**, 4052.

32 D.R. Palo, C. Erkey, *Organometallics* **2000**, *19*, 81.

33 Y. Hu, W. Chen, A.M.B. Osuna, J.A. Iggo, J. Xiao, *Chem. Commun.* **2002**, 788.

34 Z.K. Lopez-Castillo, R. Flores, I. Kani, J.P. Fackler, Jr., A. Akgerman, *Ind. Eng. Chem. Res.* **2003**, *42*, 3893.

35 Y. Hu, W. Chen, L. Xu, J. Xiao, *Organometallics* **2001**, *20*, 3206.

36 M.F. Sellin, I. Bach, J.M. Webster, F. Montilla, V. Rosa, T. Aviles, M. Poliakoff, D. J. Cole-Hamilton, *J. Chem. Soc. Dalton Trans.* **2002**, 4569.

37 (a) M.F. Sellin, P.B. Webb, D.J. Cole-Hamilton, *Chem. Commun.* **2001**, 781; (b) P.B. Webb, M.F. Sellin, T.E. Kunene, S. Williamson, A.M.Z. Slawin, D.J. Cole-Hamilton, *J. Am. Chem. Soc.* **2003**, *125*, 15577.

38 N.J. Meehan, M. Poliakoff, A.J. Sandee, J.N.H. Reek, P.C.J. Kamer, P.W. N.M. van Leeuwen, *Chem. Commun.* **2000**, 1497.

39 O. Hemminger, A. Marteel, M.R. Mason, J.A. Davies, A.R. Tadd, M.A. Abraham, *Green Chem.* **2002**, *4*, 507.

40 M. McCarthy, H. Stemmer, W. Leitner, *Green Chem.* **2002**, *4*, 501.

41 K. Wittmann, W. Wisniewski, R. Mynott, W. Leitner, C.L. Kranemann, T. Rische, P. Eilbracht, S. Kluwer, J.M. Ernsting, C.J. Elsevier, *Chem. Eur. J.* **2001**, *7*, 4584.

42 C.R. Yonker, J.C. Linehan, *J. Organomet. Chem.* **2002**, *650*, 249.

43 R.J. Sowden, M.F. Sellin, N. De Blasio, D.J. Cole-Hamilton, *Chem. Commun.* **1999**, 2511.

44 Y. Kayaki, Y. Noguchi, S. Iwasa, T. Ikariya, R. Noyori, *Chem. Commun.* **1999**, 1235.

45 L. Jia, H. Jiang, J. Li, *Green Chem.* **1999**, *1*, 91.

46 J.-C. Choi, Y. Kobayashi, T. Sakakura, *J. Org. Chem.* **2001**, 66, 5262.

47 N. Jeong, S.H. Hwang, Y. Woo, Lee, J. S. Lim, *J. Am. Chem. Soc.* **1997**, *119*, 10549.

48 J.W. Rathke, R.J. Klingler, T.R. Krause, *Organometallics* **1991**, *10*, 1350.

49 D.K. Morita, D.R. Pesiri, S.A. David, W.H. Glaze, W. Tumas, *Chem. Commun.* **1998**, 1397.

50 M.A. Carrol, A.B. Holmes, *Chem. Commun.* **1998**, 1395.

51 N. Shezad, R. S. Oakes, A. A. Clifford, C. M. Rayner, *Tetrahedron Lett.* **1999**, *40*, 2221.

52 T. Osswald, S. Schneider, S. Wang, W. Bannwarth, *Tetrahedron Lett.* **2001**, *42*, 2965.

53 B. M. Bhanage, Y. Ikushima, M. Shirai, M. Arai, *Tetrahedron Lett.* **1999**, *40*, 6427.

54 S. Fujita, K. Yuzawa, B. M. Bhanage, Y. Ikushima, M. Arai, *J. Mol. Catal. A: Chemical* **2002**, *180*, 35.

55 Y. Kayaki, Y. Noguchi, T. Ikariya, *Chem. Commun.* **2000**, 2245.

56 N. Shezad, A. A. Clifford, C. M. Rayner, *Tetrahedron Lett.* **2001**, *42*, 323.

57 N. Shezad, A. A. Clifford, C. M. Rayner, *Green Chem.* **2002**, *4*, 64.

58 (a) T. R. Early, R. S. Gordon, M. A. Carroll, A. B. Holmes, R. E. Shute, I. F. McConvey, *Chem. Commun.* **2001**, 1966; (b) R. S. Gordon, A. B. Holmes, *Chem. Commun.* **2002**, 640.

59 A. Wegner, W. Leitner, *Chem. Commun.* **1999**, 1583.

60 A. Bösmann, G. Franciò, E. Janssen, M. Solinas, W. Leitner, P. Wasserscheid, *Angew. Chem., Int. Ed.* **2001**, *40*, 2697.

61 A. Fürstner, D. Koch, K. Langemann, W. Leitner, C. Six, *Angew. Chem. Int. Ed.* **1997**, *36*, 2466.

62 F. Montilla, T. Aviles, T. Casimiro, A. A. Ricardo, M. Nunes da Ponte, *J. Organomet. Chem.* **2001**, *632*, 113.

63 F. Montilla, E. Clara, T. Aviles, T. Casimiro, A. Aguiar Ricardo, M. Nunes da Ponte, *J. Organomet. Chem.* **2001**, *626*, 227.

64 D. Ballivet-Tkatchenko, M. Picquet, M. Solinas, G. Franciò, P. Wasserscheid, W. Leitner, *Green Chem.* **2003**, *5*, 232.

65 S.-I. Fukuzawa, K. Metoki, Y. Komuro, T. Funazukuri, *Synlett* **2002**, 134.

66 J.-I. Matsuo, T. Tsuchiya, K. Odashima, S. Kobayashi, *Chem. Lett.* **2000**, 178.

67 R. Scott Oakes, T. J. Heppenstall, N. Shezad, A. A. Clifford, C. M. Rayner, *Chem. Commun.* **1999**, 1459.

68 I. Komoto, S. Kobayashi, *Chem. Commun.* **2001**, 1842.

69 E. R. Birnbaum, R. M. Le Lacheur, A. C. Horton, W. Tumas, *J. Mol. Catal. A: Chemical* **1999**, *139*, 11.

70 (a) G. Musie, M. Wei, B. Subramaniam, D. H. Busch, *Coord. Chem. Rev.* **2001**, *219–221*, 789; (b) M. Wei, G. T. Musie, D. H. Busch, B. Subramaniam, *J. Am. Chem. Soc.* **2002**, *124*, 2513.

71 N. Theyssen, W. Leitner, *Chem. Commun.* **2002**, 410.

72 F. Loeker, W. Leitner, *Chem. Eur. J.* **2000**, *6*, 2011.

73 C. Bolm, C. Palazzi, G. Franciò, W. Leitner, *Chem. Commun.* **2002**, 1588.

74 H. Jiang, L. Jia, J. Li, *Green Chem.* **2000**, *2*, 161.

75 L. Jia, H. Jiang, J. Li, *Chem. Commun.* **1999**, 985.

76 Z. Hou, B. Han, L. Gao, T. Jiang, Z. Liu, Y. Chang, X. Zhang, J. He, *New J. Chem.* **2002**, *26*, 1246.

77 U. Kreher, S. Schebesta, D. Walther, *Z. Anorg. Allg. Chem.* **1998**, *624*, 602.

78 G. R. Haas, J. W. Kolis, *Tetrahedron Lett.* **1998**, *39*, 5923.

79 F. Montilla, V. Rosa, C. Prevett, T. Aviles, M. Nunes da Ponte, D. Masi, C. Mealli, *J. Chem. Soc. Dalton Trans.* **2003**, 2170.

80 D. R. Pesiri, D. K. Morita, T. Walker, W. Tumas, *Organometallics* **1999**, *18*, 4916.

3.4
Transition Metal Catalysis using Ionic Liquids

Peter Wasserscheid

3.4.1
Ionic Liquids

Ionic liquids are characterized by the following three definition criteria: (a) they consist entirely of ions; (b) they have melting points below 100 °C; (c) they exhibit no appreciable vapor pressure below the temperature of their thermal decomposition.

As a consequence of these properties, most ions that can form ionic liquids exhibit low charge densities, resulting in low intermolecular interaction. Fig. 1 displays some of the most common ions used for the formation of ionic liquids.

However, apart from these few features common to all ionic liquids, the physical and chemical properties of these materials can cover a wide range, depending on the nature of their anion/cation combination. For example, an ionic liquid can be strongly coordinating (e.g., in the case of a chloride salt) or weakly coordinating (e.g., in the case of a hexafluorophosphate salt), depending on the nucleophilicity of their anion. Other properties, which are of great relevance to catalysis and which can be adjusted by the selection of specific combinations of ions, include solubility/polarity, acidity/basicity, viscosity, density, surface tension, heat capacity, etc. Moreover, important issues for the technical application of ionic liquids, such as price, disposal options, achievable quality, corrosion, and toxicity, are very much related to the individual ion combination under investigation.

Most ionic liquids are still based on imidazolium, pyridinium, ammonium, and phosphonium cations. Modern ionic liquid research is driven by anion develop-

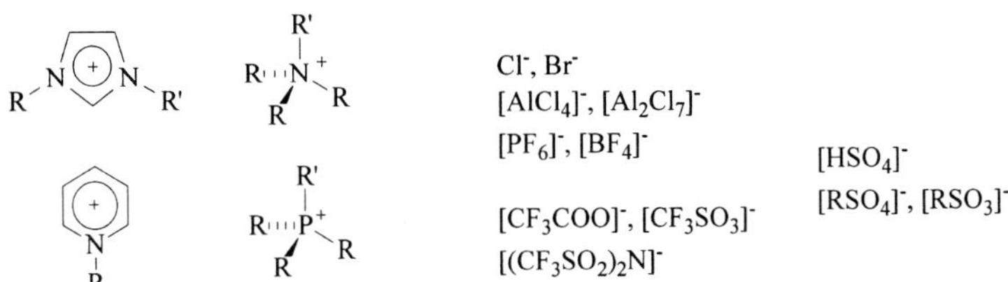

Fig. 1 Typical cations and anions for the formation of ionic liquids.

Transition Metals for Organic Synthesis, Vol. 2, 2nd Edition.
Edited by M. Beller and C. Bolm

ISBN: 3-527-30613-7

ment and by the use of functionalized derivatives of the above-mentioned cations. The aim of materials development is to provide the most suitable systems for the different areas of application that are actually discussed in the context of a technical use of ionic liquids (Fig. 2).

Toluenesulfonate [1], octylsulfate [2], and hydrogen sulfate [3] systems have recently been developed as potential solvents for synthetic applications, with some ionic liquid consumption, as heat carriers, lubricants, additives, surfactants, phase transfer catalysts, extraction solvents, solvents for extractive distillation, and antistatics. The cation and the anion of these "bulk ionic liquids" are chosen to make a relatively cheap (expected price on a multi-hundred liter scale: ca. 30 €/L), halogen-free (e.g., for easy disposal of spent ionic liquid by thermal treatment), and toxologically well-characterized liquid. Initial studies of the toxicities of several imidazolium-based ionic liquid have recently been published [4]).

Functionalized [5], fluorinated [6], deuterated [7], and chiral ionic liquids [8] are expected to play a future role as special solvents for synthetic applications with high added value and very low ionic liquid consumption as well as in analytical applications (stationary or mobile phases for chromatography, matrixes for MS, etc.), sensors, and special electrolytes. These ionic liquids are designed and optimized for the best performance in high-value-adding applications.

The wide range of ionic liquids commercially available today [9] should not lead us to forget that an ionic liquid is still a quite different product from traditional organic solvents inasmuch as it cannot be purified by distillation because of its non-volatile character. This, combined with the fact that small amounts of impurities influence the ionic liquid's properties and especially its usefulness for catalytic reactions significantly [10], makes the quality of an ionic liquid a very important consideration.

Among the potential impurities in an ionic liquid, water, halide ions, and organic starting material are of great importance for transition metal chemistry, while the color of an ionic liquid is in most applications not a critical parameter.

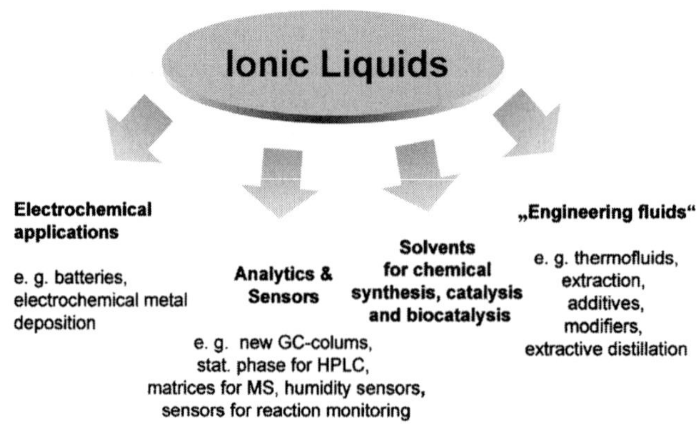

Fig. 2 Fields of applications for ionic liquids.

Without special drying procedures and completely inert handling, water is omnipresent in ionic liquids. Even the apparently hydrophobic ionic liquid $[BMIM][(CF_3SO_2)_2N]$, which has a broad miscibility gap with water, saturates under wet atmosphere to about 1.4 mass% of water, which is a significant molar amount [11]. For more hydrophilic ionic liquids, the water uptake from air can be much higher. Thus, it is important for all applications to know the amount of water present in the ionic liquid used (e.g., by determination via Karl Fischer titration). The presence of water can have a significant influence on the physico-chemical properties of the ionic liquid, its stability (a wet ionic liquid may undergo hydrolysis if its anion is not hydrolysis stable), and the reactivity of transition metal complexes dissolved in the ionic liquid.

Many ionic liquids (among them the commonly used tetrafluoroborate and hexafluorophosphate systems) are still made by halide exchange reactions. This procedure requires some know-how to avoid halide impurities in the final product. Therefore, a check with $AgNO_3$ is recommended for qualitative analysis, while titration methods and electrochemical analysis can be used to obtain quantitative information about the halide impurities. Generally, the presence of halide impurities can be detrimental for transition metal catalysis in ionic liquids, as described for example in the hydrogenation of pent-1-ene using $[Rh(nbd)(PPh_3)][PF_6]$ (where nbd=norbornadiene) [12]. In contrast, the presence of halide may be beneficial in other reactions, as is reported for the Heck reaction in $[NBu_4]Br$ using $(PPh_3)_2PdCl_2$ [13] or in the ionic liquid-mediated palladium/phosphine-catalyzed Suzuki reaction [14].

In theory, volatile impurities can easily be removed from the non-volatile ionic liquid by simple evaporation. However, it is important to know that the vapor pressure of some polar organics is significantly reduced when they are dissolved in an ionic liquid. This is particularly true for the methylimidazole that may be left in a 1-alkyl-3-methylimidazolium ionic liquid from the quaternization reaction. Because of its high boiling point (198 °C) and its strong interaction with the ionic liquid, this compound is very difficult to remove by distillation. Traces of methylimidazole in the final ionic liquid product can play an unfavorable role in many applications involving transition metal compounds. Many electrophilic catalyst complexes can coordinate to the base in an irreversible manner and be deactivated. A photometric analysis method to determine the amount of methylimidazole in an ionic liquid has been published by Holbrey, Seddon, and Wareing [15].

The limited scope of this short review does not allow us to give a detailed description of our knowledge of ionic liquid synthesis and the associated materials science. However, since this is the basis of all future developments in ionic liquid chemistry, the interested reader is strongly encouraged to study the more detailed literature covering these aspects [16].

3.4.2
Liquid-Liquid Biphasic Catalysis

Biphasic catalysis in a liquid-liquid system is an ideal approach to combine the advantages of homogeneous and heterogeneous catalysis. The reaction mixture consists of two immiscible solvents, only one of which contains the catalyst, allowing easy product separation by simple decantation. The catalyst phase can be recycled without any further treatment. However, the right combination of catalyst, catalyst solvent, and product is crucial for the success of biphasic catalysis [17]. The catalyst solvent has to provide excellent solubility for the catalyst complex without competing with the substrate of the reaction for the free coordination sites at the catalytic center.

Liquid-liquid biphasic operation is the ideal reaction mode for transition metal catalysis in ionic liquids, mainly because of the ionic liquid's exactly tuneable physico-chemical properties. The potential to enhance catalyst lifetime by recycling and the chance to improve the reaction's selectivity and the catalyst's activity by *in situ* extraction adds extra value. Finally, biphasic catalysis offers a very efficient way of reusing the relatively expensive ionic liquid itself. Thus, the ionic liquid may be seen as a capital investment for the process (in an ideal case) or at least as a "working solution", which means that only a small amount has to be replaced after a certain time of operation.

Many transition metal complexes dissolve readily in ionic liquids, which enables them to be used as solvents for transition metal catalysis. Sufficient solubility for a wide range of catalyst complexes is an obvious, but not trivial, prerequisite for a versatile solvent for homogenous catalysis. Some of the other approaches to replace traditional volatile organic solvents in transition metal catalysis by "greener" alternatives, namely the use of supercritical CO_2 or perfluorinated solvents, suffer very often from low catalyst solubility.

In the case of ionic liquids, a special ligand design is usually not necessary to get catalyst complexes dissolved in the ionic liquid. However, the application of ionic ligands can be an extremely useful tool to immobilize the catalyst in the ionic medium. In applications where the ionic catalyst layer is intensively extracted with a non-miscible solvent (i.e., under the conditions of a continuous biphasic catalysis), it is important to make sure that the amount of catalyst washed from the ionic liquid is extremely low. Fig. 3 shows a selection of ionic ligands that have been successfully used to immobilize transition metal complexes in ionic liquids.

Recently, Dupont [18], Olivier-Bourbigou and Magna [19], Sheldon [20], Gordon [21], and Wasserscheid [22] published extensive reviews presenting a comprehensive overview of transition metal catalysis involving ionic liquids. All three update earlier published reviews by Welton [23], Seddon and Holbrey [24], and Wasserscheid and Keim [25] on the same topic.

All this reviewing is a clear sign of the great research activity in the field. However, it is not the aim of this contribution to add an updated list of all relevant publications in the literature. In contrast, this chapter focusses uniquely on the application of Pd-catalyzed reactions in ionic liquids. The reader may derive, from

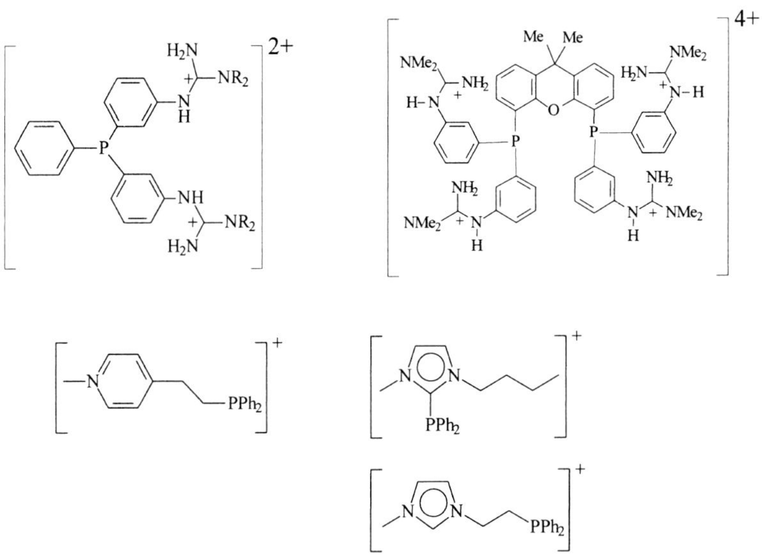

Fig. 3 Selection of ionic phosphine ligand that have been used to immobilize transition metal catalysts in ionic liquids.

this limited field, general principles for a better general understanding of the scope and limitations of using transition metal complexes in ionic liquids. In this way, the author hopes to encourage scientists working in the field of transition metal catalysis to test and further develop ionic liquids as "tool box" for their future research.

3.4.3 Pd-Catalyzed Reactions in Ionic Liquids

3.4.3.1 The Heck Reaction

The Heck reaction is the most extensively reported coupling reaction in ionic liquids so far. This is probably because of its well-understood mechanism, its industrial significance, and the compatibility of the reaction system with ionic liquids.

The use of ionic liquids as reaction media for the palladium-catalyzed Heck reaction was first described by Kaufmann et al. in 1996 [13]. His group investigated the transformation of bromobenzene with acrylic acid butyl ester to *trans*-cinnamic acid butyl ester in molten tetraalkylammonium and tetraalkylphosphonium bromide salts.

More detailed studies of the Heck reaction in low-melting salts were later presented by Herrmann and Böhm [26]. Their results indicate that ionic solvents show clear advantages over commonly used organic solvents (e.g., DMF), especially for the conversion of the commercially interesting chloroarenes. With almost all catalyst systems tested, additional activation and stabilization were observed. Molten $[NBu_4]Br$ (m.p. = 103 °C) proved to be an especially suitable reaction medium out of all the ionic solvent systems investigated.

Seddon's group described the option of carrying out Heck reactions in ionic liquids that do not completely mix with water. These authors studied different Heck reactions in the triphasic system 1-butyl-3-methylimidazolium ([BMIM])$[PF_6]$/water/hexane [27]. While the $[BMIM]_2[PdCl_4]$ catalyst used remains in the ionic liquid, the products dissolve in the organic layer. The salt formed as a by-product of the reaction ([H-base]X) is extracted into the aqueous phase.

Muzart et al. described the coupling of aryl iodides and bromides with allylic alcohols to the corresponding β-arylated carbonyl compounds [28]. Cal et al. reported the Heck coupling of substituted acrylates with bromobenzene catalyzed by Pd-benzothiazole carbene complexes in molten $[NBu_4]Br$ [29]. Similar Pd complexes in molten $[NBu_4]Br$ were later shown to be active in the Heck reaction of (*E*)-ethyl cinnamates with *p*-substituted aryl bromides and chlorides to form the corresponding trisubstituted alkene. The same solvent was found to be essential in investigations carried out by Buchmeiser et al. aimed at the Pd-catalyzed Heck coupling of arylchlorides and the amination of arylbromides [30]. The regioselective arylation of butyl vinyl ether was carried out by Xiao et al. using $Pd(OAc)_2$ as the catalyst precursor and 1,3-bis(diphenylphosphino)propane (dppp) as the ligand dissolved in $[BMIM][BF_4]$ [31].

However, as has been demonstrated by the same group [32] and Welton et al. [14], the use of imidazolium-based ionic liquids in the Pd-catalyzed Heck reaction always involves the possibility of the *in situ* formation of Pd-carbene complexes. This is because of the well-known, relatively high acidity of the H atom in the 2-position of the imidazolium ion [33]. Xiao and co-workers demonstrated that a Pd imidazolylidene complex was formed when $Pd(OAc)_2$ was heated in presence of [BMIM]Br [32]. The isolated Pd carbene complex was found to be active and stable in Heck coupling reactions. Welton et al. were later able to characterize an isolated Pd-carbene complex obtained in this way by X-ray spectroscopy [14]. The reaction pathway to form the Pd-carbene in the presence of a base is shown in Scheme 1.

It should be noted here that the abstraction of the acidic proton in the 2-position of the imidazolium ring by a base is not the only possibility to form a metal carbene complex. Cavell and co-workers observed *in situ* metal carbene complex

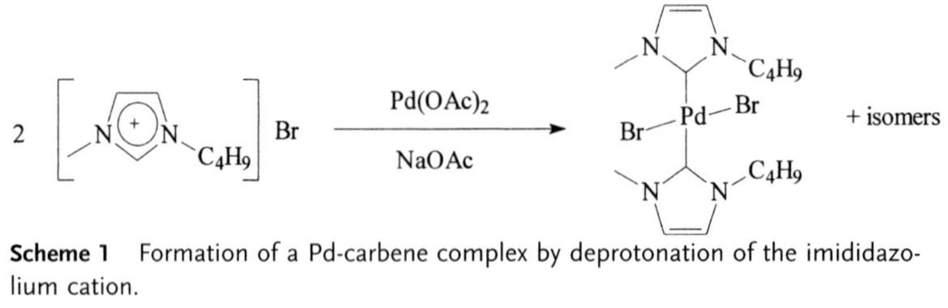

Scheme 1 Formation of a Pd-carbene complex by deprotonation of the imididazolium cation.

formation in an ionic liquid by direct oxidative addition of the imidazolium cation onto a metal center in a low oxidation state [34].

In the light of these results, it is very important to check catalytic results obtained from an imidazolium ionic liquid for a possible influence of *in situ*-formed carbene species. This can easily be carried out by also testing a given reaction in ionic liquids which do not form carbene complexes, e.g., in pyridinium- or 1,2,3-trialkylimidazolium-based ionic liquids.

Finally, some recently published Heck couplings of aryl iodides, including the use of *in situ*-formed Pd(0) nanoparticles [35], heterogeneous Pd on carbon [36], and heterogeneous Pd on silica [37], should be mentioned here. Moreover, ultrasound-assisted [38] and microwave-assisted [39] Heck reactions have been reported in several imidazolium-based ionic liquids.

3.4.3.2 Cross-Coupling Reactions

Suzuki cross-coupling reactions using $Pd(PPh_3)_4$ as the catalyst in [BMIM][BF_4] have been reported by Welton et al. (Scheme 2) [40].

The best results were achieved by pre-heating the aryl halide in the ionic liquid with the Pd complex to 110 °C. The arylboronic acid and Na_2CO_3 were added later to start the reaction. Several advantages over the reaction performed under the conventional Suzuki conditions were described. The reaction showed significantly enhanced activity in the ionic liquid, and the formation of the homo-coupling aryl by-product was suppressed.

More recent examples include the ultrasound-assisted ionic liquid-mediated Suzuki reaction [41], a polymer-supported version of the reaction [42], and the successful application of a phosphonium-based ionic liquid (tetradecyltrihexylphosphonium chloride) [43].

A number of *Stille coupling reactions* have been reported by Handy et al. [44]. Using $PdCl_2(PhCN)_2/Ph_3As/CuI$ in [BMIM][BF_4], good yields and good catalyst recyclability (up to five times) were reported for the reaction of α-iodenones with vinyl- and arylstannanes. However, the reported reaction rates were significantly lower than those obtained in NMP (Scheme 3).

Knochel et al. described Pd-catalyzed *Negishi cross-coupling reactions* between zinc organometallics and aryl iodide in [BMMIM][BF_4], e.g., the reaction for the formation of a 3-substituted cyclohexenone from 3-iodo-2-cyclohexen-1-one [45]. The reaction

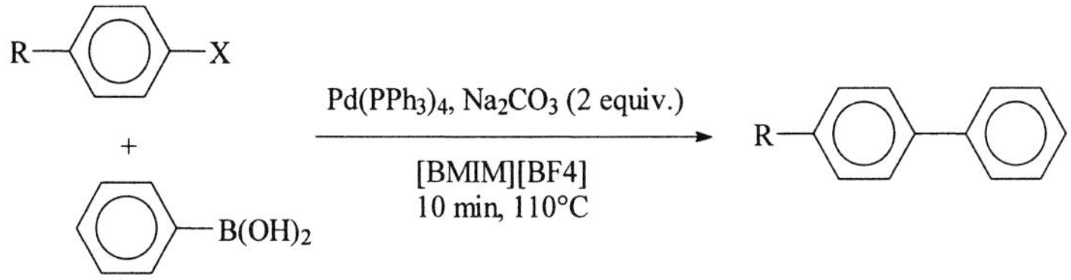

R–C₆H₄–X + C₆H₅–B(OH)₂ → R–C₆H₄–C₆H₅ ($Pd(PPh_3)_4$, Na_2CO_3 (2 equiv.); [BMIM][BF4]; 10 min, 110°C)

Scheme 2 Pd-catalyzed Suzuki cross-coupling reaction in a [BMIM][BF_4] ionic liquid.

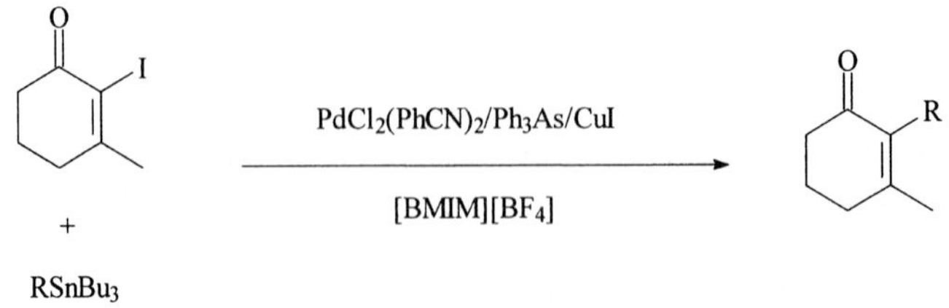

Scheme 3 Pd-catalyzed Stille coupling of α-iodoenones with vinyl and aryl stannanes in [BMIM][BF_4].

was carried out in an ionic liquid/toluene biphasic system, which allowed easy product recovery from the catalyst by decantation. However, attempts to recycle the ionic catalyst phase resulted in significant catalyst deactivation after only the third recycle.

Recently, the *Sonogashira reaction* of aryl iodides with phenylacetylene has been reported in several ionic liquids applying $(PPh_3)_2PdCl_2$ as the catalyst system without the use of a copper co-catalyst [46]. The authors report that yields using the ionic liquid [BMIM][PF_6] as the solvent for this reaction were significantly higher than those obtained using the molecular solvents toluene, THF, or DMF.

3.4.3.3
Ionic Liquid-Mediated Allylation/Trost-Tsujii Reactions

The first examples of the Pd-catalyzed allylation/Trost-Tsujii coupling reaction were reported by Bellefon and co-workers in 1999 [47]. They described the reaction of ethyl cinnamyl carbonate with ethyl acetoacetate in a methylcyclohexane/[BMIM]Cl biphasic mixture at 80 °C using $Pd(OAc)_2$/TPPTS as the catalyst system. In contrast to aqueous conditions, no cinnamyl alcohol was produced as the by-product.

The first Pd-catalyzed allylation reaction in an ambient temperature ionic liquid was published by Xiao et al. [48]. These authors investigated the $Pd(OAc)_2/PPh_3$-catalyzed allylation of 1,3-diphenylallyl acetate with a series of stabilized carbanions in the ionic liquid [BMIM][BF_4].

The enantioselective allylation of (*rac*)-(*E*)-1,3-diphenyl-3-acetoxyprop-1-ene with dimethylmalonate by a series of chiral palladium(0) ferrocenylphosphine complexes was reported in the ionic liquid [BMIM][PF_6] (Scheme 4) [49]. A significant enhancement of enantioselectivity was reported for some of the catalyst systems in the ionic liquid compared to the reaction in THF under otherwise identical conditions.

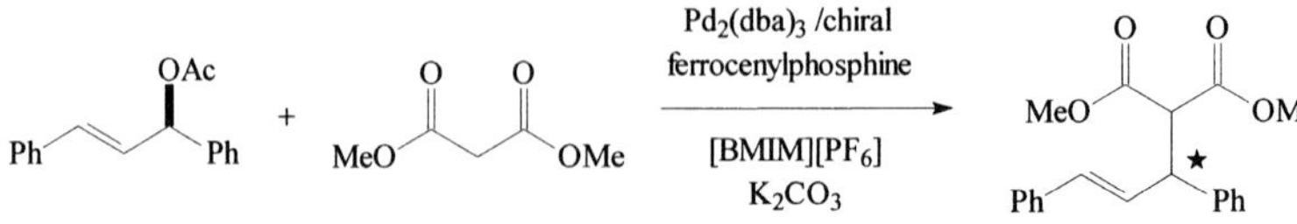

Scheme 4 Enantioselective allylation reaction in [BMIM][PF_6].

3.4.3.4
Carbonylation of Aryl Halides

The carbonylation reaction of aryl halides with a range of different alcohols using $Pd(OAc)_2/PPh_3$ and NEt_3 has been investigated in ionic liquids [50]. The reaction of bromobenzene with methanol gave, for example, an 82% yield of acetophenone after 3 h at 150 °C/30 bar CO (Scheme 5). Repetitive runs were possible. However, a significant decrease in yield was observed, with yields down to 35%, in the fourth run (4 equivalents of PPh_3). By using a larger excess of ligand (20 equivalents of PPh_3), the loss in activity during the recycling could be limited, and 74% of the original activity was still found in the fourth run. Moreover, the carbonylation of 3-alkyn-1-ols and 1-alkyn-4-ols by $Pd(OAc)_2$/2-(diphenylphosphino)pyridine in ionic liquids has been reported to quantitatively and selectively afford *exo-α*-methylene *γ*- and *δ*-lactones, respectively [51].

3.4.3.5
Pd-Catalyzed Dimerization and Polymerization

Recently, our group, in collaboration with Tkatchenko, Ballivet-Tkatchenko, and Leitner, reported the biphasic Pd-catalyzed dimerization of methyl acrylate (MA) using different tetrafluoroborate ionic liquids as the catalyst solvent, and toluene or supercritical CO_2 as the organic extraction phase (Scheme 6) [52].

It could be demonstrated that the biphasic reaction mode enabled the well-known product inhibition effect of this reaction to be overcome. Whereas in the monophasic batch reaction (with and without added IL) the reaction typically stops at a maximum MA conversion of about 80%, the continuous biphasic experiment using the ionic liquid as catalyst layer was still active after 50 h reaction time, resulting in an overall TON of 4000 mol MA converted per mol of Pd.

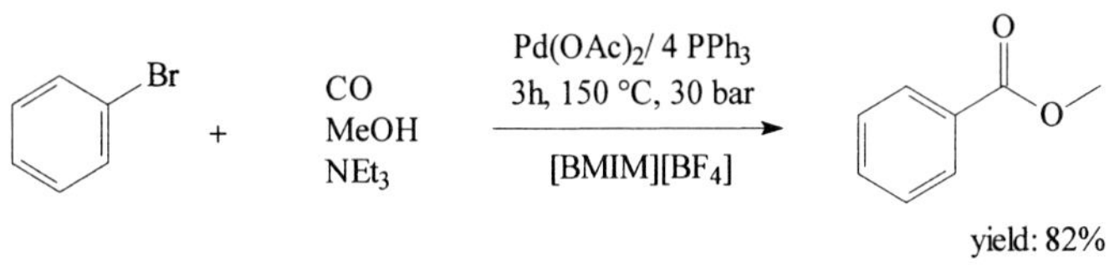

Br + CO, MeOH, NEt_3 — $Pd(OAc)_2$/ 4 PPh_3, 3h, 150 °C, 30 bar, [BMIM][BF_4] → O, O

yield: 82%

Scheme 5 Example for a Pd-catalysed carbonylation reaction in [BMIM][BF_4].

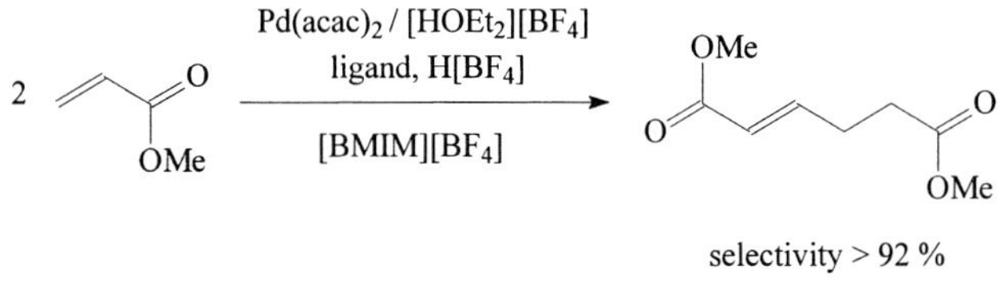

2 O, OMe — $Pd(acac)_2$ / [$HOEt_2$][BF_4], ligand, H[BF_4], [BMIM][BF_4] → OMe, O, O, OMe

selectivity > 92 %

Scheme 6 Pd-catalyzed dimerisation of methylacrylate using a tetrafluoroborate ionic liquid as catalyst solvent.

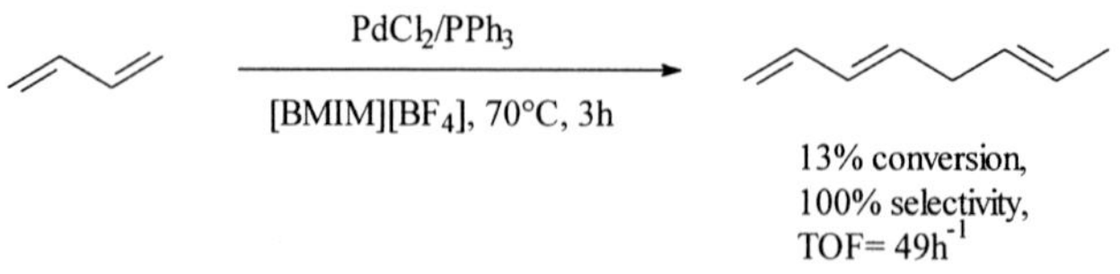

Scheme 7 Biphasic, Pd-catalyzed dimerization of butadiene in $[BMIM][BF_4]$.

Dupont and co-workers studied the Pd-catalyzed dimerization [53] of butadiene in non-chloroaluminate ionic liquids. The octatriene products are of some commercial relevance as intermediates for the synthesis of fragrances, plasticizers, and adhesives. By using $PdCl_2$ with two equivalents of the ligand PPh_3 dissolved in $[BMIM][PF_6]$, $[BMIM][BF_4]$, or $[BMIM][CF_3SO_3]$, it was possible to obtain the octatrienes with 100% selectivity (after 13% conversion) (Scheme 7) [53]. The turnover frequency (TOF) was in the range of 50 mol butadiene converted per mol catalyst per hour, which represents a substantial increase in catalyst activity in comparison to the same reaction under identical conditions in THF (TOF = 6 h^{-1}).

The formation of polyketones from styrene and CO using $[Pd(bipy)_2][PF_6]$ in several ionic liquids has been reported [54]. The ionic liquid-mediated reaction was found to be dependent on the nature of the ionic liquid's anion, with decreasing activity in the order $[(CF_3SO_2)_2N]^- > [PF_6]^- > [BF_4]^-$ for both pyridinium and imidazolium ionic liquids. Best results were obtained in a 10:1 mixture of *N*-hexylpyridinium bis(trifluorosulfon)imide and methanol, which was found to provide a much better solvent than methanol alone. The M_w/M_n values obtained were in the range 1.3–2.5, suggesting a single-site reaction even in the ionic liquid medium.

3.4.4
Conclusion

From the above-mentioned examples and from the large number of publications and reviews [18–25] covering applications of transition metal complexes in ionic liquids, it is quite obvious that this research area has become quite popular. Is it just fashionable to work in liquid salts or is there really a chance to achieve something unique?

The following aspects reflect the most important properties of ionic liquids that offer, from the author's point of view, some potential for special chemistry:

1. ***Large liquid range and non-volatile nature***
   Probably the most prominent property of an ionic liquid is its lack of vapor pressure. Transition metal catalysis in ionic liquids can particularly benefit from this on economical, environmental, and safety grounds.
   Obviously, the use of a non-volatile ionic liquid simplifies the distillative work-up of volatile products. Moreover, the application of non-volatile ionic liquids can contribute to the reduction of atmospheric pollution. This is of special rele-

vance for non-continuous reactions, where complete recovery of a volatile organic solvent is usually difficult to integrate into the process. Finally, the switch from a volatile, flammable, organic solvent to an ionic liquid may significantly improve the safety of a given process. This will be especially true in oxidation reactions where air or pure oxygen are used as oxidants, as here the use of common organic solvents is often restricted because of the potential formation of explosive mixtures of oxygen and the volatile organic solvent in the gas phase.

2. *Solubility/polarity vs nucleophilicity/coordination*

Many organic solvents applied in catalytic reactions do not behave as innocent solvents, but show significant coordination to the catalytic center. The reason why these solvents are nevertheless used in catalysis is that some polar or ionic catalyst complexes are not soluble enough in weakly coordinating organic solvents. For example, many cationic transition metal complexes are known to be excellent oligomerization catalysts [55]. However, their usually poor solubility in non-polar solvents often requires a compromise between the solvation and the coordination properties of the solvent if organic solvents are used. In order to achieve sufficient solubility of the metal complex, a solvent of higher polarity that may compete with the substrate for the coordination sites at the catalytic center is required. Consequently, in these cases, the use of an inert, weakly coordinating ionic liquid can result in a clear enhancement of catalytic activity. Ionic liquids with weakly coordinating, inert anions (e.g., $[(CF_3SO_2)2N]^-$, $[PF_6]^-$ $[PF_3(CF_3CF_2)_3]^-$ or $[BF_4]^-$) and inert cations (cations that do not coordinate to the catalyst themselves and that do not form species under the reaction conditions that coordinate to the catalyst) can be considered as relatively polar, "innocent" solvents in transition metal catalysis. In fact, it is, for example, this unique combination of high solvation power for polar catalyst complexes (polarity) and of weak coordination (nucleophilicity) that allows for the first time biphasic catalysis with highly electrophilic, cationic Ni complexes [56]. Many other catalytic applications where a rate-enhancing effect of the ionic liquid medium is claimed are very likely to have their origin in this unique solubility/polarity vs nucleophilicity/coordination properties of some ionic liquids.

3. *Tuneable acidity for catalyst activation and catalyst-supporting interaction*

Ionic liquids formed by the reaction of a halide salt with a Lewis acid (e.g., chloroaluminate or chlorostannate melts) generally act as both solvent and co-catalyst in transition metal catalysis. The reason for this is that the Lewis acidity or basicity, which is always present (at least latently), results in strong interactions with the catalyst complex. In many cases, the Lewis acidity of an ionic liquid has been used to convert the neutral catalyst precursor into the corresponding cationic active form [57–59]. Even in cases where the ionic liquid is not directly involved in creating the active catalytic species, a co-catalytic interaction between the ionic liquid solvent and the dissolved transition metal complex can take place and can result in significant catalyst activation. This type of co-catalytic influence is well known in heterogeneous catalysis, where for some reactions an acidic support activates the metal catalyst in a better way than neutral supports. In this respect, the acidic ionic liquid can be considered as a liquid, acidic support for the transition metal catalysts dissolved therein.

4. ***Electrochemistry in combination with transition metal catalysis***
   If we analyze all the different areas of application for ionic liquids it soon becomes pretty obvious that electrochemistry will establish itself as one of the important fields. Particularly for electrochemical metal deposition, the great solvation power of ionic liquids combined with their wide electrochemical windows offer unique possibilities. Despite the fact that the combination of electrochemical transformation steps and transition metal catalysis is a quite obvious challenge, the number of published examples in this area is still very limited. A first very exciting application has been described by Bedioui et al. [60]. More recent papers report the electrocatalytic cycloaddition of CO_2 to epoxide [61] and the electroreductive coupling of organic halides [62] in ionic liquids.

Lack of general understanding is still the major limitation for the further development of transition metal catalysis in ionic liquids. Obviously, it is possible for a chemical interaction between an ionic liquid solvent and a dissolved transition metal complex to be either activating or deactivating. Therefore, exact knowledge of the nature of these chemical interactions is crucial if we are to (a) derive benefit from the ionic liquid's potential to activate a catalyst and (b) avoid deactivation. Without doubt, much more fundamental work is needed and should be encouraged in order to speed up the future development of transition metal catalysis in ionic liquids. A lot of exciting chemistry is still to be done!

References

1 N. Karodia, S. Guise, C. Newlands, J.-A. Andersen, *Chem. Commun.* **1998**, 2341–2342.

2 (a) P. Wasserscheid, R. van Hal, A. Bösmann, *Green Chem.* **2002**, *4*, 400–404; (b) commercially available as ECOENG^{TM} ionic liquids from Solvent Innovation GmbH, Cologne (www. solvent-innovation.com)

3 (a) P. Wasserscheid, M. Sesing, W. Korth, *Green Chem.* **2002**, *4*, 134–138; (b) W. Keim, W. Korth, P. Wasserscheid, WO 0016902 (to BP Chemicals Limited, UK; Akzo Nobel NV; Elementis UK Limite(d) 2000 [*Chem. Abstr.* **2000**, *132*, 238691.

4 (a) J. Pernak, A. Czepukowicz, R. Pozniak, *Ind. Eng. Chem. Res.* **2001**, *40*, 2379–2383; (b) B. Jastorff, R. Stoermann, J. Ranke, K. Moelter, F. Stock, B. Oberheitmann, W. Hoffmann, J. Hoffmann, M. Nuechter, B. Ondruschka, J. Filser, *Green Chem.* **2003**, *5*, 136–142.

5 (a) A. E. Visser, R. P. Swatloski, W. M. Reichert, R. Mayton, S. Sheff, A. Wierzbicki, J. H. Davis Jr., R. D. Rogers, *Chem. Commun.* **2001**, 135–136; (b) A. E. Visser, R. P. Swatloski, W. M. Reichert, R. Mayton, S. Sheff, A. Wierzbicki, J. H. Davis Jr., R. D. Rogers, *Environ. Sci. Tech.* **2002**, *36(11)*, 2523–2529.

6 T. L. Merrigan, E. D. Bates, S. C. Dorman, J. H. Davis Jr., *Chem. Commun.* **2000**, 2051–2052.

7 C. Hardacre, J.D. Holbrey, S.E.J. McMath, *Chem. Commun.* **2001**, 367–368.

8 (a) P. Wasserscheid, A. Bösmann, C. Bolm, *Chem. Commun.* **2002**, 2000–2001; (b) M.J. Earle, P.B. McCormac, K.R. Seddon, *Green Chem.* **1999**, *1*, 23–25.

9 Some commercial suppliers are given here: (a) Solvent Innovation GmbH, Cologne (www.solvent-innovation.com); (b) Merck (www.merck.de); (b) Sachem Inc. (www.sachem.com); (c) Fluka (www.fluka.com); (d) Acros Organics (www.acros.com); ionic liquids are offered in collaboration with QUILL (www.quill.ac.uk); (e) Wako (www.wako.com).

10 K.R. Seddon, A. Stark, M.J. Torres, *Pure Appl. Chem.* **2000**, *72*, 2275–2287.

11 P. Bonhôte, A.-P. Dias, N. Papageorgiou, K. Kalyanasundaram, M. Grätzel, *Inorg. Chem.* **1996**, *35*, 1168–1178.

12 Y. Chauvin, L. Mussmann, H. Olivier, *Angew. Chem. Int. Ed. Engl.* **1995**, *34*, 2698–2700.

13 D.E. Kaufmann, M. Nouroozian, H. Henze, *Synlett* **1996**, 1091–1092.

14 C.J. Mathews, P.J. Smith, T. Welton, A.J.P. White, D.J. Williams, *Organometallics* **2001**, *20(18)*, 3848–3850.

15 J.D. Holbrey, K.R. Seddon, R. Wareing, *Green Chem.* **2001**, *3*, 33–36.

16 P. Wasserscheid, T. Welton (Eds.), Ionic Liquids in Synthesis, Wiley-VCH, **2002**, pp. 1–365.

17 B. Driessen-Hölscher, P. Wasserscheid, W. Keim, *CATTECH* **1998**, June, 47–52.

18 J. Dupont, R.F. de Souza, P.A.Z. Suarez, *Chem. Rev.* **2002**, *102*, 3667–3691.

19 H. Olivier-Bourbigou, L. Magna, *J. Mol. Catal. A: Chemical* **2002**, *182/183*, 419–437.

20 R. Sheldon, *Chem. Commun.* **2001**, 2399–2407.

21 C.M. Gordon, *Appl. Catal. A: General* **2001**, *222*, 101–117.

22 P. Wasserscheid in P. Wasserscheid, T. Welton (Eds.), Ionic Liquids in Synthesis, Wiley-VCH, **2002**, pp. 213–257.

23 T. Welton, *Chem. Rev.* **1999**, *99*, 2071–2083.

24 J.D. Holbrey, K.R. Seddon, *Clean Prod. Process.* **1999**, *1*, 223–226.

25 P. Wasserscheid, W. Keim, *Angew. Chem., Int. Ed.* **2000**, *39*, 3772–3789.

26 (a) W.A. Herrmann, V.P.W. Böhm, *J. Organomet. Chem.* **1999**, *572*, 141–145; (b) V.P.W. Böhm, W.A. Herrmann, *Chem. Eur. J.* **2000**, *6*, 1017–1025.

27 A.J. Carmichael, M.J. Earle, J.D. Holbrey, P.B. McCormac, K.R. Seddon, *Org. Lett.* **1999**, *1*, 997–1000.

28 S. Bouquillon, B. Ganchegui, B. Estrine, F. Henin, J. Muzart, *J. Organomet. Chem.* **2001**, *634*, 153–156.

29 (a) V. Calò, A. Nacci, L. Lopez, A. Napola, *Tetrahedron Lett.* **2001**, *42*, 4701–4703; (b) V. Calo, A. Nacci, A. Monopoli, L. Lopez, A. di Cosmo, *Tetrahedron* **2001**, *57*, 6071–6077.

30 J. Silberg, T. Schareina, R. Kempe, K. Wurst, M.R. Buchmeiser, *J. Organomet. Chem.* **2001**, *622*, 6–18.

31 L. Xu, W. Chen, J. Ross, J. Xiao, *Org. Lett.* **2001**, *3(2)*, 295–297.

32 L. Xu, W. Chen, J. Xiao, *Organometallics* **2000**, *19*, 1123–1127.

33 (a) A.J. Arduengo, R.L. Harlow, M. Kline, *J. Am. Chem. Soc.* **1991**, *113*, 361; (b) A.J. Arduengo, H.V.R. Dias, R.L. Harlow, *J. Am. Chem. Soc.* **1992**, *114*, 5530; (c) G.T. Cheek, J.A. Spencer, in 9th International Symposium on Molten salts (C.L. Hussey, D.S. Newman, G. Mamantov, Y. Ito, eds), The Electrochem. Soc., Inc., New York, **1994**, 426; (d) W.A. Herrmann, M. Elison, J. Fischer, C. Koecher, G.R.J. Artus, *Angew. Chem., Int. Ed. Engl.* **1995**, *34*, 2371–2374; (e) D. Bourissou, O. Guerret, F. P. Gabba, G. Bertrand, *Chem. Rev.* **2000**, *100*, 39–91.

34 D.S. McGuinness, K.J. Cavell, B.F. Yates, *Chem. Commun.* **2001**, 355–356.

35 N.A. Hamill, C. Hardacre, S.E.J. McMath, *Green Chem.* **2002**, *4*, 139–142.

36 H. Hagiwara, Y. Shimizu, T. Hoshi, T. Suzuki, M. Ando, K. Ohkubo, C. Yokoyama, *Tetrahedron Lett.* **2001**, *42(26)*, 4349–4351.

37 K. Okubo, M. Shirai, C. Yokoyama, *Tetrahedron Lett.* **2002**, *43*, 7115–7118.

38 R. R. Dashmukh, R. Rajagopal, K. V. Srinivasan, *Chem. Commun.* **2001**, 1544–1545.

39 K. S. A. Vallin, P. Emilsson, M. Larhed, A. Hallberg, *J. Org. Chem.* **2002**, *67*, 6243–6246.

40 C. J. Mathews, P. J. Smith, T. Welton, *Chem. Comm.* **2000**, 1249–1250.

41 R. Rajagopal, D. V. Jarikote, K. V. Srinivasan, *Chem. Commun.* **2002**, 616–617.

42 J. D. Revell, A. Ganesan, *Org. Lett.* **2002**, *4*, 3071–3073.

43 J. McNulty, A. Capretta, J. Wilson, J. Dyck, G. Adjybeng, A. Robertson, *Chem. Commun.* **2002**, 986–987.

44 S. T. Handy, X. Zhang, *Org. Lett.* **2001**, *3 (2)*, 233–236.

45 J. Sirieix, M. Ossberger, B. Betzemaier, P. Knochel, *Synlett* **2000**, 1613–1615.

46 T. Fukuyama, M. Shinmen, S. Nishitani, N. Sato, I. Ryu, *Org. Lett.* **2002**, *4*, 1691–1694.

47 C. de Bellefon, E. Pollet, P. Grenouillet, *J. Mol. Catal. A* **1999**, *145*, 121–126.

48 J. Ross, W. Chen, L. Xu, J. Xiao, *Organometallics* **2001**, *20*, 138–142.

49 (a) S. Toma, B. Gotov, I. Kmentová, E. Solčániová, *Green Chem.* **2000**, *2*, 149–151; (b) S. Toma, B. Gotov, I. Kmentová, E. Solčániová, *Green Chem.* **2002**, *4*, 103–106.

50 E. Mizushima, T. Hayashi, M. Tanaka, *Green Chem.* **2001**, *3*, 76–79.

51 C. S. Consorti, G. Ebeling, J. Dupont, *Tetrahedron Lett.* **2002**, *43*, 753–755.

52 (a) P. Wasserscheid, J. Zimmermann, I. Tkatchenko, S. Stutzmann, *Chem. Commun.* **2002**, 760–761; (b) M. Piquet, S. Stutzmann, I. Tkatchenko, I. Tommasi, J. Zimmermann, P. Wasserscheid, *Green Chem.* **2003**, *5*, 153–162; (c) J. Zimmermann, I. Tkatchenko, P. Wasserscheid, *Adv. Synth. Catal.* **2003**, *345*, 402–409; (d) D. Ballivet-Tkatchenko, M. Picquet, M. Solinas, G. Franchio, P. Wasserscheid, W. Leitner, *Green Chem.* **2003**, *5*, 232–235.

53 S. M. Silva, P. A. Z. Suarez, R. F. de Souza, J. Dupont, *Polym. Bull.* **1998**, *41*, 401–405.

54 (a) C. Hardacre, J. D. Holbrey, S. P. Katdare, K. R. Seddon, *Green Chem.* **2002**, *4*, 143–146; (b) M. A. Klingshirn, G. A. Broker, J. D. Holbrey, K. H. Shaughnessy, R. D. Rogers, *Chem. Commun.* **2002**, 1394–1395.

55 (a) R. B. A. Pardy, I. Tkatschenko, *J. Chem. Soc., Chem. Commun.* **1981**, 49–50; (b) J. R. Ascenso, M. A. A. F. De, C. T. Carrando, A. R. Dias, P. T. Gomes, M. F. M. Piadade, C. C. Romao, A. Revillon, I. Tkatschenko, *Polyhedron* **1989**, *8*, 2449–2457; (c) P. Grenouillet, D. Neibecker, I. Tkatschenko, *J. Organomet. Chem.* **1983**, *243*, 213–222; (d) J.-P. Gehrke, R. Taube, E. Balbolov, K. Kurtev; *J. Organomet. Chem.* **1986**, *304*, C4–C6.

56 P. Wasserscheid, C. M. Gordon, C. Hilgers, M. J. Maldoon, I. R. Dunkin, *Chem. Commun.* **2001**, 1186–1187.

57 R. T. Carlin, R. A. Osteryoung, *J. Mol. Catal.* **1990**, *63*, 125–129.

58 Y. Chauvin, S. Einloft, H. Olivier, *Ind. Eng. Chem. Res.* **1995**, *34*, 1149–1155.

59 H. Waffenschmidt, P. Wasserscheid, *J. Mol. Catal.* **2001**, *164*, 61–67.

60 L. Gaillon, F. Bedioui, *Chem. Commun.* **2001**, 1458.

61 H. Yang, Y. Gu, Y. Deng, F. Shi, *Chem. Commun.* **2002**, 274–275.

62 R. Barhdadi, C. Courtinard, J. Y. Nédèlec, M. Troupel, *Chem. Commun.* **2003**, 1434–1435.

3.5 Transition Metals in Photocatalysis

H. Hennig

3.5.1 Introduction

The advantage of photocatalysis is that the reaction conditions required when catalytic or chain processes are initiated by photons, are unusually mild. It is not surprising, therefore, that homogeneous photo-complex catalysis is a rapidly growing new field in organic synthesis, especially for the production of fine chemicals. In principle, the majority of the syntheses described in this book can be performed photocatalytically. For details, the reader is referred to review articles dealing with homogeneous photocatalysis in organic synthesis [1]. Further articles and books refer to basic principles of the photochemistry of coordination compounds and photocatalysis [2, 3].

Fig. 1 depicts the general scheme of photocatalytic processes. It is advantageous to distinguish between two types of photocatalysis, *photoinduced catalysis* and *photoassisted reactions.*

- *Photoinduced catalysis* refers to the photogeneration of a catalyst that subsequently promotes a catalyzed reaction. Photons are only required to generate the catalyst. Thus, the efficiency of such processes only depends on the activity of the catalyst produced photochemically. Therefore, high turnover of the photochemically produced catalyst is the main criterion in generating efficient photocatalytic syntheses. Quantum yields (defined as the ratio of moles of product formed to the number of photons absorbed) greater than unity may occur. The same is true for photoinduced chain reactions.
- *Photoassisted reactions* include interactions between electronically excited states or short-lived intermediates and substrate molecules leading to product formation with concomitant regeneration of the starting complex in its electronic ground state. Product quantum yields greater than unity are impossible because one photon may not initiate more than one catalytic cycle.
- *Photoinduced chain processes* resemble in part photocatalytic ones. Because such reactions offer considerable potential for the synthesis of fine chemicals they will also be included in this section.

Transition Metals for Organic Synthesis, Vol. 2, 2nd Edition.
Edited by M. Beller and C. Bolm

ISBN: 3-527-30613-7

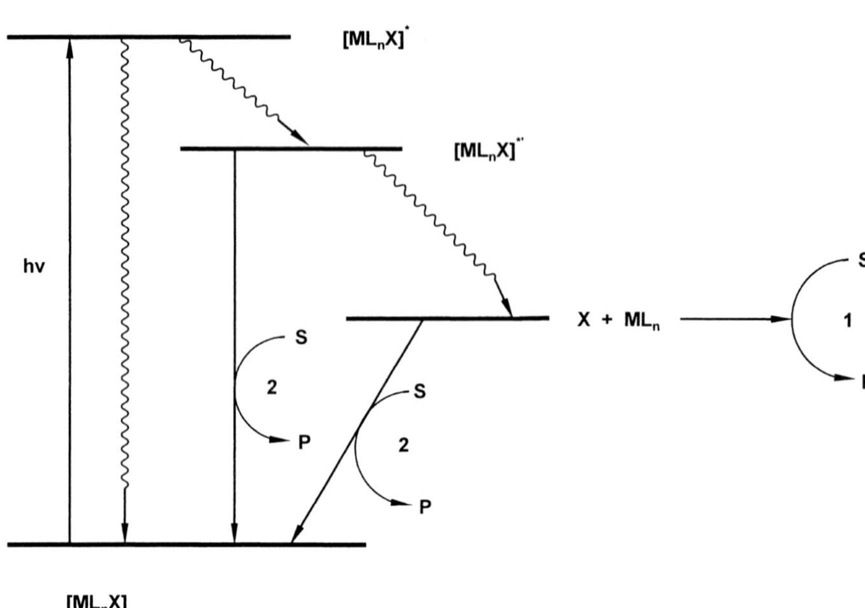

Fig. 1 Simplified Jablonski diagram of photocatalysis based on light-sensitive transition metal complexes [ML$_n$X] (1: photoinduced catalytic reaction; 2: photoassisted reaction; S: substrate; P: product).

As well as these reaction pathways, *catalyzed photolysis* might also be considered. However, since this means the catalysis of normal photochemical reactions it should not be included in this section. Although *heterogeneous photocatalysis*, particularly that based on semiconductors like TiO_2 and ZnS/CdS, has developed very rapidly during the last decade, this special field will also not be introduced. Instead, the reader is referred to some excellent papers reviewing the application of such processes in organic synthesis [4].

Fig. 2 shows the most convenient photocatalytic reaction pathways using light-sensitive transition metal complexes or organometallic compounds in organic synthesis. These include

- Photochemical generation of coordinative unsaturation (O)
- Activation of small molecules (Z) to be inserted into, e.g., C–H-, C=C-, or C≡C-bonds
- Generation of free ligands or ligand electron transfer products behaving as catalysts (A)
- Activation of coordinated ligands (X).

Fig. 2 may serve as a guide throughout this section, exemplifying how photocatalysis can be applied to such substrate conversions that are important for the syn-

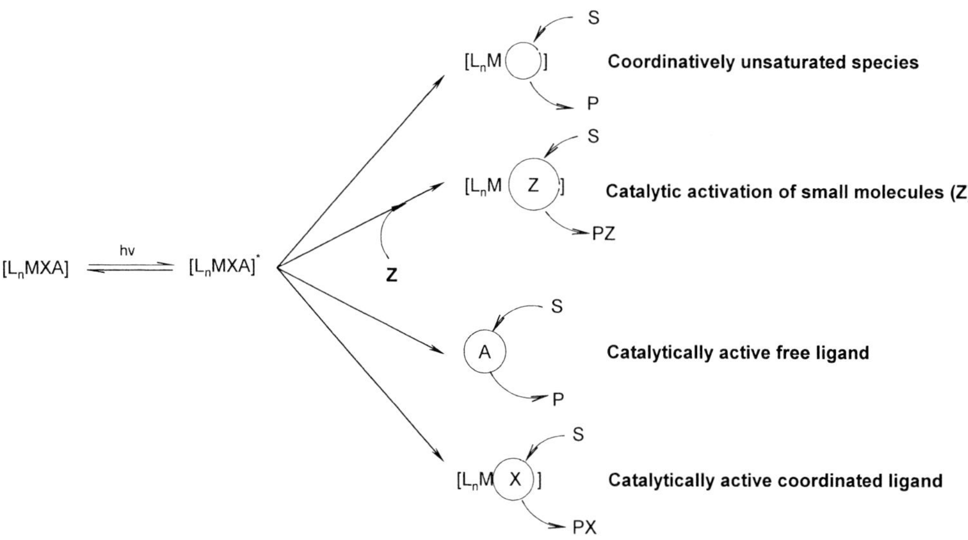

Fig. 2 General scheme of the photocatalytic activation of organic substrates (S) yielding products (P, PX, and PZ, respectively) due to the presence of light-sensitive transition metal complexes or organometallic compounds ($[L_nMX(A)]$).

thesis of fine chemicals. The given examples are selected with the aim of demonstrating the general synthetic potential behind photocatalytic reaction pathways. As a consequence, general reaction principles will preferably be introduced instead of a broad variety of various products obtainable photocatalytically.

3.5.2
Photochemical Generation of Coordinatively Unsaturated Complex Fragments

Coordinatively unsaturated species are the most common catalysts in homogeneous complex catalysis [5]. Their photochemical generation offers considerable advantage when compared with the usual thermal route because it allows for reaction tuning by light, which affords a very convenient control of the course of the reaction, for instance.

Fig. 3 illustrates very impressively how photoinduced catalytic reactions can be used in organic synthesis. It concerns the cyclotrimerization of acetylene or appropriate alkyne derivatives with diverse nitriles to the corresponding pyridines, a reaction that occurs photocatalytically in the presence of light-sensitive cobalt(I) complexes [6]. This elegant synthetic route yields various pyridine derivatives in rather high yields because of the very high catalytic turnovers. This example shows in particular the advantage of homogeneous photo-complex catalysis in the synthesis of fine chemicals. The photocatalytic process occurs under normal pressure and at room temperature. Further, a broad variety of pyridine derivatives can be obtained depending on the alkyne and nitrile synthons used. The cobalt(I)

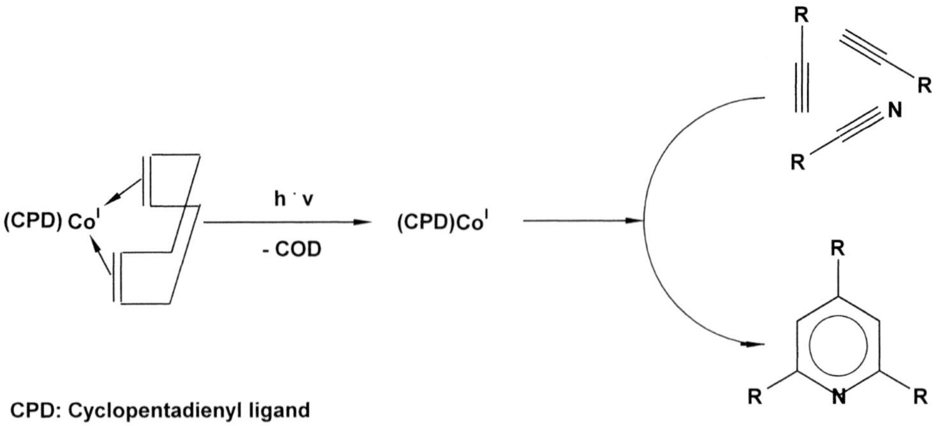

CPD: Cyclopentadienyl ligand

Fig. 3 Photocatalytic synthesis of pyridine derivatives in the presence of cobalt(I) complexes.

complexes used are light sensitive in the visible spectral region, thus allowing the use of solar energy. The catalytic efficiency is extremely high, and the entire conversion may also occur in aqueous systems.

Numerous examples of photoinduced catalytic reactions are known [1]. They are preferably applied to the activation of olefins [7], metathesis reactions [8], electron transfer processes [9], and CO insertion reactions [10], for example. Furthermore, photoinduced catalytic reactions have attracted increasing attention in the activation of small molecules such as O_2, CO, CO_2, and H_2. For example, light-sensitive metal porphyrins are useful for the selective oxygenation of terpenoid alkenes [11], as shown in Fig. 4. α-Pinene, for instance, is oxygenated selectively to *trans*-verbenol (one of the pheromones of the bark beetle, *ips typographicus*) [12]. Furthermore, considerable enantioselectivity can be achieved in the synthesis of (S)-*trans*-verbenol by using the appropriate cyclodextrin-substituted metal porphyrins [13].

3.5.3 Photochemically Generated Free Ligands as Catalysts

Free ligands or their electron transfer products photochemically generated from any transition metal complexes may also be used to initiate photoinduced catalytic processes. Fig. 5 shows how such processes may occur. It represents the photoaquation of cyanide complexes due to photochemical ligand field excitation. Particularly, octacyanomolybdate/tungstate(IV) forms cyanide with rather high quantum yields. Free cyanide ions catalyze the dimerization of heterocyclic carb-2-aldehydes to the corresponding endioles. No dark reactions occur because of the high thermodynamic complex stability of the $[Mo/W(CN)_8]^{4-}$ complexes [14].

Although it may appear that the photocatalytic synthesis of heterocyclic endioles brings no advantage when compared with the usual thermal reaction pathway,

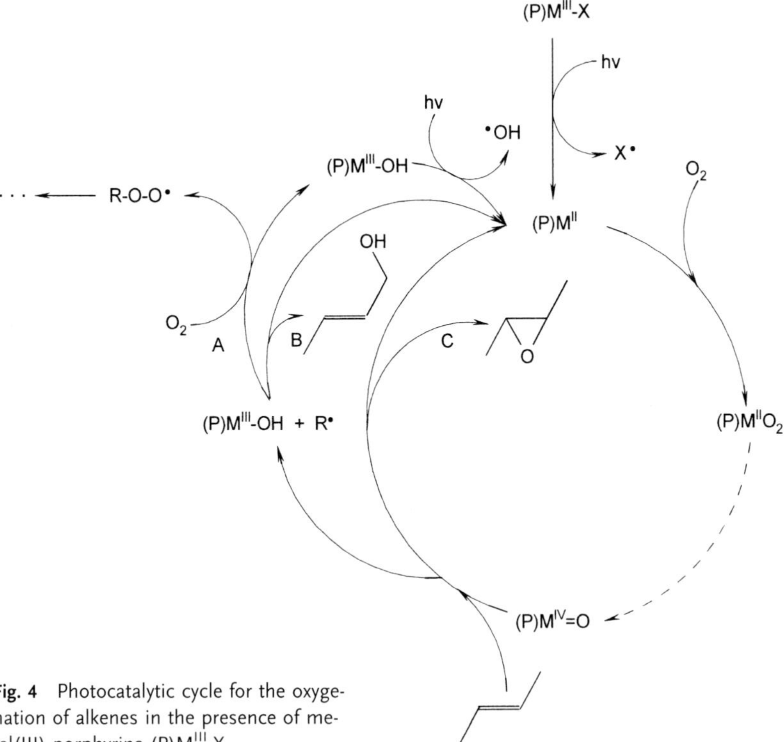

Fig. 4 Photocatalytic cycle for the oxygenation of alkenes in the presence of metal(III) porphyrins (P)M^{III}-X.

$$[Mo(CN)_8]^{4-} \underset{}{\overset{h\nu}{\rightleftharpoons}} {}^*[Mo(CN)_8]^{4-} \xrightarrow[+H_2O]{} [Mo(CN)_4O(OH)]^{3-} + 4\,CN^- \longrightarrow$$

Fig. 5 Photoinduced catalytic dimerization of heterocyclic carb-2 aldehydes to the corresponding endioles due to the photoaquation of coordinated cyanide ligands in molybdenum(IV) octacyanide.

this example confirms impressively the basic principle of photoinduced catalytic processes. The only photochemical reaction is the generation of the catalyst, here the cyanide ion, which is produced by photochemically induced dissociation from the first coordination sphere of any suitable metal complex. Incidentally, the photoinduced catalytic generation of endioles can be used in photo-imaging procedures [14].

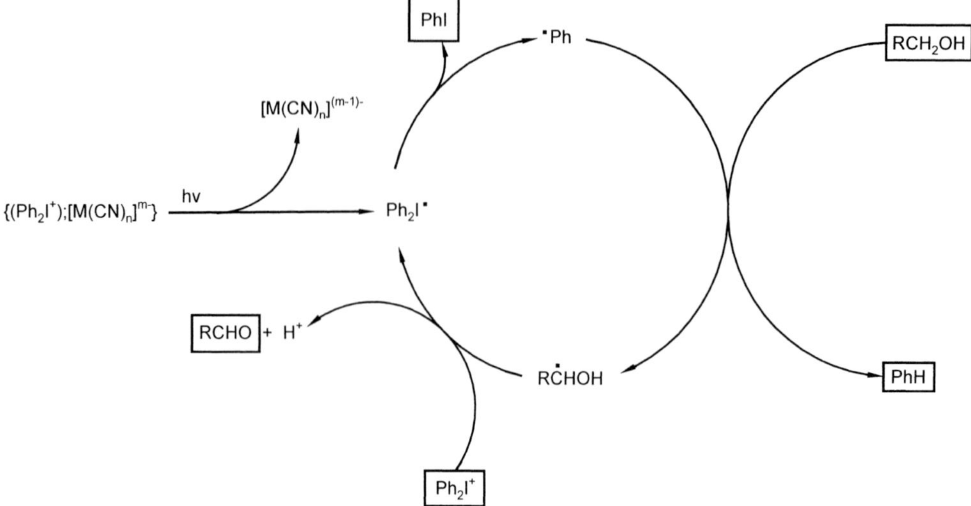

Fig. 6 Selective photoinduced chain oxidation of primary and secondary alcohols to the corresponding aldehydes or ketones caused by IPCT excitation of $\{(Ph_2I)^+;[M(CN)_n]^{m-}\}$ ion pairs in the presence of an excess of Ph_2ICl.

However, second sphere processes that may be involved in this kind of photocatalytic reactions are often much more interesting for synthetic procedures. Ion pairs are compounds distinguished by second sphere effects. Among them, donor/acceptor ion pairs are particularly interesting. Besides the usual electrostatic interactions, such ion pairs are characterized by electronic influences between the donor and the acceptor, leading to new spectroscopic transitions (IPCT) in the UV/Vis spectral region. Photochemical excitation of IPCT states may cause electron transfer processes yielding free radicals, which may be considered as possible initiators for photoinduced chain reactions. Fig. 6 shows schematically this kind of chain process used for organic syntheses [15].

Here, the photogeneration of phenyl radicals initiates a chain reaction that continues thermally because of an excess of diphenyliodonium ions reacting as strong oxidants. The advantage of such a reaction pathway is based on the exclusive oxidation of primary alcohols to the corresponding aldehydes without the formation of carboxylic acids, and further on the high stereoselectivity when allylic alcohols are considered.

Finally, coordinated ligands themselves may be photochemically converted to highly reactive species ($[L_nMX]$), as shown in Fig. 2. However, only stoichiometric reactions can be observed with $[L_nMX]$ because the regeneration of the starting complex $[L_nMA]$ cannot occur. Nitrene complexes of nickel(II) produced photochemically from the appropriate azido complexes (see Fig. 7) allow convenient N insertions into aliphatic CH or alkene C=C bonds yielding amines and aziridine derivatives, respectively [16]. Without nitrene scavengers, coordinatively unsatu-

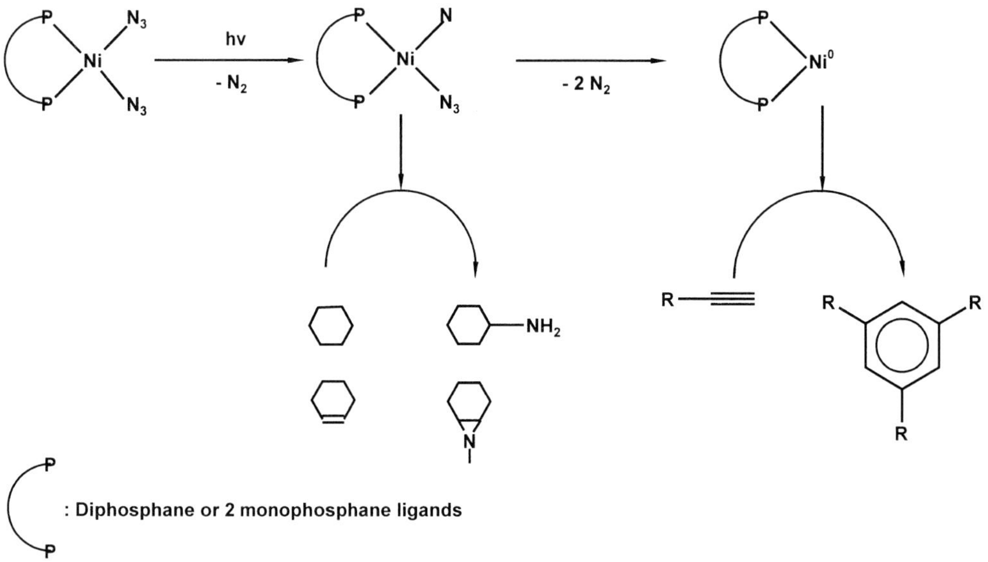

Fig. 7 N insertion into C–H or C=C bonds yielding amines and aziridines, respectively (left part); photoinduced catalytic cyclotrimerization of alkynes due to the formation of nickel(0) complex fragments (right part).

rated nickel(0) complexes are formed that are able to catalyze the cyclotrimerization of acetylene or alkyne derivatives to the corresponding benzenes.

3.5.4
Conclusions

Homogeneous photocatalysis with light-sensitive transition metal complexes or organometallic compounds allows organic syntheses at ambient temperature (or even lower) and normal pressure. Further, because the catalysts themselves are generated photochemically *in situ* from any precursor complexes, the handling of these metal complexes usually does not require any particular precautions. However, low quantum yields of the primary photoreaction, fast back electron transfer or recombination processes, and the photo-sensitivity of the products synthesized photocatalytically may diminish the efficiency of such processes considerably [3]. Heterogeneous photocatalysis based on transition metal compounds [4] or immobilized metal complexes [17] should be considered as further interesting tools in organic synthesis.

References

1 R. G. Salomon, *Tetrahedron* **1983**, *39*, 485; H. Hennig, L. Weber, R. Stich, M. Grosche, D. Rehorek, *Prog. Photochem. Photophys.* **1992**, *VI*, 167; R. G. Salomon, S. Ghosh, S. Raychaudhuri, *Adv. Chem. Ser.* **1993**, *238*, 315; J. Santamaria, C. Ferraudi in *Homogeneous Photocatalysis* (M. Chanon, ed.), Wiley, New York, USA, **1997**, Chapter 4; C. Kutal in *Homogeneous Photocatalysis* (M. Chanon, ed.), Wiley, New York, USA, **1997**, Chapter 5.

2 V. Balzani, V. Carassiti, *Photochemistry of Coordination Compounds*, Academic Press, New York, USA, **1970**; G. L. Geoffroy, M. S. Wrighton, *Organometallic Photochemistry*, Academic Press, New York, USA, **1979**; G. J. Ferraudi, *Elements of Inorganic Photochemistry*, Wiley, New York, USA, **1988**; J. Sykora, J. Sima, *Coord. Chem. Rev.* **1990**, *107*, 1; D. M. Roundhill, *Photochemistry and Photophysics of Metal Complexes*, Plenum Press, New York, USA, **1994**.

3 H. Hennig, P. Thomas, R. Wagener, D. Rehorek, K. Jurdeczka, *Z. Chem.* **1977**, *17*, 241; H. Hennig, D. Rehorek, *Photochemische und photokatalytische Reaktionen von Koordinationsverbindungen*, Teubner, Stuttgart, **1988**; N. Serpone, E. Pelizzetti (eds.), *Photocatalysis*, Wiley, New York, USA, **1989**; H. Hennig, R. Billing, *Coord. Chem. Rev.* **1993**, *125*, 89; H. Hennig, L. Weber, D. Rehorek, *Adv. Chem. Ser.* **1993**, *238*, 231; K. Kalyanasundaram, M. Grätzel (eds.), *Photosensitization and Photocatalysis Using Inorganic and Organometallic Compounds*, Kluwer, Dordrecht, Netherlands, **1993**; M. Chanon (ed.), *Homogeneous Photocatalysis*, Wiley, New York, USA, **1997**.

4 M. A. Fox in *Photocatalysis* (N. Serpone, E. Pelizzetti, eds.), Wiley, New York, USA, **1989**, Chapter 13; H. Kisch, R. Künneth, *Prog. Photochem. Photophys.*, **1991**, *IV*, 131; M. Schiavello (ed.), *Heterogeneous Photocatalysis*, Wiley, New York, USA, **1997**; H. Kisch in *Electron Transfer in Chemistry*, Vol. 4, *Heterogeneous Systems* (V. Balzani, ed.), Wiley, New York, USA, **2001**, p. 232; H. Kisch, *Adv. Photochem.* **2001**, *62*, 93; H. Kisch, W. Lindner, *Chem. unserer Zeit* **2001**, *35*, 250.

5 B. Cornils, W. A. Herrmann (eds.), *Applied Homogeneous Catalysis with Organometallic Compounds*, VCH, Weinheim, **1996**.

6 B. Heller, G. Oehme, *J. Chem. Soc., Chem. Commun.* **1995**, 179; B. Heller, D. Heller, G. Oehme, *J. Mol. Cat. A* **1996**, *110*, 211; G. Oehme, B. Heller, P. Wagler, *Energy* **1997**, *22*, 327.

7 See, for instance, U. Kölle in *Photosensitization and Photocatalysis Using Inorganic and Organometallic Compounds* (K. Kalyanasundaram, M. Grätzel, eds.), Kluwer, Dordrecht, The Netherlands, **1993**, p. 331; A. Molinari, R. Amadelli, V. Carassiti, A. Maldotti, *Eur. J. Inorg. Chem.* **2000**, 91.

8 See, for instance, T. Szymanska-Buzar, *J. Mol. Cat.* **1988**, *48*, 43; S. Pulst, F. G. Kirchbauer, B. Heller, W. Baumann, U. Rosenthal, *Angew. Chem.* **1998**, *110*, 2029, *Angew. Chem. Int. Ed. Engl.* **1998**, *37*, 1925.

9 See, for instance, M. Chanon, *Bull. Soc. Chim. Fr.* **1985**, 209.

10 See, for instance, R. H. Crabtree, in *Photosensitization and Photocatalysis Using Inorganic and Organometallic Compounds* (K. Kalyanasundaram, M. Grätzel, eds.), Kluwer, Dordrecht, The Netherlands, **1993**, p. 391.

11 L. Weber, J. Behling, G. Haufe, H. Hennig, *J. Prakt. Chem.* **1992**, *334*, 138; H. Hennig, J. Behling, R. Meusinger, L. Weber, *Chem. Ber.* **1995**, *128*, 229; H. Hennig, D. Luppa in *Peroxide Chemistry – Mechanistic and Preparative Aspects of Oxygen Transfer* (W. Adam, ed.), Wiley-VCH, Weinheim, **2000**, Chapter 12.

12 L. Weber, R. Hommel, J. Behling, G. Haufe, H. Hennig, *J. Amer. Chem. Soc.* **1994**, *116*, 2400.

13 L. Weber, I. Imiolczyk, G. Haufe, D. Rehorek, H. Hennig, *J. Chem. Soc., Chem. Commun.* **1992**, 301.

14 H. Hennig, E. Hoyer, E. Lippmann, E. Nagorsnik, P. Thomas, M. Weissenfels, *J. Inf. Recording Mater.* **1987**, *6*, 39.

15 R. Billing, D. Rehorek, H. Hennig, *Top. Curr. Chem.* **1990**, *158*, 151; H. Hennig, O. Brede, R. Billing, J. Schönewerk, *Chem. Eur. J.* **2001**, *7*, 2114.

16 H. Hennig, K. Hofbauer, K. Handke, R. Stich, *Angew. Chem.* **1997**, *109*, 373, *Angew. Chem. Int. Ed. Engl.* **1997**, *36*, 408.

17 See, for instance, A. Maldotti, A. Molinari, G. Varani, M. Lenarda, L. Storaro, F. Bigi, R. Maggi, A. Mazzacani, G. Sartori, *J. Catal.* **2002**, *209*, 3618.

3.6
Transition Metals in Radiation-Induced Reactions for Organic Synthesis: Applications of Ultrasound

Pedro Cintas

3.6.1
Sonochemistry and Metal Activation

The purpose of this chapter is to introduce readers to some of the most recent studies focused on the use of ultrasound in organometallic reactions, metal activation, and catalyst design. The reader wishing to study the use and applications of ultrasound in greater depth is referred to a series of recent monographs [1–3], while the particular field of organometallic sonochemistry has been covered in detail in two comprehensive articles [4, 5] as well as in the excellent contribution by Peters in the first edition of this book [6].

Sonochemistry deals with the chemical applications of ultrasound, i.e. frequencies beyond audible sound (10 Hz–20 kHz). Thus, sonochemistry utilizes high-power ultrasound with frequencies from 20 kHz to around 1–2 MHz, although the range 20–500 kHz is generally employed. In stark contrast with a photochemical process, acoustic radiation is a mechanical, non-quantum energy, which is transformed in part into thermal energy, and it is not absorbed by the molecules. But how does ultrasound induce chemical effects when it does not even alter the rotational or vibrational molecular states? The answer comes from a complex, nonlinear phenomenon called *acoustic cavitation*. An in-depth description lies beyond the scope of this article, but an intuitive understanding is provided by the concept of negative pressure [7]. A pressure wave (e.g., ultrasound) passing through a liquid generates a compressive (positive pressure) disturbance to its front and is followed by a decompressive phase (negative pressure) to its rear. When a sufficiently large negative pressure is applied, the distance between the molecules may exceed the critical molecular distance to hold the liquid intact, and it will break down, thereby creating voids, i.e. cavitational bubbles. Thus, cavitation involves the rapid growth and collapse of micrometer-scale bubbles in a fluid, releasing enough kinetic energy to drive the chemical reaction.

Theoretical calculations and hydrodynamic models predict that high temperatures (4500–5500 K) and rather discrete pressures (~1500–2000 atm) are produced in the cavity and at the interface of bubbles during a few nanoseconds, thereby giving rise to an almost adiabatic process. It is clear that the high-energy microenvironment provided by the cavitational event should indeed be resulting in true

Transition Metals for Organic Synthesis, Vol. 2, 2nd Edition.
Edited by M. Beller and C. Bolm

ISBN: 3-527-30613-7

chemical effects such as homolytic bond breakage, solvent pyrolysis, and generation of excited species, especially on volatile molecules capable of penetrating into the bubble. However, the collapse also causes a series of strong physical effects outside the bubble: shock waves, shear forces, microjets of liquid, and microstreaming [8]. Thus, many effects due to cavitational collapse are related to those found in mechanochemistry and tribochemistry. Accordingly, ultrasonic waves have proved to be an extremely useful means of producing highly active metal powders [4]. Ultrasonic irradiation may sweep reacted species away from the metal surface, exposing a fresh surface on which reaction can take place. This well-known cleaning effect often makes organometallic reactions possible without an inert atmosphere and with undried solvents. Moreover, metal surfaces have an enhanced susceptibility to chemical corrosion under ultrasonic irradiation [9]. The extent of erosion depends on the type of metal and can be correlated with its reactivity [10]. Soft metals (e.g., alkali metals) undergo permanent plastic deformation and can even be finely dispersed. Sonication of soft metals coated by hard oxides (Mg, Al) results in deformations which will fragment the oxide layer. Furthermore, ultrasound also reduces the adhesion of this passivating layer to the underlying metal. Some transition metals (Zn, Cu, Ni) do not undergo plastic deformations, but their surface is also activated by lowering the cohesion of the oxide coating. In addition, sonication reduces the oxygen content of the surface and the size of the particles. With hard metals possessing hard adhesive oxides (e.g. Mo, W), sonication has little or no effect.

Unlike photochemical and radiochemical reactions, for which well-defined mechanisms have been established, a similar rationale for sonochemical reactions is still a challenge. Luche has suggested an empirical classification that has found a wide range of applicability and predictability in organometallic reactions [11]. Thus, in homogeneous reactions in solution (*Type I*), cavitation can generate reactive intermediates such as radicals or radical ions, which may trigger single-electron transfer (SET) pathways in competition with a polar mechanism. Ligand-metal bond cleavage in transition metal complexes will lead to coordinatively unsaturated species or reactive complexes, as well as complete stripping off of ligands to produce amorphous metals. Such reactions will be sensitive to sonication. In heterogeneous ionic reactions, both solid-liquid and liquid-liquid (*Type II*), the ionic pathway will remain unaffected by ultrasound, although its physical effects can enhance both reaction rates and yields. However, a superior mechanical agitation other than sonication can even lead to more satisfactory results [12]. Finally, in heterogeneous radical reactions, or processes that can follow either polar or SET mechanisms (*Type III*), sonication will favor the latter, although the ratio and nature of products constitute indications of the relative importance of the two mechanisms. Most sonochemical studies involving metals belong to this category.

3.6.2
Preparation of Nanosized Materials

The extreme conditions created during the cavitational collapse along with the rapid cooling rate ($>10^9$ K/s), much greater than that obtained by conventional melting techniques (10^5–10^6 K/s), enable the preparation of nanosized amorphous particles. Some sonochemical preparations of these materials and their applications are outlined in the following paragraphs.

3.6.2.1
Metals

In addition to the synthesis of the transition metals produced from the corresponding carbonyls, such as Fe from $Fe(CO)_5$ [13], Ni from $Ni(CO)_4$ [14], and Co from $Co(CO)_3NO$ [15], other metals have also been synthesized sonochemically. Sonication of aqueous Co(II) ions and hydrazine resulted in the formation of disk-shaped cobalt nanoclusters, well suited for magnetic applications, which averaged about 100 nm in width and 15 nm in thickness [16]. Nanoparticles (50 nm diameter) of metallic copper were formed by sonochemical reduction of Cu(II) hydrazine carboxylate in aqueous solution. Sonication under Ar yields a mixture of Cu_2O and metallic copper, but the latter was exclusively obtained under a mixture of hydrogen and Ar [17]. In a further study, nanosized amorphous Cu and Cu_2O were embedded in a polyaniline matrix by irradiating a solution of Cu(II) acetate in aniline or aqueous aniline [18].

Nanoparticles of palladium clusters have been prepared by sonochemical reduction at room temperature of a mixture of $Pd(OAc)_2$ plus a surfactant (NR_4X), in THF or MeOH. The ammonium salt has not only a stabilizing effect, but also acts as a reducing agent because of the decomposition that occurs at the liquid shell surrounding the collapsing cavity, and provides reducing radicals. It is noteworthy that amorphous Pd is obtained in THF and in crystalline form in MeOH [19]. In this solvent and in higher alcohols, sonolysis of tetrachloropalladate(II) leads to Pd nanoclusters in which carbon atoms, formed by complete decomposi-

$$\mathrm{RH} \xrightarrow{))))} \text{Pyrolysis Radicals}$$

$$\mathrm{Pd(II)} + \text{Reducing Radicals} \longrightarrow \mathrm{Pd(0)}$$

$$m\mathrm{Pd(0)} \longrightarrow \mathrm{Pd}_m \text{ (clusterification)}$$

$$\mathrm{Pd}_m + \mathrm{RH} \longrightarrow \mathrm{Pd}_m\text{–RH (adsorption)}$$

$$\mathrm{Pd}_m\text{–RH} \longrightarrow (\mathrm{PdC})_m$$

Scheme 1 Formation of palladium nanoparticles by sonochemical reduction.

tion of the solvent, can diffuse (Scheme 1). The result is an interstitial solid solution of PdC_x [20]. It should also be noted that carbon-activated Pd nanoparticles have been utilized as catalysts in the Heck reaction [21].

3.6.2.2
Metallic Colloids

These are fine dispersions of metal particles of varied size (often with diameters less than 10 nm), which are held in the dispersed state by addition of a stabilizer, generally macromolecules. There have been important contributions dealing with the sonochemical formation of metallic colloids [22–25]. Noble metal nanoparticles (Pd, Ag, Au) are obtained by sonicating aqueous solutions of the corresponding salts in the presence of a surfactant. Mixtures of Au and Pd salts can also be reduced, but not simultaneously. Instead, Au(III) is reduced first, followed by Pd(II), thereby resulting in the formation of a core-shell bimetallic structure [24, 25].

Colloidal solutions of amorphous iron particles, which exhibit high magnetism, have been obtained by sonolysis of $Fe(CO)_5$ in the presence of oleic acid or polyvinylpyrrolidine [26]. Likewise, other sonochemical preparations of colloidal dispersions include metallic iron and Fe_2O_3 stabilized by large molecules [27], and colloidal cobalt in decalin stabilized by oleic acid [28]. In a recent study, coated iron nanoparticles have been formed by sonolysis of $Fe(CO)_5$ in diphenylmethane. This hydrocarbon solvent also decomposes sonochemically to produce a polymer-like solid. Accordingly, the resulting amorphous solid contains iron particles (<10 nm) and substantial amounts of C and H arising from solvent sonolysis. This material is superparamagnetic. A highly magnetic, air-stable powder was obtained by heating the pyrophoric powder at 700 °C under Ar. This substance contains Fe particles, iron carbide (Fe_3C), and small amounts of Fe_2O_3 [29].

3.6.2.3
Alloys and Binary Mixtures

Sonolysis of an equimolar mixture of $Fe(CO)_5$ and $Ni(CO)_4$ leads to a solid with a composition ($Fe_{20}Ni_{80}$) matching the ratio of the vapor pressure of the two carbonyls in the gas phase of the collapsing bubble [30]. A similar study describes the preparation of a nanophase Co/Ni [31]. Remarkably, nanosized amorphous Fe and Co as well as the amorphous alloy $Fe_{20}Ni_{80}$ with oxygen (40 atm) at room temperature, in the absence of any solvent, have been utilized for the oxidation of cycloalkanes with higher conversions (up to 57%) and selectivities than other conventional protocols (Scheme 2). The aerobic oxidation also utilizes isobutyraldehyde as co-reductant and a catalytic amount of acetic acid [32].

M50 steel powder can be obtained by sonochemical decomposition of organometallic precursors such as $Fe(CO)_5$, $V(CO)_6$, $(Et_xC_6H_{6-x})_2Cr$, and $(Et_xC_6H_{6-x})_2Mo$ in decalin [33]. The resulting amorphous powder has a porous microstructure and possesses a higher hardness than the standard M50 steel.

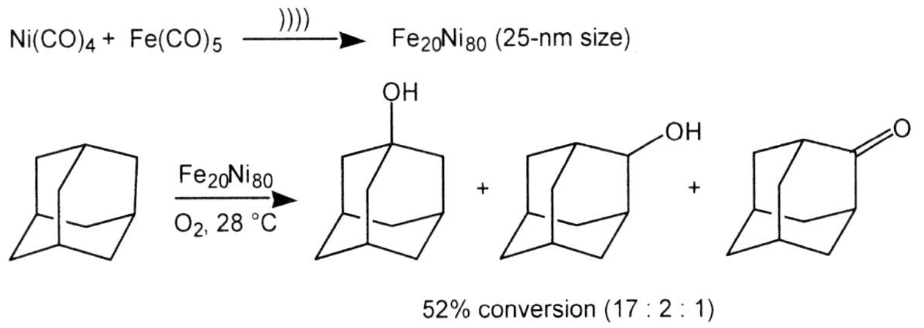

$Ni(CO)_4 + Fe(CO)_5 \xrightarrow{))))} Fe_{20}Ni_{80}$ (25-nm size)

$\xrightarrow[O_2,\ 28\ °C]{Fe_{20}Ni_{80}}$ OH + OH + O

52% conversion (17 : 2 : 1)

Scheme 2 Oxidation of cycloalkanes with a sonochemically prepared nanosized Fe/Ni alloy.

Binary mixtures of metal powders (Ni/Co, Cu/Cr, and Cu/Mo) exposed to high-intensity ultrasound in decane form intermetallic coatings. The main mechanism appears to be interparticle collision caused by the rapid movement of metal particles propelled by shock waves generated at cavitation sites [34].

The sonochemical reaction of $Fe(CO)_5$ and Et_3P, or Me_3P, produces a pyrophoric powder of amorphous iron phosphide (FeP), a low-band gap semiconductor [35]. This result contrasts with the sonolysis of $W(CO)_6$ and Ph_3P, which yields $W(CO)_5(PPh_3)$ as the main product. Both $Fe(CO)_5$ and Et_3P are volatile enough to penetrate into the cavitational bubble, whereas $W(CO)_6$ and Ph_3P are non-volatile solids that experience substitution reactions in the liquid phase.

3.6.2.4 Oxides

Sonication of a decalin solution of $Fe(CO)_5$ in air yields amorphous nanoparticles of Fe_2O_3 [36]. Ultrafine powders of Cr_2O_3 and Mn_2O_3 have been prepared by sonochemical reduction of aqueous solutions containing $(NH_4)_2Cr_2O_7$ and $KMnO_4$, respectively [37]. Ultrasound irradiation of a slurry of $Mo(CO)_6$ in decalin under air produces blue-colored $Mo_2O_5 \cdot 2H_2O$ [38]. Ultrasound also induces profound changes in both morphology and reactivity of polycrystalline MoO_3 [39]. Amorphous WO_3 can be prepared by sonicating $W(CO)_6$ in diphenylmethane under air and Ar. Further thermal treatments under Ar produce dendritic crystals of WO_2 (at 550 °C) and 50-nm rods of a WO_2-WO_3 mixture (at 1000 °C) [40].

α-Cobalt hydroxide can be prepared by sonication of an aqueous solution of $Co(NO_3)_2 \cdot 6H_2O$ and urea under Ar. Thermal decomposition, under air or Ar, produces nanometer-size cobalt oxides (Co_3O_4 and CoO) [41]. Shorter reaction times and lower temperatures, with respect to thermal processes, have been employed in the sonochemical formation of nanosized Ti(IV) oxides [42]. TiO_2 particles, generated by sonolysis of Ti(*i*-PrO)$_4$, were found to be photocatalysts superior to other commercially available samples [43]. Ultrasound also facilitates deposition of iron oxide into mesoporous titania. This catalyst has also been used for cyclohexane oxidation [44].

In an attempt to increase the pore size of zeolites, a layered zeolite precursor has been delaminated under ultrasound. The resulting material, delaminated in much the same way as the layered structure of a clay, contains catalytic sites within thin sheets. This layered structure and a typical zeolite have similar activities for *n*-decane cracking [45].

3.6.2.5
Miscellaneous Derivatives

Irradiation of $Mo(CO)_6$ and sulfur in an aromatic solvent yields amorphous MoS_2 [46], whose catalytic activity for hydrodesulfurization was found to be superior to that of commercially available ReS_2, RuS_2, and MoS_2. Colloidal CdS (< 3 nm in diameter) can be produced by sonication of Cd(II) ions in the presence of a thiol derivative [47]. Nanoparticles of CdS have been sonochemically coated onto submicron particles of silica [48]. Nanoparticles of ZnS have been prepared in a similar way [49]. Amorphous WS_2 has been prepared by ultrasound irradiation of $W(CO)_6$ with sulfur in diphenylmethane under Ar. Nanorods (3–10 nm in thickness) are obtained by further heating at 800 °C [50]. Copper and silver chalcogenides, such as Cu_3Se_2, α-Cu_2Se, and Ag_2Se have also been prepared sonochemically [51]. Ultrasound also enables the formation of nanosized and crystalline Fe, Co, and Ni monoarsenides, starting from an ethanolic solution of the corresponding metal chlorides with As and Zn metals [52].

3.6.3
Formation of Organometallic Reagents

As stated previously [6, 53], transition metal carbonyl complexes, especially those derived from Fe and Cr, are versatile reagents in organic synthesis and can easily be generated in an ultrasonic field. In pursuing such studies, a regioisomeric pair of chiral ferrilactones has been obtained by sonication of an enantiomerically pure allylic epoxide with $Fe_2(CO)_9$. Notably, the regioisomeric ratio under ultrasound changed with respect to conventional conditions. Such ferrilactones were converted into the corresponding η^4 diene complexes (Scheme 3) [54]. In a related study, ultrasound-promoted complexation of 1-azabuta-1,3-dienes with $Fe_2(CO)_9$ results in the formation of (η^4-1-azabuta-1,3-diene)tricarbonyliron complexes. These substances, which exhibit fluxional behavior, are good reagents for tricarbonyliron transfer reactions to 1,3-dienes [55].

Chromium aryl(alkoxy)carbenes react with propargylic alcohols under sonication (Ti horn, 20 kHz) to afford β-lactones in good yields. These Dötz cyclizations can also be conducted under thermal conditions, but ultrasound generally favors shorter reaction times and, moreover, proved to be more effective for obtaining the less heavily substituted β-lactones (Scheme 4) [56].

The direct insertion of metals into a C(aryl)–F bond is often sluggish, giving poor yields. However, perfluorozinc aromatics can now be obtained by direct insertion of Zn into C–F or C–Cl bonds under sonication (ultrasonic bath, 35 kHz)

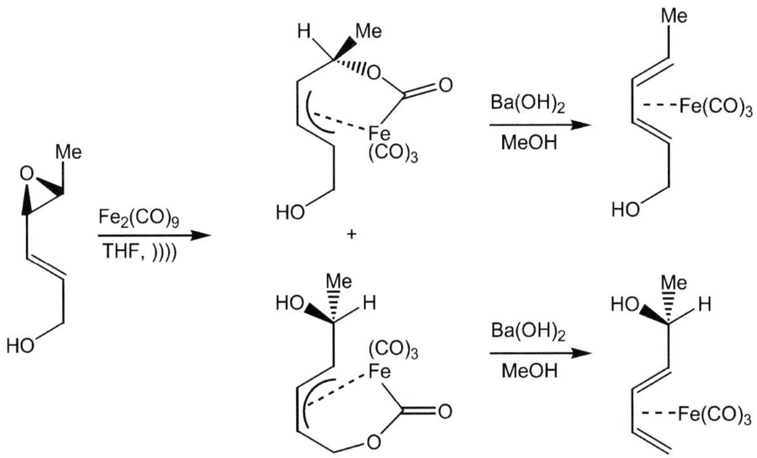

Scheme 3 Sonochemical synthesis of chiral ferrilactones.

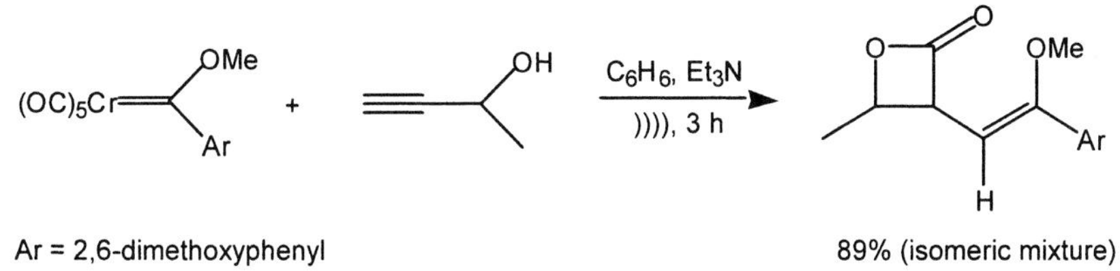

Scheme 4 Sonochemical Dötz cyclizations to β-lactones.

at room temperature [57]. The presence of metal salts (especially $SnCl_2$) as catalysts is required. Without sonication, slow transformations with poor yields take place. ^{19}F NMR provides evidence that organozinc reagents exist in equilibrium between Ar_2Zn and ArZnX species (Scheme 5). On the other hand, organozinc reagents can easily be prepared using active zinc, generated by sonoelectroreduction of zinc salts. This method also enables the preparation of other highly reactive metal powders [58].

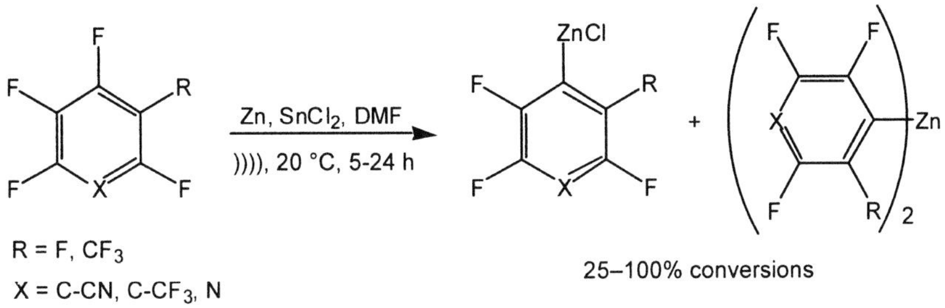

Scheme 5 Ultrasound-assisted preparation of perfluorozinc aromatics.

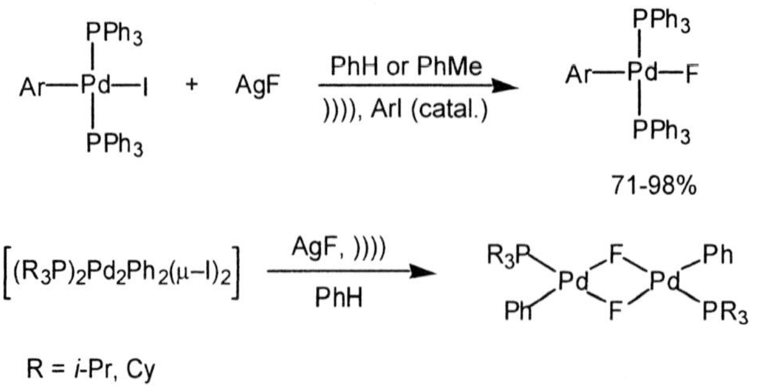

$Ar-Pd(PPh_3)_2-I$ + AgF → (PhH or PhMe;)))), ArI (catal.)) → $Ar-Pd(PPh_3)_2-F$

71-98%

$[(R_3P)_2Pd_2Ph_2(\mu\text{-}I)_2]$ → (AgF,)))); PhH) → $[(R_3P)PdPh(\mu\text{-}F)]_2$

R = *i*-Pr, Cy

Scheme 6 Preparation of palladium fluoride complexes under ultrasound.

Palladium fluoride complexes [59], which are becoming increasingly important, can be prepared by the new ultrasound-promoted I/F ligand exchange reaction of $[(Ph_3P)_2Pd(Ar)I]$ with AgF in benzene or toluene. No I/F exchange occurs without sonication. Performing this reaction in the presence of a catalytic amount of the corresponding aryl iodide (5–10%) was beneficial for the purity of the product. This ultrasonic procedure has been applied to the synthesis of the first dinuclear organopalladium μ-fluorides and their mononuclear analogs stabilized by trialkylphosphine ligands (Scheme 6) [60].

3.6.4
Bond-Forming Reactions in Organic Synthesis

Some synthetically useful organic transformations have recently been improved by sonication. Thus, Reformatsky reactions of α-bromofluoroesters have been investigated under conventional (heating plus stirring) and ultrasonic conditions. Sonication enhances metal reactivity at room temperature and increases chemical yields [61]. Likewise, perfluoroalkylation of sugar aldehydes can be effected by a sonochemical zinc-mediated Barbier-type reaction (Scheme 7) [62]. Stirring or reflux, even at 120 °C, gives no detectable product, while moderate yields are obtained

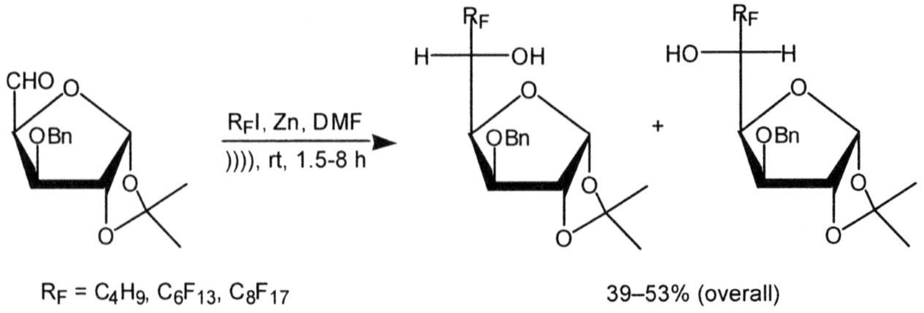

R_FI, Zn, DMF;)))), rt, 1.5-8 h

$R_F = C_4H_9, C_6F_{13}, C_8F_{17}$

39–53% (overall)

Scheme 7 Perfluoroalkylation of chiral aldehydes via a sonochemical Barbier reaction.

after sonication at room temperature. The best yields (up to 53%) are obtained with a 20 kHz probe system (120 W), although an ultrasonic cleaning bath (35 kHz) is much better (35% yield) than stirring.

Heck and Suzuki cross-coupling reactions, two of the most relevant Pd-catalyzed C–C bond-forming reactions have largely been improved by conducting these processes in ionic liquids, such as 1,3-di-*n*-butylimidazolium tetrafluoroborate or bromide, under ultrasonic irradiation. Heck reactions proceed at room temperature within 1.5–3 h in an ultrasonic bath (50 kHz) under Ar, and it is thought that the coupling occurs through the formation of Pd-biscarbene complexes and zero-valent Pd clusters stabilized by the ionic medium (Scheme 8) [63]. In fact, transmission electron microscopy (TEM) studies reveal the formation of 20-nm clusters (each containing dispersed Pd nanoparticles) under sonication. Suzuki reactions of phenylboronic acid and aryl halides have also been carried out at room temperature using an ionic liquid and methanol as cosolvent. Remarkably, no phosphine ligand is required and chlorobenzenes can also be employed [64].

The concept of emulsion electrosynthesis assisted by ultrasound has been applied to the catalytic formation of carbon-carbon bonds [65]. Here the voltammetry of aqueous vitamin B_{12} (cyanocobalamine) solutions at an electrode is modified with microscopic droplets of an organic reactant generated by applying power ultrasound. In some cases, photochemical irradiation is also required, thus representing a complex system under triple activation (electrolysis, ultrasound, and light). This protocol has been applied to bromoalkanes and activated alkenes to afford products in moderate yields (up to 50%). Thus, electroreduction of vitamin B_{12} [Co(III)L] yields the nucleophilic species Co(I)L, which reacts with the bromoalkane via a SET process. Further photolysis gives rise to an alkyl radical that adds to an α,β-unsaturated carbonyl compound (Scheme 9).

Ferrocenylalkyl amines with C_2-symmetry have been obtained via ultrasound-assisted amination of 1-ferrocenylalkyl acetates. Sonochemical activation affords greater yields than the corresponding silent reactions, although a complete diastereoselection was observed in both cases [66].

Indian authors have investigated the ultrasonically-induced transition metal-catalyzed aziridination of olefins. Bromamine-T was found to be the ideal source of nitrene, and Cu halides the best catalysts. No reaction took place at room temperature without sonication, and, interestingly, ultrasound (cleaning bath, 36 kHz, 25 °C) gave *trans*-aziridine selectively, while microwave irradiation (2.45 GHz, 30% power intensity) furnished *cis/trans* mixtures (Scheme 10) [67].

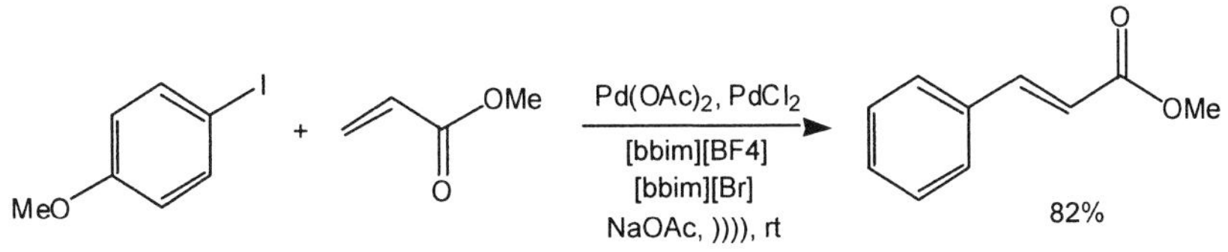

MeO–C₆H₄–I + CH₂=CH–C(O)OMe → [Pd(OAc)$_2$, PdCl$_2$ / [bbim][BF4] / [bbim][Br] / NaOAc,)))), rt] → PhCH=CH–C(O)OMe, 82%

Scheme 8 Sonochemical Heck reactions in ionic liquids.

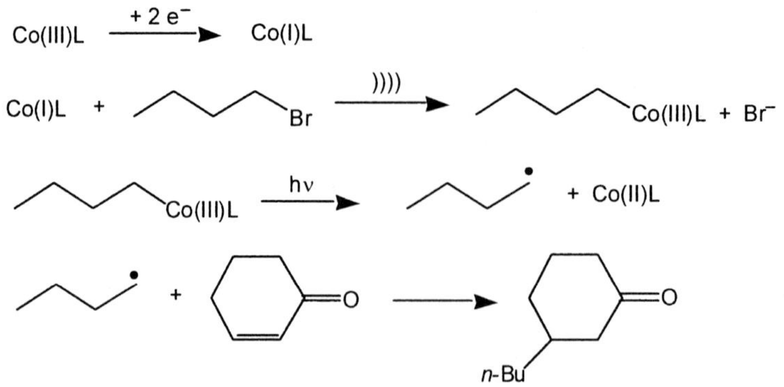

Scheme 9 Carbon-carbon bond formation via electrosynthesis assisted by ultrasound.

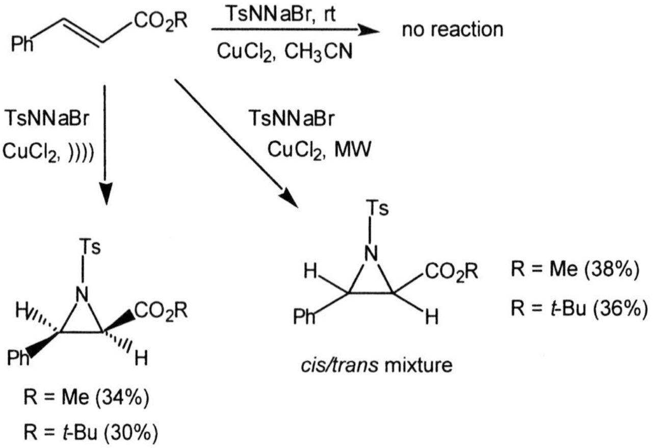

Scheme 10 Aziridination of alkenes with Bromamine-T.

3.6.5
Oxidations and Reductions

A facile and rapid oxidation of olefins to 1,2-*cis*-diols can be carried out via sonochemical activation (sonic probe) of powdered $KMnO_4$ in aqueous media. 4-Substituted styrenes are dihydroxylated in less than 20 min, while silent reactions proceed slowly (>24 h). The authors suggest that ultrasound accelerates this oxidation by decomposing the cyclic manganate diester intermediate. Under these conditions, diphenylacetylene is transformed into benzil, and terminal alkynes give mainly carboxylic acids [68].

The heterogeneous oxidation of benzyl alcohols with $KMnO_4/CuSO_4$ in CH_2Cl_2 has also been investigated under ultrasound [69]. Irradiation accelerates this transformation and minimizes overoxidation to carboxylic acid, especially when reactions are conducted under Ar. The same system has been applied to the oxidation of alkyl arenes to their corresponding carbonyl compounds.

In addition to the varied usefulness of sonochemically prepared Pd and Pt colloidal particles and catalysts (see above), the growing interest in enantioselective hydrogenations catalyzed by modified Pt catalysts has been maintained [70]. In general, rates of catalytic reactions over platinum surfaces can be greatly enhanced by ultrasound [71].

The sonochemical hydrogenation of ethyl pyruvate catalyzed by peptide-modified Pt colloids produces (*R*)-ethyl lactate as the major enantiomer with rather modest enantiomeric excesses (76–78%) [72]. The sonochemical pretreatment of a catalyst consisting of Pt/Al_2O_3 and cinchonidine and its further application to ethyl pyruvate hydrogenation in acetic acid resulted in an enhanced enantioselectivity (97% *ee*). Ultrasound also enables easy catalyst recycling. As expected, electron microscopy provided evidence of the changes in size and morphology of metal particles caused by sonication [73]. In addition to ethyl pyruvate, Pt/cinchona-catalyzed sonochemical hydrogenations have equally been applied to ketones, trifluoromethyl ketones, and unsaturated carboxylic acids, although with modest results [70, 74]. The highest enantioselection (up to 97% *ee*) has been obtained with other α-ketoesters [75].

Recent advances in sonochemical hydrogenations catalyzed by Raney nickel include an ultrasound-assisted desulfurization reaction *en route* to the marine natural product (+)-ptilocaulin [76] and extensions of the Raney-Ni/tartaric acid system, pioneered by Tai and co-workers, to varied substrates [77]. An almost complete enantiodifferentiation (up to 98% *ee*) was observed with methyl 3-cyclopropyl-3-oxopropanoate (Scheme 11) [77a]. It is believed that the enhanced enantioselectivity is caused by the ultrasonic removal of the non-enantiodifferentiating Al sites from the Ni surface. Analytical methods reveal a purer Ni surface, while the supernatant contains high Al/Ni ratios (ca. 70/30).

Boudjouk et al. have described nickel-catalyzed 1,4-additions of phenylsilane ($PhSiH_3$) to α,β-unsaturated ketones and nitriles. After hydrolysis, selective C=C reduction was achieved in high yields [78]. Activated nickel was obtained by soni-

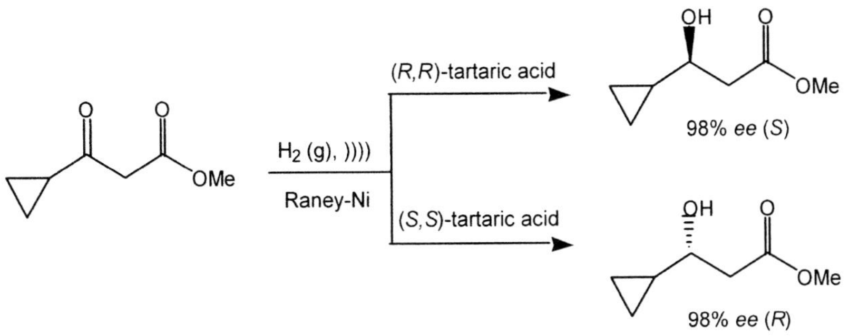

Scheme 11 Enantioselective Raney-Ni hydrogenations assisted by ultrasound.

cating Ni(II) iodide and lithium in THF solution. Commercial or unactivated nickel were ineffective.

As mentioned before [65], ultrasound emitted by a titanium horn situated in an electrochemical cell allows the formation of emulsions without stabilizing agents. Under these conditions, it is possible to carry out electrochemical reductions in aqueous media of organics hitherto thought to be essentially insoluble. Thus, a clean hydrogenation of the C-C double/triple bonds of diethyl maleate, diethyl fumarate, and diethyl acetylene dicarboxylate was observed to give isolated compounds in 50–70% yields. Reductions are commonly performed on glassy carbon electrodes, but clean reductions are also observed at a gold electrode having a large surface area [79].

3.6.6 Concluding Remarks

This account bears witness to the growing interest in sonochemistry over the last decade. Numerous results suggest the existence of species and intermediates that cannot be attained under conventional conditions. We have just started to obtain accurate information on the energy dissipation and reaction rates inside isolated microbubbles [80]. A better understanding of the kinetics and mechanisms of cavitation-based reactions could ensue. Transition metals still represent a rich and varied scenario for sonochemical reactions. Needless to say, there have also been significant and recent contributions involving alkali and alkaline-earth metals, main group elements, and lanthanides and actinides. Ultrasound and the Periodic Table will doubtless be a fruitful and long-standing research marriage.

References

1 T. J. Mason, J. P. Lorimer, *Applied Sonochemistry: Uses of Power Ultrasound in Chemistry and Processing*, Wiley-VCH, Weinheim, **2002**.

2 *Synthetic Organic Sonochemistry* (Ed.: J.-L. Luche), Plenum Press, New York, **1998**.

3 For comprehensive and updated contributions in sonochemistry, see the series *Advances in Sonochemistry* (Ed.: T. J. Mason), JAI Press Inc., Greenwich, CT, **1990**, Vol. 1; **1991**, Vol. 2; **1993**, Vol. 3; **1996**, Vol. 4; **1999**, Vol. 5.

4 J.-L. Luche, P. Cintas in *Active Metals: Preparation, Characterization, Applications* (Ed.: A. Fürstner), VCH, Weinheim, **1996**, pp. 133–190 and references therein.

5 P. Cintas, J.-L. Luche in Ref. [2], pp. 167–234 and references cited therein.

6 D. Peters in *Transition Metals for Organic Synthesis, Vol. 2* (Eds.: M. Beller, C. Bolm), Wiley-VCH, Weinheim, **1998**, pp. 420–435.

7 A. Imre, W. A. van Hook, *Chem. Soc. Rev.* **1998**, *27*, 117–123.

8 T. J. Mason, *Practical Sonochemistry. User's Guide to Applications in Chemistry and Chemical Engineering*, Ellis Horwood Ltd., Chichester, **1991**, pp. 18–30.

9 W. J. Tomlinson, *Adv. Sonochem.* **1990**, *1*, 173–195.

10 (a) B. Pugin, A. T. Turner, *Adv. Sonochem.* **1990**, *1*, 81–118; (b) K. S. Suslick, S. J. Doktycz, *Adv. Sonochem.* **1990**, *1*, 197–230; (c) O. V. Abramov, *Adv. Sonochem.* **1991**, *2*, 135–186.

11 J.-L. Luche, *Adv. Sonochem.* **1993**, *3*, 85–124.

12 Y. Kegelaers, O. Eulaerts, J. Reisse, N. Segebarth, *Eur. J. Org. Chem.* **2001**, 3683–3688.

13 K. S. Suslick, S.-B. Choe, A. A. Cichowlas, M. W. Grinstaff, *Nature* **1991**, *353*, 414–416.

14 Yu. Koltypin, G. Katabi, X. Cao, R. Prozorov, A. Gedanken, *J. Non-Crystalline Solids* **1996**, *201*, 159–162.

15 K. S. Suslick, M. Fang, T. Hyeon, A. A. Cichowlas in *Molecularly Designed Nanostructured Materials, MRS Symp. Proc. 351* (Ed.: K. E. Gonsalves), *Matl. Res. Soc., Pittsburgh*, **1994**, pp. 443–448.

16 C. P. Gibson, K. J. Putzer, *Science* **1995**, *267*, 1338–1340.

17 N. A. Dhas, C. P. Raj, A. Gedanken, *Chem. Mater.* **1998**, *10*, 1446–1452.

18 R. V. Kumar, Y. Mastai, Y. Diamant, A. Gedanken, *J. Mater. Chem.* **2001**, *11*, 1209–1213.

19 N. A. Dhas, A. Gedanken, *J. Mater. Chem.* **1998**, *8*, 445–450.

20 K. Okitsu, Y. Mizukoshi, H. Bandow, T. A. Yamamoto, Y. Nagata, Y. Maeda, *J. Phys. Chem. B* **1997**, *101*, 5470–5472.

21 N. A. Dhas, H. Cohen, A. Gedanken, *J. Phys. Chem. B* **1997**, *101*, 6834–6838.

22 F. Grieser, *Stud. Surf. Sci. Catal.* **1997**, *103*, 57–77.

23 (a) K. Okitsu, H. Bandow, Y. Maeda, Y. Nagata, *Chem. Mater.* **1996**, *8*, 315–317; (b) Y. Nagata, Y. Maeda, *Rec. Res. Devel. Pure Appl. Chem.* **1997**, *1*, 73–83.

24 Y. Mizukoshi, K. Okitsu, Y. Maeda, T. A. Yamamoto, R. Oshima, Y. Nagata, *J. Phys. Chem. B* **1997**, *101*, 7033–7037.

25 Y. Mizukoshi, T. Fujimoto, Y. Nagata, R. Oshima, Y. Maeda, *J. Phys. Chem. B* **2000**, *104*, 6028–6032.

26 K. S. Suslick, M. M. Fang, T. Hyeon, *J. Am. Chem. Soc.* **1996**, *118*, 11960–11961.

27 K. V. P. M. Shafi, S. Wizel, R. Prozorov, A. Gedanken, *Thin Solid Films* **1998**, *318*, 38–41.

28 K. V. P. M. Shafi, A. Gedanken, R. Prozorov, *Adv. Mater.* **1998**, *10*, 590–592.

29 S. I. Nikitenko, Yu. Koltypin, O. Palchik, I. Felner, X. N. Xu, A. Gedanken, *Angew. Chem.* **2001**, *113*, 4579–4581; *Angew. Chem. Int. Ed.* **2001**, *40*, 4447–4449.

30 K. V. P. M. Shafi, A. Gedanken, R. B. Goldfarb, I. Felner, *J. Appl. Phys.* **1997**, *81*, 6901–6905.

31 K. V. P. M. Shafi, A. Gedanken, R. Prozorov, *J. Mater. Chem.* **1998**, *8*, 769–773.

32 V. Kesavan, P. S. Sivanand, S. Chandrasekaran, Yu. Koltypin, A. Gedanken, *Angew. Chem.* **1999**, *111*, 3729–3730; *Angew. Chem. Int. Ed.* **1999**, *38*, 3521–3523.

33 K. E. Gonsalves, S. P. Rangarajan, A. García-Ruiz, C. C. Law, *J. Mater. Sci. Lett.* **1996**, *15*, 1261–1263.

34 D. J. Casadonte, Jr., J. D. Sweet, *J. Tribol.* **1998**, *120*, 641–643.

35 J. D. Sweet, D. J. Casadonte, Jr., *Ultrasonics Sonochem.* **2001**, *8*, 97–101.

36 X. Cao, R. Prozorov, Yu. Koltypin, G. Kataby, I. Felner, A. Gedanken, *J. Mater. Res.* **1997**, *12*, 402–406.

37 N. A. Dhas, Yu. Koltypin, A. Gedanken, *Chem. Mater.* **1997**, *9*, 3159–3163.

38 N. A. Dhas, A. Gedanken, *J. Phys. Chem. B* **1997**, *101*, 9495–9503.

39 P. Jeevanandam, Y. Diamant, M. Motiei, A. Gedanken, *Phys. Chem. Chem. Phys.* **2001**, *3*, 4107–4112.

40 Yu. Koltypin, S. I. Nikitenko, A. Gedanken, *J. Mater. Chem.* **2002**, *12*, 1107–1110.

41 P. Jeevanandam, Yu. Koltypin, A. Gedanken, Y. Mastai, *J. Mater. Chem.* **2000**, *10*, 511–514.

42 W. Huang, X. Tang, Y. Wang, Yu. Koltypin, A. Gedanken, *Chem. Commun.* **2000**, 1415–1416.

43 J. C. Yu, J. Yu, W. Ho, L. Zhang, *Chem. Commun.* **2001**, 1942–1943.

44 N. Perkas, Y. Wang, Yu. Koltypin, A. Gedanken, S. Chandrasekaran, *Chem. Commun.* **2001**, 988–989.

45 A. Corma, V. Fornes, S. B. Pergher, T. L. M. Maesen, J. G. Buglass, *Nature* **1998**, *396*, 353–356.

46 M.M. Mdleleni, T. Hyeon, K.S. Suslick, *J. Am. Chem. Soc.* **1998**, *120*, 6189–6190.

47 J.Z. Sostaric, R.A. Caruso-Hobson, P. Mulvaney, F. Grieser, *J. Chem. Soc., Faraday Trans.* **1997**, 1791–1795.

48 N.A. Dhas, A. Gedanken, *Appl. Phys. Lett.* **1998**, *72*, 2514–2516.

49 N.A. Dhas, A. Zaban, A. Gedanken, *Chem. Mater.* **1999**, *11*, 806–813.

50 S.I. Nikitenko, Yu. Koltypin, Y. Mastai, M. Koltypin, A. Gedanken, *J. Mater. Chem.* **2002**, *12*, 1450–1452.

51 T. Ohtani, T. Nonaka, M. Araki, *J. Solid State Chem.* **1998**, *138*, 131–134.

52 J. Lu, Y. Xie, X. Jiang, W. He, G. Du, *J. Mater. Chem.* **2001**, *11*, 3281–3284.

53 C.M.R. Low, *Ultrasonics Sonochem.* **1995**, *2*, S153–S163.

54 C.E. Anson, G. Dave, G.R. Stephenson, *Tetrahedron* **2000**, *56*, 2273–2281.

55 H.-J. Knölker, G. Baum, N. Foitzik, H. Goesmann, P. Gonser, P.G. Jones, H. Röttele, *Eur. J. Inorg. Chem.* **1998**, 993–1007.

56 (a) J.J. Caldwell, J.P.A. Harrity, N.M. Heron, W.J. Kerr, S. McKendry, D. Middlemiss, *Tetrahedron Lett.* **1999**, *40*, 3481–3484; (b) J.J. Caldwell, W.J. Kerr, S. Mckendry, *Tetrahedron Lett.* **1999**, *40*, 3485–3486.

57 A.O. Miller, V.I. Krasnov, D. Peters, V. E. Platonov, R. Miethchen, *Tetrahedron Lett.* **2000**, *41*, 3817–3819.

58 A. Durant, J.L. Delplancke, V. Libert, J. Reisse, *Eur. J. Org. Chem.* **1999**, 2845–2852.

59 V.V. Grushin, *Chem. Eur. J.* **2002**, *8*, 1006–1014.

60 V.V. Grushin, W.J. Marshall, *Angew. Chem.* **2002**, *114*, 4656–4659; *Angew. Chem. Int. Ed.* **2002**, *41*, 4476–4479.

61 P.L. Coe, M. Löhr, C. Rochin, *J. Chem. Soc., Perkin Trans. 1* **1998**, 2803–2812.

62 D. Peters, C. Zur, R. Miethchen, *Synthesis* **1998**, 1033–1038.

63 R.R. Deshmukh, R. Rajagopal, K.V. Srinivasan, *Chem. Commun.* **2001**, 1544–1545.

64 R. Rajagopal, D.V. Jarikote, K.V. Srinivasan, *Chem. Commun.* **2002**, 616–617.

65 T.J. Davies, C.E. Banks, B. Nuthakki, J. F. Rusling, R.R. France, J.D. Wadhawan, R.G. Compton, *Green Chem.* **2002**, *4*, 570–577.

66 M. Woltersdorf, R. Kranich, H.-G. Schmalz, *Tetrahedron* **1997**, *53*, 7219–7230.

67 B.M. Chanda, R. Vyas, A.V. Bedekar, *J. Org. Chem.* **2001**, *66*, 30–34.

68 R.S. Varma, K.P. Naicker, *Tetrahedron Lett.* **1998**, *39*, 7463–7466.

69 M. Meciarova, S. Toma, A. Heribanová, *Tetrahedron* **2000**, *56*, 8561–8566.

70 For a review on asymmetric sonochemical reactions: B. Török, K. Balázsik, K. Felföldi, M. Bartók, *Ultrasonics Sonochem.* **2001**, *8*, 191–200 and references therein.

71 S. Kelling, D.A. King, *Platinum Met. Rev.* **1998**, *42*, 8–10.

72 H. Bönnemann, G.A. Braun, *Chem. Eur. J.* **1997**, *3*, 1200–1202.

73 B. Török, K. Felföldi, G. Szakonyi, K. Balázsik, M. Bartók, *Catal. Lett.* **1998**, *52*, 81–84.

74 K. Balázsik, B. Török, K. Felföldi, M. Bartók, *Ultrasonics Sonochem.* **1999**, *5*, 149–155.

75 B. Török, K. Balázsik, G. Szöllösi, K. Felföldi, M. Bartók, *Chirality* **1999**, *11*, 470–474.

76 K. Schellhaas, H.-G. Schmalz, J.W. Batts, *Chem. Eur. J.* **1998**, *4*, 57–66.

77 (a) S. Nakagawa, T. Sugimura, A. Tai, *Chem. Lett.* **1997**, 859–860; (b) S. Nakagawa, T. Sugimura, A. Tai, *Chem. Lett.* **1998**, 1257–1258.

78 P. Boudjouk, S.-B. Choi, B.J. Hauck, A.B. Rajkumar, *Tetrahedron Lett.* **1998**, *39*, 3951–3952.

79 F. Marken, R.G. Compton, S.D. Bull, S.G. Davies, *Chem. Commun.* **1997**, 995–996.

80 Y.T. Didenko, K.S. Suslick, *Nature* **2002**, *418*, 394–397.

3.7
Applications of Microwaves

J. Lee and D.J. Hlasta

3.7.1
Introduction

The underlying mechanism in microwave-assisted organic reactions is microwave-generated heat [1]. The heat is generated by two major mechanisms. The first is a dipolar polarization mechanism in which polar molecules are heated according to their dielectric constants. One other component is solvents capabilities to absorb microwave energy and to transform it to heat. The second mechanism is a conduction mechanism in which solvents with ions heat better because of the increased collision rate.

One advantage of microwave-assisted organic synthesis is the access to fast reactions under carefully controlled reaction conditions. Higher temperatures can be applied to reactions, since the vessels are securely enclosed. These vessels allow one to obtain a pressure higher than atmospheric pressure. When higher reaction temperatures are used in microwave reaction conditions, the reactions are complete in minutes and not hours. Furthermore, the reaction temperature is reached in a matter of seconds, while typical oil baths require minutes to reach the desired temperature.

With the aid of current instrumentation, which is provided by vendors such as Personal Chemistry, CEM, and Milestone, the reactions are carried out in a safe environment [2]. In the past, microwave-assisted synthesis was performed in a typical kitchen microwave. The use of these microwaves was dangerous because of the possibility of explosion of the reaction vessels when a high temperature was desired. The kitchen microwaves produce non-homogeneous microwave field causing localized hot spots. These hot spots can promote explosions under an unsafe environment. The instruments provided by current vendors, such as Personal Chemistry, generate a homogeneous microwave field. In this way, a safer and controlled thermal reaction can be conducted. Additionally, the microwave chambers are enclosed in a safe shield, so any rupturing of vessels is contained. Some of the instruments are designed for automation, such that multiple reactions can be performed either sequentially or in parallel without any supervision. With automation, a series of reactions varying reagent stoichiometry, concentration, reaction

Transition Metals for Organic Synthesis, Vol. 2, 2nd Edition.
Edited by M. Beller and C. Bolm

ISBN: 3-527-30613-7

time, and temperature can facilitate the identification of the optimum reaction conditions.

One other advantage is that the reaction temperature and pressure are well controlled, leading to experiments that are highly reproducible. One area of needed improvement is the scale up of the reaction. Some vendors have worked on this issue. Personal Chemistry provides a contract scale-up service that uses a continuous flow reaction system [2].

Microwave-assisted reactions have found an ever-widening use in transition metal catalysis reactions. It is unclear whether microwave energy has any direct effect on metal surfaces, facilitating this type of reaction. Often, cleaner reactions are obtained in transition metal-catalyzed reactions with microwave heating. The present chapter reviews only reactions that use transition metals either as catalysts or reactants. Larhed and co-workers previously reviewed microwave-accelerated homogeneous catalysis [3].

3.7.2 C–C Bond Formation/Cross Coupling

3.7.2.1 Heck Coupling

The Heck coupling was the first microwave-assisted reaction that was directly compared with the classical thermal version of the same reaction [4]. Selected examples using palladium acetate as a catalyst are shown below (Eq. 1). The yields obtained from the microwave-assisted reaction were only slightly higher than those obtained from the thermal reaction. The most significant difference was in the time required to afford the desired product. The microwave-assisted Heck coupling took only a few minutes to obtain yields comparable with those obtained after several hours of thermal coupling conditions.

$$\text{R-C}_6\text{H}_4\text{-X} + \text{CH}_2\text{=CH-R}_2 \xrightarrow[\mu\text{W}]{\text{Pd(OAc)}_2,\ \text{DMF}} \text{R-C}_6\text{H}_4\text{-CH=CH-R}_2 \qquad (1)$$

Tab. 1 Selected results for the Heck coupling of aryl halides and triflates

R	*X*	*R_2*	*Microwave heating*		*Thermal (150–170 °C)*	
			Condition	*Yield (%)*	*Condition*	*Yield (%)*
MeO	I	CO_2Me	3.8 min, 60 W	70	300 min	68
NC	B	CO_2Me	3.8 min, 60 W	94	120 min	70
H	I	Ph	2.8 min, 90 W	87	120 min	75
Br	I	Ph	4.8 min, 60 W	63	1020 min	64
t-Bu	OTf	$(CH_2)_3CH_3$	2.8 min, 55 W	77	n.r.	n.r.

n.r.: not reported

Using the same palladium catalyst, Díaz-Ortiza and co-workers reported a Heck reaction under solvent-free conditions. The classical thermal condition without solvent produced none of the desired product (Eq. 2) [5]. However, the microwave-assisted Heck coupling provided the product with the reasonable yield of 76%.

$Pd(OAc)_2$, PPh_3, Et_3N

Classical Heating @150°C 22min No Product (2)

Microwave Heating @150°C 22 min 76% yield

In another application of the Heck coupling reaction to prepare 3-aryl-1,2-cyclohexanediones, a 10-min microwave-assisted reaction decreased the decomposition of the reactants, thus increasing the yield of the reaction (Eq. 3) [6].

OH + Ar–Br → $Pd(OAc)_2$, PPh_3, DIEA; 10 min, 40-50W, μW → OH, Ar (3)

66-69% yield

Hallberg and co-workers further took advantage of shorter reaction times to perform sequential arylations [7] and enantioselective reactions [8]. Ionic liquids have also been used to facilitate the Heck coupling [9]. Ionic liquids are particularly useful since microwaves can rapidly heat them without any significant pressure increase due to the conductive mechanism and low vapor pressure.

3.7.2.2 Stille Coupling

Few results have been reported on Stille coupling because of the popularity of other cross-coupling reactions such as the Suzuki coupling (Eq. 4) [4, 10]. The recent application of microwave-assisted Stille coupling with fluorous tin reactants promoted the facile synthesis of biaryl compounds as shown below [11, 12].

$$(C_6F_{13}CH_2CH_2)_3Sn\text{–}C_6H_5 + X\text{–}C_6H_4R_1 \xrightarrow[\mu W,\ 2\ \text{min},\ 60W]{Pd(OAc)_2,\ P(p\text{-tol})_3,\ LiCl,\ DMF} C_6H_5\text{–}C_6H_4R_1 \quad \text{47-92\% yield} \tag{4}$$

Skoda-Földes and co-workers were able to apply the microwave-assisted Stille coupling to prepare a diene that was subsequently used to form a six-membered ring [13].

3.7.2.3
Suzuki Coupling

One of the first microwave-assisted Suzuki coupling reactions was performed with KF-Al_2O_3 and without using solvent. In these reactions, biaryl compounds were obtained in excellent yield as shown below (Eq. 5) [10].

$$C_6H_5\text{–}I + (HO)_2B\text{–}C_6H_4R_1 \xrightarrow[\mu W,\ 2\text{-}15\ \text{min},\ 60\text{-}90W]{KF\text{-}Al_2O_3,\ Pd(OAc)_2} C_6H_5\text{–}C_6H_4R_1 \quad \text{58-98\% yield} \tag{5}$$

The microwave-assisted Suzuki couplings have been further applied to solid-phase reactions using palladium tetrakis(triphenylphosphine) as a catalyst. Hallberg and Schootens were able to prepare biaryl compounds in excellent yields [14, 15]. Additionally, Hallberg and co-workers applied the microwave-assisted Suzuki coupling to prepare a side chain of a protease inhibitor [16].

3.7.2.4
Sonogashira Coupling

Using aryl iodides, Kabalka and co-workers prepared aryl acetylenes using the solventless Sonogashira coupling reaction on alumina as shown below (Eq. 6) [17]. Aryl bromides and aryl chlorides did not produce the desired product. In these instances, the starting materials were recovered.

$$X\text{–}C_6H_4\text{–}I + R\text{–}C{\equiv}C\text{–}H \xrightarrow[\mu W,\ 2.5\ \text{min},\ 1000W]{Pd\text{-}CuI\text{-}PPh_3/KF\text{-}Al_2O_3} X\text{–}C_6H_4\text{–}C{\equiv}C\text{–}R \quad \text{67-94\% yield} \tag{6}$$

On the other hand, Erdélyi and Gcgoll were able to prepare aryl acetylenes from either aryl bromides or chlorides using a homogeneous reaction solution (Eq. 7)

[18]. The reaction involving either aryl bromides or aryl chlorides gave similar results to those obtained with aryl iodides.

$$\text{X-C}_6\text{H}_4\text{-Y} + \text{H-C}\equiv\text{C-Si(CH}_3)_3 \xrightarrow[\mu\text{W, 5-25 min}]{\text{Pd(PPh}_3)_2\text{Cl}_2\text{, CuI, Et}_2\text{NH, DMF}} \text{X-C}_6\text{H}_4\text{-C}\equiv\text{C-Si(CH}_3)_3 \quad (7)$$

Y = halogen; 81-99% yield

3.7.2.5
Olefin Metathesis

Controlled experiments were conducted comparing thermal heating with microwave-assisted olefin metathesis as shown below (Eq. 8) [19]. In all cases, the microwave-assisted reaction provided a shorter reaction time than the thermal reactions using the same substrate-to-solvent ratio. The microwave-assisted reaction with a **3-Ru** catalyst provided a higher yield than the reaction with a **2-Ru** catalyst.

(8)

3 mol % $Cl_2Ru(=CHPh)(P(o\text{-}Hex)_3)(P(c\text{-}Hex)_3)$ **2-Ru** or $Cl_2Ru(=CHPh)(\text{Mes-N}\frown\text{N-Mes})(P(c\text{-}Hex)_3)$ **3-Ru**

µW, 110 Watt, 15 - 60 sec

55 - 100% yield

3.7.2.6
Pauson-Khand Reaction

Various cyclopentenones were prepared from cobalt-complexed acetylenes by Evans and co-workers (Eq. 9) [20]. The authors conjectured that the rate enhancement of the Pauson-Khand reaction could be either due to the effect of the microwave energy or from “super-heating” of the media.

(9)

3.7.3
C-Heteroatom Bond Formation

3.7.3.1
Buchwald-Hartwig Reaction

The palladium-catalyzed amination of aryl bromides with various amines including an imidazole was performed using the microwave-assisted reaction (Eq. 10) [21, 22]. In these reactions, the best results were obtained with either binap [2,2′-bis(diphenylphosphino)-1,1′-binaphthyl], or dppf [1,1′-bis(diphenylphosphino) ferrocene] as ligands. Other ligands such as dppp [1,3-bis(diphenylphosphino)propane] and the monodentate ligands PPh_3 or $P(o\text{-tolyl})_3$ resulted in lower yields of the products. In all conditions, shorter reaction times of 4–16 min provided the desired products.

(10)

3.7.3.2
Aziridination of Olefins

Chanda and co-workers reported the microwave-assisted aziridination of olefins using a variety of transition metal catalysts (Eq. 11) [23]. Of the catalysts examined, copper halides gave the better yields.

(11)

Tab. 2 Aziridination of olefins with various metal halides

Entry	*Metal halide*	*Yield (%)*
1	$CuCl_2$	70
2	$NiCl_2$	60
3	$CoCl_2$	56
4	$FeCl_3$	63
5	$MnCl_2$	54
6	$SrCl_2$	40
7	$CuBr_2$	88
8	$Rh_2(OAc)_4$	30
9	No catalyst	No reaction

3.7.3.3
Other C-Heteroatom Bond Formations

Using Raney nickel, Jiang and co-workers prepared various secondary anilines from primary anilines using microwave-assisted reaction conditions as shown below (Eq. 12) [24]. The yields of the reaction ranged from 19 to 91%.

ROH
Raney Ni
μW,
30 - 60 min,
560 - 630 watts
19 - 91% yield
(12)

Combs and co-workers first demonstrated polymer-supported C-heteroatom bond formation using copper acetate as shown below (Eq. 13) [25]. The same reaction conditions were used to prepare various *N*-arylated heterocycles.

1. $4\text{-}CH_3\text{-}PhB(OH)_2$
$Cu(OAc)_2$, Mol. Sieve
Pyridine-NMP (1:1)
μW,
3 X (10 sec)
2. TFA/DCM
73 - 97% yield
(13)

Using dimethyltitanocene, Cp_2TiMe_2, Bytschkov and Doye prepared various secondary amines from alkynes and primary amines as shown below (Eq. 14) [26]. The completion of most reactions was observed in less than one tenth of the time required for reactions conventionally run in a thermal bath.

$$R_1{-}{\equiv}{-}R_2 \quad + \quad R_3{-}NH_2 \xrightarrow[\text{2. Reduction}]{\text{1. 3 mol \% }Cp_2TiMe_2\text{, various settings, }\mu W} \text{(HN}{-}R_3\text{ adduct of }R_1, R_2\text{)} \quad \text{2 - 93\% yield} \tag{14}$$

3.7.4 Synthesis of Heterocycles

3.7.4.1 Biginelli Multicomponent Condensation

The microwave-assisted library synthesis of dihydropyrimidines was demonstrated using various transition metal catalysts as shown below (Eq. 15) [27]. A variety of Lewis acids, such as $Yb(OTf)_3$, $InCl_3$, $FeCl_3$, and $LaCl_3$ were screened for the cyclization. Among the catalysts tested, $Yb(OTf)_3$ was the most effective, using the AcOH/EtOH solvent system. The cyclization produced an average yield of 52%, most products being >90% pure.

$$\text{(15)}$$

HOAc/EtOH (3:1); 10 mol % $Yb(OTf)_3$; 10 min, 120°C; μW

Ave. Yield of 52%
>90% Purity

3.7.4.2 2-Cyclobenzothiazoles via *N*-Arylimino-1,2,3-dithiazoles

2-Cyanobenzothiazoles were prepared *via* *N*-arylimino-1,2,3-dithiazoles using copper iodide (Eq. 16) [28–30]. The yields using the microwave-assisted synthesis were similar to those using conventional heating.

$$\text{(16)}$$

CuI; Pyridine

Tab. 3 Selected results for the synthesis of 2-cyclobenzothiazoles

R	*Microwave heating*			*Conventional heating*	
	Time (min)	*Watts*	*Yield (%)*	*Time (min)*	*Yield (%)*
H	10	300	69	45	67
4-F	10	300	87	60	80
4-CH_3	12	300	85	60	84
4-NO_2	10	300	65	45	68
5-CF_3	10	300	82	45	79
4,5-diF	10	300	61	45	58

3.7.4.3
Synthesis of Acridines

Using $ZnCl_2$, a variety of acridines were prepared from diarylamines and carboxylic acids as shown below (Eq. 17) [31]. Aryl carboxylic acids gave relatively lower yields than alkyl carboxylic acids.

X–(diarylamine, N–H)–Y + RCO_2H —μW, $ZnCl_2$, 5–22 min→ X–(acridine, 9-R)–Y (17)

0 to 87% yields

3.7.4.4
Dötz Benzannulation Process

Using phenyl acetylenes and a phenylchromium carbene complex, highly substituted benzenoids were prepared using microwave-assisted reaction conditions (Eq. 18) [32]. The yields of the reaction ranged from 0 to 91%.

$(OC)_5Cr$=C(OMe)Ph + phenylacetylene —μW, 5 min 130°C, CAN→ 2-phenyl-1,4-naphthoquinone (18)

0 - 91% yield

3.7.4.5 Benzofused Azoles

The microwave-assisted synthesis of 1,3-azole derivatives was accomplished on mineral supports using either $Ca(OCl)_2/Al_2O_3$ or MnO_2/SiO_2 as shown below (Eq. 19) [33]. The synthesis provided the reaction products in higher yields and better purity than those obtained from the conventional heating.

NH_2, XH; X = NH, O, S

$ArCH{=}NOH$, Al_2O_3 or ArCHO, MnO_2/SiO_2; 4-12 min, 140-226°C

Ar, N, X; 84 - 97% yield

(19)

3.7.4.6 Pyrrolidines

Highly substituted pyrrolidines were prepared from imines and methyl acrylate using a variety of Lewis acids (Eq. 20) [34]. The microwave-assisted reaction resulted in yields comparable to those obtained from a conventional reaction. Softer Lewis acids usually gave better yields than hard Lewis acids.

N, CO_2Et + CO_2Me; Et_3N, CH_3CN; MeO_2C, N, H, CO_2Et

(20)

Tab. 4 Synthesis of pyrrolidines with various Lewis acids

Lewis acid	*Condition*	*Yield (%)*	*Lewis acid*	*Condition*	*Yield (%)*
$Zn(CH_3)_2$	Microwave, 160 °C, 8 min	97	Cu(II)trifluoro-acetylacetonate	Microwave, 160 °C, 8 min	25
$ZnCl_2$	Microwave, 160 °C, 8 min	85	$MgBr_2$	Microwave, 160 °C, 8 min	24
$Yb(OTf)_3$	Microwave, 160 °C, 8 min	85	$AlCl_3$	Microwave, 160 °C, 8 min	12
LiBr	Microwave, 160 °C, 8 min	79	$TiCl_4$	Microwave, 160 °C, 8 min	0
LiBr	R.T./24 h	70	$SnCl_4$	Microwave, 160 °C, 8 min	0

3.7.5 Miscellaneous Reactions

One of the first microwave-assisted reactions that used transition metals was the oxidation of alcohols [35–39]. The oxidations were carried out in a typical kitchen microwave oven, so reproducibility of the reactions was an issue because of the differences between the microwave ovens used. But the significant results obtained in these experiments promoted the broad application of microwave technology to synthetic chemistry as described above. Microwave-induced allylic alkylations have been reported [40, 41]. Hallberg and co-workers pioneered the aminocarbonylation of aryl bromides using $Pd(OAc)_2$, and dppf [42, 43].

3.7.6 Conclusion

During recent years, there has been dramatic increase in the number of publications describing the application of microwave technology to organic reactions involving transition metals. The most common theme in these publications is a shortened reaction time at elevated temperatures. Even though there have been few reports that suggested that the microwave-assisted transition metal reaction produced better yields and cleaner products than the traditional reaction, the reasons for these results are unclear. Similar results have also been observed in non-transition metal-catalyzed reactions. Therefore, further investigation is warranted to clarify the situation.

The recent introduction of instruments that can carefully control the microwave-assisted reaction temperature and pressure allows one to reproduce reaction results consistently. Although the instruments allow one to replicate small-scale reactions, there are limitations in scaling up these reactions in a typical laboratory.

References

1 P. Lindstrom, J. Tierney, B. Wathey, J. Westman, *Tetrahedron* **2001**, *57*, 9225.

2 www.personalchemistry.com; www.cemsynthesis.com; www.milestonesci.com.

3 M. Larhed, C. Moberg, A. Hallberg, *Acc. Chem. Res.* **2002**, *35*, 717.

4 M. Larhed, A. Hallberg, *J. Org. Chem.* **1996**, *61*, 9582.

5 Á. Díaz-Ortiz, P. Prieto, E. Vázquez, *Synlett* **1997**, 269.

6 N. Gard, M. Larhed, A. Hallberg, *J. Org. Chem.* **1998**, *63*, 4158.

7 P. Nilsson, M. Larhed, A. Hallberg, *J. Am. Chem. Soc.* **2001**, *123*, 8217.

8 P. Nilsson, H. Gold, M. Larhed, A. Hallberg, *Synthesis* **2002**, *11*, 1611.

9 K. Vallin, P. Emilsson, M. Larhed, A. Hallberg, *J. Org. Chem.* **2002**, *67*, 6243.

10 D. Villemin, F. Caillot, *Tetrahedron Lett.* **2001**, *42*, 639.

11 M. Larhed, M. Hoshino, S. Hadida, D. Curran, A. Hallberg, *J. Org. Chem.* **1997**, *62*, 5583.

12 K. Olofsson, S.-Y. Kim, M. Larhed, D. Curran, A. Hallberg, *J. Org. Chem.* **1999**, *64*, 4539.

13 R. Skoda-Földes, P. Pfeiffer, J. Horváth, Z. Tuba, L. Kollár, *Steroids* **2002**, *67*, 709.
14 M. Larhed, G. Lindeberg, A. Hallberg, *Tetrahedron Lett.* **1996**, *37*, 8219.
15 C. Blettner, W. König, W. Stenzel, T. Schotten, *J. Org. Chem.* **1999**, *64*, 3885.
16 M. Alterman, H. Andersson, N. Garg, G. Ahlsén, S. Lövgren, B. Classon, U. Danielson, I. Kvarnström, L. Vrang, T. Unge, B. Samuelsson, A. Hallberg, *J. Med. Chem.* **1999**, *42*, 3835.
17 G. Kabalka, L. Wang, V. Namboodiri, R. Pagni, *Tetrahedron Lett.* **2000**, *41*, 5151.
18 M. Erdélyi, A. Gogoll, *J. Org. Chem.* **2001**, *66*, 4165.
19 K. Mayo, E. Nearhoof, J. Kiddle, *Org. Letters* **2002**, *4*, 1567.
20 M. Iqbal, N. Vyse, J. Dauvergne, P. Evans, *Tetrahedron Lett.* **2002**, *43*, 7859.
21 A. Sharifi, R. Hosseinzadeh, M. Mirzaei, *Monatsh. Chem.* **2002**, *133*, 329.
22 Y. Wan, M. Alterman, A. Hallberg, *Synthesis* **2002**, *11*, 1597.
23 B. Chanda, R. Vyas, A. Bedekar, *J. Org. Chem.* **2001**, *66*, 30.
24 Y.-L. Jiang, Y.-Q. Hu, S.-Q. Feng, J.-S. Wu, Z.-W. Wu, Y.-C. Yuan, *Synth. Commun.* **1996**, *26*, 161.
25 A. Combs, S. Saubern, M. Rafalski, P. Lam, *Tetrahedron Lett.* **1999**, *40*, 1623.
26 I. Bytschkov, S. Doye, *Eur. J. Org. Chem.* **2001**, 4411.
27 A. Stadler, C. Kappe, *J. Comb. Chem.* **2001**, *3*, 624.
28 T. Besson, M.-J. Dozias, J. Guillard, C. Rees, *J. Chem. Soc., Perkin Trans. 1* **1998**, 3925.
29 T. Besson, J. Guillard, C. Rees, *J. Chem. Soc., Perkin Trans. 1* **2000**, 563.
30 J. Guillard, T. Besson, *Tetrahedron* **1999**, *55*, 5139.
31 H. Koshima, K. Kutsunai, *Heterocycles* **2002**, *57*, 1299.
32 E. Hutchinson, W. Kerr, E. Magennis, *J. Chem. Soc., Chem. Commun.* **2002**, 2262.
33 K. Bougrin, A. Loupy, M. Soufiaoul, *Tetrahedron* **1998**, *54*, 8055.
34 H.-K. Yen, J. Matthews, J. Lee, D.J. Hlasta, "Application of Microwave Assisted Organic Reactions To Generate a Variety of Heterocycles" (ORGN-737 224th ACS National Meeting Boston, MA **2002**)
35 L. Martínez, O. García, F. Delgado, C. Alvarez, R. Patiño, *Tetrahedron Lett.* **1993**, *34*, 5293.
36 J. Gómez-Lara, R. Gutiérrez-Perez, G. Penieres-Carrillo, J. López-Cortés, A. Escudero-Salas, C. Alvarez-Toledano, *Synth. Commun.* **2000**, *30*, 2713.
37 B. Kaboudin, R. Nazari, *Synth. Commun.* **2001**, *31*, 2245.
38 A. Hajipour, S. Mallakpour, H. Backnejad, *Synth. Commun.* **2000**, *30*, 3855.
39 Q.-H. Meng, J.-C. Feng, N.-S. Bian, B. Liu, C.-C. Li, *Synth. Commun.* **1998**, *28*, 1097.
40 O. Belda, C. Moberg, *Synthesis* **2002**, *11*, 1601.
41 U. Bremberg, M. Larhed, C. Moberg, A. Hallberg, *J. Org. Chem.* **1999**, *64*, 1082.
42 Y. Wan, M. Alterman, M. Larhed, A. Hallberg, *J. Org. Chem.* **2002**, *67*, 6232.
43 N.-F. Kaiser, A. Hallberg, M. Larhed, *J. Comb. Chem.* **2002**, *4*, 109.

3.8
Transition Metal Catalysis under High Pressure in Liquid Phase

Oliver Reiser

3.8.1
Introduction

The development of efficient chemical transformations is one of the most important challenges to date. In organic synthesis, reactions are required to be highly selective as well as economically and ecologically benign, especially in large-scale productions. Since catalysts that consist of a transition metal modified by ligands are at least in theory indestructible vehicles for carrying out reactions under mild conditions with high selectivity, the development of catalytic processes plays a major research role in chemistry.

The optimization of the catalyst performance for a given reaction, which can be evaluated by its selectivity and reactivity, and the development of new reactions which can be rendered catalytic are the most important challenges in this research area. Great efforts are therefore undertaken in the development of new catalysts, mainly by designing new ligands which by coordination electronically and sterically define an environment around a metal in which a reaction can take place. However, new techniques and reaction conditions such as the application of microwaves, supercritical solvents or ionic liquids, or high pressure in liquid phases are beginning to emerge as tools contributing toward the goal of designing efficient catalytic processes.

3.8.2
General Principles of High Pressure

High-pressure chemistry [1] is usually defined as chemical processes which are carried out at pressures between 1 and 10–15 kbar in solution. The expected benefit of high pressure in most cases is to achieve acceleration of a reaction. Such high-pressure conditions are attractive from an energetic point of view, since once the pressure has been built up no additional energy is needed to maintain it. Moreover, because of the incompressibility of liquids, high pressure is a safe technique, since a sudden drop in pressure does not cause a sharp volume increase of the reaction system, in contrast to reactions carried out with compressed gases. A

Transition Metals for Organic Synthesis, Vol. 2, 2nd Edition.
Edited by M. Beller and C. Bolm

ISBN: 3-527-30613-7

general problem with the high-pressure technique is that of carrying out reactions on a larger scale, since the usual laboratory high-pressure vessels have a reaction volume between 15 and 50 cm^3. However, reaction vessels of up to 1500 L or vessels which can be run continuously way are known to be used in some industrial applications [2].

A reaction is accelerated by pressure if its volume of activation $\Delta V^\dagger$, which is defined as the differrence between the volume of the transition state and the volume of the reactants, is negative. This is generally the case in addition reactions, so that high pressure has been most widely used to accelerate cycloadditions such as Diels-Alder reactions ($\Delta V^\dagger = -20$ to -40 $cm^3 mol^{-1}$) by pressure. Moreover, dissociation reactions can be favored by pressure if charged species such as ions are formed (electrostriction). This causes an ordering of charged species (ions) and uncharged species (solvent), which results in a sharp volume decrease. Such electrostriction can be far greater ($\Delta V^\dagger$ up to -100 cm^3 mol^{-1}) than the volume decrease in addition reactions. However, this potentially useful effect has found much less application in high-pressure chemistry reactions.

3.8.3
Influence of Pressure on Rates and Selectivity in Lewis Acid-Catalyzed Cycloadditions

Since Diels-Alder reactions are activated by pressure and by Lewis acids, the combination of both has been applied in transformations which were otherwise particularly difficult to achieve. Thus, the hetero-Diels-Alder reaction between *trans*-1-methoxy-1,3-diene (**1**) with various aldehydes (**2**) can be achieved at only 10 kbar if $Eu(fod)_3$ is used as the catalyst (Scheme 1) [3]. In the absence of a Lewis acid catalyst, pressures between 15 and 25 kbar were necessary for **1** to undergo a reaction, while stronger Lewis acids such as zinc chloride, boron trifluoride etherate, or dialkoxy aluminum chloride immediately polymerized the starting diene **1**. The re-

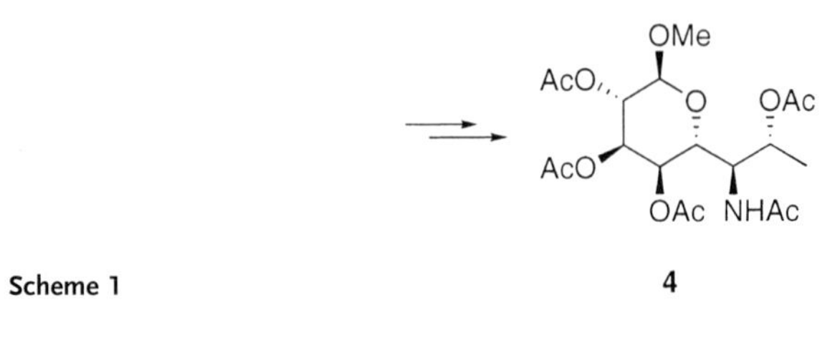

Scheme 1

action conditions described here enable amino aldehydes to be used as the heterodiene without racemization occurring, and this has been used to efficiently access amino sugars such as lincosamines **4** [4].

The combination of pressure and various Lewis acids also turned out to be beneficial in terms of rate and yield in Diels-Alder reactions between cyclopentadiene and acetoxymethylenemalonates [5]. However, no conclusive generalization for the influence of pressure and selectivity could be drawn in this study.

The intramolecular Diels-Alder reaction of **5** catalyzed by the chiral titanium complex **6** was systematically studied in a pressure range of 1 bar to 5 kbar [6]. Intriguingly, concurrent with a gradual increase in pressure, the enantioselectivity increased from 4.5% *ee* (1 kbar) to 20.4% *ee* at 5 kbar (Scheme 2). Although the increase in selectivity is relatively small, it is most important to note that the differentiation between two diastereomeric transition states leading to enantiomeric products can be improved by pressure!

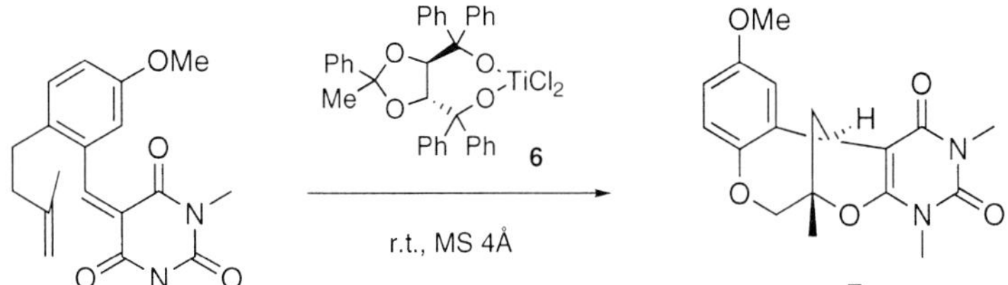

r.t., MS 4Å

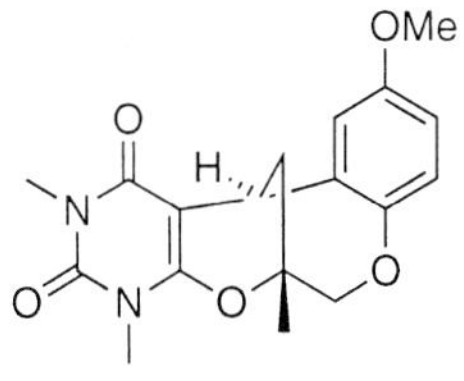

Pressure [bar]	*ee* [%]
1	4.5
1000	6.6
2000	10.1
2800	10.7
3600	16.9
5000	20.4

Scheme 2

However, these results could not be generalized for intermolecular [4+2] cycloadditions (Scheme 3) [7]. **10** was formed in 38% *ee* at normal pressure from isoprene (**8**) and the oxazolidone **9**, while the enantioselectivity decreased at 5 kbar to only 21% *ee*.

It was argued that the decreased selectivity could be due in part to a pressure-induced shift in equilibrium from the chiral catalyst **6** to the achiral catalyst precursor **11**, since this causes a decrease in molecularity. Indeed, high pressure ^{1}H NMR studies showed that the ratio of **6** to **11** decreases from 3.95 (1 bar) to 2.95 (5 kbar).

The activation of the Diels-Alder reaction between **12** and **13** by both pressure and Lewis acid catalysis was also recently investigated (Scheme 4) [8]. Such multi-activation can be beneficial, since reaction temperature and time can be reduced, as clearly demonstrated with the reaction of **13b**. However, because of competing side reactions such as polymerization, which apparently were more severe when

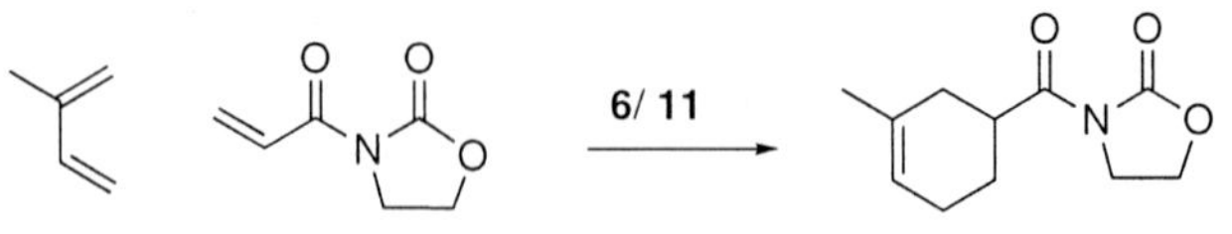

Scheme 3

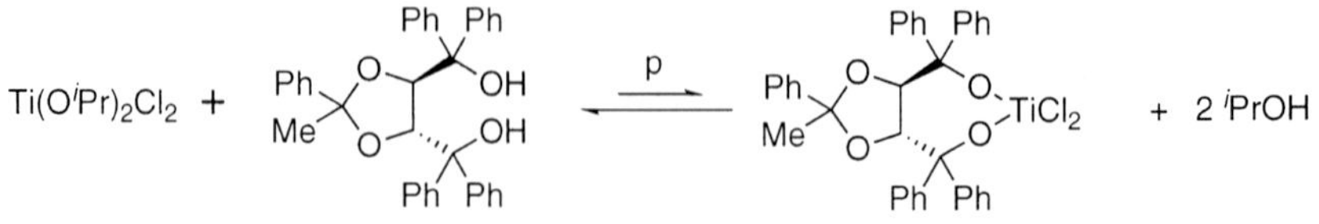

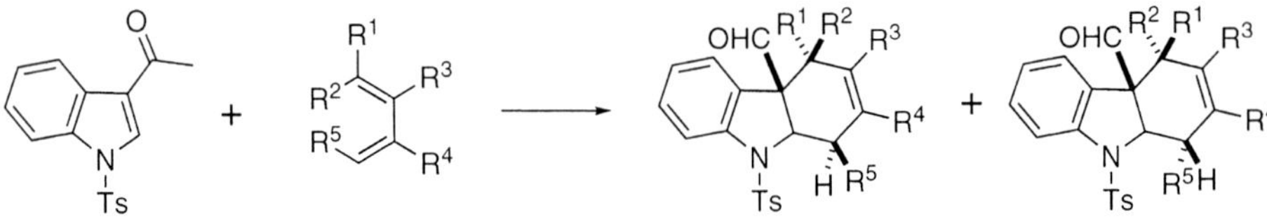

a: R^1, R^2, R^5 = H, R^3, R^4 = CH_3
b: R^1, R^2, R^5 = H, R^3, R^4 = $-(CH_2)_2^-$
c: R^2, R^3, R^5 = H, R^1 = OCH_3, R^4 = $OSiMe_3$

diene	temp (°C)	reaction time (h)	pressure (kbar)	Lewis acid	% conversion	yield **9** + **10** (%)	*endo/exo* (**9**/**10**) ratio
13a	200	216		–	85	71	
13a	50	48	16	–	93	46	
13a	25	24	16	$ZnCl_2$	96	46	
13b	195	72		–	25	25	80:20
13b	50	48	16	–	86	50	96:4
13b	25	48	16	$ZnCl_2$	100	62	>98:2
13c	195	72		–	100	57[a]	75:25
13c	45	96	12	–	100	92[a]	80:20
13c	25	24	16	$ZnCl_2$	100	decomp.	

[a] Isolated in the form of the ketone after hydrolysis of the silyl enol ether

Scheme 4

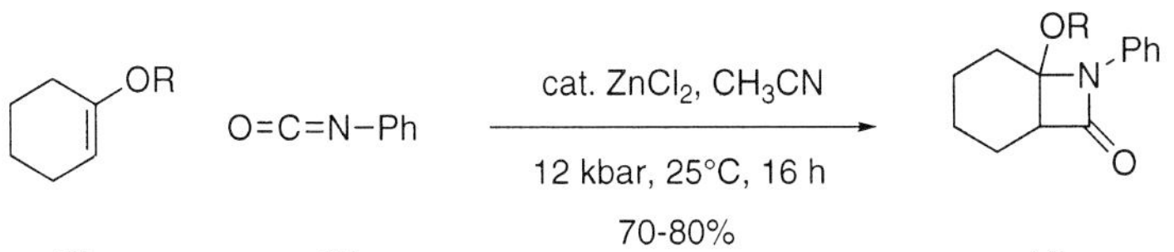

16 17 18

cat. $ZnCl_2$, CH_3CN
12 kbar, 25°C, 16 h
70-80%

Scheme 5

pressure was applied and in the presence of zinc chloride, the thermal process might still be advantageous.

The formation of lactams by [2+2]-cycloaddition of enol ethers and isocyanates proceeds even at room temperature if a combination of pressure and catalytic amounts of $ZnCl_2$ are used, as demonstrated with the synthesis of the bicyclic lactam **18**, which was derived from 1-alkoxycyclohexene **16** and phenylisocyanate **(17)** (Scheme 5) [9].

3.8.4 Nucleophilic Substitution

Nucleophilic substitution proceeding via an S_N2 pathway can be activated by pressure, as has been demonstrated in many examples. Especially, the ring opening of epoxides can be effectively initiated by pressure, but also by Lewis acid catalysis. Consequently, combining these two activation modes would be expected to lead to an even more effective way to functionalize epoxides, and indeed this strategy has been successfully applied.

The ring opening of epoxides with indole is accelerated by a combination of lanthanide catalysts and pressure, which was exploited in a synthesis of diolmycin A2 **(22)** (Scheme 6) [10]. Thus, reaction of epoxyalcohol **19** and indole **(20)** at 10 kbar in the presence of ytterbium(III) triflate and water gave rise to the adduct **21** in 51% yield. Again, the application of pressure enabled the reaction time to be drastically decreased. Subsequent debenzylation then provided the desired natural product **22**.

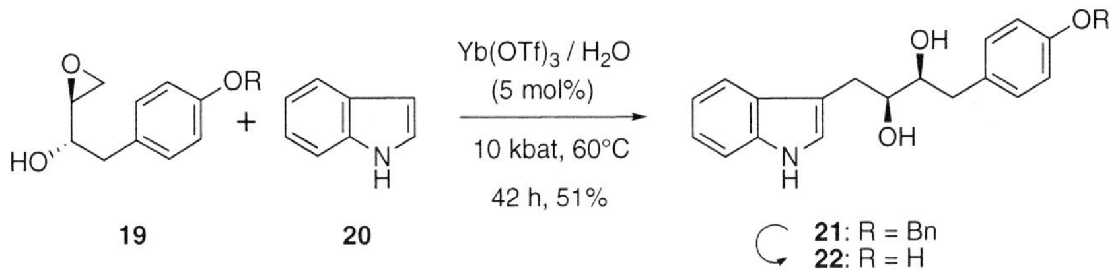

19 20

$Yb(OTf)_3$ / H_2O (5 mol%)
10 kbat, 60°C
42 h, 51%

21: R = Bn
22: R = H

Scheme 6

3.8.5
Addition of Nucleophiles to Carbonyl Compounds

The benefits of pressure or Lewis acid catalysis for the addition of nucleophiles to carbonyl compounds is also well established, e.g., in various aldol processes or allylation reactions. The combination of the two methods, however, has rarely been applied.

A very interesting example was reported with the addition of trimethylsilylcyanide to acetophenone (**23**) in the presence of the chiral titanium catalyst **25** (Scheme 7) [11]. The reaction proceeded by activation with pressure, not only with considerably improved yields but also with significantly increased enantioselectivity. The reason for the latter remains unclear, especially in light of the study of the Diels-Alder reaction between **8** and **9** carried out also with a chiral titanium catalyst (see above). Unfortunately, the exact method of preparation of **25** was not reported, which would have allowed a better comparison of these two studies. Also, it is interesting to note that the addition of TMSCN to **23** in the presence of the 3-fold or 10-fold amount of **29** proceeded with lower enantioselectivity. This could be an indication that the formation of dimers at higher concentration occurs, thereby altering the catalytic active species.

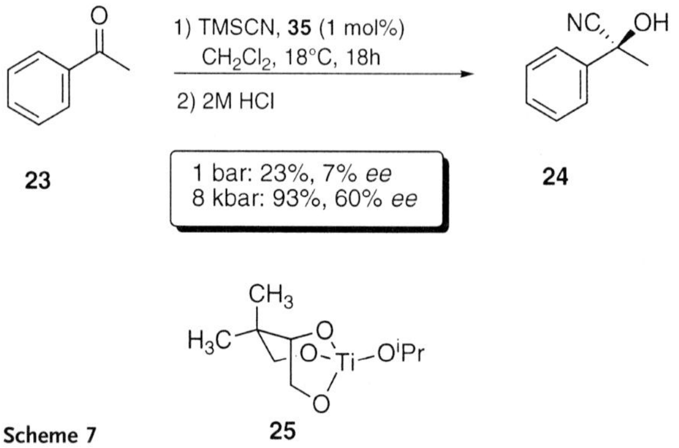

1) TMSCN, **35** (1 mol%) CH_2Cl_2, 18°C, 18h
2) 2M HCl

23 → **24** (NC, OH)

1 bar: 23%, 7% *ee*
8 kbar: 93%, 60% *ee*

CH_3, H_3C, O, O, Ti, O^iPr, O

25

Scheme 7

3.8.6
Influence of Pressure on Rates and Selectivity in Palladium-Catalyzed Cycloadditions

For the reactions discussed in the previous chapter, the catalyst simply plays the role of rendering a substrate more reactive by coordination, while the reactions can also occur – at least in principle – in the absence of the catalyst. Many transition metal-catalyzed reactions are much more complex and consist generally of more than one reaction step, each of which individually could be influenced by pressure in a positive or negative way. Consequently, to predict the net effect of pressure in such reactions is difficult, which could be the reason why a systematic investigation of pressure as a parameter in the area of palladium-catalyzed coupling reactions and [3+2]-cycloadditions was carried out only very recently.

It has been suggested that intermolecular incorporation, i.e. oxidative addition and complexation of a substrate by a metal, should be favored, intramolecular reactions, i.e. insertion, migration, and deinsertion reactions, should be invariant, and extrusion reactions, such as reductive elimination or decomplexation, should be disfavored by pressure [12]. However, decomplexation reactions are in most cases ligand exchange reactions, which can proceed in the case of coordinatively unsaturated species by an associative mechanism, and indeed there is ample evidence that ligand exchange reactions can be accelerated by pressure [13].

Thus, a rate acceleration in the reaction of iodobenzene (**26a**) and ethyl acrylate (**27**) has been observed qualitative [12]. While at room temperature under atmospheric pressure no reaction occurs, ethyl cinnamate (**28**) is obtained in high yield if a pressure of 10 kbar is applied. Surprisingly, no diarylated acrylate **29**, which arises by a second Heck reaction onto **22**, is formed at normal pressure, while at 10 kbar **26** can be obtained as the sole product. Most interestingly, however, is the observation that if bromobenzene (**26b**) is used instead of iodobenzene (**26a**), the adduct **30** was also formed, which is explained by the addition of **32** to a second molecule of ethyl acrylate (**27**) and subsequent reductive elimination and double bond isomerization (Scheme 8 and Tab. 1).

In this case, pressure seems to be able to slow down the reductive elimination leading to **28** sufficiently to make **32** accessible for further transformations. The "living nature" of similar palladium species is of great utility and has been used in intramolecular cascade cyclizations and polymerizations [14].

It was also found that the Heck reaction of certain bromoalkenes such as **34** and **35** proceeds even at 20 °C if the reaction is carried out at 10 kbar [15]. Most interesting, however, is the observation that the activated vinyl chloride would undergo a coupling reaction at a temperature as low as 60 °C, which compares favor-

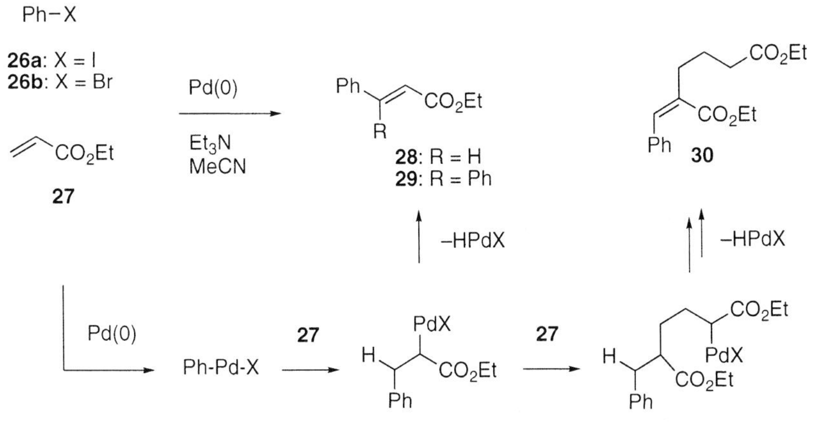

Ph–X

26a: X = I
26b: X = Br

CO_2Et (**27**)

Pd(0), Et_3N, MeCN

28: R = H
29: R = Ph

30

–HPdX

Pd(0)

Ph-Pd-X (**31**) → **27** → **32** → **27** → **33**

Scheme 8

Tab. 1 Palladium-catalyzed coupling of **26** and ethyl acrylate **27** at normal and high pressure.

PhX	*Pressure (kbar)*	*Time (h)*	*Temperature (°C)*	*Yield (%)*		
				28	***29***	***30***
26a	10^{-3}	12	25	0	0	0
26a	10	12	25	62	2	0
26a	10^{-3}	20	90	80	0	0
26a	10	26	90	54	38	0
26a	10	4	140[a)]	0	76	0
26b	10^{-3}	48	90	78	0	0
26b	10	42	90	14	41	25

a) In DMF instead of MeCn.

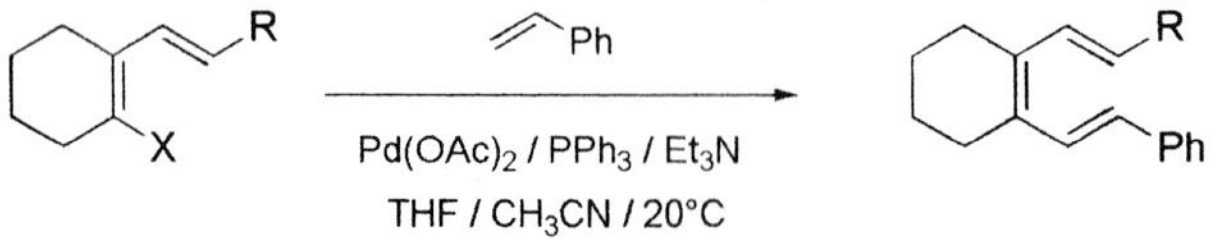

R
X
Ph
$Pd(OAc)_2$ / PPh_3 / Et_3N
THF / CH_3CN / 20°C
R
Ph

34: X = Br, R = CN
35: X = Br, R = CO_2Me
36: X = Cl, R = CO_2Me

37: R = CN
38: R = CO_2Me

Scheme 9

ably with reaction temperatures of 130 °C and above, which are required to react chloroarenes under the same reaction conditions without pressure (Scheme 9 and Tab. 2).

The cross coupling between cyclic alkenes **39** and iodobenzene **26a**, which leads to the arylated alkenes **40–42** depending on the reaction conditions, has been extensively investigated (Scheme 10). In a kinetic study [16] of the reaction between **39a** and **26a**, a 23-fold rate acceleration by increasing the pressure from 1 bar to 8 kbar was found, which corresponds to an activation volume of $\Delta V^{\dagger} = -12$ cm^3

Tab. 2 Palladium-catalyzed coupling of vinyl halides **34–36** with styrene at normal and high pressure.

Alkene	*Pressure (kbar)*	*Time (h)*	*Temperature (°C)*	*Product*	*Yield (%)*
34	10^{-3}	48	20	37	0
34	10	48	20	37	98
35	10^{-3}	48	20	38	0
35	10	48	20	38	96
36	10	72	60	38	42
36	10^{-3}	72	60	38	trace

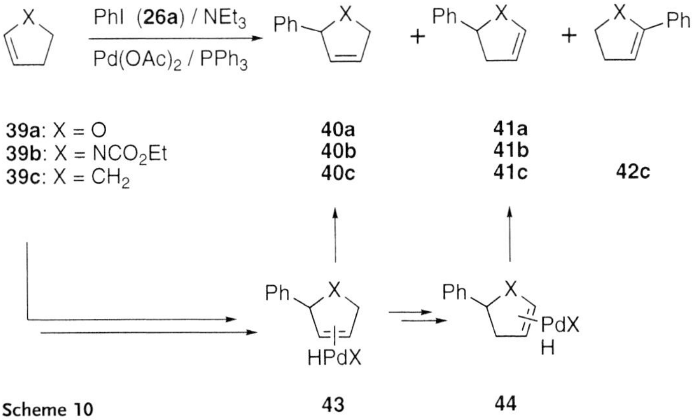

PhI (**26a**) / NEt_3

$Pd(OAc)_2$ / PPh_3

39a: X = O
39b: X = NCO_2Et
39c: X = CH_2

40a **40b** **40c**

41a **41b** **41c**

42c

HPdX

PdX H

43 **44**

Scheme 10

mol^{-1}. This study also revealed that under high pressure the rate-determining step is not the initial oxidative addition of palladium to the aryl iodide. Another effect of pressure in these coupling reactions is the dramatic increase in the lifetime of the catalyst, which is reflected in turnover numbers (TON) of up to 770000 [17]. Moreover, even in the absence of stabilizing ligands, the coupling reactions proceeded with considerable higher TON (7500) than can be reached with the catalyst $Pd(OAc)_2/PPh_3$ at normal pressure.

The phosphine ligand, however, played an important role in the regioselectivity of this reaction. While at normal pressure the ratio **41a/40a** only changed a little (95:5 with no PPh_3; 90:10 with Pd:PPh_3 1:60), at 10 kbar **40a** became the major product with an increase in phosphine (90:10 with no PPh_3; 25:75 with Pd:PPh_3 1:60) [18]. Apparently, the decomplexation of **43** by an associative substitution with PPh_3 is favored by pressure.

In agreement with this analysis, an increase in the enantioselectivity was also found in the coupling of 2,3-dihydrofuran (**39 a**) and phenyl nonaflat (PhONf) when Pd-BINAP was used as the chiral catalyst (Scheme 11). Thus, at 1 bar (*R*)-**41 a** is formed with 47% *ee*, while at 10 kbar a substantially improved selectivity of 89% *ee* is observed. Along with the increase in enantioselectivity, again the regioselectivity of the reaction had changed and **40 a** was formed to a considerable extent at high pressure. Analysing the data for regio- and enantioselectivity it becomes clear that the initial differentiation of the enantiotopic faces of **39 a** by the chiral palladium complex $PhPdL_2 \cdot ONf$ is hardly influenced by pressure. The diastereomeric intermediates **45** and **46**, however, are efficiently kinetically resolved by applying pressure: **45** undergoes associative ligand displacement more rapidly, liberating **40a**, while in **46** metal migration to **47** can take place.

A similar pressure effect on regioselectivity was reported for palladium-catalyzed [3+2]-cycloadditions (Scheme 12) [19].

Here, pressure caused a rate decrease in the TMM, contrasting with the usual rate enhancement observed in cycloadditions. The decisive effect of pressure in this study was on regioselectivity between the possible cycloadducts: while **51** is mainly formed

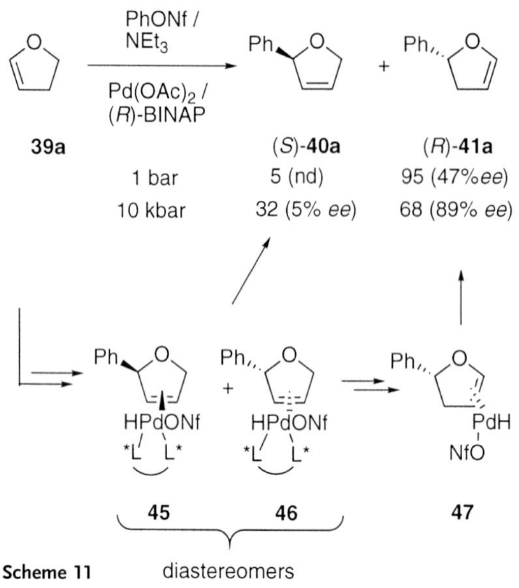

39a → PhONf / NEt_3, $Pd(OAc)_2$ / (R)-BINAP → (S)-40a + (R)-41a

	(S)-**40a**	(R)-**41a**
1 bar	5 (nd)	95 (47%ee)
10 kbar	32 (5% ee)	68 (89% ee)

HPdONf, *L L* **45**, **46** (diastereomers); PdH, NfO **47**

Scheme 11

Me_3Si, OAc **48** → Pd(0), CO_2Me, Ph, CO_2Me **49** → **50** (CO_2Me, CO_2Me, Ph) + **51**

	50		**51**
1 bar, $(Ph_3P)_4Pd$, 82%	1	:	3
10 kbar, $Pd(OAc)_2$, $(i\text{-}PrO)_3P$, 71%	only		–

Pd(0) → [**52** ⇌ **53**] (49)

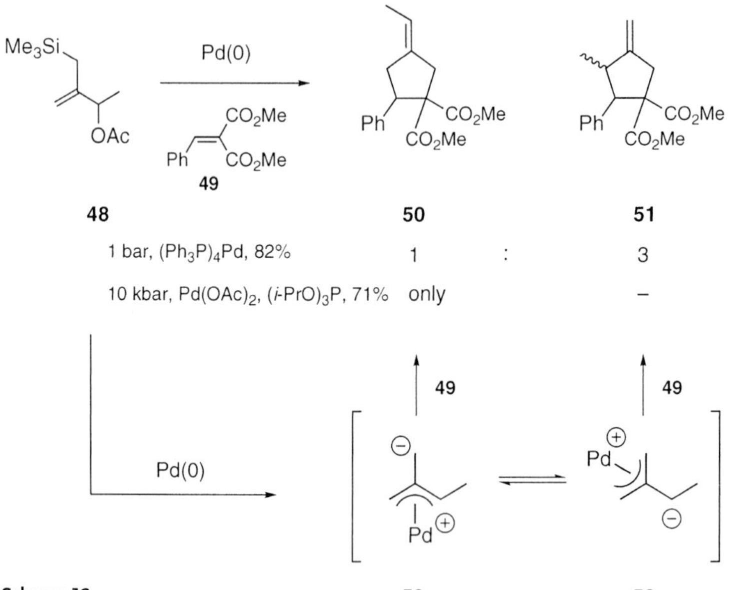

Scheme 12

at 1 bar, the only product observed at 10 kbar is **50**. A possible explanation of this dramatic change in selectivity could be the increase in rate of the bimolecular reaction of **52** and **49** to **50** compared to the unimolecular isomerization of the TMM complexes **52** and **53**. Thus, the kinetically formed complex **52** is effectively trapped under pressure by the alkene **37**.

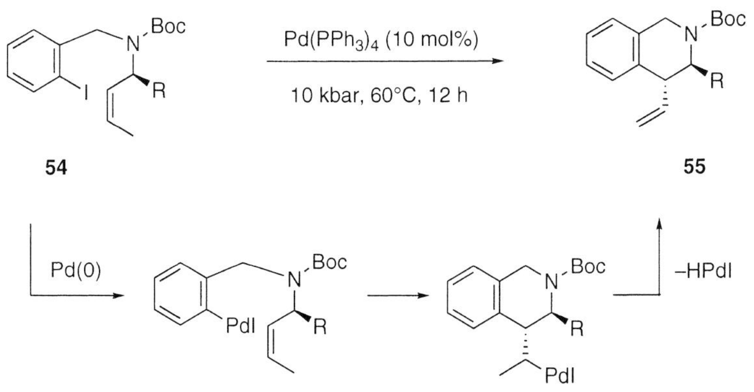

Scheme 13

The synthesis of isoquinolines by the cyclization of **54** demonstrated once more the advantageous effect that pressure could have on palladium-catalyzed coupling reactions (Scheme 13) [20]. **55** was obtained with good yield, regio- and diastereo-selectivity were obtained only if pressure was applied to the system. Especially noteworthy is the beneficial effect of pressure on an *intramolecular* reaction, since the coupling step of **56** to **57** is most likely the rate-determining step. Packing effects caused by the compact transition state and product structure, as was identified by Klärner for Diels-Alder reactions as being decisive rather than volume contractions commonly caused by an intermolecular addition [21], might very well be responsible for the observed results.

The combination of pressure and catalysis can also be used to design a new domino process. The alkenylation of aldehydes with phosphonates (Horner-Wads-

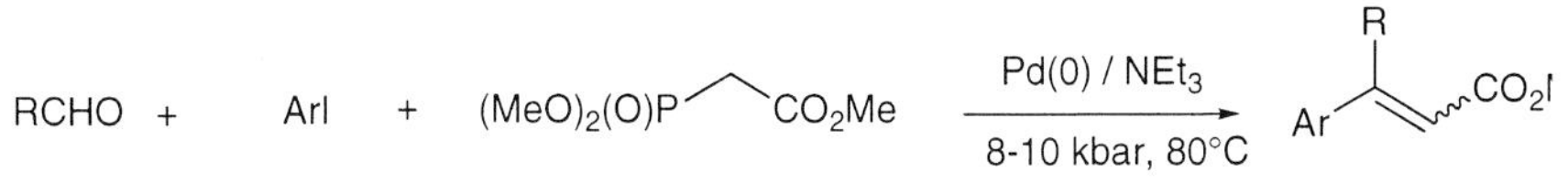

R	Ar	yield [%]	E/Z
	Ph	79	
	p-OMe-Ph	73	
p-OMe-Ph	H	80	27:73
H	p-OMe-Ph	90	70:30
Me	Ph	56	>95:<5

Scheme 14

worth-Emmons (HWE) reaction) is readily accomplished at room temperature under pressure in the presence of triethylamine as a base. These mild conditions are compatible with the Heck protocol, and consequently reacting an aldehyde, a phosphonate, and an aryliodide in the presence of palladium(0) and triethylamine under pressure leads to trisubstituted alkenes **61** (Scheme 14) [22].

Pressure proved to be beneficial not only for the alkenylation step but also for the subsequent arylation via a Heck reaction, since disubstituted alkenes are generally considerably less reactive in such coupling reactions than monosubstituted ones. However, in the case of R ≠ Ar, mixtures of (*E*)/(*Z*) isomers are formed, as was also noted in Heck reactions with cinnamic esters under normal pressure conditions [23].

3.8.7
Rhodium-Catalyzed Hydroboration

The hydroboration of alkenes is known to be activated by either pressure or catalysis. Consequently, a combination of these techniques might open the way to hydroboration of particularly unreactive substrates. Maddaluno et al. recently investigated the hydroboration of some functionalized alkenes, comparing different reagents [catecholborane (CBH) versus pinacolborane (PBH)] and activation by Wilkinson's catalyst ($RhCl[PPh_3]_3$) and pressure [24]. While bromoalkenes and allylamines were found to give the best results with CBH at ambient pressure, 2,3-dihydrofuran (**39a**) was hydroborated most effectively by PBH in the presence of the rhodium catalyst and at a pressure of 12.5 kbar (Scheme 15). No reaction took place in the absence of the catalyst at ambient pressure. Pressure alone led to the

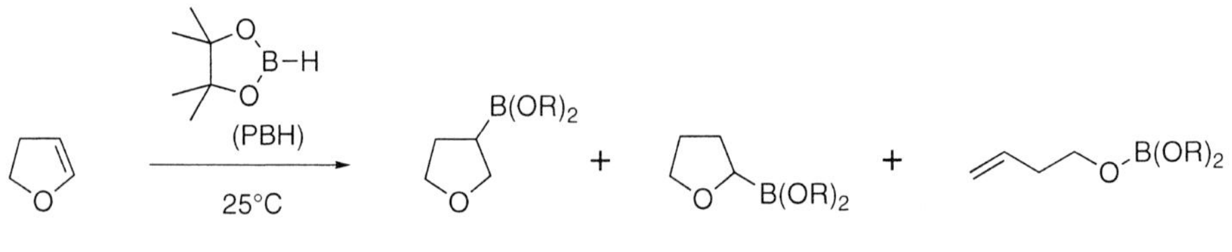

B–H
(PBH)
25°C
$B(OR)_2$
+
$B(OR)_2$
+
$B(OR)_2$
39a **62** **63** **64** 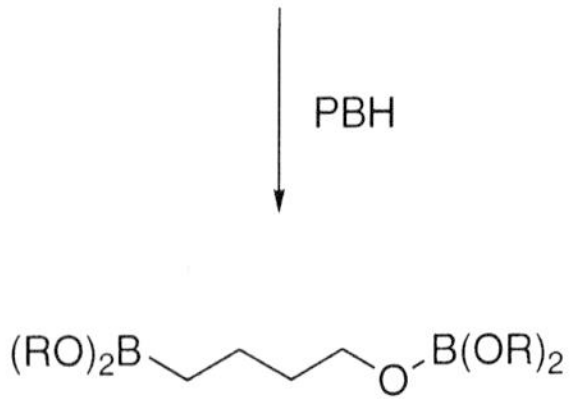
PBH
$(RO)_2B$ $B(OR)_2$
65

$RhCl(PPh_3)_3$ [mol%]	Pressure [kbar]	Time [h]	Product (Sel%) **75:76:77:78**	Yield [%]
0.5	10^{-3}	8	24:41:0:35	45
0	12.5	48	25:0:50:25	46
0.5	12.5	72	61:39:0:0	84

Scheme 15

hydroboration product **62**. However, **64** and **65**, which arise by opening of the furan ring, had also formed to a large extent. Use of Wilkinson's catalyst reversed in part the regioselectivity, leading to **63** as the major hydroboration product of 2,3-dihydrofuran (**39a**), but the ring-opening product **65** was still formed as a major by-product. When pressure and Wilkinson's catalyst were applied, the ring-opening products could be completely suppressed, and **62** and **63** could be obtained with significantly increased yields.

The application of pressure in catalysis has proved to be advantageous for a number of processes. Catalyst performance can be improved in this way, leading to higher yields, better turnover numbers and rates, and increased selectivity, demonstrating that ligand exchange on catalytic active species is facile under pressure. Nevertheless, it has also become apparent that pressure can be detrimental to catalytic processes as well, suggesting that ligand exchange can also be blocked with the application of pressure. High pressure is therefore a useful parameter to be taken into account for catalysis. However, at the current stage of development it is difficult to make general predictions, although some rules have emerged to describe the influence of pressure on transition metal-catalyzed reactions [25, 26].

Acknowledgement
This work was supported by the Fonds der Chemischen Industrie.

References

1 (a) M. Ciobanu, K. Matsumoto, *Liebigs Ann.* **1997**, in print. (b) N. S. Isaacs, *Tetrahedron* **1991**, *47*, 8463–8497. (c) J. Jurczak, B. Baranowski, *High Pressure Chemical Synthesis*; Elsevier: Amsterdam-Oxford-New York-Tokyo, **1989**. (d) K. Matsumoto, A. Sera, T. Uchida, *Synthesis* **1985**, 1–26. (e) K. Matsumoto, A. Sera, *Synthesis* **1985**, 999–1027. (f) F.-G. Klärner, *Chemie in unserer Zeit* **1989**, *23*, 53–63. (g) W. J. le Noble, *Chemie in unserer Zeit* **1983**, *17*, 152–162.

2 M. Gross, *Nachr. Chem. Tech. Lab.* **1992**, *40*, 1236–1240.

3 J. Jurczak, A. Golebiowski, T. Bauer, *Synthesis* **1985**, 928-929.

4 (a) A. Golebiowski, J. Jurczak, *Tetrahedron* **1991**, *47*, 1037–1044. (b) A. Golebiowski, J. Jurczak, *Tetrahedron* **1991**, *47*, 1045–1052.

5 N. Katagiri, N. Watanabe, C. Kaneko, *Chem. Pharm. Bull.* **1990**, *38*, 69–72.

6 L. F. Tietze, C. Ott, K. Gerke, M. Buback, *Angew. Chem.* **1993**, *105*, 1536–1538; *Angew. Chem. Int. Ed. Engl.* **1993**, *32*, 1485.

7 L. F. Tietze, C. Ott, U. Frey, *Liebigs Ann.* **1996**, 63–67.

8 I. Chataigner, E. Hess, L. Toupet, S. R. Piettre, *Org. Lett.* **2001**, *3*, 515–518.

9 R. W. M. Aben, E. P. Limburg, H. W. Scheeren, *High Pressure Res.* **1992**, *11*, 163.

10 R. W. M. Aben, E. P. Limburg, H. W. Scheeren, *High Pressure Res.* **1992**, *11*, 167–170.

11 M. C. K. Choi, S. S. Chan, K. Matsumoto, *Tetrahedron Lett.* **1997**, *38*, 6669–6672.

12 T. Sugihara, M. Takebayashi, C. Kaneko, *Tetrahedron Lett.* **1995**, *36*, 5547–5550.

13 R. v. Eldik, T. Asano, W. J. le Noble, *Chem. Rev.* **1989**, *89*, 549.

14 A. de Meijere, F. E. Meyer, *Angew. Chem.* **1994**, *106*, 2473–2506; *Angew. Chem. Int. Ed. Engl.* **1994**, *33*, 2379–2411.

15 K. Voigt, U. Schick, F. E. Meyer, A. d. Meijere, *Synlett* **1994**, 189–190.

16 S. Hillers, O. Reiser, *J. Chem. Soc. Chem. Commun.* **1996**, 2197–2198.

17 S. Hillers, S. Saratori, O. Reiser, *J. Am. Chem. Soc.* **1996**, *118*, 2077–2078.

18 S. Hillers, O. Reiser, *Tetrahedron Lett.* **1993**, *34*, 5265–5268.

19 B. M. Trost, J. R. Parquette, A. L. Marquart, *J. Am. Chem. Soc.* **1995**, *117*, 3284–3285.

20 L. F. Tietze, O. Burkhardt, M. Henrich, *Liebigs Ann.* **1997**, 1407–1413.

21 F.-G. Klärner, F. Wurche, *J. Prakt. Chem.* **2000**, *342*, 609–636.

22 K. Bodmann, S. Has-Becker, O. Reiser, *Phosphorus, Silicon and Sulfur* **1999**, *144–146*, 173–176.

23 M. Moreno-Manas, M. Perez, R. Pleixats: *Tetrahedron Lett.* **1996**, *41*, 7449–7452.

24 S. Colin, L. Vaysse-Ludot, J.-P. Lecouve, J. Maddaluno, *J. Chem. Soc. Perkin 1* **2000**, 4505–4511.

25 O. Reiser, *Rev. High Pressure Sci. Technol.* **1998**, *8*, 111–120.

26 O. Reiser, *Top. Catal.* **1998**, *5*, 105–112.

Subject Index

Numbers in front of the page numbers refer to Volumes I and II, respectively: e.g., II/254 refers to page 254 in Volume II.

a

Transition Metals for Organic Synthesis, Vol. 2, 2nd Edition.
Edited by M. Beller and C. Bolm

ISBN: 3-527-30613-7

e

f

g

m

p

u

v